MEMBRANE PATHOLOGY

Introductory Remarks

G. BIANCHI

Department of Medicine
University of Milan
Milan, Italy

E. CARAFOLI

Department of Biochemistry
Swiss Federal Institute of Technology (ETH)
Zurich 8902, Switzerland

A. SCARPA

Department of Physiology and Biophysics
Case Western Reserve University
Cleveland, Ohio 44106

This volume contains the scientific presentations and discussions of a conference sponsored by The New York Academy of Sciences held at Villa Olmo, on the shore of Lake Como in Italy on May 19–22, 1986. It was an enjoyable and unusual conference, not only because of the exquisite setting and local hospitality, but also because it permitted a relaxed and useful exchange of ideas and information among colleagues who don't often have the opportunity to meet or interact scientifically. The broadly focused topic of the meeting, membrane pathology, brought together four groups of basic scientists and a large number of clinicians and clinical scientists. In view of the increasing tendency toward specialization in medicine and the biological sciences and, consequently, the specialized nature of scientific meetings and conventions, interdisciplinary interaction, like that provided by this conference, has become increasingly rare.

Our decision to hold a conference on biological membrane pathology presented us with a dilemma: it seemed to lack originality, and yet it was long overdue. Just a random glance at recent literature in the biological sciences illustrates that during the last few years biological membrane has been recognized as playing an important, and frequently primary, role in the onset and development of disease. That the involvement of membrane in disease has now attracted so much attention is hardly surprising if one considers that membrane biochemistry and biology have developed to a point where they represent a large portion of research in biochemistry, molecular biology, physiology, and pharmacology. In arranging this conference, we hoped to emphasize the interdisciplinary character of research on biological membrane. The overall goal was to let scientists analyze a disease process, in which biological membrane's structure and/or function is or may be primarily impaired, into its most elementary components, and to allow clinical colleagues to check this analysis against the background of integrated aspects of the disease. The time appeared particularly propitious for this kind of cross-fertilization among basic scientists working in different fields and clinicians.

With this overall goal in mind the major difficulty was the choice of the disease processes to be discussed. Although our personal bias played a significant role in selecting the four diseases, the choices appear very reasonable when viewed against the following criteria: (1) the likelihood that, at the present time, the disease would benefit by detailed discussion; (2) the existence of well developed membrane-related research on the disease; and (3) the existence of some common denominator in terms of ideas, experimental approaches, or future research trends in the four diseases selected.

The volume is divided in four parts, representing four days of presentation and discussion on the four disease processes.

The first, mitochondrial disease, is an area in which impressive advances have been made in recognizing the importance of mitochondrial membrane defects as primary events in the pathogenesis of inborn disease. Comprehensive and updated information is presented on membrane defects arising from the deficiency of cytochromes or iron-sulfur clusters localized in specific tissues, or generalized in infants with congenital lactic acidosis; the defects of enzymes or cofactors for the transport of fatty acid or carnitine across the mitochondrial inner membrane; autoantigen diseases affecting the operation of the ADP/ATP carrier and other mitochondrial functions; inborn defects affecting the mitochondrial portion of the urea cycle; and mitochondrial defects in cystic fibrosis and in patients with chronic ethanol intoxication. The discussion of the specific defects is supplemented by morphological studies indicating the presence of giant mitochondria in some of these diseases and in experimental models. Genetic studies of possible defects in the systems needed to target proteins to mitochondria and newly developed ^{31}P NMR techniques permit the non invasive diagnosis of such diseases and the assessment of the efficiency of new therapeutic approaches.

Membrane defects are present in hypertension as well and affect a variety of systems or tissues. Yet the primary mechanism(s) for the onset of the disease is far from clear. The second part of this volume addresses various membrane alterations detected in hypertensive states. They include defects in Na^+ transport and Na^+ retention in a variety of cell and tissues; the appearance of endogenous changes of cytoskeleton in atheromatosis; digitalis-like factors and the natriuretic hormone; and abnormalities in intracellular calcium signalling in hypertension and in platelet membrane of hypertensive patients. These discussions have been supplemented by the results of new approaches to measure the subcellular distribution of metal ion in various tissues as affected by hypertension and by the characterization of membrane abnormalities of genetically hypertensive animal models.

Cell differentiation and cancer are traditional areas in which membrane-related research has been strong. Yet, in spite of the immense body of ongoing research, the primary events in cell differentiation and malignancy are still the object of conjecture. The third section focuses on membrane defects of cancer cells associated with or relevant to the onset of the disease that share concepts and experimental approaches with membrane defects described in the first two chapters: defects in bioenergetics and lipid composition; the role of oxy-radical; the generation of ionic signal affecting membrane transport by growth factors and oncogenes; and the intracellular signals that are amplified and/or depressed in cancer cells.

Finally, in the fourth part of this volume, diabetes is discussed as an example of a metabolic disease. While many other diseased states could have been profitably discussed, it was felt that diabetes lent itself particularly well to a study of membrane-linked factors in the development of the disease processes. Most of the membrane defects discussed in this section have clear similarities with those discussed in previous chapters such as cellular signalling during secretion, the role of altered plasma membrane ion fluxes, the control of oxidative metabolism of mitochondria, the involvement of various protein kinases, cellular Ca^{2+} homeostasis, and the occupancy and phosphorylation of receptors. The findings have been logically correlated by the presentation of morphological approaches aimed at detecting insulin packaging, and by a clinical interpretation of membrane defects on humans affected by diabetes.

We hope that the presentations contained in this volume convey the excitement and very productive scientific atmosphere of the conference. We are convinced that the contributions and discussions led to a better understanding of the basic aspects of pathological deviations and to a genuine exchange of concepts and experimental approaches between basic scientists and clinically oriented researchers. Investigators in specialized research areas will find new experiments and comprehensive reviews in their fields. We trust, however, that this volume will be even more valuable to investigators interested in other fields who plan to broaden their horizons and knowledge for educational or research purposes. They will find in the comprehensive reviews of the recent literature and in the description of new discoveries that the membrane, which lies at the heart of a very large part of pathology, can be a focal point for interactive, interdisciplinary research.

As participants in the conference at Lake Como, we learned of diseases we had not known about and came to appreciate the striking similarities between questions and approaches in fields other than our own. We benefitted from the lively discussions on the identification (or exclusion) of common mechanisms of membrane response in different diseases.

As organizers, it is a pleasure to thank the participants and the speakers who defied actual or perceived threats from terrorists and radioactive clouds to be with us; the local organizers who were able to transform four days filled with presentations and discussions into an enjoyable vacation; the various organizations and firms whose generosity made this conference possible; and The New York Academy of Sciences which, as usual, came through perfectly.

Mitochondrial Pathology: An Overview

ERNESTO CARAFOLI

Laboratory of Biochemistry
Swiss Federal Institute of Technnology (ETH)
8092 Zurich, Switzerland

Alterations of mitochondria are a common occurrence in spontaneous and experimentally induced disease processes (for a comprehensive review see Carafoli and Roman[1]). Although they are in general secondary to changes induced by the noxious agents elsewhere in the cell, in some cases primary mitochondrial alterations leading to systemic disturbances can be documented with certainty. Most are linked to genetic deficiencies in the membrane-bound or matrix-soluble enzymes (e.g., the enzymes of the oxidative and energy coupling systems or of the intramitochondrial portion of the urea cycle). However, other primitive deviations, e.g., those characterized by the production of auto-antibodies against mitochondrial components, are not genetic. The field of oncology deserves special attention even if a large portion of the observations in the literature on the relationship of mitochondria to malignancy concerns secondary phenomenology. In some rare tumors, admittedly benign in most cases, the enormous proliferation of mitochondria, unaccompanied by other obvious cellular disturbances, leaves the clear impression that mitochondria could be the primary, or at least the most conspicuous target, of the oncogenic agent(s). And it may also be mentioned that recent studies on hexokinase and its relationship to mitochondria offer new support to the idea of mitochondrial defects as the primary response of cells to malignancy-inducing agents. A comment is in order also on the matter of the mitochondrial Ca cycle, whose alterations are very common in cell pathology. Most likely, the disturbances of the cycle are not due to primary alterations of the Ca^{2+} uptake and/or the Ca^{2+} release transporters, but follow the deficiency of the plasma membrane Ca-filtering function. The result, however, will ultimately be a situation of stress for the mitochondrion, often at a stage when the remainder of the cell is still relatively unaffected. Since cell survival eventually depends on whether mitochondria will overcome the situation of stress, the deviations of the mitochondrial Ca cycle, even if not primitive, deserve a privileged place in a discussion of mitochondrial pathology.

In the early days of mitochondrial pathology, most of the reports were by necessity morphological. They described a large number of peculiar mitochondrial alterations, particularly in muscle, among them bizarre shapes of the organelles, unusual orientations of the cristae, and especially crystalloids in the intracristal space or in the space between the inner and the outer membrane. The intramitochondrial crystalloids, e.g., those of the "parking lot" type in the intracrystal space (FIGURE 1), are of obvious interest (they may correspond to the accumulation of mitochondrially made subunits of proteins assembled in collaboration with the cytoplasmic protein synthesis system) and deserve further study. However, a disproportionate emphasis has been placed in the past on their specificity. It is now known that they occasionally occur, together with the other alterations, in mitochondria of animals treated with various poisons, e.g., uncouplers of oxidative phosphorylation,[3] and even in normal mitochondria.[4] In general, their occurrence in various disease conditions involving mitochondria is now thought to reflect the very limited range of options these organelles have in responding to noxious agents of various types with structural changes. The idea that structural

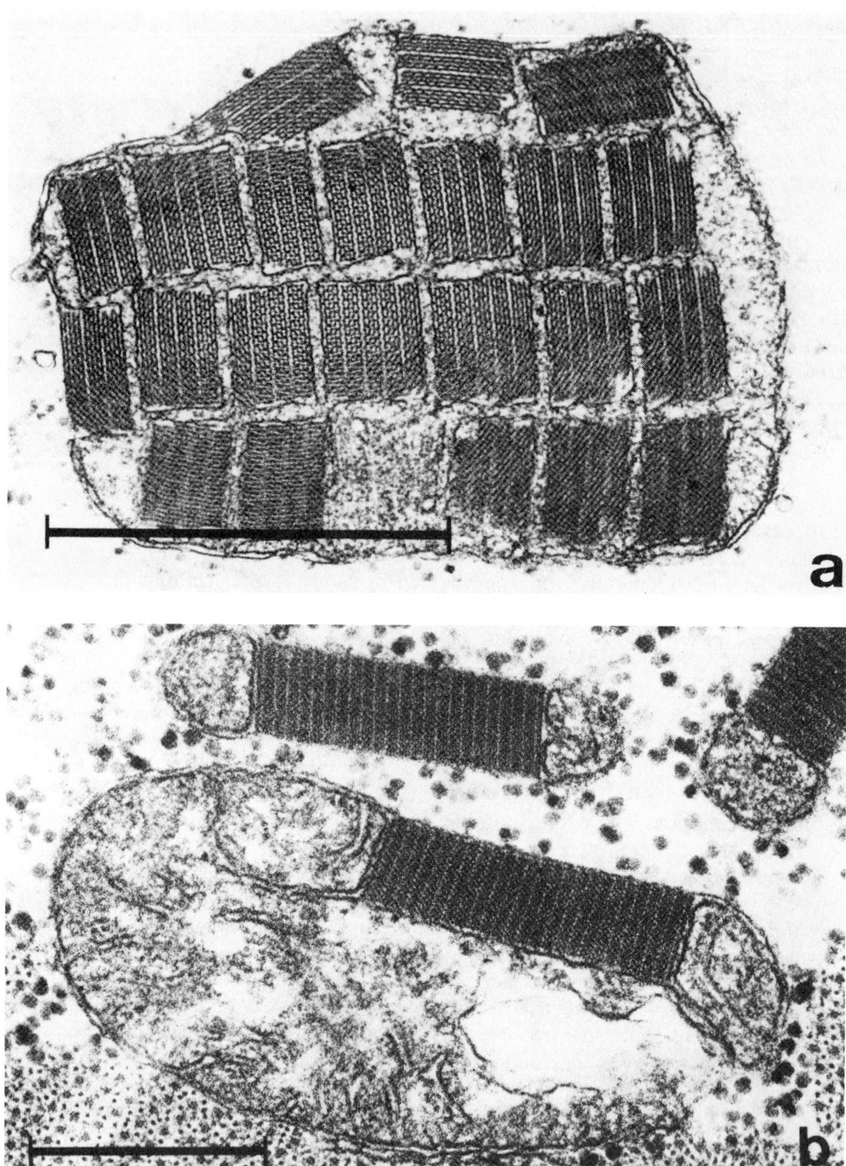

FIGURE 1. (a) Enlarged muscle mitochondrion showing "parking lot" intracristal paracrystalline inclusions. From one of the cases of mitochondrial myopathy described by J.A. Morgan-Hughes.[2] Bar represents 1 μm. (b) Enlarged muscle mitochondria from a patient with cytochrome *b* deficiency. The mitochondria have irregularly branched cristae and contain intermembranal paracrystalline inclusions. Bar represents 0.5 μm (From J.A. Morgan-Hughes.[2] With permission from Churchill-Livingstone.)

changes are specific markers for underlying biochemical defects seems rather unrealistic.

This overview will not attempt to be comprehensive, partly due to space limitations, partly due to the fact that the field of mitochondrial pathology is developing rather rapidly, and new conditions as well as areas which are now beginning to contribute to it have not yet reached the stage at which generally accepted conclusions can be discussed. The chapters that will be succinctly covered are those that have traditionally formed the backbone of research on the involvement of mitochondria in the disease process (TABLE 1).

DEFECTS OF THE RESPIRATORY CHAIN, OF THE ENZYMES INVOLVED IN THE SYNTHESIS OF ATP, AND OF SUBSTRATE TRANSPORT AND UTILIZATION

Genetic defects have now been identified in three of the four main complexes of the respiratory chain (I, III, and IV), in some of the systems for the transport of substrates, and in the energy-coupling machinery. They are rare conditions, described at most in a few dozen patients. The introduction of a modified histological trichrome stain in 1963[5] revealing reddish aggregates of mitochondria in the sub-sarcolemma spaces of muscle fibers and between myofibrils (ragged-red fibers) has greatly facilitated the qualitative recognition of these deficiencies, which almost invariably involve skeletal muscles. In the last few years, however, refined biochemical techniques have permitted the identification of the locus of these defects with relative precision. Conventionally, these conditions are clinically grouped under the heading "mitochondrial myopathies," a definition that suggests muscle specificity for the defect. They are indeed characterized by hypotonia, exercise intolerance, and muscle weakness, but it is well to remember that symptoms involving other organs, notably the central nervous system and the kidneys, are frequently also present and may even be predominant. Abnormalities of the type found in "mitochondrial myopathies" may originate directly from primary DNA defects, or be mediated by other factors; in the latter case, the mitochondrial abnormality may still dominate the picture, or be merely one of the components of conditions that have their most important expression elsewhere.

TABLE 1. Mitochondria and Pathology

Defects of the transport and utilization of substrates, of the respiratory chain, of the energy-conserving system
Defects of the mitochondrial Ca cycle, i.e., increased Ca-phosphate storage activity
Antimitochondrial antibodies
Mitochondrial involvement in the malignant transformation
Defects in the mitochondrial portion of the urea cycle
Defects of the propionate, methylmalonate, and cobalamine metabolism (e.g., the methylmalonic acidemias)

A general summary of the biochemical defects so far detected in the respiratory chain, in the transport and utilization of substrates, and in the energy-conserving systems is shown in TABLE 2 (modified from Morgan-Hughes[2]). Whereas some of the defects recorded in the table are only presumptive, others have been conclusively assigned, in some cases even to the point where single defective components of the multimolecular complexes have been identified. Given the general character of this overview, no discussion of the individual defects mentioned in the table will be presented. (Readers interested in the details of the conditions are referred to recent reviews available in the literature.[2,7–9]) However, some points that have general significance or contain particular elements of interest will be made. One that has been briefly alluded to above concerns the tissue specificity of the defects observed. Although in most cases the genetic deficiency is multi-systemic, in some cases tissue specificity indeed seems to have been documented. The specially high demands for oxidative metabolism of nervous tissue and muscle may help explain why the phenotypic manifestations in TABLE 2 frequently favor these two systems. In some cases, however, enzyme levels and/or properties have been measured and found to be

TABLE 2. Survey of the Biochemical Defects Reported in the Conditions Grouped under the Heading "Mitochondrial Myopathies" (Survey Completed in 1985)

Defects of substrate transport
carnitine deficiency
carnitine palmitoyl transferase deficiency (CPT I and II)
adenine nucleotide transport deficiency[a]
pyruvate transport deficiency[a]
Defects of substrate utilization
pyruvate dehydrogenase deficiency
dihydrolipoyl dehydrogenase deficiency
pyruvate dehydrogenase phosphatase deficiency
flavoprotein dehydrogenase deficiency
Defects of the respiratory chain
NADH-ubiquinone reductase deficiency
defects of cytochrome *b* (complex III)
defects in the region of ubiquinone (complex III)
cytochrome *c* oxidase deficiency (compex IV)
combined complex III and complex IV deficiencies
Defects of the energy-conserving systems
hypermetabolic myopathy
myopathies with "loose" respiratory control without hypermetabolism
defects in the mitochondrial ATP synthetase

[a]Defects that are only presumptive.

differentially affected in various tissues. Very likely, then, this reflects the presence of tissue-specific isozymes variously affected by the genetic deficiency.

The possibility of successful therapeutic interventions must also be mentioned in this context as a point of general interest. Traditionally, the vast majority of the genetic defects of the systems connected with the transformation of energy in mitochondria have always been considered lethal, and in any case beyond the reach of successful therapy. With the very significant advances recently made in the biochemical definition of at least some of the conditions, target-directed correction attempts have become possible, and have indeed sometimes proven successful. This has been so for the carnitine substitution therapy in some cases of carnitine deficiency[6] and for the ubiquinone substitution therapy in one case of Kearns-Sayre syndrome where a deficiency in the ubiquinone region of the respiratory chain had been documented.[10]

Another interesting general point, which is related to the lethal character of most of the deficiencies mentioned above, is the recent observation of the transient nature of a genetic defect. A certified[11] and a suspected[12] deficiency of cytochrome oxidase, which were essentially complete at birth or immediately thereafter, were found to improve spontaneously with age and to reach normalcy in two to three years. It is of interest that in one of these two cases, even when the histochemical reaction for active cytochrome oxidase was detectable only in about 5% of the muscle fibers, normal immunocytochemical reaction to the enzyme was found. This indicates that the enzyme was present, but somehow kept in an inactive state. The spontaneous remission of the symptoms, considering that the enzyme as a structural entity was apparently present in the membrane, thus becomes less difficult to rationalize. This point lends itself to another general consideration: defects of mitochondrial enzymes will have only limited chances of producing dramatically evident phenotypic changes unless they affect one or more of the rate-limiting components of the mitochondrial machinery, e.g., the respiratory chain. When the defect involves enzymes that are known to exist in large excess, e.g., cytochrome oxidase with respect to the other components of the respiratory chain, it seems reasonable to expect that evident clinical symptoms will only appear when the enzyme depletion reaches respectable proportions.

As mentioned at the opening of this chapter, no discussion of the various genetic defects listed in TABLE 2 will be presented. One exception, however, is in order for the hypermetabolic myopathy that goes under the name of Luft's disease. This is justified by the fact that, historically, this is the first mitochondrial condition where a precise biochemical defect was identified[13] and by the fact that in this case the clinical phenotype can be unusually well rationalized on the basis of the mitochondrial defect. Luft's disease, which was first described in a 35-year-old Swedish female and was subsequently observed in another patient,[14] is characterized clinically by a severe euthyroid hypermetabolism and by a great increase in the number of muscle mitochondria. Although the latter appeared enlarged, and sometimes contained bizarrely oriented cristae and paracrystalline inclusions, they apparently retained normal ability to couple the transport of electrons in the respiratory chain to the synthesis of ATP. However, whereas in normal mitochondria the rate of electron transport is greatly reduced in the absence of ADP (the so-called phenomenon of respiratory control), in the patient's muscle mitochondria the consumption of oxygen was very high even in the absence of ADP. In other words, the patient's muscle mitochondria were "loosely coupled," thus dissipating a large amount of respiratory energy as heat. The reason for the loose coupling of muscle mitochondria has not been established with certainty, and it is not even clear whether the patient's mitochondria were as loosely coupled *in vivo* as they appeared to be *in vitro*.[15] It has been proposed[16] that they lacked the ability to retain their endogenous Ca, thus abnormally activating the energy-dissipating Ca cycle (see below) whose energy-linked uptake leg would be responsible for the permanent activation of respiration in the absence of ADP.

DISTURBANCES OF MITOCHONDRIAL Ca TRANSPORT

Mitochondria contain a Ca-transporting system that consists of two portions, one used exclusively for the uptake of Ca into the organelle from the cytosol, the other used exclusively for the release of Ca from the matrix back to the cytosol.[17] The uptake route is energy dependent and is generally accepted to be energized by the electrical component of the proton motive force (about 200 mV, negative inside) created by the transport of electrons in the respiratory chain. Being energy dependent, the Ca uptake process is inhibited by reagents that interfere with the energy flow in the mitochondrion, e.g., uncouplers, but is also specifically inhibited by the polycation ruthenium red, a compound thought to interact irreversibly with a hypothetical Ca uptake carrier located in the inner membrane. The large negative electrical potential normally present in mitochondria makes the uptake route essentially irreversible, and provides the rationale for the existence of an independent Ca release route, electrically silent and thus able to operate continuously in normal mitochondria. The release route has been identified as an electroneutral Ca/Na exchanger,[18] which is particularly active in excitable tissues, but which is present in all mitochondria so far studied (in some mitochondrial types, e.g., in those of liver, a Ca/proton exchanger may also be active in the release of Ca). Taken together, the energy-driven uptake route and the Na-dependent release pathway constitute the energy-dissipating "mitochondrial Ca cycle"[17] (FIGURE 2), which probably operates at a very low rate in normal cells. The point on the overall rate of the mitochondrial Ca cycle is an important one: in the early

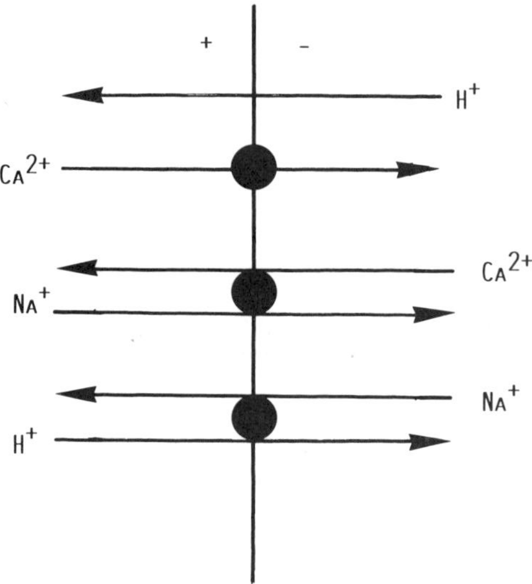

FIGURE 2. The Na-Ca cycle of mitochondria.

days of research on mitochondrial Ca transport it was generally assumed that the size of the mitochondrial Ca pool was very large, on the order of 10 nmol per mg of mitochondrial protein, and that the exchange of Ca between mitochondria and the cytosol was quantitatively important. That is, it was generally assumed that mitochondria play an important role in the regulation of the Ca concentration in the cell, possibly even contributing to the control of processes that require the rapid mobilization of large amounts of Ca such as the contraction/relaxation cycle of muscle.[19] The situation appears to have changed in the last few years, chiefly based on developments coming from two different research lines: on one hand, electron microprobe analysis work[20] has shown that the mitochondrial Ca pool *in situ* is considerably smaller than previously assumed, i.e., of the order of 1–2 nmol/mg protein. This obviously restricts considerably the quantitative contribution of mitochondria to the regulation of the cytosolic Ca homeostasis. On the other hand, the introduction of intracellular fluorescent Ca indicators[21] has shown that the ionic Ca concentration in the cytosol is less than 200 nM, i.e., 10 to 50 times lower than the concentration necessary to half-maximally activate the route for Ca uptake into mitochondria.[22] Therefore, it is now generally assumed that mitochondria, whilst not inactive in the control of cytosolic Ca in normal cells, do not play a major role in it. In fact, it now appears likely that the main function of the systems for transporting Ca in and out of mitochondria may be the regulation of the Ca-dependent enzymes in the mitochondrial matrix,[23] rather than the regulation of Ca in the cytoplasm. Naturally, this is true of cells where the ionic concentration of Ca in the cytoplasm is in the "normal" range of about 200 nM. Clearly, cells that normally experience cytosolic Ca concentrations exceeding the μM range (bone cells, Ca transporting cells, perhaps even (heart) muscle cells at the end of the activation phase) will place a greater burden on their Ca-transporting mitochondrial system, now exposed to Ca concentrations that are possibly adequate to activate its uptake portion. The overall velocity of the Ca cycle will thus presumably increase and it is also possible that a net accumulation of Ca will occur if the rate of the uptake portion of the cycle increases substantially over that of the (Na-induced) release portion, which presumably varies very little. Longer term accumulation of Ca will be aided by the simultaneous accumulation of phosphate, which will penetrate into mitochondria on an independent carrier system to precipitate the excess Ca that has been taken up. This is an old observation;[24] the phenomenon serves the useful purpose of permitting the storage of large amounts of Ca in the matrix essentially without increasing its ionic concentration to levels that would prevent the correct regulation of the intramitochondrial enzymes. In fact, the ability of mitochondria to store large amounts of precipitated Ca-phosphate salts, presumably in the form of amorphous hydroxyapatite, is a safety device of formidable importance to injured cells, which very frequently experience conspicuous increases of the level of cytosolic Ca. A common and early event in spontaneous and experimental cell pathology is indeed the decreased efficiency of the Ca filtering function of the plasma membrane,[25] resulting in the admission of abnormally high amounts of Ca into the cytoplasm. The Ca-ejecting systems of the plasma membrane, particularly the Na/Ca exchanger, will at this point increase their activity and dispose of at least a portion of the excess penetrating Ca, but it is clear that their maximal capacity will sooner or later be overcome if the injuring cause persists. The gradual increase of cytosolic Ca will then activate the energy-linked Ca uptake route and the level of cytosolic Ca will thus be maintained at a concentration that will be greater than normal, but probably not intolerably high (it is generally accepted that the set-point of external Ca in active mitochondria is probably less than 500 μM[26]). For a while, then, thanks to their ability to accumulate both Ca and phosphate, mitochondria will limit the damage linked to the situation of Ca emergency in the cytoplasm without serious impairment of their function. Since the maximal rate of the Na-induced Ca release route (and presumably also the rate of the other putative Ca-release pathways) is much slower than the maximal rate of the uptake route, once the Ca emergency has subsided (i.e., once the Ca concentration in the cytosol has been reduced to levels insufficient to activate significantly the uptake pathway), the extra Ca accumulated by mitochondria will be released at a controlled rate, compatible with the capacity of the plasma membrane exporting systems. The accumulation of large amounts of Ca and phosphate in mitochondria of injured cells has now been documented with the aid of the electron microscope in a very large number of conditions. FIGURE 3 offers one of the earliest examples,[27] i.e., the liver of a rat intoxicated with CCl_4: many mitochondrial profiles in the otherwise normal looking cells contain dense precipitates, which extraction experiments in parallel experiments have shown to be composed of Ca and phosphate. The figure also shows, in support of the discussion above, that mitochondria, despite the accumulation of the precipitates in

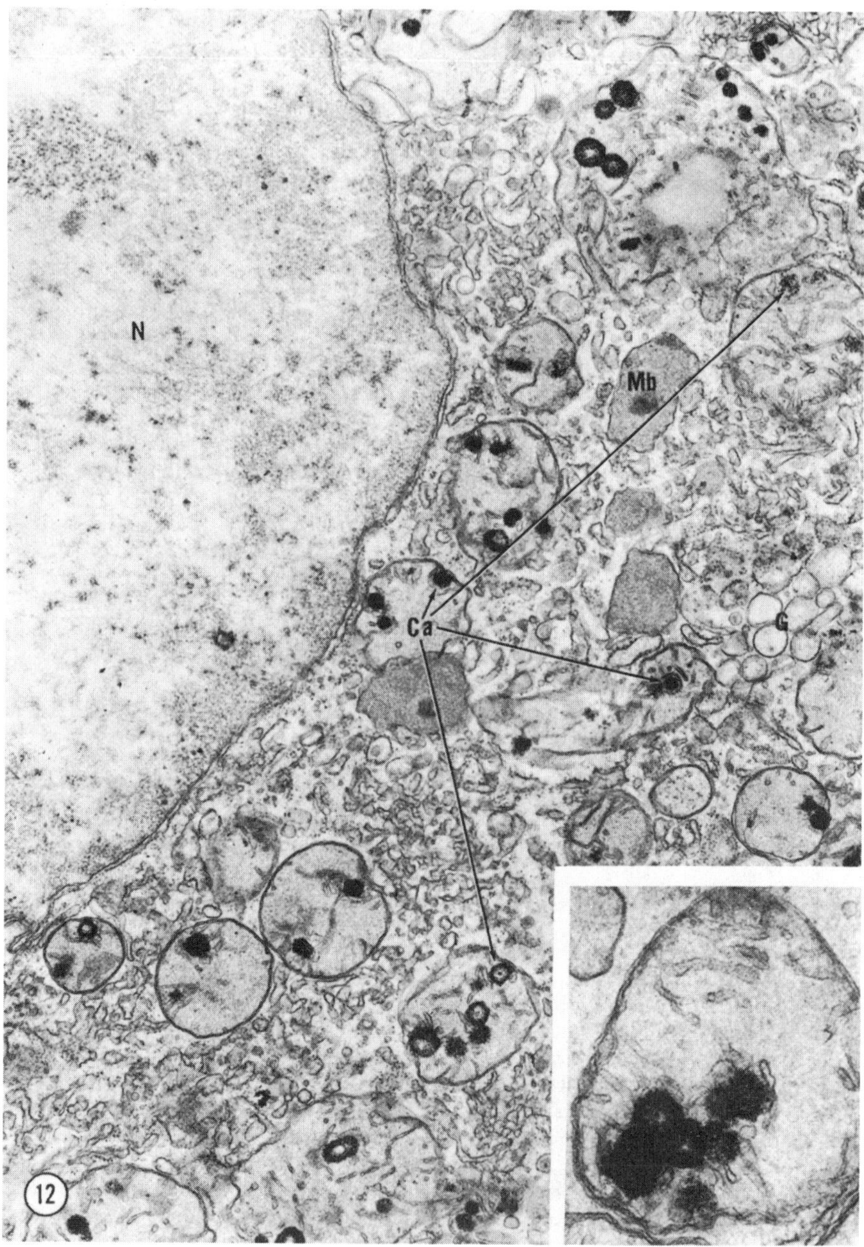

FIGURE 3. Intramitochondrial dense granules (calcium phosphate deposits) in a liver cell of a rat intoxicated with carbon-tetrachloride. The electron-transparent core seen in many of the dense masses is probably due to damage to the structure of the granule by high intensity electron beams. The mitochondrial structure, also when the organelles contain electron-opaque deposits, appears relatively well preserved. (From E.S. Reynolds.[27] With permission from the *Journal of Cell Biology*.)

their matrix, appear ultrastructurally rather well preserved. It must be understood, however, that the capacity of mitochondria to buffer Ca, however impressive, is finite and only offers the cell temporary relief. If the noxious condition does not subside, a point of no return will eventually be reached at which the amount of Ca in mitochondria will interfere with their proper functioning. In addition, the continuous uptake of Ca, which is the alternative to oxidative phosphorylation, will decrease the amount of ATP available to activate the Ca-exporting plasma membrane ATPase, resulting in ever increasing levels of Ca in the cytoplasm. FIGURE 4 summarizes these chain of events which in unfavorable cases will lead to the death of the cell. In the favorable cases, the latter event will be delayed by the mitochondria until the cell develops the ability to deal successfully with the injuring condition.

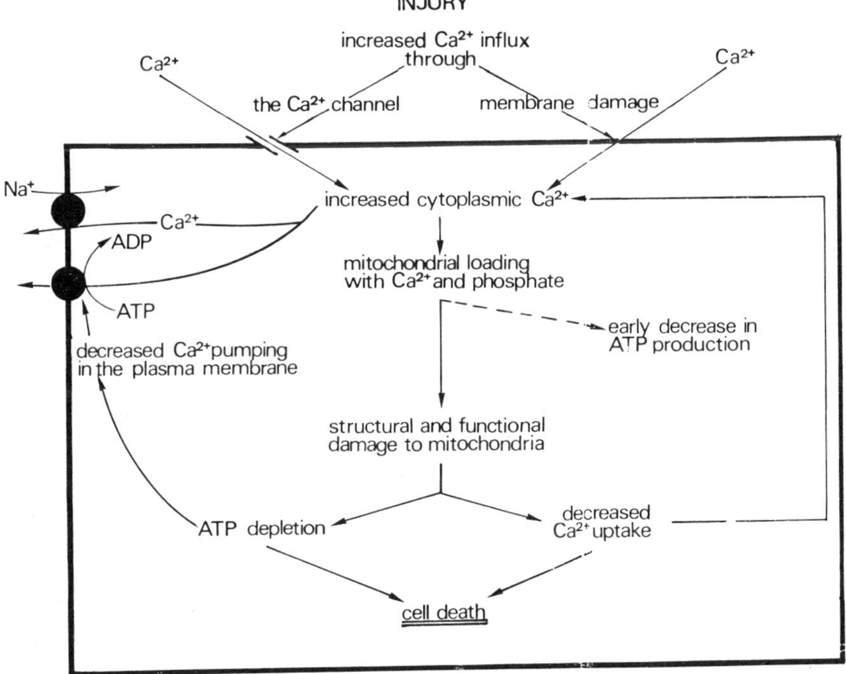

FIGURE 4. A scheme of the events following excess penetration of Ca into injured cells.

At the end of this chapter on the Ca overloading of mitochondria it is appropriate to mention same recent results on ischemic hearts that point to the importance of intramitochondrial ATP as a Ca-complexing agent.[28] That ischemia damages heart mitochondria has been known for a long time, but the novelty of the work mentioned here[28] consists in the demonstration that the drop in the level of intramitochondrial ATP induced by the condition is accompanied by the increase of the matrix free Ca. Excess intramitochondrial free Ca^{2+} is generally assumed to damage the inner membrane. Even if the mechanism of the damage is not known beyond some general statements on the ability of Ca to dehydrate the membrane, the work mentioned provides a convenient link, which could be of general significance, between noxious conditions that lead to the decrease of ATP and the impairment of mitochondrial function.

ANTI-MITOCHONDRIAL ANTIBODIES

The field of anti-mitochondrial antibodies can be traced back to the discovery in 1965[29] of a cytoplasmic antibody, which is neither organ nor species specific, in patients with primary biliary cirrhosis. The antibody was related to an antigen of mitochondrial origin,[30] but it soon became evident that the anti-mitochondrial antibodies were many and a numbering scheme from 1 to 9 (M1 to M9) is now used, the latest addition being the seventh type of anti-mitochondrial antibody, directed against the ATP/ADP translocator of the inner membrane (M7[31]). A number of methods have been used to detect the antibodies, among them immunofluorescence using unfixed cryostat sections of blocks of various human or rat tissues as antigen, complement fixation using whole or fragmented bovine heart mitochondria, and radioimmunoassays. In this overview a summary of the field will be presented: Historical reasons and the frequency of occurrence justify the particular emphasis placed on the antibodies found in patients with primary biliary cirrhosis.

Antimitochondrial antibodies are detected in the serum of a very high proportion of primary biliary cirrhosis (PBC) patients, up to about 90% using the immunofluorescence test. The antigen was originally characterized as a lipoprotein located in the inner membrane[32] and then associated by different researchers with a component loosely, and not necessarily, associated with the F_1 ATPase complex. Despite intensive studies, no definite conclusions are possible on the identity of the antigen. It can, however, be said that none of the proteins that bear known biochemical functions in mitochondria are associated with it, since none is inhibited by positive antiPBC sera. It has also been established that none of the known inhibitors of mitochondrial functions influence the antigenicity of the PBC antigen. The co-purification of the antigen with the F_1 ATPase has nevertheless directed attention to the latter and it has been tentatively proposed that the antigen may be identical with the so-called Factor B,[33] a protein that enhances the coupling between the detachable F_1 portion of the ATP synthetase and its membrane intrinsic, H^+ conducting F sector.

The PBC antibodies are not organ and species specific and have now become a standard diagnostic parameter for primary biliary cirrhosis, where they are normally detectable at titers higher than 1/100 (in less than 0.5% of normal individuals they are present at titers lower than 1/100). They are also found in a number of non-hepatic conditions such as scleroderma, Siogren's syndrome, Biermer's anemia, mixedema, anti-immune hemolytic anemia, generalized lupus erythematosus, and rheumatoid polyarthritis. In all these cases, however, the percentage of positive patients is very small (from 0.4 to 9.5%).

One point that has become clear in recent studies on the nature of the PBC antigen is that its localization is not exclusively mitochondrial. In correlating the distribution of mitochondrial markers with that of the antigen using competitive ELISA tests it has been found[31] that subcellular fractions other than mitochondria may have greater antigenic specific activity than the latter (for example kidney and heart), so that it now appears that the mitochondrial specificity may be a peculiar property of liver, which has been generally used in most of the original studies on the antigen. Similarly of interest are the reaction of PBC-positive sera with chloroplasts and the possibility (still under discussion) of reactivity with products of bacterial membranes, notably *E. coli.* The bacterial cross-reactivity, if conclusively demonstrated, would be of great general interest for the etiology of antimitochondrial antibodies in that it would suggest the possibility of foreign organisms (or of foreign toxic agents) cross-reacting with mitochondrial membranes, rendering them auto-antigenic.

The M-1 antibodies are directed against cardiolipin (diphosphatidyl glycerol), a phospholipid that is peculiarly abundant in the inner mitochondrial membrane: in fact, other eukaryotic membranes contain only traces of it. The well known antibodies of the Wassermann reaction used to diagnose syphilis are in fact anti-cardiolipin antibodies. They do not normally produce fluorescence, but complement fixation. However, the sera of some untreated cases of active secondary syphilis produce fluorescence in tissue blocks and their antibodies have thus been termed M-1F.[35]

The M-3 antibodies are found in sera of patients suffering from a pseudolupus erythematosus syndrome (PLE) induced by a drug (venocuran) once used to treat varicose veins in some countries.[36] The component of the drug responsible for the antibody induction is likely to be pyrazalone, which stimulates lymphocytes in *in vitro* tests. The antigen differs from that of PBC patients on several accounts, including the different reactivity (immunofluorescence) with various portions of the kidney tubules and the different behavior in separation experiments of mitochondrial membranes using chaotropic agents. It has been concluded that the PLE antigen is associated with the outer mitochondrial membrane.[37]

The M-4 antibody is found in sera of patients suffering from PBC and a number of other liver conditions (e.g., chronic active hepatitis).[38] Three different reaction patterns have been identified: (1) sera reacting only with the PBC antigen of the inner mitochondrial membrane; (2) sera reacting with the PBC antigen and with a trypsin-resistant outer membrane antigen; and (3) sera reacting with the PBC antigen and a trypsin-sensitive antigen present in the cytosolic fraction of fractionated tissues.

The M-5 antibody is found in the sera of patients suffering from a number of collagen disorders like lupus erythematosus, hemolytic anemia, and thrombopenic purpura.[39] It has been claimed that the antigen is associated with the outer mitochondrial membrane.

The M-6 antibody has been recently observed in the sera of some patients suffering from hepatitis induced by iproniazide treatment.[40] The mitochondrial nature of the antigen is indicated by the disappearance of the immunofluorescence (e.g., in hepatocytes) after treatment (i.e., absorption of the antibody) of the serum with liver mitochondria, but not with other liver subcellular fractions.

The recently described antibody against the mitochondrial adenine nucleotide translocator[31] is present in most PBC patients, but the interesting feature of the reaction is its organ specificity. Antibodies against the translocator isolated from liver mitochondria were detected in 13 cases (out of 13 examined) of PBC, antibodies against the translocator from heart mitochondria in 10 of the 13 cases, and antibodies against the kidney translocator in none of the 13 cases. Antibody titer (i.e., binding to the translocator) and antibody activity (i.e., inhibition of the translocation in isolated mitochondria) did not go parallel, the latter being present only in about half of the 13 PBC patients examined when liver mitochondria were used. None of the sera showed any inhibition against the transloction reaction in kidney or heart mitochondria. Also of interest is the recent observation of anti-adenine translocator antibody in the sera of 17 out of 18 patients with proven congestive cardiomyopathy.[41] The reactivity, once again, was organ specific, i.e., only visible in heart.

In closing this chapter on the antimitochondrial antibodies two general comments are in order. The first is that the antibodies apparently have no causative role in the various pathological conditions, but their frequency in some of them, particularly PBC, speaks for their relation to their etiology. The second general comment, which has already been briefly mentioned above, concerns the genesis of the antibodies. It seems probable that antigens with analogies (i.e., common epitopes) to the mitochondrial proteins may be expressed on the surface of invading microorganisms and/or that mitochondrial proteins may be modified by specific drugs to the point of becoming auto-antigenic.

MITOCHONDRIA AND TUMORS

Most of the work on the relationship between mitochondria and the malignant state carried out during the last four or five decades has been strongly influenced by the original observation by Warburg in 1930[42] that cancer cells have deficient respiratory activity and abnormally high rate of glycolysis and by his hypothesis that malignancy derives from a primary defect in cell respiration. In fact, innumerable studies carried out after Warburg's original observation have documented various deficiences of mitochondria in the cancer state. The problem is that of discriminating between changes that are secondary to defects induced by the oncogenic agent elsewhere in the cell and changes that may be primarily related to the onset of malignancy. One general difficulty, particularly in the early studies, has derived from the use of rapidly growing cancer cells, which invariably have high glycolytic activity. The difficulty has been eased somewhat by the introduction of the slowly growing Morris hepatomas, which are highly differentiated tumors, in some cases only minimally deviated from normalcy, and thus theoretically better suited to reveal a possible primary mitochondrial defect. The studies on these hepatomas have made clear that a high glycolytic rate is not a necessary marker of the malignant state, although a generally reduced number of mitochondria seems to be a consistent observation.[43] Thus, the ratio between mitochondrial and glycolytic enzymes in cancer cells seems to be shifted in favor of the latter, and this is particularly so in rapidly growing tumors. A number of differences in both the lipid and the protein portion of the mitochondrial membranes have been observed in cancer cells (see reference 43 for a comprehensive review) but their relationship to the malignant transformation remains unclear. One observation that seems to be generally accepted is that cholesterol is present in the inner membrane, which normally does not contain it. The presence of cholesterol could be responsible for the increased fragility of the membrane.

The observations on the respiratory chain in tumor cells are numerous, but it now seems clear that deficient activity, although present in some cases, is not the general rule. The same holds true for the energy-conserving systems, which seem to function normally when tested under appropriate conditions. However, a state of "loose coupling" probably reflecting the increased leakiness of the inner membrane linked to the presence of cholesterol seems to be consistently observed.

A problem that is related to the apparent preference of tumor cells for reducing pyruvate to lactate in the cytosol, rather then oxidizing it in the mitochondria, is that of the shuttle systems, which transfer the glycolytically generated reducing equivalents into the mitochondria. A once-popular view proposed that the latter (particularly the α-glycerophosphate shuttle) were deficient, so that the NADH produced by the glycolysis could not transfer its reducing equivalents to the inner mitochondrial space at a rate that could compete with lactic dehydrogenase.[44,45] This view has now lost favor, particularly after it was shown that the activity of α-glycerophosphate dehydrogenase in different strains of Ehrlich ascites cells is unrelated to the rate of lactate production. Thus, the explanation for the increased level of lactic acid in tumor cells is likely to reside elsewhere and to be related[43] to the reduced global mitochondrial activity as compared to that of glycolysis, which can indeed be normal, but is more frequently increased. As a result, the mitochondrial competition against glycolysis for

IV. Cancer Cells

G. SALVATORE, *Chair*

Poster Papers

The methyl-malonic acidemias are a group of inherited conditions (five have been identified) characterized by the accumulation of methylmalonate in blood and urine. In some patients the methylmalonyl CoA mutase was found to be practically absent from liver homogenates,[57] and in at least one patient the deficiency apparently concerned the racemase.[58] In other patients the synthesis of adenosyl cobalamine was apparently deficient, as suggested by the responsiveness of the methylmalonic acidemia to the treatment with cyanocobalamine or adenosylcobalamine, and by direct measurements of the rate of conversion of hydroxycobalamine into adenosylcobalamine in the fibroblasts of these patients.[59–62] The fibroblasts were found to be unable to oxidize propionate and methylmalonate unless high concentrations of cobalamine were added. They were, however, still able to synthesize methyl cobalamine at a normal rate, and had normal methyl malonyl CoA mutase activity in the presence of adenosyl cobalamine. The synthesis of the latter, which was clearly deficient, could be restored in the presence of ATP and a reducing system that bypassed cobalamine III reductase. In one patient, however, the synthesis of adenosyl cobalamine remained deficient also under these conditions, suggesting a defect at the level of either cobalamine III reductase or adenosyl transferase.

REFERENCES

1. Carafoli, E. & I. Roman. 1980. Molec. Aspects Med. **3:** 297–429.
2. Morgan-Hughes, J. A. 1982. *In* Skeletal Muscle Pathology. F. L. Mastaglia & J. Walton, Eds.: 309–339. Churchill-Livingstone. New York.
3. Sahgal, V., V. Subramani, R. Hughes, A. Shah & H. Singh. 1979. Acta Neuropathologica (Berlin) **46:** 177–183.
4. Hammersen, F., A. Gidlof, J. Larsson & D. H. Lewis. 1980. Acta Neuropathologica (Berlin) **49:** 35.
5. Engel, W. K. & G. G. Cunningham. 1963. Neurology **13:** 919–923.
6. Engel, A. G. 1981. *In* Disorders of Voluntary Muscle. J. Walton, Ed.: 664–711. Churchill Livingstone. New York.
7. Di Mauro, S., E. Bonilla, M. Zeviani, M. Nakagawa & D. C. DeVivo. 1985. Ann. Neurol. **17:** 609–611.
8. Sengers, R. C. A., A. M. Stadhouders & J. M. F. Trijbels. 1984. Eur. J. Pediatr. **141:** 192–207.
9. Fischer, J. C. 1985. Mitochondrial Myopathies and Respiratory Chain Defects. Ph.D. Thesis. University of Nijmegen.
10. Ogasawara, S., S. Yorifuji & Y. Nishikawa. Neurology (In press.)
11. Di Mauro, S., J. F. Nicholson, A. P. Hays, A. B. Eastwood, R. Koenigsberger & D. C. De Vivo. 1983. Ann. Neurol. **14:** 226–234.
12. Jerusalem, F., C. Angelini, A. G. Engel & R. V. Groover. 1973. Arch. Neurol. **29:** 162–169.
13. Luft, R., D. Ikkos, G. Palmieri, L. Ernster & B. Afzelius. 1962. J. Clin. Invest. **41:** 1776–1804.
14. Haydar, N. A., H. L. Conn, A. Afifi, V. Wakid, S. Ballas & K. Fanaz. 1971. Ann. Intern. Med. **74:** 548–558.
15. Edelman, N. H., T. V. Santiago, H. L. Conn. 1975. Arch. Neurol. (Chicago) **27:** 174–181.
16. Di Mauro. S. E. Bonilla, C. P. Lee, A. Schotland, A. Scarpa, H. L. Conn & B. Chance. 1976. J. Neurol. Sci. **27:** 217–232.
17. Carafoli, E. 1979. FEBS Lett. **104:** 1–5.
18. Crompton, M., M. Capano & E. Carafoli. 1976. Eur. J. Biochem. **69:** 453–462.
19. Carafoli, E. 1975. J. Molec. Cell. Cardiol. **7:** 83–89.
20. Somlyo, A. P., A. V. Somlyo, H. Shuman & M. Endo. 1982. Fed. Proc. **41:** 2883–2890.
21. Tsien, R. Y. 1981. Nature **290:** 527–528.
22. Crompton, M., E. Sigel, M. Salzmann & E. Carafoli. 1976. Eur. J. Biochem. **69:** 429–434.

23. DENTON, R. M. & J. G. MCCORMACK. 1980. Biochem. Soc. Trans. **8:** 266–268.
24. ROSSI, C. S. & A. L. LEHNINGER. 1963. Biochem. Z. **338:** 698–713.
25. SCHANNE, F. A. X., A. B. KANE, E. E. YOUNG & J. L. FARBER. Science **206:** 700–702.
26. BECKER, G. L., G. FISKUM & A. L. LEHNINGER. 1980. J. Biol. Chem. **255:** 9009–9012.
27. REYNOLDS, E. S. 1965. J. Cell Biol. **25:** 53–75.
28. TAGAWA, K., T. NISHIDA, F. WATANABE & M. KOSEKI. 1985. Molec. Physiol. **8:** 515–524.
29. WALKER, J. S., D. DONIACH, I. M. ROITT & S. SHERLOCK. 1965. Lancet **I:** 827–831.
30. BERG, P. A., D. DONIACH & I. M. ROITT. 1967. J. Exp. Med. **126:** 277–290.
31. SCHULTHEISS, H. P., P. A. BERG & M. KLINGENBERG. 1984. Clin. Exp. Immunol. **58:** 596–602.
32. BERG, P. A., I. M. ROITT, D. DONIACH & H. M. COOPER.1969. Immunology **17:** 281–293.
33. BAUM, H. & P. A. BERG. 1981. Semin. Liv. Dis. **1:** 309–321.
34. BAUM, H. & C. PALMER. 1985. Molec. Asp. Med. **8:** 201–234.
35. WRIGHT, D. J. M., D. DONIACH, M. H. LESSOF, J. L. TURK, A. S. GRIMBLE & R. D. CATTERALL. 1970. Lancet **I:** 740–744.
36. MAAS, D. & H. SCHUBOTHE. 1973. Dtsch. Med. Wochenschrift **98:** 131–139.
37. SEYERS, T. J., A. LEOUTSAKOS, P. A. BERG & H. BAUM. 1981. J. Bioenerg. Biomembr. **13:** 255–267.
38. BERG P. A., K. H. WIEDMANN, T. J. SAYERS, S. KLÖFFEL & J. LINDNER. 1980. Lancet **II:** 1329–1332.
39. LABRO, M. T., M. C. ANDRIEU, M. WEBER & J. C. HOMBERG. 1978. Clin. Exper. Immunol. **31:** 357–364.
40. HOMBERG, J. C., C. STELBY, I. ANDREIS, N. ABNAF, F. SAADOUN & J. ANDRÉ. 1982. Clin. Exp. Immunol. **47:** 93–102.
41. SCHULTHEISS, H. P., P. SCHWIMMBECK & H. D. BOLTE. 1985. Adv. Myocardiol. **6:** 311–325.
42. WARBURG, O. 1930. Metabolism of Tumors. Arnold Constable. London.
43. PEDERSEN, P. L. 1978. Proc. Exp. Tumor Res. **22:** 190–274.
44. DELBRUCH, H., H. SCHIMASSEK, K. BARTSCH & T. BUCHER. 1959. Biochem. Z. **331:** 297–311.
45. BOXER, G. E. & T. M. DEVLIN. 1961. Science **134:** 1495.
46. BUSTAMANTE, E. & P. C. PEDERSEN. 1979. Proc. Natl. Acad. Sci. USA **74:** 3735–3739.
47. ROSE, I. A. & J. V. B. WORMS. 1967. J. Biol. Chem. **242:** 1635–1645.
48. BUSTAMANTE, E. & P. L. PEDERSEN. 1980 *In* Heart Creatine Kinase, The Integration of Enzymes for Energy Distribution. W. Jacobus & J. S. Ingwall, Eds.: 147–154.
49. BUSTAMANTE, E. & P. L. PEDERSEN. 1980. Biochemistry **19:** 4972–4977.
50. TANDLER, B., R. V. P. HUTTER & R. A. ERLANDSON. 1970. Lab. Invest. **23:** 567–580.
51. TANDLER B. & F. H. SHIPKEY. 1964. J. Ultrastruct. Res. **11:** 292– .
52. SUN, C. N., H. J. WHITE & W. W. THOMPSON. 1975. Arch. Pathol. **99:** 208–214.
53. SHIH, V. E. 1978. *In* The Metabolic Basis of Inherited Disease, J. B. Stanbury, J. E. Wyngaarden & D. S. Fredrickson, Eds: 362– . MacGraw-Hill. New York.
54. KAZIRO, Y. & S. OCHOA. 1964. Adv. Enzymol. **26:** 283–305.
55. ROSENBERG, L. E. 1983. *In* The Metabolic Basis of Inherited Disease. J. B. Stanbury, J. E. Wyngaarden & D. S. Fredrickson, J. L. Goldstein, & M. S. Brown, Eds.: 474–497. MacGraw-Hill. New York.
56. HSIA, Y. E., K. J. SCULLY & L. E. ROSENBERG. 1979. Pediatr. Res. **13:** 746–751.
57. MORROW, G., L. A. BARNESS, G. L. CARDINALE, R. H. ABELES & J. G. FLAKS. 1969. Proc. Natl. Acad. Sci. USA **63:** 191.
58. KANG, E. S., P. I. SNODGRASS & P. S. GERALD. 1972. Pediat. Res. **6:** 875–879.
59. ROSENBERG, L. E., A. C. LIJEQUIST & Y. E. HSIA. 1968, Science **162:** 805–807.
60. ROSENBERG, L. E., A. C. LILJEQUIST, Y. E. HSIA & F. M. ROSENBLOOM. 1969. Biochem. Biophys. Res. Commun. **37:** 607–614.
61. MAHONEY, M. J., L. E. ROSENBERG, S. H. MUDD & B. W. UHLENDORF. 1971. Biochem. Biophys. Res. Commun. **44:** 375–381.
62. MAHONEY, M. J., A. C. HART, V. D. STEEN & L. E. ROSENBERG. 1975. Proc. Natl. Acad. Sci. USA **72:** 2799–2803.

Cytochrome Oxidase Deficiency: Clinical and Biochemical Heterogeneity[a]

SALVATORE DIMAURO,[b] MASSIMO ZEVIANI,
SERENELLA SERVIDEI,[c] EDUARDO BONILLA,
ARMAND F. MIRANDA, ALESSANDRO PRELLE,
AND ERIC A. SCHON

H. Houston Merritt Clinical Research Center
for Muscular Dystrophy and Related Diseases
College of Physicians and Surgeons
Columbia University
New York, New York 10032

In December 1984, the Biochemical Society organized a colloquium on mitochondrial myopathies, and we reviewed the already then apparent clinical and biochemical heterogeneity of cytochrome *c* oxidase (COX) deficiency.[1] During the past 15 months many new patients with COX deficiency have been reported, immunological and immunocytochemical studies have been performed on some of them, and molecular genetic investigations are under way. The new information confirms and extends the concept that COX deficiency causes diverse phenotypes.

This heterogeneity is hardly surprising when one considers the complexity of the enzyme. In mammalian tissues, including human heart (FIG. 1), thirteen different subunits can be demonstrated.[2] The three larger subunits (I–III) are encoded by mitochondrial DNA (mtDNA), while the other 10 polypeptides are encoded by nuclear DNA, synthesized on cytoplasmic ribosomes, and transported into the mitochondria where the complete enzyme is assembled as an integral component of the inner mitochondrial membrane. There is good evidence that the essential properties of the enzyme (electron transport and proton translocation) are related to the mtDNA-encoded subunits that contain the four redox centers (heme a and a_3 and two copper ions) and the proton channel.[3] The functions of the nuclear-encoded subunits have not been fully elucidated, but a regulatory role has been proposed by Kadenbach and his co-workers, who have recently shown that most if not all of these subunits occur in tissue-specific isoforms.[2]

[a]Some of the work discussed was supported by Center Grants NS 11766 from the National Institute of Neurological and Communicative Disorders and Stroke and from the Muscular Dystrophy Association. Dr. Servidei was supported by a fellowship from the Unione Italiana Lotta contro la Distrofia Muscolare (UILDM), Sezione Laziale "Giulia Testore," and Dr. Prelle by a fellowship from the Centro "Dino Ferrari," University of Milano, Italy.

[b]Address correspondence to S.D., 4-420 College of Physicians and Surgeons, 630 West 168th Street, New York, NY 10032.

[c]Present address: Clinica Neurologica, Universita' Cattolica del Sacro Cuore, Largo Gemelli 8, 00168 Roma, Italy.

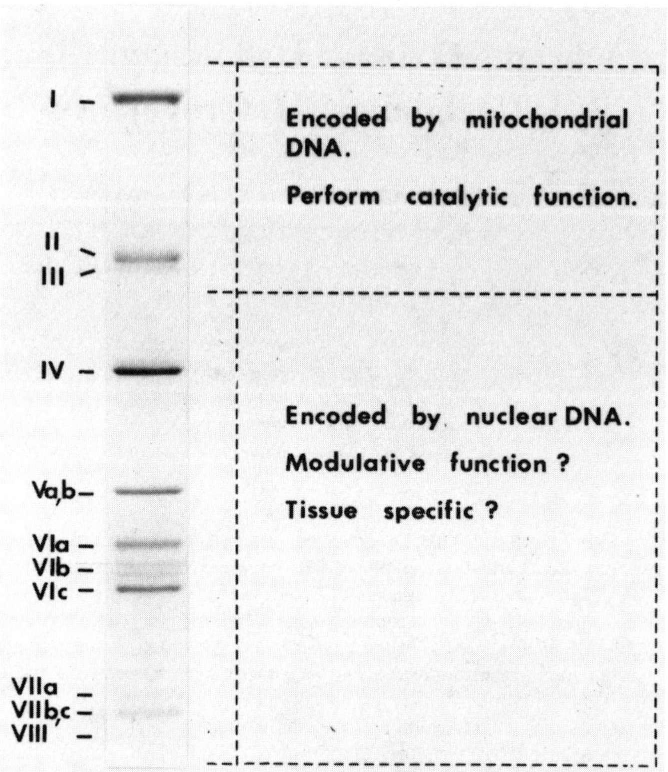

FIGURE 1. SDS-PAGE electrophoresis of COX purified from human heart.

CLINICAL HETEROGENEITY

TABLE 1 represents a tentative clinical classification of COX deficiency, based on differential tissue involvement. Three groups emerge, one in which myopathy is the main if not exclusive manifestation, another in which brain involvement predominates, and a third group comprising less well defined clinical entities associated with partial COX deficiency.

Group 1

In this group, the most common clinical picture appears to be fatal infantile myopathy with renal dysfunction: seven patients with this syndrome have been reported.[4–9] All patients but one died before 4 months of age; one survived 8 months with assisted ventilation.[9] Only two patients were overtly weak at birth[8]: in the others, hypotonia and weakness became evident in the first few weeks of life. Family history was noncontributory in three cases. In the informative families, autosomal recessive

TABLE 3. Hexokinase Activity of Liver, Regenerating Liver, and H-91 Hepatoma Cells

Homogenate	Specific Activity (units/mg protein)
Rat liver	4.6
Regenerating rat liver	10.8
H-91 hepatoma	124
+0.1 mM Glc-6-P	124
+0.6 mM Glc-6-P	124

After Bustamante & Pedersen.[48]

ADP (and phosphate) would become inefficient under conditions of high glucose utilization (i.e., high levels of glycolytic enzymes) leading to excess pyruvate production. An important point that has recently come to the forefront in this connection is the relationship of hexokinase to mitochondria. It has been shown that rapidly growing hepatomas contain much higher hexokinase activity than normal and regenerating livers and that the inhibitor glucose-6-phosphate is ineffective against the hexokinase of the rapidly growing tumors (TABLE 3). In a number of rapidly growing tumors, at variance with normal cells, a large portion of the enzyme is bound to mitochondria (TABLE 4), and the membrane-bound enzyme has higher affinity for ATP.[46–49] Recent work by Petersen and his co-workers has shown that the outer membrane protein porin may be the mitochondrial receptor for hexokinase. The association of hexokinase with mitochondria is advantageous in terms of the immediate utilization of mitochondrially synthesized ATP for the phosphorylation of glucose. These data on hexokinase are of high interest and are likely to lead to a better understanding of the complex relationship between glycolysis and respiration in tumors. Possibly, they will also lead to the identification of the mitochondrial defect most closely linked to the onset of malignancy.

The observations on the ultrastructural alterations of tumor mitochondria are also

TABLE 4. Correlation between Growth Rate and Mitochondrial Hexokinase Activity

	Growth Rate (months)	Mitochondrial Hexokinase Specific Activity
Liver	12.0	None
MH 7794A	4.5	None
MH 5123C	2.5	None
MH 44	2.0	None
MH 8995	1.7	None
MH 66	1.5	None
MH 9121F	1.0	7
MH 3924A	1.0	55
MH 9A	1.0	19
Novikoff hepatoma	0.2	280
H-91 hepatoma	0.2	330
Ehrlich cells	0.2	380
L 1210 cells	0.2	180

After Bustamante & Pedersen.[48]

very numerous, and deal with different types of alterations. A large body of information is available on the presence of inclusions in the matrix, and in the abnormal appearance of the mitochondrial profiles in the electron microscope. As already discussed in the first chapter of this overview, mitochondrial inclusions and abnormal cristae orientations are seen in many different conditions of mitochondrial injury, and thus have no specific significance to the tumor process. On a more general level, the ultrastructural literature on tumor mitochondria seems to show consistently that the organelles are smaller, contain fewer cristae, and appear somewhat swollen (however, in slowly growing hepatomas mitochondria may be normal in appearance). Again, these alterations are hardly of a specific type and are unlikely to be primarily related to the oncogenic agent. The situation may be different in some rare tumors of salivary glands, Whartin's tumor, and oncocytoma. Both tumors are composed of large

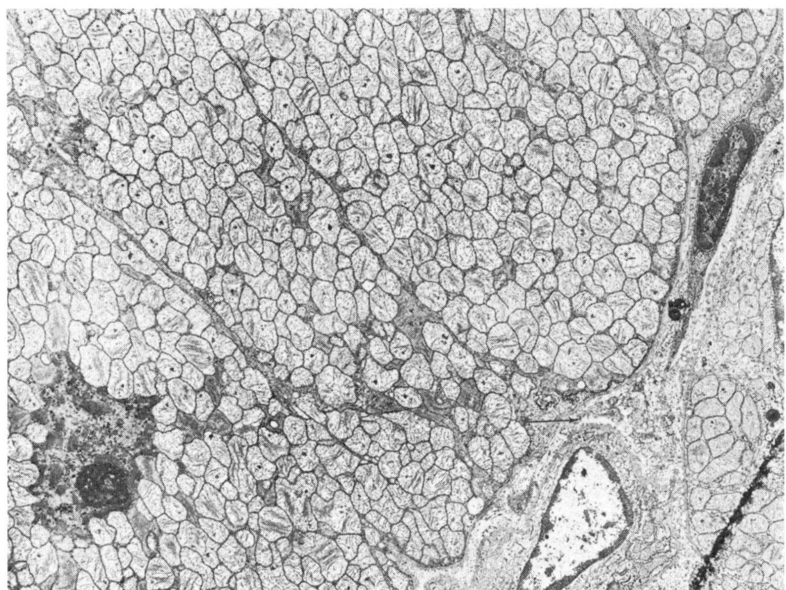

FIGURE 5. Mitochondria pack the entire cytoplasm of oncocytes. They frequently show centrally located stacks of cristae. (From C.N. Sun *et al.*[52] With permission from *Archives of Pathology.*)

epithelial cells (the oncocytes), which appear strongly granular and acidophilic in the light microscope. This is due to the fact that their body is literally filled with mitochondria (FIGURE 5), which are so densely packed that hardly anything else is visible in the cytoplasm.[50,51] Although some of the mitochondria may show signs of degeneration, most appear normal, at least in the typical oncocytes. Oncocytomas and Whartin's tumor, then, are apparently different from other tumors where a less developed (and structurally deficient) mitochondrial system seems to be the rule. However, with few exceptions they are benign tumors and oncocytes not very different from those described in FIGURE 5 are sometimes encountered in normal salivary glands. It must also be noted that not all oncocytes have normal looking mitochondria: in some the cristae appear stacked in the center of the organelle or pushed to its periphery by large deposits of glycogen. It is clear however, that the ultrastructural findings on mitochondria dominate the picture in these cells, and it thus seems permissible to consider them directly (and primarily) involved in their transformation. In fact, it has been speculated that in oncocytes excess mitochondria would accumulate in response to a putative primitive mitochondrial defect.[52] Oncocytomas would thus be "intracellular" tumors, characterized by the uncontrolled proliferation of mitochondria.[54]

DEFECTS OF MISCELLANEOUS MITOCHONDRIAL ENZYMES UNRELATED TO THE MAIN ENERGY CONSERVATION SYSTEM

Of the many ancillary systems present in mitochondria, two, the urea cycle and the propionate metabolizing pathway, appear to be of particular importance of the subject matter of this overview since inborn defects of their enzymes have been repeatedly documented. They will thus be succinctly discussed in this closing chapter.

The Urea Cycle

In mammals, urea constitutes the largest portion of the nitrogen excreted in the urine. Urea is synthesized in a cycle of reactions that is compartmentalized between the matrix of mitochondria and the cytoplasm. Of its five steps, two are mitochondrial: carbamoyl-phosphate synthetase, which catalyzes the ATP-dependent condensation of CO_2 and NH_3 into carbamoyl-phosphate ($H_2N\text{-}CO\text{-}PO_3H_2$), and ornithine trans-carbamylase, which synthesizes citrulline from ornithine and carbamoyl phosphate. Although a limited urea cycle activity is found also in brain and kidney cells, the tissue responsible for most of the urea formation is liver: in fact, the urea synthesis in kidney and brain is dependent on the import of carbamoyl phosphate into kidney and brain mitochondria from the blood, since these mitochondrial types do not contain carbamoyl phosphate synthetase. The cycle collects into the molecule of urea two amino groups, one derived from aspartate in the cytoplasm, the other from the oxidative deamination of glutamate by glutamate dehydrogenase in the mitochondrial matrix. Glutamate in turn carries the amino group donated to it by the transamination of 11 of the 20 amino acids derived from the catabolism of proteins. The transaminations take place in the cytosol, and the resulting glutamate is then transported into mitochondria on a specific carrier in exchange for OH^-. The raison d'etre of the cycle is naturally the protection of tissues from the toxic action of ammonia. Since the latter is particularly harmful to the nervous system, the predominant nervous symptomatology of hyperammonenias deriving from the malfunction of the cycle is easily understood.

The mitochondrial carbamoyl-phosphate synthetase, unlike the cytoplasmic isozyme, requires acetyl-glutamate as an allosteric activator that facilitates the binding of the two ATP molecules used in the reaction. Acetyl-glutamate is synthesized by acetyl-glutamate synthetase from acetyl CoA and glutamate. Since the enzyme, which is specifically activated by arginine, is located in mitochondria, it is also frequently considered as one of the enzymes of the urea cycle.

Inborn defects have been described in both carbamoyl phosphate synthetase and ornithine trans-carbamylase (a few dozen patients, see reference 53 for a recent review). The prevailing symptoms are nervous, among them vomiting, lethargy, coma, severe mental retardation in the more protracted cases, and are initiated and/or increased by protein feeding. Hyperammonemia has been the rule in both defects and very low levels of carbamoyl phosphate synthetase and ornithine transcarbamylase, respectively, have been found in liver.

Propionate, Methylmalonate, and Cobalamine Metabolism

Propionyl CoA deriving from the β-oxidation of fatty acids with odd carbon atoms numbers is transformed into succinyl CoA through a pathway in which methylmalonyl CoA is an intermediate (FIGURE 6). Propionyl CoA also arises from the catabolism of valine, isoleucine, methionine, and threonine, and from the cleavage of the side chain of cholesterol. Of the three enzymatic steps of the pathway, the first requires biotin, Mg^{2+}, and ATP, and produces only the D-isomer of methylmalonyl CoA. After racemization in the second step of the pathway, methyl-malonyl CoA is isomerized to succinyl CoA in the third reaction, which requires adenosyl cobalamine.[56] Adenosyl cobalamine itself is synthesized from cobalamine through a series of reactions that are

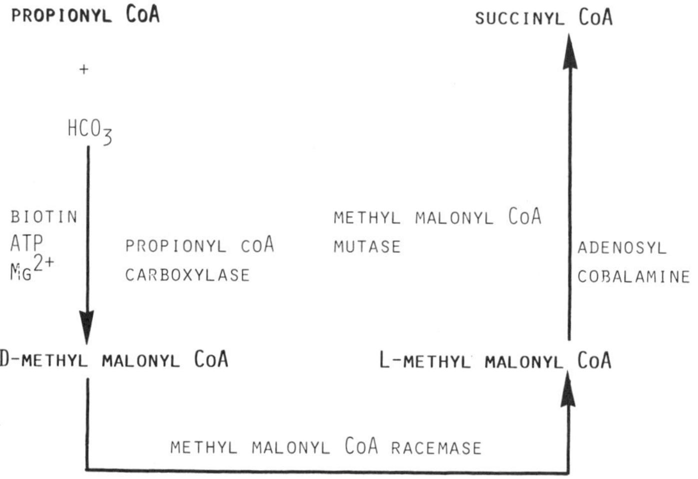

FIGURE 6. Propionate metabolism in mitochondria.

also located in mitochondria, two reductases and a transferase that converts cobalamine I into adenosyl cobalamine.[54]

Inborn errors have been described in the three enzymes of the propionyl CoA-succinyl CoA pathway proper, and in the three enzymes of the sequence of reactions that lead to the synthesis of adenosyl cobalamine.[55] In addition, one defect involves ketothiolase, the enzyme that produces propionyl CoA from L-methylacetoacetyl CoA, an intermediate in the isoleucine catabolism pathway. The defect has not been conclusively documented with enzyme activity measurements, but is strongly suggested by the state of ketoacidosis, with excretion in the urine of large quantities of metabolites of the isoleucine-propionyl CoA route. Propionyl CoA carboxylase has been suggested to be deficient in several children who suffer from ketoacidosis and severe nervous symptoms, and accumulate in the blood and in urine propionate and intermediates of the catabolism routes of leucine, valine, methionine, and threonine. In one case[56] the carboxylase defect was demonstrated directly in the leukocytes of the patient.

- HOECHST PHARMA AG, SWITZERLAND
- ITALIAN NATIONAL RESEARCH—C.N.R. ITALIA
- MERCK, SHARP & DOHME RESEARCH LABORATORIES, ITALIA
- MERCK, SHARP & DOHME RESEARCH LABORATORIES, U.S.A.
- SANDOZ, U.S.A.
- SCHERING AG, F.R.G.
- SCHERING-PLOUGH CORPORATION
- SMITH KLINE & FRENCH LABORATORIES, U.S.A.
- SQUIBB CORPORATION
- S.P.A.—ITALIA
- STUART PHARMACEUTICALS/DIVISION OF ICI AMERICAS, INC.

Financial assistance was received from:

- FARMITALIA—CARLO ERBA
- FIDIA FARMACEUTICI, ITALIA
- SIGMA-TAU FARMACEUTICI, ITALIA
- SQUIBB—ITALIA

ADDITIONAL CONTRIBUTIONS HAVE BEEN RECEIVED FROM

- BAYER AG/MILES, U.S.A.
- BRISTOL MYERS, U.S.A.
- DEPARTMENT OF THE NAVY—OFFICE OF NAVAL RESEARCH (GRANT N00014-86-G-0077)
- EASTMAN KODAK CORPORATION
- GÖDECKE AG, F.R.G.

ANNALS OF THE NEW YORK ACADEMY OF SCIENCES
Volume 488

MEMBRANE PATHOLOGY

Edited by G. Bianchi, E. Carafoli, and A. Scarpa

The New York Academy of Sciences
New York, New York
1986

Library of Congress Cataloging-in-Publication Data

Membrane pathology.

(Annals of the New York Academy of Sciences, ISSN 0077-8923 ; v. 488)
Result of a conference entitled Conference on Biological Membrane Pathology, held in Como, Italy, May 19–22, 1986, and sponsored by the New York Academy of Sciences.
Includes bibliographies and index.
1. Cell membranes—Abnormalities—Congresses. 2. Mitochondrial membranes—Abnormalities—Congresses. 3. Diabetes—Congresses. 4. Hypertension—Congresses. 5. Pathology, Cellular—Congresses. I. Bianchi, G. II. Carafoli, Ernesto. III. Scarpa, Antonio, 1942– . IV. New York Academy of Sciences. V. Conference on Biological Membrane Pathology (1986 : Como, Italy) VI. Series. [DNLM: 1. Cell Membrane—pathology—congresses. W1 AN626YL v.488/ QH 601 M53254 1986]
Q11.N5 vol. 488 500 s 87-1652
[RB152] [616.07]

(*Cover:* Watercolor of the Villa Olmo, Como, Italy by L. Orci.)

SP
Printed in the United States of America
ISBN 0-89766-371-3 (cloth)
ISBN 0-89766-372-1 (paper)
ISSN 0077-8923

inheritance appeared likely because twin sisters of one patient[4] and a brother of another[7] died in infancy with similar symptoms, and two patients were siblings.[8]

Myopathy without renal dysfunction was seen in two children: one died of respiratory failure at 6 weeks of age,[10] the other died at age 7 months after 5 months of supported ventilation.[11]

The association of myopathy and cardiopathy was described in two patients.[12,13] We have studied a third patient, a girl with severe congenital weakness, cardiopathy, and lactic acidosis.[14] Although her strength seemed to increase gradually and the lactic acidosis improved, she died of heart failure at 7 months.

Myopathy and liver disease were reported in the same family, but not in the same patient: two siblings had severe myopathy and respiratory insufficiency without liver dysfunction, while their second cousin died of hepatic failure at 9 months of age.[15]

The clinical and laboratory features of the patients described above are summarized in TABLE 2: the unifying elements are the COX-deficient myopathy and the fatal outcome in infancy.

Three unrelated children also had life-threatening myopathy in infancy, but improved spontaneously[16–18]: The two older patients[16,17] are now virtually normal. This benign infantile mitochondrial myopathy has been ascribed to a reversible defect of muscle COX in two patients[17,18]; in one patient,[16] the enzyme defect was not documented but was strongly suggested by the similarities of clinical course and muscle biopsy (TABLE 3).

Group 2

Of the second group of disorders, dominated by involvement of the central nervous system (TABLE 1), subacute necrotizing encephalomyelopathy (Leigh's syndrome) is the most common. Leigh's syndrome is characterized by respiratory abnormalities, weak cry, poor feeding, impaired vision and hearing, ataxia, weakness and hypotonia, intellectual deterioration, and seizures. Onset may be in early infancy, but some patients develop normally for several months. The characteristic neuropathological lesions are focal, bilateral, and symmetrical necrotic lesions extending from the thalamus to the pons and involving the inferior olives and the posterior columns of the spinal cord. Microscopically, these "spongiform" lesions show demyelination, vascular

TABLE 1. Cytochrome Oxidase Deficiency: Clinical Presentations

Syndromes dominated by muscle involvement
Fatal infantile myopathy
Fatal infantile myopathy and renal dysfunction (De Toni-Fanconi-Debre syndrome)
Myopathy and cardiopathy
Myopathy and liver disease in second cousins
Benign infantile mitochondrial myopathy (MLG disease)
Syndromes dominated by brain involvement
Subacute necrotizing encephalomyelopathy (Leigh's syndrome)
Progressive sclerosing poliodystrophy of childhood (Alpers' disease)
Trichopoliodystrophy (Menkes' disease)
Other
Encephalomyopathies of adult onset
Progressive external ophthalmoplegia (?)

TABLE 2. Clinical and Laboratory Features of 12 Patients with Fatal Infantile COX Deficiency of Muscle[a]

Features	Patient Number											
	1	2	3	4	5[b]	6[b]	7	8	9	10[c]	11	12
Reference	4	5	6	7	8	8	9	10	11	15	12	13
Sex	M	M	F	M	F	F	F	M	M	F	F	F
Age at admission	7 wk	4 wk	8 wk	6 wk	Birth	Birth	3 mo	Birth	2 mo	3 mo	5 mo	4 mo
Age at death	13 wk	14 wk	16 wk	15 wk	7 wk	7 wk	8 mo	6 wk	7 mo	4½ mo	5 mo	8 mo
Generalized weakness	+	+	+	+	+	+	+	+	+	+	+	+
Hypotonia	+	+	+	+	+	+	+	+	+	+	+	+
Hyporeflexia	+	+	+	+	+	+	+	+	+	+	−	+
Respiratory failure	+	+	+	+	+	+	+	+	+	+	+	?
De Toni-Fanconi-Debre syndrome	+	+	+	+	+	+	+	−	−	−	−	−
Cardiopathy	−	−	−	−	−	−	−	−	−	−	+	+
Lactic acidosis	+	+	+	+	+	+	?	+	+	+	+	+
Family history	+	−	−	+	+	+	−	+	−	+	+	+

[a]After DiMauro *et al.*[42]
[b]Siblings.
[c]A second cousin had cytochrome *c* oxidase deficiency of the liver.

proliferation, and astrocytosis. COX deficiency was reported in four patients,[19–21] and we have confirmed it in five more.[22] However, it is important to keep in mind that COX deficiency is only one of the biochemical causes of Leigh's syndrome. Disorders of pyruvate metabolism (pyruvate dehydrogenase or pyruvate carboxylase deficiencies) have been documented in some patients, and in others the biochemical defect remains unknown.

Another severe brain disorder of infancy or childhood with "spongy degeneration" of the nervous system is progressive sclerosing poliodystrophy, or Alpers' disease.[23]

TABLE 3. Clinical and Laboratory Features of 3 Patients with Benign Infantile Myopathy due to Reversible COX Deficiency[a]

Features	Patient Number (Ref.)		
	1 (16)	2 (17)	3 (18)
Sex, age at first admission	F; 6 wk	M, 2 wk	M; 6 wk
Feeding difficulties, gavage	+	+	+
Weak cry	+	+	+
Respiratory difficulties	+	+	+
Assisted ventilation	+	−	+
Diffuse weakness	+	+	+
Hypotonia	+	+	+
Hypo- or areflexia	+	+	+
Macroglossia	+	+	−
Hepatomegaly	+	−	+
Spontaneous improvement	+	+	+
Mild residual weakness (age)	22 mo	33 mo	?
Lactic acidosis	?	+	+
Lactic acidosis resolving at age	?	12 mo	6 mo
Family history	−	−	−
Muscle biopsy	7 wk: MLG (most ff) 22 mo: MLG (4% ff)	1 mo: MLG (mild, most ff) 7 mo: MLG (severe, 50%) 23 mo: normal	4 mo: MLG (most ff) 11 mo: MLG (few ff) + fibrosis

[a]The enzyme defect is only presumptive in case 1. MLG = mitochondria, lipid, and glycogen accumulation.

COX deficiency of muscle was described in one patient with this condition.[23] Trichopoliodystrophy (Menkes' disease) is an X-linked recessive disorder characterized by early infantile onset of seizures, developmental regression, hair abnormalities (pili torti), tortuous arteries, fragile bones, hypopigmentation, and temperature instability. These diverse symptoms and signs have been attributed to secondary deficiencies of copper-dependent enzymes, one of which is COX. The primary defect involves intestinal transport of copper, as reflected by low levels of serum copper and ceruloplasmin. A defect of COX was reported in muscle, brain, and liver mitochondria in one patient,[24] but this may vary, because we found normal COX activity in muscle from two other patients.

Group 3

The third group in TABLE 1 includes clinically heterogeneous disorders apparently due to partial COX deficiency. We have studied three adults with slowly progressive symptoms and signs of muscle and central nervous system involvement ("mitochondrial encephalomyopathy"), in whom muscle COX activity was about half normal while other mitochondrial enzymes were spared.[25,26] Two of these patients also had signs of peripheral axonal neuropathy.[25]

Complete lack of histochemically demonstrable COX in individual fibers was reported in 15 patients with progressive external ophthalmoplegia and mitochondrial myopathy.[27,28] Three had the characteristic symptoms and signs of the Kearns-Sayre syndrome[29]; the others had only ophthalmoplegia and weakness. Biochemical studies showed a partial defect of COX activity and of cytochrome aa_3 concentration only in one patient with Kearns-Sayre syndrome. The pathogenic significance of this observation is uncertain because scattered COX-deficient fibers can be seen in patients with morphological abnormalities of muscle mitochondria ("ragged-red fibers") with or without ophthalmoplegia,[26] and also in patients with clearly defined respiratory chain defects, such as complex I deficiency.[30]

BIOCHEMICAL HETEROGENEITY

Differential Tissue Involvement

The existence of multiple tissue-specific isozymes of COX, proposed by Kadenbach and co-workers several years ago[31,32] has been confirmed by data from human pathology. In agreement with clinical and pathological observations in cases with fatal infantile myopathy, COX deficiency was confined to skeletal muscle, sparing heart, liver, and brain.[4,5,7,11] Accordingly, reduced-minus-oxidized cytochrome spectra showed that the cytochrome aa_3 peak was lacking in mitochondria isolated from muscle,[4,5,7] but present in heart mitochondria.[11] A partial defect of COX in kidney was demonstrated by both enzyme assay[5,7] and spectral analysis[7] in two patients with myopathy and renal dysfunction. Of the three cases with myopathy and cardiopathy, the heart was studied in two: COX activity was, surprisingly, normal in one[12] but markedly decreased (12% of normal) in the other.[14]

In Leigh's syndrome, COX deficiency appears to be generalized, including cultured fibroblasts, which may provide a useful tool for prenatal diagnosis, but the liver was spared in two of seven patients.[19,22]

Differential tissue involvement can also be demonstrated in muscle biopsies by histochemistry. In patients with fatal infantile myopathy we found that the enzyme stain was virtually absent in extrafusal fibers, but present in intrafusal fibers of the muscle spindle and in smooth muscle cells of blood vessels.[9] Similar observations were made in early biopsies from children with benign infantile myopathy (FIG. 2).

Reversibility of COX Deficiency

The spontaneous clinical improvement in the benign infantile myopathy correlates with a return of COX activity in muscle, which can be demonstrated histochemically (FIG. 3) and biochemically (FIG. 4). In one of the two children so studied, less than 5% of the fibers stained for COX in a first biopsy taken at 1 month of age, about 60% in a second biopsy at 7 months of age, and all the fibers stained intensely in a third biopsy

obtained at age 3 years[17] (FIG. 3). Accordingly, COX activity increased from 8 to 47 to 175% of the normal mean in the three biopsies (FIG. 4). In the other patient[18] the return of muscle COX activity seemed slower: COX increased from 11% of the normal mean at 4 months to 57% at 11 months. The apparently delayed biochemical recovery in this child could explain the slower pace of clinical improvement and the presence of fibrosis in the second biopsy.[18]

The basis for the reversibility of the enzyme defect is unknown, but two hypotheses can be considered. (1) If the genetic error affects mtDNA, the return of the enzyme could be ascribed to a gradual "selection" of fibers containing predominantly wild-type mitochondria over those that contain a majority of mutant mitochondria and are presumably not viable.[17] In favor of this hypothesis is the apparently "all-or-none"

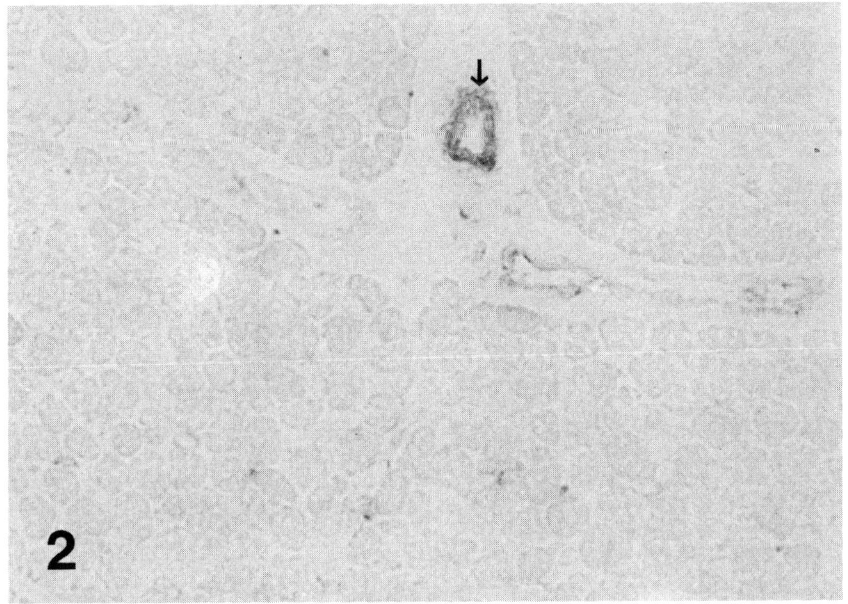

FIGURE 2. Histochemical reaction for COX activity in a frozen section of muscle from a patient with the benign form of COX deficiency. There is no stain in muscle fibers but activity can be seen in the walls of an intramuscular arteriole (arrow). ×250. Figure reduction, 90%.

involvement of individual fibers, with increasing number of fibers becoming reactive rather than all fibers showing a gradual increase of COX reactivity (FIG. 3). There is also a "precedent" for this mechanism in human pathology: bone marrow cells of patients with abnormal chloramphenicol (CAF) sensitivity who have recovered from CAF-induced anemia may become less sensitive than normal cells to the effects of the drug *in vitro*. It was suggested that the initial exposure may have destroyed those cells that contain predominantly abnormally CAF-sensitive mitochondria leaving behind cells that contain predominantly or exclusively normal mitochondria.[33] Against this hypothesis of a mtDNA mutation is the lack of any evidence of maternal transmission in the families of the three patients. In addition, the exclusive involvement of muscle seems incompatible with the general assumption that mtDNA is identical in all cells of

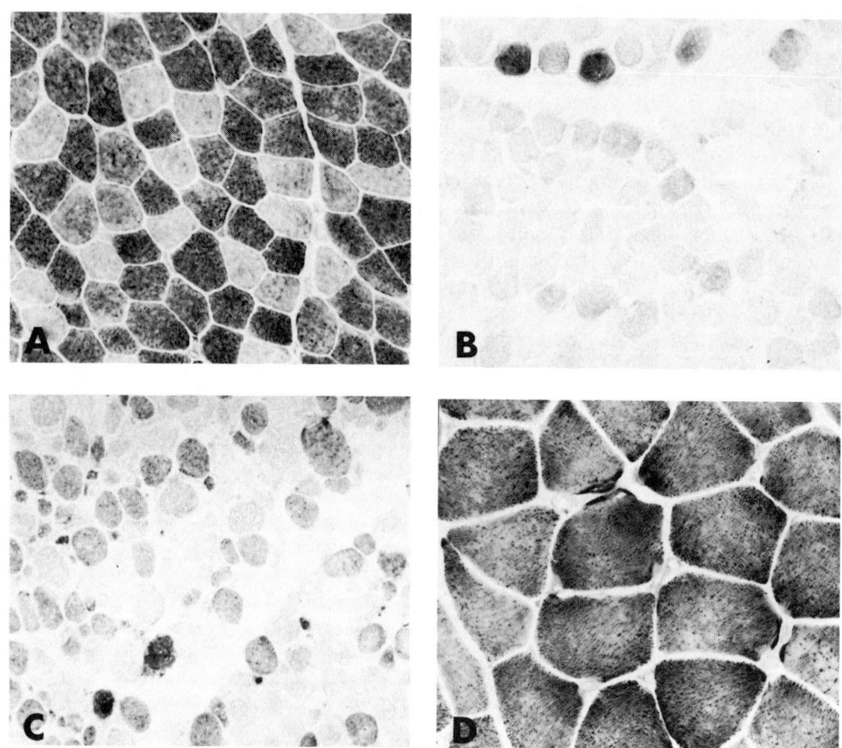

FIGURE 3. Histochemical reaction for COX activity in muscle of a patient with benign COX deficiency (B, C, D) compared with muscle from a normal infant (A). At 1 month of age (B), only a few fibers are reactive for COX. The proportion of reactive fibers is increased at 7 months of age (C). At 36 months of age (D), the staining appears slightly increased in intensity compared with that in normal muscle (A). ×400. Figure reduction, 65%. (From DiMauro *et al.*[17] With permission from the Little Brown & Co.).

an organism.[34] (2) If the mutation involves nuclear DNA, it could affect a subunit of COX that is not only muscle-specific but developmentally regulated. Expression of immunologically different isoproteins during development was demonstrated for multiple COX subunits by comparing fetal and adult rat tissues.[2] There is also evidence that developmentally controlled COX isozymes occur in human tissues: a monoclonal antibody against subunit IV of human heart COX reacted differently with mature as compared to immature cultured or regenerating human muscle.[35] COX deficiencies due to mutations of genes encoding "fetal" or "neonatal" muscle COX isozymes would be spontaneously corrected when the "mature" isozyme begins to be expressed. The reverse situation has been amply documented in human muscle pathology: mutations of "mature" muscle-specific isozymes, such as phosphorylase or phosphoglycerate mutase, are not expressed in immature cultured or regenerating muscle.[36] For these cytosolic enzymes, the transition from "fetal" to mature isozyme pattern occurs well before birth, around the fourth month of gestation. To explain the return of COX in children with benign infantile myopathy, one would have to assume that the transition from fetal to mature isozyme pattern for COX and, perhaps, other mitochondrial enzymes occurs later than for cytosolic enzymes, around the neonatal period. Another possibility is that, in addition to the fetal and adult isoforms, there may be a temporary, intermediate, "neonatal" isozyme of COX, as shown for myosin.[37] These questions will be answered only when we have a better understanding of the developmental control of COX.

Immunological Studies

To investigate whether COX deficiency was due to lack of enzyme protein or to the presence of immunologically reactive but enzymatically inactive protein, we have conducted immunological studies in muscle biopsies and postmortem tissues from patients with different clinical forms of COX deficiency. Polyclonal antibodies obtained against the COX holoenzyme purified from normal human heart[11] were used for immunocytochemistry and immunotitration by enzyme-linked immunosorbent assay (ELISA). The biochemical heterogeneity of COX deficiency extended to the presence or absence of immunologically cross-reacting material (CRM). Patients with fatal infantile myopathy, with or without renal dysfunction, showed markedly decreased CRM in muscle extracts or in isolated mitochondria by ELISA.[9,11] This was confirmed by immunocytochemistry of frozen sections in which the immunoperoxidase reaction was evaluated semiquantitatively with a microscope photometer[11] The intrafusal fibers of the muscle spindle, which showed normal histochemical reaction for COX, were also positive by immunocytochemistry.

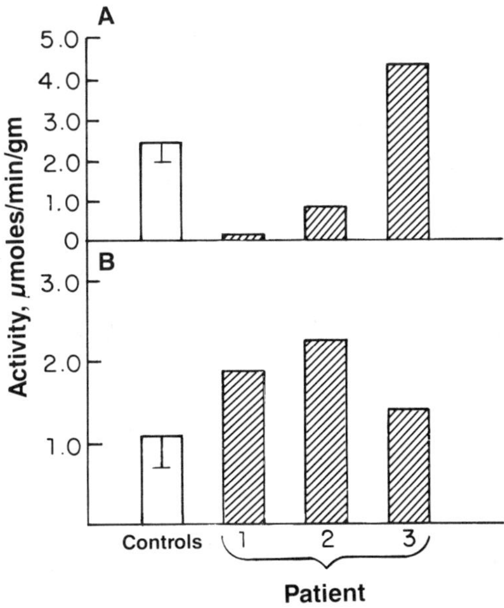

FIGURE 4. Schematic representation of COX activity (A) and succinate-cytochrome *c* reductase activity (B) in three muscle biopsies obtained at 1 month (1), 7 months (2), and 36 months (3) of age from a patient with benign COX deficiency. Normal values are the mean of 31 controls for COX and 21 for succinate-cytochrome *c* reductase; bars represent S.D.

In contrast, children with benign infantile myopathy due to reversible COX deficiency have normal amount of CRM in muscle, even in biopsies taken early in the course of the disease and showing very little COX activity. This was documented by immunocytochemistry in one case (FIG. 5), and by both immunocytochemistry and immunotitration in another.[18] The presence of CRM in muscle biopsies of infants with the benign myopathy may have important prognostic value: in one of our patients, the presence of normal CRM in a biopsy obtained at 4 months of age, when he was very sick and not yet clearly improving, encouraged us to continue assisted ventilation.[18]

In Leigh's syndrome, we found normal or only moderately decreased CRM in all organs studied, including brain, muscle, liver, and kidney.[22]

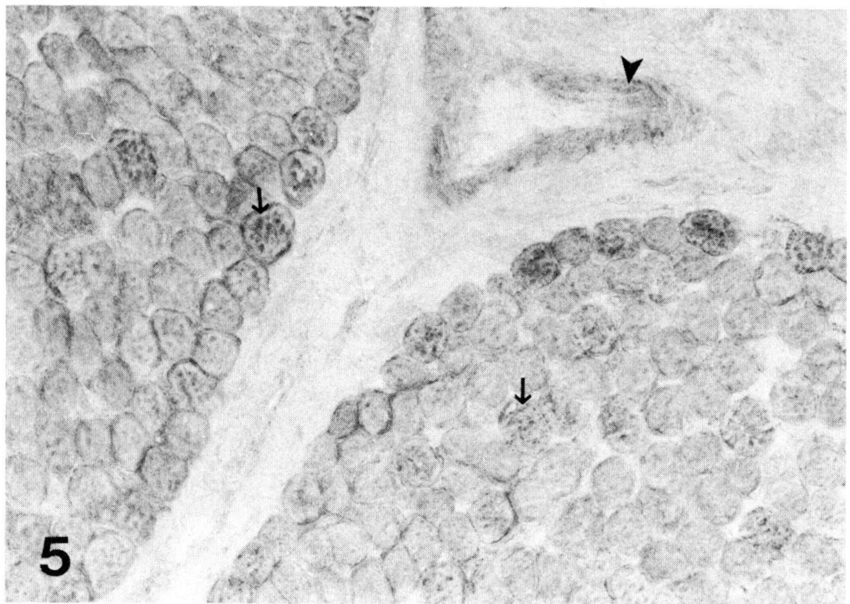

FIGURE 5. Binding of antibodies against human heart COX in a frozen section of muscle from a patient with benign COX deficiency using immunoperoxidase stain. There is binding of antibodies in most muscle fibers (arrows) and in the arterial walls (arrow heads). ×250. Figure reduction, 90%.

In one adult patient with encephalomyopathy and partial muscle COX deficiency, there was a clear decrease of CRM by immunocytochemical stain and by ELISA.[26]

In agreement with data of Muller-Hocker *et al.*,[38] we also found decreased immunoreactivity in the isolated histochemically COX-negative fibers seen in patients with mitochondrial myopathies and ragged-red fibers. Different results were obtained by Johnson *et al.*[39] who, using antibodies against rat COX, reported that all histochemically COX-negative fibers in their patients with progressive ophthalmoplegia were positive by immunocytochemistry. This apparent discrepancy may simply reflect the difficulty of assessing the immunocytochemical reaction quantitatively.

ANNALS OF THE NEW YORK ACADEMY OF SCIENCES

Volume 488
December 31, 1986

MEMBRANE PATHOLOGY[a]

Editors and Conference Organizers
G. BIANCHI, E. CARAFOLI, AND A. SCARPA

CONTENTS

[a]This volume is the result of a conference entitled Conference on Biological Membrane Pathology, held in Como, Italy on May 19–22, 1986 and sponsored by the New York Academy of Sciences.

II. Hypertension

A. P. SOMLYO, *Chair*

II. Diabetes

A. I. WINEGARD, *Chair*

Subunit Composition of Mutant Enzymes

Immunoenzymatic stain (Western blot) of SDS-polyacrylamide gel electropherograms of purified human COX showed that, as expected from data in the literature,[32] all 13 subunits reacted with antiserum to the holoenzyme. We used this technique to investigate the possibility that COX deficiency in some patients might be due to lack of one or more subunits. In two patients with Leigh's syndrome, Western blots of brain mitochondrial extracts showed normal patterns and apparently normal amounts of all subunits.[22]

The same technique applied to muscle mitochondria from a child with fatal infantile myopathy failed to show detectable bands, in keeping with the marked decrease of CRM. However, immunoprecipitation of the same mitochondrial extracts followed by SDS-polyacrylamide gel electrophoresis showed a faint but normal subunit pattern.[11] These results do not exclude defective synthesis of one or more muscle-specific subunits, because low concentrations of individual subunits may well impair the assembly of the holoenzyme.

MOLECULAR GENETICS

The clinical and biochemical heterogeneity of COX deficiency illustrated above suggests that different molecular defects may be involved. We have begun a long-range project aimed at elucidating the different genetic defects at the molecular level using the tools of recombinant DNA technology.

We decided to restrict our investigation to the nuclear-encoded subunits for three reasons: The sequence of human mtDNA is already known, including the genes for COX I–III,[40] there is no evidence so far of maternal inheritance for any of the known COX deficiencies, and conversely, the selective involvement of one or more tissues suggests that some of the mutations may affect nuclear-encoded subunits that confer tissue specificity to COX.[2] The general strategy has been to use antibodies against individual subunits to screen cDNA expression libraries from human liver. Besides monoclonal antibodies against subunit IV,[35] we have obtained polyclonal antibodies against subunit V and VII extracted from polyacrylamide gels. Using these antibodies, we have already isolated and sequenced a full-length cDNA clone encoding subunit IV[41] and have also isolated a clone encoding subunit Vb. Utilizing these and other nuclear genes as probes, we will attempt to identify the molecular lesions. Initially, this will be done by a combination of Southern and Northern analysis in conjunction with *in situ* hybridization, but ultimately we may have to identify the mutant genes by cloning and sequencing them directly.

We anticipate that these studies will not only contribute to a better understanding of pathogenic mechanisms, but will also help understand basic biological processes, such as the molecular basis of tissue specificity and the coordination between mitochondrial and nuclear gene expression.

ACKNOWLEDGMENT

We are grateful to Ms. Mary Tortorelis for typing the manuscript.

REFERENCES

1. DiMauro, S., M. Zeviani, E. Bonilla, N. Bresolin, M. Nakagawa, A. F. Miranda & M. Moggio. 1985. Cytochrome c oxidase deficiency. Biochem. Soc. Trans. **13**(4): 651–653.
2. Kuhn-Nentwig, L. & B. Kadenbach. 1985. Isolation and properties of cytochrome c oxidase from rat liver and quantification of immunological differences between isozymes from various rat tissues with subunit-specific antisera. Eur. J. Biochem. **149:** 147–158.
3. Capaldi, R. A., F. Malatesta & V. M. Darley-Usmar. 1983. Structure of cytochrome c oxidase. Biochim. Biophys. Acta **726:** 135–148.
4. Van Biervliet, J. P. A. M., L. Bruinvis, D. Ketting, P. K. DeBree, E. V. Heiden, S. K. Wadman, J. L. Willems, H. Bookelman, U. V. Haelst & L. A. Monnens. 1977. Hereditary mitochondrial myopathy with lactic acidemia, a DeToni-Fanconi-Debre syndrome, and a defective respiratory chain in voluntary striated muscles. Pediat. Res. **11:** 1088–1093.
5. DiMauro, S., J. R. Mendell, A. Sahenk, D. Bachman, A. Scarpa, R. M. Scofield & C. Reiner. 1980. Fatal infantile mitochondrial myopathy and renal dysfunction due to cytochrome c oxidase deficiency. Neurology **30:** 795–804.
6. Heiman-Patterson, T. D., E. Bonilla, S. DiMauro, J. Foreman & D. L. Schotland. 1982. Cytochrome c oxidase deficiency in a floppy infant. Neurology **32:** 898–900.
7. Minchom, P. E., R. L. Dormer, I. A. Hughes, D. Stansbie, A. R. Cross, G. A. F. Hendry, O. T. G. Jones, M. A. Johnson, H. S. A. Sherratt & D. M. Turnbull. 1983. Fatal infantile mitochondrial myopathy due to cytochrome c oxidase deficiency. J. Neurol. Sci. **60:** 453–463.
8. Muller-Hocker, J., D. Pongratz, T. Deufel, J. M. F. Trijbels, W. Endres & G. Hubner. 1983. Fatal lipid storage myopathy with deficiency of cytochrome c oxidase and carnitine. Virchows Arch. **399:** 11–23.
9. Zeviani, M., I. Nonaka, E. Bonilla, E. Okino, M. Moggio, S. Jones & S. DiMauro. 1985. Fatal infantile mitochondrial myopathy and renal dysfunction caused by cytochrome c oxidase deficiency: Immunological studies in a new patient. Ann. Neurol. **17:** 414–417.
10. Trijbels, F., R. Sengers, L. Monnens, A. Janssen, T. Willem, H. Terlaak & H. Stadhouders. 1983. A patient with lactic acidaemia and cytochrome oxidase deficiency. J. Inher. Metab. Dis. **6** (Suppl. 2): 127–128.
11. Bresolin, N., M. Zeviani, E. Bonilla, R. H. Miller, R. W. Leech, S. Shanske, M. Nakagawa & S. DiMauro. 1985. Fatal infantile cytochrome c oxidase deficiency: decrease of immunologically detectable enzyme in muscle. Neurology **35:** 802–812.
12. Rimoldi, M., E. Bottacchi, L. Rossi, F. Cornelio, G. Uziel & S. DiDonato. 1982. Cytochrome c oxidase deficiency in muscles of a floppy infant without mitochondrial myopathy. J. Neurol. **277:** 201–207.
13. Sengers, R. C. A., J. M. F. Trijbels, J. A. J. M. Bakkeren, W. Ruitenbeek, A. J. M. Janssen, A. M. Stadhouders & H. J. Laak. 1984. Deficiency of cytochromes b and aa_3 in muscle from a floppy infant with cytochrome oxidase deficiency. Eur. J. Pediat. **141:** 178–180.
14. Zeviani, M., D. H. Van Dyke, S. Servidei, S. C. Bauserman, E. Bonilla, E. T. Beaumont, J. Sharda, K. Vanderlaan & S. DiMauro. 1986. Myopathy and fatal cardiopathy due to cytochrome c oxidase deficiency. Arch. Neurol. (In press).
15. Boustany, R. N., J. R. Aprille, J. Halperin, H. Levy & G. R. DeLong. 1983. Mitochondrial cytochrome deficiency presenting as a myopathy with hypotonia, external ophthalmoplegia, and lactic acidosis in an infant and as a fatal hepatopathy in a second cousin. Ann. Neurol. **14:** 462–470.
16. Jerusalem, F., C. Angelini, A. G. Engel & R. V. Groover. 1973. Mitochondria-lipid-glycogen (MLG) disease of muscle. Arch. Neurol. **29:** 162–169.
17. DiMauro, S., J. F. Nicholson, A. P. Hays, A. B. Eastwood, A. Papadimitriou, R. Koenigsberger & D. C. DeVivo. 1983. Benign infantile mitochondrial myopathy due to reversible cytochrome c oxidase deficiency. Ann. Neurol. **14:** 226–234.
18. Zeviani, M., P. Peterson, S. Servidei, E. Bonilla & S. DiMauro. 1987. Benign reversible muscle cytochrome c oxidase deficiency: A second case. Neurology (In press.)

19. WILLEMS, J. L., L. A. W. MONNENS, J. M. F. TRIJBELS, J. H. VEERKAMP, A. E. F. H. MEYER, K. VAN DAM & U. VAN HAELST. 1977. Leigh's encephalomyelopathy in a patient with cytochrome c oxidase deficiency in muscle tissue. Pediatrics **60:** 850–857.
20. MIYABAYASHI, S., K. NARISAWA, K. TADA, K. SAKAI, K. KOBAYASHI & Y. KOBAYASHI. 1983. Two siblings with cytochrome c oxidase deficiency. J. Inher. Metab. Dis. **6:** 121–122.
21. HOGANSON, G. E., D. J. PAULSON, R. CHUN, R. L. SUFIT & A. L. SHUG. 1984. Deficiency of muscle cytochrome c oxidase in Leigh's disease. Pediatr. Res. **18:** 222 A (abstract).
22. SERVIDEI, S., M. ZEVIANI, S. DIMAURO, M. DIROCCO, D. C. DEVIVO, S. DIDONATO, G. UZIEL, K. E. BERRY, M. G. NORMAN, G. HOGANSON, S. D. JOHNSEN & P. C. JOHNSON. 1986. Cytochrome oxidase deficiency in Leigh's syndrome. Ann. Neurol. **20:** 400 (abstract).
23. PRICK, M. J. J., F. J. M. GABREELS, J. M. F. TRIJBELS, A. J. M. JANSSEN, R. LECOULTRE, K. VAN DAM, H. H. J. JASPER, E. J. EBELS & A. A. W. OP DE COUL. 1983. Progressive poliodystrophy (Alpers' disease) with a defect in cytochrome aa_3 in muscle: a report of two unrelated patients. Clin. Neurol. Neurosurg. **85:** 57–70.
24. FRENCH, J. H., E. S. SHERARD, H. LUBELL, M. BROTZ & C. L. MOORE. 1972. Trichopoliodystrophy. I. Report of a case and biochemical studies. Arch. Neurol. **26:** 229–244.
25. PEZESHKPOUR, G., C. KRARUP, F. BUCHTHAL, S. DIMAURO & F. MCBURNEY. 1984. Involvement of peripheral nerve in mitochondrial disease. Neurology **34:** 182 (abstract).
26. SERVIDEI, S., R. P. LAZARO, E. BONILLA, K. D. BARRON, M. ZEVIANI & S. DIMAURO. 1987. Mitochondrial encephalomyopathy and partial cytochrome c oxidase deficiency. Neurology. (In press.)
27. JOHNSON, M. A., D. M. TURNBULL, D. J. DICK & H. S. A. SHERRATT. 1983. A partial deficiency of cytochrome c oxidase in chronic progressive external ophthalmoplegia. J. Neurol. Sci. **60:** 31–53.
28. MULLER-HOCKER, J., D. PONGRATZ & G. HUBNER. 1983. Focal deficiency of cytochrome c oxidase in skeletal muscle of patients with progressive external ophthalmoplegia. Virchows Arch. **402:** 61–71.
29. ROWLAND, L. P., A. P. HAYS, S. DIMAURO, D. C. DEVIVO & M. BEHRENS. 1983. Diverse clinical disorders associated with morphological abnormalities of mitochondria. *In* Mitochondrial Pathology in Muscle Diseases. G. Scarlato & C. Cerri, Eds.: 140–158. Piccin Medical Books. Padova, Italy.
30. MORGAN-HUGHES, J. A. & D. N. LANDON. 1983. Mitochondrial respiratory chain deficiencies in man. Some histochemical and fine-structural observations. *In* Mitochondrial Pathology in Muscle Diseases. G. Scarlato & C. Cerri, Eds.: 19–37. Piccin Medical Books. Padova, Italy.
31. KADENBACH, B., R. HARTMANN, R. GLANVILLE & G. BUSE. 1982. Tissue-specific genes code for polypeptide VIa of bovine liver and heart cytochrome c oxidase. FEBS Lett. **138:** 236–238.
32. JARAUSCH, J. & B. KADENBACH. 1982. Tissue-specificity overrides species-specificity in cytoplasmic cytochrome c oxidase polypeptides. Hoppe-Seyler's Z. Physiol. Chem. **363:** 1133–1140.
33. FINE, P. E. M. 1978. Mitochondrial inheritance and disease. Lancet **2:** 659–662.
34. GILES, R. E., H. BLANC, H. M. CANN & D. C. WALLACE. 1980. Maternal inheritance of human mitochondrial DNA. Proc. Natl. Acad. Sci USA **77:** 6715–6719.
35. MIRANDA, A. F., M. NAKAGAWA, E. BONILLA & S. DIMAURO. 1985. Subunit IV of cytochrome c oxidase (COX): Evidence for developmental regulation in skeletal muscle. J. Cell Biol. **101:** 449a (abstract).
36. MIRANDA, A. F. & T. MONGINI. 1986. Diseased muscle in tissue culture. *In* Myology. A. G. Engel & B. Q. Banker, Eds.: 1123–1149. McGraw-Hill. New York.
37. LOWEY, S. 1986. The structure of vertebrate muscle myosin. *In* Myology. A. G. Engel & B. Q. Banker, Eds.: 563–587. McGraw-Hill. New York.
38. MULLER-HOCKER, J. S. STUNKEL, D. PONGRATZ & G. HUBNER. 1985. Focal deficiency of cytochrome c oxidase and of mitochondrial ATPase with histochemical evidence of loosely coupled oxidative phosphorylation in a mitochondrial myopathy of a patient with bilateral ptosis. J. Neurol. **69:** 27–36.
39. JOHNSON, M. A., B. KADENBACH, L. KUHN-NENTWIG, J. J. FULTHORPE, H. S. A. SHER-

RAT & D. M. TURNBULL. 1985. Immunocytochemical studies of cytochrome oxidase deficiencies in human skeletal muscle. Biochem. Soc. Trans. **13:** 729.

40. ANDERSON, S., A. T. BANKIER, B. G. BARRELL, M. H. L. DE BRUIJN, A. R. COULSON, J. DROUIN, I. C. EPERON, D. P. NIERLICH, B. A. ROE, F. SANGER, P. H. SCHREIER, A. J. H. SMITH, R. STADEN & I. G. YOUNG. 1981. Sequence and organization of the human mitochondrial genome. Nature **290:** 457–465.
41. ZEVIANI, M., J. HERBERT, A. SHERBANY, S. DIMAURO & E. A. SCHON. 1986. Isolation of a cDNA clone encoding subunit IV of cytochrome c oxidase. Neurology **36** (Suppl. 1): 78 (Abstract).
42. DIMAURO S, E. BONILLA, M. ZEVIANI, M. NAKAGAWA & D. C. DEVIVO. 1985. Mitochondrial myopathies. Ann. Neurol. **17:** 521–538.

DISCUSSION OF THE PAPER

B. CHANCE (*University of Pennsylvania, Philadelphia, PA*): We noted no loss of subunits in your elegant gels. Was there spectroscopic evidence of loss of hemes or copper atoms in the respiratory chain?

S. DI MAURO (*Columbia University, New York, NY*): Reduced-minus-oxidized spectra of isolated muscle mitochondria show a complete lack of the peak corresponding to cytochrome aa_3. However, more sophisticated techniques for measuring hemes and copper directly in tissues could be usefully applied to the study of these disorders. In fact, in collaboration with Dr. Scarpa in your laboratory, we did apply such techniques to show a partial defect of cytochrome *c* oxidase in the kidney of a patient with fatal infantile cytochrome oxidase deficiency and renal dysfunction.

S. CORVERA (*University of Massachusetts, Worcester, MA*): In your immunoprecipitation studies of cytochrome oxidase in diseased tissues, how can you distinguish whether the immunoprecipitated enzyme resides in normal or abnormal fibers?

DI MAURO: The tissues that we have used for immunoprecipitation have a great predominance of abnormal fibers, so the contribution of enzyme derived from normal fibers must be negligible.

L. E. ROSENBERG (*Yale University, New Haven, CT*): Would you comment on the Mendelian inheritance of either the severe or benign forms of human cytochrome oxidase deficiency?

DI MAURO: In the malignant form of cytochrome oxidase–deficiency myopathy, autosomal recessive inheritance is suggested by the presence of affected siblings and the occurrence of consanguineity in some families. The three children with benign cytochrome oxidase deficiency were the children of non consanguineous parents.

Mitochondrial Myopathies Involving the Respiratory Chain: A Biochemical Analysis[a]

S. TAKAMIYA, W. YANAMURA, AND R. A. CAPALDI

Institute of Molecular Biology
University of Oregon
Eugene, Oregon 97403

N. G. KENNAWAY AND R. BART

Department of Medical Genetics
Oregon Health Sciences University
Portland, Oregon 97201

R. C. A. SENGERS, J. M. F. TRIJBELS, AND W. RUITENBEEK

Department of Pediatrics
University of Nijmegen
Nijmegen, The Netherlands

INTRODUCTION

Mitochondrial myopathies are relatively rare diseases in which defective mitochondrial metabolism results in muscle weakness and/or exercise intolerance and in some cases also in neurological, renal, or cardiac dysfunctions. Lactic acidosis is a common symptom of this class of disease but there is no consistent clinical presentation, with disorders not only involving different tissues, but being congenital or late onset as well as progressive or occasionally reversible (see Reference 1 for review). When examined by electron microscopy, the mitochondria of muscle samples from patients with mitochondrial myopathies are often seen in large aggregates under the sarcolemma. They are usually enlarged, the cristae may be increased or form unusual shapes, and there may be crystalline inclusions in the matrix or intracristal spaces.

As pointed out by DiMauro *et al.*,[1] mitochondrial myopathies can be divided into three groups at the biochemical level, based on the area of mitochondrial metabolism affected, i.e. substrate utilization, electron transfer or energy coupling, and ATP synthesis.

We have focused our attention on defects of the electron transfer chain, our studies being directed towards an understanding of the molecular basis of mitochondrial myopathies, as well as at identifying novel features of the biogenesis and developmental regulation of the respiratory chain proteins.

[a]Supported by the Muscular Dystrophy Association (N.G.K. and R.A.C.) and the National Institutes of Health (Grant HL 22050 to R.A.C.).

[b]Address correspondence to R.A.C.

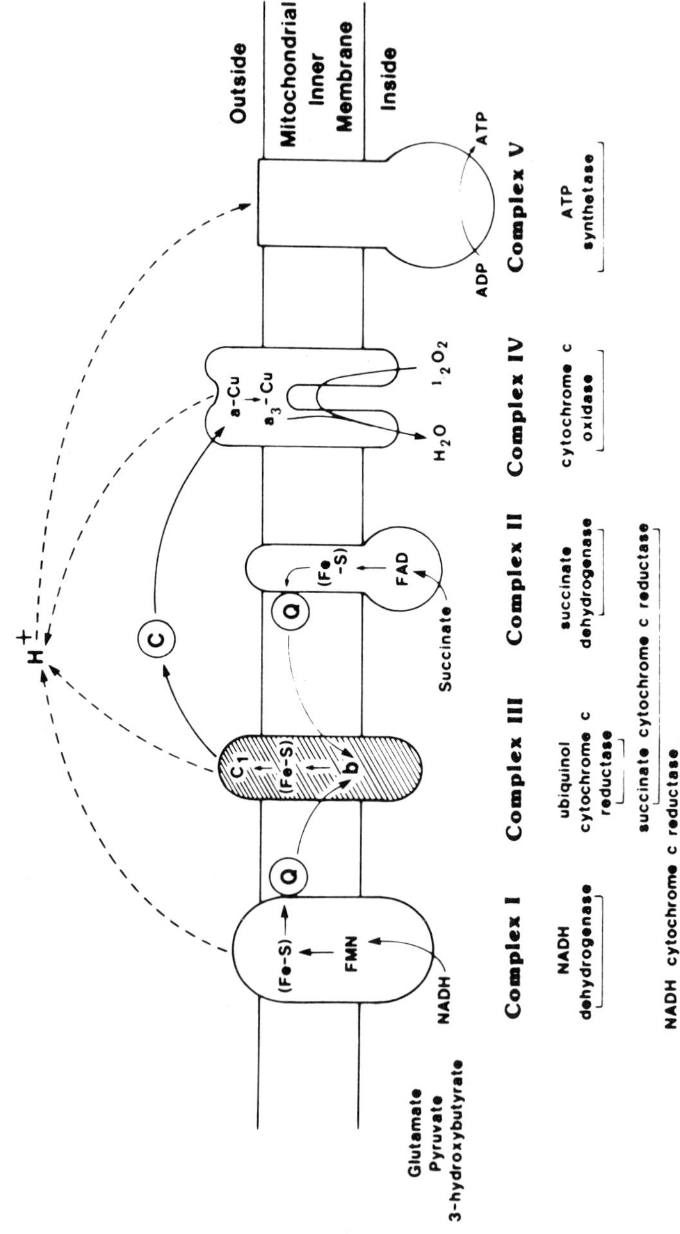

FIGURE 1. Schematic of the mitochondrial respiratory chain. (From Kennaway *et al.*[6] With permission from *Pediatric Research*.)

Complexity of the Mitochondrial Respiratory Chain

The mitochondrial respiratory chain is divided into four functional units or complexes, each an intrinsic protein complex of the mitochondrial inner membrane. The distribution of prosthetic groups between these complexes, i.e. complex I or NADH coenzyme Q reductase, complex II or succinate coenzyme Q reductase, complex III or reduced CoQ cytochrome *c* reductase, and complex IV or cytochrome *c* oxidase, is shown in FIGURE 1. Coenzyme Q and cytochrome *c* are essential additional components of the electron transport chain, acting as small mobile electron carriers between the larger complexes. The approximate molecular weights of complexes I through IV are 900,000, 200,000, 300,000 and 200,000, respectively.[2] Each is a multipolypeptide complex as shown in FIGURE 2 with complex I having more than 30 different component polypeptides and with complex II, the simplest structurally, having five different "subunits." Complexes, I, III, and IV represent the three coupling sites at which the free energy available from the redox reactions is converted into a transmembrane proton gradient. This gradient is used by the ATP synthase to drive ATP synthesis.[2]

Localization of Respiratory Chain Defects

Defects have been described involving complexes I,[3–5] III,[5–9] and IV.[10–12] In addition, several patients have been reported with a combined defect of complex IV (cytochrome *c* oxidase) with complex III[13] or complex I.[14] We have examined several patients in our laboratories: one, with an isolated complex III defect in skeletal muscle[6] and another, with a combined defect of complexes III and IV are described below. Patient 1 had lactic acidosis and progressive exercise intolerance from the age of 9–10 years. Patient 2 died at 5 months of age with generalized weakness, lactic acidosis, and cardiorespiratory insufficiency.[13] TABLE 1 lists the cytochrome content of muscle samples from these two patients. In patient 1, the content of *b* heme is very low but other cytochromes are present in normal amounts. In patient 2, the content of aa_3 and *b* hemes are all low with some reduction also in the amount of $c + c_1$. TABLE 2 lists the electron transfer activities showing that for patient 1 the defect is limited to complex III. This is in contrast to patient 2 where both complexes III and IV are involved.

For both patients, the polypeptide composition of the defective respiratory chain complexes has been examined by immunoblotting with antibodies raised against the various components of beef heart mitochondria. FIGURE 3(a and b) shows immunoblotting of purified mitochondria from patient 1 with antibody raised against complex III isolated from beef heart. FIGURE 3c shows data for patient 2. In both cases, several components of complex III including core proteins (I and II), non-heme iron protein FeS, and polypeptide VI (in patient 1) are greatly diminished (along with cytochrome *b* spectrally). Cytochrome c_1, in contrast (polypeptide IV), is present in close to normal amounts in patient 1 (FIG. 3a) and in significant amounts in patient 2 (FIG. 3c). This cytochrome c_1 is present as the mature polypeptide ($M_r \simeq 31{,}000$) and not as the precursor (made larger by an N terminal leader sequence) that is processed by a protease at the matrix side of the mitochondrial inner membrane.[15]

FIGURE 4 shows immunoblotting of mitochondria from patient 2 with anti-cytochrome *c* oxidase antibodies. Polypeptides Mt_{II}, STA, and ASA (see figure legend for the nomenclature of cytochrome *c* oxidase subunits) are present only in much reduced amounts. Polypeptide C_{IV} appears to be present in somewhat higher amounts than background, with this polypeptide incorporated as a mature subunit and not as the unprocessed precursor protein. These immunological studies show that there is

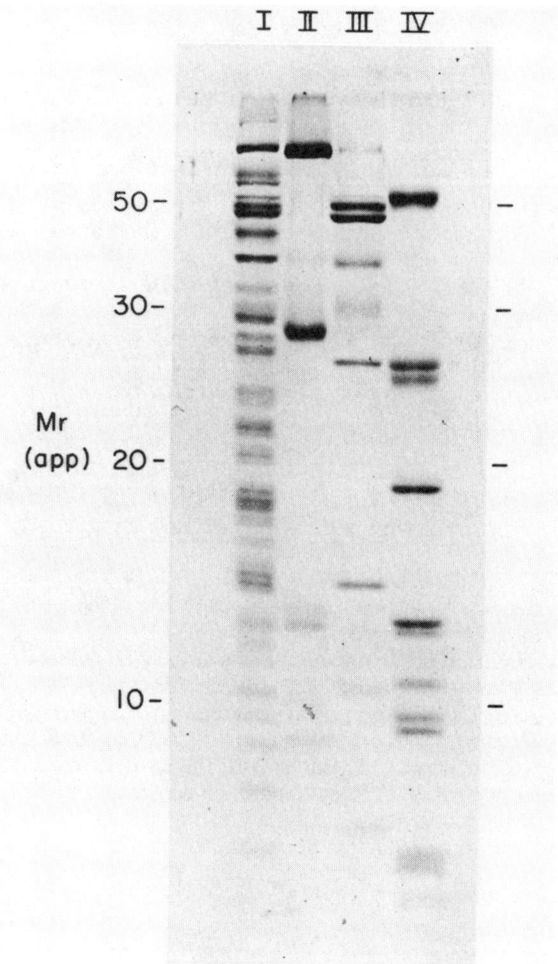

FIGURE 2. Sodium dodecyl sulphate polyacrylamide gel electrophoresis of purified electron transfer complexes. I, NADH coenzyme Q reductase; II, succinate coenzyme Q reductase; III, reduced coenzyme Q cytochrome *c* reductase; IV, cytochrome *c* oxidase.

TABLE 1. Content of Cytochromes in Skeletal Muscle Mitochondria

Patient	b	$c + c_1$ (pmol/mg mitochondrial protein)	aa_3	Reference
1	100	660	300	6
2	49	382	30	13
Controls	630	660	500	6
	312 ± 69	576 ± 113	431 ± 113	13

TABLE 2. Enzymatic Activities in Muscle Mitochondria

Activity	Patient 1	Patient 2	Controls
	(nmol/min/mg mitochondrial protein at 37°C)		
ATPase	2000	–	1200
Cytochrome *c* oxidase	1222	38[a]	945, 646[a]
NADH cytochrome *c* reductase	0	–	136
Succinate dehydrogenase	57	–	94
Succinate cytochrome *c* reductase	8	<10[a]	340, 197[a]

[a]Activities measured at 30°C.

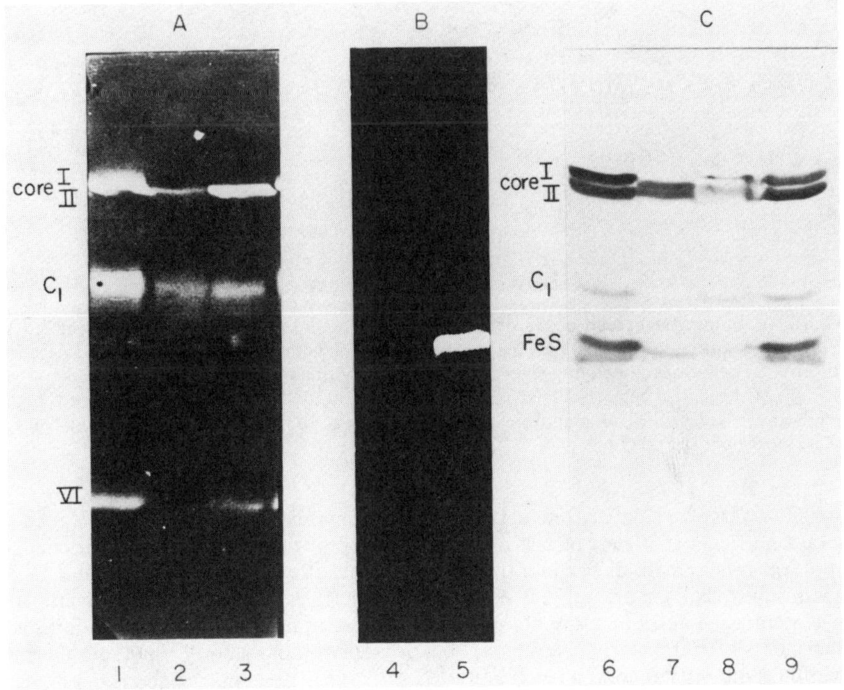

FIGURE 3. Binding of anticomplex III antibodies to mitochondrial samples. (A and B) Reaction with mitochondria from patient 1: (Lane 1) 2 mg isolated complex III from beef heart mitochondria; (Lane 2) 80 μg patient mitochondria; (Lane 3) 80 μg control human muscle mitochondria; (Lane 4) 80 μg control human muscle mitochondria; and (Lane 5) 80 μg patient muscle mitochondria. In this experiment, polypeptides transferred onto nitrocellulose were reacted with antibody made against holo complex III (Panel A) and the non-heme iron sulfur protein (Panel B) and the reaction visualized by using fluorescein-labeled, anti-rabbit antibody. (C) Reaction with mitochondria from patient 2: (Lane 6) 10 μg beef heart mitochondria; (Lane 7) 80 μg control muscle mitochondria; (Lane 8) 80 μg patient muscle mitochondria; and (Lane 9) 3 μg isolated Complex III from beef heart mitochondria. In this experiment, a mixture of antibodies against holoenzyme, core proteins and non-heme iron sulfur protein were used and visualized by using peroxide-labeled, anti-rabbit antibody.

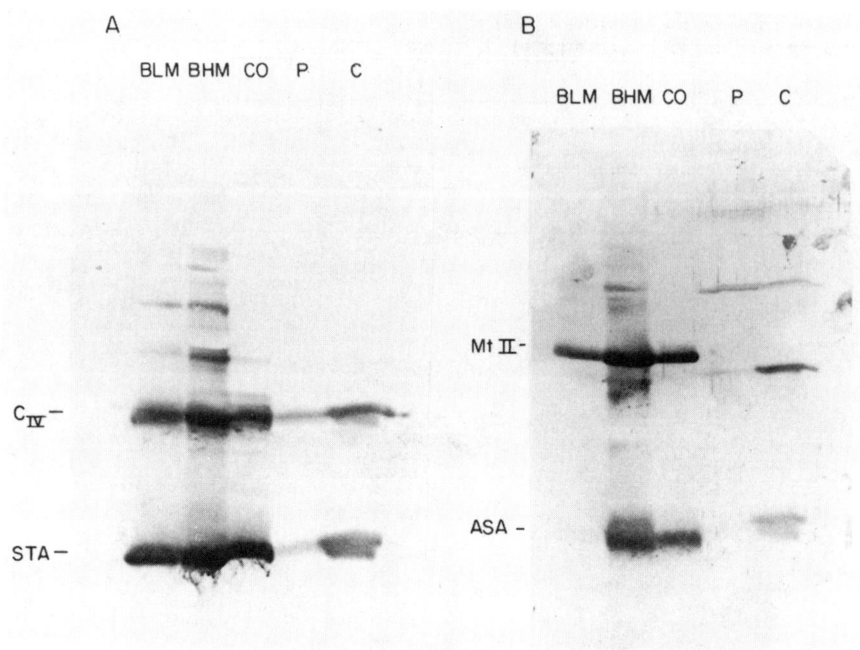

FIGURE 4. Binding of cytochrome *c* oxidase antibodies with mitochondria from patient 2. (A) Reaction with antibodies against purified C_{IV} and STA. The nomenclature for subunits is detailed in FIGURE 6 and the accompanying legend. (B) Reaction with antibodies against purified Mt_{II} and ASA: BLM, beef liver mitochondria; BHM, beef heart mitochondria; CO, isolated beef heart cytochrome *c* oxidase; P, 160 μg muscle mitochondrial fraction from patient 2; C, 160 μg control muscle mitochondrial fraction.

deficient assembly of defective electron transfer complexes, implying that the genetic alteration affects synthesis of one of the components without which association into complexes is not possible. The fact that cytochrome c_1 in complex III and subunit IV of cytochrome *c* oxidase are present could mean that these components insert into the inner membrane early in assembly, perhaps as the template for binding of additional components. Alternatively, they may be degraded more slowly than other non-assembled subunits of complexes III and IV.

Biosynthesis of the Respiratory Chain

An intriguing aspect of the mitochondrial electron transport chain is its mode of biosynthesis, involving two genomes; one in the mitochondrion, the other in the nucleus.[16] FIGURE 5 summarizes schematically the different steps in synthesis and assembly of cytochrome *c* oxidase. The model equally applies to other electron transport complexes and the ATP synthase. One or more of the polypeptides of complexes I, III, IV (cytochrome *c* oxidase) and the ATP synthase is coded for on mitochondrial (mt) DNA and is synthesized on ribosomes inside the mitochondrion. The available evidence is that the mtDNA is transcribed as a polycistronic message, which is then processed to give mRNA against the different reading frames.[17] These are translated on mitochondrial ribosomes close to or associated with the mitochondrial inner membrane.

The majority of the electron transport chain components are synthesized on cytoplasmic ribosomes. This synthesis occurs on free ribosomes; the proteins being made as precursors with an N terminal extension.[15] These leader sequences have been shown to direct components to the mitochondrion and facilitate their incorporation in the correct compartment, i.e. matrix space, inner membrane, intracristal space, or outer membrane.[18]

Respiratory chain defects causing mitochondrial myopathies could be due to lesions in mtDNA or in the nuclear genome. In the case of patient 2, the fact that complex III and cytochrome *c* oxidase are both defective is particularly interesting

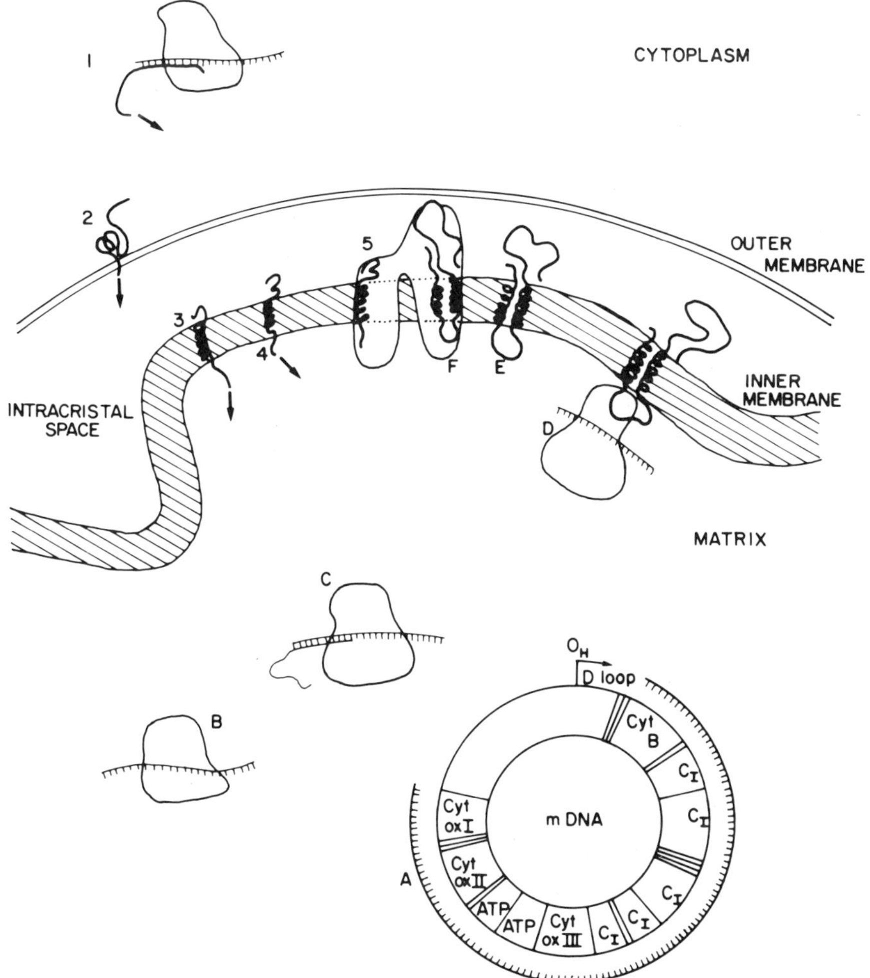

FIGURE 5. Schematic of the synthesis and assembly of the respiratory chain complexes.

because the synthesis of these two complexes appears to be linked only at the levels of the translation of mtDNA and processing of the polycistronic messenger RNA. In patient 2 the genetic defect probably involves a nuclear-coded protein required in these events.

Cytoplasmically made proteins that affect the coordinated synthesis of complex III and cytochrome *c* oxidase in yeast have been identified genetically.[19]

Therapy of Patients with Mitochondrial Myopathies

Patient 1 is now a 19-year-old female. The diagnosis of mitochondrial myopathy with lactic acidosis was made from muscle biopsy tissue.[6,20] The bioenergetic capacity of muscle in this patient was also studied using phosphorous 31 nuclear magnetic

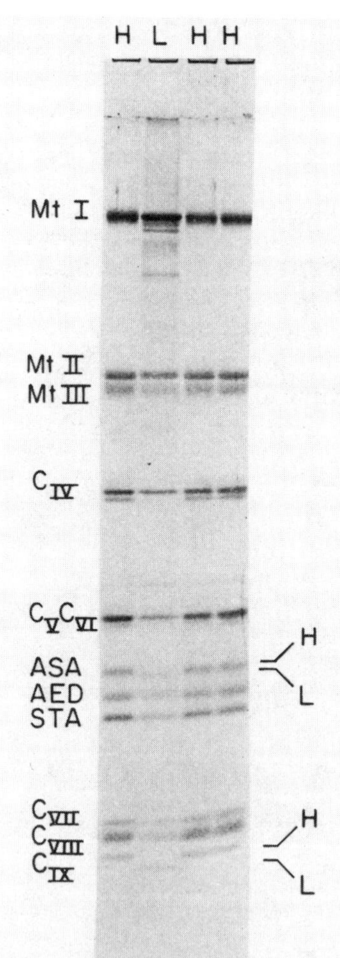

FIGURE 6. Sodium dodecyl sulfate polyacrylamide gel electrophoresis of cytochrome *c* oxidase from beef heart and beef liver: H, heart; L, liver. The mitochondrially coded polypeptides are labeled Mt_{I-III}, the nuclear coded polypeptides C_{IV-IX} to correspond with those of lower eukaryotes such as yeast. The remaining polypeptides are labeled ASA, AED, and STA corresponding to their N terminal sequences in beef heart.

resonance.[21] The ratio of phosphocreatine to inorganic phosphate concentration (PCr/P_i) proved to be only 15% of normal at rest and very little exercise was needed to decrease it to the lowest tolerable limit of 1.0. After mild exercise, recovery of PCr/P_i ratios to already low pre-exercise values proceeded very slowly (20 min or more compared with 3–4 min in normal individuals). This patient was treated with reagents to bypass complex III, the most successful being menadione (vitamin K_3), which is readily reduced by coenzyme Q, and ascorbate (vitamin C), which may accept electrons from menadione and then reduce cytochrome *c* (see FIG. 1 for the rationale).[21] There was dramatic improvement in exercise capacity as measured by nuclear magnetic resonance and by the capability of the patient to undertake a more vigorous daily routine.

Tissue Specificity of Mitochondrial Myopathies

Clinical observations of patient 1 suggested that the complex III defect was limited to muscle. Recent studies on cultured skin fibroblasts, peripheral blood leukocytes, and transformed lymphoid cells indicate that the defect is not expressed in these cells.[22] This tissue specificity is a striking and unexpected finding and has been established now for defects in complex III and in cytochrome *c* oxidase.[6,11,22] Direct biochemical evidence for tissue specific forms of complex III is lacking and is only newly available for cytochrome *c* oxidase.[23,24] FIGURE 6 shows the polypeptide composition of cytochrome *c* oxidase from beef heart and beef liver. The enzyme contains 13 different polypeptides as isolated; three coded for in the mitochondria (Mt_{I-III}) and ten coded for by nuclear genes. Six of the nuclear coded polypeptides have counterparts in lower eukaryotes such as yeast (C_{IV-IX}): four do not and these are labeled by their N terminal sequences. From FIGURE 6 it is evident that two of the polypeptides, C_{IX} and ASA, migrate differently in the two tissues. FIGURE 4 provides immunological evidence of the tissue specificity of polypeptide ASA. Antibodies against this polypeptide from beef heart muscle react with human skeletal muscle but not beef liver. Kadenbach and colleagues have suggested that all of the cytoplasmically made subunits of cytochrome oxidase occur in tissue specific as well as adult and fetal forms.[24] The presence of developmentally regulated isoenzyme forms could explain the reversibility of the benign form of cytochrome *c* oxidase deficiency,[25] if the genetic defects(s) occurred only in the immature isoenzyme form.

REFERENCES

1. DIMAURO, S., E. BONILLA, M. ZEVIANI, M. NAKAGAWA & D. C. DEVIVO. 1985. Ann. Neurol. **17:** 521–538.
2. CAPALDI, R. A. 1982. Biochim. Biophys. Acta **694:** 291–306.
3. MORGAN-HUGHES, J. A., D. N. LANDON, J. M. LAND & J. B. CLARK. 1979. J. Neurol. Sci. **43:** 27–46.
4. LAND, J. M., J. A. MORGAN-HUGHES & J. B. CLARK. 1981. J. Neurol. Sci. **50:** 1–13.
5. MORGAN-HUGHES, J. A., D. J. HAYES, J. B. CLARK, D. N. LANDON, M. SWASH, R. J. STARK & P. RUDGE. 1982. Brain **105:** 553–582.
6. KENNAWAY, N. G., N. R. BUIST, V. M. DARLEY-USMAR, A. PAPADIMITRIOU, S. DIMAURO, R. I. KELLEY, R. A. CAPALDI, N. K. BLANK & A. D'AGOSTINO. 1984. Pediatric Res. **18:** 992–999.
7. MORGAN-HUGHES, J. A., P. DARVENIZA, S. N. KAHN, D. N. LANDON, R. M. SHERRATT, J. M. LAND & J. B. CLARK. 1977. Brain **100:** 617–624.

8. PAPADIMITRIOU, A., H. B. NEUSTEIN, S. DIMAURO, R. STANTON & N. BRESOLIN. 1984. Pediatric Res. **18:** 1023–1028.
9. SENGERS, R. A. C., J. C. FISCHER, J. M. F. TRIJBELS, W. RUITENBEEK, A. M. STADHOUDERS, H. J. TER LAAK & H. H. J. JASPER. 1983. Eur. J. Pediatr. **140:** 332–337.
10. ZEVIANI, M., I. NONAKA, E. BONILLA, E. OKINO, M. MOGGIO, S. JONES & S. DIMAURO. 1985. Ann. Neurol. **17:** 414–417.
11. BRESOLIN, N., M. ZEVIANI, E. BONILLA, R. H. MILLER, R. W. LEECH, S. SHANSKE, M. NAKAGAWA & S. DIMAURO. 1985. Neurology **35:** 802–812.
12. DIMAURO, S., J. R. MENDELL, Z. SAHENK, D. BACHMAN, A. SCARPA, R. M. SCOFIELD & C. REINER. 1980. Neurology **30:** 795–804.
13. SENGERS, R. A. C., J. M. F. TRIJBELS, J. A. J. N. BAKKERSEN, W. RUITENBEEK, J. C. FISCHER, A. J. M. JANSSEN, A. M. STADHOUDERS & H. J. TER LAAK. 1984. Eur. J. Pediatr. **141:** 178–180.
14. SHERRATT, H. S. A., N. E. F. CARTLIDGE, M. A. JOHNSON & D. M. TURNBULL. 1984. J. Inher. Metab. Des. **7** (suppl. 2): 107–108.
15. SCHATZ, G. & R. A. BUTOW. 1983. Cell **32:** 316–318.
16. CLAYTON, D. A. 1984. Ann. Rev. Biochem. **53:** 573–594.
17. ALONI, Y. & G. ATTARDI. 1971. Proc. Natl. Acad. Sci. USA **68:** 1757–1761.
18. MAARSE, A. C., A. P. G. M. VAN LOON, H. RIEZMAN, I. GREGOR, G. SCHATZ & L. A. GRIVELL. 1984. EMBO J. **3:** 2831–2837.
19. DIECKMANN, C. L., L. K. PAPE & A. TZAGOLOFF. 1982. Proc. Natl. Acad. Sci. USA **79:** 1805–1809.
20. DARLEY-USMAR, V. M., N. G. KENNAWAY, N. R. M. BUIST & R. A. CAPALDI. 1982. Proc. Natl. Acad. Sci. USA **80:** 5103–5106.
21. ELEFF, S., N. G. KENNAWAY, N. R. M. BUIST, V. M. DARLEY-USMAR, R. A. CAPALDI, W. J. BANKS & B. CHANCE. 1984. Proc. Natl. Acad. Sci. USA **81:** 3529–3533.
22. DARLEY-USMAR, V. M., M. WATANABE, Y. UCHIYAMA, I. KONDO, N. G. KENNAWAY, L. GROHNKE & H. HAMAGUCHI. 1983. Clin. Chim. Acta. (In press.)
23. KADENBACH, B., R. HARTMANN, R. GLANVILLE & G. BUSE. 1982. FEBS Lett. **138:** 236–238.
24. MERLE, P. & B. KADENBACH. 1982. Eur. J. Biochem. **125:** 239–244.
25. DIMAURO, S., J. F. NICHOLSON, A. P. HAYES, A. B. EASTWOOD, A. PAPADIMITRIOU, R. KOENIGSBERGER & D. C. DEVIVO. 1983. Ann. Neurol. **14:** 226–234.

DISCUSSION OF THE PAPER

G. F. AZZONE (*University of Padova, Padova*): The replacement effect following administration of menadione may be counterbalanced by the fact that menadione causes oxidation of NADH and this in turn is followed by an increase of membrane permeability for protons. It would be nice to have one effect i.e., restoration of electron transport, without alteration of the membrane.

B. CHANCE (*University of Pennsylvania, Philadelphia, PA*):The fact that menadione administration doesn't cause complete recovery may suggest that it has some side effect. Of course, there is no proof that such a side effect occurs *in vivo.* The suggestion for the side effects originates mainly from the analysis of the effect of menadione *in vitro.*

H. R. SCHOLTE (*Erasmus University, Rotterdam*): Maybe supplementation with CoQ could be of help in patients with site 2 defects. Was that tried?

KENNAWAY: We have not tried CoQ although we have certainly considered doing this.

S. DiMauro (*Columbia University, New York, NY*): In your second patient, there was a cardiomyopathy. Were heart mitochondria altered either morphologically or biochemically?

Kennaway: Unfortunately, heart tissue was not available for biochemical studies.

Thierauch: Is there evidence of increased occurrence of free radical production in mitochondria of patients with mitochondrial diseases?

Kennaway: There is no evidence that I am aware of.

E. Rubin (*Hahnemann University, Philadelphia, PA*): If there is a problem with subunit assembly, isn't it possible that the primary defect is in the assembly mechanism, for example, a docking protein, rather than in subunit synthesis?

Kennaway: I can't rule out that possibility.

The ADP/ATP Carrier as a Mitochondrial Auto-antigen—Facts and Perspectives

H.-P. SCHULTHEISS, K. SCHULZE, U. KÜHL,
G. ULRICH, AND M. KLINGENBERG

Department of Internal Medicine
Klinikum Grosshadern
University of Munich
D-8000 Munich 70, Federal Republic of Germany

INTRODUCTION

Autoantibodies directed against a wide range of tissue antigens have been found in the sera of patients with different autoimmune diseases.[1–5] The causes of autoimmune diseases are in most instances unknown. However, one popular notion is that many autoimmune diseases may be caused directly or indirectly by viral infections.[6,7] Viruses may induce autoimmune processes by altering the host's immune system, by causing the release or expression of sequestered antigens, or through antigenic determinants shared by the virus and the host cells.

Recently, we showed that the sera of patients with dilated cardiomyopathy—a suspected virus-induced autoimmune disease—contain circulating autoantibodies directed against the ADP/ATP carrier.[8,9] As the inner mitochondrial membrane is *a priori* impermeant to hydrophilic metabolites, the ADP/ATP carrier facilitates the transport of ATP to the cytosol with its energy-consuming processes and the return of ADP to the inner mitochondrial space for regeneration by oxidative phosphorylation.[10,11]

Of great metabolic significance is the differentiation between ADP and ATP in the transport by the energization of the inner mitochondrial membrane, which makes the exchange strongly asymmetric.[12,13] By prefering uptake of ADP and release of ATP this energization of the membrane modulates the ADP/ATP carrier activity so as to correspond exactly to the requirements of ATP production in the cell.

In this paper we discuss the immunochemical characteristics of the ADP/ATP carrier and the specificity of the autoantibodies directed against this protein in patients with autoimmune diseases. For elucidating the role of the autoantibodies against the ADP/ATP carrier in the pathogenesis of heart muscle diseases, we studied the hearts of guinea pigs immunized with the isolated carrier protein. The subcellular distribution of the adenine nucleotides was measured, as an inhibition of the carrier should change the intracellular energy balance by diminishing the ATP efflux from the mitochondria into the cytosol. Based on these results we shall discuss our working hypothesis that autoimmunity to intracellular proteins, such as the ADP/ATP carrier, may contribute to the pathophysiology of autoimmune diseases.

ORGAN- AND CONFORMATION-SPECIFICITY OF THE ADP/ATP CARRIER

To characterize the targets of the autoantibodies specific for the CAT-protein complex, different methods like radioimmunoassay,[14] immunoprecipitation,[17] and Western blotting[18] were used. To test the possible effect of the antibodies on the nucleotide transport *in vitro* the exchange rate of mitochondria was determined by the inhibitor stop method combined with the back-exchange.[12,19]

The specificity of the antibodies raised against the ADP/ATP carrier was tested by the immunoblot technique. In Western blots of mitochondrial proteins from heart, liver, and kidney with the anti-ADP/ATP carrier antibodies, an antibody binding was only seen on the 30,000 MW band, which corresponds with the isolated ADP/ATP carrier (FIG. 1). From the weak antibody binding to the kidney protein and the failing binding to the liver protein an organ specificity of the translocator protein can be inferred (FIG. 1). For a further immunochemical characterization of the ADP/ATP carrier, we elaborated a solid-phase double-antibody immunoradiometric assay (IRMA).[14] First the surface of the plastic microtiter well is coated with the isolated ADP/ATP carrier protein in the form of the carboxyatractylate (CAT)-protein complex,[15,16] and residual binding sites on the plastic are blocked with an irrelevant protein such as albumin. Then the plate is washed and incubated with antiserum to

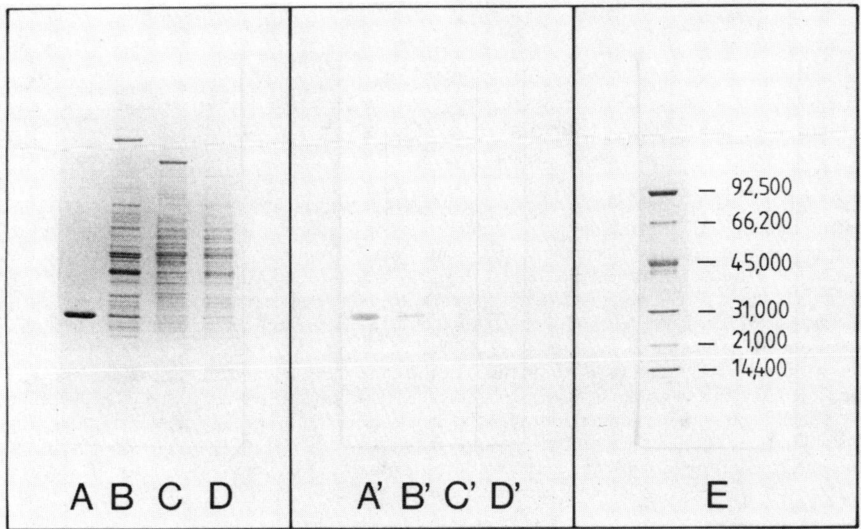

FIGURE 1. Detection of antibodies against the ADP/ATP carried by the immunoblot technique. The isolated ADP/ATP carrier (A) and total mitochondrial proteins from heart (B), liver (C), and kidney (D) as well as marker proteins (E) were separated on SDS-polyacrylamide slab gels. One section of the gel was stained with Coomassie blue (A–D, E), the other section was electrophoretically blotted on a nitrocellulose sheet and incubated with the anti-ADP/ATP carrier antiserum and then with horseradish peroxidase-conjugated anti-IgG (A′, B′, C′, D′).

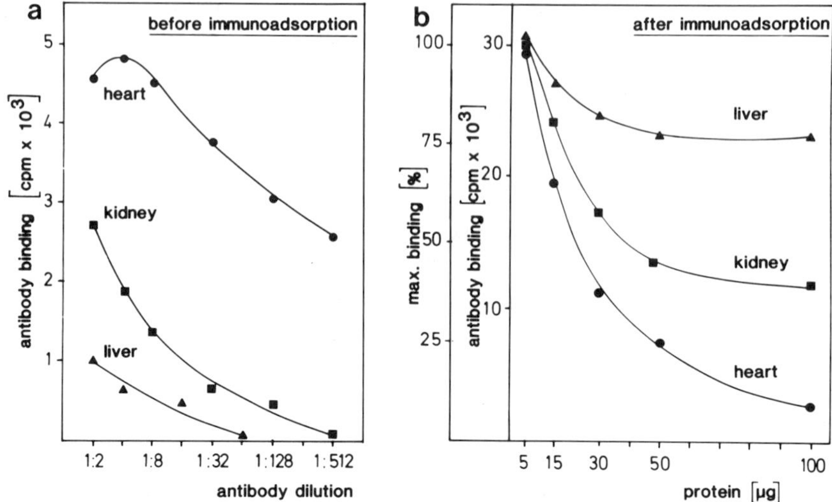

FIGURE 2. The binding of the heart carboxyatractylate protein antiserum to the ADP/ATP carrier isolated from heart, kidney, and liver (a) and the residual anti-heart protein activity of the antiserum after immunoadsorption on the isolated carrier proteins from heart, kidney and liver (b). (a) 90 μl antiserum (serial dilution 1:2 to 1:512) were incubated in microtiter wells precoated with the isolated carrier proteins from heart (●), kidney (■), and liver (▲) overnight at 4°C. After incubation with ^{125}I-protein A (80 μl, 100,000 cpm/well) the wells were counted in a gamma spectrometer. All samples were tested in duplicates. Values for radioactivity are experimental results minus background (cpm, determined in the presence of equal amounts of non immuneserum). (b) Increasing amounts of the isolated carboxyatractylate-protein complexes (5–100 μg) were incubated with a fixed amount of isolated IgG of the anti-heart protein antiserum. After "precipitation" of the antigen-antibody complexes with protein A bearing *Staphylococcus aureus* (Cowan I) the residual antibody activity of the supernatant was determined in the IRMA with carboxyatractylate-protein heart precoated wells.

allow the antibodies to bind to the antigen. After further washing, a second antibody (^{125}I-protein A) is incubated in the wells to bind to the first antibody. The amount of bound ^{125}I-protein A is determined by a gamma spectrometer.

In IRMA the anti–heart protein antibody showed the highest binding activity with the heart protein (FIG. 2a). However, cross-reacting antibodies reacting with kidney and liver protein were found (FIG. 2a). The maximum binding of the anti-heart antibody to the kidney protein was about 50% and to the liver protein about 20% of the binding to the heart protein. In contrast, anti-kidney antibody showed the highest binding activity with the kidney protein, but cross-reacting antibodies with the heart and liver protein were seen.

The cross-reactivity between the CAT-protein from heart, kidney, and liver was also confirmed by immunoadsorption studies. The residual activity of a fixed amount of anti-heart antibodies after incubation with increasing amounts of isolated CAT-protein complexes from heart, kidney, and liver (plateau binding value) was measured in IRMA with CAT-protein coated wells (FIG. 2b). While the residual activity of the anti-heart antibody was only about 10% after immunoadsorption of the heart protein, the residual activity after immunoadsorption on the kidney (approximately 45%) and on the liver protein (80%) was significantly higher. As both proteins from kidney and liver were not able to displace anti-heart antibodies completely, organ-specific antigenic determinants have to be postulated, although an incomplete cross-reaction between the three organs exist.[37] This partial cross-reactivity could also be demonstrated by immunoadsorption studies on mitochondria or by crossed immunoelectrophoresis.[14,20]

In an attempt to gain further evidence of an organ-specificity of the ADP/ATP carrier, we could show that isolated immunoglobulins (IgG) against the heart protein could inhibit the nucleotide transport from heart mitochondria, while the nucleotide transport from kidney and liver mitochondria was not influenced (FIG. 3). A removal of the inhibitory activity of the heart antibodies by adsorption with kidney or liver mitochondria was not possible. However, preadsorption of the antibodies on heart mitochondria eliminated the inhibitory activity.

The specificity of the antibody-mediated inhibition of the adenine nucleotide transport could also be demonstrated in the reconstituted system.[21] Depending on the antibody concentration, a nearly complete inhibition of the transport activity was obtained (FIG. 4).

It had been postulated that there is a single binding center for the substrates and both inhibitors—carboxyatractylate (CAT) and bongkrekate (BKA).[22] Therefore the binding of the specific ligand CAT should also be blocked by the antibodies. As shown in FIG. 5, after the preincubation of heart mitochondria with isolated IgG directed against the CAT-protein from heart the [^{3}H]CAT binding on heart mitochondria was significantly decreased, while the ligand binding on liver and kidney mitochondria was not affected by the antibody. Also, the BKA-binding to heart mitochondria was also decreased, indicating antibody-mediated fixation of the carrier protein in the c-

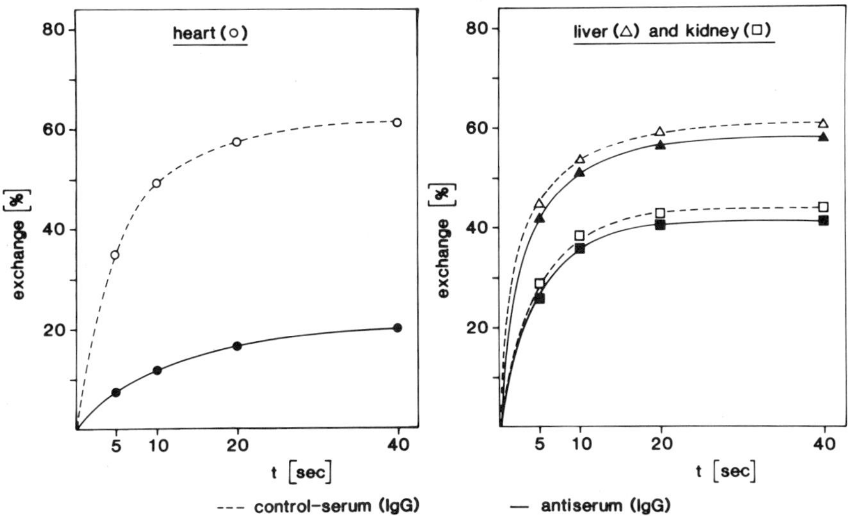

FIGURE 3. Influence of the anti-heart CAT-protein antiserum on the nucleotide transport from heart (○), liver (△), and kidney (□) mitochondria. Isolated and preloaded ([^{14}C]ADP) mitochondria were incubated with equilibrated antiserum (isolated IgG) for 30 min at 4°C. Afterwards the ADP/ATP exchange rate was determined by the inhibitor stop method combined with the back-exchange (methodological details[14,19]). Exchange time was 5 sec, 10 sec, 20 sec, and 40 sec at 4°C for heart and kidney mitochondria and at 8°C for liver mitochondria.

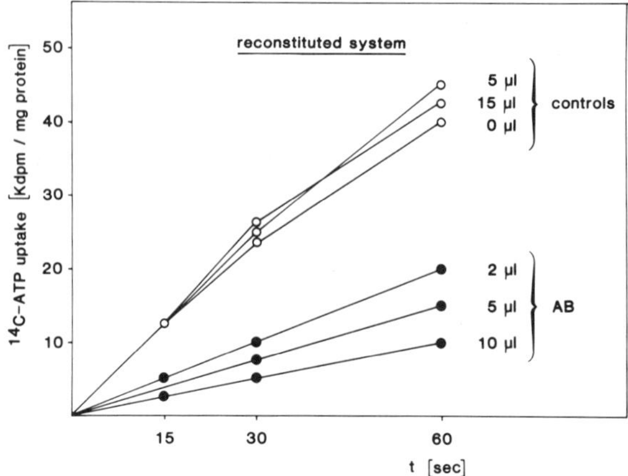

FIGURE 4. Effect of the anti-heart CAT protein antiserum on the reconstituted nucleotide transport.[21] The antibody concentration-dependent inhibition was determined by the forward exchange ([^{14}C]ADP uptake).[19] After incubation of the liposomes for 30 min with 2 μl, 5 μl, and 10 μl immune serum (●) (AB) or 0 μl, 5 μl, and 15 μl control serum (○) (controls) the exchange experiments were done simultaneously.

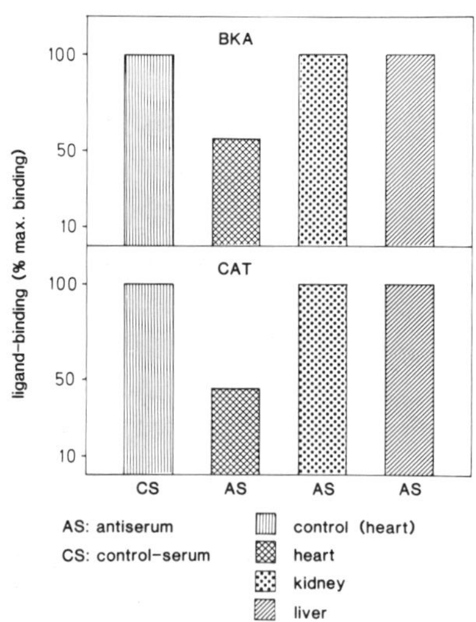

FIGURE 5. Binding of [^{3}H]carboxyatractylate (CAT) and Bongkrekat (BKA) to mitochondria isolated from heart, kidney, and liver. Mitochondria were preincubated with isolated and equilibrated immunoglobins from control and heart anti-CAT protein complex serum.

conformation. The inhibition of the CAT-binding by the antibody was concentration dependent (FIG. 6).

Based on these results, one could speculate that the substrate/ligand binding site of the ADP/ATP carrier is organ specific. As the antibodies, which inhibit the nucleotide transport and the ligand binding are not only organ specific but also conformation specific (see above), it seems likely that the antibodies recognize topographic antigenic determinants consisting of residues on different fragments, widely separated in the sequence, but brought into close proximity by the folding of the native protein.[27]

Polyclonal antisera to any large antigen contain a large number of antibody populations, each specific for a precise topographical domain on the surface of the antigen.[24,25] Some of the antibodies are directed to a domain that contains specific amino acid sequences as the major contributor to the binding energy. However, the majority of polyclonal antibodies is specific for domains constructed from the three-dimensional folding.[24,28]

The ADP/ATP carrier exists in two translocational conformations, the "c"-state

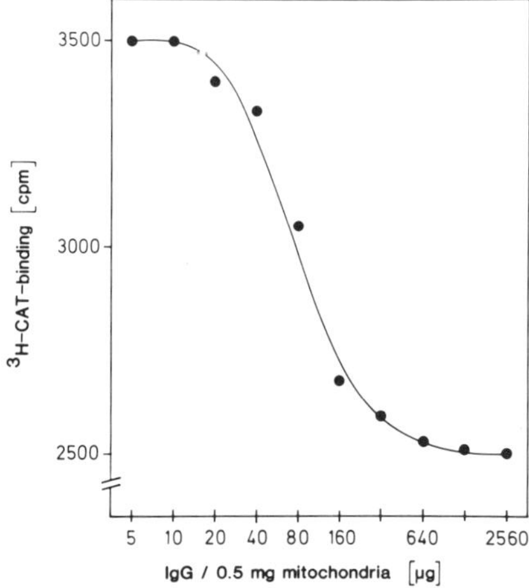

FIGURE 6. Binding of [^{3}H]carboxyatractylate to heart mitochondria. Anti-CAT protein complex antibody concentration-dependent inhibition of the [^{3}H]carboxyatractylate binding. Mitochondria (0.5 mg) were preincubated with increasing amounts of isolated IgGs (5–2,560 μg).

and the "m"-state. These are obtained in a fixed form by binding with carboxyatractylate (CAT) (= "c"-state) or with Bongkrekat (BKA) (= "m"-state).[10] This conformation specificity of the ADP/ATP carrier was also demonstrated by immunochemical studies (FIG. 7). Thus the binding activity of the antibodies against the CAT-protein was higher to the CAT-protein than to the BKA-protein (FIG. 7a). However, after denaturation of the CAT- and BKA-protein the binding activity of the anti-CAT-protein antibody was the same for both proteins reflecting identical antigenic determinants of the denatured state of both protein complexes (FIG. 7a). The different antibody binding to the two native conformational states of the protein, however, implicates a role in antigenicity for amino acids that are brought together by the three-dimensional folding of the molecule. This could be confirmed by immunoadsorption studies on the isolated CAT- and BKA-protein complexes. As shown in FIG. 7b, more than 95% of the anti-CAT-protein antibodies were adsorbed by the CAT-protein, while the BKA-protein adsorbed only about 60%. These results indicating conformation specific antibodies were also confirmed by immunoadsorption studies on intact mitochondria (c-conformation) and submitochondrial particles (m-conformation).

The conformation specificity of the anti-CAT-protein antibodies is also expressed by the fact that the transport was inhibited only in mitochondria (for the "c"-side) but not in submitochondrial particles (for the "m"-side). These data confirming earlier results[29] indicate that the substrate-/ligand-binding center located in the central channel between the two subunits of the carrier dimer changes strongly its "configuration" between the "c"- and "m"-state. Although the center has a common core sharing

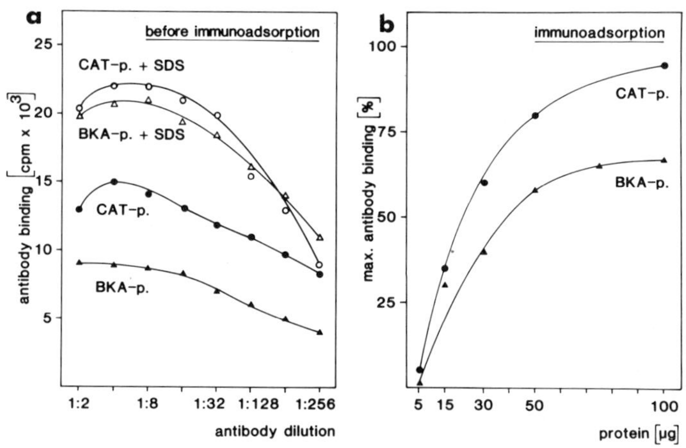

FIGURE 7. (a) The binding of the heart-carboxyatractylate protein antiserum to the carboxyatractylate protein complex (CAT-p) (●) and the Bongkrekat protein complex (BKA-p) (▲) in its native form and after denaturation by SDS (+SDS) (○, △), For details see legend to FIGURE 2a. (b) Immunoadsorption of carboxyatractylate protein antiserum on the isolated CAT-p (●) and BKA-p (▲). For details see legend to FIGURE 2B.

the binding both in the "c"- and "m"-state,[26] the side-specific inhibition of the nucleotide-transport indicates topographic differences among the two states along the channel either directed to the "c"-side or the "m"-side.

AUTO-ANTIBODIES AGAINST THE ADP/ATP CARRIER IN DILATED CARDIOMYOPATHY

Sera of 32 patients with proven dilated cardiomyopathy (DCM) were tested. The diagnosis of DCM was established according to the WHO/ISFC task force as a chronic disorder of heart muscle of unknown cause or association.[30] Cases of specific heart muscle diseases or secondary cardiomyopathies were excluded. Diagnosis was principally based on cardiac catheterization data and the morphological analysis of endomyocardial biopsies, which were compatible with DCM.[31] The hemodynamic characteristics are given in TABLE 1. Sera from patients with proven coronary heart disease (CHD) ($N = 20$), suspected alcoholic heart disease (AHD) ($N = 8$), hyper-

TABLE 1. Cardiological Characteristics of Patients with Dilated Cardiomyopathy

	Age (years)	Sex	EF (%)	EDV (ml)	CI ($1/min/m^2$)
DCM group I	42	3 f	19	330	1.9
EF < 40%	±17	9 m	±10	±99	±0.4
DCM group II	44	5 f	62	155	2.7
EF > 40%	±10	15 m	±12	±45	±0.6

Abbreviations: DCM: dilated cardiomyopathy, EF: ejection fraction, EDV: end diastolic volume, CI: cardiac index, f: female, m: male.

trophic obstructive cardiomyopathy (HOCM) ($N = 6$), and from healthy blood donors (controls) ($N = 30$) were used as controls.

As shown in FIG. 8a the autoantibodies directed against the ADP/ATP carrier were significantly elevated in the sera from 24 of 32 DCM patients tested. In comparison sera from patients with CHD, AHD, or HOCM were within the control range. Regression analysis showed a weak but statistically significant relation between the antibody titer and the hemodynamic function. Those patients with an ejection fraction higher than 40% (group II) had a significantly lower antibody titer than those with an ejection fraction of less than 40% (group I) (FIG. 8b).

Since the ADP/ATP carrier seems to exhibit organ-specific antigenic determinants, we expected that the DCM autoantibodies are also organ specific. The data shown in FIG. 9a for group I (EF < 40%) demonstrate a significantly lower binding activity of the autoantibodies to the liver protein in comparison to the heart protein

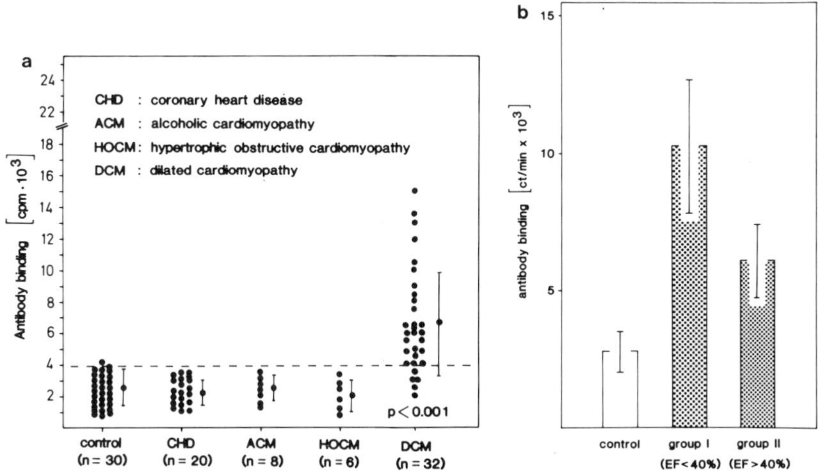

FIGURE 8. (a) The binding of autoantibodies to the ADP/ATP carrier from heart in IRMA from patients with dilated cardiomyopathy (DCM), coronary heart disease (CHD), suspected alcoholic cardiomyopathy (ACM), hypertrophic obstructive cardiomyopathy (HOCM), and 30 healthy blood donors (controls). For methodological details see legend to FIGURE 2a. (b) Autoantibody activity expressed in counts per minute (cpm) from patients with DCM in correlation to the hemodynamic function: group I = ejection fraction (EF) <40%; group II = EF > 40%. Control = 30 healthy blood donors.

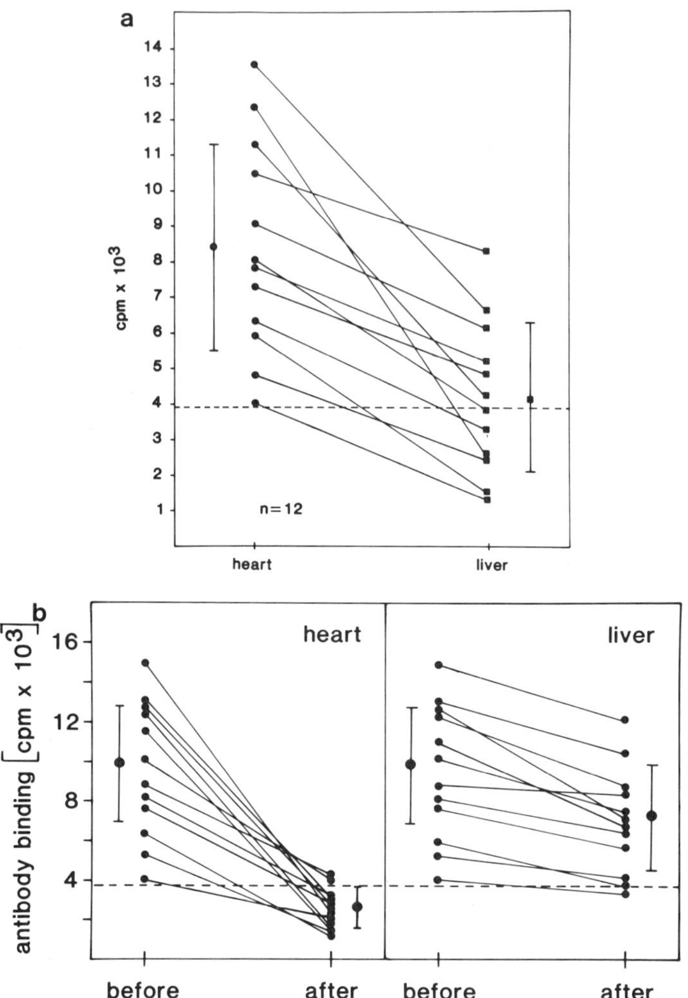

FIGURE 9. The binding of the autoantibodies from patients with dilated cardiomyopathy (group I) to the isolated ADP/ATP carrier from heart (●) and liver (■) (a). Binding activity of the autoantibodies to the heart protein before and after immunoadsorption on the isolated ADP/ATP carrier proteins from heart and liver (b). For methodological details see legend to FIGURE 2b.

($p < 0.01$). To exclude the possibility that the suspected organ specificity is only the result of a lower affinity of the antibodies to the antigenic determinants of the liver protein, immunoadsorption studies were performed. For that purpose the interaction of the autoantibodies with the isolated ADP/ATP carrier protein from heart and liver was examined. The maximum amount of an antibody bound by a given immunoadsorbant (plateau binding value) was determined by titrating a fixed amount of antiserum with increasing amounts of antigen. As shown for the group I sera, the ADP/ATP carrier from heart adsorbed most of the autoantibodies, while the residual activity of the antiserum after immunoadsorption on the liver protein was significantly higher ($p < 0.001$) (FIG. 9b). These results were also confirmed by immunoadsorption studies on intact mitochondria from heart and liver (data not shown).

The ability of the antiserum to inhibit nucleotide exchange activity was studied by the inhibitor-stop method combined with the back-exchange.[19] None of the control immunoglobulin preparations produced a blockade of the exchange rate. However, about 60% of the immunoglobulin preparations from DCM-patients tested caused a decrease in the exchange rate (FIG. 10A). To exclude artifactual causes of the inhibitory effect, the tests were performed with the purified IgG fraction of the sera. Secondly, the inhibition of the exchange rate was abolished after precipitation of the antibodies by Sepharose-protein A. Lastly, the preimmunoadsorption of the antibodies on isolated heart mitochondria prevented the inhibition of the ADP/ATP carrier.

Regression analysis revealed a significant relation between the nucleotide exchange rate inhibition and the clinical class (FIG. 10B). However, no apparent correlation was found between the anticatalytic activity and the IRMA binding properties. The adenine nucleotide transport was also measured on liver and kidney mitochondria. Under these conditions none of the DCM sera (isolated IgG) inhibited the nucleotide transport. Again these results demonstrate the organ specificity not only of the autoantibodies but also of the adenine nucleotide translocator itself.

ANTIBODY-MEDIATED DISTURBANCE OF THE CELLULAR ENERGY METABOLISM

The autoantibodies reacting with the ADP/ATP carrier are directed against various antigenic parts of the protein. However, for a possible pathophysiological role

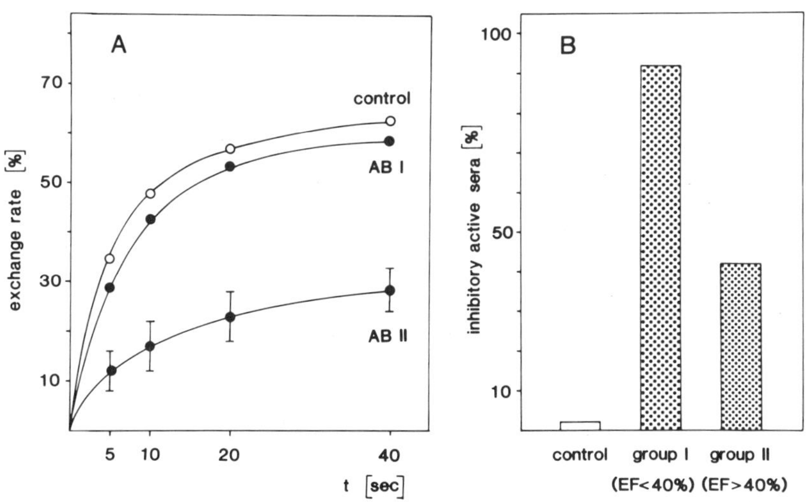

FIGURE 10. Effect of the autoantibodies from patients with dilated cardiomyopathy on the nucleotide transport (A) and the correlation of the inhibitory activity of the antisera to the hemodynamic function: group I = ejection fraction (EF) <40%; group II = EF > 40%. For methodological details see legend to FIGURE 3.

of the antibodies only those that bind to the substrate binding site or inhibit the translocation-step are of interest. Suspecting a pathogenetic role, one has to ask whether these immunoinactivating antibodies can act as antagonists not only *in vitro* but also *in vivo* by blocking the binding of the "natural ligand" with the consequence of an inhibition of the physiological effect?

Autoantibodies to cell membrane receptors have been documented in a number of disease states in man and there is good evidence for a pathogenic role for at least some of these autoantibodies.[1–5] However, up to now it is not clear whether autoantibodies against intracellular antigens may also have a pathophysiological meaning. To answer the question, if anti-ADP/ATP carrier antibodies do not only act *in vitro* but also influence the adenine nucleotide transport *in vivo,* we established an animal model: guinea pigs were immunized with the isolated ADP/ATP carrier from heart. After five booster injections in monthly intervals, the sera were tested for their anti-ADP/ATP carrier activity. The antibodies were characterized by Western blotting and by measuring the nucleotide transport.

In those animals, which produced significant titers of immunoinactivating antibodies, an intracellular action of the antibodies *in vivo* was demonstrated by measurements of the cytosolic and mitochondrial adenine nucleotide concentrations. These were evaluated in hemoglobin-free, perfused working guinea pig hearts[32,33] using density gradient centrifugation of the myocardial homogenates in non-aqueous media.[34,35] Perfusion experiments were terminated by stop-freezing the myocardium between aluminum blocks precooled in liquid nitrogen. Since at the time of freeze-clamping the myocardium, the heart preparations were in a steady state with respect to hemodynamic function and oxidative metabolism, the obtained compartmentation of adenine nucleotides appears to reflect an equilibrium between the disposal of ATP by extra mitochondrial sites and the rate of oxidative phosphorylation of ADP in the mitochondria.

Control animals showed a normal intracellular distribution of ATP (FIG. 11a). These values are in good agreement with those reported by Soboll[35] and Geisbühler.[36] However, the concentration of ATP in the hearts of guinea pigs immunized with the ADP/ATP carrier was lower in the cytosol, while the concentration was significantly increased in the mitochondria (FIG. 11a). Besides this remarkable shift of the cytosolic and mitochondrial ATP concentration, the asymmetric distribution of ATP between the cytosolic and mitochondrial compartment was reversed in the immunized animals. Consequently, the ATP/ADP ratio was significantly lower in the cytosol, while in the mitochondria, a substantial increase was observed (FIG. 11b). By calculating the subcellular phosphorylation potentials, the difference between the cytosolic and mitochondrial phosphorylation potentials was significantly decreased in the immunized animals (FIG. 11c).

These results show a significant decrease of myocardial high energy phosphates in the immunized guinea pig hearts in comparison to the control hearts, although the myocardial function was the same in both groups. Since it was shown that the ratio ATP/ADP is a suitable parameter for the evaluation of the energy state or the balance between mitochondrial phosphorylation and ATP consumption in the cytosol, our findings indicate an inhibition of the nucleotide transport in the hearts of the immunized animals. This becomes even more evident by the drastically lowered phosphorylation potential difference (Δ G) between the cytosolic and mitochondrial ATP and the parallel increase of the mitochondrial ATP/ADP ratio. Similar results were described when carboxylatractyloside is added to the perfused liver.[38] This clearly demonstrates that the diminished difference of the cytosolic and mitochondrial ADP/ATP ratio is indeed caused by the impaired function of the ADP/ATP carrier.

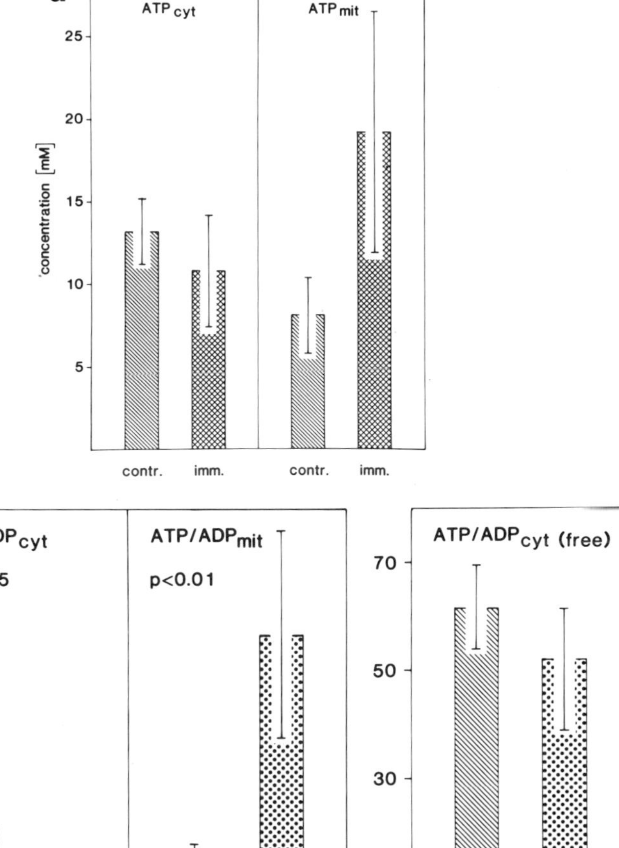

FIGURE 11. Subcellular distribution of myocardial high energy phosphates in perfused working guinea-pig hearts stimulated by 0.08 μM noradrenalin. Data for cytosol (cyt) and mitochondria (mit) are given as concentrations (mM), $x \pm$ SEM, $N = 5$ for controls, $N = 4$ for immunized animals. Ratios of free cytosolic ATP/ADP or (ATP)/(ADP) $\times$ (P_i) were calculated from the mass action ratio of the creatine kinase reaction (ATP)/(ADP) $-$ (creatine phosphate) $\times$ (H^+)/((creatine) $\times$ K_{CPK}). G(kJ/mol) = difference between concentration terms of cytosolic and mitochondrial phosphorylation potentials. Statistical significance between controls and immunized hearts using the Student *t*-test for unpaired samples are indicated as $p < 0.005$ or $p < 0.01$. Fractionation of the tissue into cytosol and mitochondria was accomplished utilizing density gradient certrifugation in non-aqueous media according to Elbers *et al.*[34] Lyophilized tissue was homogenized in heptane/carbon tetrachloride ($d = 1.23$ g/ml) and subsequently fractionated in a heptan/carbon tetrachloride density gradient ($d = 1.29$–1.38 g/ml). Centrifugation (27,000 $\times$ *g*, 4 hr) yielded eight fractions, each contained different portions of cytosolic and mitochondrial protein. In each fraction the cytosolic and mitochondrial marker enzymes, the high energy phosphates as well as the protein contents were determined. Subcellular concentrations were obtained assuming 3.8 μl water per mg cytosolic protein and 1.8 μl water per mg protein in heart mitochondria, respectively.[52,57] The equilibrium constant of the creatine kinase reaction was $K_{CPK} = 2.04 \times 10^{-9}$.[58]

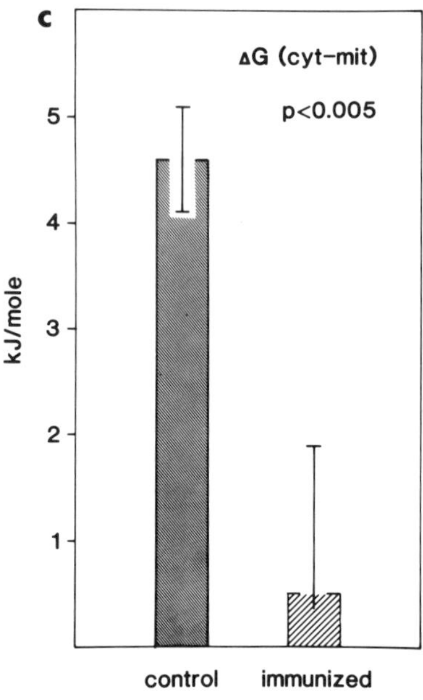

FIGURE 11. *Continued*

IMMUNOLOGICAL CROSS-REACTIVITY BETWEEN THE ADP/ATP CARRIER AND A CELL-SURFACE PROTEIN OF CARDIAC MYOCYTES

The data presented suggest that autoimmunity to the ADP/ATP carrier may contribute to the pathophysiology of autoimmune diseases by causing an autoantibody-mediated imbalance between energy delivery and demand. As a prerequisite for an internalization of antibodies into myocardial cells one has to postulate a binding-site at the cell surface. First, evidence for a cross-reactivity between the ADP/ATP carrier and a cell surface protein was obtained by indirect immunofluorescence. Incubation of frozen sections of heart tissue with anti-ADP/ATP carrier antibodies showed beside an intracellular staining an antibody binding to the plasma membrane (FIG. 12a). This was confirmed by a positive staining of isolated cardiac myocytes showing a sarcolemmal immunofluorescence (FIG. 12b). After neutralization of the anti-ADP/ATP carrier antibodies by preadsorption with the isolated ADP/ATP carrier the intracellular staining and the staining of the cell surface disappeared. A time- and concentration-dependent binding of antibodies to the cell surface was also demonstrated in a radioimmuno binding assay (FIG. 13a). Isolated cardiac myocytes were incubated with anti-ADP/ATP carrier antibodies (IgG). Bound antibodies were detected by iodinated protein A.[39,40] To exclude an unspecific binding of the antibodies to the cell surface, the antiserum was preincubated with the isolated ADP/ATP carrier. As shown in FIGURE 13b, there was a concentration-dependent decrease of the antibody binding.

Immunofluorescence data and the results from the radioimmunobinding assay are further supported by immunoblot analysis and immunoprecipitation of plasma membranes isolated from cardiac myocyte (FIG. 14). Membranes were obtained by a modified method of Cates and Holland,[59] using sucrose density gradient centrifugation

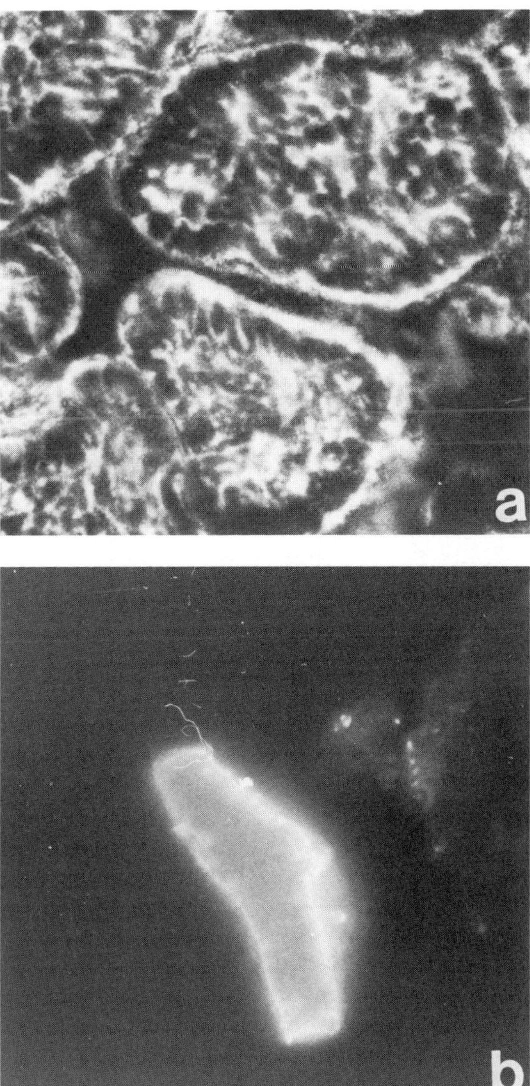

FIGURE 12. Immunofluorescence localization of anti-ADP/ATP carrier antibody binding. (a) Frozen sections (4 μm) of guinea pig heart were covered with anti-ADP/ATP carrier antiserum diluted 1:10 for 30 min. The sections were then washed in phosphate-buffered saline (PBS) and covered with FITC-conjugated anti-rabbit IgG. (b) Isolated cardiac myocytes[60] were incubated with the anti-ADP/ATP carrier antiserum and stained by FITC-conjugated anti-rabbit IgG.

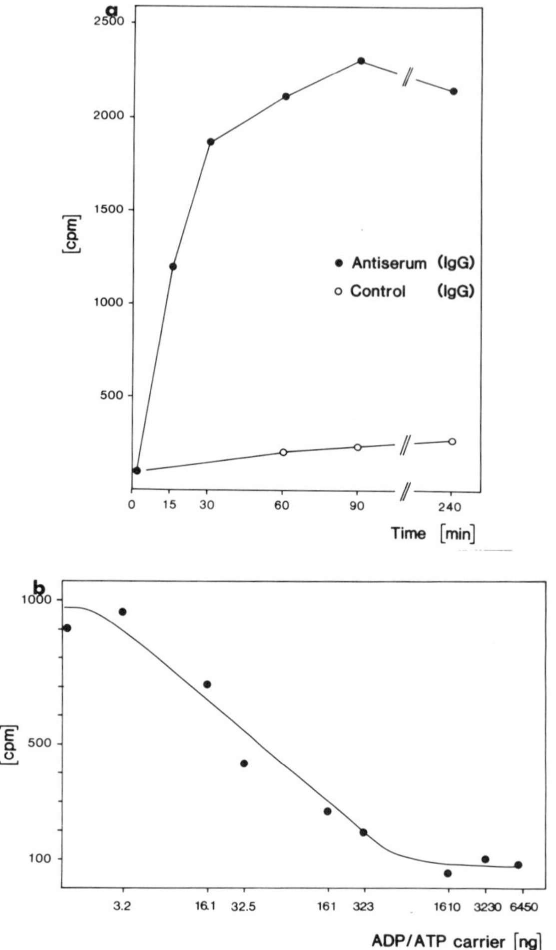

FIGURE 13. Binding of antibodies to the ADP/ATP carrier is isolated cardiac myocytes (a) and inhibition of the antibody binding by preadsorption of the antibodies on the isolated ADP/ATP carrier protein (b). Cardiac myocytes in Krebs-phosphate buffer + 0.5% BSA were incubated with anti-ADP/ATP carrier antibodies, washed, and the bound IgGs were detected using [^{125}I]-protein A 20,000 cpm/200,000 cells, specific activity 0.7 μCi/μg).

of homogenized heart muscle tissue after high salt extraction of contractile elements. In the Western blot these antibodies directed against the ADP/ATP carrier recognized three proteins of the membrane fraction (FIG. 14c). The main component detected by the antibodies is a 32 K protein. This protein could be identified as the ADP/ATP carrier of the mitochondria (FIG. 14d), contaminating our plasma membrane fraction. The other two proteins, having molecular weights of 47 K and 29 K, were not found in mitochondrial membranes. By immunoprecipitation experiments it could be shown that both proteins belong to the plasma membrane (FIG. 14c). For that purpose the cell surface of isolated adult living cardiac myocytes was iodinated by the lactoperoxidase-glucoseoxidase method. In the following immunoprecipitation of detergent-solubilized plasma membranes again three proteins were recognized by the anti-ADP/ATP carrier antibody. However, only the 47 K and 29 K proteins were iodinated demonstrating that both proteins belong to the cell surface (FIG. 14e). Accordingly the intracellular 32 K component was not detected by autoradiography, however, after staining of the gels with $AgNO_3$.

In conclusion these results demonstrate a cross-reactivity between the ADP/ATP carrier of the inner mitochondrial membrane and two proteins of the cell surface. These two proteins might serve as receptors for "receptor-mediated endocytosis," by which the antibodies against the ADP/ATP carrier are taken up by the cells. Recent data[41] indicate that the protein with a molecular weight of 47 K is the subunit of the connexon, a channel-forming protein located within the gap-junctional region, while the 29 K protein is a degradation product of the 47 K protein.[42]

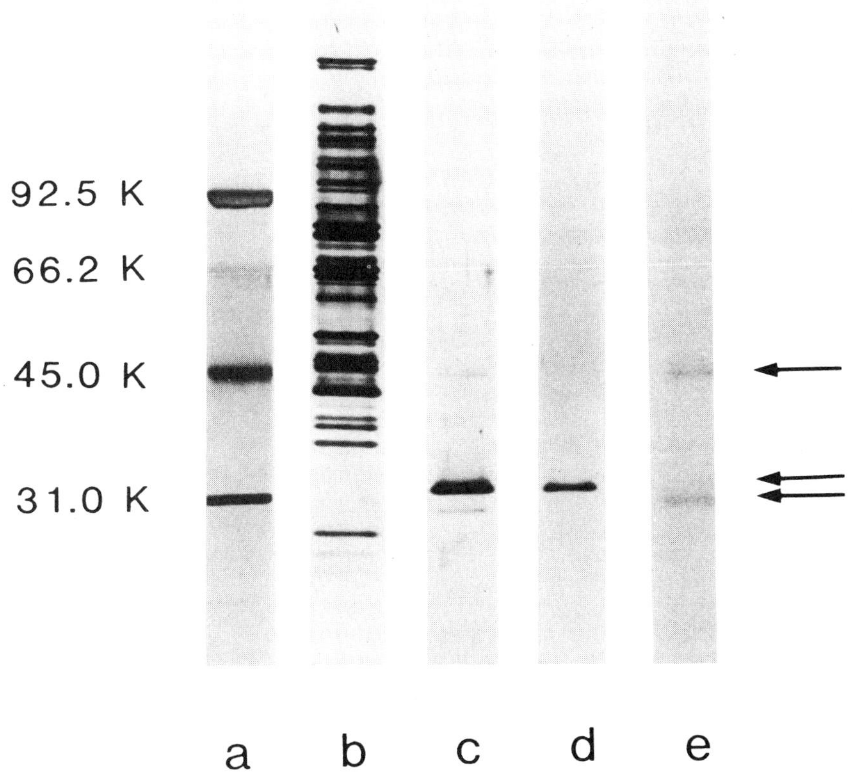

FIGURE 14. Immunoblot analysis and immunoprecipitation experiments. Samples were resolved in 12.5% SDS-polyacrylamide gels and transferred to nitrocellulose; antibody binding was determined using peroxidase-conjugated anti-rabbit IgG. (Lane a) molecular weight standard, (b) plasma membrane fraction stained with $AgNO_3$ after transfer to nitrocellulose, (c) the positions of the 47K, 32K, and 29K membrane proteins recognized by immunoblot technique with anti-ADP/ATP carrier antibodies, (d) purified ADP/ATP carrier, and (e) autoradiography of immunoprecipitated and iodinated cell surface proteins. The cell surface proteins were precipitated with the anti-ADP/ATP carrier antiserum (isolated IgG).

CONCLUSIONS

After the identification of the ADP/ATP carrier as an autoantigen in different autoimmune diseases,[8,9,43,44] the precise immunochemical characterization of the antigen will provide important clues regarding the cause of autoimmune diseases such as dilated cardiomyopathy. The present results confirm and extend earlier observations on the organ and conformation specificity of this protein.[14,20,29,45] The organ specificity shown by the radioimmunoassay and by immunoadsorption studies was also substantiated by the tissue specificity of the inhibitory activity of the antibodies, as demonstrated both with the experimental sera and with autoantibodies from patients suffering from dilated cardiomyopathy. Part of the antibodies also inhibited the binding of the specific ligand carboxyatractylate to the carrier protein. Also this inhibition of the ligand binding was organ specific.

Based on these results, one could suspect that the substrate/ligand binding center itself is organ specific. However, this seems surprising, since one would expect an invariant substrate binding site of the carrier in the various organs, as it was shown recently for the NADPH binding region of enzymes possessing dehydrogenase activities.[46] Therefore one can assume that the specific inhibition of the ligand binding is caused by an antibody-induced conformational change of the binding site and not by a direct blockade of the ligand-binding center through the antibodies. The organ-specific inhibitory effect of the antibodies indicates the existence of organ-specific topographic regions along the central channel between the two subunits of the carrier dimer. However, it cannot totally be excluded that the binding center itself is also tissue specific. To answer this question definitely, monoclonal antibodies will be necessary, since the polyclonality of both autoantibodies and experimental sera prevents the exact standardization of methodology. Nevertheless, this remarkable organ specificity of the ADP/ATP carrier reflects isoenzyme distribution in various organs, which may be adapted to the specific requirements in different tissues.

Beside the principle importance of the organ specificity for the understanding of basic regulatory mechanisms of differentiated organisms, this aspect has also become particularly significant for autoimmune processes.[8,44] In relation to the pathogenetic etiology of these diseases, one can assume that the causes for the production of these highly specific autoantibodies are limited to the local milieu. This observation is not in accord with previous theories suggesting that the immunologic abnormalities in autoimmune diseases result from a generalized defect in immune regulation. In particular, the high concentration and restricted specificities of the anti-ADP/ATP autoantibodies argue strongly against a random polyclonal activation of lymphocytes.

The specificity of the autoantibodies for the ADP/ATP carrier and their functional activity as inhibitors *in vitro* suggest that they are mediators of autoimmune diseases at the molecular level. Inhibition of the ADP/ATP carrier protein *in vivo* could upset the balance between energy delivery and demand in the cell. In experimental animals, evidence is obtained for a diminished activity of the nucleotide transport *in vivo* after immunization of guinea pigs with the isolated carrier protein. The decreased transport activity is followed by an increase of the mitochondrial ADP/ATP ratio and a parallel lowering of the cytosolic ADP/ATP ratio. Based on the functional and regulatory characteristics of the ADP/ATP carrier[10,11,53] this indicates a stimulation of the remaining active carrier proteins, thereby partly compensating for the inhibitory action of the antibodies by an increased net transfer of ATP to the cytosol.

There are several ways in which antibodies against the ADP/ATP carrier could alter the function of this protein. For example, the antibodies might inhibit carrier function by antibody binding to the protein. This inhibition might be due to a binding of the antibody direct to the active site of the translocator or near the active site thereby sterically hindering the translocation process. Another possibility to be considered is that the antibodies react with the primary translation product,[54,55] hindering the incorporation of the carrier protein into the inner mitochondrial membrane. Both possibilities are suggested by immunohistological and immunoelectronmicroscopical results, showing an immunoglobulin binding to the mitochondrial membrane in the immunized animals (Schultheiss and Herzog, in preparation). Finally, the anti-carrier antibodies might cause an antigenic modulation of the protein increasing carrier degradation, as it was shown for other receptors in autoimmune diseases.[4,47–49]

However, for an intracellular action of the antibodies the internalization of the antibodies is an prerequisite. Up to now there was no experimental proof for a specific uptake of autoantibodies by cells and an intracellular action of these antibodies—a surprising fact considering the enormous variety of autoantibodies against mitochondrial, nuclear, cytoplasmic, and other intracellular antigenic determinants found in autoimmune diseases. However, some authors discussed the possibility of intracellular acting autoantibodies.[50,51,56] The mechanism proposed for the antibody uptake was Fc-receptor mediated endocytosis.[56] The data presented here demonstrate the existence of a cell surface protein, which cross-reacts with the ADP/ATP carrier of the inner mitochondrial membrane. Immunoblot and immunoprecipitation experiments make it quite likely that this cross-reacting protein on the cell surface is the connexon, a channel-forming protein within the gap junctional region. These results were recently confirmed by immunoelectronmicroscopical studies showing a specific binding of the anti-ADP/ATP carrier antibodies to the gap junctions (H.P. Schultheiss und V. Herzog, in preparation). By the specific binding of the antibodies to the connexon one can suspect that this cell surface protein is the receptor for a "receptor-mediated endocytosis" of the anti-ADP/ATP carrier antibodies.

Regardless of the mechanism of the intracellular action of the autoantibodies directed against the ADP/ATP carrier, the data indicate that autoimmunity to this translocator protein may cause an antibody-mediated imbalance between energy delivery and demand. The clinical consequences and theoretical implications of the antibodies may be important. Are they the factor in heart muscle diseases of unknown cause that is a prominent feature of dilated cardiomyopathy? Considering the enormous variety of human autoimmune diseases of which etiologies are still unknown, it is interesting to speculate that many of these could be caused by a failure of specialized or differentiated function of a cell due to energy deficiency, which is caused by intracellularly acting autoantibodies directed against the ADP/ATP carrier.

REFERENCES

1. VENTER, J. C. & L. C. HARRISON. 1980. Science **207:** 1361–1363.
2. DOBERSEN, M. J., J. E. SCHARFF, F. GINSBERG-FELLNER & A. L. NOTKINS. 1980. N. Engl. J. Med. **303:** 1493–1498.
3. MARON, R., D. ELIAS, B. M. DE JONGH, G. J. BRUINING, J. J. VAN ROOD, Y. SHECHTER & I. R. COHEN. 1983. Nature **303:** 817–818.
4. DRACHMAN, D. B., R. N. ADAMS, L. F. JOSIFEK & S. G. SELF. 1982. N. Engl. J. Med. **307:** 769–775.
5. HASPEL, M. V., T. ODNODERA, B. S. PRABHAKAR, P. R. MCCLINTOCK, K. ESSANI, U. R. RAY, S. YAGIHASHI & A. L. NOTKINS. 1983. Nature **304:** 73–76.
6. SMITH H. R. & A. D. STEINBERG. 1983. Ann. Rev. Immunol. **1:** 175–210.
7. SHOENFELD, Y. & R. S. SCHWARTZ. 1984. N. Engl. J. Med. **311:** 1019–1029.
8. SCHULTHEISS, H. P. & H. D. BOLTE. 1985. J. Mol. Cell. Cardiol. **17:** 603–617.
9. SCHULTHEISS, H. P., P. SCHWIMMBECK, H. D. BOLTE & M. KLINGENBERG. 1985. *In*

Advances in Myocardiology. N.S. Dhalla & D.J. Hearse, Eds.: 311–327. Plenum Publ. Corp. New York.
10. KLINGENBERG, M. 1976. *In* The Enzymes of Biological Membranes. A. Martonosi, Ed. **3:** 383–438. Plenum Publ. Corp. New York.
11. KLINGENBERG, M. & H. W. HELDT. 1982. *In* Metabolic Compartmentation. H. Sies, Ed.: 101–122. Academic Press. London.
12. PFAFF, E. & M. KLINGENBERG. 1968. Eur. J. Biochem. **6:** 66–79.
13. KLINGENBERG, M. 1980. J. Membr. Biol. **56:** 97–105.
14. SCHULTHEISS, H. P. & M. KLINGENBERG. 1984. Eur. J. Biochem. **143:** 599–605.
15. RICCIO, P., H. AQUILA & M. KLINGENBERG. 1975. FEBS Lett. **56:** 129–132.
16. RICCIO, P., H. AQUILA & M. KLINGENBERG. 1975. FEBS Lett. **56:** 133–138.
17. KESSLER, S. W. 1981. Methods Enzymol. **73:** 442–457.
18. TOWBIN, H., T. STAEHELIN & J. GORDON. 1979. Proc. Natl. Acad. Sci. USA **76:** 4350–4354.
19. PALMIERE, F. & M. KLINGENBERG. 1971. Methods Enzymol. **56:** 279–301.
20. SCHULTHEISS, H. P. & M. KLINGENBERG. 1985. Arch. Biochem. Biophys. 239–279.
21. KRÄMER, R. & M. KLINGENBERG. 1979. Biochemistry **18:** 4209–4215.
22. ERDELT, H., M. J. WEIDEMANN, M. BUCHHOLZ & M. KLINGENBERG. 1972. Eur. J. Biochem. **30:** 107–122.
23. SCATCHARD, G. 1949. Ann. N.Y. Acad. Sci. **51:** 660–672.
24. EAST, I. J., J. G. R. HURRELL, P. E. E. TODD & S. J. LEACH. 1982. J. Biol. Chem. **257:** 3199–3202.
25. EAST, I. J., P. E. TODD & S. J. LEACH. 1980. Molec. Immunol. **17:** 519–525.
26. KLINGENBERG, M. & M. APPEL. 1980. FEBS Lett. **119:** 195–199.
27. BERZOFSKY, J. A., G. K. BUCKENMEYER, G. HICKS, F. R. N. GURD, R. J. FELDMANN & J. MINNA. 1982. J. Biol. Chem. **257:** 3189–3198.
28. URBANSKI, G. J. & E. MARGOLIASH. 1977. J. Immunol. **118:** 1170–1180.
29. BUCHANAN, B. B., W. EIERMANN, P. RICCIO, H. AQUILA & M. KLINGENBERG. 1976. Proc. Natl. Acad. Sci. USA **73:** 2280–2284.
30. Report of a WHO Expert Committee. 1984. *In* Cardiomyopathies. World Health Organization Technical Report Series **697:** 7–68. World Health Organization. Geneva.
31. OLSEN, E. G. J. 1979. Am. Heart J. **98:** 385–392.
32. BÜNGER, R., O. SOMMER, G. WALTER, H. STIEGLER & E. GERLACH. 1979. Pflügers Arch. **380:** 259–266.
33. BECKER, B. F. & E. GERLACH. 1984. Klin. Wochenschr. **62** (Suppl. II): 58–66.
34. ELBERS, R., H. W. HELDT, P. SCHMUCKER, S. SOBOLL & H. WIESE. 1974. Hoppe-Seyler's Z. Physiol. Chem. **355:** 378–393.
35. KLINGENBERG, M. & H. W. HELDT. 1982. *In* Metabolic Compartmentation. H. Sies, Ed.: 101–122. Academic Press. London.
36. GEISBUHLER, T., R. A. ALTSCHULD, R. W. TREWYN, A. Z. ANSEL, K. LAMKA & G. P. BRIERLY. 1984. Circ. Res. **54:** 536–546.
37. BERZOFSKY, J. A. & A. N. SCHECHTER. 1981. Molec. Immunol. **18:** 751–763.
38. SOBOLL, S., R. SCHOLZ & H. W. HELDT. 1978. Eur. J. Biochem. **87:** 377–390.
39. KÜHL, U., G. ULRICH & H. P. SCHULTHEISS. 1986. Eur. Heart J. (In press.)
40. ULRICH, G., U. KÜHL, J. JANDA & H. P. SCHULTHEISS. 1986. Eur. J. Heart. (In press.)
41. KÜHL, U., G. ULRICH, E. PAWELS & H. P. SCHULTHEISS. 1986. EMBO J. (In press.)
42. MANJUNATH, C. K. & E. PAGE. 1985. Am. Physiol. Society: H783–H791.
43. SCHULTHEISS, H. P., P. A. BERG, & M. KLINGENBERG. 1983. Clin. Exp. Immunol. **54:** 648–654.
44. SCHULTHEISS, H. P., P. A. BERG & M. KLINGENBERG. 1984. Clin. Exp. Immunol. **58:** 596–602.
45. EIERMANN, W., H. AQUILA & M. KLINGENBERG. 1977. FEBS Lett. **74:** 209–214.
46. KATIYAR, S. S. & J. W. PORTER. 1983. Biochemistry **80:** 1221–1223.
47. APPEL, S. H., R. ANWYL, M. W. MCADAMS & S. ELIAS. 1977. Proc. Natl. Acad. Sci. USA **74:** 2130–2134.
48. FIGURA, VON, K., V. GIESELMANN & A. HASILIK. 1984. EMBO J **3:** 1281–1286.
49. GRUNFELD, C. 1984. Proc. Natl. Acad. Sci. USA **81:** 2508–2511.
50. BOTTAZZO, G. F. 1984. Diabetologica **26:** 241–249.

51. MATHEWS, M. B. & R. M. BERNSTEIN. 1983. Nature **304:** 177–179.
52. SOBOLL, S., R. SCHOLZ, M. FREISL, R. ELBERS & H. W. HELDT. 1976. *In* Use of Isolated Liver Cells and Kidney Tubules in Metabolic Studies. J.M. Tager, H.D. Söling & J. R. Williamson, Eds.: 29–40. North Holland Publishing Company. Amsterdam.
53. KLINGENBERG, M. & H. ROTTENBERG. 1977. Eur. J. Biochem. **73:** 125–130.
54. HACKENBERG, H., P. RICCIO & M. KLINGENBERG. 1978. Eur. J. Biochem. **88:** 373–378.
55. ZIMMERMANN R. & W. NEUPERT. 1980. Eur. J. Biochem. **109:** 217–229.
56. ALARCON-SEGOVIA, D., A. RUIZ-ARGUELLES & E. FISHBEIN. 1978. Nature **271:** 67–69.
57. KAUPPINEN, R. A., J. K. HILTUNEN & I. E. HASSINEN. 1980. FEBS Lett. **112:** 273–276.
58. ALTSCHULD, R. A. & G. P. BRIERLEY. 1977. J. Mol. Cell Cardiol. **9:** 875–896.
59. CATES, G. A. & P. C. HOLLAND. 1978. Biochem. J. **174:** 873–881.
60. ULRICH, G., U. KÜHL & H. P. SCHULTHEISS. 1986. J. Mol. Cell Cardiol. (In press.)

DISCUSSION OF THE PAPER

L. H. OPIE (*University of Capetown, Capetown, South Africa*): Is the effect on the heart really organ specific? From your data it seemed more organ selective. Could some of the impaired functions of skeletal muscle in patients with cardiomyopathy be explained by interaction of circulating antibodies with the skeletal muscle ADP-ATP transporter?

SCHULTHEISS: We could show that the ADP/ATP carrier proteins from heart, kidney, and liver partially cross-react. However, there are organ-specific determinants, as we could show by immunoadsorption studies. In this connection it is important that the antibody-mediated inhibition of the translocator was always organ specific: that is, with an antibody directed against the heart protein one can only inhibit the nucleotide transport of heart mitochondria and not of liver or kidney mitochondria.

To your second question: up to now we did not look for an impaired function of skeletal muscle mitochondria in dilated cardiomyopathy.

L. E. ROSENBERG (*York University, New Haven, CT*): Do the anti-ADP/ATP carrier antibodies interfere with proper insertion of the newly synthesized ADP/ADP carrier into the inner mitochondrial membrane?

SCHULTHEISS: We have not done these experiments yet. There are some preliminary results, reported by Neupert and co-workers, that might indicate that this is a possible mechanism.

E. CARAFOLI (*Swiss Federal Institute of Technology (ETH), Zurich*): You have reported high titers of anti-adenine nucleotide translocase in primary biliary cirrhosis patients. Considering that the antigen in this condition does not seem to be any of the major proteins of the respiratory chain or of the energy-conserving machinery, what else do you think your antibody recognizes in addition to the ATP/ADP translocase?

SCHULTHEISS: We could show recently that the antibodies directed against the ADP/ATP carrier bind specifically to the cell surface indicating that there are cross-reacting proteins on the cell surface. Recent results indicate that the connexon located within the gap junctions binds the anti-ADP/ATP carrier antibodies. However at this moment we can't say whether the cell-surface protein or the ADP/ATP carrier is the primary antigen against which the auto-aggression is directed.

I. M. ARIAS (*Columbia University, New York*): The primary cellular site of the lesion in PBC is not well understood. Application of your studies to PBC may hopefully elucidate which cells (hepatocyte or cells of the biliary tract) is the primary site.

SCHULTHEISS: Up to now we did not differentiate between the different cells in the liver. We can only say that we find autoantibodies against the ADP/ATP carrier isolated from liver in patients suffering from PBC. These antibodies are also organ specific and inhibit the nucleotide transport from liver mitochondira *in vitro.*

B. CHANCE (*University of Pennsylvania, Philadelphia, PA*): The cardiac respiratory inhibition you showed is so small in relation to the tremendous bioenergetic reserve that it is confusing that a cardiomyopathy is observed. Perhaps some other factor is required for the development of the disease?

SCHULTHEISS: Up to now we did not measure the respiratory activity in the immunized animals. What we could show is a significant increase of the mitochondrial ATP concentration and a slight decrease of the cytosolic ATP concentration. Parallel to that there is an increased lactate release. The question as to whether the inhibition of the nucleotide transport *in vivo* is or is not the main factor for the development of the disease can't be answered yet. The aim of this contribution was to show that there is an antibody-mediated disturbance of the myocardial cellular metabolism and that this new mechanism might be of pathophysiological importance in this and other diseases.

Q. AL-AWQATI (*Columbia University, New York, NY*): (1) After the antibody is internalized by endocytosis how does it reach the mitochondria? (2) You don't think it enters the cell by focal cell lysis?

SCHULTHEISS: (1) To my knowledge—up to now—there are no experimental data that could explain this phenomenon. However, there seems to be a second possibility to explain how the antibody could act intracellularly; it could bind to the protein before this is integrated into the inner mitochondrial membrane hindering the assembly in becoming an active functioning transport system.

(2) Based on our results we can say that the antibodies can and do enter the cells without local cell lysis. Morphologically, based on electron microscopy criteria, the cells are not destroyed. Functionally, we obtain a time- and concentration-dependent reduction of the contraction velocity; however, the isolated cardiac myocytes still function regularly. Finally, we only see intracellular antibodies located in endocytotic vesicles.

P. COLEMAN (*New York University, New York, NY*): You briefly mentioned that the etiology of dilated cardiomyopathy involves a primary viral infection. Could you briefly describe what is known about the viral agent and the specificity of the target organs for the suspected virus?

SCHULTHEISS: In most cases virus infection in myocarditis and dilated cardiomyopathy is induced by Coxsackie B viruses. Recently it was shown by hybridization techniques that viral RNA can be found in the myocardium of patients suffering from dilated cardiomyopathy. In addition, there are experimental studies that show the development of a dilated cardiomyopathy after virus infection (for example with Coxsakie B3-Nancy).

Studies on Giant Mitochondria

BERNARD TANDLER[a] AND CHARLES L. HOPPEL

Department of Oral Biology
School of Dentistry
Departments of Pharmacology and Medicine
School of Medicine
Case Western Reserve University
VA Medical Center
Cleveland, Ohio 44106

Mitochondria, which play a cardinal role in the economy of the cell, are excellent morphological indicators of the state of health of cells. One of the ways in which these organelles respond to metabolic injury is to increase in size, often to a point where they merit the appellation of giant mitochondria or megamitochondria. Although quite large, even gigantic, mitochondria may occur in certain normal cells,[1] it is those organelles that have become greatly enlarged as a result of pathological processes in typical cells such as hepatocytes or cardiomyocytes that are of concern in this review.

Giant mitochondria have been reported to occur in variety of human diseases and in liver cells of aged humans[2–16] (TABLE 1). Such mitochondria are usually observed retrospectively in fixed specimens and, because of their source, rarely are available for experimentation. Obviously, giant mitochondria that are experimentally induced in laboratory animals are indispensable to the study of altered mitochondrial morphology and function. Giant mitochondria have been experimentally produced in rodents, principally in their hepatocytes (but in other animals and cell types as well), by nutritional manipulation[17–28] (TABLE 2), pharmacological agents[29–49] (TABLE 3), ethanol intoxication[50–64] (TABLE 4), and hormones[65–68] (TABLE 5).

Giant mitochondria in liver cells have one of three basic morphologies, depending on the means whereby they are induced. Hepatic megamitochondria occurring in mice fed a riboflavin-deficient diet may be up to 10 μm in diameter[17] (FIG. 1). They tend to be spherical and are very similar to, although considerably larger than, normal-sized mitochondria in the same cells. Within a unit cross-section area, the number and distribution of cristae are the same regardless of organelle size, although some megamitochondria may contain a stack of cristae. Intramitochondrial dense granules are scattered throughout the matrix of the giant mitochondria, which is of normal density (FIG. 2). In contrast, giant hepatic mitochondria (greater than 10 μm in diameter) resulting from feeding mice or rats nialamide,[43] ethidium bromide,[40,41] or the copper-chelating agent, cuprizone[29–33] (FIG. 3), or from injection with triamcinolone,[68] have short cristae that are restricted to the organelle periphery (FIG. 4); the matrix, which is greatly expanded, is of near normal density but lacks membranes. It should be noted that these seemingly diametrically opposed arrangements of cristae are not invariably related to a particular method of megamitochondrial induction. When galactoflavin, a reversible antagonist of riboflavin, is added to a riboflavin-deficient diet, the appearance of symptoms of ariboflavinosis is greatly accelerated, with giant mitochondria being observed in mice as early as 13 days after initiation of the diet,

[a]Address all correspondence to Dr. Bernard Tandler, School of Dentistry, Case Western Reserve University, Cleveland, Ohio 44106.

TABLE 1. Giant Mitochondria in Human Disease States and Aging

Condition	References
Hepatic porphyria	2
Cerebral cortex in microcephaly	3
Kidney diseases	4–6
Reye's syndrome	7
Erythroleukemia	8
Megakaryoblastic leukemia	9
Gilbert's disease	10
Systemic scleroderma	11
Heart diseases	12
Pleomorphic adenoma	13
Diabetes	14
Aging	15,16

TABLE 2. Giant Mitochondria Induced by Nutritional Deficiency

Condition	Species	Tissue	Duration	References
Riboflavin deficiency	Mouse	Hepatocyte	42 d	17,18
	Rat	Hepatocyte	82 d	18,19
	Mouse	Renal tubule	5–7 wk	20
Galactoflavin supplementation	Mouse	Hepatocyte	21–22 d	21
Vitamin D deficiency	Rat	Hepatocyte	21 d	22
Partial starvation	Rat	Hepatocyte	6 wk	23
Starvation	Rat	Hepatocyte	9 d	24
Protein deficiency	Rat	Hepatocyte	13 wk	25
	Chick	Hepatocyte	1 d	26
Iron deficiency	Rat	Hepatocyte	3–8 wk	27
Necrogenic diet	Rat	Hepatocyte	19 d	28

TABLE 3. Drug-Induced Giant Mitochondria in Liver

Condition	Duration	Reference
Cuprizone	1–3 wk	29
	9–12 d	30
	5–6 d	31,32
	10 d	33
Penicillamine	3 d	34
Diethyldithiocarbamate	9–18 d	35,36
Chloramphenicol	10 d	37
	5 d	38
	9–11 d	39
Ethidium bromide	5 d	40,41
Concanavalin A	1 d	42
Nialamide	7–10 d	43,44
Isoniazid	7–10 d	44
Iproniazid	7–10 d	44
Hydrazine	7–10 d	45
γ-Benzene hexachloride	6–10 mo	46
n-Propyl alcohol	5–13 wk	47
n-Butyl alcohol	5–13 wk	47
l-Octadecanol	3–5 wk	48
Amanitotoxin	4 d	49

TABLE 4. Ethyl Alcohol-Induced Giant Mitochondria in Hepatocytes

Species	Reference
Human	50–61
Monkey	62
Baboon	63
Rat	64

TABLE 5. Hormone-Induced Giant Mitochondria in Hepatocytes

Condition	Species	Reference
Oral contraceptives	Human	65
Pregnancy	Human	66
Anabolic steroids	Sheep	67
Triamcinolone	Mouse	68

rather than after the 42 days required for their appearance in simple riboflavin deficiency.[21] In contrast to the enlarged mitochondria caused by simple deficiency in which the cristae have an ostensibly normal distribution, the giant mitochondria produced by galactoflavin supplementation contain cristae that are largely restricted to the periphery (FIG. 5). The latter organelles frequently have a stack of closely packed cristae, a configuration that also occurs in hepatic megamitochondria in rats

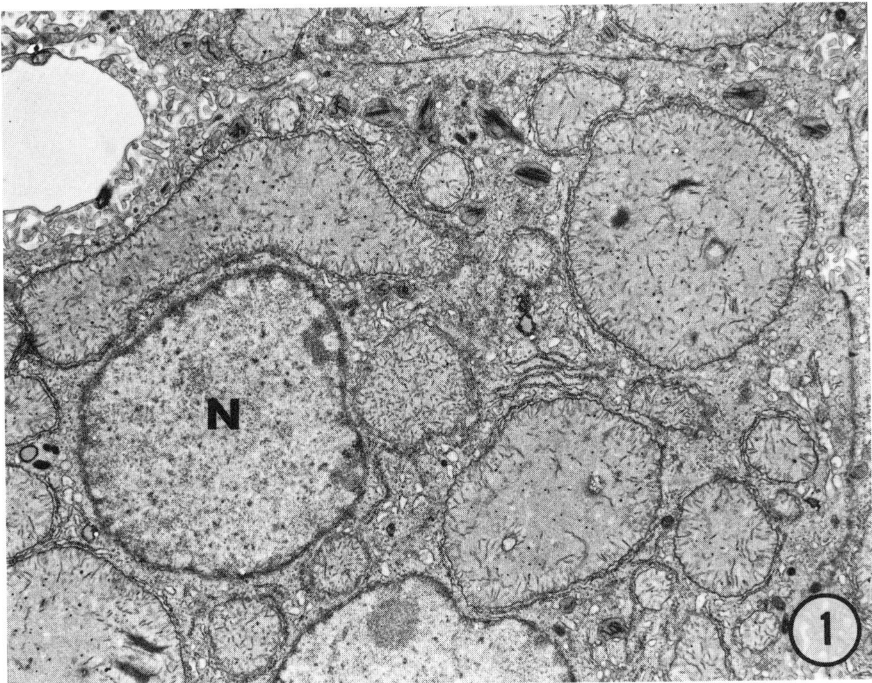

FIGURE 1. Survey electron micrograph of hepatocyte in a mouse fed a riboflavin-deficient diet for 7 wk. Some mitochondria are as large as the nucleus (N). × 7,000.

fed a simple riboflavin-deficient diet for prolonged periods.[19] Wakabayashi *et al.*[69] have further shown that the appearance of the cristae in cuprizone-intoxicated mice depends on the length of time the animals are fed the experimental diet: after 7–8 days, the megamitochondria are cristae-enriched, whereas after 15–16 days the cristae are drastically decreased in number and occupy a peripheral position. When an alternative copper chelator, namely, D-penicillamine, is used, megamitochondria can occur in just

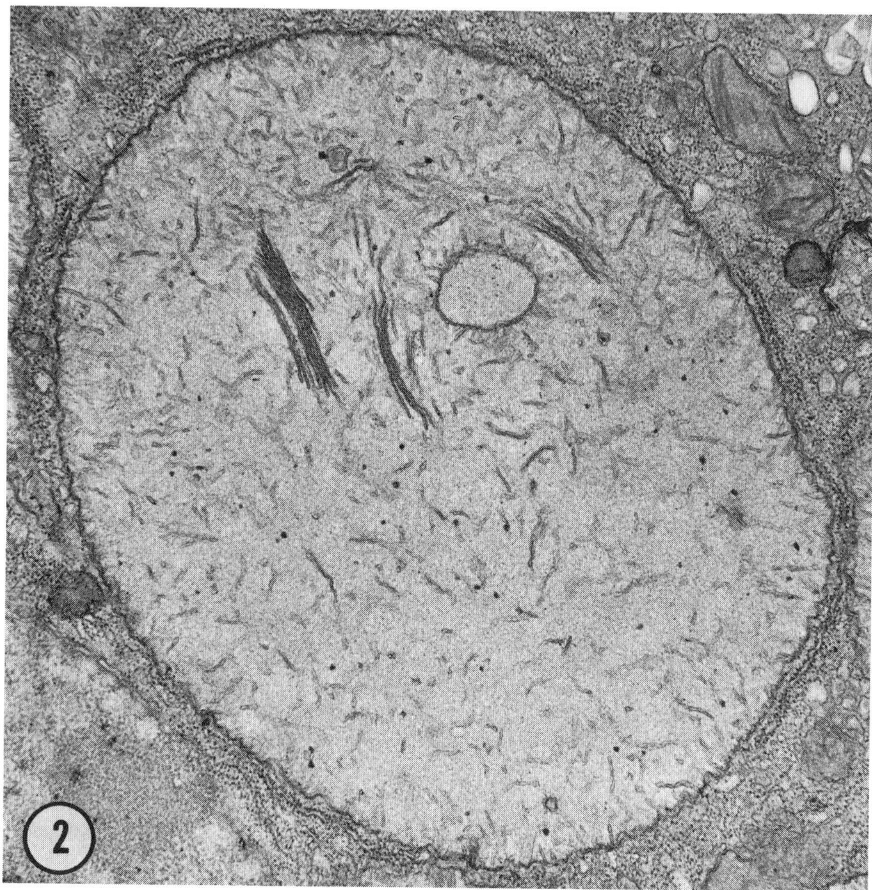

FIGURE 2. A megamitochondrion in a hepatocyte of a 7-wk riboflavin-deficient mouse. This giant organelle resembles its normal-sized counterparts except for several stacks of cristae. × 24,800.

three days; in most the cristae are restricted to the periphery, but some of these organelles have cristae that are uniformly distributed throughout the inner compartment.[34]

A third morphology can be obtained in mouse liver megamitochondria by simply adding ethanol, propanol, or butanol to the drinking water. Such mitochondria are characterized by a total absence of cristae.[47,48,57] Similar cristae-free giant mitochon-

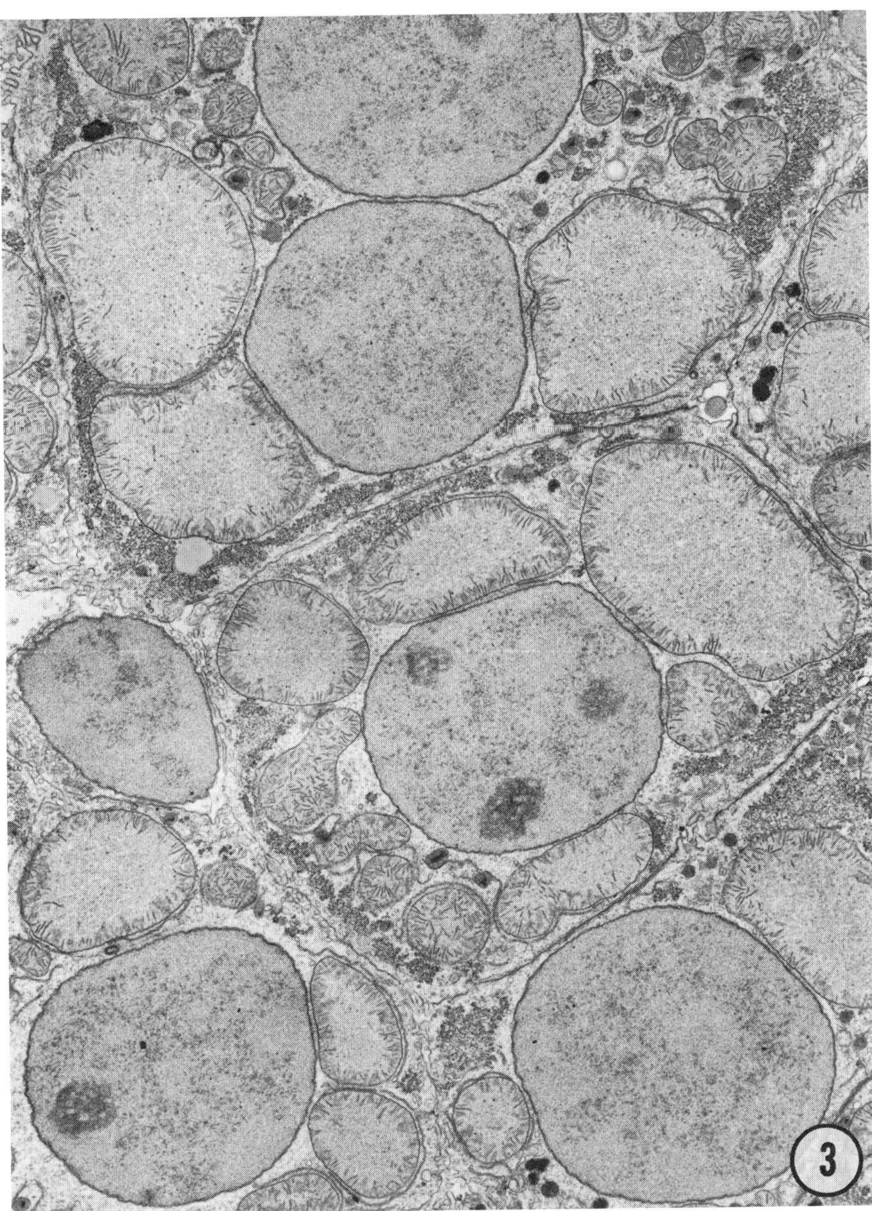

FIGURE 3. Hepatocytes in a mouse fed a cuprizone-supplemented diet for 9 dy. Giant mitochondria surround the nuclei. × 5,000.

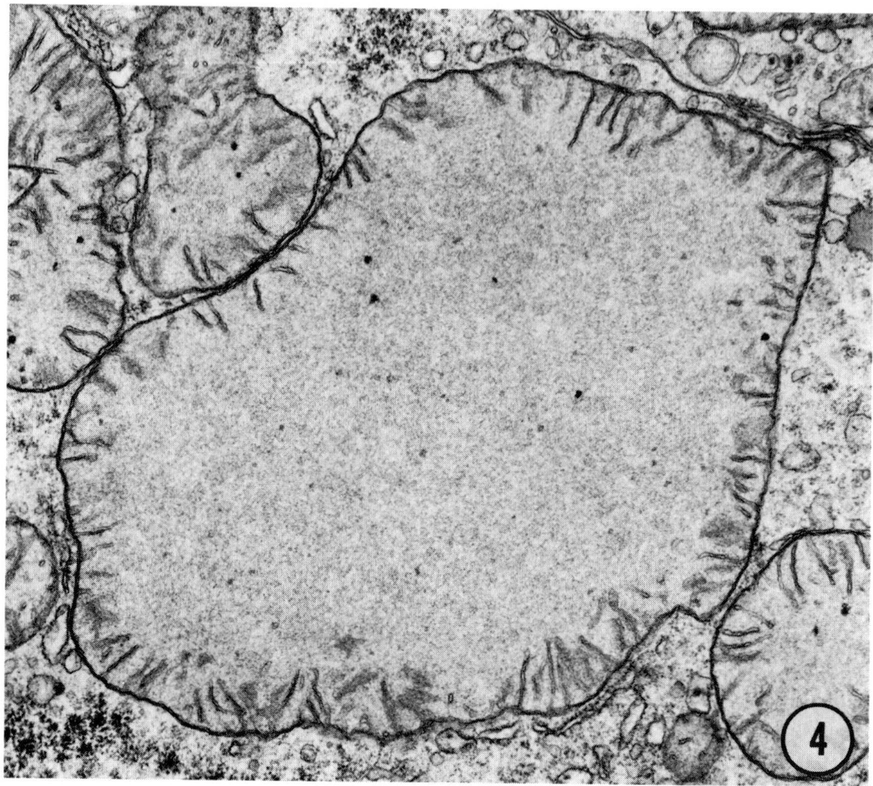

FIGURE 4. A hepatic megamitochondrion in a mouse fed a cuprizone-supplemented diet for 9 dy. The cristae are short and confined to the organelle periphery; the matrix is structureless. × 21,600.

dria have been noted in hepatic cells of monkeys subjected to chronic ethanol administration[62] and in kidney proximal tubule cells of patients with nephrotic syndrome.[4]

Giant mitochondria of diverse origins may contain a variety of inclusions. The stacks of cristae engendered in rat hepatic mitochondria by riboflavin deficiency may become dilated to form a string of vacuoles.[19] Random vacuoles may be present in enlarged mitochondria in adrenocortical cells of rats receiving aminoglutethimide.[70] Relatively large, amorphous dense bodies that are greater in size than typical intramitochondrial dense granules have been observed in giant mitochondria in ariboflavinosis,[17,19] alcoholism,[52,55] systemic scleroderma,[11] Gilbert's syndrome,[10] cuprizone intoxication,[31] and various kidney diseases. In the riboflavin-deficient mouse liver, some of these dense inclusions appear to consist of a tangled skein of dense filaments; such inclusions are readily distinguishable from myelin figures, which may be present in the matrix of the same mitochondria.[17] Myelin figures also may occur in certain stages of cuprizone intoxication.[31] It is common to find one or more filamentous paracrystalline inclusions in giant mitochondria, particularly those resulting from alcoholism (cf. reviews by Spycher and Ruttner[71] and by Bhagwat and Ross[72]). The dimensions and spacing of the paracrystalline filaments in enlarged mitochondria from various sources have been reviewed by Haust.[73] The nature of these filaments is for the most part unknown, but, through the use of several proteases, Schaff *et al.*[74] were able to digest them from thin sections of liver biopsies from patients with Gilbert's syndrome, showing that at least in this case the inclusions are proteinaceous and acidic. In a variation on the usual structure of filamentous paracrystals, those in giant hepatic mitochondria in griseofulvin-fed mice consist of parallel filaments, each of which is partly surrounded by a semicircular density, with all such densities facing in the same direction within a given organelle.[75] In the proximal convoluted tubule cells of a patient with chronic glomerulonephritis, giant mitochondria contained clusters of short,

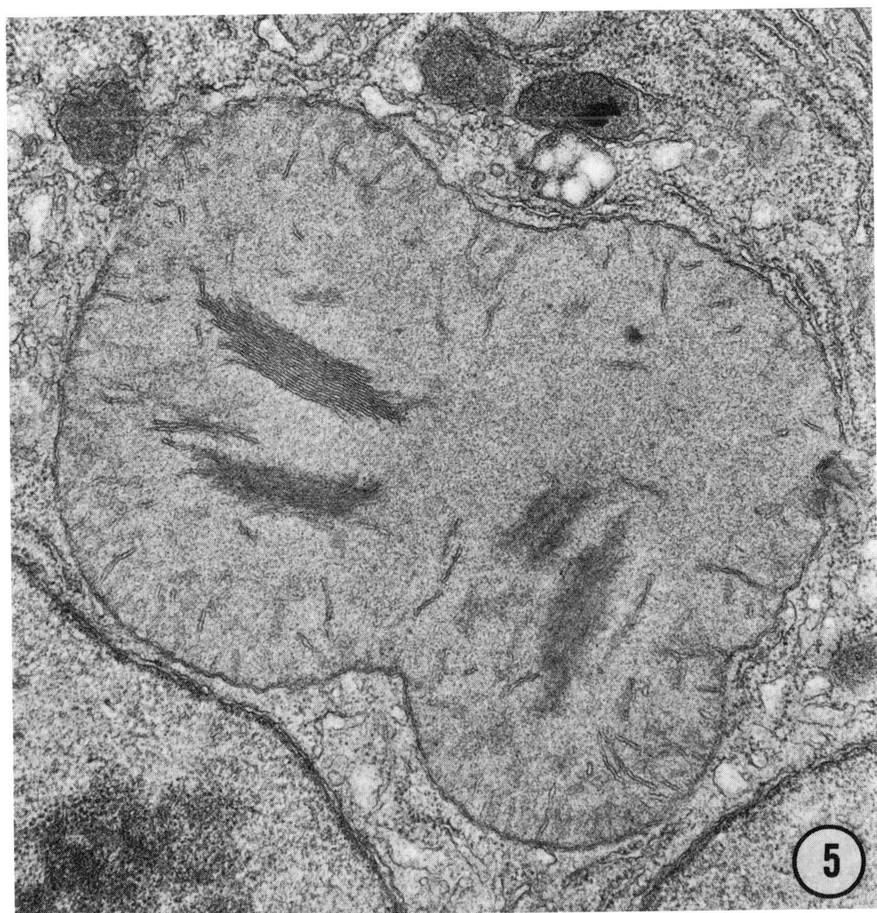

FIGURE 5. An enlarged mitochondrion in a hepatocyte from a mouse fed a riboflavin-deficient, galactoflavin-supplemented diet for 3 wk. Cristae for the most part are short and peripheral, but several stacks of these membranes are present in the matrix. × 25,400.

randomly oriented filaments.[76] Large mitochondria in the thick limb of Henle in rats often have bundles of closely packed filaments that show a regular periodicity.[77] In addition to intramatrical filaments, giant mitochondria sometimes contain filaments within an expanded crista.[13,25] These filaments have a helical configuration and are about 4 nm thick. They are of unknown nature. Finally, expanded cristae in giant mitochondria in human myocardium have been found to contain depositis of β-glycogen,[12] as do occasional cristae in large hepatic mitochondria of riboflavin-deficient mouse liver.[17]

Some controversy exists as to the way in which megamitochondria form. It seems obvious that they can arise by two different mechanisms: either normal-sized mitochondria grow or they fuse to form larger versions of themselves. Morphometric analyses of hepatocytes of rats subjected to total[24] or partial chronic starvation[23] or to vitamin D deficiency[22] or to injection of ethidium bromide[41] have shown that as the mitochondria increase in size there is a corresponding decrease in the number of mitochondria per cell. If only a few mitochondria are selected for unbridled growth, then the remaining normal-sized organelles would have to be eliminated, presumably by autophagy, in order to effect the observed reduction in mitochondrial number. In temporal studies of megamitochondrial formation, no increase in mitochondrial autophagy has been noted.[17,19,21] The inescapable conclusion is that the smaller mitochondria fuse.[17] However, the total mitochondrial volume per cell is increased during the formation of the large mitochondria,[24] indicating that true mitochondrial growth accompanies the fusion process.

In normal hepatocytes, mitochondria never touch each other—they are seemingly repelled, probably because of their surface charge. During the genesis of megamitochondria, as exemplified by those resulting from riboflavin deficiency,[17] mitochondria closely approach one another and a protrusion from one organelle may extend deep into a corresponding invagination in the surface of a neighboring organelle. In such manner, large aggregates consisting of interlocked mitochondria are formed. These appressed mitochondria frequently are linked by myelin figures, which may indicate membrane flux and fusion.[17] A similar mechanism for formation of megamitochondria by treatment of cuprizone or diethyldithiocarbamate has been proposed by Wakabayashi and co-workers,[36,78] but they also suggest that giant mitochondria with medial septa are indicative of a fusion process. However, Tandler *et al.*[79] have adduced evidence that large mitochondria with a membranous partition are in fact dividing, a supposition that is strongly borne out by observation of certain invertebrate cells in which the number of mitochondria is increasing at an extremely rapid pace.[80]

The ability to induce experimentally giant mitochondria in a readily accessible organ such as the rodent liver presents the biochemist with the opportunity to study these intriguing organelles in functional terms. A paradox becomes evident, however, when publications dealing with this area are inspected. Despite the fact that the *in situ* hepatocytes contain megamitochondria that may exceed 10 μm in diameter, the pellets of isolated so-called giant mitochondria rarely contain organelles that are more than 2 μm in diameter; in most instances the isolated experimental mitochondria are virtually identical to those in pellets from controls.[81–83] Giant mitochondria are not broken by the homogenization procedure per se, since they are present in whole, unfractionated homogenates.[83] Examination of the nuclear fraction of cells that contained giant mitochondria has shown that these organelles sediment in the nuclear fraction as a consequence of their large size and characteristic density.[81] It may be concluded that most of the biochemical studies that purport to deal with megamitochondrial function actually have been carried out on normal- or nearly normal-sized organelles.

Only Wagner and Rafael[39] have provided electron microscopic evidence that they were dealing with isolated megamitochondria. They found that giant mitochondria

TABLE 6. Hepatic Mitochondrial Oxidative Metabolism

Condition	Species	Pyruvate (+ malate)	Glutamate	α-Keto Glutarate	Fatty Acids	β-Hydroxy Butyrate	Succinate (+ Rotenone)	Reference
Riboflavin deficiency	Mouse	↓	↓	↓	↓	↓	↓	88
+ galactoflavin	Mouse	↓	↓	↓	↓	↓	↓	88
Riboflavin deficiency	Rat			↓		↓	↓	89
	Rat		↓				↓	90
	Rat	0	0		↓	↓	↓	91
	Rat				↓			92
	Rat		↓		↓			93
+ galactoflavin	Rat		0			↓	0	90
	Rat	↓	↓	↓		↓	↓	94
Cuprizone	Mouse	↓	↓	↓	↓	↓	↓	81
	Mouse						0	39
	Mouse						0	33
Penicillamine	Mouse	0	0	0	0		0	34
Chloramphenicol	Mouse			↓	↓		↓	39,95
Hydrazine	Mouse		↓				↓	45
Isonicotinic acid derivatives	Mouse						0	44
α, α′ Dipyridyl	Mouse						0	96

The data from each paper are normalized to the controls by showing 0, no change; and ↓, decreased activity.

induced by chloramphenicol injection showed the following decreases (compared to controls) in enzyme activities: cytochrome oxidase, 80%, ATP synthetase, 50%; cytochrome aa_3, 75%; cytochrome *b*, 20%; and the antimycin sensitivity of succinate-cytochrome *c* reductase, 60%. These results may be a reflection of the restriction by the chloramphenicol of the protein-synthesizing system of hepatic mitochondria; no comparable reduction of inner membrane enzyme activities was found in cuprizone-induced megamitochondria.

Some workers have attempted to isolate single giant mitochondria either by use of a suction capillary on a disrupted piece of liver tissue or by use of differential centrifugation. Using the first method, Suchy and Cooper[84] have studied single megamitochondria in an oxygen electrode; they concluded that such mitochondria have normal respiratory properties. Giant mitochondria lend themselves to impalement with microelectrodes—studies using this technique have measured megamitochondrial membrane potentials and resistances as well as the protonmotive force.[85–87]

TABLE 7. Hepatic Cytochrome Content and Oxidase Activity

Condition	Species	Cytochromes			Cytochrome Oxidase	Monoamine Oxidase	References
		b	cc_1	aa_3			
Riboflavin	Rat	0	↓	0			97
	Rat					↓	98
Cuprizone	Mouse	0	0	0	0	0	83
	Mouse				0	↓	99
	Mouse	0	0	0			33
	Mouse	↓	↓	0	0		39
Diethyldithiocarbamate	Mouse	0	0	0	0	0	36,82
Chloramphenicol	Mouse	↓		↓	↓		39
	Mouse	↓	0	↓			100
Hydrazine	Mouse				0	↓	45
Isonicotic acid derivatives	Mouse				0	↓	44

The data from each paper are normalized to the controls by showing 0, no change; and ↓, decreased content or activity.

No ultrastructural evidence was presented in these studies on single mitochondria that the structures in question were indeed intact giant mitochondria.

Bearing in mind the caveat that most studies on alleged giant mitochondria actually were carried out on pellets of normal-sized organelles from the affected cells, certain alterations in mitochondrial oxidative metabolism have been recorded. These are summarized in TABLE 6.[33,34,39,44,45,81,88–96] Based on the available data, it is obvious that mitochondria derived from livers that exhibited megamitochondria (regardless of mode of induction) showed lessened oxidative capacity for most substrates tested. The lack of oxidative changes after administration of penicillamine probably reflects the paucity of hepatocytic alterations in this treatment.[34] As shown in TABLE 7[33,34,39,44,45,82,83,97–100] the copper-chelating agents, cuprizone and DDC, do not lead to significant changes in cytochrome content or cytochrome oxidase activity in hepatic mitochondria. In contrast, chloramphenicol, an inhibitor of protein synthesis by mitochondria, adversely affects various cytochromes, which are generated in part by these organelles. Wakabayashi *et al.*[44] originally suggested that there is a relationship between reduced monoamine oxidase activity and the production of giant mitochondria in liver, but further work by these authors has failed to support this contention.[45]

Mitochondria are DNA-containing organelles. When a liver mitochondrion undergoes a thousand-fold increase in volume (in riboflavin deficiency the diameter of a mitochondrion may increase from 1 μm to 10 μm), is there a concomitant increase in DNA content? If there is a parallel increase, are there a thousand separate molecules of DNA per megamitochondrion? The answer to such questions can be ascertained only if highly purified fractions of giant mitochondria are used as the starting material for isolation of m-DNA. The only study that attempted to examine m-DNA from megamitochondria did not document the nature of the starting mitochondrial pellet.[101] Although relatively small alterations were found in the m-DNA, it is not certain whether these altered molecules were derived from megamitochondria or from their normal-sized counterparts.

CONCLUSION

Giant mitochondria, which occasionally occur in a variety of disease states, can be induced in laboratory animals by many different experimental means. Critical questions concerning the biochemistry and physiology of these enlarged organelles can be addressed only if a methodology is developed that permits isolation of structurally and metabolically intact megamitochondria in pure yield. With such a methodology, the composition of membranes, the complement of enzymes and their activities, the DNA status, all can be determined, as well as the metabolic abilities of these mitochondria. Such studies should provide important clues to the mechanism of mitochondrial giantism and to whether this propensity for enlargment is centered in a single metabolic alteration, independent of the means used to evoke this response.

REFERENCES

1. MUNN, E. A. 1974. The Structure of Mitochondria. Academic Press. London.
2. JEAN, G., G. LAMBERTENGHI & T. RANZI. 1968. Ultrastructural study of the liver in hepatic porphyria. J. Clin. Pathol. **21:** 501–507
3. SUZUKI, K. & I. RAPIN. 1969. Giant neuronal mitochondria in an infant with microcephaly and seizure disorder. Arch. Neurol. **20:** 62–72.
4. THOENES, W. 1966. Über matrixreiche Riesenmitochondrien. Elektronemikroskopische Beobachtung an Tubulusepithel der menschlichen Niere bei nephrotischem Syndrom. Z. Zellforsch. **75:** 422–433.
5. SCHUURMANS STEKHOVEN, J. H. & U. J. v. HAELST. 1970. Matrixreiche Riesenmitochondrien in den Zellen der proximalen Nierentubuli. Beobachtungen bei 4 Patienten mit einem Nierenleiden. Virchows Arch. Abt B. Zellpathol. **5:** 105–112.
6. SUZUKI, T., M. FURUSATO, S. TAKASAKI & E. ISHIKAWA. 1975. Giant mitochondria in the epithelial cells of the proximal convoluted tubules of diseased human kidneys. Lab. Invest.**33:** 578–590.
7. PARTIN, J. C., W. K. SCHUBERT & J. S. PARTIN. 1971. Mitochondrial ultrastructure in Reye's syndrome (encephalopathy and fatty degeneration of the viscera). N. Engl. J. Med. **285:** 1339–1343.
8. GHADIALLY, F. N. & L. F. SKINNIDER. 1974. Giant mitochondria in erythroleukaemia. J. Pathol. **114:** 113–117.
9. ENOMOTO, Y. & Y. WATANABE. 1985. Giant mitochondria with filamentous structures in bone marrow macrophages. J. Electron Microsc. **34:** 25–32.

10. Slabodsky-Brousse, N., G. Feldmann, J. Brousse & P. Dreyfus. 1974. Étude stéréologigue de la fréquence des mitochondries géantes hépatiques dans la maladie de Gilbert. Comparaison avec le sujet normal. Biol. Gastroenterol. (Paris) **7:** 179–186.
11. Feldmann, G., M. Maurice, J. M. Husson, J. N. Fiessinger, J. P. Camilleri, J. P. Benhamou & E. Housset. 1977. Hepatocyte giant mitochondria: An almost constant lesion in systemic scleroderma. Virchows Arch. A Pathol. Anat. Histol. **374:** 215–227.
12. Kraus, B. & H. Cain. 1980. Giant mitochondria in the human myocardium—morphogenesis and fate. Virchows Arch. B Cell Pathol. **33:** 77–89.
13. Tandler, B. & R. A. Erlandson. 1983. Giant mitochondria in a pleomorphic adenoma of the submandibular gland. Ultrastruct. Pathol. **4:** 85–96.
14. Petersen, P. 1977. Ultrastructure of periportal and centrilobular hepatocytes in human fatty liver of various aetiology. Acta Pathol. Microbiol. Scand. Sect. A **85:** 421–427.
15. Sato, T. & H. Tauchi. 1975. The formation of enlarged and giant mitochondria in the aging process of human hepatic cells. Acta Pathol. Jpn. **25:** 403–412.
16. Wilson, P. D. & L. M. Franks. 1975. The effect of age on mitochondrial ultrastructure. Gerontologia **21:** 81–94.
17. Tandler, B., R. A. Erlandson & E. L. Wynder. 1968. Riboflavin and mouse hepatic cell structure and function. I. Ultrastructural alterations in simple deficiency. Am. J. Pathol. **52:** 69–95.
18. Tandler, B. & C. L. Hoppel. 1972. Effects of riboflavin deficiency on liver cells. Meth. Achiev. Exp. Pathol. **6:** 25–48.
19. Tandler, B. & C. L. Hoppel. 1980. Ultrastructural effects of riboflavin deficiency on rat hepatic mitochondria. Anat. Rec. **196:** 183–190.
20. Kobayashi, K., T. Yamamoto & T. Omae. 1980. Ultrastructural changes of renal tubules in the riboflavin deficient mouse. Virchows Arch. B Cell Pathol. **34:** 99–109.
21. Tandler, B. & C. L. Hoppel. 1974. Ultrastructural effects of dietary galactoflavin on mouse hepatocytes. Exp. Mol. Pathol. **21:** 88–101.
22. Riede, U. N., U. Leibundgut & H. P. Rohr. 1973. Ultrastruktureller und morphometrischer Nachweis einer durch Vitamin-D-Mangel induzierten Störung im oxidativen Stoffwechsel. Ein ultrastrukturell-morphometrische Analyse der Ratten-Leberparenchymzelle. Beitr. Pathol. **150:** 378–388.
23. Riede, U. N., J. Hodel, C. V. Matt, Y. Rasser & H. P. Rohr. 1973. Einfluss der Hungers auf die quantitative Cytoarchitektur der Rattenleberzelle. II. Chronischer partieller Hunger. Beitr. Pathol. **150:** 246–260.
24. Rohr, H. P. & U. N. Riede. 1973. Experimental metabolic disorders and the subcellular reaction pattern. Curr. Top. Pathol. **58:** 1–48.
25. Svoboda, D. & J. Higginson. 1964. Ultrastructural changes produced by protein and related deficiencies in the rat liver. Am. J. Pathol. **45:** 353–379.
26. Wyllie, L., N. Marchi & J. Verne. 1978. Modification des mitochondries dans les hépatocytes en culture en fonction de la teneur en protéines du milieu. C. R. Soc. Biol. **172:** 231–235.
27. Dallman, P. R. & J. R. Goodman. 1971. The effects of iron deficiency on the hepatocyte: a biochemical and ultrastructural study. J. Cell Biol. **48:** 79–90.
28. Svoboda, D. J. & J. Higginson. 1963. Ultrastructural hepatic changes in rats on a necrogenic diet. Am. J. Pathol **43:** 477–495.
29. Suzuki, K. 1969. Giant hepatic mitochodria: Production in mice fed with cuprizone. Science **163:** 81–82.
30. Tandler, B. & C. L. Hoppel. 1973. Division of giant mitochondria during recovery from cuprizone intoxication. J. Cell Biol. **56:** 266–272.
31. Wakabayashi, T., M. Asano & C. Kurono. 1975. Mechanism of the formation of megamitochondria induced by copper-chelating agents. I. On the formation process of megamitochondria in cuprisone-treated mouse liver. Acta Pathol Jpn.**25:** 15–37.
32. Wakabayashi, T., M. Asano, K. Ishikawa & H. Kishimoto. 1977. Cuprizone-induced megamitochondrial formation and membrane fusion. J. Electron Microsc. **26:** 137–140.
33. Flatmark, T., H. Kryvi & A. Tangeras. 1980. Induction of megamitochondria by cuprizone (biscyclohexanone oxaldihydrazone). Evidence for an inhibition of the mitochondrial division process. Eur. J. Cell Biol. **23:** 141–148.

34. WALTER, R. J., B. TANDLER & C. L. HOPPEL. 1980. Ultrastructural and biochemical effects of D-penicillamine on mouse hepatocytes. Anat. Rec. **197:** 289–295.

35. ASANO, M. & T. WAKABAYASHI. 1974. Induction of giant mitochondria in mouse hepatocytes by diethyldithiocarbamate (DDC). J. Electron Microsc. **23:** 189–191.

36. ASANO, M., C. KURONO & T. WAKABAYASHI. 1974. On the formation process and biochemical properties of diethyldithiocarbamate (DDC)-induced megamitochondria. Nagoya Med. J. **19:** 139–145.

37. ALBRING, M., K. RADSAK & W. THOENES. 1975. Chloramphenicol-induced giant hepatic mitochondria. Naturwissenschaften **62:** 43–44.

38. OBERHOLZER, M. & H. P. ROHR. 1975. Ultrastructural-morphometric studies on the rat and liver and adrenocortical cell mitochondria after chloramphenicol administration. Pathol. Eur. **10:** 29–36.

39. WAGNER, T. & J. RAFAEL. 1977. Biochemical properties of liver megamitochondria induced by chloramphenicol or cuprizone. Exp. Cell Res. **107:** 1–13.

40. ALBRING, M., K. RADSAK & W. THOENES. 1973. Giant mitochondria. II. Induction of matrix enriched megamitochondria in mouse liver parenchymal cells by ethidium bromide. Virchows Arch. Abt. B Zellpathol. **14:** 373–377.

41. ROHR, H. P., M. WACKER & A. v. PEIN. 1976. Ethidium bromide and hepatic mitochondrial structure in mice. A morphometric analysis. Pathol. Eur. **11:** 129–135.

42. NOPANITAYA, W. & M. L. TYAN. 1975. Giant mitochondria in hepatocytes following concanavalin A administration. Proc. 33rd Annu. Meeting Electron Microsc. Soc. Am. 310–311.

43. ASANO, M., T. WAKABAYASHI, K. ISHIKAWA & H. KISHIMOTO. 1977. Induction of megamitochondria in mouse hepatocytes by nialamide. J. Electron Microsc. **26:** 141–144.

44. WAKABAYASHI, T., M. ASANO & S. KAWAMOTO. 1979. Induction of megamitochondria in the mouse liver by isonicotinic acid derivatives. Exp. Mol. Pathol. **31:** 387–399.

45. WAKABAYASHI, T., M. HORIUCHI, M. SAKAGUCHI, H. ONDA & K. MISAWA. 1983. Induction of megamitochondria in the mouse and rat livers by hydrazine. Exp. Mol. Pathol. **39:** 139–153.

46. WATARI, N. 1973. Ultrastructural alterations of the mouse liver after the prolonged administration of BHC. J. Clin. Electron Microsc. **5:** 1449–1456.

47. WAKABAYASHI, T., M. HORIUCHI, M. SAKAGUCHI, H. ONDA & M. IIJIMA. 1984. Induction of megamitochondria in the rat liver by n-propyl alcohol and n-butyl alcohol. Acta Pathol. Jpn. **34:** 471–480.

48. WAKABAYASHI, T., M. HORIUCHI, K. ADACHI & T. KOYAMA. 1984. Induction of megamitochondria by *l*-octadecanol J. Electron Microsc. **33:** 236–238.

49. PANNER, B. J. & R. J. HANSS. 1969. Hepatic injury in mushroom poisoning. Electron microscopic observations on two nonfatal cases. Arch. Pathol. **87:** 35–45.

50. SOGA, J., I. TERADA, M. KASUKAWA & F. KOIZUMI. 1970. Ultrastructural changes of cirrhotic liver associated with chronic alcoholism. A case report. Acta Pathol. Jpn. **20:** 339–356.

51. OUDEA, M. C., A. N. LAUNAY, S. QUÉNÉHERVÉ & P. OUDEA. 1970. The hepatic lesions produced in the rat by chronic alcoholic intoxication. Histological, ultrastructural and biochemical observations. Rev. Eur. Etudes Clin. Biol. **15:** 748–764.

52. KIESSLING, K. -H. & L. PILSTRÖM. 1971. Ethanol and the human liver. Structural and metabolic changes in liver mitochondria. Cytobiologie **4:** 339–348.

53. ALBOT, G. & M. PARTURIER-ALBOT. 1972. Les mégamitochondries de l'hépatocyte. Ann. Gastroenterol. Hepatol. **8:** 1–12.

54. MA, M. H. 1972. Ultrastructural pathologic findings of the human hepatocyte. I. Alcoholic liver disease. Arch. Pathol **94:** 554–571.

55. HORVATH, E., K. KOVACS & R. C. ROSS. 1973. Alcoholic liver lesion. Frequency and diagnostic value of fine structural alterations in hepatocytes. Beitr. Pathol. **148:** 67–85.

56. BIANCHI, L., K. WINCKLER, M. MITHATSCH & H. P. ROHR. 1973. Mallory bodies and giant mitochondria—two different structures in liver biopsies from alcoholics. Beitr. Pathol. **150:** 298–310.

57. RUBIN, E. & C. S. LIEBER. 1967. Experimental alcohol hepatic injury in man: ultrastructural changes. Fed. Proc. **26:** 1458–1467.
58. RUBIN, E. & C. S. LIEBER. 1974. Fatty liver, alcoholic hepatitis and cirrhosis produced by alcohol in primates. N. Engl. J. Med. **290:** 128–135.
59. PETERSEN, P. 1977. Abnormal mitochondria in hepatocytes in human fatty liver. Acta Pathol. Microbiol. Scand. Sect. A **85:** 413–420.
60. STEWARD, R. V. & H. P. DINCSOY. 1982. The significance of giant mitochondria in liver biopsies as observed by light microscopy. Am. J. Clin. Pathol. **78:** 293–298.
61. LEO, M. A., M. ARAI, M. SATO & C. S. LIEBER. 1982. Hepatotoxicity of vitamin A and ethanol in the rat. Gastroenterology **82:** 194–205.
62. FRENCH, S. W., B. H. RUEBNER, E. MEZEY, T. TAMURA & C. H. HALSTEAD. 1983. Effect of chronic ethanol feeding on hepatic metochondria in the monkey. Hepatology **3:** 34–40.
63. ARAI, M., M. A. LEO, M. NAKANO, E. R. GORDON & C. S. LIEBER. 1984. Biochemical and morphological alterations of baboon hepatic mitochondria after chronic ethanol consumption. Hepatology **4:** 165–174.
64. KOCH, O. R., L. L. ROATTA DE DE CONTI, L. PACHECO BOLAÑOS, G. BUGARI & A. O. M. STOPPANI. 1977. Alteraciones morfofuncionales de las mitocondrias hepaticas en ratas tratadas cronicamente con alcohol. Regresion con la abstinencia. Medicina (Buenos Aires) **37:** 21–32.
65. COSSEL, L., H. BRAULKE & W. STORCH. 1979. Intrazisternales Hyalin und Riesenmitochondrien in Hepatozyten und orale Kontrazeptiva. Zbl. Allg. Pathol. Pathol. Anat. **123:** 318–336.
66. GONZÁLEZ-ANGULO, A., R. AZNAR-RAMOS, H. MÁRQUEZ-MONTER, G. BIERZWINSKY & J. MARTÍNEZ-MANAUTOU. 1970. The ultrastructure of liver cells in women under steroid therapy. I. Normal pregnancy and trophoblastic growth. Acta Endocrinol. **65:** 193–206.
67. REID, I. M., I. A. DONALDSON & R. J. HEITZMAN. 1978. Effects of anabolic steroids on liver cell ultrastructure in sheep. Vet. Pathol. **15:** 753–762.
68. LUDATSCHER, R. & M. SILBERMANN. 1976. Effect of triamcinolone on the ultrastructure of mouse hepatocytes. Acta Anat. **96:** 176–187.
69. WAKABAYASHI, T., M. HORIUCHI, S. KAWAMOTO & H. ONDA. 1984. Membrane fusion in mitochondria. II. An ultrastructural interpretation of cristae-enriched megamitochondria induced by cuprizone. Acta Pathol. Jpn. **34:** 481–488.
70. ITOH, G. 1977. Elektronenmikroskopische und elektronenmikroskopish-histochemische Untersuchungen über Nebennierenrinde der Ratte mit Aminoglutäthimid (Elipten, CIBA) und über ihren Reparatprozess. Acta Pathol. Jpn. **27:** 75–91.
71. SPYCHER, M. A. & J. R. RÜTTNER. 1968. Kristalloide Einschlüsse in menschlichen Lebermitochondrien. Virchows Arch. Abt. B Zellpathol. **1:** 211–221.
72. BHAGWAT, A. G. & R. C. ROSS. 1971. Hepatic intramitochondrial crystalloids. Arch. Pathol. **91:** 70–77.
73. HAUST, M. D. 1968. Crystalloid structures of hepatic mitochondria in children with heparitin sulphate mucopolysaccharidosis (Sanfilippo type). Exp. Mol. Pathol. **8:** 123–134.
74. SCHAFF, Z., K. LAPIS & J. ANDRÉ. 1974. Effect of proteolytic digestion on mitochondrial crystalline inclusions in Gilbert's syndrome. J. Microscopie **20:** 265–270.
75. SHAPIRO, S. H., Z. WESSELY & J. V. KLAVINS. 1980. Hepatocyte mitochondrial alterations in griseofulvin fed mice. Ann. Clin. Lab. Sci. **10:** 45–53.
76. DATSIS, S. A. G. 1973. Intramitochondrial filamentous inclusions in chronic glomerulonephritis. Beitr. Pathol. **149:** 396–407.
77. SUZUKI, T. & F. K. MOSTOFI. 1967. Intramitochondrial filamentous bodies in the thick limb of Henle of the rat kidney. J. Cell Biol. **33:** 605–624.
78. WAKABAYASHI, T., M. ASANO & C. KURONO. 1974. Some aspects of mitochondria having a septum. J. Electron Microsc. **23:** 247–254.
79. TANDLER, B., R. A. ERLANDSON, A. L. SMITH & E. L. WYNDER. 1969. Riboflavin and mouse hepatic cell structure and function. II. Division of mitochondria during recovery from simple deficiency. J. Cell Biol. **41:** 477–493.

80. LARSEN, W. J. 1970. Genesis of mitochondria in insect fat body. J. Cell. Biol. **47:** 373–383.
81. HOPPEL, C. L. & B. TANDLER. 1973. Biochemical effects of cuprizone on mouse liver and heart mitochondria. Biochem. Pharmacol. **22:** 2311–2318.
82. ASANO, M., C. KURONO & T. WAKABAYASHI. 1974. On the formation process and biochemical properties of diethyldithiocarbamate (DDC)-induced megamitochondria. II. Some biochemical properties of DDC-induced megamitochondria. Nagoya Med. J. **19:** 147–153.
83. WAKABAYASHI, T., M. ASANO, C. KURONO & T. OZAWA. 1975. Mechanism of the formation of megamitochondria induced by copper-chelating agents. II. Isolation and some properties of megamitochondria from the cuprizone-treated mouse liver. Acta Pathol. Jpn. **25:** 39–49.
84. SUCHY, J. & C. COOPER. 1974. Isolation and respiratory measurements on a single large mitochondrion. Exp. Cell Res. **88:** 198–202.
85. MALOFF, B. L., S. P. SCORDILIS, C. REYNOLDS & H. TEDESCHI. 1978. Membrane potentials and resistances of giant mitochondria. Metabolic dependence and the effects of valinomycin. J. Cell Biol. **78:** 199–213.
86. MALOFF, B. L., S. P. SCORDILIS & H. TEDESCHI. 1978. Assays of the metabolic viability of single giant mitochondria. Experiments with intact and impaled mitochondria. J. Cell Biol. **78:** 214–226.
87. CAMPO, M. L. & H. TEDESCHI. 1984. Protonmotive force in giant mitochondria. Eur. J. Biochem. **141:** 5–7.
88. HOPPEL, C. L. & B. TANDLER. 1975. Riboflavin and mouse hepatic cell structure and function. Mitochondrial oxidative metabolism in severe deficiency states. J. Nutr. **105:** 562–570.
89. BURCH, H. B., F. E. HUNTER, JR., A. M. COMBS & B. A. SCHUTZ. 1960. Oxidative enzymes and phosphorylation in hepatic mitochondria from riboflavin-deficient rats. J. Biol. Chem. **235:** 1540–1544.
90. BEYER, R. E., S. L. LAMBERG & M. A. NEYMAN. 1961. The effect of riboflavin deficiency and galactoflavin feeding on oxidative phosphorylation and related reactions in rat liver mitochondria. Can. J. Biochem. Physiol. **39:** 73–88.
91. HOPPEL, C., J. P. DIMARCO & B. TANDLER. 1979. Riboflavin and rat hepatic cell structure and function. Mitochondrial oxidative metabolism in deficiency states. J. Biol. Chem. **254:** 4164–4170.
92. SAKURAI, T., S. MIYAZAWA, S. FURUTA & T. HASHIMOTO. 1982. Riboflavin deficiency and β-oxidation systems in rat liver. Lipids **17:** 598–604.
93. OLPIN, S. E. & C. J. BATES. 1982. Lipid metabolism in riboflavin-deficient rats. 2. Mitochondrial fatty acid oxidation and microsomal desaturation pathway. Br. J. Nutr. **47:** 589–596.
94. ZAMAN, Z. & R. L. VERWILGHEN. 1975. Effects of riboflavin deficiency on oxidative phosphorylation, flavin enzymes and coenzymes in rat liver. Biochem. Biophys. Res. Commun. **67:** 1192–1198.
95. WAGNER, T. & J. RAFAEL. 1975. ATPase complex and oxidative phosphorylation in chloramphenicol-induced megamitochondria from mouse liver. Biochim. Biophys. Acta **408:** 284–296.
96. RIFKIN, R. J. & P. A. GAHAGAN-CHASE. 1970. Morphologic and biochemical effects of chelating agent, α,α'-dipyridyl, on kidney and liver in rats. Lab. Invest. **23:** 480–488.
97. REITH, A., D. BRDICZKA, J. NOLTE & H. W. STAUDTE. 1973. The inner membrane of mitochondria under influence of triiodothyronine and riboflavin deficiency in rat heart muscle and liver. A quantitative electronmicroscopical and biochemical study. Exp. Cell Res. **77:** 1–14.
98. KWATRA, M. & T. L. SOURKES. 1980. Monoamine oxidase A and B activities in liver of riboflavin-deficient rats. Biochem. Pharmacol. **29:** 2693–2694.
99. WAKABAYASHI, T., M. ASANO, K. ISHIKAWA & H. KISHIMOTO. 1978. Mechanism of the formation of megamitochondria by copper-chelating agents. V. Further studies on isolated megamitochondria. Acta Pathol. Jpn. **28:** 215–223.
100. SOWERS, A. E. & C. R. HACKENBROCK. 1981. Alterations in density and size distribution

of intramembrane particles in the inner membrane of mitochondria from chloramphenicol-fed mice. Eur. J. Cell Biol **24:** 101–107.

101. GUERINEAU, M., S. GUERINEAU & C. GOSSE. 1974. Abnormal mitochondrial DNA molecules in megamitochondria from cuprizone-treated rats. Eur. J. Biochem. **47:** 313–319.

DISCUSSION OF THE PAPER

R. WATTIAUX (*Laboratoire de Chimie Physiologique, Namur, Belgium*): When you observe giant mitochondria in the liver, are they present in each hepatocyte?

TANDLER: The distribution of giant mitochondria in liver cells depends on the means whereby they are produced. In riboflavin deficiency or in cuprizone intoxication, giant mitochondria are present in varying numbers in every hepatocyte regardless of location within the lobule. In contrast, penicillamine produces a few megamitochondria in scattered cells, whereas most cells lack such organelles.

M. MILANICK (*Yale University, New Haven, CT*): Have you attempted to induce mitochondria fusion or aggregation *in vitro* with some of your drug treatments, both by treatment of cells or of an isolated mitochondria fraction?

TANDLER: We have tried to add galactoflavin to the culture medium of rat cells, but it does not seem to have any effect. Neither does cuprizone. Protein-deficient culture medium or addition of dexamethasone produces some enlargement of the mitochondria, but these do not qualify as megamitochondria.

T. GALEOTTI (*Catholic University, Rome*): Do you know whether giant mitochondria are less resistant to lipid peroxidation? Has anybody measured the content of Cu-Zn superoxide dismutase?

TANDLER: Such studies have not been done.

A. SCARPA (*Case Western Reserve University, Cleveland, OH*): Dr. Tandler, one question that should be addressed is whether or not cells with giant mitochondria have ATP/ADP ratios comparable to those of control cells. The mitochondrial volume-to-surface ratio of those mitochondria may be unfavorable to some oxidative-phosphorylation reactions.

TANDLER: Since giant mitochondria have not been isolated, we can't compare their ATP/ADP ratios. This study has not, to my knowledge, been carried out on tissues or tissue slices of affected organs. The normal-sized mitochondria isolated from livers exhibiting giant mitochondria have lowered ATP/ADP and respiratory control ratios.

E. CARAFOLI (*Swiss Federal Institute of Technology, Zurich*): (1) One of the treatments that produce giant mitochondria is iproniazid intoxication, which also induces anti-mitochondrial antibodies. Do you know of any studies on anti-mitochondrial antibodies in conditions known to induce giant mitochondria?

(2) Cuprizone is a copper chelator that presumably induces copper deficiency. Menke's disease is characterized by copper malabsorption and thus, presumably, by copper deficiency. Are you aware of any findings of giant mitochondria in Menke's disease?

(3) You mentioned charge repulsion as a mechanism to prevent mitochondrial aggregation and then, eventually, fusion. One way to decrease mitochondrial charge is by uncoupling them, i.e., by collapsing their proton motive force. What is the degree of coupling of isolated giant mitochondria?

TANDLER: (1) No studies of this sort have been done.

(2) Although cuprizone is a copper chelator, its mode of action may be via another

mechanism. When mice are fed a diet that contains both cuprizone and excess copper, their hepatocytes simultaneously exhibit megamitochondria and the morphological signs of copper toxicity. Therefore, it is not surprising that giant mitochondria have not been reported in Menke's syndrome.

(3) Since giant mitochondria have not been isolated, we don't know the degree of coupling. However, small mitochondria obtained from livers with megamitochondria are poorly coupled.

G. F. AZZONE (*University of Padova, Padova*): If formation of giant mitochondria is due to fusion of smaller mitochondria one would expect constancy of specific activity of the various mitochondrial enzymatic and transport reactions. Is there any indication that this is the case? For example, which are the properties of the giant mitochondria isolated by H. Tedeschi?

TANDLER: Once small mitochondria have fused, they may undergo a variety of functional alterations, possibly because of a markedly changed surface-to-volume ratio. In any event, comparison of giant vs. normal mitochondria requires that each be available in pure yield. This is not yet possible in the case of giant mitochondria. Tedeschi and co-workers have impaled megamitochondria with microelectrodes and measured several physiological parameters. Again, comparisons between giant and normal organelles cannot be drawn, since normal-sized mitochondria are too small to be impaled.

Cystic Fibrosis—A Lethal Exocrinopathy with Altered Mitochondrial Calcium Metabolism

ROBERT J. FEIGAL[a,b] AND BURTON L. SHAPIRO[a,c]

Departments of [a]Oral Biology and [b]Pediatric Dentistry
School of Dentistry
Laboratory Medicine and [c]Pathology
Medical School
University of Minnesota
Minneapolis, Minnesota 55455

INTRODUCTION

Cystic fibrosis (CF) is a lethal genetic disease. It is one of the leading causes of death in the first decade of life.[1] While aggressive symptomatic care has extended the average life span dramatically, premature death is inevitable from the chronic lung and digestive anomalies. The protein responsible for these symptoms is unknown.

One in 20 Caucasians is a carrier for this autosomal recessive disease.[2] Carrier status of both parents, undetectable by any available test, becomes apparent only after the birth of an affected child.

Research goals in the CF field include: (1) identification of the altered gene product thus increasing possibilities for more direct treatment, (2) development of a carrier detection test allowing for accurate genetic counseling, (3) chromosome mapping of the CF gene, and (4) understanding its pathogenesis.

Our laboratory is one of several that recognized an alteration of cellular calcium (Ca^{2+}) related to CF.[3–7] We have followed our initial observations with investigations that show clear evidence of mitochondrial changes.[8–11] This paper presents background on the disease itself and traces the path of inquiry that leads us to conclusions regarding the fundamental nature of mitochondrial alterations in CF. The results presented here summarize our investigations concerning mitochondrial function in CF.

CYSTIC FIBROSIS

CF is characterized by chronic pulmonary obstruction and infection, gastrointestinal malabsorption due to pancreatic insufficiency, and increased sweat salinity.[12] The confluence of the first two of these major clinical signs was first reported as a distinct disease entity by Fanconi in 1936.[13] Research over the ensuing 50 years has been productive in defining the clinical aspects of CF, but pathogenesis continues to be an enigma. After the first report that changes in sweat salinity[14] were coupled with exocrine secretory abnormalities, CF was accepted as a generalized exocrine dysfunction syndrome.[2] The observation of increased sweat chloride[14] led to the diagnostic test for CF—sweat chloride measurement.

Farber[15] made the early observation that accumulation and apparent obstruction by mucous secretions in the respiratory and gastrointestinal tracts are prominent pathologic features of the disease. The thick and tenacious mucus is accepted as an underlying mechanism responsible for the pancreatic insufficiency and chronic respiratory pathology.[16] Early microobstruction of pancreatic ducts leads to cystic degeneration of the exocrine portion of the pancreas resulting in malabsorption of fat-soluble nutrients. Respiratory function is compromised by obstruction of bronchioles by viscous mucus. The partially or completely obstructed alveolar areas are prone to infections, obstructive emphysema, bronchiectasis, irreversible alveolar fibrosis, and acute respiratory failure. The exact nature of the mucus changes is still being explored.

Reports of altered exocrine secretions are numerous and have recently been reviewed.[17,18] Of interest to this discussion are observations that increased concentrations of Ca^{2+} are present in several secretions in CF, i.e., saliva, tears, seminal fluid, and cerumen.[19]

Treatment has focused on the respiratory and digestive symptoms. Daily physical therapy to remove mucus, aerosols to change mucus viscosity, and antibiotics to combat infections are cornerstones of CF therapy.[16] To enable more normal digestive and absorptive functions, pancreatic enzymes are administered orally. Such aggressive therapy has proven effective in extending the life span of patients with CF. Prior to the advent of antibiotic therapy, few patients survived infancy. More recent data from the CF Patient Registry show median survival age to be approximately 10 years in 1966 and approximately 20 years in 1980.[20]

Progress in treatment has been limited by the lack of information on molecular pathogenesis. There is no clear understanding of the metabolic events that link the unknown altered gene product with the myriad symptoms observable by the clinician.

Recently, the gene responsible for CF has been shown to be on chromosome 7.[21,22] Despite this progress, the protein alteration is still a mystery. Research goals include identification of the gene product and development of an accurate and efficient test to determine carrier status. Both goals will aid in the solution of problems caused by this major genetic disease.

Cell Calcium and CF

Earliest indications that cellular Ca^{2+} is not normal in CF cells are found in a preliminary study by Baur *et al.*[23] Using X-ray spectroscopy, three samples (one each from CF, carrier, and control fibroblast lines) were examined. CF fibroblasts showed a greater Ca^{2+} peak. Soon after that first report, our laboratory and others observed increased cell Ca^{2+} in cultured fibroblasts,[3] parotid acinar cells,[6] and leukocytes[5] from CF patients as compared with control cells. Our laboratory has explored this observation in greater detail and has expanded upon our early report that Ca^{2+} was greater in CF than in normal skin fibroblasts.[4]

CF is a generalized exocrinopathy. No animal model exists. Since fibroblasts are secretory cells, are easily obtained, and maintain a stable genotype in culture, we have used skin fibroblasts as a model for the study of CF. Skin fibroblasts from patients with CF, heterozygotes (HZ), and controls were used in matched pair experiments.[4] These experiments were designed to compare CF with control cell strains and HZ with their adult control cell strains. Therefore, experiments were performed with strains that were matched for age, sex, passage number, day of experiment, and multiwell culture tray. By careful matching, we controlled for subtle day-to-day culture variability and unknown extraneous variables.

Using whole cells in monolayers, our first studies compared ^{45}Ca influx over time. Methods for isotope-exchange experiments have been described.[24] Following preincubation with appropriate buffers containing Ca^{2+} and phosphate, cells were exposed to the same buffer containing ^{45}Ca. After exposure to ^{45}Ca, cells were washed with ice-cold lanthanum chloride ($LaCl_3$) (5 mM) in order to remove extracellular bound ^{45}Ca. Equilibration of ^{45}Ca influx occurred after one hour. Measurements on numerous cell strains at the 75-minute time point revealed a larger intracellular pool size in CF fibroblasts compared with controls ($p < 0.005$) (FIGURE 1).

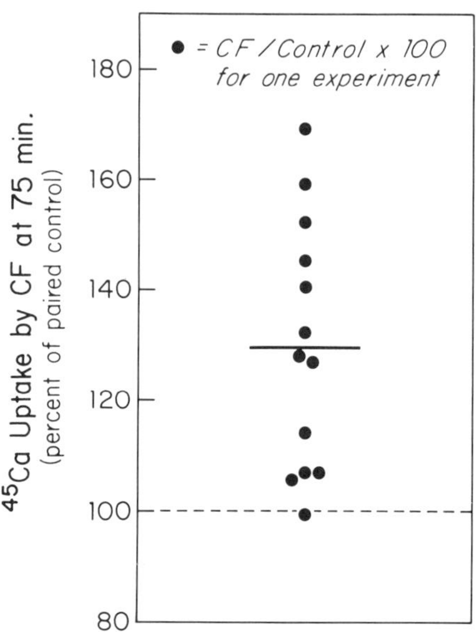

FIGURE 1. ^{45}Ca influx by CF fibroblasts compared with respective ages and passage-matched controls. ^{45}Ca exchange into fibroblast monolayers was measured at 75 minutes. After ^{45}Ca uptake, the cells were washed with $LaCl_3$. Each data point represents the influx by one CF cell strain as a percentage of the influx by one control cell strain matched for age, passage number, and experimental tray.

Because these Ca^{2+} observations seemed potentially far-reaching, comparisons have been repeated: (1) using other cells from additional donors, (2) using lymphocytes as well as fibroblasts, (3) using an independent method for Ca^{2+} measurement and (4) most importantly, using HZ donors and their matched controls.

Lymphocytes obtained from CF donors and matched controls exhibit increased cellular Ca^{2+} (TABLE 1) ($p < 0.005$).[25] Measured by atomic absorption spectrophotometry, these findings support the earlier ^{45}Ca studies. Fibroblasts in culture and lymphocytes isolated from fresh blood both exhibit markedly increased cellular Ca^{2+}.

In the first of many studies using cells from obligate HZ and their matched controls, we showed a difference in cellular Ca^{2+} in cultured fibroblasts.[4] This was considered strong evidence that a Ca^{2+} anomaly was related to the CF gene, not secondary to the disease process or to treatment. Cells donated by healthy, symptom-

TABLE 1. Calcium Content of Fibroblasts and Lymphocytes Assayed by Atomic Absorption ($\mu g\ Ca^{2+}/10^7$ cells)

Fibroblasts		Lymphocytes	
CF	Control	CF	Control
19.00 ± 2.2 $N = 5$	10.80 ± 1.9 $N = 5$	4.07 ± 0.44 $N = 10$	1.80 ± 0.22 $N = 6$
$p < 0.01$		$p < 0.005$	

free carriers of the CF gene consistently differed in ^{45}Ca influx from cells donated by matched controls. These studies have been repeated in our laboratory with similar results (FIGURE 2).

More recent data on cells from greater numbers of HZ donors confirm our original observations. Using ^{45}Ca exchange methods, whole cell Ca^{2+} is increased in HZ cells when compared with matched controls. This difference is present when preincubation of the cells is done with the Ca^{2+} concentration at 2.5 mM and the phosphate concentration at 3.0 mM. Atomic absorption spectrophotometry of cells under the same incubation conditions gave values for $\mu g\ Ca^{2+}$ per mg cell protein of HZ $= 2.10 \pm 0.09$ ($N = 5$) and controls $= 1.70 \pm 0.10$ ($N = 4$), $p < 0.025$.[26]

While the difference in intracellular Ca^{2+} is small, it is consistent. The Ca^{2+} pool size differences in HZ cells, obtained from individuals who would be undetectable as carriers had they not had a CF child, suggests that altered Ca^{2+} metabolism is related to the basic gene defect in CF. To pursue the metabolic defect that would lead to altered intracellular calcium, more specificity was required.

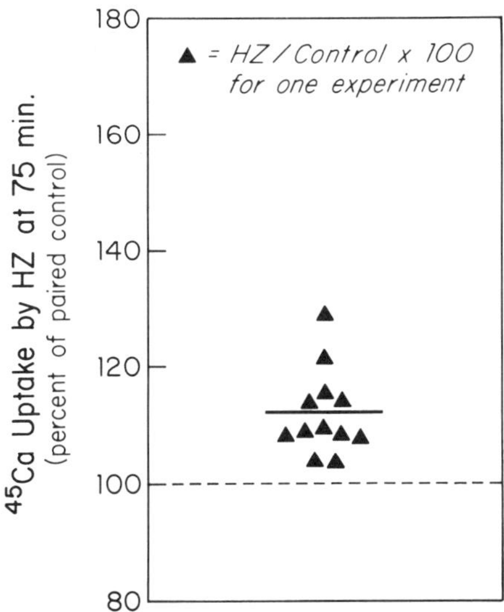

FIGURE 2. ^{45}Ca influx by heterozygous fibroblasts compared with respective age and passage matched controls. Same conditions as in FIGURE 1.

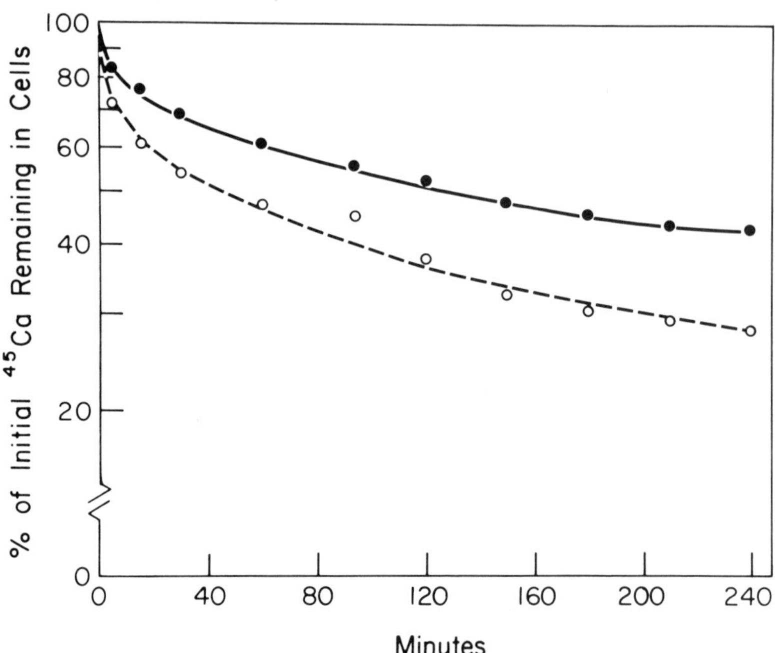

FIGURE 3. ^{45}Ca efflux from CF (●) and control (○) fibroblasts. A graph of data from a representative experiment. Cells in suspension were prelabeled with ^{45}Ca for 90 minutes in Krebs Ringer bicarbonate buffer containing 2.5 mM Ca^{2+} and 2.0 mM phosphate, then washed and resuspended in the same buffer without ^{45}Ca for efflux measurements. Each point represents the percent of total (0-time) ^{45}Ca remaining in the cells. Graphical analysis of these experiments yields values for Ca^{2+} compartment sizes and efflux rate constants as shown in TABLE 2.

MITOCHONDRIAL CALCIUM AND CF

Two early pieces of evidence suggested mitochondria as the location of the altered Ca^{2+} pool in CF cells. The first was graphical analysis of ^{45}Ca efflux from preloaded cells. Methods previously described were followed.[24] These studies of tracer equilibration out of cells into buffer containing cold Ca^{2+} suggested that an intracellular Ca^{2+} pool with slow efflux was the site of the CF alteration[25] (FIGURE 3 and TABLE 2).

TABLE 2. Calcium Pool Sizes and Efflux Rate Constants (From Graphical Analysis of ^{45}Ca Efflux from Whole Cells)

	Pool Size (nmoles Ca^{2+}/mg protein)		Efflux Rate Constant (nmoles/mg protein · min) / (nmoles/mg protein)	
	Cytosolic	Mitochondrial	From Cytosol	From Mitochondria
CF	144 ± 20	2353 ± 499	$2.52 \times 10^{-2} \pm 0.27 \times 10^{-2}$	$1.78 \times 10^{-3} \pm 0.39 \times 10^{-3}$
Control	126 ± 17	1291 ± 247	$2.77 \times 10^{-2} \pm 0.91 \times 10^{-2}$	$2.40 \times 10^{-3} \pm 0.84 \times 10^{-3}$
	$p < 0.05$	$p < 0.025$	$p > 0.3$	$p > 0.1$

Other measurements on Ca^{2+} binding by crude cell homogenates in the presence and absence of ruthenium red suggested a mitochondrial anomaly (TABLE 3). ^{45}Ca binding differences between CF and control cell homogenates was entirely accounted for by the ruthenium red sensitive portion of ^{45}Ca uptake or binding, implicating mitochondria.[8]

Cell fractionation studies followed. Our findings show that mitochondria are, at least in part, responsible for the increased Ca^{2+} in CF cells. Studies with isolated mitochondria-rich subcellular fractions show a marked increase in Ca^{2+} accumulation over time in CF mitochondria.[8,10]

Mitochondria-rich fractions were isolated by differential centrifugation in 0.25 M sucrose/Tris, pH 7.4, containing 1 mM EDTA following usual procedures. Isolated mitochondrial fractions (total protein = 500 μg) were washed twice in EDTA-free 0.25 M sucrose/Tris, pH 7.4, then suspended in solution A (Tris-HCl 33.6 μmol, KCl 24 μmol, $MgCl_2$ 3 μmol and sucrose 25 μmol per 400 μl, pH 7.4). A 100-μl aliquot of mitochondrial suspension was delivered to each experimental tube containing 400 μl solution A at 20°C with air as the gas phase. After a 10-minute incubation, 100 μl of

TABLE 3. Calcium Uptake by Cell Homogenates with and without Ruthenium Red (nmoles Ca^{2+}/mg protein at 8 minutes)

	CF			Control		
Experiment	Total	With RR	MT[a]	Total	With RR	MT[a]
1	91.0	6.6	84.4	76.9	7.1	69.8
2	123.9	10.7	113.2	108.6	9.6	99.0
3	106.2	7.2	99.0	95.7	12.2	83.5
4	119.3	10.8	108.5	92.5	9.9	82.6
Mean	110.1	8.8	101.3	93.4	9.7	83.7

CF compared to control	matched pair *p* value
Total uptake	$p < 0.01$
Mitochondrial uptake	$p < 0.005$

[a]Assumed mitochondrial uptake = Total uptake minus uptake under the influence of ruthenium red.

solution B (sodium succinate 6 μmol, phosphate 1.8 μmol, and ATP 0.45 μmol per 100 μl, pH 7.4) were added, and 60 seconds later 100 nmol ^{45}Ca were added in a 100-μl volume. At each sampling time point, a 100-μl aliquot was removed from the buffer and delivered to a microcentrifuge tube (500 μl) containing 100 μl of quench solution (20 mM EGTA and 10 μM ruthenium red). The quenching tubes were centrifuged at 12,000 *g* for 4 minutes to pellet the mitochondria and a 100-μl sample of the supernatant was taken for ^{45}Ca determination.

CF mitochondria accumulated more Ca^{2+} than control mitochondria (FIGURE 4). At 30 minutes, the Ca^{2+} uptake by CF mitochondria was about 78% more than in controls (10^{-9} moles Ca^{2+} per mg mitochondrial protein: CF = 122.7 ± 10.7, control = 69.1 ± 6.0, $N = 15$ experiments). This difference was highly significant ($p < 0.005$). Others have repeated these studies using similar methods and have found comparable results.[27]

Mitochondrial Ca^{2+} differences have been measured by other techniques as well. Using ^{45}Ca and a standard filtration protocol, CF mitochondrial uptake of Ca^{2+} was shown to be increased over control values with malate-glutamate as the respiratory substrate.[10] ATP and phosphate were present in the buffer. Antimycin A, an inhibitor of electron transport, at 2×10^{-4} M reduced CF and control uptake of Ca^{2+} to the

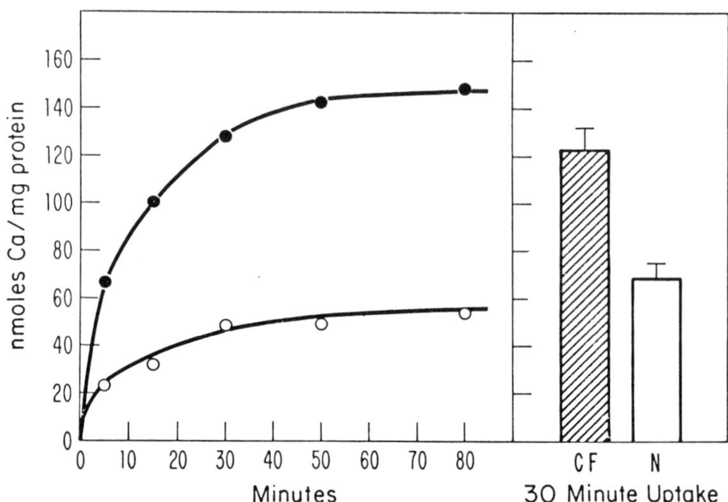

FIGURE 4. Ca^{2+} uptake by mitochondria isolated from CF and control fibroblasts. The left side of the figure is the time course of Ca^{2+} uptake per mg mitochondrial protein by CF (●) and control (O) mitochondria. Each point represents the mean of duplicate determinations on two CF and two control mitochondrial preparations. The right side of the figure gives the mean and S.E.M. of Ca^{2+} uptake by CF (filled bar) and control (open bar) mitochondria. The bars represent 15 experiments in which the 30 minute sample was measured. $p < 0.0005$. The buffer for these experiments contained 100 μM Ca^{2+}.

same low value. Oligomycin, which inhibits ATP hydrolysis, reduced uptake in both CF and control mitochondria, at 8×10^{-5} M, yet a difference between genotypes remained (FIGURE 5). The inhibitor studies suggested that the Ca^{2+} uptake alteration in CF was dependent on electron transport activity and not on differences in ATP hydrolysis.

Kinetic analysis of Ca^{2+} uptake by mitochondria at 10 seconds with varying concentrations of free Ca^{2+} revealed no significant differences in affinity of the carrier for Ca^{2+} between CF and control[10] (FIGURE 6).

Intracellular Ca^{2+} determinations using the indicator, Quin 2, in whole cells may be supportive of our mitochondrial data. Measuring cytoplasmic Ca^{2+} of cells preloaded with Quin 2, no change in cytoplasmic Ca^{2+} in CF cells has been observed.[28-30] This is not surprising. Our hypothesis of Ca^{2+} differences in CF only requires transient and local alterations of Ca^{2+}. Clearly, generalized changes in cytoplasmic free Ca^{2+} of any magnitude would not be compatible with cell viability.

The Quin 2 studies have included cell Ca^{2+} measurements after cell treatment with the mitochondrial uncoupler, carbonyl cyanide *p*-trifluoromethoxyphenyl hydrazone (FCCP), to estimate FCCP-releasable (mitochondrial) Ca^{2+}. Dearborn *et al.*[29] reported a marked increase in the mitochondrial Ca^{2+} pool in CF lymphocytes. Other similar studies do not report an intracellular mitochondrial pool difference. Grinstein *et al.*[28] were unable to make accurate measurements of FCCP-releasable Ca^{2+} stores. Suter *et al.*[30] reported no difference between CF and control neutrophils in a ionomycin releasable Ca^{2+} pool. This group did not use FCCP to specifically release mitochondrial Ca^{2+}. It is also important to note that mature neutrophils do not contain an appreciable number of active mitochondria.[31]

Measurements of mitochondrial Ca^{2+} by atomic absorption spectroscopy of isolated mitochondrial fractions confirm an increased capacity for Ca^{2+} within CF mitochondria. In this study, mitochondria were incubated for one hour in buffer high in Ca^{2+} and phosphate. The resultant measurement must be considered nonphysiologic as it reflects a "massive loading" environment. Nonetheless, statistically different Ca^{2+} values were found when comparing CF and control mitochondria. After contact for one hour with 1 mM Ca^{2+}, the CF levels were (μg Ca^{2+}/mg protein): CF = 6.0 ± 0.08 ($N = 3$) and control = 4.18 ± 0.04 ($N = 3$), ($p < 0.025$).[32]

Recent work in our laboratory suggests that the Ca^{2+} uptake difference is independent of phosphate content of the buffer. Incubation media free of phosphate were used. The buffer contained 130 mM KCl, 10 mM HEPES buffer (pH 7.2), 5 mM succinate, 25 nmole ^{45}Ca, and 2 μg rotenone per mg mitochondrial protein; final volume 945 μl, at 20°C. Uptake started at the addition of 200 μl mitochondrial suspension (approximately 0.50 mg). At specified time points, duplicate 100 μl samples were filtered through 0.45 μm Millipore filters and washed. Total Ca^{2+} uptake was less in all measured samples than our previous studies because of the lack of phosphate. Nonetheless, mitochondria from CF cells repeatedly accumulated more Ca^{2+} than did mitochondria from matched control cells ($p < 0.02$) (FIGURE 7).

Presently, efforts are underway to compare Ca^{2+} efflux rates between CF and control mitochondria. Initial measurements would suggest that nonstimulated efflux of Ca^{2+} is greater in CF mitochondria than that seen from control mitochondria (FIGURE 7). Attempts to stimulate Ca^{2+} efflux from mitochondria have just begun. Sodium and *t*-butyl hydroperoxide (BuOOH) have been used. Each of these potential stimulating methods seems to increase the rate of Ca^{2+} from normal mitochondria, while both Na^+ and BuOOH seem to slow efflux from CF mitochondria. More data are necessary before conclusions can be drawn on efflux differences.

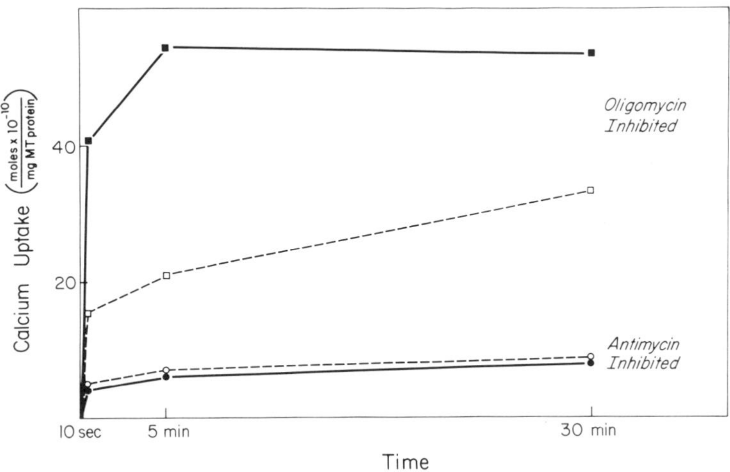

FIGURE 5. The time course of Ca^{2+} uptake by CF (■, ●) and control (□, O) mitochondria under the influence of inhibitors. Each point represents the mean of duplicate determinations on one cell strain. Ca^{2+} uptake was measured in KCl-Tris buffer at pH 7.4, 20°C. Malate and glutamate (5 mM), ATP (1 mM), and phosphate (3 mM) were added prior to Ca^{2+} uptake. Ca^{2+} concentration was 200 μM.

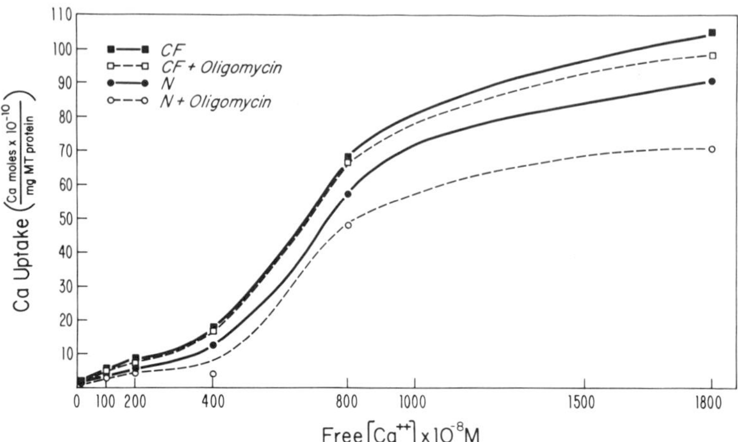

FIGURE 6. Kinetic analysis of Ca^{2+} uptake by mitochondria at 10 seconds. Free Ca^{2+} ion concentrations were controlled by the use of nitrilotriacetic acid/Ca^{2+} buffer within the same solutions as for FIG. 5. Each point represents the mean of duplicate measurements on five cell strains of CF uninhibited (■), CF inhibited by oligomycin, 8×10^{-5} M (□), control uninhibited (●), and control inhibited by oligomycin, 8×10^{-5} M (○).

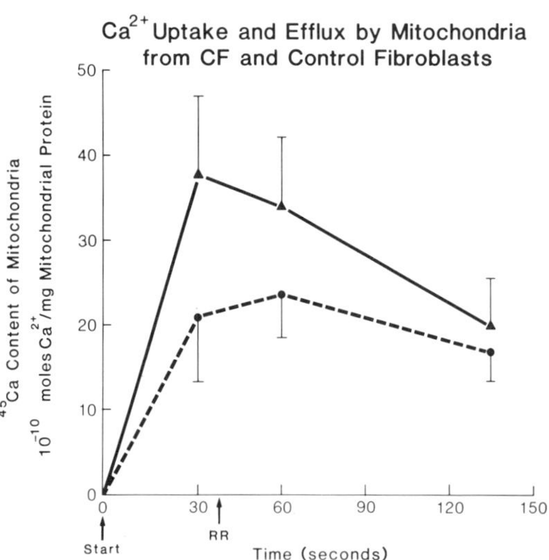

FIGURE 7. Ca^{2+} uptake and efflux by mitochondria from CF and control fibroblasts. Mitochondria (0.5 mg in 200 μl of buffer) was added to 800 μl of KCl-Hepes containing 30 nmoles ^{45}Ca, 5 mM succinate, and 5 μM rotenone. Samples were taken when indicated, filtered, and washed to determine ^{45}Ca content as previously described. Ca^{2+} uptake was stopped by the addition of 1 nmole ruthenium red at the time indicated (RR). The graph represents mean and S.E.M. for seven experiments each measuring one CF (▲) and one control (●) cell strain simultaneously.

Other Mitochondrial Changes

Further studies of mitochondrial function in CF have been accomplished in an attempt to determine mechanisms responsible for the altered Ca^{2+}. Oxygen consumption rates measured on CF, HZ, and control cells show increased respiration in both HZ and CF cell strains.[8,9]

The altered oxygen consumption rate was differentially affected by rotenone, an inhibitor of Site 1 of the mitochondrial electron transport chain[9] (FIGURE 8). Several possible mechanisms exist for this increased oxygen consumption: (1) Mitochondrial respiration in CF and HZ cells may be relatively more uncoupled. (2) Under the stress of *in vitro* study, CF cells may respond with greater respiration. (3) Finally, metabolic events responsible for electron flow to the respiratory chain may be altered. Because of the differential response to rotenone, our laboratory focused efforts on the enzyme complex NADH dehydrogenase to explore the respiratory differences.

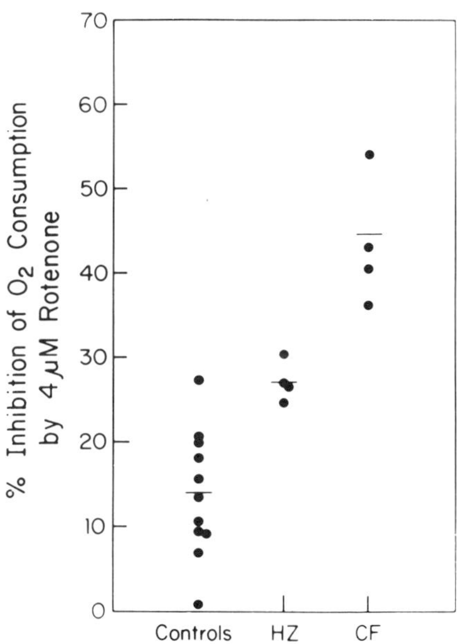

FIGURE 8. Differential effect of rotenone on oxygen consumption by whole fibroblasts from CF, HZ, and control donors. Oxygen consumption rate by fibroblasts was measured before and after the addition of 4 μM rotenone. The change in oxygen consumption was recorded as percentage of initial oxygen consumption rate. Each point represents inhibition measured on one cell strain.[9]

Examination of mitochondrial NADH dehydrogenase (NADH:[acceptor] oxidoreductase, E.C. 1.6.99.3) from isolated mitochondria-rich fractions from cultured skin fibroblasts has revealed enzyme kinetics that differ in the three CF genotypes. pH optima were different among the genotypes[9,11] (FIGURE 9). Apparent binding of the substrate to the enzyme (K_m[NADH]) ranged from 10.9–16.1 μM ($N = 7$) for CF and from 20.9–26.3 μM ($N = 5$) for HZ. With three exceptions, K_m for controls ($N = 12$) ranged from 31.8–42.8 μM (FIGURE 10). K_m of the three exceptional controls were in the HZ range. Two of the three are identical twins. pH optima of enzyme from these three strains did not differ from that of known HZ.[11]

These data correlating two kinetic parameters of an enzyme and the three CF genotypes suggest an association between the CF gene and mitochondrial NADH dehydrogenase.

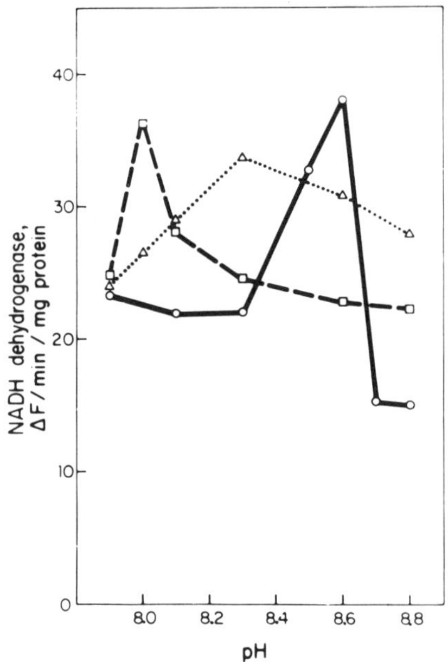

FIGURE 9. NADH dehydrogenase pH optima differences. NADH dehydrogenase activity in isolated mitochondrial preparations from CF (○), HZ (△), and control (□) cell strains was assayed at various pH values. Mitochondrial fractions were frozen and thawed three times to ensure membrane disruption prior to enzyme assay. The control cell strain had its highest activity at pH 8.0, whereas the CF cell strain has its highest activity at pH 8.6. The HZ cell strain showed a broad peak with its optimum at pH 8.3.

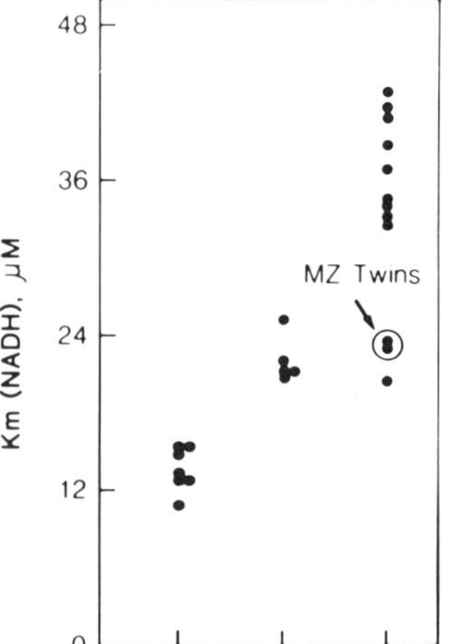

FIGURE 10. Apparent K_m (NADH) of NADH:potassium ferricyanide oxidoreductase. Each dot represents the average of replicate estimates for an individual. pH optima of preparations from the three subjects whose K_m falls in the HZ range were no different from known HZ.

To examine the possibility of a generalized abnormality of the inner membrane of mitochondria, to which NADH dehydrogenase is bound, we examined succinate dehydrogenase (SDH). No differences could be found between CF and control values for SDH.[11]

A recent working hypothesis links several diverse observations in CF cells through the effects of pyridine nucleotide oxidation-reduction (redox) state. Abnormal Ca^{2+} levels in CF mitochondria and altered kinetics of NADH dehydrogenase may be a part of an overall cellular redox change in CF cells. Good evidence exists linking relative redox levels to Ca^{2+} movement at mitochondria,[33–35] plasma membrane,[36] and endoplasmic reticulum.[37] The relationship between mitochondrial Ca^{2+} changes in CF and mitochondrial NADH dehydrogenase alterations related to CF is the most easily envisioned of the possible redox-Ca^{2+} relationships. Other connections can be made between NADH/NAD^+ ratios and other membrane systems. Of interest are reports that plasma membrane Ca^{2+}-ATPase from CF cells shows decreased activity compared with controls.[38] One can hypothesize a link from altered NADH/NAD^+ in mitochondria to altered plasma membrane Ca^{2+}-ATPase. Plasma membrane Ca^{2+}-ATPase activity is affected by oxidation state of glutathione which, in turn, is in balance with cytoplasmic and mitochondrial redox pairs.

This hypothesis should be examined because it offers a potential unifying basis for diverse observations in CF, i.e., increased mitochondrial Ca^{2+} accumulation and decreased plasma membrane Ca^{2+}-ATPase activity levels.

Initial work in our laboratory on whole cell levels of NAD^+ and NADH suggests a shift in pyridine nucleotide redox state in CF. Using freeze-stop methods[39] with harvested fibroblast cells, NADH was extracted with 0.5 N alcoholic KOH and NAD^+ extracted with 0.6 N perchloric acid. Assays were done fluorimetrically using methods published by Klingenberg.[39] Comparing five CF and control experiment pairs, the NADH/NAD^+ ratio appears to be increased in CF (FIGURE 11).

Additional evidence of oxidative changes in CF has been found by Dr. Elli Kohen.[40] Using microspectrofluorometry on CF and normal fibroblasts, Dr. Kohen has found a relative uncoupling of CF mitochondria. NAD(P)H transients measured in individual cells after microinjection of metabolic substrates suggest that mitochondrial control of extra mitochondrial pathways is less tight in CF cells. Additionally, NAD(P)H response is greater in CF cells than in controls when mitochondria are damaged by anthrolin or 2,4-dinitrophenol.

DISCUSSION

Several lines of evidence argue for altered mitochondrial function in CF. Changes in Ca^{2+} uptake, oxygen consumption, and mitochondrial NADH dehydrogenase kinetics have been reported. Newer evidence of Ca^{2+} efflux changes by mitochondria, response to stimuli measured by microspectrofluorometry, and redox state of pyridine nucleotides lends more support to the hypothesis that mitochondria are affected in this disease.

More powerful data in support of this hypothesis can be found in those areas where differences are observable between carriers for the CF gene and controls. These observations in otherwise normal adults are strongly suggestive that mitochondrial differences are closely related to the CF gene. Parameters shown to be altered in carriers include Ca^{2+} uptake by whole cells, oxygen consumption by whole cells, and mitochondrial NADH dehydrogenase kinetics.

What these mitochondrial changes mean to the clinical aspects of CF is open to speculation. We have previously hypothesized a relationship between altered Ca^{2+} movement at mitochondria and altered secretions in CF.[25] It has been demonstrated that secretory granules in exocrine acinar cells accumulate and export significant amounts of Ca^{2+} in conjunction with their protein and glycoprotein products.[41,42] It would seem that increased Ca^{2+} stores may result in increased secretory Ca^{2+}. Several

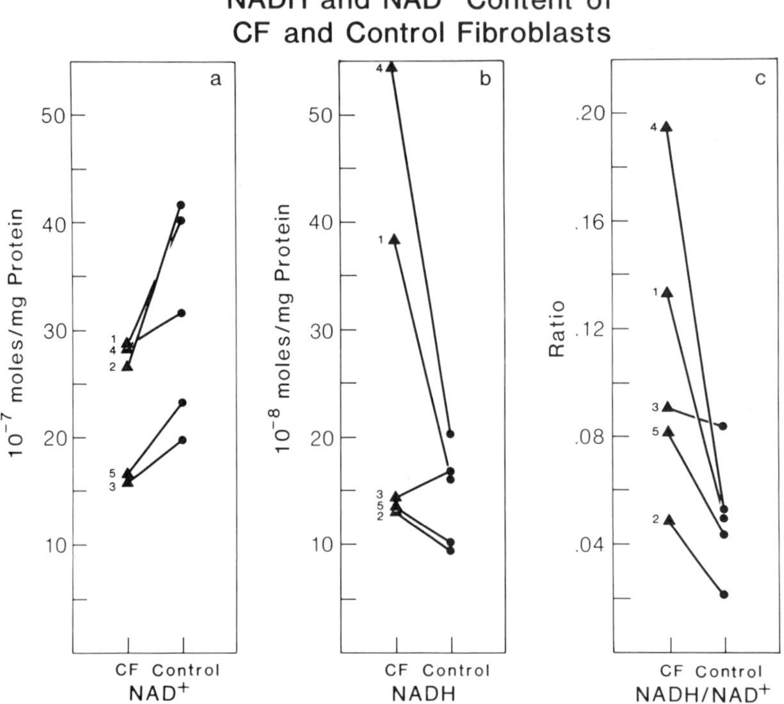

FIGURE 11. NADH and NAD^+ content of CF and control fibroblasts. Confluent fibroblast cultures were harvested by scraping, then washed, pelleted, and quickly frozen. Extraction of NADH and NAD^+ were done as previously described.[36] Each point on the graph represents the mean of three determinations on one CF (▲) or control (●) pooled cell population. No two pooled cell populations were composed of identical cell strains. Four CF and five control strains were used. Each experimental pair is joined by a solid line and the five experiments numbered 1–5. (a) NAD^+ concentration (10^{-7} moles/mg protein), (b) NADH concentration (10^{-8} moles/mg protein); (c) $NADH/NAD^+$ ratio comparison.

groups have demonstrated that increased Ca^{2+} associated with exocrine secretions leads to decreased solubility or increased viscosity of the secretions.[43–45] Some have proposed that these less soluble secretions are the basis for microobstruction of acini and small ducts that occur in many exocrine glands and that these widespread obstructions lead to consequent pathology[44,46,47] in CF. Other related hypotheses have been suggested for a Ca^{2+} relationship to CF pathology.[48]

The actual import eventually given to the mitochondrial changes in CF will have to await further study. Mechanisms responsible for the observed changes are not yet delineated. The part these alterations play in pathogenesis of CF is unknown. Nonetheless, significant evidence has accumulated that mitochondria are changed in this disease.

REFERENCES

1. U.S. Vital Statistics. 1984.
2. NADLER, H. L., G. J. S. RAO & L. M. TAUSSIG. 1978. Cystic Fibrosis. *In* The Metabolic Basis of Inherited Disease. 4th edit. J. B. Stanbury, J. B. Wyngaarden & D. S. Fredrickson, Eds.: 1683–1710. McGraw-Hill. New York.
3. SHAPIRO, B. L., R. J. FEIGAL, N. J. LAIBLE, M. H. BIROS & W. J. WARWICK. 1978. Doubling time, α-aminoisobutyrate transport and calcium exchange in cultured fibroblasts from cystic fibrosis and control subjects. Clin. Chim. Acta **82:** 125–131.
4. FEIGAL, R. J. & B. L. SHAPIRO. 1979. Altered intracellular calcium in fibroblasts from patients with cystic fibrosis and heterozygotes. Pediat. Res. **13:** 764–768.
5. BANSCHBACH, M. W., A. G. KARAM, P. K. LOVE & B. C. HILMAN. 1978. Cystic fibrosis serum promotes (^{45}Ca) uptake by normal human leukocytes. Biochem. Biophys. Res. Commun. **84:** 922–927.
6. MANGOS, J. A. & W. H. DONNELLY. 1981. Isolated parotid acinar cells from patients with cystic fibrosis. Morphology and composition. J. Dent. Res. **60:** 19–26.
7. ROOMANS, G. M., O. CEDER & H. KOLLBERG. 1981. Electrolyte redistribution in cystic fibrosis fibroblasts studied by electron probe X-ray microanalysis. Ultrastruct. Pathol. **2:** 53–58.
8. FEIGAL, R. J. & B. L. SHAPIRO. 1979. Mitochondrial calcium uptake and oxygen consumption in cystic fibrosis. Nature **278:** 276–277.
9. SHAPIRO, B. L., R. J. FEIGAL & L. F. H. LAM. 1979. Mitochondrial NADH dehydrogenase in cystic fibrosis. Proc. Natl. Acad. Sci. USA **76:** 2979–2983.
10. FEIGAL, R. J., M. S. TOMCZYK & B. L. SHAPIRO. 1982. The calcium abnormality in cystic fibrosis mitochondria: Relative role of respiration and ATP hydrolysis. Life Sci. **30:** 93–98.
11. SHAPIRO, B. L., L. F. H. LAM & R. J. FEIGAL. 1982. Mitochondrial NADH dehydrogenase in cystic fibrosis: Enzyme kinetics in cultured fibroblasts. Am. J. Human Genetics **34:** 846–852.
12. DAVIS, P. B. & P. A. DI SANT'AGNESE. 1980. A review. Cystic fibrosis at forty—quo vadis? Pediat. Res. **14:** 83–87.
13. FANCONI, G., E. UELINGER & C. KNAUER. 1936. Das coeliak-syndrome bei angeborener zystischer pankreasfibromatose and brochiectasien. Wien. Med. Wochenschr. **86:** 753–756.
14. DI SANT'AGNESE, P. A., R. C. DARLING, G. A. PERERA & E. SHEA. 1953. Abnormal electrolyte composition of sweat in cystic fibrosis of the pancreas. Pediatrics **12:** 549–563.
15. FARBER, S. 1944. Pancreatic function and disease in early life. V. Pathologic changes associated with pancreatic insufficiency in early life. Arch. Pathol. **37:** 238–250.
16. LOBECK, C. 1972. Cystic Fibrosis. *In* The Metabolic Basis of Inherited Disease. 3rd edit. J. B. Stanbury & J. B. Wyngaarden & D. S. Fredrickson, Eds.: 1605–1628. McGraw-Hill. New York.
17. DEARBORN, D. G. 1976. Water and electrolytes of exocrine secretions. *In* Cystic Fibrosis: Projections into the Future. J. A. Mangos & R. C. Talamo, Eds.: 179–191. Intercontinental Medical Book Corp. New York.
18. BOAT, T. F. & D. G. DEARBORN. 1984. Etiology and Pathogenesis. *In* Cystic Fibrosis. L. M. Taussig, Ed.: 25–84. Thieme-Stratton, Inc. New York.
19. KATZ, S., M. H. SCHONI & M. A. BRIDGES. 1984. The calcium hypothesis of cystic fibrosis. Cell Calcium **5:** 421–440.

20. WOOD, R. E. 1984. Prognosis. *In* Cystic Fibrosis. L. M. Taussig, Ed.: 434–460. Thieme-Stratton, Inc. New York.
21. KNOWLTON, R. G., O. COHEN-HAGUENAUER, N. V. CONG, J. FREZAL, V. A. BROWN, D. BARKER, J. C. BRAMAN, J. W. SCHUMM, L. C. TSUI, M. BUCHWALD & H. DONIS-KELLER. 1985. A polymorphic DNA marker linked to cystic fibrosis is located on chromosome 7. Nature **318:** 380–382.
22. WAINWRIGHT, B. J., P. J. SCAMBLER, J. SCHMIDTKE, E. A. WATSON, H. Y. LAW, M. FARRALL, H. J. COOKE, H. EIBERG & R. WILLIAMSON. 1985. Localization of cystic fibrosis locus to human chromosome 7 cen-q22. Nature **318:** 384–385.
23. BAUR, P. S., W. E. BOLTON & S. C. BARRANCO. 1976. Electron microscopy and microchemical analysis of cystic fibrosis diploid fibroblasts in vitro. Texas Rep. Biol. Med. **34:** 113–134.
24. BORLE, A. B. 1975. Methods for assessing hormone effects on calcium fluxes in vitro. Methods Enzymol. **39:** 513–573.
25. SHAPIRO, B. L., R. J. FEIGAL & L. F-H. LAM. 1980. Intracellular calcium and cystic fibrosis. *In* Perspectives in Cystic Fibrosis. J. M. Sturgess, Ed.: 15–28. Imperial Press. Ontario.
26. SHAPIRO, B. L. & L. F-H. LAM. (Personal communication.)
27. VON RUECKER, A., R. BERTELE & K. HARMS. 1984. Calcium metabolism and cystic fibrosis: Mitochondrial abnormalities suggest a modification of the mitochondrial membrane. Pediat. Res. **18:** 594–599.
28. GRINSTEIN, S., B. ELDER, C. A. CLARKE & M. BUCHWALD. 1984. Is cytoplasmic Ca^{2+} in lymphocytes elevated in cystic fibrosis? Biochim. Biophys. Acta **769:** 270–274.
29. DEARBORN, D. G., R. L. WALLER & W. J. BRATTIN. 1984. Cytosolic free Ca^{2+} concentration and intracellular Ca^{2+} distribution in CF lymphocytes. *In* Cystic Fibrosis: Horizons. E. Lawson, Ed.: 402. Wiley & Sons. Chichester.
30. SUTER, S., P. D. LEW, J. BALLAMAN & F. A. WALDVOGEL. 1985. Intracellular calcium handling in cystic fibrosis: Normal cytosol calcium and intracellular stores in neutrophils. Pediat. Res. **19:** 346–348.
31. KLEBANOFF, S. J. & R. A. CLARK. 1978. The Neutrophil: Function and Clinical Disorders. Elsevier/North-Holland Biomedical Press. New York.
32. SHAPIRO, B. L. & L. F-H. LAM. 1985. Mitochondrial calcium in fibroblasts from subjects with cystic fibrosis and in older cells. Am. J. Hum. Genet. **37** (Suppl.): A17 (abstract).
33. LEHNINGER, A. L., A. VERCESI & E. A. BABABUNMI. 1978. Regulation of Ca^{2+} release from mitochondria by the oxidation-reduction state of pyridine nucleotides. Proc. Natl. Acad. Sci. USA **75:** 1690–1694.
34. LÖTSCHER, H. R., K. H. WINTERHALTER, E. CARAFOLI & C. RICHTER. 1979. Hydroperoxides can modulate the redox state of pyridine nucleotides and the calcium balance in rat liver mitochondria. Proc. Natl. Acad. Sci. USA **76:** 4340–4344.
35. RICHTER, C., K. H. WINTERHALTER, S. BAUMHÜTTER, H. R. LÖTSCHER & B. MOSER. 1983. ADP-ribosylation in inner membrane of rat liver mitochondria. Proc. Natl. Acad. Sci. USA **80:** 3188–3192.
36. BELLOMO, G., F. MIRABELLI, P. RICHELMI & S. ORRENIUS. 1983. Critical role of sulfhydryl group(s) in ATP-dependent Ca^{2+} sequestration by the plasma membrane fraction from rat liver. FEBS Lett. **163:** 136–139.
37. JONES, D. P., H. THOR, M. T. SMITH, S. A. JEWELL & S. ORRENIUS. 1983. Inhibition of ATP-dependent microsomal Ca^{2+} sequestration during oxidative stress and its prevention by glutathione. J. Biol. Chem. **258:** 6390–6393.
38. KATZ, S. & A. TWUM-AMPOFO. 1980. (Mg^{2+} + Ca^{2+})-ATPase activity in plasma membrane enriched preparations of human skin fibroblasts: Decreased activity in fibroblasts derived from cystic fibrosis patients. Clin. Chim. Acta **100:** 245–252.
39. KLINGENBERG, M. 1974. Nicotinamide-Adenine Dinucleotides. *In* Methods of Enzymatic Analysis. 2nd English edit. H. U. Bergmeyer, Ed.: 2045–2072. Academic Press. New York.
40. KOHEN, E. (personal communication) University of Miami. Department of Biology.
41. WALLACH, D. & M. SCHRAMM. 1971. Calcium and the exportable protein in rat parotid gland. Parallel subcellular distribution and concomitant secretion. Eur. J. Biochem. **21:** 433–437.

42. ARGENT, B. E., R. M. CASE & T. SCHRATCHERD. 1973. Amylase secretion by the perfused cat pancreas in relation to the secretion of calcium and other electrolytes and as influenced by the external ionic environment. J. Physiol. **230:** 575–593.
43. BLOMFIELD, J., K. L. WARTON & J. M. BROWN. 1973. Flow rate and inorganic components of submandibular saliva in cystic fibrosis. Arch. Dis. Child. **48:** 267–274.
44. GUGLER, E., J. C. PALLAVICINI, H. SWERDLOW & P. A. DI SANT'AGNESE. 1967. Role of calcium in submaxillary saliva of patients with cystic fibrosis. J. Pediat. **71:** 585–592.
45. FORSTNER, J. F. & G. G. FORSTNER. 1976. Effects of calcium on intestinal mucin: Implications for cystic fibrosis. Pediat. Res. **10:** 609–613.
46. GIBSON, L. E., W. J. MATTHEWS, JR., P. T. MINIHAN & J. A. PATTI. 1971. Relating mucus, calcium, and sweat in a new concept of cystic fibrosis. Pediat. **48:** 695–710.
47. BLOMFIELD, J., A. R. RUSH, H. M. ALLARS & J. M. BROWN. 1976. Parotid gland function in children with cystic fibrosis and child control subjects. Pediat. Res. **6:** 574–578.
48. SORSCHER, E. J. & J. L. BRESLOW. 1982. Cystic fibrosis: A disorder of calcium-stimulated secretion and transepithelial sodium transport? Lancet **I:** 368–370.

DISCUSSION OF THE PAPER

B. CHANCE (*University of Pennsylvania, Philadelphia, PA*): (1) The pH profiles are so sharp (0.2–0.4 pH) that they are hard to explain on enzyme chemistry, can you explain?

(2) The differences in K_m NADH may be small compared to the [NADH] in the mitochondria matrix. Please comment!

FEIGAL: For both questions I can only report that differences among the three genotypes have been reproducible on numerous cell strains of the three genotypes. The import we can give these findings to cell physiology is unknown at this point. The consistency of the differences among the genotypes is of interest genetically.

G. F. AZZONE (*University of Padova, Padova*): None of the alterations of the mitochondrial properties mentioned in your talk can explain an increase in role or in extent of Ca^{2+} accumulation in mitochondria. In particular I see no possible effect of changes of activity of respiratory chain components. To modify Ca^{2+} distribution it is necessary to alter the activity of the pathways for Ca^{2+} transport, i.e., influx and efflux pathways.

FEIGAL: I agree with the comments within this question. We are presently directing our attention to the Ca^{2+} efflux pathway. An alteration in efflux during non stimulated or stimulated conditions may help in answering the question of why Ca^{2+} accumulation is changed in CF.

The observations on oxygen consumption rate and NADH dehydrogenase were presented to show that several mitochondrial changes are apparent in CF and, more importantly, to show that the carriers for CF have altered mitochondrial parameters. There are no proven causal relationships among these observations yet the number of mitochondrial changes observed argue for a role of mitochondria in CF.

E. CARAFOLI (*Swiss Federal Institute of Technology, Zurich*): Do you have any measurements of phosphate content in cystic fibrosis mitochondria? If phosphate increased it could help explain the augmented Ca^{2+} uptake by CF mitochondria.

FEIGAL: No measurement of phosphate content of cystic fibrosis mitochondria has been made. An early observation that may pertain to this question is that increasing phosphate content of buffer overlying whole cell monolayers results in amplified differences in ^{45}Ca influx between cystic fibrosis and control or heterozygotes and their control cells.

S. CORVERA (*University of Massachusetts, Worcester, MA*): Are general growth characteristics of skin fibroblasts in culture similar in normal and cystic fibrosis?

FEIGAL: Yes, growth characteristics such as doubling time, time to confluence, and cell number per unit area are identical at early passage number. We routinely do our assays at passage 10 or below. Work by Shapiro has shown some changes in cell doubling time in cystic fibrosis cells at higher passages (15 and above). This remains a mystery but should not affect our data taken at low passages.

P. NICHOLLS (*Brock University, St. Catherine's, Ont.*): The high frequency of the cystic fibrosis gene in the population suggests a "balanced polymorphism" with some selective advantage to the heterozygote. Is there any evidence for such an advantage (e.g., a decreased susceptibility to another disease)?

FEIGAL: Population genetics theory would agree that a heterozygote advantage must exist or may have existed at one time. To date, no specific disease has been shown to be of decreased susceptibility in heterozygotes. Others have speculated on such an advantage (see reference by Taussig, L.M. in *Cystic Fibrosis*, 1984, Thieme-Stratton, Inc., New York).

Targeting of Nuclear-Encoded Proteins to the Mitochondrial Matrix: Implications for Human Genetic Defects

LEON E. ROSENBERG, WAYNE A. FENTON, ARTHUR L. HORWICH, FRANTISEK KALOUSEK, AND JAN P. KRAUS

Department of Human Genetics
Yale University School of Medicine
New Haven, Connecticut 06510

Many inherited metabolic disorders in man result from markedly impaired activity of single enzymes localized to mitochondria. Although these specific mitochondrial enzyme deficiencies involve a number of metabolic pathways including glycolysis (pyruvate dehydrogenase deficiency), porphyrin biosynthesis (ferrochelatase deficiency), and electron transfer (cytochrome oxidase deficiency), most affect the pathways by which amino acids are catabolized—either those routes responsible for transamination (ornithine aminotransferase deficiency) or ammonia detoxification (deficiency of N-acetylglutamate synthestase, carbamyl phosphate synthetase, and ornithine transcarbamylase) or the degradation and/or interconversion of the carbon skeleton (deficiency of branched chain ketoacid dehydrogenase, medium chain acyl CoA dehydrogenase, isovaleryl CoA dehydrogenase, propionyl CoA carboxylase, and methylmalonyl CoA mutase).

The vast majority of mitochondrial proteins—including all those mentioned above—share several crucial features: they are coded for by nuclear genes, synthesized on cytoplasmic polyribosomes, and subsequently imported by mitochondria. It follows, therefore, that defective import of such proteins will likely be among the several general molecular mechanisms ultimately shown to be responsible for mitochondrial enzyme deficiency, along with the more common defects as failure to be synthesized, inability to bind substrates or cofactors, or accelerated degradation. It was to test this prediction of defective import that we began, nearly a decade ago, to try to understand how a few nuclear-coded mitochondrial matrix enzymes in mammalian cells get from where they are made to where they function. It is not our intent here to review the entire field of biogenesis of mitochondrial proteins in eukaryotic organisms. This is a rapidly advancing area of research, too broad to be covered in these pages. We will review the progress we have made and direct the interested reader to more comprehensive recent reviews.[1,2]

GENERAL STRATEGY AND EARLY FINDINGS

Model Enzymes

We have used three mitochondrial matrix enzymes as our models, each of which has been purified to homogeneity in our laboratory and against each of which polyclonal antibodies of high titer have been raised in rabbits. Ornithine transcarbamylase (OTC), the first discovered mitochondrial component of the urea cycle, is encoded by a gene on the short arm of the X chromosome, It is a trimer of identical subunits (α_3), each of 36 kDa. The two other enzymes are part of the propionate pathway by which the carbon skeletons of several essential amino acids and odd chain fatty acids are ultimately supplied to the tricarboxylic acid cycle. Propionyl CoA carboxylase (PCC) is a biotin-dependent enzyme composed of two non-identical subunits—α of ~70 kDa and β of ~54 kDa. This holoenzyme probably has a quaternary structure of $\alpha_4\beta_4$, the biotin-binding site residing on the α subunit. Methylmalonyl CoA mutase (MUT) is a dimer of identical subunits (α_2) of ~77kDa; it requires adenosylcobalamin as a cofactor.

Subunits Synthesized as Larger Precursors

After demonstrating that intact mitochondria are unable to import homogeneous, assembled OTC, we decided to determine the nature of the primary translation product for this enzyme using, as an experimental protocol, cell-free protein synthesis in a reticulocyte lysate system programmed with rat liver polysomal RNA, specific immunoprecipitation, and polyacrylamide gel electrophoresis. We found that OTC was synthesized as a single polypeptide about 4 kDa larger than its mitochondrial subunit homologue[3] (FIG. 1). Moreover, incubation of this larger putative precursor polypeptide with intact rat liver mitochondria led to its internalization and proteolytic cleavage to a size identical to that of the mature mitochondrial OTC subunit (36 kDa).[4] Subsequent experiments showed that this bonafide precursor polypeptide, designated pOTC, consisted of a cleavable 4 kDa leader peptide joined to the mature OTC sequence at its amino terminus, and that the amino-terminal leader was absolutely required for mitochondrial uptake.[5] Similar experiments were carried out for the MUT subunit and for the α and β subunits of PCC. In each case, these polypeptides were synthesized as larger precursors (pMUT, pαPCC, pβPCC) that could be taken up by mitochondria *in vitro* and processed to their respective mature-sized counterparts[6,7] (FIG. 1). Interestingly, analysis of the mobility of each of these precursors on polyacrylamide gels suggests that the estimated size of the amino-terminal leader peptide was distinct for each precursor examined: ~3 kDa for pMUT; ~4 kDa for pOTC; ~4.5 kDa for pαPCC; and ~7 kDa for pβPCC.[5,7]

Energy-Dependent Uptake, Enzymatic Cleavage, and Assembly

Having demonstrated that mitochondrial uptake of several precursor polypeptides depends upon the presence of amino-terminal leader peptides, which are cleaved after entry, we next examined three key features of this import pathway. First, we demonstrated that a variety of inhibitors of electron transport (rotenone, antimycin, cyanide) and several uncouplers (2,4-dinitrophenol, valinomycin, carbonyl cyanide p-trifluoromethoxyphenylhydrazone) impaired markedly the posttranslational uptake and conversion of pOTC to mature OTC.[8] Similar findings were obtained with pMUT and pβPCC.[6,7] Significantly, atractyloside (an inhibitor of adenine nucleotide transport) failed to impair conversion of pOTC to OTC suggesting that maintenance of the inner membrane potential (rather than the supply of ATP) is the predominant source of the energy required for translocation. Second, we prepared a series of submitochondrial fractions and used them to show that the protease responsible for cleaving pOTC to its mature species is a divalent cation-dependent protein localized to the mitochondrial matrix.[9] Third, we demonstrated that assembly of mature OTC subunits into the active trimeric form occurred only after cleavage of the leader peptide.[10]

Confirmatory Data with Intact Cells

The availability of reagents such as dinitrophenol and rhodamine 6G, which interfere with mitochondrial energy metabolism both *in vitro* and in intact cultured cells, allowed us to test certain features of the import pathway in intact cell systems. When cells were pulse labeled in the absence of inhibitors, only mature-sized subunits

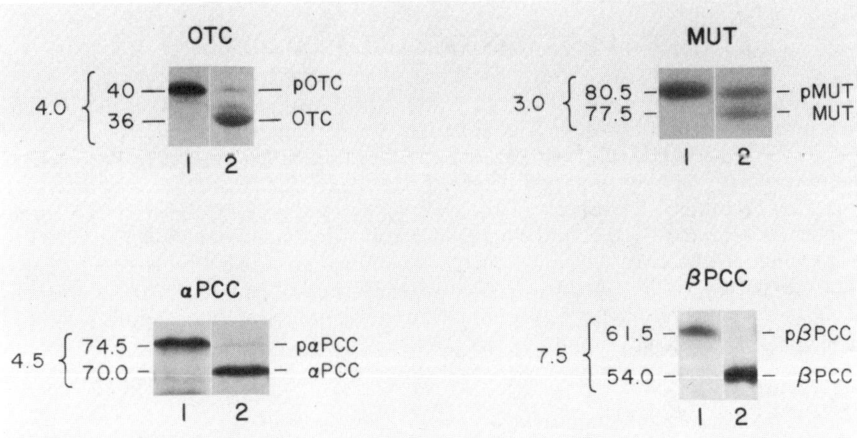

FIGURE 1. *In vitro* translation of precursors and posttranslational processing by intact mitochondria. Portions of fluorographs of SDS-polyacrylamide gels showing products specifically immunoprecipitated with polyclonal antisera raised against ornithine transcarbamylase (OTC), methylmalonyl CoA mutase (MUT), the α subunit of propionyl CoA carboxylase (αPCC), or the β subunit of propionyl CoA carboxylase (βPCC), respectively, are shown. In each lane marked 1 is shown the primary translation product isolated from a cell-free rabbit reticulocyte lysate system programmed with rat liver polysomal RNA. In each lane marked 2 is shown the products isolated from the mitochondrial pellet after incubation of the translation mixture with intact rat liver mitochondria for 30–60 min. The numbers to the left of each pair of lines denote the size of the respective species in kDa. The abbreviation pOTC, pMUT, etc. refer to the precursors for the various subunits. For details see references 4, 6, and 7.

for OTC and MUT were observed. When pulsed in the presence of dinitrophenol, however, only pOTC and pMUT were apparent. A pulse in the presence of inhibitor followed by a chase in its absence allowed us to determine a $t_{1/2}$ for the conversion of the precursors to mature subunits in the range of 2–4 min, indicating that mitochondrial uptake and processing occurs rapidly after synthesis.[7,11]

Model of Biogenesis

Based on these findings and the results of others, we proposed the multi-step model shown in FIGURE 2. It posits that the mitochondrial proteins we have studied are synthesized on cytoplasmic polyribosomes as larger precursors bearing amino terminal leader peptides that function as address signals. Once the precursors are recognized by mitochondria (presumably via outer membrane receptors), they are transported across the outer and inner membranes by a process that requires an intact membrane potential. Then, the leader peptide is cleaved by a divalent cation-dependent matrix protease. Finally, the mature subunit(s) assemble to their active conformation. It should be emphasized that similar models of import of mitochondrial proteins have been proposed by many groups working with model proteins from *Saccharomyces cerevisiae* and *Neurospora crassa,* as well as with mammalian cells.[1,2] Thus, it seems certain that the general features of import of proteins to the internal subcompartments of the mitochondrion have been conserved throughout evolution.

INFORMATION FROM CLONED GENES

To define further the structure and function of the amino-terminal leader peptides so crucial to the most mitochondrial proteins, we have cloned in succession cDNAs for rat OTC,[12,13] human OTC,[14] and rat βPCC.[15] Our experimental strategy has been the same for each, namely: isolation of enriched mRNA for each species by polysome immunoadsorption[15,16] synthesis of the corresponding double-stranded cDNA; insertion of cDNA into pBR 322 and cloning in *E. coli;* identification of bonafide clones by a combination of colony hybridization, blot hybridization, and hybrid-selected translation; confirmation by matching nucleotide sequence of cDNAs with amino acid sequences of the respective pure proteins; and formation of full length cDNAs by molecule construction techniques when necessary.

Amino Acid Sequence of Leader Peptides

We deduced first the complete primary structure of human pOTC from the nucleotide sequence of the cloned cDNA.[14] The amino terminal leader peptide was shown to contain 32 amino acid residues, which is in excellent agreement with prior work suggesting its mass to be ~4 kDa (FIG. 3). The amino acid composition of the

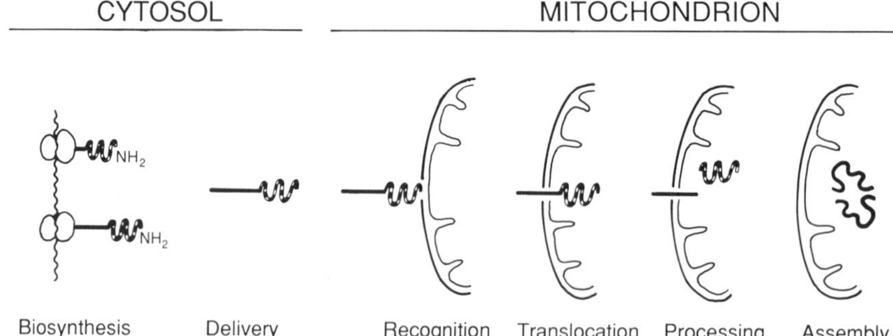

FIGURE 2. Model of biogenesis of nuclear-encoded mitochondrial matrix enzymes. The amino terminal leader peptide is denoted by the wavy, hatched line; the mature portion by the solid, straight line.

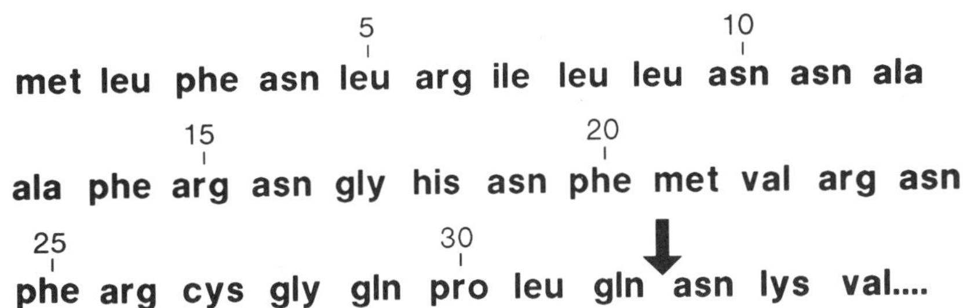

FIGURE 3. Complete amino acid sequence of the human pre-ornithine transcarbamylase (pOTC)leader peptide. The vertical arrow denotes the site of cleavage between the leader and mature portions of pOTC. See reference 14 for details.

leader is interesting in that it contains four arginines (at position 6, 15, 23, and 26) but no glutamates or aspartates. This suggests that the leader carries a net positive charge. Cleavage of the leader peptide from the mature portion occurs between glutamine and asparagine residues. Rat pOTC was also predicted from its corresponding cDNA to have a 32 amino acid long leader peptide.[13] The rat and human pOTC leader sequences are only 69% homologous compared to 93% homology in their respective mature sequences. In addition to arginine residues at positions 6, 15, 23, and 26 (identical to human), the rat pOTC leader has three lysine residues at positions 11, 16, and 29. Thus, it is even more basic in composition than its human homologue. The precise length of the βPCC leader peptide has not been determined. It is at least 38 residues long and contains 5 arginines. Once again, it is devoid of acidic residues.

Leader Peptide Sufficient to Direct Imports

All our previous work had indicated that the amino-terminal leader peptides characteristic of mitochondrial precursors were necessary for import. To determine if they were sufficient as well as necessary, we asked whether such a leader could "misdirect" a cytosolic protein to the mitochondria. Thus, we fused the nucleotide sequence encoding the human OTC leader with a cDNA encoding the cytosolic enzyme, dihydrofolate reductase (DHFR).[17] This fusion gene, along with required SV40 regulatory elements and a selectable marker, was used to transfect a mutant Chinese hamster ovary cell line devoid of endogenous DHFR. In pulse-labeled transformants, the predicted chimeric protein was detected, as were its expected mitochondrial cleavage products. Further, immunofluorescent staining of transformants with anti-DHFR antiserum revealed a pattern characteristic of mitochondrial localization. Finally, when the chimeric precursor was synthesized in a cell-free system and incubated posttranslationally with rat liver mitochondria, it was imported by mitochondria and cleaved to a product that was protected from the proteolytic action of tryspin. Thus, both in intact cells and *in vitro,* the human OTC leader sequence was sufficient to direct import of DHFR.

Critical Regions and Residues in the Human OTC Leader Peptide

We have produced a series of deletions as well as numerous site-directed and random amino acid substitutions in the human OTC leader peptide to obtain additional detailed information about functionally important features of this sequence.[18,19] Analysis of the deletions (FIG. 4) revealed that, whereas neither the extreme amino-terminal nor carboxy-terminal portions of the leader is absolutely required for mitochondrial uptake and proteolytic processing, the mid-portion of the leader is critical.[19] When codons 8–22 were deleted, no uptake or processing was noted. All other deletions resulted in much less extreme impairment of import. Next, we assessed the role of arginine residues in the leader. Two preliminary experiments had suggested that one or more of these residues was important. In one, intact transfected cells expressing the human OTC gene were incubated with pulse-labeled in medium in which arginine was replaced by its less charged analogue, canavanine. The substituted precursor failed to be imported and cleaved.[9] In the other, arginine residues at positions 15, 23, and 26 were replaced at once in the altered leader by charge-neutral glycines.[18] The triply substituted precursor failed to be taken up and cleaved by mitochondria *in vitro*. This result could be interpreted as evidence that the net positive charge conferred on the leader by these arginine residues plays some important role in uptake and processing. It could, alternatively, indicate the existence of some important secondary structural feature that is lost in the triply substituted mutant. Thus, we proceeded to assess the role of each of the four arginine residues individually (FIG. 5).[19] When the arginine residue at position 6 or position 26 was substituted by glycine, no significant effect on import and cleavage *in vitro* was noted. Substitution of glycine for arginine at position 15 reduced processing by about 50%. Arginine at position 23 was, without question, the most crucial. When it was replaced by glycine, no processing of the precursor was observed. We next substituted individually three other amino acids, asparagine, alanine, and lysine, for arginine at position 23. The asparagine-substituted

Construct	Amino Acid Sequence of Leader Peptide (10, 20, 30)	↓	Import/Cleavage of Precursor by Intact Mitochondria (%)
wt	MLFNLRILLNNAAFRNGHNFMVRNFRCGQPLQ	NK	100
d2-3	M NLRILLNNAAFRNGHNFMVRNFRCGQPLQ	NK	100
d2-7	M LLNNAAFRNGHNFMVRNFRCGQPLQ	NK	45
d2-12	M AFRNGHNFMVRNFRCGQPLQ	NK	20
d8-22	MLFNLRI RNFRCGQPLQ	NK	0
di 1	MLFNLRI AAAI RNFRCGQPLQ	NK	20
d26-31	MLFNLRILLNNAAFRNGHNFMVRNF Q	NK	90

FIGURE 4. Schematic analysis of the effect of various deletions within the human pre-ornithine transcarbamylase (pOTC) leader peptide of subsequent uptake and processing by intact mitochondria. The wild-type (wt) leader peptide sequence is shown in single letter code in the top line. The sequences of deletion constructs are shown below, and these constructs are denoted by numbering the codons that were deleted, i.e., d2–3 denotes deletion of codons 2 and 3; d2–12 denotes deletion of codons 2 through 12. Product di 1 is a deletion-insertion derivative of d8–22 containing four inserted codons as shown. The vertical arrow identifies the site of cleavage between the leader and mature portions. The % figures shown in the far right lane were obtained from densitometric tracings of the polyacrylamide gels, setting the wild-type value at 100%. See reference 19 for details.

Substitution	Amino Acid Sequence of Leader Peptide: 6	15	23	26	Import/Cleavage of Precursor by Intact Mitochondria (%)
none	R	R	R	R	100
arg^{6} → gly	G	R	R	R	80
arg^{15} → gly	R	G	R	R	40
arg^{23} → gly	R	R	G	R	0
arg^{26} → gly	R	R	R	G	100
arg^{15,26} → gly	R	G	R	G	15
arg^{15,23,26} → gly	R	G	G	G	0

FIGURE 5. Effect of substitutions of glycine residues for arginines in the human pre-ornithine transcarbamylase (pOTC) leader peptide. Substitutions are shown in the far left column with the schematic representations of the leader residues in the center. The right column shows % of uptake and processing of each construct compared to the value for the unsubstituted leader, which was set at 100%. These values were obtained from densitometric tracing of polyacrylamide gels. R denotes arginine; G denotes glycine. See references 18 and 19 for details.

precursor was taken up and cleaved about 20% as efficiently as wild-type; the alanine-substituted form about 50% as well, and the lysine-substituted molecule indistinguishably from wild-type. Clearly, a positive charge at position 23 is favored (as shown by the results with arginine and lysine) but is not absolutely required (as shown by the ability of asparagine and alanine to retain appreciable function). Given additional information from theoretical predictions that glycine is an α-helix–breaking residue and that the order of increasing α-helix potential of the residues substituted here is gly $<$ asn $\ll$ arg $<$ lys $<$ ala, we proposed that residue 23 participates in (and perhaps is the center of) a short stretch of α-helix required in some, as yet unexplained, way for uptake and processing of the OTC precursor.[19] This putative helix does not appear to be amphipathic because substitution of arginine for the residues at positions 21 or 24 (singly or in combination) does not interfere with uptake and cleavage (unpublished results). The extent of this helical region, the means by which it promotes uptake and cleavage, and its relationship to the functional role of other positively charged residues are not known currently. It would not be surprising to find that different structural features of the leader have unique, discrete functions in the multi-step pathway of import and processing.

IMPLICATIONS FOR GENETIC DEFECTS

We are unaware of a single documented example of an inherited deficiency of a mammalian mitochondrial protein caused by defective transport from cytosol to mitochondrion. From the foregoing, however, it should be clear that many mutational events could alter the structure and function of these crucial and cleavable leader peptides in ways that would prevent the proteins from reaching their appropriate destinations. Leader deletions might be detectable by looking for size differences on gel electrophoresis. Discovery of point mutations will require systematic study, first with the kinds of *in vitro* reconstitution protocols discussed here and subsequently by molecular genetic analysis of cloned genes. Identification and characterization of such natural mutations will undoubtedly add valuable information to our ultimate understanding of how proteins are targeted to mitochondria and other cellular destinations.

REFERENCES

1. ADES, I. Z. 1972. Transport of newly synthesized proteins into mitochondria—a review. Molec. Cell. Biochem. **43:** 113–127.
2. HAY, R., P. BOHNI & S. GASSER. 1984. How mitochondria import proteins. Biochim. Biophys. Acta **779:** 65–87.
3. CONBOY, J. G., F. KALOUSEK & L. E. ROSENBERG. 1979. In vitro synthesis of a putative precursor of mitochondrial ornithine transcarbamylase. Proc. Natl Acad. Sci. USA **76:** 5724–5779.
4. CONBOY, J. & L. E. ROSENBERG. 1981. Posttranslational uptake and processing of *in vitro* synthesized ornithine transcarbamylase precursor by isolated rat liver mitochondria. Proc. Natl. Acad. Sci. USA **78:** 3073–3077.
5. KRAUS, J P., J. G. CONBOY & L. E. ROSENBERG. 1981. Pre-ornithine transcarbamylase: Properties of the cytoplasmic precursor of a mitochondrial matrix enzyme. J. Biol. Chem. **256:** 10739–10742.
6. KRAUS, J. P., F. KALOUSEK & L. E. ROSENBERG. 1983. Biosynthesis and mitochondrial processing of the β subunit of propionyl CoA carboxylase from rat liver. J. Biol. Chem. **258:** 7245–7248.
7. FENTON, W. A., A. M. HACK, D. HELFGOTT & L. E. ROSENBERG. 1984. Biogenesis of the mitochondrial enzyme methylmalonyl CoA mutase: Synthesis and processing of a precursor in a cell-free system and in cultured cells. J. Biol. Chem. **259:** 6616–6621.
8. KOLANSKY, D. M., J. G. CONBOY, W. A. FENTON & L. E. ROSENBERG. 1982. Energy dependent translocation of the precursor of ornithine transcarbamylase by isolated rat liver mitochondria. J. Biol. Chem. **257:** 8467–8471.
9. CONBOY, J. G., W. A. FENTON & L. E. ROSENBERG. 1982. Processing of pre-ornithine transcarbamylase requires a zinc-dependent protease localized to the mitochondrial matrix. Biochem. Biophys. Res. Commun. **105:** 1–7.
10. KALOUSEK, F., M. D. ORSULAK & L. E. ROSENBERG. 1984. Newly processed ornithine transcarbamylase subunits are assembled to trimers in rat liver mitochondria. J. Biol. Chem. **259:** 5392–5395.
11. HORWICH, A. L., W. A. FENTON, F. A. FIRGAIRA, J. E. FOX, D. KOLANSKY, I. S. MELLMAN & L. E. ROSENBERG. 1985. Expression of amplified DNA sequences for ornithine transcarbamylase in HeLa cells: Arginine residues may be required for mitochondrial import of enzyme precursor. J. Cell Biol. **100:** 1515–1521.
12. HORWICH, A. L., J. P. KRAUS, K. WILLIAMS, F. KALOUSEK, W. KONIGSBERG & L. E. ROSENBERG. 1983. Molecular cloning of cDNA coding for rat ornithine transcarbamoylase. Proc. Natl. Acad. Sci. USA **80:** 4258–4262.
13. KRAUS, J. P., P. E. HODGES, C. L. WILLIAMSON, A. L. HORWICH, F. KALOUSEK, K. R. WILLIAMS & L. E. ROSENBERG. 1985. A cDNA clone for the precursor of rat mitochondrial ornithine trancarbamylase: comparison of rat and human leader sequences and conservation of catalytic sites. Nucl. Acids Res. **13:** 943–952.
14. HORWICH, A. L., W. A. FENTON, K. R. WILLIAMS, F. KALOUSEK, J. P. KRAUS, R. F. DOOLITTLE, W. KONIGSBERG & L. E. ROSENBERG. 1984. Structure and expression of a complementary DNA for the nuclear coded precursor of human mitochondrial ornithine transcarbamylase. Science **224:** 1068–1074.
15. KRAUS, J. P., C. L. WILLIAMSON, F. A. FIRGAIRA, T. L. YANG-FENG, M. MUNKE, U. FRANCKE & L. E. ROSENBERG. 1986. Cloning and screening with nanogram amounts of immunopurified messenger RNAs: cDNA cloning and chromosomal mapping of cysta-

thionine β-synthase and the β-subunit of propionyl CoA carboxylase. Proc. Natl. Acad. Sci. USA **83:** 2047–2051.

16. KRAUS, J. P. & L. E. ROSENBERG. 1982. Purification of low-abundance messenger RNAs from rat liver by polysome immunoadsorption. Proc. Natl. Acad. Sci. USA **79:** 4015–4019.
17. HORWICH, A. L., F. KALOUSEK, I. MELLMAN & L. E. ROSENBERG. 1985. A leader peptide is sufficient to direct mitochondrial import of a chimeric protein. EMBO J. **4:** 1129–1135.
18. HORWICH, A. L., F. KALOUSEK & L. E. ROSENBERG. 1985. Arginine in the leader peptide is required for both import and proteolytic cleavage of a mitochondrial precursor. Proc. Natl. Acad. Sci. USA **82:** 4930–4933.
19. HORWICH, A. L., F. KALOUSEK, W. A. FENTON, R. A. POLLOCK & L. E. ROSENBERG. 1986. Targeting of pre-ornithine transcarbamylase to mitochondria: Definition of critical regions and residues in the leader peptide. Cell **44:** 451–459.

DISCUSSION OF THE PAPER

P. COLEMAN (*New York University, New York, NY*): Since the OTC leader peptide is required for mitochondrial uptake of the enzyme, it seems obvious to ask: (1) Whether mitochondria take up the leader peptide by itself? and, if yes, (2) Can one employ this leader peptide, appropriately coupled, to allow mitochondria to import anything one wishes?

ROSENBERG: We have recently synthesized the OTC leader peptide and are examining currently its ability to be taken up by mitochondria or, alternatively, to compete for uptake of *bonafide* precursor proteins. Theoretically, if leader peptides can be imported alone, then coupling them with other materials could lead to their import by mitochondria.

M. KLINGENBERG (*Institute for Physical Biochemistry, Munich*): Are there defects known for mitochondrial proteins that can be attributed to change in the leader sequence, with a consequent alteration of the insertion or sorting process? For example, accumulation of precursors in cystosol or in a semicomplete insertion stage.

ROSENBERG: There is, as of now, no well-documented example of a deficiency of a mitochondria enzyme caused by failure of import. I have no doubt, however, that such abnormalities occur and will be observed.

R. WATTIAUX (*Laboratoire de Chimie Physiologique, Namur, Belgium*): Is the leader peptide required to target proteins in the correct submitochondrial compartment (outer membrane, intermembrane space, inner mitochondrial membrane, matrix)?

ROSENBERG: To date, cleavable leader peptides have not been observed for imported outer membrane mitochondrial proteins. Proteins destined for the intermembrane space, the inner membrane, and the matrix generally contain cleavable leaders.

E. RUBIN (*Hahnemann University, Philadelphia, PA*): What about proteins synthesized by mitochondria? Do they have leader sequences or other recognition mechanisms?

ROSENBERG: I do not believe that proteins synthesized in the mitochondrion have been shown to contain leader peptides. I have not seen any schemes explaining proper targeting of such mitochondrial subunits.

C. LEIFER (*Temple University, Philadelphia, PA*): Can the process of endocytosis provide an alternative to the transport system involving a leader attached to a protein or polypeptide?

ROSENBERG: It is entirely possible that endocytosis of mitochondrial membranes bearing precursor protein-receptor complexes could be involved with import. We have no knowledge of the precise means of translocation.

S. CORVERA (*University of Massachusetts, Worcester, MA*): Are extramitochondrially synthesized mitochondrial proteins sorted through the Golgi before targeting?

ROSENBERG: There is no evidence for involvement at the Golgi in the sorting at mitochondrial proteins.

Inborn Defects of the Mitochondrial Portion of the Urea Cycle[a]

J. P. COLOMBO, C. BACHMANN,
AND AURELIA SCHRÄMMLI

Department of Clinical Chemistry
Inselspital
University of Berne
CH-3010 Berne, Switzerland

THE UREA CYCLE

The main purpose of the urea cycle is to eliminate nitrogen in the form of ammonia, from the body. Ammonia is produced by several reactions in the catabolism of proteins and amino acids. It is toxic to the nervous system and is detoxified through incorporation into urea. Urea contains two nitrogens atoms, one from ammonia and the other from aspartate. Urea synthesis requires the stoichiometric production of carbamylphosphate and aspartate and the presence of ornithine. Carbamylphosphate is formed in the mitochondria from ammonia, ATP, and bicarbonate. It condenses then with ornithine to citrulline, which leaves the mitochondrion, and in a further reaction is metabolized via argininosuccinic acid to arginine. Arginase splits urea from this amino acid leading again to ornithine which is then recycled within the mitochondrion. As a consequence the overall pathway of urea formation involves enzyme reactions in both the cytosol and the mitochondria, which are controlled by the concentration of substrates, activators, and inhibitors. In addition, compartmentation of the urea cycle enzymes necessitates transport of the intermediates across the mitochondrial membrane. Hence, many factors may contribute to the regulation of urea synthesis under different metabolic conditions.

Congenital defects involving intramitochondrial enzymes are carbamoyl phosphate synthetase (CPS), ornithine carbamyl transferase (OCT), and N-acetylglutamine synthetase (NAGS) deficiency. The latter enzyme is not directly a part of the cycle but is essential for its function.

When urea synthesis is impaired due to congenital enzyme deficiency, hyperammonemia and other manifestations ensue. This occurs mainly in the newborn period or infancy. Some of the patients show an acute clinical course that without appropriate treatment, leads to psychomotor retardation or death. As a rule the clinical symptoms arise or are precipitated by an increased protein intake or by protein catabolism. The severity varies greatly depending on the genetic variant of the enzyme defect.

MITOCHONDRIAL SUBSTRATE TRANSPORT

During the operation of the urea cycle ornithine enters the mitochondrion whereas citrulline must leave it. The transport of ornithine into the mitochondrion is therefore an important regulatory step.

[a]This project was supported by the Swiss National Research Foundation (grant 3.910.0.85).

Several mechanisms of transport have been proposed. Gamble and Lehninger were the first to show an energized phosphate-dependent, unidirectional entry of ornithine into the mitochondrion and a separate, respiratory-independent uniporter for citrulline.[1] However, dependence on respiratory energy and phosphate flux for ornithine uptake into the mitochondrion was questioned.[2-4] In more recent studies Bradford and McGiven proposed two ornithine transport systems: an ornithine/H^+ antiporter[2] and an ornithine/citrulline exchange system.[5]

The first proposed system, which is independent of citrulline transfer, is involved in the catabolism of ornithine via the ornithine-δ-amino-transferase (OAT) reaction whereas the other system would be required for the function of the urea cycle. Recently Hommes *et al.* found no evidence for an ornithine/citrulline exchange.[6]

In our own studies using the method of centrifugal filtration[7] we demonstrated an ornithine/citrulline antiporter, thus confirming the findings of Bradford and McGiven (FIG. 1). The transport rate of ornithine was estimated to 5,[1] 8.3,[2] 38,[6] and 10.2 (our observations) nmol · min^{-1} · mg^{-1} mitochondrial protein. The apparent K_m for ornithine was found to be 1.0 mM[2] and 1.3 mM.[6]

MITOCHONDRIAL ORNITHINE METABOLISM

In the cytosol ornithine is formed from arginine by arginase or by glycine transaminidase. Once transported into the mitochondrion, ornithine is available for two metabolic reactions. It may be transaminated to Δ^1-pyrroline-5-carboxylate and glutamate in the OAT reaction or react with carbamoylphosphate (CP) in the OCT reaction to produce citrulline in the urea cycle.

The relative rates of metabolic fluxes through the OAT and OCT pathway in the liver cell are difficult to outline. It seems that the major function of OAT is to maintain a constant level of ornithine,[2] which could arise from dietary arginine since an excess level of ornithine is inhibitory to OCT.[9]

Patients with OAT deficiency may accumulate ornithine produced from arginine in the cytosol. Thus minor changes in the intracellular ornithine pool may lead to major changes in the plasma concentration of ornithine with 10- to 20-fold increase in this disorder. This could be an indication that the amount of ornithine metabolized through the OAT pathway *in vivo* is considerable.

Comparing the kinetic data of OAT (rat) (K_m ornithine 4 mM,[2] 7.2 mM[10]; enzyme activity 25 nmol · min^{-1} · mg^{-1} mitochondrial protein) to that of OCT[9] (human) (K_m ornithine 0.47–0.78 mM; enzyme activity 6.2 μmol · min^{-1} · mg^{-1} protein) ornithine is preferentially metabolized in the urea cycle pathway. Extrapolation of the above kinetic and transport data to the *in vivo* situation suggests that the availability of ornithine influences greatly the rate of citrulline synthesis per se or stimulates CPS[11,12] and transamination. In enzyme defects of the urea cycle ornithine may become rate limiting and is needed for therapy.

DEFECTS OF ORNITHINE METABOLISM

Hyperornithinemias

Hyperornithinemia is the major feature of two distinct congenital defects of ornithine metabolism.

Hyperornithinemia with gyrate atrophy of the choroid and retina usually leads to blindness in the fourth decade of life. It can be attributed to a defect of OAT. In patients the enzyme deficiency has been documented in skin fibroblasts, liver biopsy, hair roots, stimulated lymphocytes, and primary muscle cell cultures. The skin fibroblasts showed 0–2% of the normal OAT activity.

Another entity affecting ornithine metabolism is the hyperammonemia, hyperornithinemia, and homocitrullinuria syndrome (HHH-syndrome).[13] A normal activity of

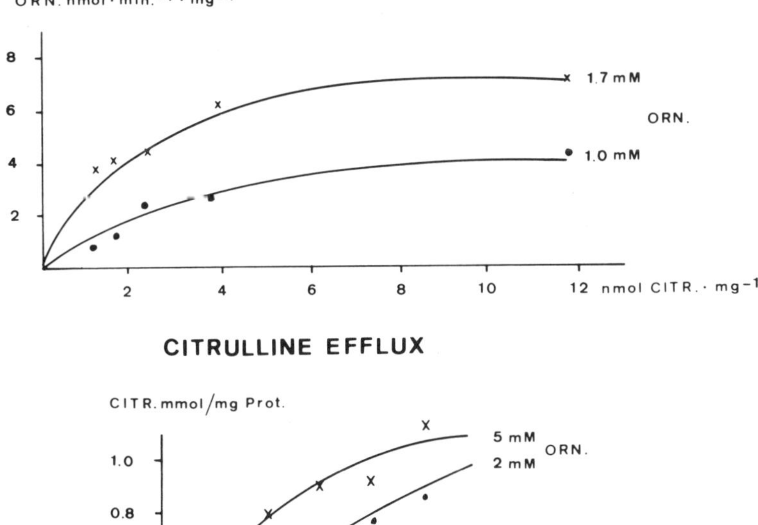

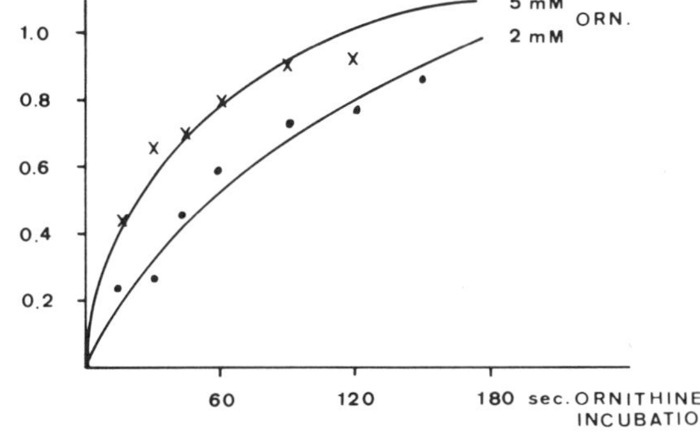

FIGURE 1. (Top) Ornithine uptake of isolated rat liver mitochondria at different citrulline concentrations. (Bottom) Citrulline efflux from isolated rat liver mitochondria preloaded with 2 and 5 mM ornithine. The centrifugal filtration technique according to Palmieri and Klingenberg was used.[7]

OAT but a partial deficiency of ornithine decarboxylase in skin fibroblasts was reported in those patients. There is a variable degree of mental retardation. Hyperammonemia can be controlled by a low protein diet and an improvement of the clinical condition occurs after ornithine or arginine treatment. In view of the favorable amino acid substitution and the findings of defective incorporation of the ornithine skeleton into the proteins by fibroblasts of those patients, a deficit of ornithine transport across the mitochondrial membrane was postulated.[8] The exact nature of those defects still awaits a clear proof.[14,15]

INTRAMITOCHONDRIAL UREA CYCLE DEFECTS

OCT Deficiency

The most frequent intramitochondrial urea cycle enzyme defect is OCT deficiency. OCT deficiency is an X-linked autosomal disorder characterized by protein intolerance, hyperammonemia, and mental retardation. The OCT gene is located on the short arm band 21.1 of the X chromosome.[16] Lethal outcome in the newborn period of the affected male hemizygotes is common. The symptomatology in these cases resembles CPS deficiency. The clinical and biochemical manifestation in the heterozygote female may vary depending on the genetic heterogeneity and the random X chromosomal inactivation.[17] Several polymorphisms at the OCT gene locus[18] have been demonstrated.

OCT is trimer, whose subunits are synthesized in the cytosol, coupled to a precursor, transported, and reassembled in the mitochondrion after cleavage of the precursor protein.[19] The enzyme is present in liver, intestinal mucosa, brain, leukocytes, amniotic cells, and serum. The liver and the intestinal enzyme are probably of the same genetic origin.[20,21] Several kinetic mutants of the enzyme such as increased affinity for ornithine, decreased affinity for a CP, and pH dependency were reported. Patients with neonatal onset show practically no immunoreactive cross-reacting material with bovine OCT in their liver biopsies.[22]

OCT deficiency leads to an accumulation of CP. Under these conditions CP may be exported out of the mitochondrion into the cytosol and in the presence of aspartate leads to an increased pyrimidine synthesis. The hypothesis that in rat liver the glutamine-dependent CPS II localized in the cytoplasm is the exclusive source of pyrimidine synthesis has been questioned.[11,23,24] It apparently contributes to it mainly under physiological and not hyperammonemic conditions.[25]

In hyperammonemic rats with portacaval shunt, we observed an increased activity of aspartate transcarbamylase.[26]

Patients with OCT deficiency show increased orotic aciduria in addition to hyperammonemia.[27] Orotic aciduria is used as a major diagnostic parameter in connection with the elevated plasma ammonia levels. Normal or even low concentrations of citrulline, arginine, and argininosuccinic acid in plasma and urine make the diagnosis of OCT deficiency very probable. It can be confirmed by enzyme determination in liver tissue. Heterozygotes may be asymptomatic or according to the degree of lyonization, exhibit 10–50% of normal OCT activity wich clinical symptoms of variable severity. Asymptomatic heterozygotes may be detected by measuring orotic acid excretion in urine after a protein load (1 g/kg body weight).[28] However, failure of the test has been reported.[29] New mutation in the parental germ cells must then be considered. A better way of detecting asymptomatic heterozygote females in OCT pedigrees is the use of cDNA probes with restriction fragment length polymorphism techniques.[18]

CPS Deficiency

CPS I (ammonia) catalyzes the formation of CP from ammonia, 2 ATP, and HCO_3^- (FIG. 2). Its mitochondrial matrix concentration is about 0.4–1.0 mmol/l, which amounts to 15–25% of the total mitochondrial protein.[30–32]

CPS is activated in an allosteric manner by N-acetyl-glutamate (NAG) leading to conformational changes of the CPS molecule. The average CPS activity in rat liver *in vivo* amounts to 743 ± 128 μmol · hr^{-1} · g^{-1} wet weight.[33]

This corresponds to a calculated CP synthesis of 206 nmol · min^{-1} · mg^{-1} mitochondrial protein, which fits very well with the experimental data of 150 nmol · min^{-1} · mg^{-1}.[32] In human liver wedge biopsy an activity of 279 ± 65 μmol · hr^{-1} · g^{-1} wet weight was reported.[33] The enzyme is present as a dimer with a molecular mass of about 160,000 per subunit. Like OCT, it is synthesized in the cytosol as a larger

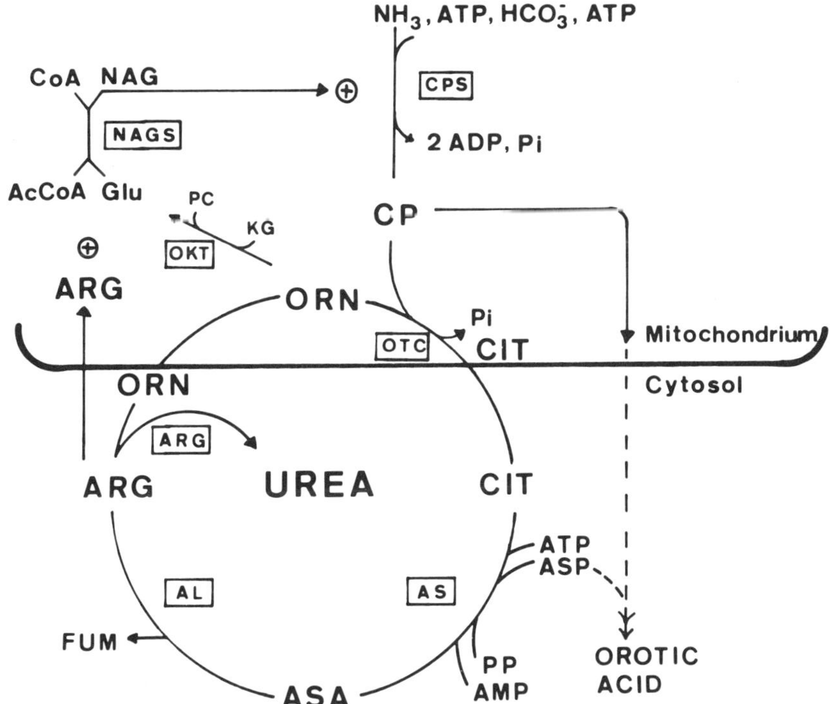

FIGURE 2. Urea cycle. NAG = N-acetylglutamate; NAGS = N-acetylglutamate synthetase; OKT = ornithine-α-ketotransaminase; CPS = carbamylphosphate synthetase; CP = carbamylphosphate; CIT = citrulline; AS = argininosuccinate synthetase; ASP = aspartate; ASA = argininosuccinate; AL = argininosuccinate lyase; ARG = arginine; ARG = arginase; ORN = ornithine

precursor molecule (p-CPS, MW 165,000), transported into the mitochondrion, and processed to the mature enzyme.[34,35] At normal concentrations of ammonia in the liver (<1 mM) CPS is the rate-limiting reaction of urea synthesis, whereas at high ammonia levels the argininosuccinate synthetase reaction becomes rate limiting.

Most cases of CSP deficiency have been observed in the newborn period with less than 10% enzyme activity in liver biopsy tissue. Beside hyperammonemia these patients exhibit attacks of vomiting, seizures, irritability, somnolence, and lethargy progressing to coma. The symptoms are precipitated by an increased intake of protein. Without appropriate treatment aiming to augment the disposal of nitrogen, such as peritoneal dialysis and sodium benzoate infusions, the babies die within the first year of life. A less severe form with a residual enzyme activity between 20–30% and a later onset around one year of age exists. Children who survive have mental retardation. For the description of the detailed clinical symptomatology see references 36 and 37.

Patients with hyperammonemia but normal or subnormal plasma levels of citrulline, arginine, absent urinary excretion of argininosuccinic acid, organic acids, and orotic acid must be considered as having CPS deficiency, which must be confirmed by enzyme measurements.

Long term treatment of CPS and OCT deficiency is identical.[38] It includes low natural protein and essential amino acids. Except for argininemia, urea cycle disorders lead to an insufficienct production of arginine. This latter amino acid is needed for the synthesis of protein, creatine, and ornithine. In addition it is an intramitochondrial activator of NAGS. As a consequence, arginine supply is prerequisite in defects of carbamyl phosphate synthetase and OCT deficiency. Citrulline can be used instead of arginine in equimolar amounts. The advantage is that 1 mol of nitrogen will be fixed for argininosuccinate formation per 1 mol of citrulline supplied and consequently be released as urea.

N-ACETYLGLUTAMATE SYNTHETASE DEFICIENCY

NAG Synthesis

An enzyme also involved in mitochondrial disorders of urea synthesis is NAGS. The enzyme catalyzes the synthesis of NAG from acetyl-CoA and glutamate. Most studies have been made on the rat and mouse enzyme. We have purified the human enzyme and developed a method to measure it.[39,40]

The enzyme is activated by arginine, disclosing thus a positive feedback mechanism of urea synthesis. Kinetic properties in the absence of arginine are compatible with a rapid equilibrium random bibi mechanism. The K_m of the forward reaction of the human amino acyltransferase for glutamate and acetyl CoA are 8.1 and 4.4 mmol/l, respectively.

At a concentration of 0.7 mmol/l of acetyl CoA, the level present in the mitochondrion (0.6–0.8 mmol/l),[41] the K_m for glutamate was 8.7 mmol/l, similar to the concentration found in rat liver mitochondria (3–15 mmol/l). With a highly purified rat liver enzyme Sonoda and Tatibana found a K_m for acetyl CoA of 0.7 mmol/l.[42] Small changes in the concentration of the two substrates might therefore markedly affect the initial velocity of NAGS.

The regulation of the mitochondrial NAG concentration depends on the rate of its synthesis and degradation. Since NAG is mainly degraded by amino acylase in the cytosol it must cross the mitchondrial membrane.[43] The rate of efflux is about 0.05 nmol $\cdot$ min^{-1} $\cdot$ mg^{-1} mitochondrial protein in energized mitochondria.[44] The activity of NAGS in human liver amounts to 0.6 nmol $\cdot$ min^{-1} $\cdot$ mg^{-1} mitochondrial protein.[40] Thus about 12 times more NAG accumulates in the mitochondrion by unit time as can be transported out of the mitochondria. Reported concentrations of 0.71 $\pm$ 0.04 (0.2–2.0) nmol $\cdot$ mg^{-1} mitochondrial protein are sufficient to fully activate CPS (Ka NAG 0.1 mmol).[45]

There is also a low product inhibition of NAG by acting on NAGS. The accumulated NAG can also be split in a reverse reaction.[46] This reverse activity of NAGS in human liver is very low (about 0.0041 nmol $\cdot$ min^{-1} $\cdot$ mg^{-1} mitochondrial protein, which is about 10% of the rate of efflux of NAG from the mitochondrion. If these *in vitro* data can be extrapolated to the *in vivo* situation, NAGS synthesis is very sensitively regulated and therefore influences considerably the flux through the CPS reaction.

NAGS DEFICIENCY

We have observed two cases of NAGS deficiency.[47,48] We are not aware of other case reports up to now. The first case was closely followed because two siblings died in the neonatal period. In the patient who was clinically well, a plasma ammonia concentration of 160–240 μmol/l (normal <130 μmol/l)[49] during the first three days of life was measured. Plasma amino acid pattern was unspecific, orotic acid excretion normal. NAGS activity in a liver biopsy was undetectable (normal 34–203 nmol $\cdot$ min^{-1} $\cdot$ mg^{-1} protein).[47] CPS with 1.13 μmol $\cdot$ hr^{-1} $\cdot$ mg^{-1} protein was in the normal range (0.66–2.1 μmol $\cdot$ hr^{-1} $\cdot$ mg^{-1} protein).

A second patient was hospitalized at three days of life because of somnolence, vomiting, reduced muscle tone and peripheral circulation. The plasma ammonia concentration was 711 μmol/l (normal <104 μmol/l). Glutamine and alanine concentration in plasma were increased, citrulline, ornithine, and arginine low. The urinary excretion of orotic and organic acids was normal.

The child died at the age of 8 days despite the treatment with sodium benzoate, arginine, and peritoneal dialysis. NAGS acitivity in the postmortem liver biopsy was not detectable.

In the first patient carbamylglutamate (CG) was used in addition to arginine and sodium benzoate to control hyperammonemia. Brown *et al.* had advocated the use of CG and arginine in the treatment hepatic coma.[50]

Administration of CG decreased ammonia levels to <100 μmol/l. Despite a continuous arginine supplement, when CG was discontinued for 24 hr, hyperammonemia ensued. Resumption of CG treatment (1,800 mg/day) lead to normal ammonia levels.

Meijer *et al.* have shown that in mitochondria in rats fed a standard diet CG was more active in stimulating citrulline synthesis than NAG itself.[44] In a low mitochondrial energy state CG more easily penetrates the mitochondrial membrane than NAG. Furthermore CG protects rats more efficiently from ammonia intoxication than NAG since it is less rapidly degraded by amino acylases present in the cytosol.[51,52] On the basis of the facilitated entry and despite a lower affinity of CPS for CG ($K_a = 4$ mmol/l) than for NAG (0.1–0.2 mmol/l), stimulation of CPS must take place at high doses of CG.[44] This may explain in part the clinical improvement in our patient.

The congenital deficiencies of the mitochondrial enzymes of the urea cycle lead to severe clinical disorders. All of them in their homozygote state have low to zero activity of the corresponding urea cycle enzymes in liver tissue. Due to the compartmentation of the urea cycle, membrane transport of intermediates must also be considered. However, it has not yet been assessed to what extent, besides the enzyme defects, impaired membrane transport contributes to the pathogenesis of these disorders.

REFERENCES

1. Gamble, J. G. & A. L. Lehninger. 1973. J. Biol. Chem. **248:** 610–616.
2. MGivan, J. D., N. M. Bradford & A. D. Beavies. 1977. Biochem. J. **162:** 147–156.
3. Bryla, J. & E. J. Harris. 1976. FEBS Lett. **72:** 331–336.
4. Aronson, D. L. & J. J. Diwan. 1981. Biochemistry **20:** 7064–7068.
5. Bradford, N. M. & J. D. McGivan. 1980. FEBS Lett. **113:** 294–298.

6. HOMMES, F. A., L. KITCHINGS & A. G. ELLER. 1983. Biochem. Med. **30:** 313–321.
7. PALMIERI, F. & M. KLINGENBERG. 1979. Direct methods for measuring metabolite transport and distribution in mitochondria. Methods Enzymol. **56:** 279–301.
8. SHIH, V. E. 1981. Enzyme **26:** 254–258.
9. SNODGRASS, P. J. 1968. Biochemistry **7:** 3047–3051.
10. YIP, M. C. M. & R. K. COLLINS. 1971. Enzyme **12:** 187–200.
11. COHEN, N. S., C. W. CHEUNG & L. RAIJMAN. 1980. J. Biol. Chem. **255:** 10248–10255.
12. COHEN, N. S., C. W. CHEUNG, F. S. KYAN, E. E. JONES & L. RAIJMAN. 1982. J. Biol. Chem. **257:** 6898–6907.
13. SHIH, V. E., M. L. EFRAN & H. W. MOSER. 1969. Am. J. Dis. Child **117:** 83–92.
14. HOMMES, C., K. HO, R. A. ROESEL & M. E. CORYELL. 1982. J. Inher. Metab. Dis. **5:** 41–47
15. GRAY, R. G. F., S. E. HILL & R. J. POLLITT. 1983. J. Inher. Metab. Dis. **6:** 143–148.
16. LINDGREN, V., B. DE MARTINVILLE, A. L. HORWICH, L. E. ROSENBERG & UTA FRANCKE. 1984. Science **26:** 698–700.
17. ROWE, P. C., S. L. NEWMAN & S. W. BRUSILOW. 1986. New Engl. J. Med. **314:** 541–547.
18. ROZEN, R., J. FOX, W. A. FENTON, A. L. HORWICH & L. E. ROSENBERG. 1985. Nature **313:** 815–817.
19. HORWICH, A. L., F. KALOUSEK & L. E. ROSENBERG. 1985. Proc. Natl. Acad. Sci. USA **82:** 4930–4933.
20. NOGATA, N., I. AKABOSHI, J. E. F. YAMAMATO, I. MATSUDA & T. KATSUKI. 1982. Adv. Exper. Med. Bio. **153:** 47–52.
21. WRAIGHT, C, K. LINGELBACH & N. HOOGENRAAD. 1985. Eur. J. Biochem. **153:** 239–242.
22. FRANÇOIS, B., P. BRIAND & L. CATHELINEAU. 1982. Adv. Exper. Med. Biol. **153:** 53–62.
23. BOURGET, P. A. & G. C. TREMBLAY. 1972. Biochem. Biophys. Res. Commun. **46:** 752–756.
24. NATALE, P. J. & G. C. TREMBLAY. 1974. Arch. Biochem Biophys. **162:** 357–368.
25. PAUSCH, J., J. RASENACK, D. HÄUSSINGER & W. GEROK. 1985. Eur. J. Biochem. **150:** 189–194.
26. COLOMBO, J. P., J. BERÜTER, C. BACHMANN & E. PEHEIM. 1977. Enzyme **22:** 391–398.
27. BACHMANN, C. & J. P. COLOMBO. 1980. Eur. J. Pediat. **134:** 109–113.
28. BRUSILOW, S. W. & D. L. VALLE. 1985. Ped. Res. **19:** 244A.
29. BECROFT, D. M. O., D. M. J. BARRY, D. R. WEBSTER & H. A. SIMMONDS. 1984. J. Inher. Metab. Dis. **7:** 157–159.
30. RAIJMAN, L. & M. E. JONES. 1976. Arch. Biochem. Biophys. **175:** 270–278.
31. LASTY, C. J. 1978. Eur. J. Biochem. **85:** 373–383.
32. COHEN, N. S., C. W. CHEUNG, F. S. KYAN, E. E. JONES & L. RAIJMAN. 1982. J. Biol. Chem. **257:** 6898–6907.
33. NUZUM, C. T. & P. J. SNODGRASS. 1976. Multiple assays of the five urea cycle enzymes in human liver homogenates. *In* The Urea Cycle. S. Grisolia, R. Baguena & F. Mayer, Eds.: 325–349. John Wiley. New York.
34. MORI, M., S. MIVRA, T. MORITA & M. TATIBANA. 1982. Adv. Exp. Med. Biol. **153:** 267–276.
35. BHAT, N. K. & N. G. AVADHANI. 1985. Biochemistry **24:** 8107–8113.
36. BATSHAW, M. L. 1984. Curr. Prob. Ped. **XIV:** 11.
37. NYHAN, W. L. 1984. Abnormalities in amino acid metabolism in clinical medicine. Appleton-Century-Crofts. Norwalk, CT.
38. BACHMANN, C. 1984. Enzyme **32:** 56–64.
39. BACHMANN, C., S. KRÄHENBÜHL & J. P. COLOMBO. 1982. Biochem. J. **205:** 123–127.
40. COLOMBO, J. P., S. KRÄHENBÜHL, C. BACHMANN & P. AEBERHARD. 1982. J. Clin. Chem. Clin. Biochem. **20:** 325–329.
41. SIESS, E. A., D. G. BROCKS, H. H. LATTKE & O. H. WIELAND. 1977. Biochem. J. **166:** 225–235.
42. SONODA, T. & M. TATIBANA. 1983. J. Biol. Chem. **258:** 9839–9844.
43. REGLERO, A., J. RIROS, J. MENDELSON, R. WALLACE & S. GRISOLIA. 1977. FEBS Lett. **81:** 13–17.
44. MEIJER, A. J., G. M. VAN WOERKOM, R. J. A. WANDERS & C. LOF. 1982. Eur. J. Biochem. **124:** 325–330.

45. LOF, C., M. COHEN, L. P. VERMEULEN, C. W. T. VAN ROERMUND, R. J. A. WANDERS & A. J. MEIJER. 1983. Eur. J. Biochem. **135:** 251–258.
46. GRÜNIG, A. 1985. Dissertation. Universität Bern.
47. BACHMANN, C., S. KRÄHENBÜHL, J. P. COLOMBO, G. SCHUBIGER, K. H. JAGGI & O. TÖNZ. 1981. N. Engl. J. Med. **304:** 543.
48. BACHMANN, C., M. BRANDIS, E. WEISSBARTH-RIEDEL, R. BURGHARD & J. P. COLOMBO. (Submitted for publication.)
49. COLOMBO, J. P., E. PEHEIM, R. KRETSCHMER, H. DAUWALDER & D. SIDIROPOULOS. 1984. Clin. Chim. Acta **138:** 283–291.
50. BROWN, R., M. MONNING, M. HELP & S. GRISOLIA. 1958. Lancet **I:** 591.
51. KIM, S., W. K. PAIK & P. P. COHEN. 1972. Proc. Natl. Acad. Sci. USA **69:** 3530–3533.
52. RUBIO, V. & S. GRISOLIA. 1981. Nature **292:** 496.

DISCUSSION OF THE PAPER

E. CARAFOLI (*Swiss Federal Institute of Technology, Zurich*): N-acetyl glutamate synthetase is a matrix enzyme and is activated by arginine. How is arginine transported into mitochondria?

J. P. COLOMBO (*University of Berne, Berne, Switz.*): Arginine is taken up by mitochondria. However, arginine transport studies are hampered by the presence of arginase, which sticks to the mitochondria and is difficult to remove without damaging them. So arginine transport may end up to be ornithine transport. I am not aware of recent studies overcoming this and clearly demonstrating the mechanisms of arginine transport.

Transport and Function of Carnitine: Relevance to Carnitine-Deficient Diseases

NORIS SILIPRANDI

Istituto Chimica Biologica
Università di Padova
Centro Studio Fisiologia Mitocondriale CNR
Via Marzolo 3
35131 Padua, Italy

Several clinical syndromes associated with a lack of carnitine have been described in recent years and classified as either primary or secondary carnitine deficiencies.[1] The primary ones include myopathic and systemic syndromes, while the secondary constitute a variety of pathologies that are consequences of genetic or acquired metabolic defects.[1]

MYOPATHIC CARNITINE DEFICIENCY

This is the best known primary carnitine deficiency, first described by Engel and Angelini[2] and so far recognized in about 30 patients. It is characterized by a pronounced weakness of muscles, coupled with the infiltration of triacylglycerols within and between muscle fibers. In all cases the carnitine content of muscle, but not of liver or blood, is decreased and occasionally is absent altogether. The oxidation of ^{14}C-labeled long chain fatty acids by muscle homogenates is impaired, but restored upon carnitine addition. Carnitine administration generally results in a substantial improvement of the clinical symptoms, in spite of the fact that the amount of carnitine in muscles is hardly altered. This myopathy has been interpreted as due to an altered transport of carnitine into muscles.[2,3]

SYSTEMIC CARNITINE DEFICIENCY

Originally described by Karpati *et al.*[4] and then thoroughly studied by Rebouche and Engel,[5] the syndrome is characterized by recurrent episodes of acute encephalopathy, progressive muscular weakness, and steatosis. In some patients a dilative cardiomyopathy has also been found. The carnitine level is reduced not only in skeletal muscle, but also in heart, liver, and blood. A rapid depletion of glycogen stores, hypoglycemia, and lactate acidosis is often encountered and can be adequately explained by an underutilization of fatty acids with compensatory increase in glycolysis. The etiology is still obscure. A block in carnitine biosynthesis (as postulated by Karpati *et al.*[4]) though not excluded, is not considered to be the primary defect by Rebouche and Engel,[5] who suggest additionally an abnormal renal handling of carnitine, together with an altered cellular transport.

SECONDARY CARNITINE DEFICIENCY

An up-to-date list of all metabolic disorders associated with secondary carnitine deficiency has recently been made by Engel and Rebouche,[1] who also propose an interesting rationale for the underlying mechanism. A common feature is, or seems to be, an excessive intramitochondrial production of intermediates capable of reacting with CoA, thus creating an abnormally high acyl CoA:CoA ratio. The increase in this ratio may be prevented, or moderated, by the available free carnitine with formation of acyl carnitine; these may be eliminated from the cell and excreted in the urine. This loss of carnitine may create, according to Chalmers *et al.*[6] an increased requirement for free carnitine and thereby a "secondary deficiency." For this reason "carnitine insufficiency" is a more appropriate term than "carnitine deficiency" for defining the actual state of carnitine balance in this group of disorders.[6] Whereas primary carnitine deficiencies are characterized by a decrease of both free and total carnitine, secondary carnitine deficiencies might be expected to exhibit a low level of free carnitine and an above normal amount of esterified carnitine. Unfortunately, conclusive data about the

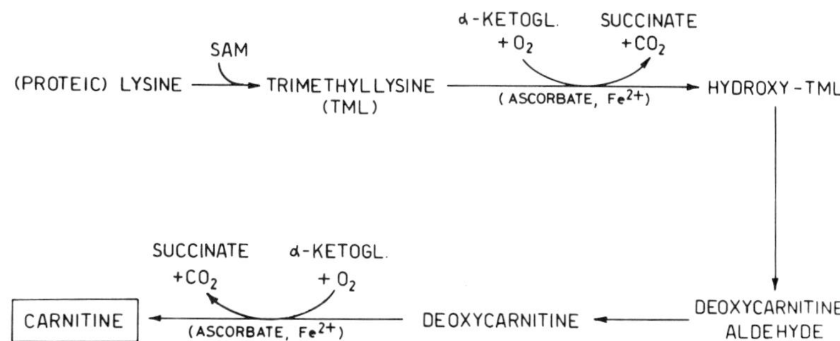

FIGURE 1. Schematic pathway of carnitine biosynthesis. (SAM = adenosylmethionine).

concentration of free and total carnitine in tissues and blood in these disorders are not available to test this theory.

The following survey of carnitine biosynthesis, transport, and function may serve as a guide in clarifying the pathogenesis of carnitine deficiency diseases and carnitine insufficiency, as well as in understanding some of the strange effects of carnitine on membrane stability. A complete review of carnitine biochemistry may be found in the excellent review by Bremer.[7]

CARNITINE BIOSYNTHESIS

The pathway of carnitine synthesis in mammals and microorganisms is summarized in FIGURE 1. ϵ-N-trimethyllysine (TML) is the common precursor of carnitine in both species. However, while microorganisms (*Neurospora crassa*) obtain TML by methylation of free lysine (as catalyzed by methylase I), mammals derive TML destined for carnitine synthesis by methylation of proteic lysine (as catalyzed by protein methylase III).[8] In all animal tissues TML is transformed into deoxycarnitine, the immediate precursor of carnitine, but liver, brain, and (in humans) kidney are also capable of hydroxylating deoxycarnitine to carnitine.[9] The enzyme catalyzing this last step (deoxycarnitine hydroxylase) is missing in the remaining tissues, which are thus entirely dependent on carnitine uptake from the blood.

It has been estimated that the amount of carnitine synthesized in 24 hours by a rat is around 20 μmoles/kg[10] and from 3 to 28 μmoles/kg by a dog.[11] According to Rebouche[12] the TML released from methylated proteins is sufficient to support carnitine biosynthesis in the rat. The largest amount by far of proteic TML is present in muscles.[8] This lends weight to the idea suggested by Rebouche[12] that, although a part of the TML is metabolized by the muscles themselves, some is also taken up by kidney and converted into carnitine. Evidence for this is that (1) in man at least, the kidneys have considerable β-hydroxylase activity, much higher than that found in other tissues[13] and (2) kidneys have a high capacity to take up TML from blood.[12]

As will be discussed below in more detail, deoxycarnitine formed in muscles from TML is certainly released into the blood, converted into carnitine in liver and kidney and then returned back to the muscles (FIG. 2). So far, information for quantitating TML and deoxycarnitine in tissues, blood, and urine is lacking, considerably hindering

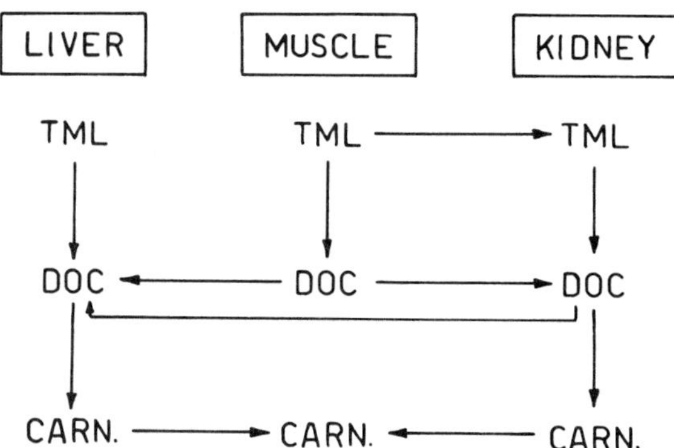

FIGURE 2. Transport of trimethyllysine (TML), deoxycarnitine (DOC), and carnitine (CARN) among tissues.

our understanding of the mechanism of the pathogenesis of primary carnitine deficiencies.

The dependence of both TML and deoxycarnitine hydroxylase on ascorbate and Fe^{2+} (FIG. 1) accounts for the considerable decrease in deoxycarnitine and carnitine concentration in the tissues and blood of ascorbate-deficient guinea pigs[14] and in iron-deficient rat pups.[15]

CARNITINE TRANSPORT

The concentration of carnitine in tissues is generally related to their capacity to oxidize fatty acids (e.g. skeletal and cardiac muscle[16]). It appears paradoxical therefore that the tissues richest in carnitine are unable to synthesize it and must take it up from blood. Since the concentration in blood is considerably lower, active transport might be thought to occur. However such transport has yet to be experimentally demonstrated. Thus inhibitors of both glycolysis and oxidative phosphorylation only partially inhibit carnitine transport into either cardiac or skeletal muscle[17,18] and Vary and Neely[19] have reported that carnitine uptake by perfused heart is insensitive to anoxia. An alternative mechanism is exchange transport and this must be seriously considered because of the results of Mølstad,[20] showing that the efflux of endogenous carnitine from heart cells may occur in exhange for external carnitine or carnitine analogues. In experiments with rat heart slices we have provided further evidence that the transport of carnitine, its esters or analogues, occurs in an exchange-diffusion process by showing that its efflux is insensitive to metabolic inhibitors (F^-, CN^-, azide, etc.), but sensitive to thiol reagents (N-ethylmaleimide or Mersalyl).[21] The rapid exchange between tissue carnitine and medium acetyl carnitine occurs with an apparent K_m of 13 μM, indicating that, *in vivo*, muscles may readily take up acetyl carnitine from the blood without altering the endogenous pool of total carnitine. D-carnitine (the non-physiological isomer) is also able to exchange with endogenous L-carnitine, though with an apparent higher K_m. This may account for the finding of Paulson and Shug[22] that administration of D-carnitine induces a depletion of L-carnitine.

More significant is the exchange between extracellular deoxycarnitine and intracellular carnitine.[23] This exchange may occur bidirectionally in a ratio close to 1:1, but at a higher rate when deoxycarnitine is the internal partner, thus reflecting the physiological condition whereby muscles export deoxycarnitine and import carnitine. Since the concentration gradients of deoxycarnitine and carnitine between rat heart and blood are 21 and 16, respectively, it is reasonable to assume that the free energy from the downhill export of deoxycarnitine may drive the concurrent uphill import of carnitine.

Preliminary *in vivo* experiments show that carnitine administration (500 mg/kg) to rats induces a transient but significant depletion of deoxycarnitine in tissues, particularly in kidney, whereas administration of equivalent amounts of deoxycarnitine induces an analogous depletion in carnitine.

This latter exchange may be relevant to our understanding of the primary carnitine deficiencies and particularly the myopathic form. This latter syndrome could be due either to an alteration of the sarcolemma carrier responsible for the exchange or to a deficient synthesis of deoxycarnitine in muscles.

CARNITINE FUNCTION

The main function of carnitine is the transport of long chain fatty acids from the extramitochondrial space, where they are activated as acyl CoA, into the mitochondrial matrix space, where they are oxidized. It is true that a fraction of these fatty acids diffuses passively across the inner mitochondrial membrane and, together with those generated *in situ* by the activity of phospholipases, is activated in the matrix and thereafter directly oxidized.[24] However, transport of the major part requires the carnitine translocating system. This system is made up of two membrane bound CoA:carnitine acyl transferases (one on the cytosolic and the other on the matrix surface of the inner mitochondrial membrane) interconnected by the carnitine translocase. By this system, long chain acyl carnitines enter the mitochondria in exchange for free carnitine.[7]

Any condition that impairs the carnitine-dependent transport leads to an accumulation of triacylglycerols (FIG. 3). Such lipid storage spontaneously occurs in myopathic carnitine deficiency (see above) and may be induced in liver by administration

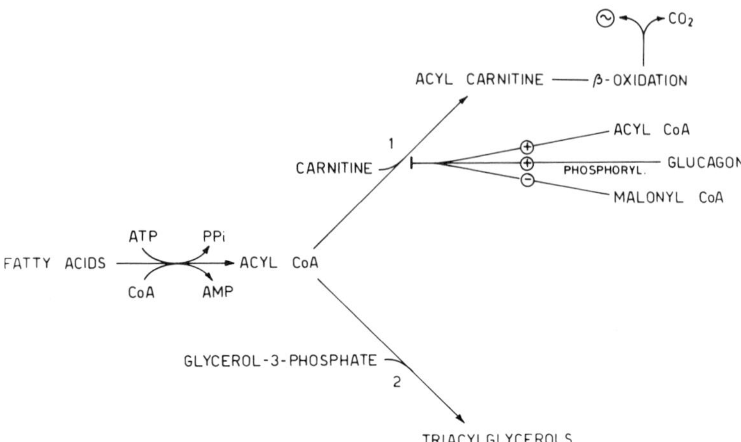

FIGURE 3. Utilization of long chain fatty acids into (1) oxidative energy (~) yielding pathway, dependent on CoA:carnitine acyl transferase activity, stimulated (+) by long chain acyl CoA, or by phosphorylation of the enzyme promoted by glucagon and inhibited (−) by malonyl CoA and (2) esterification in triacylglycerols, dependent on CoA:glycerol-3-phosphate acyl transferase activity.

to experimental animals of aminocarnitine, a potent inhibitor of carnitine palmitoyl transferase.[25]

The carnitine-dependent transport allows the controlled access of long chain fatty acids to the intramitochondrial oxidation site. For instance malonyl CoA, the first committed intermediate in the conversion of glucose to lipids, inhibits the cytosolic CoA:carnitine palmitoyl transferase,[26] thus depressing the oxidation of fatty acids. On the other hand, an excess of long chain acyl CoA increases the activity of the same transferase, thus enhancing fatty acid oxidation.[27] The activity of hepatic CoA:carnitine palmitoyl transferase is moreover under the control of glucagon, which activates the enzyme by promoting its phosphorylation through a cAMP-dependent protein kinase.[28] Therefore the rate of fatty acid oxidation is under both dietary and hormonal control.

Through its effect on CoA:carnitine palmitoyl transferase, carnitine decreases the amount of acyl CoA and at the same time renders free CoA available for CoA requiring processes. By removing long chain acyl CoA molecules, carnitine prevents the many adverse effects of these amphipathic intermediates.[28] Thus carnitine protects liver mitochondria exposed to submicellar concentrations of long chain fatty acids by preventing the inhibition of these thioesters on both adenylate[29] and dicarboxylate[30] transport.

Long chain acyl CoA may also be able to deform intracellular membranes by interpolation into the lipid bilayer; thereby membranes become more susceptible to the damaging action of noxious agents or conditions. Indeed we have provided evidence that liver mitochondria from starved rats, when exposed to hydroperoxides, damaging concentrations of Ca^{2+} and phosphate, or simply kept at room temperature, undergo a rapid deenergization, which is consistently delayed by carnitine.[31] Our explanation is that carnitine, by removing part of the membrane-bound acyl CoA, makes mitochondria less vulnerable to the abovementioned agents. The same mechansim has also been invoked to interpret the protective action of carnitine against the damaging action of halothane on liver mitochondria.[32] It seems that by a still unknown mechanism, halothane may induce an abnormal formation of long chain acyl CoA, which is abolished by carnitine. Conceivably, a non-specific protective action of carnitine against a variety of noxious agents (which even include ammonium acetate[33]) might be due to a stabilization of biological membranes by removal of the induced excess of acyl CoA. This function of carnitine may be important in heart anoxia or ischemia where both free fatty acids and acyl CoA accumulate.[24]

The efficacy of carnitine in patients with a variety of disorders of organic acid metabolism[34] may also be ascribed to the removal of accumulated acyl CoA with generation of the corresponding acyl carnitine and their elimination in the urine. This process, implying an overutilization of free carnitine, may lead to a secondary carnitine deficiency. As outlined by Kim *et al.*[35] an accumulation of organic anions and their CoA esters, as it occurs in propionic and isolaveric acidemia, methylmalonic aciduria, and medium chain acyl CoA dehydrogenase deficiencies, can produce encephalopathy. Hence the possibility of eliminating these metabolites through the formation of the corresponding acyl carnitines seems relevant from a therapeutic point of view. The continuous replenishment of free CoA pool resulting from the CoA:carnitine transferase catalyzed reactions, is also necessary for the maintenance of α-ketoglutarate dehydrogenase activity,[36] a necessary condition for pyruvate, fatty acid, and branched chain α-keto acid oxidation.

In this context it is worthy of mention that acyl carnitines, which unlike the corresponding acyl CoA can diffuse across cellular membranes, are not necessarily eliminated in the urine but may also be redistributed in the tissues. Being readily oxidizable, this extra acyl carnitine may be of considerable value under particular conditions.

A typical example is offered by the results obtained some years ago with aged kidney mitochondria, incapable of oxidizing either endogenous fatty acids or added acetate.[37] Addition of catalytic amounts of acetyl carnitine to these deenergized mitochondria induced an immediate oxidation of the substrates. It was assumed that the minute amount of added acetyl carnitine was sufficient to supply the initial amount of ATP needed for the activation of the substrates and for starting their oxidation. Once started, the process proceeded autocatalytically (FIG. 4). In this regard acyl carnitine may be considered a readily metabolizable acyl reservoir, analogous to the phosphate reserve of phosphocreatine.

By the same mechanism (illustrated in the scheme of FIG. 4) acetyl carnitine may remove oxaloacetate, which accumulated either when citrate synthase is inhibited or

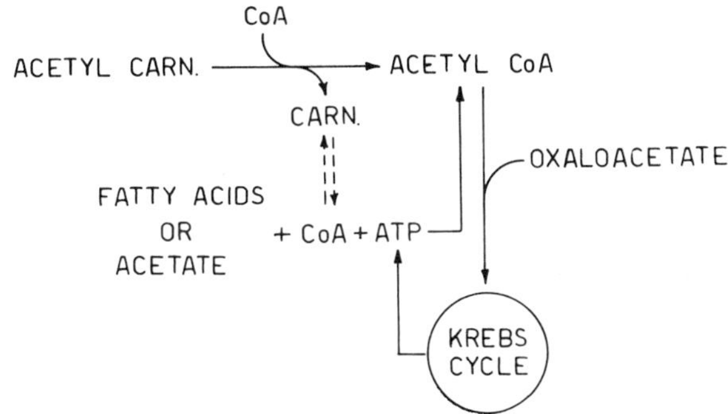

FIGURE 4. Sparking action of acetyl carnitine on fatty acids and acetate utilization in de-energized rat kidney mitochondria.

acetyl CoA is deficient. In its turn accumulation of oxaloacetate inhibits succinate dehydrogenase, thus further hindering the flux in the Krebs cycle. A situation of this nature seems to occur in shocked patients, characterized by a high succinate/fumarate ratio in blood.[38] Administration of acetyl carnitine to these patients results in a significant decrease of this ratio, coupled with an improvement in their clinical condition.

Thus a redistribution of acyl carnitine and particularly acetyl carnitine between tissues might contribute to the adaptation of metabolic processes to particular conditions.

REFERENCES

1. ENGEL, A. G. & C. J. REBOUCHE. 1984. Carnitine metabolism and inborn errors. J. Inher. Metab. Dis. 7(Suppl. 1): 38–43.
2. ENGEL, A. G. & C. ANGELINI. 1973. Carnitine deficiency of human skeletal muscle with associated lipid storage myopathy: a new syndrome. Science **179:** 899–902.
3. REBOUCHE, C. J. & A. G. ENGEL. 1984. Kinetic compartmental analysis of carnitine metabolism in the human carnitine deficiency syndromes: evidence for alterations in tissue carnitine transport. J. Clin. Invest. **73:** 857–867.
4. KARPATI, G., S. CARPENTER, A. G. ENGEL, G. WATTERS, J. ALLEN, S. ROTHMAN, G. KLASSEN & O. A. MAMMER. 1975. The syndrome of systemic carnitine deficiency: clinical, morphological biochemical and pathophysiologic features. Neurology **25:** 16–24.
5. REBOUCHE, C. J. & A. G. ENGEL. 1984. Carnitine biosynthesis and metabolism in the primary carnitine deficiency syndromes. *In* Neuromuscular Disease. G. Serratrice, C. Desnuelle, J. L. Gastaut, & J. F. Pellisier, Eds.: 95–99. Raven Press. New York.
6. CHALMERS, R. A., C. R. ROE, B. M. TRACEY, T. E. STACEY, C. L. HOPPEL & D. S. MILLINGTON. 1983. Secondary carnitine insufficiency in disorders of organic acid metabolism; modulation of acyl-CoA/CoA ratios by L-carnitine in vivo. Biochem. Soc. Trans. **11:** 724–725.
7. BREMER, J. 1983. Carnitine: metabolism and functions. Physiol. Rev. **63:** 1420–1480.
8. PAIK, W. K. & S. KIM. 1975. Protein methylation: chemical, enzymological, and biological significance. Adv. Enzymol. **42:** 227–286.
9. REBOUCHE, C. J. & A. G. ENGEL. 1980. Tissue distribution of carnitine biosynthetic enzymes in man. 1980. Biochim. Biophys. Acta **630:** 22–29.
10. CEDERBLAD, G. & S. LINDSTEDT. 1976. Metabolism of labeled carnitine in the rat. Arch. Biochem. Biophys. **175:** 173–180.
11. REBOUCHE, C. J. & A. G. ENGLE. 1983. Kinetic compartmental analysis of carnitine metabolism in the dog. Arch. Biochem. Biophys. **220:** 60–70.
12. REBOUCHE, C. J. 1982. Sites and regulation of carnitine biosynthesis in mammals. Fed. Proc. **41:** 2848–2852.
13. REBOUCHE, C. J. 1980. Comparative aspects of carnitine biosynthesis in microorganisms and mammals with attention to carnitine biosynthesis in man. *In* Carnitine Biosynthesis, Metabolism and Functions. R. A. Frenkel & J. D. McGarry, Eds.: 57–67. Academic Press. New York.
14. CIMAN, M., L. SARTORELLI, G. MANTOVANI & N. SILIPRANDI. 1985. Carnitine and deoxycarnitine in ascorbate deficient guinea pigs. IRCS Med. Sci. **13:** 717.
15. BARTHOLMEY, S. J. & A. R. SHERMAN. 1985. Carnitine levels in iron-deficient rat pups. J. Nutr. **115:** 138–145.
16. BOHMER, T. & P. MØLSTAD. 1980. Carnitine transport across the plasma membrane. *In* Carnitine Biosynthesis, Metabolism and Functions. R. A. Frenkel & J. D. McGarry, Eds.: 73–89. Academic Press. New York.
17. REBOUCHE, C. J. 1977. Carnitine movement across muscle cell membranes. Studies in isolated rat muscle. Biochim. Biophys. Acta **471:** 145–155.

18. BAHL, J. J., T. R. NAVIN & R. BRESSLER. 1980. Carnitine uptake and stimulation of carnitine uptake in the isolated beating adult rat heart myocytes. *In* Carnitine Biosynthesis, Metabolism and Functions. R. A. Frenkel & J. D. McGarry, Eds.: 91–112. Academic Press. New York.
19. VARY, T. C. & J. R. NEELY. 1982. Characterization of carnitine transport in isolated perfused adult rat hearts. Am. J. Physiol. **242:** H585–H592.
20. MØLSTAD, P. 1980. The efflux of L-carnitine from cells in culture (CCL 27). Biochim. Biophys. Acta **597:** 166–173.
21. SARTORELLI, L., M. CIMAN & N. SILIPRANDI. 1985. Carnitine transport in rat heart slices. I. The action of thiol reagents on the acetylcarnitine/carnitine exchange. Ital. J. Biochem. **34:** 275–281.
22. PAULSON, D. J. & A. L. SHUG. 1981. Tissue specific depletion of L-carnitine in rat heart and skeletal muscle by D-carnitine. Life Sci. **28:** 2931–2938.
23. SARTORELLI, L., M. CIMAN, G. MANTOVANI & N. SILIPRANDI. 1985. Carnitine transport in rat heart slices. II. The carnitine/deoxycarnitine antiport. Ital. J. Biochem. **34:** 282–287.
24. SILIPRANDI, N., F. DI LISA, C. R. ROSSI & A. TONINELLO. 1984. Overview of lipid metabolism. *In* Myocardial Ischemia and Lipid Metabolism. R. Ferrari, A. M. Katz, A. Shug & O. Visioli, Eds.: 1–13. Plenum Press. New York.
25. JENKINS, D. L. & O. W. GRIFFITH. 1985. DL-Aminocarnitine and acetyl-DL-aminocarnitine. Potent inhibitors of carnitine acyltransferases and hepatic triglyceride catabolism. J. Biol. Chem. **260:** 14748–14755.
26. MILLS, S. E., W. FOSTER & J. D. MCGARRY. 1983. Interaction of malonyl-CoA and related compounds with mitochondria from different rat tissues. Biochem J. **214:** 83–91.
27. BREMER, J., G. WOLDEGIORGIS, K. SCHALINSKE & E. SHRAGO. 1985. Carnitine palmitoyltransferase. Activation by palmitoyl-CoA and inactivation by malonyl-CoA. Biochim. Biophys. Acta **833:** 9–16.
28. BRECHER, P. 1983. The interaction of long-chain acyl CoA with membranes. Mol. Cell. Biochem. **57:** 3–15.
29. PANDE, S. V. & M. C. BLANCHAER. 1971. Reversible inhibition of mitochondrial adenosine diphosphate phosphorylation by long chain acyl coenzyme A esters. J. Biol. Chem. **246:** 402–411.
30. SILIPRANDI, N., F. DI LISA & L. SARTORELLI. 1985. Transport and function of carnitine in cardiac muscle. *In* Membranes and Muscle. M. C. Berman, W. Gevers & L. H. Opie, Eds.: 105–119. ICSU Press. Oxford.
31. DI LISA, F., V. BOBYLEVA-GUARRIERO, P. JOCELYN, A. TONINELLO & N. SILIPRANDI. 1985. Stabilising action of carnitine on energy linked processes in rat liver mitochondria. Biochem. Biophys. Res. Commun. **131:** 968–973.
32. TONINELLO, A., D. BRANCA, G. SCUTARI, N. SILIPRANDI, E. VINCENTI & GIRON. 1986. L-carnitine effect on halothane-treated mitochondria. Biochem. Biophys. Res. Commun. **139:** 303–307.
33. O'CONNOR, J. E., M. COSTELL & S. GRISOLIA. 1984. Protective effect of L-carnitine on hyperammonemia. FEBS Lett. **166:** 331–334.
34. DI DONATO, S., M. RIMOLDI, B. GARAVAGLIA & G. UZIEL. 1984. Propionylcarnitine excretion in propionic and methylmalonic acidurias: a cause of carnitine deficiency. Clin. Chim. Acta **139:** 13–21.
35. KIM, C. S., D. R. DORGAN & C. R. ROE. 1984. L-carnitine: therapeutic strategy for metabolic encephalopathy. Brain Res. **310:** 149–153.
36. HÜLSMANN, W. C., D. SILIPRANDI, M. CIMAN & N. SILIPRANDI. 1964. Effect of carnitine on the oxidation of α-oxoglutarate to succinate in the presence of acetoacetate or pyruvate. Biochim. Biophys. Acta **93:** 166–168.
37. SILIPRANDI, N., D. SILIPRANDI & M. CIMAN. 1965. Stimulation of oxidation of mitochondrial fatty acids and of acetate by acetylcarnitine. Biochem. J. **96:** 777–780.
38. GASPARETTO, A., G. G. CORBUCCI & N. SILIPRANDI. 1986. Action of acetyl carnitine on patients with circulatory shock. Clin. Pharm. Research. **4:** 76–78.

DISCUSSION OF THE PAPER

S. ZIERZ (*Universität-Nervenklinik, Bonn*): Is there any positive evidence for the existence of a primary carnitine deficiency? Or aren't all so far reported cases of primary carnitine deficiency just secondary to another metabolic/enzyme defect that is or was not yet identified or looked for?

N. SILIPRANDI (*University of Padova, Padova, Italy*): Although the mechanism of the primary carnitine deficiency is still obscure, these pathologies have in common a decreased content of total carnitine in tissues and decreased excretion of total carnitine in urine. Secondary carnitine deficiencies, on the contrary, are characterized by an opposite condition: higher esterified carnitine/free carnitine ratio in tissues and increased excretion of total carnitine in urine.

In other words, in primary carnitine deficiencies carnitine is insufficient (defective synthesis or impaired transport) to support normal metabolism. In secondary carnitine deficiencies the normal carnitine available is insufficient to eliminate the excess acyl CoA produced as a consequence of other metabolic defects.

E. RUBIN (*Hahnemann University, Philadelphia, PA*): The dissipation of the proton motive force by halothane may depend on the partition of halothane into the membrane, which is time-dependent. If carnitine changes the partition coefficient for halothane, this would explain the protection against halothane-induced dissipation of the energy gradient.

SILIPRANDI: Carnitine does not modify the partition coefficient for halothane, but prevents the formation of acyl CoA secondarily induced by halothane with a still unknown mechanism.

M. KLINGENBERG (*University of Munich, Munich*): Which protein is the major source of trimethyllysine (TML)? A possible candidate could also be the ADP/ATP carrier, which contains one residue of TML. Due to its great abundance it could also contribute to the TML level and the control of carnitine formation.

SILIPRANDI: The information is very interesting and useful, still myosin seems to be, at last, quantitatively the principal source of trimethyllysine.

Effects of Ethanol on the Structure and Function of Rat Liver Mitochondrial and Microsomal Membranes

THEODORE F. TARASCHI, WILLIAM S. THAYER,[b]
JOHN S. ELLINGSON, AND EMANUEL RUBIN

Department of Pathology
Jefferson Medical College
Philadelphia, Pennsylvania 19107
and
[b]*Department of Pathology and Laboratory Medicine*
Hahnemann University School of Medicine
Philadelphia, Pennsylvania 19102

Chronic ethanol ingestion leads to serious degenerative and inflammatory diseases of the major body organs, particularly the liver.[1] A major morphologic feature associated with alcohol-induced injury consists of alterations in membrane-bounded hepatic organelles, especially mitochondria.[2–6] Mitochondria become enlarged and exhibit unusual shapes, disoriented cristae,[2–6] and in certain species (e.g. primates) paracrystalline inclusions.[2,3,7] Chronic ethanol abuse also produces many functional aberrations in liver mitochondria.[8–10] An altered redox state in the hepatocyte or the primary metabolite of ethanol, namely acetaldehyde, have been suggested as the cause of the widespread manifestations of ethanol-induced maladies in the liver. However, experimental evidence to support these hypotheses is not conclusive. Our group has been pursuing an alternative mechanistic pathway by which ethanol could exert its deleterious effects. Ethanol, similar to anesthetics, may act through its incorporation into biological membranes. Chronic ethanol ingestion by rats leads to alterations in the fatty acid composition of the phospholipids of subcellular organelle membranes obtained from the brain, blood, pancreas, and liver.[11] Another common result of chronic ethanol ingestion is the development of membrane resistance (tolerance) to disordering by physiologically relevant levels of ethanol *in vitro.* The addition of 50–100 mM ethanol to biological membranes derived from the organs of normal animals causes significant molecular disordering (fluidization) of the lipid hydrocarbon chains. By contrast, membranes from ethanol-fed animals are resistant to molecular disordering by the same concentration of ethanol.[12] The major structural and functional alterations in rat liver membranes will be summarized and the molecular basis for these alterations discussed.

ALTERATIONS IN PROTEIN CONTENT AND FUNCTION

Oxidative Phosphorylation

Cederbaum *et al.*[8] demonstrated that mitochondria isolated from rats chronically fed ethanol have decreased rates of respiration and ATP synthesis compared to

[a]Supported by U.S. Public Health Service Grants AA3442, AA5662, and AA88. T.F.T. is the recipient of a Research Scientist Development Award from the National Institute on Alcohol Abuse and Alcoholism.

untreated rats. Subsequently, Thayer and Rubin[13] reported that liver submitochondrial particles isolated from ethanol-fed rats exhibit decreased respiratory rates under coupled or uncoupled conditions. P/O ratios were decreased 10–15% with all substrates under phosphorylating conditions. However, the steady-state rate of ATP synthesis was found to be decreased by ~40% with all substrates as a result of the

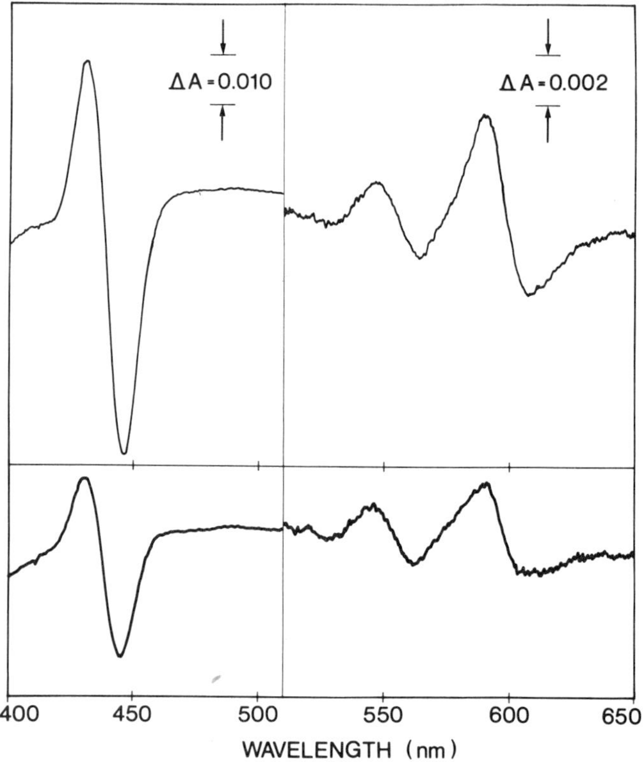

FIGURE 1. Spectrophotometric determination of enzymatically functional cytochrome oxidase content in alcoholic and control membranes. Difference absorbance spectra (dithionite-reduced plus CO minus reduced) were recorded for rat liver submitochondrial particles (2.0 mg/ml) from a control (*top*) and ethanol-fed rat (*bottom*). Comparison of the spectral changes at wavelengths characteristic of the cyt a_3^{2+}-CO complex, 429 minus 445 nm and 590 minus 605 nm, indicates that the functional cytochrome oxidase content of membranes from the alcoholic rat is about half that of the control. Similar results were obtained from reduced minus oxidized difference spectra of several preparations. (From Thayer & Rubin.[14] With permission from *Journal of Biological Chemistry*.)

combined effects on respiration and phosphorylation produced by chronic ethanol feeding.[13] It appears that the rate of ATP synthesis during oxidative phosphorylation in submitochondrial particles is mainly limited by the rate of electron transfer through the respiratory chain. Consequently, we sought to identify alterations in the content and activity of the respiratory components that might be responsible for the attenuating effects of chronic ethanol feeding on oxidative phosphorylation.

Cytochromes

We have previously analyzed the cytochrome content of mitochondrial membranes from ethanol-fed rats by optical spectroscopy,[14] employing the characteristic heme absorbance differences between the reduced and oxidized states. Significant decreases (~50%) in cytochrome oxidase (FIG. 1) and cytochrome *b* content were observed in isolated hepatocytes, intact mitochondria, and submitochondrial particles following chronic ethanol feeding. The content of cytochrome *c* was unaffected by chronic ethanol consumption, a finding reflected in a doubling of the cytochrome *c*/cytochrome oxidase ratio in the mitochondrial membranes from ethanol-fed animals. Interestingly, even though the amount of cytochrome oxidase in the mitochondrial membrane was only half of that in the control membranes, the remaining enzyme appeared normal enzymatically. Thus, although chronic ethanol consumption decreased the membrane content of cytochrome oxidase, its kinetic properties were unaltered. In analyzing these findings, we were intrigued by the fact that although the content of cytochrome oxidase that was measured spectrophotometrically decreased by about half, the amount of total mitochondrial protein was unchanged, and no compensatory increase in other mitochondrial proteins was detected. In addressing this problem, we recognized that we had been measuring functional cytochrome oxidase by spectrophotometric and enzymatic criteria rather than the physical content of the protein or apoprotein moiety. To determine the latter, an immunochemical assay was developed that utilized anticytochrome oxidase antiserum.[15] Cytochrome oxidase protein in submitochondrial particles was estimated by immunoinhibition titrations (FIG. 2). Triton extracts of submitochondrial particles from control and ethanol-fed rats containing the antigen, cytochrome oxidase, were incubated overnight with the rabbit antiserum. The immunoprecipitate that formed was removed by centrifugation and the amount of cytochrome oxidase remaining in the supernatant was determined by measuring its activity. The inhibition patterns obtained in either fixed membrane protein or fixed antiserum titrations were quantitatively similar for both types of submitochondrial particles, indicating a similar amount of immunologic reactivity per mg protein. Thus, the protein moiety of cytochrome oxidase recognized by the antiserum was present at similar amounts in both membranes. This finding suggested that an inactive form of cytochrome oxidase was also present in the mitochondrial membranes of ethanol-fed rats, since cytochrome oxidase biochemical reactivity was one half that of controls but immunologic activity was unchanged. The inactive form of cytochrome oxidase appears to be devoid of heme, since this was undetected spectrophotometrically. The origin of this defect in protein assembly caused by chronic ethanol consumption may lie in an inability of heme *a* to become attached to the oxidase moiety or a deactivation of functional oxidase in the mitochondrion.

Dehydrogenases

In addition to the effects on the mitochondrial cytochromes, chronic ethanol consumption affects the membrane content of proteins associated with NADH dehydrogenase. Since this enzyme contains several iron-sulfur clusters, which are paramagnetic in their reduced state, we measured mitochondrial enzyme content of NADH dehydrogenase by electron spin resonance. In NADH-reduced submitochondrial particles, clusters N-2, N-3, and N-4 were decreased by 20–30% following chronic ethanol treatment.[16] By contrast, chronic ethanol treatment did not affect the membrane content of iron-sulfur clusters associated with succinate dehydrogenase.

ATPase Activity

Studies of ATP-dependent energy-linked functions such as $^{32}P_i$-ATP exchange and reverse electron transfer (e.g. reduction of NAD^+ by succinate) indicated that chronic

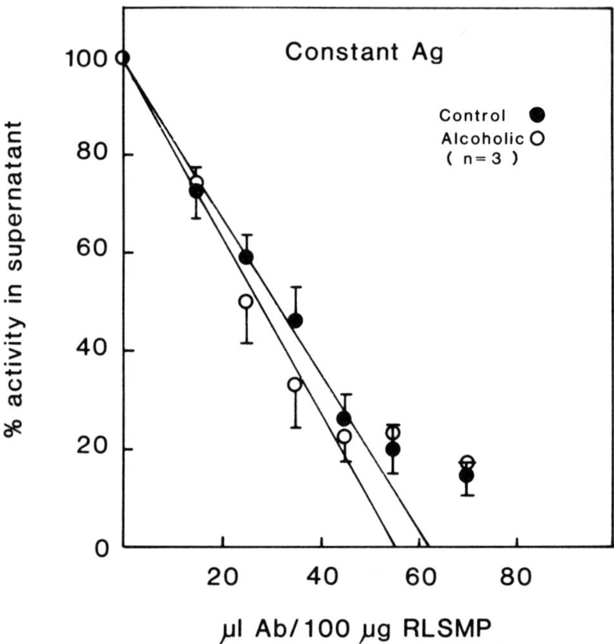

FIGURE 2. Immunochemical determination of physical cytochrome oxidase content in alcoholic and control membranes. Immunoinhibition titrations were conducted with varying amounts of anti-rat liver cytochrome oxidase serum using a fixed amount of submitochondrial particles (0.1 mg), solubilized in 0.5% Triton X-100, in a final volume of 0.10 ml. Following centrifugation to move the immunoprecipitate, cytochrome oxidase activity remaining in the supernatants was measured. Symbols indicate mean and S.D. relative activity for three different pairs of submitochondrial particle preparations. Absolute specific activities were control: 6970, alcoholic: 4050 natoms 0/min/mg membrane protein. The equivalent amount of antiserum necessary to inhibit the respective activity of each preparation by 50% of its original value is indicative of quantitatively similar cytochrome oxidase immunologic reactivity per mg membrane protein. (From Thayer & Rubin. With permission from *Biochimica Biophysica Acta.*)

ethanol treatment caused a 10–15% decrease in coupling efficiency.[13] ATPase activity measured directly under uncoupled conditions was decreased by 30% in preparations from ethanol-fed rats.[13] This finding suggests that the content of the ATPase protein is also substantially reduced in the mitochondrial membranes following chronic ethanol treatment.

ALTERATIONS OF LIVER MEMBRANE STRUCTURE AND COMPOSITION BY CHRONIC ETHANOL CONSUMPTION

Mitochondria

The addition of ethanol *in vitro* to liver mitochondrial membranes from untreated rats causes significant molecular disordering (fluidization) of the lipid hydrocarbon chains. By contrast, liver mitochondria from ethanol-fed rats are resistant to molecular disordering by the same concentration of ethanol.[17] The specific structural and chemical changes in the membranes that are responsible for the acquired tolerance to ethanol have yet been identified. Alterations in fatty acid composition have been evoked as being responsible, but no consistent pattern of change has emerged, and reported changes are small. In the mitochondrial membrane, the only noteworthy change is a significant decrease in 18:2 fatty acids following ethanol feeding.[18] We have shown that membrane tolerance to disordering is associated with a decrease in the partition coefficient of ethanol and other lipophilic molecules into the membrane.[19] An increased cholesterol content in the membranes from ethanol-fed animals has also been suggested to play a major role in the development of membrane tolerance,[20] but in many instances, the cholesterol content of the membranes is unchanged[21] or even decreased [22] following chronic ethanol treatment. In addition, mitochondrial membranes, which do not contain cholesterol, develop membrane tolerance and recombined membranes prepared from red blood cell and microsomal phospholipids from ethanol-fed animals also exhibit membrane tolerance.[23] Accordingly, new approaches and methodologies are necessary to determine the molecular basis of membrane tolerance.

Microsomes

The majority of the physicochemical and compositional studies that have been undertaken to investigate the effects of chronic ethanol consumption on membrane structure, composition, and function have been directed toward alterations that occur following a long period (e.g. > 35 days) of ethanol treatment. Few studies have been directed toward investigating the loss of membrane tolerance after withdrawal from ethanol feeding, i.e. the reestablishment of a normal membrane.

In our animal model, rats ingest 14–16 g of ethanol/kg body weight for 35 days in a nutritionally adequate diet, while pair-fed controls consume the same diet, except that carbohydrate isocalorically replaces ethanol. We have investigated the structural properties of liver microsomes obtained from rats that had been chronically administered ethanol for 35 days and then withdrawn from ethanol for 1–10 days.[23] In addition, using electron spin resonance, high pressure liquid chromatography (HPLC), and membrane reconstitution techniques, the membrane component(s) responsible for membrane tolerance in the microsomal membrane are identified for the first time.

Response to Withdrawal from Ethanol

The effect of ethanol *in vitro* on the membrane molecular order parameter was assessed by ESR for liver microsomes isolated from untreated rats, rats chronically fed ethanol, and ethanol-fed rats withdrawn from ethanol (FIG. 3). The molecular order parameter S, derived from the motionally sensitive spectral hyperfine splittings, provides a measure of the lipid hydrocarbon chain molecular order (fluidity). The results were obtained from microsomes labeled with the phospholipid probe 1-

palmitoyl-2-{12-[β-(4′,4′-dimethyloxazolidinyl-N-oxyl)]stearoyl}phosphatidylcholine (PtdCho 12) (FIG. 3). Similar results were obtained with microsomes labeled with 12-[β-(4′,4′-dimethyloxazolidinyl-N-oxyl)]stearic acid (Ste 12). In agreement with an earlier report from this laboratory,[24] the addition of various amounts of ethanol *in vitro* (50, 75, and 100 mM) caused a decrease in the order parameter, from 0.372 to 0.356. Such behavior is indicative of significant membrane molecular disordering. Molecular disordering, characterized by an increase in the spectral splitting $2T_{\perp}$, reflects the greater angular displacement of the lipid fatty acid chains bearing the spin-label group from the long molecular axis of the lipid in the membrane bilayer. Conversely, the

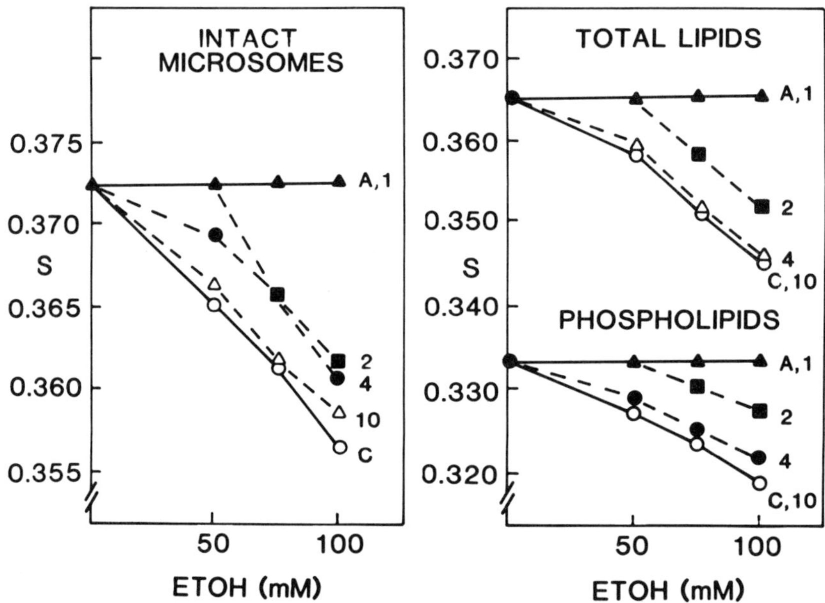

FIGURE 3. The effect of ethanol *in vitro* on the molecular order parameter *S* obtained for PtdCho 12-labeled intact rat liver microsomes and multilayer vesicles prepared from extracted microsomal total lipids and phospholipids. Microsomes were obtained from untreated rats (curve C, ○), alcoholic rats (curve A, ▲), and alcoholic rats withdrawn from ethanol feeding for 1 day (curve 1, ▲), 2 days (curve 2, ■), 4 days (curve 4, ●), and 10 days (curve 10, △). Each point represents the mean obtained from four pairs of animals. Errors in each point were never greater than 0.75%.

order parameters determined for microsomes obtained from ethanol-fed rats were unchanged over the same range of ethanol concentrations. This resistance to disordering by ethanol has been reported for microsomes from ethanol-fed rats labeled with fatty acid or phospholipid spin probes.[24] No differences in baseline-order parameters (no ethanol added) were observed for microsomes obtained from control or ethanol-treated animals. We found that a period of 28–35 days was required for all of the animals treated chronically with ethanol to display membrane tolerance.

Following one day of ethanol withdrawal, the microsomes remained totally resistant to disordering by ethanol (up to 100 mM) *in vitro* (FIG. 3). After only two days of ethanol withdrawal, the microsomes were disordered by the addition of 75 mM ethanol (FIG. 3), as evidenced by a decrease in the order parameter. The extent of membrane disordering after two days of withdrawal was not as large (ΔS = 2.9%) as in the control membranes (ΔS = 4.6%) between 0 and 100 mM ethanol. Microsomes obtained from ethanol-fed animals that had been withdrawn from ethanol for four days had order-parameter profiles more similar to the controls. By 10 days of withdrawal, the order-parameter profiles were identical to those of the controls.

To locate the origin of the rapid loss of membrane tolerance, total lipid extracts were prepared from the microsomes used in the experiments described above. The results are summarized in FIG. 3. Multilayer vesicles prepared from the control lipids were disordered by the addition of increasing amounts of ethanol (0 mM, S = 0.365; 75 mM, S = 0.358). Following four days of withdrawal, the order-parameter profile could be superimposed on that of the control lipids (FIG. 3). Between 0 and 100 mM ethanol, the bilayers from microsomal lipids obtained from rats withdrawn for four days experienced the same decrease in the order parameters as did bilayers from microsomes of control rats. This recovery was more rapid than in intact microsomes, which required a longer period of withdrawal (4–10 days) to display the full loss of membrane tolerance (FIG. 3).

To determine whether membrane resistance to disordering by ethanol is a characteristic of the phospholipids only, bilayers prepared from microsomal phospholipids were examined for membrane tolerance. Bilayers prepared from phospholipids of control microsomes were considerably disordered by ethanol *in vitro*. The baseline order parameters obtained from the bilayers comprised solely of the phospholipids (S = 0.333) were lower than those from the intact microsomes (S = 0.372) and from the total lipid extracts (S = 0.365). Bilayers prepared from microsomal phospholipids from ethanol-fed rats were resistant to disordering by ethanol. Furthermore, as was observed in the intact membranes and the total lipid extracts, resistance to disordering by ethanol was lost in the phospholipid bilayers from animals withdrawn for two days. In a fashion similar to that of the total lipid extacts, four days of ethanol withdrawal were required for the order parameter profiles of those membranes to display disordering to the same extent of those from control phospholipids (FIG. 3).

There was no difference in the cholesterol:phospholipid ratios between control and ethanol-treated animals.

The Origin of Membrane Tolerance in Rat Liver Microsomes

We embarked upon a study to assess by ESR the ability of individual liver microsomal phospholipids from ethanol-fed rats to promote membrane tolerance in membrane vesicles composed of recombined individual phospholipids of the microsomal membrane.[25] Microsomal membrane phospholipids were separated into classes by preparative HPLC. The membrane composition obtained with this procedure was 22 mole % phosphatidylethanolamine (PE), 8.5% phosphatidylinositol (PI), 4.0% phosphatidylserine (PS), and 65.5% of a phosphatidylcholine (PC)-sphingomyelin fraction. No significant differences in the molar proportions of individual phospholipid classes were observed as a result of chronic ethanol feeding.

In vesicles composed of varying mole fractions of microsomal phospholipids (before separation by HPLC) obtained from the livers of untreated and ethanol-fed rats, we determined that only 30–40 mole % of the phospholipids derived from the ethanol-fed animals was necessary to render the vesicles tolerant to ethanol-induced disordering.

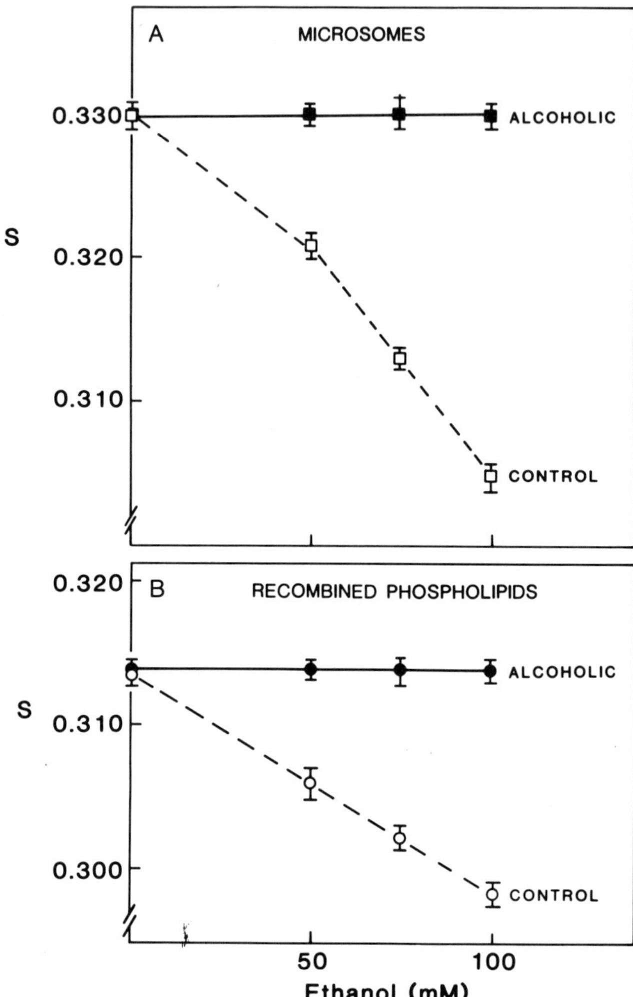

FIGURE 4. Typical order parameter profiles showing the effect of ethanol *in vitro* on the molecular order parameters, *S*, obtained by ESR at 37°C (A) intact rat liver microsomes and (B) multilamellar vesicles prepared from recombined microsomal phospholipids that had been separated into individual classes by HPLC and labeled with 12 doxyl stearic acid. Each phospholipid class in the recombined vesicles was present in its naturally occurring mole fraction. Microsomes were obtained from untreated (control) rats (□----□) and rats chronically consuming ethanol (alcoholic) (■—■). The total control (untreated) (○----○) or alcoholic (ethanol-fed) (●—●) microsomal phospholipids used to prepare the reconstituted vesicles were isolated from the same microsomal preparations used to obtain the data in part A. Points and bars represent the mean + or − SD from four pairs of animals.

Thus, the phospholipids obtained from the ethanol-fed rats contained a component potent enough to confer membrane tolerance, even when diluted with a larger amount of phospholipids from untreated rats.

Vesicles were then prepared by recombining the HPLC-separated microsomal phospholipids in their naturally occurring molar proportions. Reconstituted membranes composed of phospholipids from untreated rats were disordered by the addition of ethanol, as evidenced by a significant decrease (4.6%) in the order parameter, from 0.314 to 0.299 (FIG. 4B). On the other hand, vesicles composed of microsomal phospholipids from ethanol-treated rats showed no change in the order parameter over the same range of ethanol concentrations (FIG. 4B). As in the intact membrane, chronic ethanol treatment did not alter the baseline order parameters of the vesicles prepared from the phospholipids derived from the ethanol-fed rats. The response of the reconstituted vesicles to ethanol, *in vitro,* qualitatively paralleled that of the intact microsomal membranes, thereby providing the opportunity to identify the phospholipid(s) responsible for conferring membrane tolerance.

In an attempt to identify the component responsible for membrane tolerance, vesicles were made by recombining either all the individual liver microsomal phospholipids from the untreated animals or all the individual phospholipids from the ethanol-fed animals (in the same molar proportions as found in the microsomal membrane), except that in each preparation one different phospholipid class was omitted. The missing phospholipid in the untreated preparation was then replaced by the corresponding phospholipid from the ethanol-fed rats. Similarly, in the preparation of phospholipids from the ethanol-fed rats, a phospholipid was deleted and replaced by one from the control rats. Recombined membranes prepared from the phospholipids of ethanol-fed rats retained their resistance to fluidization by ethanol after substitution of PC (66.5%), PE (21%), PI (8.5%), or (PS 4.0%) from untreated rats (FIG. 5). Recombined "untreated" membranes were still fluidized by ethanol after substitution of PC, PE, or PS from the ethanol-fed animals. By contrast, when PI (8.5%) from treated rats was substituted in the membranes made from phospholipids from untreated rats, the membranes were rendered resistant to disordering by ethanol (FIG. 5). To determine whether the effect of PI is peculiar to rat hepatic microsomes, we prepared vesicles that had the same proportions of phospholipid classes as the rat microsomal membrane, but were made from bovine liver PC , PE, and PI, and bovine brain PS and sphingomyelin. These reconstituted vesicles were disordered by ethanol to the same extent as vesicles made from the microsomal phospholipids of untreated animals ($S = 0.316$, 0 mM; $S = 0.301$, 100 mM; $\%\Delta S = 4.6$). Vesicles in which bovine liver PI was replaced with microsomal PI from untreated rats were also disordered to the same extent. However, when the bovine liver PI was replaced by PI from the ethanol-fed rats, the bovine vesicles became resistant to disordering. Thus, 8.5% microsomal PI from the treated animals also conferred membrane tolerance to vesicles composed of 91.5% phospholipids from another animal species and another organ. The minimum amount of PI from ethanol-fed animals needed to confer tolerance to either microsomal phospholipids or bovine liver phospholipids was subsequently determined to be only 2.5%. Interestingly, when twice the naturally occurring amount of the microsomal PS (i.e. 8.0% instead of 4.0%) from ethanol-fed rats was substituted into the vesicles made from phospholipids from untreated rats or bovine standard lipids, they were rendered tolerant.

We have very recently performed similar experiments on mitochondrial inner membranes obtained from the livers of ethanol-fed rats. Cardiolipin, in its naturally occurring amount, was the only phospholipid that, on its own, rendered the reconstituted lipid bilayers from untreated rats tolerant to disordering by ethanol.[26] Thus, the

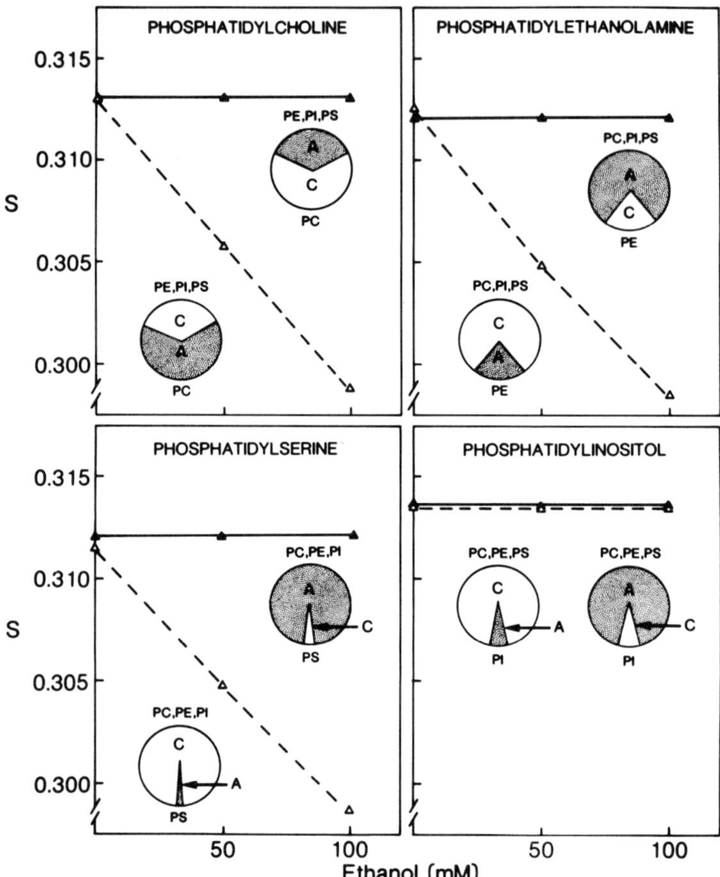

FIGURE 5. The capacity of individual microsomal phospholipid classes isolated from ethanol-fed rats to produce membrane tolerance in vesicles composed of microsomal phospholipids from untreated rats. The phospholipid compositions of each of the reconstituted membrane systems are indicated in the pie charts. Shaded areas indicate the mole fraction of phospholipid(s) (A) from ethanol-fed rats and unshaded areas the mole fraction of phospholipid(s) (C) from untreated, control animals. ESR order parameters in the absence and presence of ethanol for recombined phospholipid vesicles comprised of three phospholipid classes from ethanol-fed rats and one phospholipid class from untreated rats are shown as (▲—▲) and vesicles comprised of three phospholipid classes from untreated rats and one phospholipid class from ethanol-fed rats (Δ----Δ). The same results were obtained for four other sets of pair-fed animals. The error in each point was never greater than 0.75%.

basis for acquired membrane tolerance in liver mitochondria and microsomes appears to be chronic ethanol-induced alterations in the acidic phospholipids.

DISCUSSION

The sequence of molecular events involved in the assembly of cytochrome oxidase and other mitochondrial proteins is fairly well understood. The cytoplasmically synthesized subunits of ATPase and probably of cytochrome oxidase are transported into the membrane as precursor proteins in an energy-dependent reaction prior to processing and assembly with the mitochondrially synthesized subunits.[27,28] The subunits of cytochrome oxidase from both the mitochondrial and cytoplasmic translation systems are assembled in a specific order within the mitochondria.[29] It also appears that heme *a* promotes association of specific cytochrome oxidase subunits during assembly.[31] The role of membrane lipids in the insertion of the subunits into the membrane and in heme attachment is unknown. Therefore, at this time it is difficult to assess how the alterations in lipid composition and structure produced by chronic ethanol feeding might modulate the assembly of functional cytochrome oxidase.

Membrane tolerance does not solely result from ethanol-induced differences in the molecular order of the individual microsomal phospholipids. Using phospholipid spin probes, we found no differences in baseline molecular order (no ethanol present) among PC, PE, and phosphatidic acid in microsomes from untreated or ethanol-fed rats.[24] Changes in cholesterol content are not responsible for the development of membrane tolerance in rat liver mitochondria, microsomes, or erythrocytes. This conclusion is supported by the fact that no changes in cholesterol:phospholipid ratio are found in microsomes or erythocytes before or after withdrawal from ethanol.[23]

Since tolerance is observed in phospholipids extracted from three different membranes, changes in protein-lipid interactions cannot be implicated as the cause of tolerance. However, such interactions may be involved to some degree, since loss of tolerance in reconstituted bilayers is demonstrable after shorter periods of time than in intact membranes. The considerably lower baseline order parameters obtained for the phospholipid bilayers, compared to the intact membranes (FIG. 3), reflects greater "fluidity" in the absence of lipid-protein interactions and cholesterol, both of which tend to order the membrane. We have previously shown[19] that acquired membrane tolerance is associated with a decreased partitioning of ethanol and halothane into the alcoholic membranes. The decreased partition of ethanol into the alcoholic membrane may result from alterations in phospholipid molecular species, which might confer different molecular shapes to the phospholipids. Such alterations, which would go undetected using standard analytical techniques, could result in different packing properties of the membrane lipids, resulting in decreased partitioning of ethanol, and hence membrane tolerance. These alterations in lipid molecular species and packing properties would not necessarily cause any detectable change in membrane molecular order (fluidity) measured by ESR.

Interestingly, PI, which is a minor component of the microsomes, uniquely confers membrane tolerance, whereas PC or PE alone, which are present in much greater amounts, are without effect. Even more surprising is the finding that as little as 2.5% content of PI from ethanol-fed animals renders the membrane resistant to fluidization. The ability of PI to promote membrane tolerance is not restricted to microsomal phospholipids; the inclusion of as little as 2.5% of PI from the ethanol-fed animals into vesicles of bovine phospholipids of brain and liver has the same effect. The observation that vesicles prepared from phospholipids of ethanol-fed animals, in which PI from untreated animals replaced the PI from treated animals, remain resistant to disordering by ethanol suggests that minor ethanol-induced modifications in the other microsomal phospholipids make them weak promoters of tolerance, and that they can confer tolerance to reconstituted membranes when they comprise 90% of the vesicle phospholipids. Thus, while PI is a very strong promoter of membrane tolerance, combinations of other microsomal phospholipids (e.g. PC + PE) from ethanol-fed rats, when present in very high concentrations, can also produce this effect.

In our studies, PI isolated from ethanol-fed rats exhibited a small decrease in the level of arachidonic acid and a slight increase in that of oleic acid. We do not know whether such minor ethanol-induced alterations in fatty acid composition are related to the capacity of alcoholic PI to confer membrane tolerance. Chronic ethanol treatment may also cause changes in the phospholipid molecular species, although such alterations have not yet been sought. It may be that chemical alterations that occur in PI as a result of ethanol feeding change the packing properties of the microsomal phospholipids, such that the partition of ethanol into the membrane is reduced. The identification of the chemical and structural alterations in PI and cardiolipin caused by chronic ethanol feeding is being actively pursued in our laboratory.

REFERENCES

1. MAJCHROWICZ, E. & E. P. NOBEL. 1978. Biochemistry and Pharmacology of Ethanol. Plenum Press. New York.
2. RUBIN, E. & C. S. LIEBER. 1973. Science **182:** 712–713.
3. RUBIN, E. & C. S. LIEBER. 1974. N. Engl. J. Med. **290:** 128–135.
4. LIEBER, C. S., D. P. JONES & L. M. DE CARLI. 1965. J. Clin. Invest. **44:** 1009–1021.
5. Rubin. E. & C. S. Lieber. 1967. Gastroenterology **52:** 1–13.
6. RUBIN, E. & C. S. LIEBER. 1968. N. Engl. J. Med. **278:** 869–876.
7. RUBIN, E. & C. S. LIEBER. 1975. Clin. Gastroenterol. **4:** 247–272.
8. CEDERBAUM, A. I., C. S. LIEBER, A. TOTH, D. S. BEATTIE & E. RUBIN. 1973. J. Biol. Chem. **248:** 4977–4986.
9. GORDON, E. R. 1973. J. Biol. Chem. **248:** 8271–8280.
10. CEDERBAUM, A. I., C. S. LIEBER & E. RUBIN. 1974. Arch. Biochem. Biophys. **165:** 560–569.
11. TARASCHI, T. F. & E. RUBIN. 1985. Lab. Invest. **52:** 120–128.
12. CHIN, J. H. & D. B. GOLDSTEIN. 1977. Science **196:** 684–686.
13. THAYER, W. S. & E. RUBIN. 1979. J. Biol. Chem. **254:** 7717–7723.
14. THAYER, W. S. & E. RUBIN. 1981. J. Biol. Chem. **256:** 6090–6097.
15. THAYER, W. S. & E. RUBIN. 1986. Biochim. Biophys. Acta **849:** 366–373.
16. THAYER, W. S., T. OHNISHI & E. RUBIN. 1980. Biochim. Biophys. Acta **591:** 22–36.
17. WARING, A. J., H. ROTTENBERG, T. OHNISHI & E. RUBIN. 1982. Arch. Biochem. Biophys. **216:** 51–61.
18. WARING, A. J., H. ROTTENBERG, T. OHNISHI & E. RUBIN. 1981. Proc. Natl. Acad. Sci. USA **78:** 2582–2586.
19. ROTTENBERG, H., A. WARING & E. RUBIN. 1981. Science **213:** 583–585.
20. CHIN, J. H., L. M. PARSON & D. B. GOLDSTEIN. 1978. Biochim. Biophys. Acta **513:** 358–363.
21. ALLING, C., S. LILIJEQUIST & J. ENGEL. 1982. Med. Biol. **60:** 149–153.
22. HARRIS, R. A., D. M. BAXTER, M. A. MITCHELL & R. J. HITZEMANN. 1984. Mol. Pharmacol. **25:** 401–407.
23. TARASCHI, T. F., J. S. ELLINGSON, A. WU, R. ZIMMERMAN & E. RUBIN. 1986. Proc. Natl. Acad. Sci. USA **83:** 3669–3673.
24. TARASCHI, T. F., A. WU & E. RUBIN. 1985. Biochemistry **24:** 7096–7101.
25. TARASCHI, T. F., J. S. ELLINGSON, A. WU, R. ZIMMERMAN & E. RUBIN. 1986. Proc. Natl. Acad. Sci. USA. (In press.)
26. ELLINGSON, J. S., T. F. TARASCHI, A. WU, R. ZIMMERMAN & E. RUBIN. 1986. Unpublished observation.

27. ADES, I. Z. 1982. Mol. Cell. Biochem. **43:** 113–127.
28. HARMEY, M. A. & W. NEUPERT. 1985. *In* The Enzymes of Biological Membranes. A. Martonosi, Ed. **4:** 431–464. Plenum Press. New York.
29. WIELBURSKI, A. & D. B. NELSON. 1983. Biochem. J. **212:** 829–834.
30. WIELBURSKI, A. & D. B. NELSON. 1984. FEBS Lett. **177:** 291–294.

DISCUSSION OF THE PAPER

R. WATTIAUX: (*Laboratoire de Chimie Physiologique, Namur, Belgium*): Microsomes are very heterogeneous. Do you think that the effects you observe on microsomes mainly concern plasma membrane fragments, endoplasmic reticulum membranes, or Golgi?

RUBIN: We have observed the same changes in hepatic mitochondria, pancreatic plasma membranes, synaptosomes, and erythrocytes. It seems that resistance to fluidization, i.e., membrane tolerance, is produced in all membranes by chronic ethanol ingestion.

S. DIMAURO (*Columbia University, New York*): (1) Did you conduct studies similar to these in muscle of your experimental animals?

(2) Did you study fatty acid oxidation by mitochondria in your volunteers or experimental animals?

RUBIN: (1) We are actively pursuing studies of muscle in animals and in muscle biopsies from human alcoholics.

(2) Some years ago we showed that hepatic mitochondria from rats chemically fed ethanol display decreased fatty acids oxidation.

P. CHATELAIN (*Labaz-Sanofi, Brussels*): Do you have any modification of cholesterol content in mitochondria and microsomes after ethanol chronic ingestion?

RUBIN: Mitochondrial inner membrane contains little, if any, cholesterol. Moreover, recombined phospholipid membranes contain no cholesterol. We have measured cholesterol content of hepatic microsomes and found no change.

H.-P. SCHULTHEISS (*University of Munich, Munich*): Did you also find the changes of lipid composition in heart mitochondria? Is this reversible if you stop the addition of ethanol to the animals?

RUBIN: We have not studied heart mitochondria. Changes in hepatic mitochondria are, indeed, reversible.

G. AZZONE (*University of Padova, Padova*): Administration of ethanol results in an increase of reducing power, which means increase of NADH and NADPH. But in turn this may result in the increased extent of saturation of fatty acids. The increase in the order parameter and in membrane rigidity may be simply a consequence of the presence of larger amounts of saturated fatty acids in the bilayer.

RUBIN: The recombined membranes are made exclusively from extracted phospholipids. Neutral lipids and cholesterol are not present; yet the difference between the membranes from control and alcoholic animals is still observed.

E. CARAFOLI (*Swiss Federal Institute of Technology, Zurich*): Do you know whether it is the heme that is not assembled in 50% of cytochrome oxidase or whether there is a deficiency in one or more of the enzymes of the porphyrin biosynthesis pathway, e.g., ferrochelatase?

RUBIN: The total heme content of the hepatocyte is actually increased, owing to an increase in microsomal P450. Thus there is no defect in heme synthesis. We have measured several enzymes involved in heme synthesis and found no decrease.

Phosphorus Magnetic Resonance Spectroscopy Studies of the Role of Mitochondria in the Disease Process[a]

B. CHANCE, J. S. LEIGH, D. S. SMITH, S. NIOKA, AND B. J. CLARK

Department of Biochemistry and Biophysics
University of Pennsylvania
Philadelphia, Pennsylvania 19104

INTRODUCTION

I take pleasure in presenting a new and unusually fruitful technique for the *in vivo* study of mitochondrial disease. In the past six years this noninvasive technique has involved a wide number of adults and neonates having mitochondrial disease. The scope of the project has been far reaching and thus it is possible to include cardiac diseases, vascular effects of diabetes, and neonatal cancer. Because of the limitations of time, these topics are touched upon only briefly here, but they serve as examples of a unifying theory of nuclear magnetic resonance (NMR) studies, which we have recently published[1] and presaged by a number of preliminary notes.[2–4]

This presentation will be focused upon the applications of P MRS (phosphorus magnetic resonance spectroscopy) as this reports directly the consequences of the impact of mitochondrial disease upon readily measured NMR parameters. Other nuclei, for example ^{1}H and ^{13}C, report metabolites whose levels are regulated by the mitochondrial activity. Thus, primary events of diseases are best determined at the level of the high and low energy phosphate compounds.

BASIC PRINCIPLES OF P MRS STUDY OF METABOLIC DISEASE

The Phosphate Potential

What does P MRS measure and where does it do it? FIGURE 1 indicates the where and what of P MRS studies and emphasizes that the components of the highly viscous mitochondrial matrix space are largely, if not completely, NMR "invisible." However, the phosphorylation potential ($ATP/ADP \times P_i$)[5,6] is communicated through the adenine nucleotide translocase, which serves to export ATP and to import ADP; the mitochondrial contents of these substances are largely regulated by the property of their transporters. However, inorganic phosphate is transported by various pathways, one of which is shown here to be specific for inorganic phosphate but others involving carboxylic acids may also be functional.

[a]Supported by National Institutes of Health Grants NS 22881, HR 34004, RR 02305, HL 18798, HL 31934 and the Benjamin Franklin Partnership's Advanced Technology Center of Southeastern Pennsylvania.

The export of ATP into the cytosol affords an immediate conversion of incoming creatine to phosphocreatine (PCr) for the export to more remote cellular uses of phosphocreatine accompanied by the import of creatine. These substances, including inorganic phosphate (P_i) are small, readily diffusible, and often at high concentrations (tens of millimolar). The energy utilizing portion of the cell, whether it be pumps for the brain or myofibrils for the muscle, breaks down its local pool of ATP into ADP+P_i;

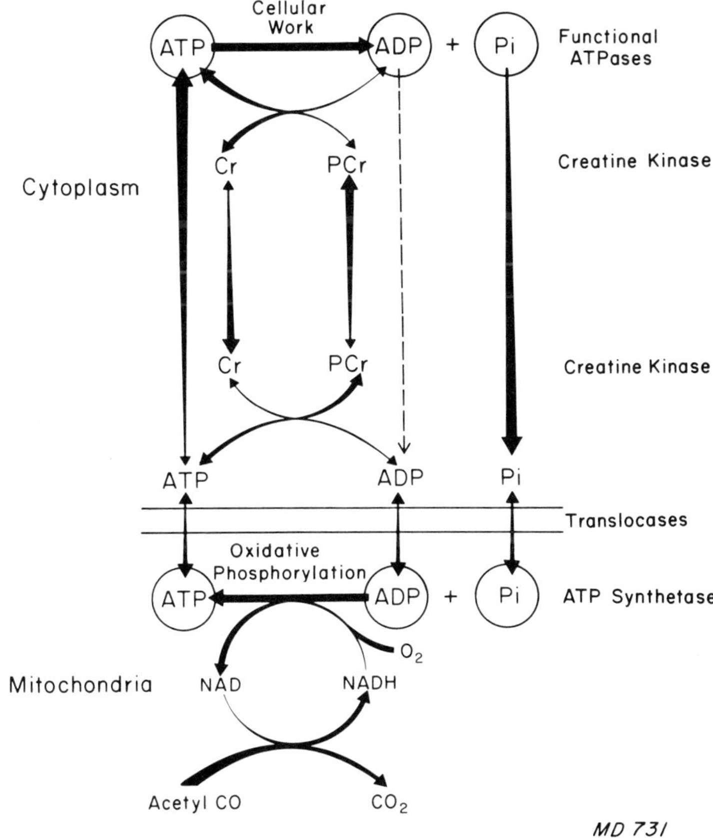

FIGURE 1. Mitochondrial/cytosolic relationships of particular interest in P MRS. The important diffusables are PCr, Cr, and P_i, while the adenine nucleotides may, under appropriate conditions, be localized near the site of ATPase and ATP synthetase activities. P MRS reports the cytosolic phosphate potential.

the latter is exported directly to the mitochondria, and the former is rephosphorylated by the highly diffusible phosphocreatine.[7]

It is these highly mobile, readily diffusible components, PCr and P_i, that are readily detected by P MRS. The cytosolic pool of magnesium-bound ATP can also be determined, but since ATP concentration is usually kept constant by the activity of creatine kinase, it gives little direct information regarding mitochondrial disease, except in its severest cases when ATP will vanish due to conversion of ADP and AMP and subsequently to nonphosphorylated products. Creatine is not detected by phosphorus NMR and some attempts to measure it by proton NMR have been made, although only in a preliminary fashion in animal studies. Nevertheless, we employ the approximation that the creatine and [P_i] are produced in equal amounts from PCr.

The Reaction Velocity in Metabolic Disease

While the "quality" of mitochondrial function is addressed by the phosphate potential, ATP/ADP$\times P_i$, the value approaches its theoretical maximum only as the cell activity tends toward zero, a state that is scarcely achieved in resting muscle (state 4), and not achieved in any of the vital organs—upon which life depends (state 3). However, a key parameter for evaluating mitochondrial health and disease is the velocity of oxidative metabolism in the steady state; by steady state we mean that the velocity of ATP synthesis is equal to the velocity of ATP breakdown. Claude Bernard first realized that the "milieu interieur fixé" is essential for life of the cell.[8]

The velocity of oxidative metabolism as identified by the equation:

$$3\,\mathrm{ADP} + 3\,\mathrm{P_i} + \mathrm{NADH} + \mathrm{H^+} + \tfrac{1}{2}\,\mathrm{O_2} = 3\,\mathrm{ATP} + \mathrm{NAD^+} + \mathrm{H_2O}, \qquad (1)$$

and is here assumed to be a reaction in which all substrates react independently in such a way that the overall velocity of oxidative metabolism could be controlled by their concentrations in relation to their Michaelis-Menten affinity.[9] While Michaelis himself did not envisage this "steady state" representation of his simple theory, Chance and Higgins made the necessary extension some time ago.[10] Thus, the simplified equation is

$$\frac{V}{V_m} = \frac{1}{1 + \dfrac{K_1}{\mathrm{ADP}} + \dfrac{K_2}{\mathrm{P_i}} + \dfrac{K_3}{\mathrm{O_2}} + \dfrac{K_4}{\mathrm{NADH}}}. \qquad (2)$$

It can be seen that if any one of the concentrations of the several substrates is equal to its K_m value, and all but one of the others are large, then control by that one component will reach only 50% of V_m. Thus, our rationale is to measure the V_m of systems affected with mitochondrial diseases, hypertension, or diabetes and to determine the extent to which V_m has been compromised. In this way, we propose to diagnose diseases of these three categories and, in a number of cases, propose therapies that can be followed continuously and noninvasively by the simple character of the P MRS examination.

As the equation stands above, only P_i is directly determined by P MRS. ADP can be determined by substituting for its value the parameters of the creatine kinase equilibrium, the value of K_1 as determined *in vitro,* the ATP concentration of the tissue (as determined by analytical biochemistry) and for a constant pH, the term, K_1/ADP becomes 0.60 $\times$ PCr/P_i; making the assumption that PCr breaks down into equal amounts of creatine and P_i. Furthermore, since the K_m for ADP is low compared to that of P_i, ADP is in control of mitochondrial respiration *in vivo.*[11] While P_i may control, it is supplementary to that of ADP (except in special cases of hypophosphatemia, as for example in PFK deficiency).[12]

The enzyme which is involved in the equation (**1**) above is found in the mitochondrion. As shown in FIGURE 1, this involves the translocase and the respiratory chain. Since no distinct rate-limiting step exists in the normally complemented mitochondrion, the deletion of any component in the sequential path from ADP to ATP via the respiratory chain and the associated ion and charge separations across the membrane will all be surveyed, PCr/P_i being directly compromised by a deletion of any one in this series of over 10 multi-subunit enzyme systems. It is for this reason, that the P MRS has very important diagnostic values.

If, however, the mitochondrial chain is functioning unsatisfactorily, where a genetic deficiency in the Embden-Myerhoff system or in fatty acid metabolism occurs, for example, then the NADH level will suffer and may be less than its K_m, causing only a fraction of the mitochondrial capability (V_m) to be available (Eq. (**2**)). Similar remarks occur in hypertension where vascular disease may fail to deliver adequate oxygen to the heart, the brain, or skeletal tissues as caused by or exacerbated by diabetes. Thus, we can evaluate the interplay of all substrates and enzymes involved directly in oxidative metabolism, and the supply of substrates to the mitochondrial system via the Embden-Meyerhoff or fatty acid pathways.

Validation of the Equation

Validation of the simple equation for ADP control in the presence of excess phosphate, substrate, and oxygen, has been done in studies of voluntary exercise of human skeletal tissue. V may be varied from less than 10% of maximum (resting value) up to 60 to 80% of maximum by voluntary exercise, and therefore the equation above can be tested, taking note that a steady state of metabolism must be achieved. The transient method employed by Radda gives a non-steady state evaluation of mitochondrial function.[13]

$$\frac{V}{V_m} = \frac{1}{1 + \dfrac{0.60}{P_i/PCr}}. \qquad (\mathbf{3})$$

The value of 0.60 ± 0.06 is obtained for a young experienced athlete training for the America's Cup competition. The V_m is 52 Joules/min characterizing his "mitochondrial respiratory system."

A MITOCHONDRIAL BYPASS

While our studies have embraced numerous examples of increases of mitochondrial activity in adult humans due to appropriate training regimens, this volume is focused upon pathological consequences of the loss of mitochondrial activities. We shall present a striking example of the use of P MRS to study a cytochrome *b*–deficient patient reported earlier in this volume by Dr. Nancy Kennaway together with Dr. Neil Buist who made this patient accessible to us on three occasions over the past four years. More details on the case history of the invasive study and other parameters are given in their report.[14,15] As indicated there, nearly all of the complex III activity had been deleted by a genetic defect that became prominent at age 9 and has continued up to the present age of 20 of the patient. The disease appears to be confined to the skeletal tissues and data presented from studies of the arm at rest and following mild exercise.

An Eel Model

In order to indicate the state of metabolism of a tissue devoid of mitochondria, we have employed the well known organ of the electric eel (*Electrophrous electricus*) studied in tissue slices in collaboration with others.[16] Independent studies indicate that

95% of the energy of the electric organ is derived from glycolysis and 5% from respiration. This organ serves as a good model for a patient where only 3% of the respiratory activity was observed to be retained in the excised tissue. FIGURE 2 displays the P MRS signature of the eel tissue approximately four hours after excision, and shows very large, highly significant P_i and PCr peaks, the ratio of which is slightly graded upon (the half-life of the excised slice is six hours) and the initial value of PCr/P_i is less than 3. The phosphocreatine values are in good agreement with those determined by analytical biochemistry.[17]

DIAGNOSIS OF A METABOLIC DISEASE

High phosphate and low PCr values would be expected to be diagnostic of deficient oxidative metabolism. In both the electric organ and in the human skeletal tissue, the only workload imposed is the maintenance of the tissue at rest. As mentioned above, the capability of intact mitochondria to generate a high phosphate potential at rest is expressed by $ATP/ADP \times P_i$ ratios of greater than 10^4 corresponding to PCr/P_i values of approximately 30.[18] FIGURE 3 illustrates a cytochrome *b*–deficient patient's arm (left), which shows a PCr/P_i value of only 1.6 (whether calculated by peak height or area determinations).

A Therapy

Based upon earlier work indicating that menadione (K_3) and ascorbate (C) can bypass the Antimycin-a block in mitochondria,[15] we have provided the patient with

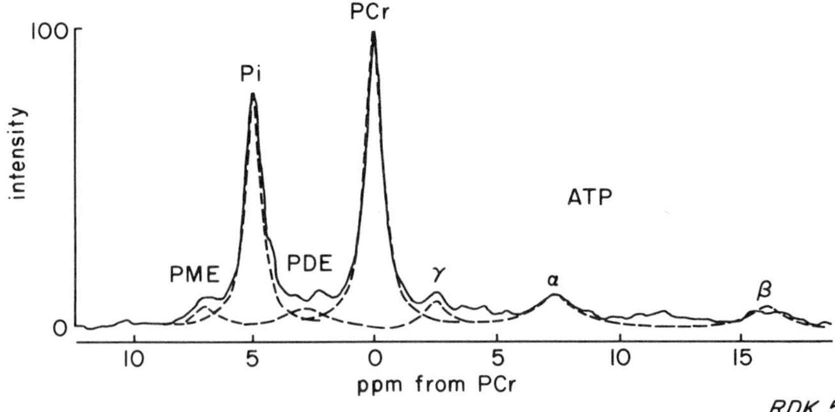

FIGURE 2. NMR spectrum of the organ of the electric eel (*Electrophrous electricus*) as an example of the phosphate potential of a predominantly glycolytic organ. The PCr/P_i ratio is 1.5 similar to that in the arm of a cytochrome *b*–deficient patient. The initial value of PCr/P_i of the organ extrapolated to time 0 is less than 3.

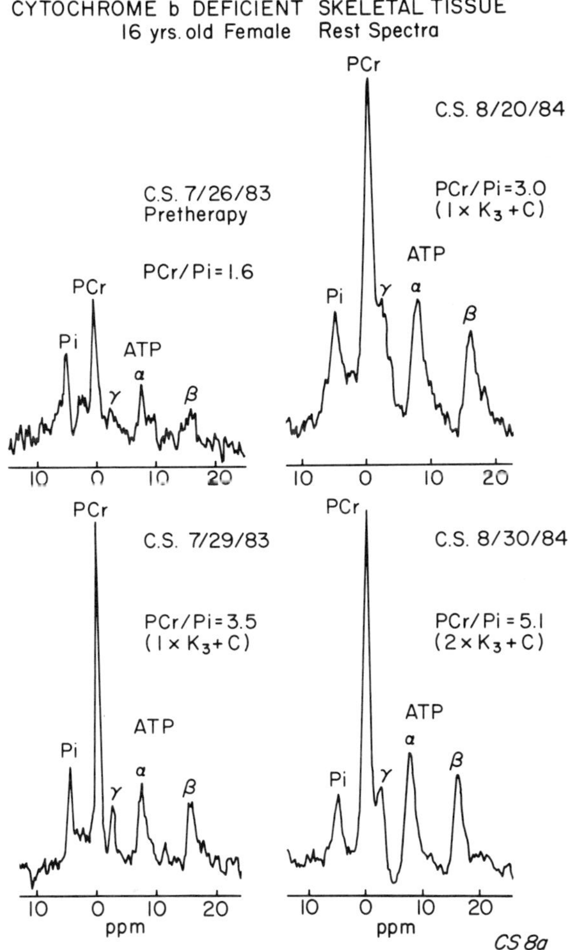

FIGURE 3. A series of P MRS studies of a cytochrome *b*–deficient muscle (upper left) with no therapy, immediately below and to the right with successive therapeutic treatments over the course of a year.

high levels of vitamins K_3 and C in order to determine whether the block could be bypassed *in vivo.* In fact, the PCr/P_i value rose to 3.5, was stabilized over a year at value of 3, and rose further to 5 and higher with a double the dose of vitamin K_3 a year later. This dose-response profile (FIGURE 4) shows a linear relationship for the three doses available. Furthermore, extrapolation to zero $C + K_3$ suggests that a significant activity was provided by what we postulate to be endogenous levels of $C + K_3$.

An Evaluation

FIGURE 5 indicates the rise of the calculated value of V_m from no dose to the single and double doses of the previous FIGURE 4 together with a profile for a normal control.

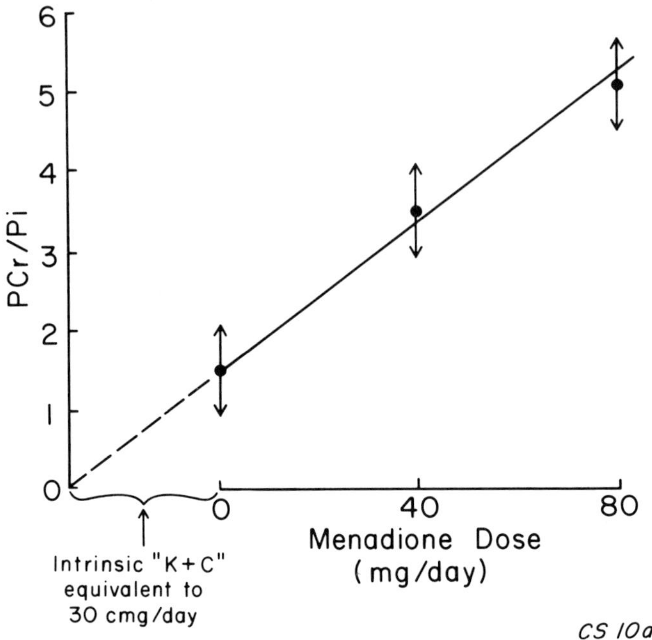

FIGURE 4. Dose/response profile for vitamins K_3 + C therapy of cytochrome *b*–deficient arm showing the lack of saturation with respect to the therapy. An extrapolation to 0 suggests a significant endogenous content of vitamin C + K_3.

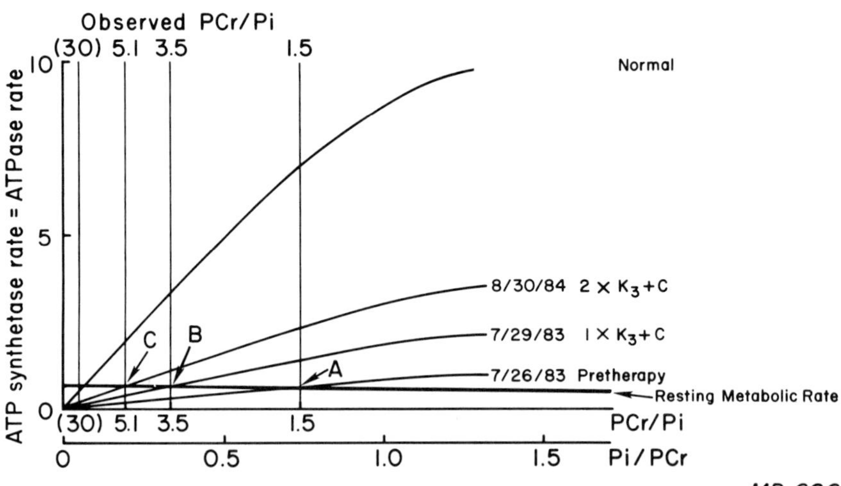

FIGURE 5. The use of the Michaelis-Menten equation to calculate the fraction of V_m available as a function of therapeutic dose. The fraction of V_{max} available is increased regularly with the increase of dose. At the metabolic operating point, P_i/PCr significantly decreases from A to B to C. Moreover, approximately 40% activity is recovered with respect to the normal control.

The results are consistent with FIGURE 3 above, no "saturation" of the effect of therapy has been obtained, and the interpretation of FIGURE 6 indicates that only about 40% of the total activity has been reconstituted.

A Progonosis

As shown by Dr. Kennaway, recovery from exercise with the mitochondrial bypass in operation is biphasic. A portion of the rise is 20 times faster than that of the untreated patient, but a second portion rises much more slowly to the initial value. We interpret this to mean that the size of the K_3/C pool available to mitochondrial oxidations following exercise is insufficient to afford a complete accelerated recovery.

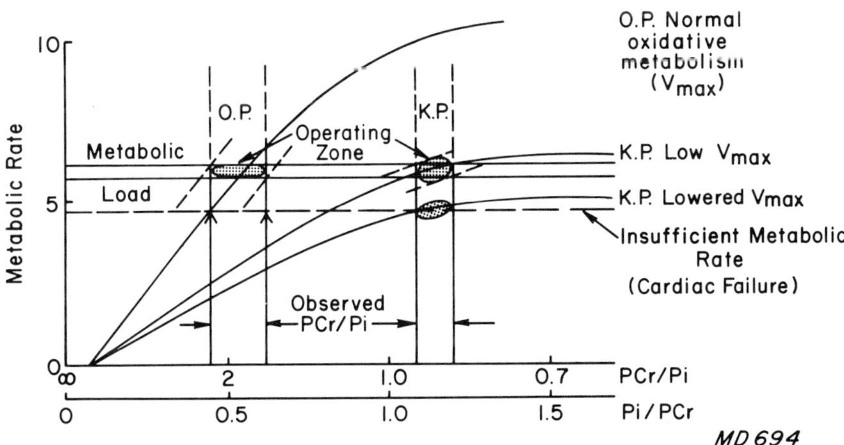

FIGURE 6. Michaelis-Menten analysis of experimental data from hypertrophic neonatal heart observed to have a P_i/PCr of 1.1 and estimated to be operating at 80% of V_{max}. A symptom-free sibling was observed to have a P_i/PCr of 0.5 and was in stable condition. A postulated lowering of V_{max} at the same metabolic load places the operating point beyond the V_{max} capability and the heart would unstabilize. Subsequently, cardiac failure occurred, in this case suggested by the approximation of the metabolic load to the V_{max} of the hypertrophic heart.

In order to increase the pool size available, the therapy has been changed from the water-soluble K_3 to the lipid-soluble K_1; the restoration of V_m was as effective as the single dose of $K_3 + C_1$; K_1 having been given at the same dosage. We endeavor to monitor the patient at least once a year and possibly every six months to ensure a stable restitution of a larger percentage of V_m, as is possible within the safe dosage for the patient.

CARDIOMYOPATHY IN NEONATES

The hypertrophic neonatal heart fills the chest cavity and is readily accessible to NMR studies through the thin sternum, particularly in ill, underweight babies having a cardiomyopathy. This pathological condition provides excellent NMR signals (ungated), characteristic of cardiac tissue without the need for lengthy studies involving various imaging or pseudo imaging techniques. With neonates, particularly those suffering heart disease, no "stress test" is possible to explore the V/V_m profiles to calculate the deficit of V_m in cardiomyopathy. Such data have been determined from animal models of adult dogs and neonate lambs indicating that the normal beagle dog heart operates at normal workloads with a low P_i/PCr value, often 0.2. We have proposed that a multiplicity of control mechanisms afford a remarkable homeostatis of PCr/P_i including microvascular control,[1] and as Denton and others point out, substrate control as well.[19,20]

Diagnosis of Two Cases of Cardiomyopathy

The first case of cardiomyopathy we have studied was referred to us by Dr. Hans Bode of MGH[21] and the NMR spectrum shows a pronounced P_i peak in the spectrum and a high P_i/PCr, atypical of age-matched controls and of a sibling who was asymptomatic.

The Michaelis-Menten analysis,[9] similar to that employed for the cytochrome *b*–deficient patient, is illustrated in FIGURE 7. In this case the operating point obtained

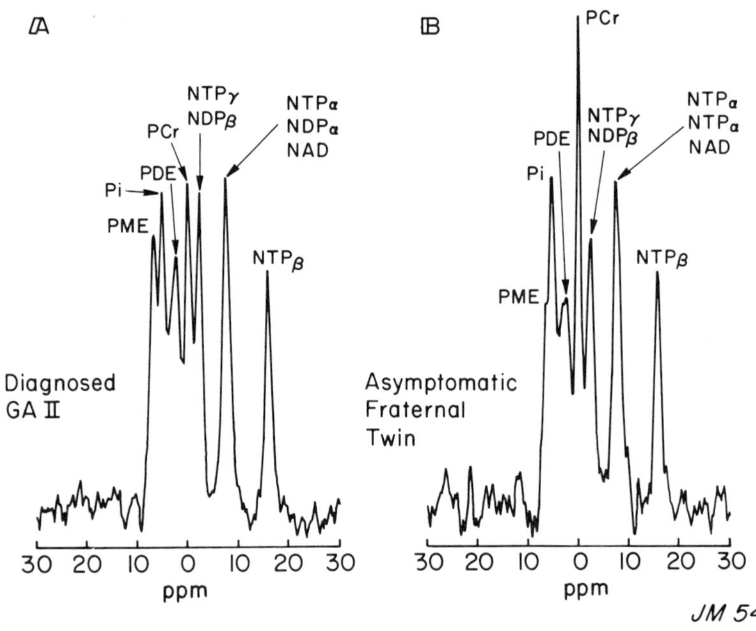

FIGURE 7. Comparison of MRS from diagnosed glutaric urea type II (A) with spectra of an asymptomatic fraternal twin. The PCr/P_i ratio of the patient with cardiac hypertrophy is inferior to that of the normal twin. The time course of variations of PCr with therapy are indicated in FIGURE 8.

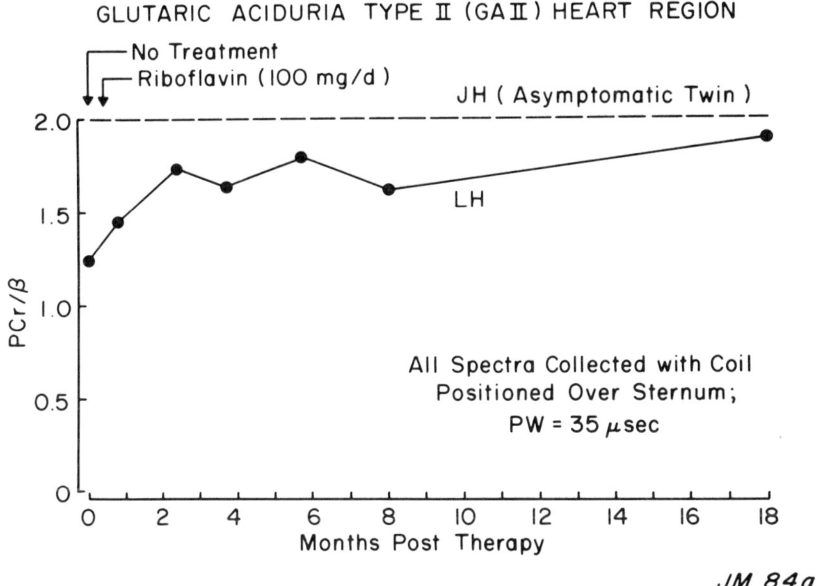

FIGURE 8. Riboflavin therapy of suspected deficiency of FAD in glutaryl CoA dehydrogenase. The progress of the disease has been followed by PCr/ATP_B measurements over 18 months using as an asymptomate, the twin's initial value.

with P_i/PCr value of greater than 1 is indicated to be very near the V_{max} available as compared with the sibling whose P_i/PCr value was less than half that of the diseased heart and whose operating point, based on the assumption of the same workload corresponded to a V/V_m of roughly 0.5. A further decrease of P_i/PCr indicated in the diagram as a supposed progress of the disease further raises V/V_m to the point where little further reserve capacity of the cardiac tissue is available. A month after the P MRS study, cardiac failure occurred. The sibling whose P_i/PCr was 0.5 has remained stable.

The second case is glutaric aciduria type II, a hypertrophic cardiomyopathy, studied in collaboration with Dr. Richard Kelley and John Maris.[22] As in the previous study, an asymptomatic was also studied (FIGURE 8). The spectra was taken in the same manner as those in the previous study but with a somewhat improved technique (two years had elapsed), showing a higher P_i and a lower PCr in the diseased cardiac tissue than in the normal twin. A dosage of 100 mg per day of riboflavin, to a point where riboflavin sufficed as judged from urine analysis, leads to an improvement of the PCr/P_i ratio towards that of the sibling over the roughly 18 months of observation (FIGURE 8).

An Interpretation

In this case of glutaric aciduria Type II, a preliminary diagnosis suggested a deficiency of the prosthetic group of the glutaryl CoA dehydrogenase leading to a deficiency of citric acid cycle components and a substrate-deprived cardiac tissue. In this case, regulation of the V_m is by the NADH term of the general equation for sufficient P_i and oxygen:

$$\frac{V}{V_m} = \frac{1}{1 + \frac{0.06}{P_i/PCr} + \frac{K_4}{NADH}}. \tag{4}$$

Increased activity of the delivery of substrate would increase the PCr as observed in FIGURE 8 below.

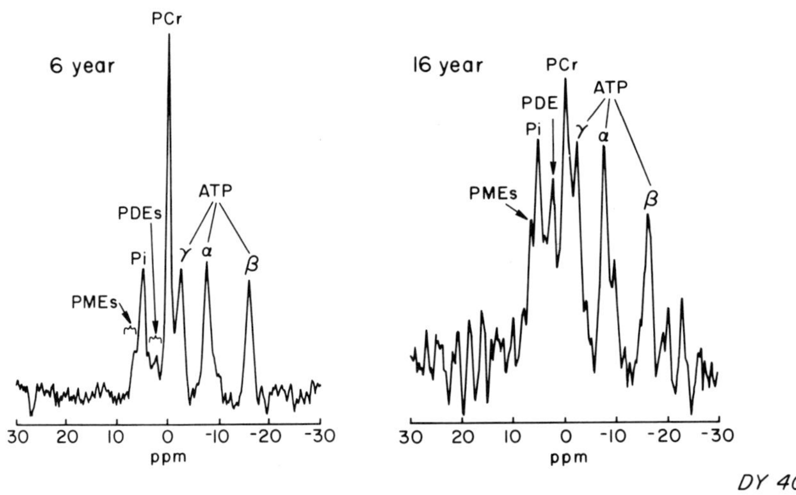

FIGURE 9. Illustrating the lowering of PCr and the rise of P_i in two cases of Duchenne's dystrophy as studied by P MRS of two patients with diagnosed Duchenne's dystrophy of 6 and 15 years. Normal values of PCr/P_i for the 15-year-old patient are approximately twice as high as that for the 6-year-old patient.

Therapy

In view of the glutaric acid urea, the lack of FAD at glutaryl CoA dehydrogenase was suspected and 100 mg/day supplement of riboflavin was given with the progressive improvement of PCr (referenced to ATP as shown in FIGURE 8).

DUCHENNES DYSTROPHY

In ongoing studies of Duchenne's dystrophy, we have observed an increasing P_i/PCr ratio between 6 to 16 years (FIGURE 9). Furthermore, the P_i/PCr value at the youngest age measured (6 year old) is half that of the age-matched normal control.[23]

The P MRS indication of bioenergetic insufficiency is attributable to damage to the mitochondria or to an insufficiency of oxygen or NADH. The progression of the energy deficiency with age is an interesting possibility that merits further study.

HYPOXIA

Equally important to the study and diagnosis of metabolic disease is the use of P MRS in the evaluation of tissue hypoxia where the factor controlling the availability of V_m is oxygen delivery

$$\frac{V}{V_m} = \frac{1}{1 + \frac{0.06}{P_i/PCr} + \frac{K_3}{O_2}}. \tag{5}$$

This equation has been particularly effective in the study of pre- and full-term neonates and detailed examples of this are presented elsewhere.[24]

SUMMARY

The incisive detection of bioenergetic insufficiency in an organ of known workload by P MRS is noninvasive and nondestructive, and in some cases the portion of the organ involved can be determined, particularly if both PCr and ATP are depleted. The fractional loss of ATP and hence the relative volumes of viable and "metabolically dead" tissue are thereby evaluated.[25] In addition, the value of P MRS in following a therapy complements its value in diagnosis as this has been demonstrated in cases followed over 6 months to three years.

The fact that deficiencies of the enzymes and substrates of oxidative metabolism can be detected by P MRS affords a global overview of energy metabolism that can be a key to rapid diagnosis. The distinction of the enzyme and/or substrate deficiency, while not directly indicated by steady state P MRS, can be identified by use of the "Crossover Theorem"[26] and its impact upon blood and tissue levels of substrates (including oxygen). In the case of neonatal systemic hypoxia, there is no doubt about which of the equations applies, and similarly in metabolic disease, a glutaric acid urea is a direct consequence of the crossover response of metabolism and signifies that an enzyme deficiency may be involved.

Furthermore, the clinical danger of a high P_i/PCr value is clarified by our observations, both from the animal models and from the theory, the high values; i.e. 2 and over, suggest work stresses near the capability of oxidative metabolism and imminent failure of the negative feedback afforded by metabolic regulation, particularly ADP control of oxidative metabolism. This control is lost because of the fall of phosphocreatine to the point where creatine kinase is no longer in equilibrium, leading to the loss of ATP and its conversion to large amounts of ADP and its breakdown products. ATP then stimulates glycolysis and results in a massive lactic acidosis. At the same time, the low thermodynamic capability of glycolytic metabolism is unable to prevent irreversible ion disequilibration, water movements, edema, and eventually rupture of the cell membrane. The pathway of resynthesis of ATP is then tortuous, particularly as AMP is deaminated and adenosine is converted eventually to hypoxanthine. Thus, NMR reports that metabolic control is operating in the region where homeostasis of biochemical parameters is feasible. It further reports regions where the metabolic control is susceptible to failure and most aggressive clinical care is required.

ACKNOWLEDGMENT

Phospho-Energetics Inc. provided the NMR equipment used in these studies.

REFERENCES

1. CHANCE, B., J. S. LEIGH, B. J. CLARK, J. MARIS, J. KENT, S. NIOKA & D. SMITH. 1985. Proc. Natl. Acad. Sci. USA **82:** 8384–8388.
2. CHANCE, B., J. S. LEIGH, JR. & Z. ARGOV. 1985. *In* News of Metabolic Research. J. S. Leigh, Jr., Ed. **1**(3): 25–27. University of Pennsylvania. Philadelphia.
3. CHANCE, B. & J. S. LEIGH, JR. 1985. *In* News of Metabolic Research. J. S. Leigh, Jr., Ed. **1**(4): 26–33. University of Pennsylvania. Philadelphia.
4. CHANCE, B., J. S. LEIGH & S. NIOKA. 1985. *In* News of Metabolic Research. J. S. Leigh, Jr., Ed. **2**(1): 26–31. University of Pennsylvania. Philadelphia.
5. CHANCE, B. & G. HOLLUNGER. 1960. Nature **185:** 666–672.
6. KLINGINBERG, M. & P. SCHOLLMEYER. 1961. Biochem. Biophys. Res. Commun. **4:** 5.
7. MEYER, R. A., M. KUSHMERIC & T. BROWN. 1982. Am. J. Physiol. **11:** C1–C11.
8. BERNARD, C. 1878. Les Phenomenes de la Vie. Vols. 1 & 2. Paris.
9. MICHAELIS, L. & M. MENTEN. 1913. Biochem. Z. **49:** 333.
10. CHANCE, B. & J. HIGGINS. 1952. Arch. Biochem. Biophys. **41:** 432–441.
11. CHANCE, B. & G. WILLIAMS. 1955. J. Biol. Chem. **217:** 383–393.
12. CHANCE, B., S. ELEFF, W. BANK, J. S. LEIGH, JR. & R. WARNELL. 1982. Proc. Natl. Acad. Sci. USA **79:** 7714–7718.
13. ARNOLD, D. L., P. M. MATTHEWS & G. K. RADDA. 1984. Mag. Res. Med. **1:** 307–315.
14. CAPALDI, R. A., S. TAKAMIYA, W. YANAMURA, N. KENNAWAY & N. BUIST. 1986. Ann. N.Y. Acad. Sci. (This volume.)
15. ELEFF, S., N. G. KENNAWAY, N. R. M. BUIST, V. M. DARLEY-USMAR, R. A. CAPALDI, W. J. BANK & B. CHANCE. 1984. Proc. Natl. Acad. Sci. USA **81:**3529–3533.
16. CHANCE, B., R. D. KENYES & P. MUELLER. 1986. Soc. Mag. Res. Med. (Montreal). **4:** 1370–1371.
17. MAITRA, P. K., A. GHOSH, B. SCHOENER & B. CHANCE. 1964. Biochim. Biophys. Acta **88:** 112–119.
18. GYULAI, L., Z. ROTH, J. S. LEIGH, JR. & B. CHANCE. 1985. J. Biol. Chem. **260:** 3947–3954.
19. DENTON, R. M. & J. G. MCCORMACK. 1981. Clin. Sci. **61:** 135–140.
20. BALABAN, R. S., H. L. KANTOR, L. A. KATZ & R. W. BRIGGS. 1986. Science **232:** 1121–1123.
21. WHITMAN, G. J. R., B. CHANCE, H. BODE, J. MARIS. J. HASELGROVE, R. KELLEY, B. J. CLARK & A. H. HARKEN. 1985. J. Am. Col. Cardiology **5:** 745–749.
22. KELLEY, R., B. J. CLARK, J. M. MARIS & B. CHANCE. 1986. P NMR Spectroscopy of Pediatric Cardiomyopathy.
23. YOUNKIN, D. P., P. BERMAN, J. SLADKY, C. CHEE, W. BANK & B. CHANCE. 1986. ^{31}P NMR Studies in Duchenne Muscular Dystrophy: Age Related Metabolic Changes. Neurology. (In press.)
24. LIEN, R. I., T. SINNWELL, B. CHANCE & M. DELIVORIA-PAPADOPOULOS. 1986. Soc. Mag. Res. Med. (Montreal) **2:** 293–294.
25. CHANCE, B., S. NIOKA, D. SMITH & J. S. LEIGH, JR.. 1986. Soc. Mag. Res. Med. (Montreal) **4:** 1372–1373.
26. CHANCE, B., G. R. WILLIAMS, W. F. HOLMES & J. HIGGINS. 1955. J. Biol. Chem. **217:** 439–451.

DISCUSSION OF THE PAPER

S. DiMauro (*Columbia University, New York*): (1) Did your NMR data in Duchenne patients suggest a primary mitochondrial alteration in Duchenne dystrophy?

(2) Did you study very young patients?

(3) Did you study obligate carriers?

Chance: Your first and third questions are closely linked; and at present we have neither a carrier study nor a reason to suppose a mitochondrial change on the primary event but it surely is a topic of intense research. Our youngest patient is six years. The PCr/Pi versus age profile gives 5 at this point, about half the normal control value, suggesting detectability at younger ages.

M. Klingenberg (*University of Munich, Munich*): (1) What are the causes for the instability observed in heart if the P_i/PCr ratio reaches more than 0.5? Is the high P_i content partially accumulated into mitochondria and does it impair substrate uptake into mitochondria?

(2) Is, in contrast to heart with high degree of aerobic metabolism, the high P_i/PCr ratio in the electric organ more easily reversed, since the primary glycolytic phosphorylation is not sensitive to the complications of mitochondrial function?

Chance: (1) At P_i/PCr of 1–2 the V/V_m is 70–80% and ADP control is "loose" and glycolysis is stimulated, an unstable state where increased heart work can lead to heart failure.

(2) The phosphate potential of glycolysis of the electric organ "run down" continuously *in vitro;* the Na^+ leak at rest always exceeds the ATP synthesis *in vitro.* The *in vivo* state needs to be studied.

J. P. Colombo (*University of Berne, Berne, Switz.*): From your ^{31}P NMR data on impaired oxidative metabolism can you, e.g. in the case of cytochrome oxidase deficiency, draw a conclusion on the intracellular pH?

Chance: In cytochrome *b* deficiency, significant lactate was found in the blood (~10 mM). However, the muscle did not show a significant acidosis, possibly due to enhanced lactate efflux.

G. Azzone (*University of Padova, Padova, Italy*): To what extent does your conclusion depend on the relative volumes of the various cellular compartments?

Chance: Edema would dilute the cytosol with the total moles of, for example, phosphocreatine would be unaltered. We have measured relaxation times of PCr and P_i in normal and in hypermetabolic stress and found no significant change.

N. Siliprandi (*University of Padova, Padova, Italy*): Do you have some information on the PCr content in muscles of trained sprinters and long distance runners at rest?

Chance: Yes, our assay of white fibers shows the expected larger PCr over that of the red muscles. We are studying exercise protocols that will better differentiate the two fiber types.

Biological Membranes in Hypertension: Is Control of Intracellular Calcium and Other Ions Mediated by a Membrane Defect?

L. H. OPIE[a] AND D. A. DAVEY

Hypertension Clinic and Heart Research Unit
Department of Medicine;
Reproductive Medicine Research Unit
Department of Obstetrics and Gynecology;
University of Cape Town
Observatory, South Africa

IS THERE A GENERALIZED MEMBRANE DEFECT IN HYPERTENSION?

A prominent pathophysiological alteration in systemic hypertension is the increase in peripheral arterial tone and peripheral vascular resistance. Could this change be the ultimate result of a widespread membrane defect as proposed by several groups including Postnov and Orlov[1] or due to an inherited abnormality of membrane sodium transport, as proposed by Garay *et al.*?[2–4] These apparently simple questions seem to require complex answers, especially because hypertension is multifactorial in origin and not just a straightforward disease; the etiology of hypertension may reflect different genetic mechanisms in some patients, while in others environmental factors may dominate. A further complexity is the important difference between a marker and a mechanism (FIG. 1). The various abnormalities of ion transport and contents found in a variety of models of hypertension could be either genetic markers[5,6] of subpopulations of hypertensives or alternatively part of the basic mechanism involved in hypertension. As most studies on ion content and transport have been done on blood cells, a further pertinent issue is the relation of findings on those cells to those in vascular smooth muscle cells, which are the ultimate target of the hypertensive abnormalities.

A generalized membrane defect could be expected to lead to high-sodium, low-potassium, and high-calcium intracellular changes by virtue of "leaks" across the membrane. Alternatively, there could be specific defects, for example of sodium transport, including the sodium pump and the sodium countertransport systems.[4] The question of a generalized membrane defect could in part be answered if values of sodium, potassium, and calcium in vascular smooth muscle were known. Vascular cells are difficult to analyze and we have little firm information on the concentration of these ions in vascular cells in human or experimental hypertension. There is even less information about the ionized values (i.e. the intracellular activity) of these cations. However, several other types of non-vascular cells have been analyzed including white cells, red cells, and platelets (TABLES 1–4), and it is such tissues that have been most widely studied.

[a]Address correspondence to: L.H.O., Heart Research Unit, University of Cape Town, Medical School, Observatory 7925, South Africa.

THE ULTIMATE TARGET—VASCULAR SMOOTH MUSCLE

The thus far consistent finding of an increased ionized calcium, usually measured indirectly by Quin 2, in blood cells of hypertensive animals or man (TABLE 1) is a useful lead, without directly proving that there is an increased cytosolic calcium available to the contractile apparatus of vascular smooth muscle cells. Only one report examines ionized calcium in vascular smooth muscle, and cultured cells were used that need not necessarily reflect the situation *in situ*. If the data available can be interpreted as showing a real increase of cytosolic calcium for contractile purposes, then there must

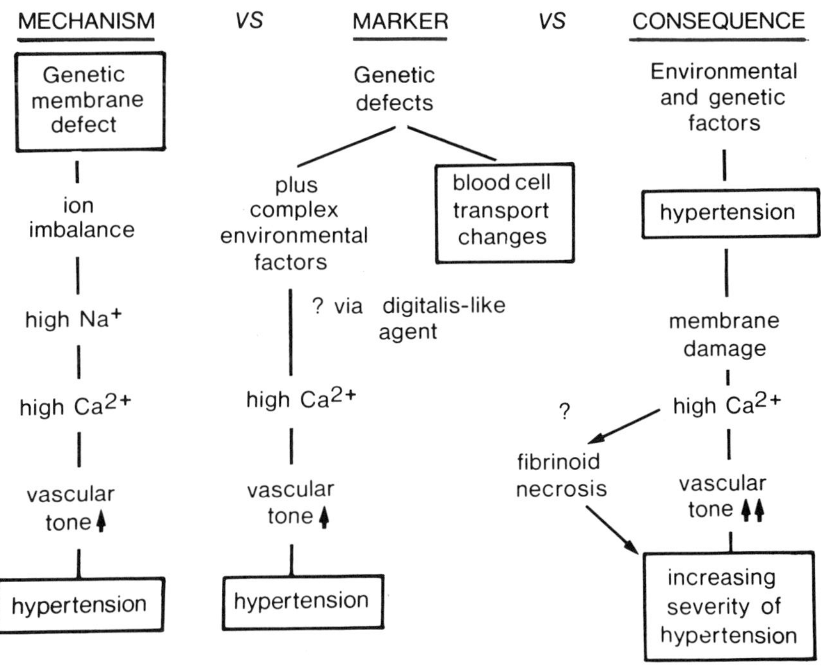

FIGURE 1. Proposed schema for multifactorial events underlying development of hypertension.

be (1) increased entry of calcium by channels or exchange mechanisms (Na/Ca), (2) downhill leaks from without to within the vascular cell, or (3) a decreased exit of calcium by depressed activity of the Ca^{2+}-ATPase pump of the sarcolemma or the Na/Ca exchanger. Of the above, only the Ca^{2+}-ATPase has been directly measured. In patients, there is some indirect evidence for increased calcium entry mediated by the calcium channel.[7] It is postulated that an increased vascular smooth muscle calcium could be associated with an increased internal sodium (the two being interrelated by the Na/Ca exchanger.[8] However, there are few direct measurements of both sodium and calcium ions on vascular smooth muscle cells (TABLE 1).

TABLE 1. Calcium Levels (Free or Ionized) or Indices of Calcium Metabolism in Tissues from Hypertensive Animals or Man

Tissue	Source of Tissue	Change Found	Interpretation/Comment	Reference
Free Ca (Quin 2)				
Platelets	Man	+	Rises with BP and falls with therapy by beta-blocker, Ca antagonists, or diuretics	Erne *et al.*[52]
Platelets	Man	+	External Ca required to show increase versus normals	Le Quan Sang *et al.*[5]
Platelets	Man	+	Slight rise of Ca with BP rise (with wide overlap)	Lenz *et al.*[51]
Lymphocytes	Milan HR	+	Associated with low cell Na	Bruschi *et al.*[60]
Free Ca (ionized)				
RBC	Man	+	Associated with low cell Na	Zidek *et al.*[11]
VSM, RBC	SHR	+	Elevated Ca but not Na disappears on culture and therefore due to "humoral" factors	Losse *et al.*[61]
RBC	Man	+	Ca changes associated with hypertension, Na changes ? genetic	Zidek *et al.*[62]
Ca ATPase				
RBC	Milan HR	–	Explains high Ca	Vezzoli *et al.*[63]
Arteries	SHR	–	Explains high Ca	Kwan and Grover[64]
Platelets	Man	–	Blunted calmodulin stimulation	Resink *et al.*[65]
Ca binding/accumulation				
RBC	SHR	–	Inside-out membrane vesicles	Devynck *et al.*[66]
Heart	SHR prehypertensive	–	Differences vs. control eliminated by calmodulin	David-Dufilho *et al.*[38]
RBC, heart	SHR	–	Found in prehypertensive rats ? genetic	Robinson[7]
Ca uptake				
Brain synaptosome, platelets	SHR	+	Increased Na^+ permeability in primary hypertension "type of generalized membrane pathology"	Postnov *et al.*[67] and Orlov and Postnov[68]

NOTE: (+) = increase and (–) = decrease.

Calcium and Vascular Tone

The significance of even small degrees of elevation of cytosolic calcium, achieved by any of several mechanisms to be considered, is that an increased vascular smooth muscle tone should follow. Increased entry of vascular smooth muscle calcium levels could be achieved by an enhanced slow inward calcium current, either activated by depolarization or by receptors such as those of the α-receptor. Secondly, enhanced entry of calcium could be achieved by the activity of the Na/Ca exchanger, as discussed later. Thirdly, increased entry could result from non-specific leaks. The role of the Na/Ca exchanger has been emphasized by Blaustein.[8,9] Because of the large concentration gradient between extra- and intracellular calcium in vascular smooth muscle (TABLE 2), a small "leak" (genetic, hypertension-induced, or exaggerated by hypertension) could lead to a critically important increase in tone. However, the properties of the membrane itself with its channels and pumps could be completely unaltered, with enhanced calcium entry or decreased calcium exit achieved by altered regulation by altered activity of the normal regulatory mechanisms.

TABLE 2. Concentrations of Sodium and Calcium in Smooth Muscle Cells

Ion	Extracellular Value (ionized) (mmol/l)	Smooth Muscle Preparation	Value (mmol/l)	Reference
Na^+	140	Mesenteric artery	11	Mulvany *et al.*[45]
		Cultured aortic cells[a]		
		Normal rats	3	Losse *et al.*[61]
		SHR rats	15	Losse *et al.*[61]
		Taenia coli	22	Aaronson and Van Breemen[41]
Ca^{2+}	1.25	Cultured aortic cells[a]		
		Normal rats	0.016	Losse *et al.*[61]
		SHR rats	0.121	Losse *et al.*[61]

[a]May not reflect physiological conditions.[69]
SHR = spontaneously hypertensive rat.

SODIUM AND POTASSIUM CONTENTS OF BLOOD CELLS

The sodium concentration of blood cells is an important parameter because of the hypothesis that cellular sodium overload is indirectly related to excess dietary sodium; for example, Blaustein[8,9] stresses the links between an increased intracellular sodium and calcium by the Na/Ca exchanger. In many studies, blood cell sodium is increased (TABLE 3) yet there are sufficiently many opposing results with decreased values to question any conclusions based on sodium content alone. Some of the conflict may be explained by diverse methodologies used, for example the new "*in vivo*" method of Simon and Conklin,[10] using rapid separation techniques, is in contrast to most other methods requiring prolonged treatment of cells before analysis. Nonetheless, the data of Zidek *et al.*,[11] showing a reduced intracellular sodium activity that normalized with antihypertensive therapy, were obtained with apparently standard methods. Thus it is difficult to base any hypothesis solely on the conflicting data presented in TABLE 3. If the Na/Ca exchanger were a critical link in the chain of development of hypertension, then the cellular sodium must be increased, at least in vascular smooth muscle. It is,

TABLE 3. Ionic Composition (Na, K, Cl) of Tissues from Hypertensive Animals or Man

Tissue	Source	Na^+	K^+	Cl^-	Comment	Reference
Sodium increased						
Artery, renal	Man	+	0	?	Higher water content; not expressed in concentration	Tobian and Binion[70]
Artery, caudal	Rat	+	?	?	Decreased transmembrane K^+ (inferred) gradient leads to depolarized muscle responding more readily to NE	Hermsmeyer[50]
WBC	Various	+	?	?	"Convincing evidence" for sodium pump abnormality	Hilton[15]
RBC	SHR	+	Modest –	?	"Enhanced membrane permeability"	Postnov *et al.*[67]
Platelets	Man	?	?	?	Platelet Ca^{2+} rises with BP and falls with therapy by beta-blocker, Ca antagonist, or diuretic	Erne *et al.*[52]
VSM, RBC	SHR	+	?	?	Elevated Ca disappears on culture and therefore regarded as due to "humoral factors"	Losse *et al.*[61]
RBC	Man	+	0	?	Decreased Na efflux	Lijnen *et al.*[71]
RBC	Man, FH	+	?	?	Related to low Na^+ efflux	Gudmundsson *et al.*[72]
RBC	Man, FH	+	?	?	Related to enhanced Li efflux	Clegg *et al.*[55]

TABLE 3. *Continued*

Sodium decreased						
Proximal tubular cells	Milan HR	[SB	0	–	"One cannot expect to draw a simple unifying hypothesis"	Thurau *et al.*[73]
RBC	Milan HR	–	?	?	Erythrocytic volume decreased	Ferrari *et al.*[57]
RBC	Man	–	+	?	New "*in vivo*" method with rapid separation	Simon and Conklin[10]
RBC	Man	–	0	?	Na activity rose with treatment of hypertension	Zidek *et al.*[11]
Sodium unchanged						
Aorta	Rat	0	0	0	Associated with increased K turnover	Jones[74]
RBC	Man, FH	0	–	?	Na not genetic marker of hypertension	Semplicini *et al.*[75]

SHR = spontaneously hypertensive rat; WBC = white blood cells; RBC = red blood cells; ? = no data; + = increase; – = decrease; 0 = no data; BP = blood pressure; HR = hypertensive rat; and FH = family history of hypertension.

however, important to look at the sodium value of blood cells in a dynamic way, as reflecting the balance of the effects of pumps, exchange mechanisms, and leaks.

The potassium concentration of blood cells is seldom measured. As in the case of sodium, it is the kinetic approach to the handling of potassium that is more likely to be of interest. In two studies, the potassium content was unchanged and in a third it increased (TABLE 3). In arterial muscle, the intracellular potassium activity has been inferred from changes in the transmembrane resting potential. Thus, in the spontaneously hypertensive rat, there may be an abnormally low intracellular potassium in the arterial muscles.[12] As a result, it is proposed that there is a smaller potassium electrochemical gradient causing the resting membrane potential to be less electronegative. In response, there is thought to be an increased electrogenic ion transport (enhanced sodium pump activity) that results in an ultimate resting membrane potential not much altered.[12] Such "compensated defects" are difficult to disprove.

Membrane Fluidity

In red blood cells, a direct relation between ion transport and membrane fluidity has been found. Cholesterol enrichment of red cell membranes leads to inhibition of sodium-potassium cotransport.[13] The data of Heagerty *et al.*[14] suggest that dietary manipulation of membrane fluidity could likewise lead to altered sodium transport rates in the white cells.

SODIUM AND POTASSIUM TRANSPORT SYSTEMS IN BLOOD CELLS

Whereas the sodium pump is energy-dependent, a number of other non–energy-dependent systems help to regulate intracellular sodium and potassium of isolated blood cells.[3,15] These include the sodium-lithium countertransporter,[16] the sodium-potassium cotransporter,[2] the sodium-sodium system,[17] and (in intestinal cells) the sodium-glucose coporter. As there is no evidence that any of these play a role in ion regulation in vascular smooth muscle, a consideration of their possible abnormalities in hypertension may be of more interest as a marker rather than as a fundamental mechanism; several ways of viewing this problem are shown in FIGURE 1.

Sodium-Potassium Cotransport

Garay and Meyer[18] and several others have found a decreased Na/K cotransport system in red blood cells from hypertensive rats or man (TABLE 4). When conditions thought to be closer to the physiological were employed, cotransport was increased.[19] In normotensive patients with diabetes mellitus, sodium-potassium transport is also decreased, showing the lack of specificity of this finding for hypertensives.[20] In Milan hypertensive rats, high cotransport values are linked with low intracellular sodium values in red cells, and a similar finding occurs in the offspring of hypertensive patients.[21] The data of Canessa[22] suggest that more subtle kinetic parameters of cotransport activity such as K_m and V_{max} in response to calcium, need measurement and that cotransport in some hypertensives is sensitive to the sodium content of the diet. In this complex area, the techniques used to measure the transport systems may be critical.[19] Thus data without full kinetic parameters such as those of our group in Cape Town[23] may have somewhat limited significance. Nevertheless it is of interest that cotransport, measured in a standard way, varied much more between monozygotic twins than between hypertensives and non-hypertensives (FIG. 2). Complex genetic changes cannot be excluded, nor can genetic polymorphism. For example, Garay *et al.*[4] were able to show a subgroup of patients with altered cotransport values.

Renal Tubular Cells

Renal alterations may be critical in causing excess sodium reabsorption to help maintain or initiate a state of generalized sodium overload. Bianchi's group[21] has compared the properties of renal tubular cells and red blood cells in their Milan hypertensive rat model. In both types of cells, decreased total sodium content was coupled with increased cotransport activity; possibly the latter change was significant in "sensitizing" the kidney of hypertensive rats to sodium loading. Similar detailed studies of sodium handling by renal tubular cells in other models and in man are required.

Ethnic Differences in Hypertension

One hypothesis is that hypertension in black ethnic groups may be more severe than in Caucasians, and is less amenable to therapy by β-adrenergic blockade, suggesting that a "low renin" type of hypertension prevails. Hence the role of the circulating sodium pump inhibitor could also be more prominent[24] so that greater ionic changes could perhaps be expected in red cells of black hypertensives. Etkin *et al.*,[25] measuring passive influx of sodium in ouabain-inhibited cells, found (unexpectedly) that white but not black American hypertensives had increased permeability of red cells to sodium. Davidson *et al.*[23] found that cotransport, studied without kinetic parameters, was no different between white and black South African hypertensives. Canessa *et al.*[26] studied red cell countertransport and cotransport in black American hypertensives and suggested differences from whites although there were no direct comparisons. Of related interest is the decreased ouabain-sensitive rubidium uptake (taken as an index of sodium pump activity) in normotensive British blacks; hypertensives were not reported.[27] It therefore seems that further strict comparisons between white and black populations are urgently needed to settle this important question.

SODIUM-LITHIUM COUNTERTRANSPORT

Sodium-lithium countertransport, studied on red cells, may reflect Na/Na exchange or Na/H exchange. Sodium-lithium countertransport may also be increased in some human hypertensives,[26] a change that may be associated with increased sodium-potassium cotransport.[4] If there were a sodium-proton exchange in vascular smooth muscle (not yet established), sodium ion activity could increase inside vascular smooth muscle cells[28] resulting in an increased internal sodium (see section on Na/Ca exchange).

MEMBRANE LEAKS

Besides the cotransport, countertransport, and sodium-potassium pump systems, sodium may leak into blood cells through the membrane. Red cells of different species have widely different contributions of the leak system, cotransport, and sodium pump

TABLE 4. Indices of Ionic Fluxes (Na, K, and related) in Tissues from Hypertensive Animals or Man

Tissue	Source of Tissue	Ions Involved	Change Found	Interpretation/Comment	Reference
Cotransport decreased					
RBC	Man	Na^+ K^+	—	Proposed genetic marker of hypertension	Garay *et al.*[2] and Meyer *et al.*[76]
RBC	Man	Na^+ efflux	Variable-	Big variation between two laboratories ? genetic polymorphism	Dagher and Canessa[56]
RBC	Man	Na^+ efflux	—	Family history of hypertension—big fall in cotransport	Lijnen *et al.*[71]
RBC	Man	Na^+	Increased $(Na^+)_i$ for ½ max	Abnormal cotransport in 46/65 patients	Garay *et al.*[3]
RBC	Man/Black	Na^+/K^+	Decreased	Reduction in V_{max} of Na and K cotransport	Canessa *et al.*[26]
Cotransport increased or unchanged					
RBC	Man	K^+ influx	Increased	"Conditions closer to physiological" than Garay *et al.*	Talib *et al.*[19]
RBC	Man	Na^+ K^+	Slight −, not significant	Changes in hypertension not as important as genetic changes	Davidson *et al.*[23]
RBC	Man	Na^+	Increased	Furosemide-sensitive Na^+ efflux	Stokes *et al.*[17]
Na efflux decreased or unchanged					
WBC	Man	Na^+	Decreased	In FH normotensives. Abnormality not direct cause of hypertension.	Heagerty *et al.*[14]
Thymocytes	SHR	Na^+	None, only with age	No effect of high salt diet on total or ouabain-sensitive Na^+ efflux. "Altered membrane transport may not participate directly in pathogenesis of hypertension."	Swales *et al.*[6]
RBC	Man, FH	Na^+	Decreased	Efflux increases to normal with high salt diet	Gudmundsson *et al.*[72]
RBC	Man	Na^+	Decreased	Ratio Na^+/K^+ flux reduced	Garay and Meyer[18]
Na influx increased or unchanged					
RBC	Man, FH	Na^+	Unchanged	Normal cell permeability to Na_i, no change with high salt	Gudmundsson *et al.*[72]
RBC	Man White/black	Na^+	Increased in white; unchanged in black	Probable genetic difference; ouabain-inhibition, therefore RBC Na-permeability measured	Etkin *et al.*[25]
RBC	Man	Na^+	Increased	Both furosemide-sensitive and resistant components increased	Stokes *et al.*[17]

TABLE 4. *Continued*

Tissue	Source of Tissue	Ions Involved	Change Found	Interpretation/Comment	Reference
Na pump enhanced					
Heart	SHR pre-hypertensive	Na^+/K^+ ATPase	Enhanced	In prehypertensive stage abnormal	David-Dufilho *et al.*[38]
RBC	Man	Rubidium	Enhanced	Also in some normotensives with FH	Woods *et al.*[27]
RBC	Man	Na^+	Enhanced	No positive evidence for effect of dietary Na^+ on pump	Stokes *et al.*[17]
RBC	Man	Na^+/K^+	Pump enhanced ?	Only indirect evidence (Na increased, K decreased)	Simon and Conklin[10]
Na pump inhibition					
RBC	Man, FH	Na^+	Ouabain-sensitive efflux down	In normotensive FH; genetic defect	Heagerty *et al.*[5]
RBC	Man, also FH	Na^+/K^+ ATPase	Pump inhibition	In 40% of patients with FH of hypertension; correlates with inhibition of ouabain binding	De The *et al.*[77]
Na pump unchanged					
RBC	Man	Rubidium	None	Ouabain-resistant rubidium flux	Lijnen *et al.*[71]
WBC	Man	Na^+	Falls only with age	Altered Na^+ transport related to age not BP	Swales *et al.*[6]
Na/Li increased					
RBC	Man	Na^+/Li^+	Increased	Also in FH	Canessa *et al.*[26]
RBC	Man	Na^+/Li^+	Increased	Increased in a subgroup of hypertensives	Garay *et al.*[3]
RBC	Man	Li^+	Increased	Increased in hypertensives with FH	Clegg *et al.*[55]
Na/Li unchanged					
RBC	Man/Black	Li^+	No change	See Caucasians, Canessa *et al.*[26]	Canessa *et al.*[26]
RBC	Man	Li^+	No change	Not direct marker of high BP; population heterogeneity	Dagher and Canessa[56]
K efflux					
Aorta	SHR	K^+	Enhanced	"Decreased Ca for stabilization" invoked	Jones[74]
Femoral					
K influx					
RBC	Man	K^+	Enhanced	Reduction in Na^+/K^+ flux ratio in essential hypertension	Garay and Meyer[18]

[a]For further references, see TABLE 3 of Garay *et al.*[3] SHR = spontaneously hypertensive rat; FH = family history of hypertension; VSM = vascular smooth muscle; RBC = red blood cell; + = increase; − = decrease; 0 = no change; ? = no data.

activity to the overall regulation of ion flux.[29] Increased sodium "leaks" occur in the red cells of 15–30 percent of human hypertensives.[30] In response to a membrane leak for sodium, varying degrees of compensatory mechanisms, such as an increased sodium pump or cotransport system, may cause the ultimate cellular sodium value to be increased, unchanged, or decreased.

THE SODIUM PUMP AND MEMBRANE CHARACTERISTICS

Because of the early studies emphasizing an increased red cell sodium in essential hypertension, it became reasonable to look for an underlying alteration of sodium pump function (i.e. the Na/K ATPase = sodium pump = electrogenic sodium pump).

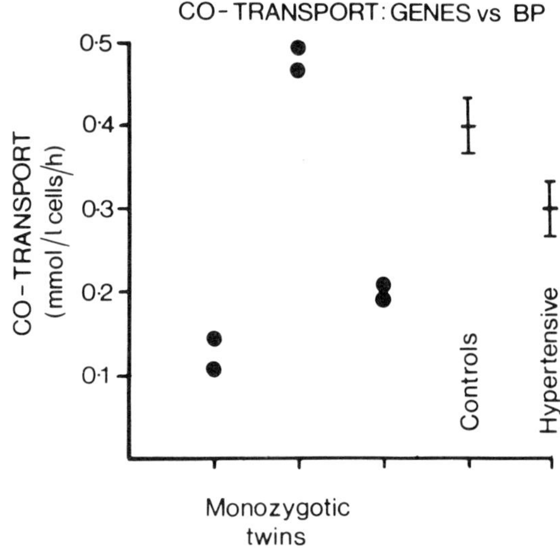

FIGURE 2. Cotransport activity in three pairs of monozygotic twins compared with means ± SEM for controls and hypertensive subjects. (From Davidson *et al.*[23] With permission from *British Medical Journal.*)

The classic inhibitor of the sodium pump is digitalis. Several authors have suggested that sodium pump activity is inhibited in tissues from hypertensive patients or rats[31] (TABLE 1) and that there is a circulating digitalis-like compound and/or natriuretic hormone in the blood of patients with essential hypertension.[8,9,24,32,33] Meyer *et al.*[34] find an inhibitor, called endaline, in about one third of patients with sustained moderate hypertension. Haupert *et al.*[35] and Carilli *et al.*[36] have identified a hypothalamic inhibitory factor (HIF), which inhibits the sodium pump in nanomolar concentrations, a finding that fits well with the hypotheses of Blaustein[8] and of Haddy and Overbeck.[37]

Other workers have found enhanced pump activity.[17,27,38] These opposite findings do not necessarily exclude a role for abnormal functioning of the sodium pump in hypertension, but rather argue against a primary membrane defect localized to the sodium pump.

A recent and highly interesting finding is that the sodium pump activity of husbands correlates with that of wives, irrespective of the blood pressure showing the dominant importance of an environmental factor.[39] One such possible factor could be the sodium intake, because a high sodium intake may call for the circulating digitalis-like factor with consequent sodium pump inhibition and a corresponding increase in the intracellular sodium level.[34]

Consequences of Pump Inhibition

The two major mechanisms whereby sodium pump inhibition could lead to vasoconstriction are by the activity of the Na/Ca exchanger and depolarization with a consequent increase in calcium influx.

SODIUM-CALCIUM EXCHANGE IN VASCULAR SMOOTH MUSCLE

Critical to links between intracellular sodium and vascular smooth muscle tone is the concept of the Na/Ca exchange mechanism (TABLE 5). This exchanger, well established in the case of the myocardium, is becoming established for vascular smooth muscle, and is apparently absent in blood cells. In vascular smooth muscle, some authors[8] emphasize the likelihood of the Na/Ca exchange, especially in large vessels like the aorta.[40] Other authors[12] conclude that the Na/Ca exchange is not well established, especially in the small resistance arteries or in intestinal smooth muscle.[41] Four arguments against the sodium-calcium countertransport are: (1) in vascular muscle there is no contraction on transfer to a low sodium solution, in contrast to the heart;[12] (2) there is no hyperpolarization as might be expected on the assumption that three or more sodium ions exchange for each calcium ion;[12,42] (3) there is no increase of labeled calcium uptake during exposure to sodium-free solution;[43] and (4) sodium pump inhibition in rat mesenteric artery gives rise to a depolarization that calcium antagonists inhibit.[44] Mulvany *et al.*[45] show that a combination of low external sodium and inhibition of the sodium pump by ouabain is required to raise the tension of rat mesenteric small arteries; their interpretation is that the Na/Ca exchange only plays a role in extreme conditions.

There is now direct and satisfactory evidence for Na/Ca exchange in that sodium-loaded vesicles from rat mesenteric arterial sarcolemma do take up calcium.[46] The rates seem low, supporting the conclusion that Na/Ca exchange is not as important in vascular smooth muscle as in the heart.[47] In rat uterine smooth muscle, the flux of calcium by the Na/Ca exchange is much less than that by the energy-dependent calcium ATPase.[40] Nevertheless, the Na/Ca exchanger is of sufficient magnitude to help maintain intracellular calcium homeostasis especially during muscular relaxation.[9]

MEMBRANE DEPOLARIZATION AND CALCIUM INFLUX

Another theory is that inhibition of the sodium pump leads to membrane depolarization because the sodium pump is electrogenic. Alternatively, a primary defect in membrane depolarization could lead to a secondary increase in sodium pump activity.[49]

Membrane depolarization could theoretically, if severe enough, reach the threshold

TABLE 5. Evidence for and against Role of Na/Ca Exchange in the Increased Calcium Level of Vascular Smooth Muscle

Observation	Tissue	Source	Interpretation/comment	Reference
Na pump inhibition gives depolarization, which calcium antagonists inhibit	Small arteries	Rat mesenteric	Depolarizing effect of Na pump may be more important than Na/Ca exchange	Mulvany[44]
Na-loaded sarcolemmal vesicles take up Ca	Arteries	Rat mesenteric	Direct evidence for Na/Ca exchange; rate similar in SHR to controls; low rate when compared to heart	Matlib *et al.*[46]
Na-dependent Ca uptake in vesicles from enriched sarcolemmal preparation	Uterine smooth muscle	Rat myometrium	Direct evidence for Na/Ca exchange; max value 2 μmol/g vs. 40–70 μmol/g for ATP-dependent Ca uptake	Grover *et al.*[48]
In Na-free solution, there is no increased uptake of labeled Ca or increased contraction nor any hyperpolarization	Arterial	Various	Indirect evidence against Na/Ca exchange	Droogmans and Casteels[43] Hermsmeyer[12]

for activation for the opening of the slow calcium channel, which is −30 to −40 mV. This degree of depolarization seems unlikely to be a primary defect in hypertension since the resting membrane potential of the caudal artery is about −55 mV in both normotensive and hypertensive rats, provided that external potassium is normal.[50] However, added sodium pump inhibition may be able to depolarize arterial tissue sufficiently for calcium ions to enter via the slow calcium channel.[44]

Vascular "Responsiveness" and Na/Ca Exchange

Whereas vascular tone seems to be related to free cytosolic calcium levels in smooth muscle cells, "responsiveness" is the change in tension resulting from a given neural or humoral stimulus. The resting membrane potential may help to determine "responsiveness." Thus a vascular cell with an "inhibited" sodium pump and a decreased membrane potential may constrict more vigorously in response to adrenergic stimulation. This "myogenic" result of sodium pump inhibitors is in contrast to their "neurogenic" effects, whereby an enhanced intracellular calcium in nerve triggers the release of norepinephrine. The relative roles of "myogenic" versus "neurogenic" effects may vary according to vessel size, with the myogenic mechanism more prominent in the larger vessels such as the aorta and the neurogenic effect more prominent in the resistance arterioles.[44]

CALCIUM ABNORMALITIES: A CONSISTENT FINDING?

In contrast to the diverse findings in relation to intracellular sodium concentrations and indices of sodium and potassium flux across the cell membrane, a thus far consistent observation has been the increase of intracellular free calcium in a variety of tissues (TABLE 1). In some studies, the degree of elevation of platelet calcium is only modest with a wide overlap.[51] The ideal measurement, not yet made, would be an intracellular ion-sensitive electrode to detect true intracellular activity of blood cells.

Robinson[7] thoroughly reviews the relationship between calcium abnormalities and hypertension. An interesting clinical observation is that verapamil decreased blood pressure more than did sodium nitroprusside in severe hypertensives. Assuming that top doses of both agents were employed, the "excess hypotensive" effect of verapamil could be plotted against the BP; the higher the BP, the more the excess effect. This observation agrees with the powerful antiarrhythmic effect of nifedipine in hypertensives and the modest or absent hypotensive effect in normals. A simple explanation for such findings can be proposed provided that it be accepted that the sodium pump is inhibited, which leads to depolarization of vascular smooth muscle[44] thereby promoting calcium entry through voltage-operated channels, the latter would be blocked by calcium antagonists.[51] These observations would explain why calcium antagonists and not captopril decrease platelet calcium levels,[51] but do not explain why β-blockers and diuretics do likewise.[52]

DIETARY LIPIDS AND MEMBRANE PROPERTIES

Relevant to the present paper are correlations between the lipid composition of the diet and membrane properties. Recently Heagerty *et al.*[14] showed that a safflower seed diet increased membrane linoleic acid with a significant increase of the sodium pump rate in leukocytes, a decrease in blood pressure, and a fall in the intracellular free calcium. These important data seem to be the first link suggesting that dietary induced changes in membrane properties can directly influence the cellular ionic events associated with hypertension. The finding that extreme dietary changes can alter blood pressure and membrane characteristics as well as sodium pump activity, does not necessarily prove a direct causal relationship. However, these data emphasize the potential importance of environmental factors. Likewise the inverse relation between the levels of circulating high-density lipoproteins and calcium influx into red cells[53] also stresses that dietary factors can influence membrane properties and ion fluxes.

PROBLEMS WITH THE PRIMARY "MEMBRANE THEORY"

An important question is whether hypertension could cause the changes in intracellular calcium via a non-specific mechanism. That is, ionic changes in hypertension could be the "marker" or the "innocent bystander"[6] and not the "mechanism" or the "culprit." Some effect of hypertension itself may increase free calcium at least in the platelets and in the red cells of patients.[11,52] Hypertension itself could mechanically damage the sarcolemmal membranes to give rise to a high intracellular calcium (FIG. 1) that would explain the relationship between the height of hypertension and the effect of a variety of therapies on intracellular calcium, as shown by Buhler's group.[52] The cellular mechanism may involve altered turnover of membrane inositol lipids.[54] Antihypertensive therapy by a loop diuretic decreases intracellular calcium activity of red cells and at the same time increases the sodium activity, which leads to the conclusion that at least some of the membrane abnormalities must result from, rather than be a cause of the hypertension.

The absence of consistent data that ionic alterations occur early in the course of hypertension and are genetically linked, can be explained by arguing that membrane alterations are not required for hypertension to develop; however in hypertensives, there are a variety of changes in pump and transport functions that may be variously linked to different populations of hypertensives.[3]

WHAT IS LIKELY: HYPERTENSION AS A MULTIFACTORIAL DISEASE THAT MAY HAVE HIGHLY COMPLEX CHANGES IN ION FLUXES

The apparent conflicts in the literature may be resolved by the proposal that hypertension is a multifactorial condition, in part genetic and the genetic component is in turn also highly complex, varying from population to population.[3,30,55,56] That early pre-hypertensive genetic effects in regulation of ion flux could occur in experimental models and in some patients has been suggested by several studies.[21,26] Abnormalities in the red cell structural components may be found in the stem cells of Milan hypertensive rats as well as in proximal tubular cells.[57,58] The differences between Milan hypertensive rats and spontaneously hypertensive rats suggest different genetic defects in different animal models, and raises the possibility of multiple genetic defects in man, together with varying gene penetration and polymorphism, plus a prominent hypertension-induced or hypertension-related change in intracellular calcium concentration.[52,54]

CONCLUSIONS

In the broadest sense of the term, biological membranes are likely to play an important role in the causation of hypertension because it is increasingly agreed that intracellular calcium in the vascular tissue is a probable and important event in hypertension, which in itself indicates a loss of the normal ionic gradient (extracellular to intracellular). This alteration in calcium is frequently linked with an enhanced intracellular sodium content or activity, but here the evidence for a direct link is less clear and frequently based on indirect findings in white or red blood cells. Some workers feel that monovalent cation changes are a "marker" rather than a "mechanism" in hypertension. Varying degrees of compensatory mechanisms may exist to result in normal, increased, or decreased sodium levels in blood cells of patients with hypertension. The usual supposition is that the sodium pump is inhibited, for example by a circulating digitalis-like factor, the "natriuretic hormone" or "endaline" and/or the "hypothalamic inhibitory factor." The opposing studies that show an increased pump activity could be explained by postulating that there is compensatory hyperactivity, for example in response to an increased sodium leak in a subgroup of hypertensive patients. New studies emphasize the complex inheritance of a variety of transport abnormalities, differently linked to hypertension in different subpopulations. Thus a simple "primary membrane hypothesis" becomes unlikely. There are, however, consistent findings concerning the elevation of internal calcium in blood cells including platelets.

How this critical elevation of intracellular calcium concentration comes about is still open to question. The fact that it can be normalized by therapy with a variety of agents suggests that an elevated blood pressure level leads to an abnormality of the calcium level, not the other way round. In hypertension, the complex interrelation of environmental and genetic factors makes it difficult to exclude a summation of "minor" membrane defects or "localized" membrane defects located, for example, at the level of the renal tubules, or a "minor" genetic defect in either renal tubules or vascular smooth muscle that could become a "major" functional defect when the blood pressure rises.

At present, a reasonable hypothesis is that hypertension is caused by a multiplicity of factors including genetic and dietary abnormalities, which ultimately cause an increased intracellular calcium in vascular smooth muscle cells. In some hypertensives, an enhanced extracellular volume calls forth a circulating digitalis-like factor that inhibits sodium pump activity with varying compensatory cellular transport mechanisms ultimately causing the intracellular sodium level to be unchanged, increased or decreased. As the accuracy of techniques for measurement of true cytosolic calcium concentrations and transients in vascular smooth muscle cells develops, there will be an increasing focus on the ultimate regulation of the biochemical events directly governing vascular smooth muscle and the consequent development of hypertension.

REFERENCES

1. POSTNOV V. V. & S. N. ORLOV. 1984. Cell membrane alteration as a source of primary hypertension. J. Hypert. **2:** 1–6.
2. GARAY, R. P., G. DAGHER, M. G. PERNOLLET, M. A. DEVYNCK & P. MEYER. 1980. Inherited defect in Na^+, K^+ cotransport system in erythrocytes from essential hypertensive patients. Nature **284:** 281–283.
3. GARAY, R. P., C. NAZARET, P. HANNAERT & M. PRICE. 1983. Abnormal Na^+, K^+

cotransport function in a group of patients with essential hypertension. Eur. J. Clin. Invest. **13:** 311–320.
4. GARAY, R., C. ROSATI & P. MEYER. 1986. Na^+ transport in primary hypertension. Ann. N.Y. Acad. Sci. This volume.
5. HEAGERTY, A. M., M. MILNER, R. F. BING, H. THURSTON & J. D. SWALES. 1982. Leucocyte membrane sodium transport in normotensive populations: dissociation of abnormalities of sodium efflux from raised blood pressure. Lancet **ii:** 894–896.
6. SWALES, J. D., R. F. BING, R. BRADLAUGH, A. EL-ASHRY, N. GODFREY, A. M. HEAGERTY & H. THURSTON. 1984. Cell membrane handling of sodium, sodium balance and blood pressure. J. Cardiovasc. Pharmacol. **6:** S42-S48.
7. ROBINSON. B. F. 1984. Altered calcium handling as a cause of primary hypertension. J. Hypert. **2:** 453–460.
8. BLAUSTEIN, M. P. 1977. Sodium ions, calcium ions, blood pressure regulation and hypertension: a reassessment and a hypothesis. Am. J. Physiol. **232:** C165-C173.
9. BLAUSTEIN, M. P., J. M. HAMLYN & T. ASHIDA. 1986. Sodium ions, calcium ions, natriuretic hormone and hypertension. Ann. N.Y. Acad. Sci. This volume.
10. SIMON, G. & D. J. CONKLIN. 1986. In vivo erythrocyte sodium concentration in human hypertension is reduced, not increased. J. Hypert. **4:** 71–75.
11. ZIDEK, W., H. LOSSE, W. SCHMIDT, K. J. FEHSKE & H. VETTER. 1984. Intracellular electrolytes during antihypertensive treatment with a loop diuretic. J. Hypert. **2:** 393–395.
12. HERMSMEYER K. 1984. Altered arterial muscle ion transport mechanism in the spontaneously hypertensive rat. J. Cardiovasc. Pharmacol. **6:** S10–S15.
13. COOPER, R. A. 1977. Abnormalities of cell-membrane fluidity in the pathogenesis of disease. New Engl. J. Med. **297:** 371–377.
14. HEAGERTY, A. M., J. D. OLLERENSHAW, J. JACKSON, R. F. BING, H. THURSTON & J. D. SWALES. 1985. Linoleic acid, leucocyte sodium transport, free calcium and blood pressure (abstr). Brit. Heart J. **54:** 635–636.
15. HILTON, P. J. 1986. Cellular sodium transport in essential hypertension. New Engl. J. Med. **314:** 222–229.
16. TOSTESON, D. C. 1981. Cation countertransport and cotransport in human red cells. Fed. Proc. **40:** 1429–1433.
17. STOKES, G. S., J. C. MONAGHAN, A. T. MIDDLETON, M. SHIRLOW & J. F. MARWOOD. 1986. Effects of dietary sodium deprivation on erythrocyte sodium concentration and cation transport in normotensive and untreated hypertensive subjects. J. Hypert. **4:** 35–38.
18. GARAY, R. P. & P. MEYER. 1979. A new test showing abnormal net Na^+ and K^+ fluxes in erythrocytes from essential hypertensive patients. Lancet **i:** 349–353.
19. TALIB, H. K. B., A. R. CHIPPERFIELD & P. F. SEMPLE. 1984. Potassium influx into erythrocytes in essential hypertension. J. Hypert. **2:** 405–409.
20. WILKINS, M. R., P. E. JENNINGS, M. J. WEST, M. J. KENDALL & A. H. BARNETT. 1985. Dissociation of changes in sodium transport in erythrocytes from changes in blood pressure. J. Hypert. **3** (Suppl 3): S21–S23.
21. BIANCHI, G., G. FERRARI, S. SALARDI, B. BARBER, L. TORIELLI, M. FERRANDI & D. CUSI. 1986. Membrane abnormalities, hypertension and genetic links in MHS. Ann. N.Y. Acad. Sci. This volume.
22. CANESSA, M. 1986. Pathophysiology of the Na exchange and Na-K-Cl cotransport in essential hypertension. Ann. N.Y. Acad. Sci. This volume.
23. DAVIDSON, J. S., L. H. OPIE & B. KEDING. 1982. Sodium-potassium cotransport activity as genetic marker in essential hypertension. Brit. Med. J. **284:** 539–541.
24. HADDY, F. J. & M. B. PAMNANI. 1983. The role of a humoral sodium-potassium pump inhibitor in low-renin hypertension. Fed. Proc. **42:** 2673–2680.
25. ETKIN, N. L., J. R. MAHONEY, M. W. FORSTHOEFEL, J. R. ECKMAN, J. D. MCSWIGAN, R. F. GILLUM & J. W. EATON. 1982. Racial differences in hypertension-associated red cell sodium permeability. Nature **297:** 588–589.
26. CANESSA, M., N. ADRAGNA, H. S. SOLOMON, T. M. CONNOLLY & D. C. TOSTESON. 1980. Increased sodium-lithium countertransport in red cells of patients with essential hypertension. New Engl. J. Med. **382:** 772–776.

27. WOODS, K. L., D. G. BEEVERS & M. WEST. 1981. Familial abnormality of erythrocyte cation transport in essential hypertension. Brit. Med. J. **282:** 1186–1188.
28. MAHNENSMITH, R. L. & P. S. ARONSON. 1985. The plasma membrane sodium-hydrogen exchanger and its role in physiological and pathophysiological processes. Circ. Res. **56:** 773–788.
29. MILANICK, M. A. & J. F. HOFFMAN. 1986. Ion transport and volume regulation in red blood cells. Ann. N.Y. Acad. Sci. This volume.
30. GARAY, R. P. & C. NAZARET. 1985. Na^+ leak in erythrocytes from essential hypertensive patients. Clin. Sci. **69:** 613–624.
31. RASMUSSEN, H. 1983. Cellular calcium metabolism. Ann. Int. Med. **98:** 809–816.
32. MACGREGOR, G. A. & H. E. DE WARDENER. 1984. A circulating sodium transport inhibitor and essential hypertension. J. Cardiovasc. Pharmacol. **6:** S55–S60.
33. DE WARDENER, H. E. & E. M. CLARKSON. 1985. Concept of natriuretic hormone. Physiol. Rev. **65:** 658–759.
34. MEYER, P., J. F. CLOIX & M. A. DEVYNCK. 1986. Chemical and clinical studies of endogenous digitalis-like factor in hypertension. Ann. N.Y. Acad. Sci. This volume.
35. HAUPERT, G. T. JR., C. T. CARILLI & L. C. CANTLEY. 1984. Hypothalamic sodium-transport inhibitor is a high-affinity reversible inhibitor of Na^+-K^+-ATPase. Am. J. Physiol. **247:** F919–F924.
36. CARILLI, C. T., M. BERNE, L. C. CANTLEY & G. T. HAUPERT, JR. 1985. Hypothalamic factor inhibits the (Na,K)ATPase from the extracellular surface. Mechanism of inhibition. J. Biol. Chem. **260:** 1027–1031.
37. HADDY, F. J. & H. W. OVERBECK. 1976. The role of humoral factors in volume expanded hypertension. Life Sci. **19:** 935–948.
38. DAVID-DUFILHO, M., M. CIRILLO, J. P. BEUGRAS & M. A. DEVYNCK. 1984. Active Na^+ and Ca^{2+} transport by cardiac sarcolemmal membranes of young genetically hypertensive rats. J. Hypert. **2:** 431 (abstr).
39. CUSI, D., G. TRIPODI, E. ALBERGHINI, E. NIUTTA, C. BARLASSINA, E. FOSSALI, F. DOSSI & G. BIANCHI. 1986. Hereditability of sodium transport systems and hypertension. Ann. N.Y. Acad. Sci. This volume.
40. OZAKI, H. & N. URAKAWA. 1981. Involvement of a Na-Ca exchange mechanism in contraction induced by low-Na solution in isolated guinea-pig aorta. Pflug. Arch. **390:** 107–112.
41. AARONSON, P. & C. VAN BREEMEN. 1981. Effects of sodium gradient manipulation upon cellular calcium, ^{45}Ca fluxes and cellular sodium in the guinea-pig taenia coli. J. Physiol. **319:** 443–461.
42. PITTS, B. J. R. 1979. Stoichiometry of sodium-calcium exchange in cardiac sarcolemmal vesicles. Coupling to the sodium pump. J. Biol. Chem. **254:** 6232–6235.
43. DROOGMANS, G. & R. CASTEELS. 1979. Sodium and calcium interactions in vascular smooth muscle cells of the rabbit ear artery. J. Gen. Physiol. **74:** 57–70.
44. MULVANY, M. J. 1985. Changes in sodium pump activity and vascular contraction. J. Hypert. **3:** 429–436.
45. MULVANY, M. J., C. AALKJAER & T. T. PETERSEN. 1984. Intracellular sodium, membrane potential, and contractility of rat mesenteric small arteries. Circ. Res. **54:** 740–749.
46. MATLIB, M. A., A. SCHWARTZ & Y. YAMORI. 1985. A Na^+-Ca^{2+} exchange process in isolated sarcolemmal membranes of mesenteric arteries from WKY and SHR rats. Am. J. Physiol. **249:** C166–C172.
47. BRADING, A. F. & T. W. LATEGAN. 1985. Na-Ca exchange in vascular smooth muscle. J. Hypert. **3:** 109–116.
48. GROVER, A. K., C. Y. KWAN, P. K. RANGACHARI & E. E. DANIEL. 1983. Na-Ca exchange in a smooth muscle plasma membrane-enriched fraction. Am. J. Physiol. **244:** C158–C165.
49. POSTNOV, Y. V., G. M. KRAVTSOV & I. Y. POSTNOV, JR 1985. Evidence of lowered plasma membrane potential in different cell types in primary hypertension. J. Hypert. **3** (Suppl 3): S9–S11.
50. HERMSMEYER. K. 1976. Electrogenesis of increased norepinephrine sensitivity of arterial vascular muscle in hypertension. Circ. Res. **38:** 362–367.
51. LENZ, T., H. HALLER, M. LUDERSDORF, A. KRIBBEN, M. THIEDE, A. DISTLER & T.

PHILLIPP. 1985. Free intracellular calcium in essential hypertension. Effects of nifedipine and captopril. J. Hypert. **3** (Suppl 3): S13–S15.

52. ERNE, P., P. BOLLI, E. BURGISSER & F. R. BUHLER. 1984. Correlation of platelet calcium with blood pressure. Effect of antihypertensive therapy. New Engl. J. Med. **310:** 1084–1088.
53. STIMPEL, M. L., NEYSES, R. LOCHER, M. KNORR & W. VETTER. 1985. High density lipoproteins—modulators of the calcium channel? J. Hypert. **3** (Suppl 3): S49–S51.
54. BUHLER, F. R., T. J. RESINK, D. DIMITROV, A. ZSCHAUER & P. ERNE. 1986. Membrane abnormalities in platelets of patients with essential hypertension. Ann. N.Y. Acad. Sci. This volume.
55. CLEGG, G., D. B. MORGAN & C. DAVIDSON. 1982. The heterogeneity of essential hypertension. Relation between lithium efflux and sodium content of erythrocytes and a family history of hypertension. Lancet **ii:** 891–894.
56. DAGHER, G. & M. CANESSA. 1984. The Li-Na exchange in red cells of essential hypertensive patients with low Na-K cotransport. J. Hypert. **2:** 195–201.
57. FERRARI, P., D. CUSI, B. BARBER, C. BARLASSINA, G. VEZZOLI, L. DUSSI, E. MINOTTI & G. BIANCHI. 1982. Erythrocyte membrane and renal function in relation to hypertension in rats of the Milan hypertensive strain. Clin. Sci. **63:** 61s–64s.
58. BIANCHI, G., P. FERRARI, D. TRIZIO, M. FERRANDI, L. TORIELLI, B. R. BARBER & E. POLLI. 1985. Red blood cell abnormalities and spontaneous hypertension in the rat. A genetically determined link. Hypertension **7:** 319–325.
59. LE QUAN SANG, K. H., P. BENLIAN, K. CHAHNAZ, T. MONTENAY-GARESTIER, P. MEYER & M. A. DEVYNCK. 1985. Platelet cytosolic free calcium concentration in primary hypertension. J. Hypert **3** (Suppl 3): S33–S36.
60. BRUSCHI, G., M. E. BRUSCHI, G. ORLANDINI, A. CAVATORTA, A. BORGHETTI, M. FERRANDI & G. BIANCHI. 1985. Why is $[Ca^{2+}]_i$ increased in blood cells in primary hypertension J. Hypert. **3** (Suppl 3): S45–S47.
61. LOSSE, H., W. ZIDEK & H. VETTER. 1984. Intracellular sodium and calcium in vascular smooth muscle of spontaneously hypertensive rats. J. Cardiovasc. Pharmacol. **6:** S32–S34.
62. ZIDEK, W., H. VETTER, K.-G. DORST, H. ZUMKLEY & H. LOSSE. 1982. Intracellular Na^+ and Ca^{2+} activities in essential hypertension. Clin. Sci. **63:** 41s–43s.
63. VEZZOLI, G., A. A. ELLI, G. TRIPODI, G. BIANCHI & E. CARAFOLI. 1985. Calcium ATPase in erythrocytes of spontaneously hypertensive rats of the Milan strain. J. Hypert. **3:** 645–648.
64. KWAN, C. Y. & A. K. GROVER. 1983. Membrane abnormalities occur in vascular smooth muscle but not in non-vascular smooth muscle from rats with deoxycorticosterone-salt induced hypertension. J. Hypert. **1:** 257–265.
65. RESINK, T. J., V. A. TKACHUK, P. ERNE & F. R. BUHLER. 1985. Platelet membrane Ca^{2+}-ATPase: blunted calmodulin-stimulation in essential hypertension. J. Hypert. **3** (Suppl 3): S37–S40.
66. DEVYNCK, M. A., M. G. PERNOLLET, A. M. NUNEZ & P. MEYER. 1981. Analysis of calcium handling in erythrocyte membranes of genetically hypertensive rats. Hypertension **3:** 397–403.
67. POSTNOV, Y. V., S. N. ORLOV, G. M. KRAVTSOV & P. V. GULAK. 1984. Calcium transport and protein content in cell plasma membranes of spontaneously hypertensive rats. J. Cardiovasc. Pharmacol. **6:** S21–S27.
68. ORLOV, S. N. & Y. V. POSTNOV. 1982. Ca^{2+} binding and membrane fluidity in essential and renal hypertension. Clin. Sci. **63:** 281–284.
69. DUHM, J. 1984. *In* Discussion of paper by Losse *et al.* J. Cardiovasc. Pharmacol **6:** S34.
70. TOBIAN, L. & J. T. BINION. 1952. Tissue cations and water in arterial hypertension. Circulation **5:** 754–758.
71. LIJNEN, P., J.-R. M'BUYAMBA-KABANGU, R. H. FAGARD, D. R. GROESENEKEN, J. A. STAESSEN & A. K. AMERY. 1984. Intracellular concentration and transmembrane fluxes of sodium and potassium in erythrocytes of white normal male subjects with and without a family history of hypertension. J. Hypert. **2:** 25–30.
72. GUDMUNDSSON, O., O. ANDERSSON, H. HERLITZ, O. JONSSON, J. NAUCLER, J. WIKSTRAND & G. BERGLUND. 1984. Blood pressure, intraerythrocyte content, and transmembrane

fluxes of sodium during normal and high salt intake in subjects with and without a family history of hypertension: evidence against a sodium transport inhibitor: J. Cardiovasc. Pharmacol. **6:** S35–S41.

73. THURAU, K., F. BECK, M. BORST, A. DORGE, R. RICK & G. BIANCHI. 1984. Intracellular electrolyte composition in various experimental models of hypertension: an electron microprobe study. J. Cardiovasc. Pharmacol. **6:** S28–S31.
74. JONES, A. W. 1973. Altered ion transplant in vascular smooth muscle from spontaneously hypertensive rats. Circ. Res. **33:** 563–572.
75. SEMPLICINI, A., G. B. AMBROSIO, E. RIGON, L. DISSEGNA, S. ZAMBONI, G. P. ROSSI, A. C. PESSINA & C. DAL PALU.1985. Intra-erythrocytic sodium content in normotensive offspring of normotensive and hypertensive subjects: an epidemiological study. J. Hypert. **3** (Suppl 3): S61–S63.
76. MEYER, P., R. P. GARAY, C. NAZARET, G. DAGHER, M. BELLET, M. BROYER & J. FEINGOLD. 1981. Inheritance of abnormal erythrocyte cation transport in essential hypertension. Brit. Med. J. **282:** 1114–1117.
77. DE THE, H., M.-A. DEVYNCK, J. ROSENFELD, M.-G. PERNOLLET, J.-L. ELGHOZI & P. MEYER. 1984. Plasma sodium pump inhibitor in essential hypertension and normotensive subjects with hypertensive heredity. J. Cardiovasc. Pharmacol. **6:** S49–S54.

Ion Transport and Volume Regulation in Red Blood Cells[a]

MARK A. MILANICK[b] AND JOSEPH F. HOFFMAN

Department of Physiology
Yale University School of Medicine
New Haven, Connecticut 06510

INTRODUCTION

A pump/leak model has been proposed to explain how cell volume can be maintained in the steady state.[1] This model has been quantitatively tested in sheep red blood cells and provides a reasonable fit to the data. The fact that the cation content of sheep red cells is apparently controlled by a single gene suggests that one gene may control both the number of transporters and the amount of leak. We thought that another test of the pump/leak model was to examine the effect of adaptative responses of red cells on the relationship between membrane transport properties and cell volume. For this, we chose to look at the red cells from rats on different K diets because a decrease in dietary K is known to alter plasma [K]. Alterations in red cell membrane transport activity have been reported when the plasma [K] falls. Some reports ascribe most of the change to a change in Na/K pump activity[2-5] (or Na,K-ATPase activity) others to cotransport (Na/K/C1) activity changes.[6] If transport activity changes, does cell volume change in a way consistent with the pump/leak model?

The pump/leak model is evidently limited since it only qualitatively rather than quantitatively predicts the flux parameters in red cells from species other than sheep. Furthermore, since the model was first proposed the contribution of other ion transport pathways to the unidirectional fluxes has now been established. Thus we also wanted to extend the pump/leak model to include these other transport systems. We hoped to be able to determine whether Na/K/Cl cotransport plays a significant role in red cells at their normal steady-state volume.

The final section of this paper illustrates some of the information that emerges from the steady-state flux equations.

PUMP/LEAK EQUATIONS

A basic tacit assumption in this formulation is that the steady-state cell volume is determined by the sum, $Na_i + K_i$. Rats on a low K diet are known to have low plasma [K]. The previous equations[1] are not in a format in which it is easy to determine the effect of K_o on volume because the parameter, β, the ratio of the pump to leak flux for K, is an implicit function of K_o. The leak flux is always linearly dependent on K_o. The pump dependence on K_o is complex but for saturating values of K_o the pump flux is independent of K_o. For simplicity, the equations are solved in terms of the pump flux

[a]Supported in part by National Institutes of Health grants HL 09906 and AM 17433.
[b]Present address is Department of Physiology, University of Missouri School of Medicine, Medical Sciences Building, Columbia, MO 65212.

and the leak rate constants so the only implicit K_o dependence, if any, is the pump flux dependence on K_o.

The four equations that are used to derive the volume, Na_i, K_i, and Cl_i from the external ion concentrations and flux values are given in TABLE 1. Equations 1 and 2 represent electroneutrality and osmotic balance, respectively. Equations 3 and 4 represent the steady-state assumption that there is no net flux of Na nor K. The leak fluxes are assumed (or defined) to follow the Ussing flux ratio. For red cells the membrane potential, E_m, is determined by the chloride ratio Cl_i/Cl_o.[7] If P is the net pump influx for K, N is the pump stoichiometry for net Na to net K, and ${}^ik^L_K$ and ${}^ik^L_{Na}$

TABLE 1. Pump/Leak Equations

$$Na_i + K_i - Cl_i + \frac{ZX}{V} = 0 \quad (1)$$

$$Na_i + K_i + Cl_i + \frac{X}{V} = Na_o + K_o + Cl_o \quad (2)$$

K leak influx + net pump K influx = K leak efflux (3)

Na leak influx = Na leak efflux + net pump Na efflux (4)

$$\frac{X}{V} = \frac{2\,(-\,Cl_o + \{Cl_o[Cl_o + (z^2 - 1)(P)(N/{}^ik^L_{Na} - 1/{}^ik^L_K)]\}^{1/2})}{[z^2 - 1]} \quad (5)$$

$$Na_i = \frac{-zCl_o + \{Cl_o[Cl_o + (z^2 - 1)(P)(N/{}^ik^L_{Na} - 1/{}^ik^L_K) + z(Na_o - NP/{}^ik^L_{Na})(1 + z)]\}^{1/2}}{(1 - z)(1 + [K_o + P/{}^ik^L_K]/[Na_o - NP/{}^ik^L_{Na}])} \quad (6)$$

$$K_i = \frac{-zCl_o \pm \{Cl_o[Cl_o + (z^2 - 1)(P)(N/{}^ik^L_{Na} - 1/{}^ik^L_K) + z(Na_o - NP/{}^ik^L_{Na})(1 + z)]\}^{1/2}}{(1 - z)(1 + [Na_o - NP/{}^ik^L_{Na}[K_o + P/{}^ik^L_K])} \quad (7)$$

$$Cl_i = \frac{zCl_o + \{Cl_o[Cl_o + (z^2 - 1)(P)(N/{}^ik^L_{Na} - 1/{}^ik^L_K)]\}^{1/2}}{z + 1} \quad (8)$$

Symbols: z = equivalent valence of impermeant solutes, X = total intracellular content of impermeant solutes, P = net pump K influx, N = pump Na/K stoichiometry, ${}^ik^L_K$ = leak rate constant for K, and ${}^ik^L_{Na}$ = leak rate constant for Na.

are the leak rate constants for K and Na influx, respectively, then

$$1 + \frac{P}{{}^ik^L_K\,K_o} = \frac{K_iCl_i}{K_oCl_o} \quad (9)$$

and

$$1 = \frac{Na_iCl_i}{Na_oCl_o} + \frac{NP}{{}^ik^L_{Na}\,Na_o}. \quad (10)$$

These equations do not include explicitly the dependence of pump flux on Na and K. Equations 9 and 10 can be solved for K_i and Na_i to obtain equations of the form:

$$K_i = Cl_o/Cl_i\,f \text{ and } Na_i = Cl_o/Cl_i\,g\,, \quad (11)$$

where

$$f = K_o + P/{}^{i}k^{L}_{K} \text{ and } g = Na_o - NP/{}^{i}k^{L}_{Na} . \quad \textbf{(12)}$$

The result of solving equations 1, 2, 11, and 12 for Na_i, K_i, Cl_i, and volume are shown in the bottom of TABLE 1 (Equations 5–8).

Note that the equation for volume behaves as expected: an increase in the pump flux leads to a decrease in volume, an increase in the leak rate constant for Na causes an increase in cell volume (due to an increase in Na_i), an increase in the leak rate constant for K causes a decrease in volume (due to K_i loss), and an increase in N causes a decrease in volume (as the pump extrudes more Na). However, the volume and ion concentrations are determined solely by the ratio of the pump flux to leak rate constants for Na and for K (and the pump stoichiometry), which is equivalent to the result of Tosteson and Hoffman[1] and that the volume was determined by the ratio of K pump flux to K leak flux and the ratio of the leaks for Na and K (and the pump stoichiometry). Note that when $z = -1$ there is no quadratic equation, so the volume equation reduces to

$$\frac{X}{V} = Cl_o - (f + g) = P\left(\frac{N}{{}^{i}k^{L}_{Na}} - \frac{1}{{}^{i}k^{L}_{K}}\right). \quad \textbf{(13)}$$

In this case it is obvious that in order to obtain positive values of volume one needs N ${}^{i}k^{L}_{K} > {}^{i}k^{L}_{Na}$. The same constraint holds for the more general equation ($z \neq -1$) in order to obtain a positive, real value of volume.

This derivation ignores HCO_3^--dependent Na fluxes; yet the HCO_3^--dependent Na flux is presumably a substantial portion of the Na influx *in vivo.*[8] Nevertheless, the HCO_3^--dependent Na pathway probably behaves as a leak flux in the sense the it follows the ratio Na_iCl_i/Na_oCl_o (since it represents $NaCO_3$/Cl exchange and the $NaCO_3^-$ concentration is not saturating). The Na flux in 150 mM NaCl is a bit smaller than in 120 mM NaCl 30 mM HCO_3, but the effect is probably negligible given the errors in the other determinations.

Inclusion of Na/K/Cl Cotransport and Other Mediated Pathways

Because several additional cation transport pathways have been more recently characterized it is important to extend the pump/leak model to include these pathways to evaluate their possible role in cell volume regulation. The previous equation 5 defining volume can be extended to include additional pathways in the steady-state flux equations. For example the K steady-state equation 3 including Na/K/Cl cotransport, C, is

K leak influx + net K pump influx = K leak efflux + net K cotransport efflux, **(14)**

or

$$K_i\, Cl_i = Cl_o\, [K_o + P/{}^{i}k^{L}_{K} - C/{}^{i}k^{L}_{K}] , \quad \textbf{(15)}$$

which can be re-arranged to

$$K_i = Cl_o/Cl_i\, f' , \quad \textbf{(11')}$$

where

$$f' = K_o + \frac{\text{the sum of the net mediated K efflux pathways}}{{}^{i}k_K^L}. \quad (12')$$

For convenience, we take the net efflux mediated by the cotransport pathway as positive. Also, in this formulation, the process need not necessarily be passive; one can mathematically choose C so that the system works uphill. Furthermore, since in red cells the Cl leak fluxes are much larger than known Na/K cotransport mediated fluxes, we assume Cl is distributed at equilibrium, thus providing a measure of E_m. Because of this, the results presented here do not depend upon whether the pathway is Na/K cotransport or Na/K/Cl cotransport.

TABLE 2. Effect of K Diet on Rat Blood Parameters

	Low K Diet	Normal Diet	High K Diet	p^a
Plasma [K] (mM)	1.8 ± 0.5	4.0 ± 0.1	5.3 ± 0.7	<0.01
$[Na_{cell}]$ (mM)	7.0 ± 1.3	5.5 ± 1.0	5.6 ± 1.4	<0.01
Cell volume (u^3)	53 ± 5	58 ± 6	60 ± 4	~0.05
Blood pH	7.48 ± 0.07	7.42 ± 0.05	7.43 ± 0.05	N.S.
Pump K influx	10.8 ± 3.2	8 ± 1.3	7.5 ± 0.7	<0.05
Cotransport K influx	3.3 ± 1.2	1.4 ± 0.4	1.3 ± 0.3	<0.02
Leak K influx	0.58 ± .12	0.45 ± .06	0.49 ± .06	<0.05
Pump Na efflux	4.8	4.3	—	
Cotransport Na efflux	3.6	1.1	1.6	
Leak Na efflux	4.0	1.2	1.4	

Units: The values presented are the mean ± standard deviation of triplicate samples. The Na and K fluxes were measured by standard techniques,[24] though on different days. Similar results were obtained in three other experiments. The flux media contained 140 mM NaCl, 2.5 mM KCl, 20 mM HEPES pH 7.4. Plasma K was measured with a flame photometer; units are mM. Cell Na content was determined on fresh cells washed with isotonic $MgCl_2$; units are mmoles/l cell H_2O. Cell volume was determined from measurement of hematocrit and number of cells as determined with a Coulter counter; units are u^3. K influxes are expressed as mmoles K/(kg Hb × hr). Na fluxes are expressed as mmoles Na/(kg wet cells × hr). The pump flux was taken as that portion of the flux inhibited by 5 mM ouabain. The cotransport flux was taken as that portion of the flux inhibited by 100 μM bumetanide in the presence of 5 mM ouabain. The leak flux was taken to be the flux in the presence of 5 mM ouabain and 100 μM bumetanide.

[a]Paired *t*-test between cells from rats on low K diet and normal K diet.

EFFECT OF K DIET DEPLETION ON RAT RED CELL TRANSPORT

The results presented in TABLE 2 provide a comparison of the blood of rats placed on both a low and high K diet for 8 weeks with blood from control rats on a normal K diet. As expected the low K diet resulted in a two-fold decrease in plasma [K] and a slight increase in $[Na]_i$. The red cells from the rats on the low K diet may have had a slightly smaller cell volume ($p \sim 0.05$). The ouabain-insensitive, bumetanide-sensitive K influx was increased in red cells from rats on a low K diet compared to red cells from rats on a normal or high K diet; there was also a small increase in the ouabain-sensitive K influx but no significant change in the ouabain-insensitive, bumetanide-insensitive K influx. The bumetanide-sensitive Na efflux was also increased.

The "leak" fluxes as measured in NO_3 and bumetanide were compared with those measured in Cl and bumetanide; the K influx was about one third less in the latter medium (0.67 ± 0.09 vs. 0.45 ± 0.04 mmoles K/kg Hb × hr). Thus, the Cl-dependent pathway and the bumetanide-sensitive pathway are different. In rat red cells we define the bumetanide-sensitive pathway as a measure of cotransport (assuming complete and specific inhibition by bumetanide) and the bumetanide-insensitive pathway as a measure of the leak.

Because Na_i was higher in cells from rats on the low K diet, the increased ouabain- and bumetanide-sensitive fluxes could be due to this increased Na_i. As shown in TABLE 3 the ouabain-sensitive K influx is very similar from all three types of cells if Na_i is increased to 50 mM. This suggests that the V_{max} of the pump is the same in all the cells and thus that the number of pumps is probably the same. The increased bumetanide-sensitive flux in the red cells from rats on the low K diet is not due to the incrased Na_i; the flux is still high even when Na_i = 50 mM.

The bumetanide-sensitive flux in rat red cells is volume dependent (TABLE 3).[6] However, a relatively large change in the volume of cells from rats on the normal diet

TABLE 3. Effects of Na_i and Volume on K Influx Pathways

	Na_i = 50 mM Pump K Influx	Cotransport	Leak
Low K diet	8.3	2.1	0.43
Normal K diet	8.6	0.44	0.31
High K diet	8.9	0.32	0.28
Low K diet			
isotonic	10	3.2	0.8
hypotonic	10.8	1.5	1.2
High K diet			
isotonic	6.5	1.8	0.6
hypertonic	7.1	2.6	0.5

The flux values are the mean of triplicate measurements (the values ranged less than ±10% of the mean). The Na_i and volume variation experiments were done on different days, but similar results to those presented were also obtained in other experiments of each type. Na_i was varied by the nystatin procedure. When the osmolarity was varied, water was first added to reduce the osmolarity to 250 mOsm and then sucrose was added to increase the osmolarity to 350 mOsm. Flux units: mmole K/(kg Hb × hr). Other details as in TABLE 2.

was required in order to see a bumetanide-sensitive flux of the same size as that seen in the cells from rats on the low K diet. Also, a substantial swelling of the cells from rats on the low K diet was required in order to reduce the bumetanide-sensitive flux to the value seen in control cells.

Thus we have found that rats on a low K diet have an increased red cell Na content, a decreased plasma K concentration, and an increase in bumetanide-sensitive K influx. After these results were obtained, Duhm, Gobel, and Beck[6] reported similar findings. In agreement with their results and in contrast to previous reports,[2–5] we observed only a small increase in the ouabain-sensitive flux of K and Na (TABLE 2). Furthermore, we found that when the red cells were treated to have the same, high concentration of Na, there was no significant increase in the ouabain-sensitive K influx. Quantitatively, our results differed from those of Duhm, Gobel, and Beck[6] in several respects. We observed a smaller increase in the bumetanide-sensitive flux but we did not observe a significant change in cell volume or in the K leak. Also, we did not observe as much of a volume dependence of the bumetanide-sensitive flux. We doubt that cell shrinkage is sufficient to account for all the increase in bumetanide-sensitive flux seen in the results presented in TABLE 2. We have also shown that an increase in Na_i does not increase the bumetanide-sensitive K influx in control rats.

Effects of K_o on Volume and Na_i

If the pump flux is independent of K_o then volume and Cl_i are independent of K_o. This is evident from equations 5 and 8 in TABLE 1. The sum of Na_i and K_i is independent of K_o.

However, Na_i does depend upon K_o (equation 6). This result is easily derived from the steady-state flux equation for Na, equation 4; the equivalent equation 10,

$$1 = \frac{Na_i\,Cl_i}{Na_o\,Cl_o} + \frac{NP}{{}^ik^L_{Na}\,Na_o} \tag{10}$$

can be rewritten as

$$NP/{}^ik^L_{Na} = Na_o - Na_i\,Cl_i/Cl_o\,. \tag{16}$$

Suppose K_o changes from K_o' to K_o'', then, because $Na_o + K_o + Cl_o$ is fixed, Na_o must change from Na_o' to Na_o''. Thus

$$NP''/{}^ik''^L_{Na} = Na_o'' - Na_i''\,Cl_i/Cl_o \tag{17}$$

Assume that neither the pump nor leak flux change. ($P' = P''$ and ${}^ik'^L_{Na} = {}^ik''^L_{Na}$.) Thus,

$$Na_i'' = Na_i + \frac{Cl_o\,[Na_o'' - Na_o']}{Cl_i}\,. \tag{18}$$

Thus, a decrease in K_o (at constant extracellular osmolarity) will lead to an increase in $[Na_i]$ even if the pump flux is constant. The decrease in plasma K and increase in $[Na_i]$ would evidently represent compensatory changes with respect to the pump. On the other hand, both the decrease in plasma K and the increase in Na_i would tend to increase the driving force for exit via an Na/K/Cl cotransport system that would shrink the cells, as observed by Duhm, Gobel, and Beck.[6] In our studies, $[Na_i]$ increased less than expected, suggesting that there is some net cotransport exit. We cannot rule out the possibility of a slight change of volume as well.

Hypokalemia is a symptom in many diseases and can presumably be caused by a variety of different defects. In all cases of potassium depletion, however, there is an increase in $[Na]_i$.[9] The above equation suggests that, in some cases, this increase in Na_i may occur without any changes in red cell membrane properties. In Bartter's syndrome, though the red cells have an increase in Na_i, there appears to be no change in cell volume[10] and perhaps no change in membrane transport properties.[a] The fact that in about 12 hours of K replacement, plasma K, and cell Na returned to normal suggests that the circulating red cell had normal transport components. Red cells from

[a]The pump flux was increased when measured *in vitro;* this was attributed to the increase in Na_i. However, since the *in vitro* assay included 5 mM K_o for both the patient and the control red cells, it may not be fair to conclude that the patient red cells had an increased pump flux *in vivo* since the plasma K was low.

hypokalemia rats also return to normal cell Na if incubated for 4 hr in normal K media.[6]

Comparison of Predicted and Observed Transport Parameters

The steady-state flux equations can be solved for the cation flux parameters as a function of the ion concentrations. These equations are useful to assess the adequacy of the pump/lead model (TABLE 4). Indeed, the original paper compared the predicted values of α and β from ion concentrations with the values from the fluxes. The equations for Na_i and volume are very sensitive to the flux values. For example, the measured values of α and β for sheep red cells are within 20% of the values predicted by theory (TABLES III and VII in Tosteson & Hoffman[1]) yet if one calculates the predicted Na_i from equation 17 in Tosteson & Hoffman,[1] it is off by almost three fold. The predicted and calculated values of α, β, and γ for rat red cells are presented in TABLE 5. The values are within a factor of 2. This difference between the ratio from the ion concentration and from the transport parameter is probably due to the difficulty of fully inhibiting the pump in rat red cells. Ouabain is a poor inhibitor of rodent red cell pumps. Since the K leak influx is only a small fraction of the total K influx, the residual pump flux may lead to a large error in the determination of the leak flux for K (estimated as the ouabain- and bumetanide-insensitive fluxes). For example, if it is assumed that ouabain only inhibits 95% of the pump flux (at $K_o = 2$ mM), then the leak flux should be revised to include 5% of the ouabain-sensitive flux. The value thus revised is close to the value predicted from the ion concentrations.

In most red cells that have been studied, the bumetanide-sensitive pathway makes a small, if not negligible contribution to the net cation fluxes at or near physiological conditions and thus contributes little to the steady-state volume. However, it seems important to take account of this flux (presumably Na/K/Cl cotransport) when checking to see how well the (extended) pump/leak model works since the bumetanide-sensitive flux is often a significant part of the ouabain-insensitive unidirectional flux (especially for K). A better measure of the leak flux is often the bumetanide- and ouabain-insensitive flux, provided that these substances fully inhibit their targeted fluxes. In human red cells, the ratio of the ouabain-sensitive K flux to the ouabain-insensitive K flux is less than the value predicted from the K and Cl gradients. The ratio of ouabain-sensitive K flux to the ouabain- and bumetanide-insensitive K flux is closer to the predicted value, but still significantly different (data not shown). If the ouabain- and bumetanide-insensitive K flux were also to include a contribution from a

TABLE 4. Transport Parameters

$$\alpha' = \frac{\text{leak Na influx}}{\text{leak K influx} \times \text{N}} = \frac{{}^{i}k^{L}_{Na}}{{}^{i}k^{L}_{K}} = \frac{K_iCl_i - K_oCl_o}{Na_oCl_o - Na_iCl_i}$$

$$\beta' = \frac{\text{mediated K pathway net influxes}}{\text{rate constant for K leak}} = \frac{P}{{}^{i}k^{L}_{K}} = \frac{K_iCl_i - K_oCl_o}{Cl_o}$$

$$\gamma = \frac{\text{mediated Na pathway net influxes}}{\text{rate constant for Na leak}} = \frac{NP}{{}^{i}k^{L}_{Na}} = \frac{Na_oCl_o - Na_iCl_i}{Cl_o}$$

Note that the definition of β' is slightly different from β as used by Tosteson and Hoffman.[1] β' is the ratio of pump K influx to K leak rate constant whereas β is the ratio of pump K influx to K leak influx. Note also that the three parameters are not independent, e.g., $\alpha' = \beta'/\gamma$.

TABLE 5. Comparison of Ion Ratios and Transport Parameters

Parameter	Ion Ratio	Transport Parameter
β'	$\frac{K_iCl_i}{Cl_o} - K_o = 90$	$P/ik_K^L = 44$
γ	$Na_o - \frac{Na_iCl_i}{Cl_o} = 143$	$NP/ik_{Na}^L = 77$

mediated K/K exchange pathway, the value would then be even closer to the predicted value.

Note that a K/Cl cotransport pathway may behave as a K leak when the concentrations of both ions are less than saturating. This probably explains why the pump/leak model worked so well for LK sheep cells, which are now known to have a Cl-dependent K transport system,[11] though the fit still requires experimental verification.

K_i/K_o RATIO AS A PREDICTOR OF TRANSPORT PATHWAYS

For a Na/K pump/leak system, K must be accumulated since the pump mediates Na/K exchange. On the other hand, a strictly Na/K/Cl cotransport/leak system (which appears to require some as yet unspecified energy source[12] to drive net cation efflux) both K and Na must be extruded, i.e., both intracellular concentrations must be less than the concentration expected from passive distribution. When K is exactly passively distributed, either the K leak is much larger than any mediated pathway or the fluxes via the mediated pathway exactly cancel, e.g. pump K influx equals cotransport K efflux.

Though red cells from some sheep and cattle have low K_i, the K_i is considerably greater than that predicted from a strictly passive distribution, consistent with a relatively large portion of the mediated K flux being due to the Na/K pump. On the other hand, in red cells from carnivores, e.g., dogs and ferrets, K is essentially passively distributed, consistent with either no active K flux or a mediated downhill flux exactly balanced by pump-mediated K influx. In dog red cells, the former is apparently the case as there is no ouabain-sensitive K influx. Qualitatively these cells are thought to maintain their volume by extruding Na via a Na/Ca exchange system; the Ca gradient is maintained by a Ca pump. The equations derived here allow a quantitative assessment of such a regulatory process.

In recent years, the dog red cell cation pathways that respond to osmotic and volume changes have received the most attention.[13] The difficulties associated with estimation of Na/Ca exchange and Ca pump activity under physiological or steady state conditions have been avoided by studying the characteristics of these pathways in non–steady states where the fluxes became magnified. Nevertheless, it is of interest to extrapolate from the currently available results to determine if the ratio of Na/Ca exchange to Na leak fit with the predictions from the ion concentration ratios. By assuming that 4 Na ions are extruded for each Ca ion that enters, the net Na efflux via Na/Ca exchange can be estimated from the Ca influx. (Note that the stoichiometry of Na/Ca exchange in dog red cells apparently varies greatly depending upon conditions.) The Ca influx has not been measured under physiological Ca_o and Na_o, but its value can be estimated from the value of $K_{1/2}$ for Ca_o when $Na_o = 0$ ($K_{1/2}$ is 100 μM,

FIGURE 2 of Parker[13]) and the value of an inhibitory constant, K_I, for Na_o inhibition of Ca influx at $Ca_o = 40\ \mu M$ is 5–10 mM.[b] At 5 mM Ca_o and 140 mM Na_o the Ca influx via Na/Ca exchange would be about one eighth of V_{max}, i.e., 1 mmole/l cells hr. Thus, the Na efflux via Na/Ca exchange at normal plasma Na and Ca concentration would be 4 mmoles Na/(1 cells × hr). The Na leak rate constant is ~0.07 hr^{-1} (FIGURE 7 of Parker[13]). The transport parameter, γ, representing the ratio, net Na efflux/$^{i}k^{L}_{Na}$) is estimated to be 4/0.07 = 57. The steady-state Na flux equation predicts that this parameter, γ, is equal to $Na_o - Na_i\ Cl_i/Cl_o = 50$.[c] Since a number of assumptions were required to estimate Na/Ca exchange the agreement between the flux parameters and the ion concentrations may be fortuitous. Obviously, a more rigorous test should include the direct determination of the Na/Cl flux under physiological conditions. Such information and application of the Na steady-state equation[c] should help clarify the role of Na/Ca exchange for dog red cell volume regulation. Lack of agreement between the transport parameters and the ion ratios would not eliminate an alternative Na/Ca exchange hypothesis, that volume changes *in vivo* turn on Na/Ca exchange that leads to volume correction and turn off of Na/Ca exchange (FIGURE 4 of Parker[14]).

Ferret red cells have very large bumetanide-sensitive chloride-dependent Na and K fluxes.[15] The K flux requires Na and the Na flux requires K, so this appears to be a Na+K cotransport system. Our present understanding of Na+K cotransport in other systems is that it is passive and downhill. If this is the case in ferrets, then this system cannot be responsible for maintaining the *in vivo* steady-state cell volume. Energy is required to maintain a steady volume; if cation leaks exist one needs to extrude net (Na+K). Since the cotransport flux in ferrets is inhibited in ATP-depleted cells,[16] it is possible that this system could operate uphill. It should be pointed out though, that since K is passively distributed, and the bumetanide-insensitive fluxes are small, that the bumetanide-sensitive Na + K net flux must be nearly 0 *in vivo* so even an active bumetanide-sensitive Na+K cotransport does not appear to occur. Indeed Flatman[15] reports little or no net bumetanide-sensitive K movements under plasma ion concentrations of Na and K. In red cells such as those of the ferret, a K extrusion mechanism is not alone sufficient because even when a K_i is maintained at 0 (and Na passively distributed) there is not enough cation extrusion to provide for proper volume regulation. Thus, for cell volume control the ferret red cell requires some form of active extrusion of Na. One possibility is active bumetanide-sensitive Na (or Na + Cl) extrusion. Preliminary results[16] suggest that vanadate, an inhibitor of several types of cation transport systems, cannot inhibit a Na efflux pathway in these cells. In addition vanadate also inhibits uphill Ca extrusion and Ca stimulated ATPase activity,[16a] suggesting that ferret red cells have a Ca pump. Does a Na/Ca exchange system operating in opposition to a Ca pump provide a mechanism for maintaining steady-state cell volume in the ferret? Preliminary evidence[16a] suggests that ferret red cells do in fact possess a Na/Ca exchange system. For example, Ca_o stimulates Na efflux and

[b]The inhibitory constant, K_I, for Na_o is estimated from the concentration dependence of Na inhibition of Ca entry at 40 μM (FIGURE 3 of Parker[13]), where K_I is defined by

$$V = \frac{V_{max}\,Ca_o}{K_{1/2}\left(1 + \dfrac{Na_o}{K_I}\right) + Ca_o},$$

which assumes simple competition between Na_o and Ca_o.

[c]Na/Ca exchange mediated net Na efflux + Na leak efflux = Na leak influx.

Ca influx can be inhibited either by replacing Li_o with Na_o or by replacing of Na_i with Li_i. Furthermore, in the absence of Na_o, the cells can accumulate up to 500 μmoles Ca_i/l · cells from an initial value of about 10 μmoles Ca_i/1 cells in 20 min when Ca_o is 50 μM. The Na/Ca stoichiometry in different experiments varied from 3.5–4.5 and was relatively insensitive to changes in cell volume. Thus, the amount of Na/Ca exchange that occurs at normal (plasma) concentrations of Na and Ca can be estimated to be four times the Ca influx when Ca_o is 2 mM and Na_o is 140 mM. Measurements of Ca influx varied between 0.5–0.7 mM Ca/l · cells × hr (4 experiments) under these conditions. The Na leak was ~0.05 hr^{-1} so the transport parameter, γ, = 4 × (0.5 − 0.7)/0.05 = 40 to 56. The ion concentrations predict a value of 40. Thus, Na efflux, driven out by Na/Ca exchange, appears to be sufficient to balance the influx of Na through the leak. Under the same conditions vanadate-inhibitable Ca efflux operates to keep Ca_i very low in normal cells, since Ca_i accumulates in the presence of vanadate but not in its absence.[d] These preliminary results[16a] point to a common mechanism for steady-state volume control in ferret and dog red cells at their normal equilibrium cell volumes *in vivo*. When either cell type is perturbed away from its normal steady-state volume then volume-dependent pathways that are bumetanide-sensitive (Na/K cotransport) or amiloride-sensitive (Na/H exchange) appear to become activated.

Transport Responses and Cell Volume Regulation

Some types of red blood cells when placed in anisotonic media are known[17] to regulate the return of their volume back to normal. Although some kind of pump is needed to balance opposing leaks in order to maintain the normal steady-state cell volume and ion distribution, various ion exchange systems, cotransport systems and channels have been implicated as the pathways activated in various cell types during acute volume adjustments under volume-perturbed conditions. This division of labor appears to allow for efficient energy utilization during unstressed times and for rapidity of response to stress. In this way, the regulation of cell volume is analogous to the regulation of the membrane potential (E_m) in excitable cells since the pumps are necessary to maintain the ion gradients that create the E_m. The rapid changes in E_m in response to stimuli are mediated by other systems, usually channels. A problem with this analogy is that there are well documented physiological stimuli that cause changes in E_m and responses to changes in E_m whereas in mammals, the osmolarity of the plasma is well controlled so that *in vitro* cell volume regulation in response to osmotic stress does not seem to be the physiological role of cell volume–dependent transport pathways in blood cells.[18] In some respects the analogy seems particularly apt in the case of lymphocytes since the volume-responsive ion transport pathways are time and volume dependent.[19] In other cells, it appears that the fluxes stimulated by placement in an anisotonic media persist at an elevated level, even after the cells have returned to their normal volume (in an anisotonic medium).[20–22]

The pump/leak model and the steady-state flux equations discussed above can be used as a preliminary tool to evaluate hypotheses concerning the extent of involvement of the different flux pathways in the original steady-state cell volume and the steady-state cell volume assumed after recovery from the perturbed state. It is

[d]The Ca leak is small in ferret red cells; there is almost no increase in Ca influx when Ca_o is raised from 2 to 100 mM.

important to note that the equations presented here only apply to the steady state and it is often difficult to be certain that cells after returning to nearly normal volume are in fact in the steady state. This is apparent for instance when ouabain is added to cells that have volume readjusted in anisotonic media because the response is often only a slow change of cell volume even though in the long term cell swelling would occur. For this reason cells in an anisotonic media with ouabain probably do not achieve a true steady state near their original or readjusted cell volume. It seems plausible, however, that cells in an anisotonic media without ouabain may reach a new steady state. Consider the response of amphiuma red cells to hypotonic stess.[20] In the absence of ouabain, after several hours, there is almost no further change in cell volume and Na_i and K_i. In contrast, when ouabain is present, though there is only a small change in cell volume with time after the cell has returned to normal volume, Na_i and K_i continue to change (FIGURE 2 of Seibens & Kregenow[20]).

When duck red cells are swollen by placement in a hypotonic medium (e.g., 88 mM Na_o) the cells undergo a volume regulatory decrease by activating a K/Cl cotransport pathway.[21] Is the activation of this pathway sufficient to result in the final steady state or must any other transport pathways (pump or leak) also change? Consider the steady-state flux equation 3 for Na: Na leak influx = Na leak efflux + net Na pump efflux, which is equivalent to

$$Na_i = \frac{Cl_o}{Cl_i}\left(Na_o - \frac{NP}{{}^{i}k^{L}_{Na}}\right). \tag{19}$$

At isotonicity, steady state Na_i is low and Na_o (150 mM) is only slightly larger than the value of the fraction, $NP/{}^{i}k^{L}_{Na}$, whose value is approximately 140 mM. If $NP/{}^{i}k^{L}_{Na}$ does not change when the NaCl concentration is reduced, then the steady-state equation predicts a negative Na_i since the lower Na_o (e.g., 88 mM) is less than the value of $NA/{}^{i}k^{L}_{Na}$ or 140 mM. Thus at least one Na transport pathway must change.

A similar type of analysis can be applied to amphiuma red cells: hypotonic stess activates a K efflux pathway (probably K/H exchange[17]) but must also alter the ratio $NP/{}^{i}k^{L}_{Na}$ if Na_i is to remain positive. Seibens and Kregenow[20] report that the pump flux increases about three fold; this only worsens the situation since one predicts from the steady-state equation a decrease in $NP/{}^{i}k^{L}_{Na}$ as Na_o decreases. However, since there is a four-fold increase in Na influx under this condition and since the Na concentration has decreased by about one third, the Na influx rate constant must increase about six fold[20]; thus $NP/{}^{i}k^{L}_{Na}$ decreased by one half compared to a one-third decrease in Na_o. The steady-state flux equation can be used to determine what types of changes must occur between an old and new steady state. Unfortunately, this theory can provide no information on why or how the increase in pump flux and large increase in leak flux take place. As pointed out by Siebens and Kregenow,[20] both responses are unexpected from the changes in ion concentrations since Na_i actually decreases as the pump flux increases and Na_o decreases as the Na influx increases.

Volume regulatory decrease to a steady-state cell volume during hyposomotic stress apparently requires not only the activation of a K loss pathway but also requires adaptive changes in Na fluxes. Thus it may be that the inability of many types of mammalian red cells to volume regulate reflects the lack of an adaptive response in the cell's Na transport pathways. In this regard, it is interesting to note that human red cells containing hemoglobin CC can volume regulate[23]; analysis by means of the Na steady-state equation indicates that this process involves an adaptation of the Na pathways in addition to the activation of a K pathway.

REFERENCES

1. Tosteson, D. C. & J. F. Hoffman. 1960. Regulation of cell volume by active cation transport in high and low potassium sheep red cells. J. Gen. Physiol. **44:** 169–194.
2. Chan, P. C. & W. R. Sanslone. 1969. The influence of a low-potassium diet on rat-erythrocyte membrane adenosine triphosphatase. Archiv. Biochem. Biophys. **134:** 48–52.
3. Cumberbatch, M. & D. B. Morgan. 1983. Erythrocyte sodium and potassium in patients with hypokalaemia. Clin. Sci. **64:** 167–176.
4. Lamb, J. F., J. F. Aiton & P. Ogden. 1982. K^+ depletion and Na^+ pump density. Nature **295:** 717–718.
5. Rubython, E. J. & D. B. Morgan. 1983. Effect of hypokalaemia on the ouabain-sensitive sodium transport and the ouabain-binding capacity in human erythrocytes. Clin. Sci. **64:** 177–182.
6. Duhm, J., B. O. Gobel & F.-X. Beck. 1983. Sodium and potassium ion transport accelerations in erythrocytes of DOC, DOC-salt, one clip-two kidney and spontaneously hypertensive rats. The role of hypokalemia and cell volume. Hypertension **5:** 642–652.
7. Hoffman, J. F. & P. C. Laris. 1974. Determination of membrane potentials in human and *Amphiuma* red blood cells by means of a fluorescent probe. J. Physiol. **239:** 519–552.
8. Funder, J. & J. O. Weith. 1967. Effects of some monovalent anions on fluxes of Na and K, and on glucose metabolism of ouabain treated red cells. Acta Physiol. Scand. **71:** 168–185.
9. Leaf, A. & R. S. Cotran. 1985. *In* Renal Pathophysiology. 3rd edit. Oxford University Press. New York.
10. Korff, J. M., A. W. Siebens & J. R. Gill, Jr. 1984. Correction of hypokalemia corrects the abnormalities in erythrocyte sodium transport in Bartter's Syndrome. J. Clin. Invest. **74:** 1724–1729.
11. Dunham, P. B. & J. C. Ellory. 1981. Passive potassium transport in low potassium sheep red cells: dependence upon cell volume and chloride. J. Physiol. (Lond.) **317:** 511–539.
12. Hoffman, J. F. & F. M. Kregenow. 1966. The characterization of new energy dependent cation transport processes in red blood cells. Ann. N.Y. Acad. Sci. **137:** 566–576.
13. Parker, J. C. 1978. Sodium and calcium movements in dog red blood cells. J. Gen. Physiol. **71:** 1–17.
14. Parker, J. C. 1977. Solute and water transport in dog and cat red blood cells. *In* Membrane Transport in Red Cells. J. C. Ellory & V. L. Lew, Eds.: 427–465. Academic Press. New York.
15. Flatman, P. W. 1983. Sodium and potassium transport in ferret red cells. J. Physiol. (Lond.) **341:** 545–557.
16. Mercer, R. W. & J. F. Hoffman. 1985. Bumetanide-sensitive Na/K cotransport in ferret red blood cells. Biophys. J. **47:**157a.

16a. Milanick, M. & J. F. Hoffman. 1986. Na/Ca exchange and Ca pump fluxes in ferret red blood cells. J. Gen. Physiol. (In press.)

17. Siebens, A. 1985. Cellular volume control. *In* The Kidney: Physiology and Pathophysiology. D. W. Seldin & G. Giebisch, Eds.: 91–115. Raven Press. New York.
18. Haas, M. & T. J. McManus. 1985. Effects of norepinephrine on swelling-induced potassium transport in duck red cells: evidence against a volume-regulatory decrease under physiological conditions. J. Gen. Physiol. **85:** 649–667.
19. Sarkadi, B., E. Mack & A. Rothstein. 1984. Ionic events during the volume response of human peripheral blood lymphocytes to hypotonic media II. Volume- and time-dependent activation and inactivation of ion transport pathways. J. Gen. Physiol. **83:** 513–527.
20. Seibens, A. W. & F. M. Kregenow. 1985. Volume-regulatory responses of *Amphiuma* red cells in anisotonic media. The effect of amiloride. J. Gen. Physiol. **86:** 527–564.
21. Kregenow, F. M. 1971. The response of duck erythrocytes to non-hemolytic hypotonic media: evidence for a volume controlling mechanism. J. Gen. Physiol. **58:** 396–412.
22. McManus, T. J. & W. F. Schmidt, III. 1978. Ion and co-ion transport in avian red cells. *In* Membrane Transport Processes. J. F. Hoffman, Ed. **1:** 79–106. Raven Press. New York.

23. BRUGNARA, C., A. S. KOPIN, H. F. BUNN, & D. C. TOSTESON. 1985. Regulation of cation content and cell volume in hemoglobin erythrocytes from patients with homozygous hemoglobin C disease. J. Clin. Invest. **75:** 1608–1617.

DISCUSSION OF THE PAPER

A. P. SOMLYO (*University of Pennsylvania, Philadelphia, PA*): Do cells swell in vanadate?

(2) Do ATPases provide energy or volume regulation?

MILANICK (*Yale University, New Haven, CT*): We have incubated red cells in vanadate and also in the absence of Ca for three hours and have only observed a slight swelling. However, because of the low net cation fluxes of red cells this experiment is not definitive. Indeed, it is difficult to demonstrate that human red cells swell when incubated with ouabain for 3–5 hours.

E. RUBIN (*Hahnemann University, Philadelphia, PA*): Your data indicate adaptive responses in nucleated red blood cells in hypertension. Why study non-nucleated red blood cells (human) in hypertension, when the target cell, the vascular smooth muscle cell, is nucleated and capable of adaptive responses?

MILANICK: Non-nucleated red cells have not been shown to volume regulate when osmotically shocked, i.e., reach a new steady state. It may be that this response takes a long time. Nevertheless, Prof. Bianchi and others have shown that non-nucleated cells do have volume-dependent pathways. If it is a volume-dependent pathway that is an early defect in hypertension, and if the defect is present in the target cell and in the red cell, then study of the red cell may help in detecting the defect.

G. BIANCHI (*University of Milan, Milan*): When considering the role of cell volume on ion transport regulation in relation to hypertension we should also include the role of long term determinants of cell volume regulation. In fact, as Dr. Ferrari is showing in her poster, the lower volume of RBC of MHS persists also in resealed ghosts placed in a solution of different osmotic pressure. This means that the difference in volume is caused by a primary change of the membrane that, by itself, may effect volume and ion transport.

MILANICK: I agree. When the red cell behaves as an ideal osmometer, the immediate determinants of cell volume are presumably the ion contents, which are controlled by the ion fluxes. I certainly agree that, in many cases, the intriguing problem is the long term regulation of the ion fluxes and, in part, the influence of plasma, cytosolic, and cytoskeletal factors.

CANESSA (*?*): Dr. Bianchi: in relation to your question, an important factor determining mechanisms of volume regulation is the cell-age dependence of the transporters.

MILANICK: I agree; age can certainly affect, directly or indirectly, ion transport and therefore cell volume.

Na^+ Transport in Primary Hypertension

RICARDO GARAY, CLELIA ROSATI,
AND PHILIPPE MEYER

Inserm U7
Hôpital Necker
161, rue de Sèvres
Paris 75015 France

INTRODUCTION

One major environmental factor contributing to the onset of high blood pressure is excess Na^+ intake.[1] In recent years, the considerable progress in the understanding of Na^+ metabolism at a molecular level has contributed to the elucidation of this causal relationship.[2–13]

MOLECULAR MECHANISMS OF Na^+ TRANSPORT ACROSS CELL MEMBRANES

Knowledge of Na^+ metabolism at a molecular level is not so advanced as that of oxygen, sugar, or lipid metabolism; this may be related to the fact that all known functional proteins of Na^+ metabolism are membrane transport proteins. They are therefore difficult to extract, purify, and almost impossible to crystallize.

The human red cell has been widely used for molecular studies of Na^+ transport. FIGURE 1 shows three different Na^+ transport systems that have been well characterized in these cells: a ouabain-sensitive Na^+,K^+-pump, which catalyzes the exchange of internal Na^+ for external K^+ coupled to the hydrolysis of ATP, thus generating electrochemical gradients of Na^+ and K^+ across the cell membrane; a furosemide (or bumetanide)-sensitive Na^+,K^+ cotransport system, which catalyzes inward and outward fluxes of Na^+ and K^+; and a Na^+,Na^+ countertransport system, which catalyzes a one-to-one exchange of internal for external Na^+ (Li^+ and perhaps H^+ may replace Na^+ in this system).

Unfortunately, Na^+ transport in other cells is not so well known as in the human erythrocyte. FIGURE 1 shows that the above three erythrocyte Na^+ transport systems are also present in different segments of the nephron: a ouabain-sensitive Na^+,K^+-pump is present in basolateral membranes all along the nephron, a furosemide (or bumetanide)-sensitive Na^+,K^+-cotransport system is located in the luminal side of Henle's loop cells, and Na^+,Na^+-countertransport appears to be the same as the Na^+:H^+ exchange catalyzing a fraction of luminal Na^+ reabsorption in the proximal tubule. Conversely, the amiloride-sensitive Na^+ channel of distal and collecting tubules does not appear to be represented in the human red cell membrane. It is important to note that some other Na^+ transport systems, such as Na^+:Ca^{2+} exchange, that are present in other cells appear to be lacking in human red cell membranes.

Cell Na^+ content of non-epithelial cells (like vascular smooth muscle cells) is the result of the activity of all membrane Na^+ transport systems (and channels). FIGURE 2 shows two different situations: (1) the stationary (basal) cell Na^+ content, which

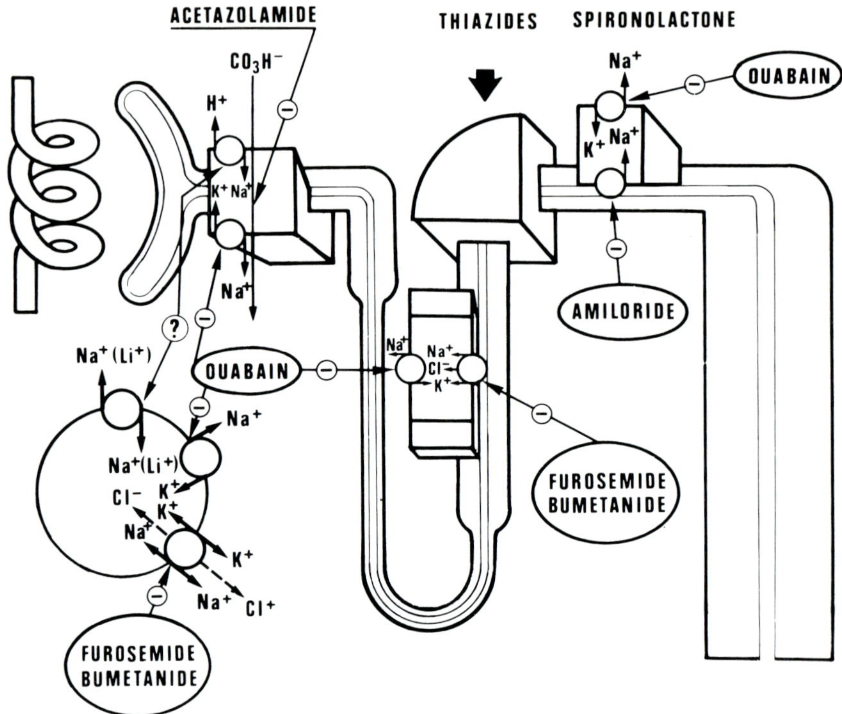

FIGURE 1. Na^+ transport systems in human red cells and kidney. Three erythrocyte Na^+ transport systems [the ouabain-sensitive Na^+,K^+-pump, the furosemide (or bumetanide)-sensitive Na^+,K^+ cotransport system, and the Na^+,Na^+ countertransport system] appear to be representative of those present in different segments of the nephron.

generally speaking depends on the balance between Na^+ entry by passive Na^+ permeability (membrane Na^+ leak) and Na^+ extrusion by the Na^+,K^+-pump and (2) cell Na^+-load, which is normally extruded by the Na^+,K^+-pump and the Na^+,K^+-cotransport system.

NATRIURETIC HORMONES

Excess Na^+ intake tends to induce volume expansion, which may be rapidly compensated by the transient secretion of at least two natriuretic hormones, the atrial natriuretic factor (ANF), which is a peptide present in specific granules of atrial cardiocytes[14] and the endogenous "ouabain-like" factor(s) (OLF), of unknown chemical structure and site of production. The site of action of ANF is the renal glomeruli where it increases glomerular filtration rate.[15] The mechanism of this effect may involve an increase in renal blood flow and in the area of glomerular filtration (through an indirect action on mesangial cells). At the molecular level the interaction of ANF with specific receptors in vascular smooth muscle cells provokes a cascade of events, stimulation of cyclic GMP production,[16] decrease in cytosolic free Ca^{2+} content,[17] and vasorelaxation.

The second natriuretic hormone, OLF, seems to inhibit the Na^+,K^+-pump.[4,5] At the tubular level, this results in a natriuretic effect by decrease of renal Na^+reabsorption. However, a similar pump inhibition at the vascular wall may transiently increase cell Na^+ content. FIGURE 2 shows that a normal membrane Na^+ transport function may rapidly ensure the extrusion of such cell Na^+ load.

STABLE Na^+ TRANSPORT ABNORMALITIES IN PRIMARY HYPERTENSION

In the past ten years, the extensive investigation of Na^+ transport mechanisms in circulating cells of humans and rats with primary hypertension strongly suggested that different genetic abnormalities can be associated with high blood pressure.

Studies in Human Hypertensives

FIGURE 3 shows four different and stable Na^+ transport abnormalities found in erythrocytes from essential hypertensive patients. These may induce transitory (or compensatory) changes in the same or in other Na^+ transport systems. The four stable Na^+ transport abnormalities can be summarized as follows.

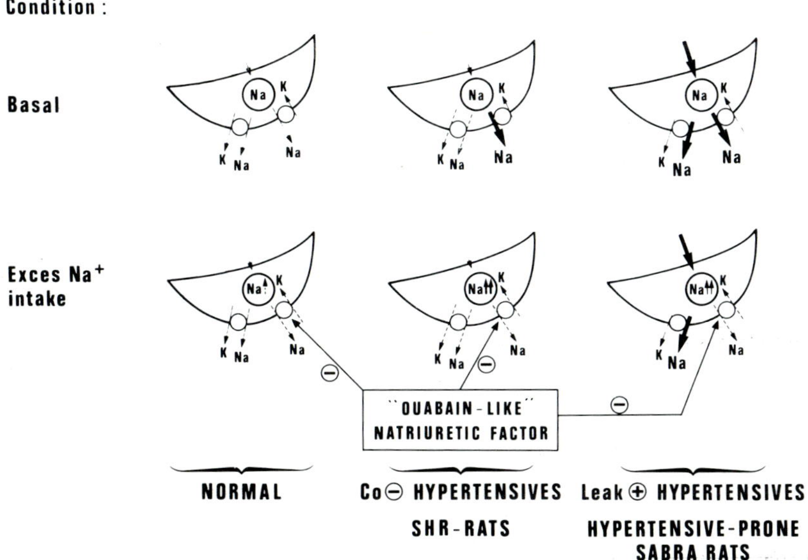

FIGURE 2. Abnormal cell Na^+ homeostasis under excess Na^+ intake in some forms of primary hypertension. A key step in the development of primary hypertension is the disturbance in cell Na^+ homeostasis induced by the interaction of "ouabain-like" factors with vascular cells having genetic abnormalities in membrane Na^+ transport.

Increased Passive Entry of Na^+ [Leak(+)]

This first Na^+ transport abnormality was suspected by Wessels *et al.* in 1967, measuring unidirectional ^{22}Na influx and recently confirmed by more specific studies[10,11,13] in about 15 to 30% of human hypertensives (Leak(+) hypertensives). A leak(+) abnormality may be compensated by an increase in the maximal rate of the Na^+,K^+ pump and the Na^+,K^+ cotransport system.[11] Indeed, internal Na^+ contents (and blood pressure levels) in Leak(+) hypertensives are inversely correlated to these compensatory phenomena.[11]

Decreased Apparent Affinity of the Na^+,K^+ Cotransport System for Internal Na^+ [Co(−)]

About 30 to 40% of human hypertensives (Co(−) hypertensives) exhibited a decreased apparent affinity of the Na^+,K^+ cotransport system for internal Na^+.[6,8] This

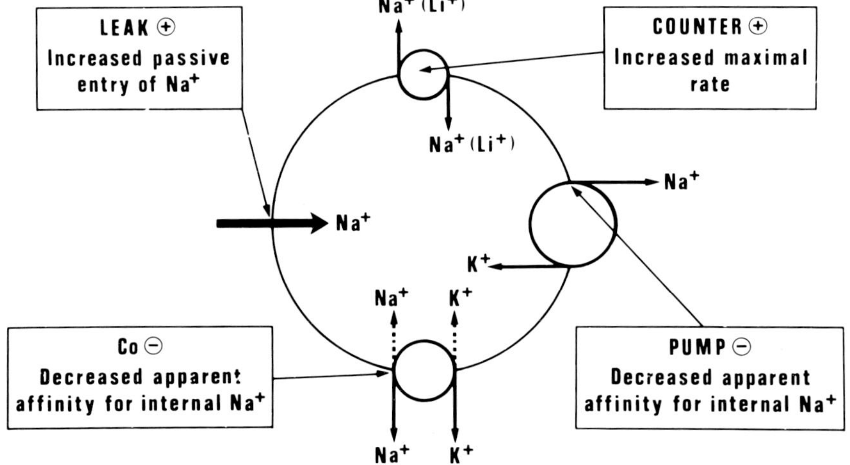

FIGURE 3. Stable Na^+ transport abnormalities in erythrocytes from essential hypertensive patients.

results in a decreased ability of the Na^+,K^+ cotransport system to extrude a cell Na^+ load. The maximal rate of the Na^+,K^+ cotransport system may be decreased, normal, or increased as a function of the severity of high blood pressure and several other factors besides hypertension.[6] In addition, most Co(−) hypertensives exhibit a normal interaction with external K^+.[18]

Increased Maximal Rate of Na^+:Li^+ Countertransport [Counter(+)]

Firstly described by Canessa *et al.* in 1980, this abnormality is present in about 20 to 40% of human hypertensives (Counter(+) hypertensives) and is frequently associated with an increased maximal rate of outward Na^+,K^+ cotransport.[6,19]

Decreased Apparent Affinity of the Na^+,K^+ Pump for Internal Na^+ [Pump(−)]

We recently observed a fourth stable Na^+ transport abnormality, decreased apparent affinity of the Na^+,K^+ pump for internal Na^+ in about 5–10% of human hypertensives.[20] This abnormality appears to be compensated by an increase in the maximal rates of the Na^+,K^+ pump and Na^+, K^+ cotransport system.

Studies in Genetically Hypertensive Rats

Studies of Na^+ transport systems in erythrocytes from genetically hypertensive rats showed varied results. Spontaneously Hypertensive Rats of the Okamoto strain (SHR) had a stable Co(−) abnormality (apparent affinity of the Na^+,K^+ cotransport system for internal Na^+ reduced by 50% in SHR rats aged from 2 to 26 weeks). Sabra Rats of the Hypertensive-prone substrain (HPS) had a stable Leak(+) abnormality that is compensated by increased maximal rates of the Na^+,K^+ pump and the Na^+,K^+ cotransport system.[3] Milano Hypertensive Rats exhibited increased maximal rates of the Na^+,K^+ cotransport system.[8]

PHYSIOPATHOLOGICAL IMPLICATIONS

The physiological relevance of each Na^+ transport system differs from cell to cell. It appears therefore that the physiopathological disturbances induced by each Na^+ transport abnormality may differ from organ to organ. For instance, Weder has shown that increased erythrocyte Na^+:Li^+ countertransport in human hypertensives is correlated with lower renal fractional lithium clearance.[21] This suggests that Counter(+) hypertensives have increased proximal tubular sodium reabsorption and thus their target organ is the kidney. Increased sodium reabsorption may induce volume expansion, permanent secretion of OLF, inhibition of the pump in vascular cells, and hypertension.[5]

In Co(−) hypertensives (and SHR rats) the target cells may be noradrenergic neurons. Indeed, a Co(−) abnormality in noradrenergic endings may lead to an abnormal recovery of cell Na^+ content after an action potential, thus explaining the increased noradrenergic activity of SHR and of Co(−) hypertensives.[13] In addition, the transitory secretion of endogenous OLF following excess Na^+ intake may unmask the Co(−) defect in vascular smooth muscle cells leading to a more pronounced increase in cell Na^+ content (FIGURE 2).[3]

Leak(+) hypertensives and HPS rats may present a similar disturbance in Na^+ handling by vascular smooth muscle cells than Co(−) hypertensives and SHR rats (FIGURE 2).[3]

The Link between Abnormal Cell Na^+ Handling and Hypertension

The final step in the sequence of events leading to primary hypertension involves cytosolic free Ca^{2+} content[2] and catecholamines. In the vascular wall, pump inhibition may partially depolarize cell membranes thus opening potential-dependent Ca^{2+} channels,[22] increase catecholamine output in noradrenergic endings[23] thus opening receptor-dependent Ca^{2+} channels, and stimulate Ca^{2+} influx through Na^+:Ca^{2+} exchange.[2] The final consequence is an increase in both cytosolic free Ca^{2+} content and peripheral arterial resistance.

CLINICAL ASPECTS

It is still too early to draw the clinical and therapeutical characteristics of the different subgroups of hypertensives. Nevertheless preliminary studies suggest that Co(−) hypertensives present labile or moderate hypertension[6,13,19] with elevated plasma norepinephrine.[13] Conversely, Counter(+) hypertensives exhibit more severe hypertension and increased renin activity.[24]

PHARMACOLOGICAL ASPECTS

The above studies suggested that a key step in the development of some forms of primary hypertension is the disturbance in cell Na^+ homeostasis induced by the interaction of "ouabain-like" factors with vascular cells having genetic abnormalities in membrane Na^+ transport. We therefore proposed that drugs that may "protect" the digitalis receptor site of vascular cells against pump inhibition are of potential interest as a new approach to the treatment of high blood pressure.[25]

Several arguments have suggested to us that canrenone, an antihypertensive drug, may interfere with the above mechanism. Canrenone is the active metabolite of

TABLE 1. Na^+,K^+-Pump in Human Erythrocytes Preincubated with Canrenone

Preincubation (4 hr)	Na^+,K^+-Pump Activity[a]		
	Control	42.5 nM Ouabain	% Inhibition
—	2310 ± 15	2075 ± 30	10.2%
1 μM Canrenone	2370 ± 10	2190 ± 10	7.6%
10 μM Canrenone	2365 ± 15	2265 ± 30	4.2%

Values in this table are given as mean ± range of duplicates. A similar result was obtained in two further experiments.

[a]Ouabain-sensitive Na efflux, in μmoles $(1 \cdot \text{cells} \times \text{hr})^{-1}$.

spironolactone. Its action mechanism was supposed to involve a competition with aldosterone for a common cytosolic receptor in distal and collecting tubules of the nephron. However, several authors have found that, in addition to this mechanism, canrenone is able to interact with the Na^+,K^+ pump *in vitro*.[25,26] This interaction is the one expected for a partial agonist at the digitalis receptor site. Under basal conditions, canrenone slightly inhibits Na^+,K^+-pump activity. This effect could explain the previous observation that canrenone administration (as K-canrenoate) potentiates the inotropic effect of digitalis in dogs.[27] If the pump is blocked by high doses of ouabain, canrenone is able to restimulate it. This effect is likely correlated with the ability of canrenone to protect against digitalis-induced cardiac toxicity[28] and to reverse the inhibitory effect of digoxin on basal and furosemide-stimulated renin secretion.[29]

It is important to note that the above experiments *in vitro* required canrenone concentrations one order of magnitude higher than those observed in plasma of treated patients (which are of about 8–10 μM[30]). We therefore investigated if preincubation with canrenone may elicit effects *in vitro* at pharmacological doses. TABLE 1 shows the effect of a 4 hr preincubation of human red cells with pharmacological doses of canrenone. It can be seen that the ability of low doses of ouabain to inhibit the pump is reduced by about one fourth and one half at canrenone concentrations of 1 and 10 μM, respectively. Under those conditions canrenone behaves like an antagonist at the digitalis receptor site.

Interestingly canrenone is able to partially antagonize *in vivo* the secondary effects of endogenous "ouabain-like" factors on cell Na^+ handling and blood pressure in rats with reduced renal mass under excess Na^+ intake.[31] These results suggest that chronic administration of canrenone to subjects with primary hypertension may induce a lowering of blood pressure by antagonism with endogenous "ouabain-like" factors at the vascular wall.

ACKNOWLEDGMENTS

We are greatly indebted to G. Dagher, M. De Mendonça, C. Nazaret, P. Hannaert, M.L. Grichois, M. Price, and J. Diez who worked in our laboratory in the past eight years.

REFERENCES

1. Dahl, L. K. 1977. Salt intake and hypertension. *In* Hypertension. J. Genest, E. Koiw & O. Kuchel, Eds.: 548–559. McGraw-Hill. New York.
2. Blaustein, M. 1977. Sodium ions, calcium ions, blood pressure regulation and hypertension: A reassessment and a hypothesis. Am. J. Physiol. **232:** C165–C173.
3. DeMendonca, M., A. Knorr, M. L. Grichois, D. Ben-Ishay, R. Garay & P. Meyer. 1982. Erythrocytic ion transport systems in primary and secondary hypertension of the rat. Kidney Int. **21** (Suppl. 11): S69–S75.
4. Buckalew, V. M. & K. A. Gruber. 1983. Natriuretic hormone. *In* The Kidney in Liver Disease. M. Epstein, Ed.: 479–499. Elsevier Biomedical. New York.
5. DeWardener, H. E. & G. A. MacGregor. 1983. The natriuretic hormone and its possible relationship to hypertension. *In* Hypertension. J. Genest, O. Kuchel, P. Hamet & M. Cantin, Eds.: 84–95. McGraw Hill. New York.
6. Garay, R. P., C. Nazaret, P. Hannaert & M. Price. 1983. Abnormal Na^+,K^+ cotransport function in a group of patients with essential hypertension. Eur. J. Clin Invest 13:311–320.
7. Canessa, M., Spalvins, N. Adragna & B. Falkner. 1984. Red cell sodium countertransport and cotransport in normotensive and hypertensive blacks. Hypertension **6:** 344–351.
8. Cusi, D., C. Barlassina, P. Ferrari, M. Ferrandi & G. Bianchi. 1984. Cation transport abnormalities in human and rat essential hypertension. *In* Topics in Pathophysiology of Hypertension. H. Villarreal & M. P. Sambhi, Eds.: 136–146. Martinus-Nijhoff Publishers. Boston.
9. Postnov, Y. V. & S. N. Orlov. 1985. Ion transport across plasma membrane in primary hypertension. Physiol. Rev. **65:** 904–945.
10. Wessels, F. & H. Zumkley. 1985. New aspects concerning the ^{22}Na influx into red cells in essential hypertension. Klin. Wochenschr. **63**(Suppl III): 38–41.
11. Garay, R. P. & C. Nazaret. 1985. Na^+ leak in erythrocytes from essential hypertensive patients. Clin. Sci. **69:** 613–624.
12. Behr, J., H. Witzgall, R. Lorenz, P. C. Weber & J. Duhm. 1985. Red cell Na^+-K^+ transport in various forms of human hypertension. Role of cardiovascular risk factors and plasma potassium. Klin. Wochenschr. **63**(Suppl III): 63–65.
13. Mongeau, J. G. 1985. Erythrocyte cation fluxes in essential hypertension of children and adolescents. Int. J. Ped. Nephrol. **6:** 41–46.
14. De Bold, A. 1985. Atrial natriuretic factor: A hormone produced by the heart. Science **230:** 767–770.
15. Atlas, S. A., H. D. Kleinert, M. J. Camargo, A. Januszewicz, J. E. Sealey, J. H. Laragh, J. W. Schilling, J. A. Lewicki, L. K. Johnson & T. Maack. 1984.

Purification, sequencing and synthesis of natriuretic and vasoactive rat atrial peptide. Nature **309:** 717–719.

16. HAMET, P., J. TREMBLAY, S. PANG, R. GARCIA, G. THIBAULT, J. GUTKOWSKA, M. CANTIN & J. GENEST. 1984. Effect of native and synthetic Atrial Natriuretic Factor on cyclic GMP. Biochem. Biophys. Res. Commun. **123:** 515–527.
17. GARAY, R., P. HANNAERT, F. RODRIGUE, B. DUNHAM, P. MARCHE, J. GENEST, P. BRAQUET, C. BIANCHI, M. CANTIN & P. MEYER. 1985. Atrial natriuretic factor inhibits Ca-dependent, K-fluxes in cultured vascular smooth muscle cells. J. Hypertension **3**(suppl 3): S297–S298.
18. PRICE, M., P. HANNAERT, G. DAGHER & R. P. GARAY. 1984. The interaction of internal Na^+ and external K^+ with the erythrocyte Na^+, K^+ cotransport system in essential hypertension. Hypertension **6:** 352–359.
19. ADRAGNA, N., M. CANESSA, H. SOLOMON, E. SLATER & D. C. TOSTESON. 1982. Red cell lithium-sodium countertransport and sodium-potassium cotransport in patients with essential hypertension. Hypertension **4:** 795–804.
20. GARAY, R., J. DIEZ & P. BRAQUET. 1985. Na,K-pump in essential hypertension. *In* The Sodium Pump. I. Glynn & C. Ellory, Eds.: 657–662. The Company of Biologists Limited. Cambridge, UK.
21. WEDER, A. B. 1986. Red cell lithium countertransport and renal lithium clearance in hypertension. N. Eng. J. Med. **314:** 198–201.
22. MULVANY, M. J. 1985. Changes in sodium pump activity and vascular contraction. J. Hypertension **3:** 429–436.
23. NAKAZATO, Y., A. OHGA & Y. ONODA. 1978. The effect of ouabain on noradrenaline output from peripheral adrenergic neurons of isolated guinea pig vas deferens. J. Physiol. (Lond.) **278:** 45–54.
24. BRUGNARA, C., R. CORROCHER, L. FORONI, M. STEINMAYR, F. BONFANTI & G. DE SANDRE. 1983. Lithium-sodium countertransport in erythrocytes of normal and hypertensive subjects: relationship with age and plasma renin activity. Hypertension **5:** 529–534.
25. GARAY, R. P., J. DIEZ, C. NAZARET, G. DAGHER & J. P. ABITBOL. 1985. The interaction of canrenone with the Na,K-pump in human red blood cells. Naunyn-Schmiedeberg's Arch. Pharmacol. **329:** 311–315.
26. FINOTTI, P. & P. PALATINI. 1981. Canrenone as a partial agonist at the digitalis receptor site of sodium-potassium activated adenosine triphosphatase. J. Pharmacol. Exp. Ther. **217:** 784–790
27. MARCHETTI, G., E. VITOLO, G. F. DIFRANCESCO, B. CAVALLARO & C. SPONZILLI. 1983. Positive inotropic effect of K-Canrenoate: an investigation in anaesthetized dogs untreated or treated with digitalis. Arch. Int. Pharmacodyn. Ther. **266:** 250–263.
28. YEH, B. K., B. N. CHIANG & P. K. SUNG. 1976. Antiarrhythmic activity of K-Canrenoate in man. Am. Heart J. **92:** 308–314.
29. FINOTTI, P. & A. ANTONELLO. 1982. Canrenoate reversal of inhibitory effects of digoxin on basal and furosemide-stimulated renin secretion. Clin. Pharmacol. Ther. **32:** 1–6.
30. MUGGE, A., W. SCHMITZ & H. SCHOLZZ. 1984. Negative inotropic effects of aldosterone antagonists in isolated human and guinea-pig ventricular heart muscle. Klin. Wochenschr. **62:** 717–723.
31. DEMENDONCA, M., M. L. GRICHOIS, M. G. PERNOLLET, B. THORMANN, P. MEYER, M. A. DEVYNCK & R. P. GARAY. 1985. Hypotensive action of canrenone in a model of hypertension where a ouabain-like factor is present. J. Hypertension. **3** (suppl. 3): 573–575.

DISCUSSION OF THE PAPER

Q. AL-AWQATI (*Columbia University, New York, NY*): Does canrenone displace labeled ouabain from membranes and if so what is the EC_{50}?

GARAY: Erdmann has shown that canrenone displaces [^{3}H]ouabain from heart cell membranes. Finotti and Palatini showed the same in purified Na^+,K^+-ATPase from guinea pig brain. The EC_{50} was between 10^{-5} and 10^{-4} M.

G. BIANCHI (*University of Milan, Milan*): Did you try other diuretics to see whether for the same increase in Na^+ excretion the fall in blood pressure was greater with canrenone?

GARAY: We compared the diuretic effect of canrenone in rats with reduced renal mass with that of furosemide and thiazides. In contrast with those diuretics, the diuretic effect of canrenone is slight and only lasts 24 hr. Blood pressure fall is not correlated with such diuretic effects.

L. H. OPIE (*University of Cape Town, Observatory, South Africa*): (1) Did the leak abnormalities in hypertensions change with therapy?

(2) Was there an overlap between leak and cotransport abnormalities in hypertensive patients?

GARAY: (1) We studied the effect of a new diuretic (cicletanine) on ion transport in 55 hypertensives treated during two years. None of the four abnormalities (Pump(−), Leak(+), Co(−), or Counter(+)) was changed by this treatment. I do not know what happens with other antihypertensive drugs.

(2) In our first studies we only observed patients with one (or none) stable abnormality. However, in recent studies we have found some patients presenting two abnormalities. We also found a patient (and only one) with three abnormalities (Co(−), Pump(−), and Leak(+)).

The Cytoskeleton of Rat Aortic Smooth Muscle Cells

Normal Conditions, Experimental Intimal Thickening, and Tissue Culture

G. GABBIANI

Department of Pathology
University of Geneva
1211 Geneva 4
Switzerland

It is presently well accepted that smooth muscle cells (SMC) are the major contributor to the formation of atheromatous plaque.[1–4] However, little is known about morphologic and/or biochemical differences between SMC of the media and atheromatous SMC.[1,2,5] The cytoskeletal features of SMC in the arterial wall have been clarified only recently:[6–10] practically all SMC contain intermediate filaments (IF) composed of vimentin and many (more than 50%) contain IF composed of desmin. However, vascular SMC are characterized by the prevalence of a special α-actin isotype.[7,9] It has been shown that the large majority of rat aortic SMC located in intimal thickening after endothelial injury (the most commonly used model for the atheromatous plaque) contain IF composed only of vimentin.[8] This suggests that changes in cytoskeletal elements may represent useful markers of SMC adaptation during experimental and possibly human atheromatosis. We have therefore compared the cytoskeletal features of rat aortic SMC during normal conditions, experimental intimal thickening, and *in vitro* culture.[11,12] Our results show that the modifications of cytoskeletal elements during intimal thickening or tissue culture furnish new useful information on SMC reaction to stimuli inducing their replication and movement.

NORMAL CONDITIONS AND ISOLATED CELLS

Total tissues were treated as previously described.[12] Cells were isolated by means of enzymatic digestion.[13] Vimentin- and desmin-positive cells were identified by means of indirect immunofluorescent staining using affinity-purified polyclonal antibodies against vimentin purified from eye lens and desmin purified from chicken gizzard.[12] For the biochemical evaluation of total actin we first determined the quantities of DNA and protein per mg of tissue in the different specimens;[12] then, we evaluated actin as percentage of total protein by densitometric analysis of sodium dodecyl sulfate polyacrylamide gels. These data, correlated with the estimations of DNA and total proteins per mg of tissue, allowed us to calculate the quantities of actin per cell.[12] The proportion of actin isoforms was calculated in two-dimensional gels according to Quitschke and Schechter.[14] The values were obtained as percentages of total actin in the same gel.

Results using total tissues or isolated cells were similar and hence will be reported together. The percentage of vimentin-positive and vimentin plus desmin–positive cells was 30 and 70, respectively. The content of actin per cell was about 20 pg and the proportions of α-, β-, and γ-actin isoforms were 58%, 29%, and 13%, respectively.

INTIMAL THICKENING AFTER ENDOTHELIAL INJURY

Fifteen days after balloon-induced endothelial injury, the percentage of vimentin-positive cells in the intimal thickening was 79 and the percentage of vimentin plus desmin-positive cells was 21. Total actin per cell decreased to 13.62 pg and the proportions of actin isoforms was 35% for α, 45% for β, and 20% for γ. Seventy-five days after endothelial injury, when the lesion was reendothelialized and SMC were not replicating, the percentage of vimentin-positive cells was 48 and the percentage of vimentin plus desmin–positive cells 58. The content of actin per cell was 18 pg and the proportions of different actin isoforms were 55% for α, 30% for β, and 15% for γ, practically the same as that of normal cells.

PRIMARY CONFLUENT CELLS

SMC cultured in 10% fetal calf serum reached confluence about 7 days after seeding. At that time the percentage of vimentin-positive cells was about 52 and the percentage of vimentin plus desmin–positive cells about 48. The content of actin per cell was about 15 pg and the proportions of actin isoform were 34% for α, 43% for β, and 23% for γ.

DISCUSSION

Our results show that in experimental intimal thickening observed in the rat aorta 15 days after balloon-induced endothelial injury, distinct cytoskeletal changes appear in SMC compared with normal media SMC. These consist of increased amount of cells containing only vimentin and decreased amount of cells containing in addition desmin, decreased actin content per cell, and a switch in pattern of actin isoform expression from a predominant α-smooth muscle type to a predominant β-type and an increased γ-type. These changes are present also in primary cultures of SMC in the presence of 10% FCS and are similar to those previously reported in human aortic atheromatous plaques.[11] All the described intimal changes regress by 75 days after injury when endothelial repair is complete. Thus, at least in our experimental conditions, it appears that absence of endothelium is important in order to maintain SMC cytoskeletal changes and after proliferation and migration into the intima, SMC can evolve *in situ* toward a mature type of cell morphologically and biochemically similar to the SMC of the normal media. The finding of a molecular switch of actin expression in rat SMC during intimal thickening and primary cultures provides a new biochemical marker allowing a better definition of the phenotype of these cells and suggests that cultured SMC may represent a useful experimental model for atheromatous SMC.

In conclusion, it appears that evaluation of cytoskeletal changes of SMC under various conditions furnishes useful information as to the expression of SMC phenotype during experimental and human atheromatosis. It is conceivable that a better understanding of the mechanisms leading to the described cytoskeletal alterations will help in the clarification of the molecular phenomena resulting in the production of experimental and human atheromatous plaques.

REFERENCES

1. BENDITT, E. P. & A. M. GOWN. 1980. Atheroma: the artery wall and the environment. Int. Rev. Exp. Pathol. **21:** 55–118.
2. HAUST, M. D. 1974. Reaction patterns of intimal mesenchyme to injury and repair in atherosclerosis. Adv. Exp. Med. Biol. **43:** 35–57.
3. ROSS, R. & J. A. GLOMSET. 1976. The pathogenesis of atherosclerosis (first of two parts). N. Engl. J. Med. **295:** 369–377.
4. ROSS, R. & J. A. GLOMSET. 1976. The pathogenesis of atherosclerosis (second of two parts). N. Engl. J. Med. **295:** 420–425.
5. CHAMLEY-CAMPBELL, J., G. R. CAMPBELL & R. ROSS. 1979. The smooth muscle cell in culture. Physiol. Rev. **59:** 1–61.
6. FRANK, E. D. & L. WARREN. 1981. Aortic smooth muscle cells contain vimentin instead of desmin. Proc. Natl. Acad. Sci. USA **78:** 3020–3024.
7. GABBIANI, G., E. SCHMID, S. WINTER, C. CHAPONNIER, C. DE CHASTONAY, J. VANDEKERCKHOVE, K. WEBER & W. W. FRANKE. 1981. Vascular smooth muscle cells differ from other smooth muscle cells: predominance of vimentin filaments and a specific α-type actin. Proc. Natl. Acad. Sci. USA **78:** 298–302.
8. GABBIANI, G., E. RUNGGER-BRÄNDLE, C. DE CHASTONAY & W. W. FRANKE. 1982. Vimentin-containing smooth muscle cells in aortic intimal thickening after endothelial injury. Lab. Invest. **47:** 265–269.
9. SCHMID, E., M. OSBORN, E. RUNGGER-BRÄNDLE, G. GABBIANI, K. WEBER & W. W. FRANKE. 1982. Distribution of vimentin and desmin filaments in smooth muscle tissue of mammalian and avian aorta. Exp. Cell Res. **137:** 329–340.
10. TRAVO, P., K. WEBER & M. OSBORN. 1982. Co-existence of vimentin and desmin type intermediate filaments in a subpopulation of adult rat vascular smooth muscle cells growing in primary culture. Exp. Cell Res. **139:** 87–94.
11. GABBIANI, G., O. KOCHER, W. S. BLOOM, J. VANDEKERCKHOVE & K. WEBER. 1984. Actin expression in smooth muscle cells of rat aortic intimal thickening and human atheromatous plaque. J. Clin. Invest. **73:** 148–152.
12. KOCHER, O., O. SKALLI, W. S. BLOOM & G. GABBIANI. 1984. Cytoskeleton of rat aortic smooth muscle cells. Normal conditions and experimental intimal thickening. Lab. Invest. **50:** 645–652.
13. IVES, H. E., G. S. SCHULTZ, R. E. GALARDY & J. D. JAMIESON. 1978. Preparation of fractional smooth muscle cells from the rabbit aorta. J. Exp. Med. **148:** 1400–1413.
14. QUITSCHKE, W. & N. SCHECHTER. 1982. A noncomputerized scanning method for determining relative protein quantities and synthesis rates on two-dimensional electrophoretic gels. Anal. Biochem. **124:** 231–238.

Sodium/Calcium Exchange in Vascular Smooth Muscle:

A Link between Sodium Metabolism and Hypertension[a]

MORDECAI P. BLAUSTEIN,[b] TERUNAO ASHIDA,
WILLIAM F. GOLDMAN, W. GIL WIER,
AND JOHN M. HAMLYN

*Departments of Physiology and Medicine
and the Hypertension Center
University of Maryland School of Medicine
Baltimore, Maryland 21201*

SALT AND HYPERTENSION

The role of sodium in the etiology of many forms of hypertension, including essential hypertension, is well documented. A direct association is recognized in hypermineralocorticoidism (Conn's Syndrome)[1] and in renal parenchymal diseases that compromise the ability to excrete a salt load.[2] Similarly, indirect evidence indicates that the incidence of hypertension in various populations is directly related to the levels of Na intake in the populations.[3] In societies in which lifelong intake of Na is very low, blood pressure (BP) does not increase with age, and hypertension is virtually absent.[3] Furthermore, a salt-restricted diet or treatment with natriuretic agents may be effective blood pressure-lowering therapy for a large majority of patients with primary (essential) hypertension.[4,5] Also, in at least some instances, it has been possible to demonstrate a direct relationship between short term manipulation of dietary Na and BP.[6]

Perhaps even more direct is the evidence that Na metabolism is altered in many individuals with essential hypertension and some of their first degree relatives. For example, many normotensive offspring of hypertensive parents have a reduced ability to excrete a salt load.[7] Also, fractional clearance of Li by the kidneys, an indication of reduced proximal tubular clearance of Na, is reduced in hypertensive individuals and some of their first degree relatives.[8,9] Many hypertensive patients also have elevated intracellular Na concentrations, $[Na^+]_i$, in their erythrocytes and leukocytes, as a result of reduced Na extrusion.[10]

These numerous observations have provided the rationale for the search to identify the links between Na metabolism and vascular smooth muscle (VSM) contraction that may play a role in the etiology of hypertension.

[a]Supported by grants from the American Heart Association (W.G.W.), the Muscular Dystrophy Association (M.P.B.), the National Institutes of Health (AM-32276 to M.P.B. and HL-29473 to W.G.W.), the Veterans Administration (Dr. Bruce P. Hamilton), the Searle Pharmaceutical Company and the Upjohn Company, and by fellowships from the AHA-Maryland Affiliate (W.F.C.) and Eli Lilly and Company (T.A.). W.G.W. is an Established Investigator of the AHA.

[b]Address correspondence to M.P.B., Department of Physiology, University of Maryland School of Medicine, 655 West Baltimore Street, Baltimore, MD 21201.

CALCIUM AND THE REGULATION OF CONTRACTILITY IN VASCULAR SMOOTH MUSCLE

The elevated BP of chronic essential hypertension and most other forms of hypertension is characterized by increased peripheral vascular resistance (TPR) with normal cardiac output (CO). The increased TPR reflects a functional alteration of VSM that may be manifest by increased reactivity to vasopressor agents including norepinephrine, angiotensin II, and vasopressin.[11] The generalized nature of this increased reactivity makes it unlikely that the primary mechanism involves the agonist receptors *per se*. A more plausible possibility is that the mechanisms that link receptor activation to contraction (in the "final common path") are altered in hypertension: either the sensitivity of the contractile elements to Ca or the availability of Ca for contraction is modified. At present, there is no evidence that the sensitivity of the contractile elements to Ca is altered, whereas there is a known link between Na

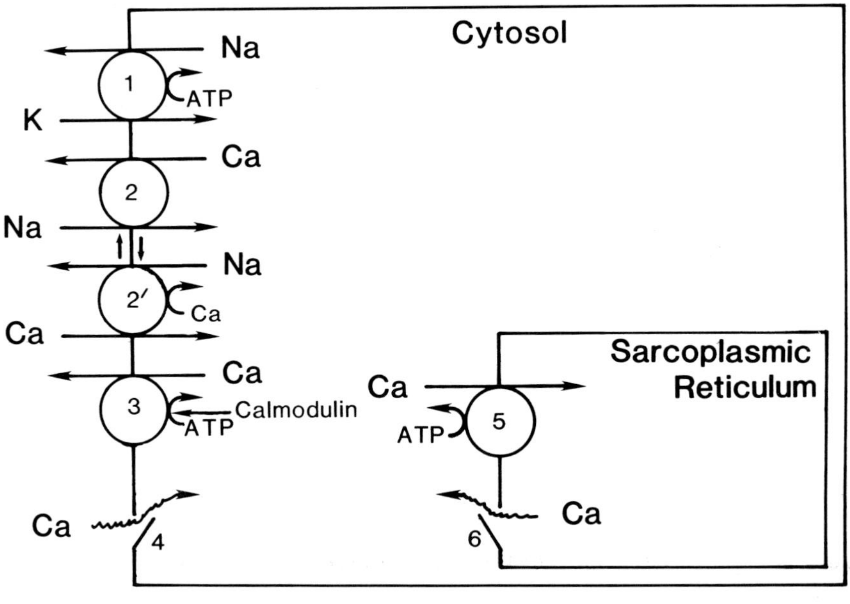

FIGURE 1. Diagram of a vascular smooth muscle cell showing the plasma membrane and sarcoplasmic reticulum (SR) mechanisms that play a role in the regulation of free Ca^{2+} in the cytosol. (1) Ouabain-sensitive, ATP-driven Na pump; (2) Plasma membrane Na/Ca exchange system operating in the "forward (Ca efflux) mode" (2), and in the "reverse (Ca influx) mode" (2′); (3) ATP-driven, calmodulin-regulated plasma membrane Ca pump; (4) Ca-selective channels (both voltage-gated and receptor-operated); (5) ATP-driven, calmodulin-insensitive SR Ca pump; and (6) Ca channels involved in Ca release from the SR. The reverse mode Na/Ca exchange (2′) is activated by (non-transported) intracellular free Ca^{2+} in the dynamic physiological concentration range (10^{-7} to 10^{-6} M).[38] Although not indicated here, the Na/Ca exchange is also modulated by (internal) ATP, which affects the cation affinities.[32] Note that reverse mode Na/Ca exchange (2′) and Ca channels (4) are parallel Ca entry mechanisms, while forward mode Na/Ca exchange (2) and the ATP-driven, calmodulin-sensitive Ca pump (3) are parallel Ca extrusion mechanisms.

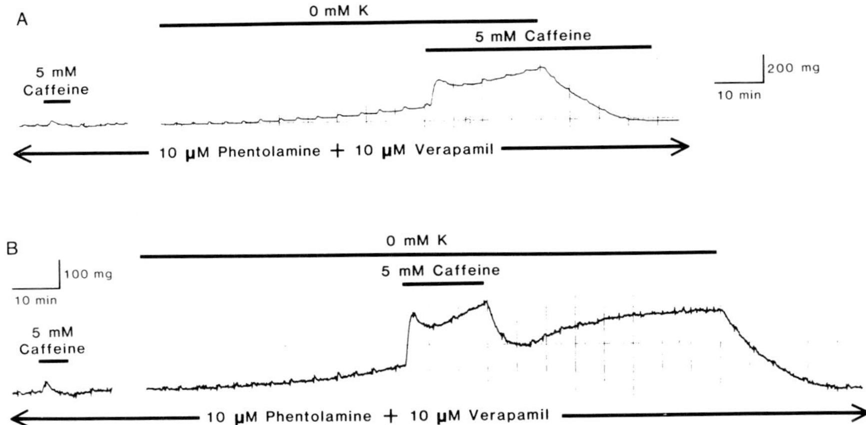

FIGURE 2. Effect of K-free media on rat aortic ring resting tension and the contractile response to 5 mM caffeine. Small rings of rat thoracic aorta (2–2.5 mm long, 1–1.5 mm diameter) were suspended in water-jacketed tissue chambers (volume = 0.5 ml) and superfused with well-oxygenated physiological salt solution (PSS) at a rate of 2 ml/min at 37 °C. The main cations in the standard PSS were (in mM): 139.2 Na; 4.7 K, 1.8 Ca, and 1.2 Mg. All ion substitutions involved isosmotic cation replacement; e.g., in K-free solution, the K was replaced by Na. After equilibration in PSS (1–1.5 hr), isometric tension was continuously recorded on a strip chart recorder. The periods of K-free (0 mM K) superfusion and applications of caffeine are indicated by the bars. The superfusion fluid contained 10 μM phentolamine and 10 μM verapamil. There was a 45 min gap between the left-hand and the right-hand records in A, and a 50 min gap between the left-hand and right-hand records in (B). K-free superfusions were started 20 min before the right-hand record in (A), and 22 min before the right-hand record in (B). Resting tension = 500 mg. Similar methods were used for the experiments of FIGURES 3, 5, 6, and 8–11.

metabolism and Ca regulation in VSM cells, namely, Na/Ca exchange.[12–14] With these considerations in mind, we will focus on Ca metabolism in VSM cells.

Na/Ca EXCHANGE AND THE REGULATION OF CELL Ca IN VASCULAR SMOOTH MUSCLE: EFFECTS OF Na PUMP INHIBITION

Contraction of VSM is normally triggered by a rise in the cytoplasmic free Ca level, $[Ca^{2+}]_i$, as a result of Ca entry from the extracellular fluid and the release of Ca from the sarcoplasmic reticulum, SR.[15] Relaxation is promoted by resequestration of Ca in the SR and by extrusion of Ca across the plasma membrane. Two types of mechanisms may be involved in Ca entry (FIG. 1): Ca-selective channels (both agonist receptor–operated and voltage-gated) and Na/Ca exchange. Likewise, two types of mechanisms may be involved in Ca extrusion (FIG. 1): an ATP-driven Ca pump (mediated by a calmodulin-modulated, Ca-dependent ATPase[16]) and Na/Ca exchange. The Na/Ca exchange can mediate either net Ca entry or exit, depending upon the prevailing Na electrochemical gradient across the plasmalemma, $\Delta\bar{\mu}_{Na}$. In this article, we describe recent studies on the properties of the Na/Ca exchange system in VSM, a system that has been the subject of considerable controversy in recent years.[14,17,18] Furthermore, since the Na/Ca exchange operates in parallel with the Ca channels and ATP-driven Ca pump (FIG. 1), we will consider the relative physiological roles of these various Ca transport mechanisms.

Small rings of rat thoracic aorta and bovine tail artery were used for these experiments, in which contractile tension was employed as a measure of the amount of Ca delivered to the contractile apparatus. Contractions were induced in a variety of ways: (1) with caffeine, which effects Ca release from the SR and inhibits resequestration; (2) with norepinephrine (NE), which promotes Ca influx through receptor-operated (and, perhaps, voltage-gated) Ca channels and Ca release from the SR; (3) by raising the external K concentration, $[K^+]_o$, which depolarizes the VSM cells thereby opening voltage-gated Ca channels; (4) by strophanthidin or by removal of external K, to inhibit the Na pump, and raise $[Na^+]_i$; and (5) by lowering the external Na concentration, $[Na^+]_o$, as one means of reducing $\Delta\bar{\mu}_{Na}$. (Note that strophanthidin and changes in $[K^+]_o$ will also affect $\Delta\bar{\mu}_{Na}$.)

FIGURE 2 shows data from an experiment in which the effect of external K removal was tested on the contraction of rat aorta. K removal not only increased tonic ("unstimulated") tension, but also increased the amplitudes of the contractions induced by caffeine and (not shown) NE. The rise in tonic tension and the response to caffeine were observed in the presence of an α-receptor blocker (10 μM phentolamine) and a Ca channel blocker (10 μM diltiazem, nitrendipine, or, as shown here, verapamil). The enhanced NE response also occurred in the presence of the Ca channel blockers. All of the effects of external K removal were dependent upon external Ca, and were reversible: for example, re-introduction of K caused prompt relaxation, either in the presence or absence of caffeine.

As illustrated in FIGURE 3, effects similar to those seen with external K removal were observed when the rat aortic rings were treated with strophanthidin: the responses to caffeine and NE were enhanced. Since rat tissues are relatively insensitive to

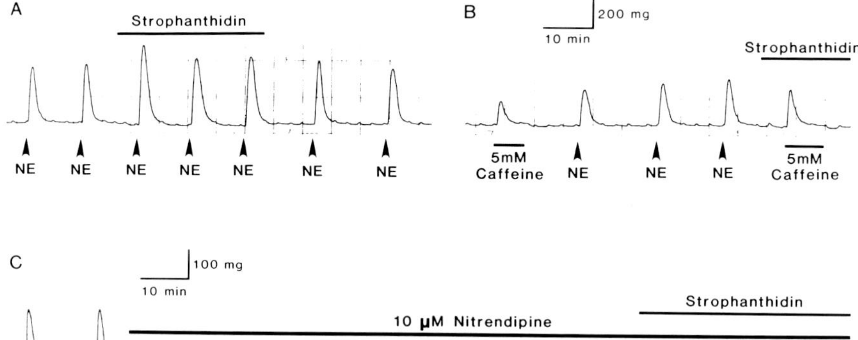

FIGURE 3. Effect of strophanthidin (5×10^{-5} M) on the contractile response of rat aorta to 6×10^{-8} M NE (A and C) and 5 mM caffeine (B and C). A and B are data from the same aortic ring; there was a 20 min gap between A and B. Nitrendipine (10 μM) was present during the period indicated by the bar in C.

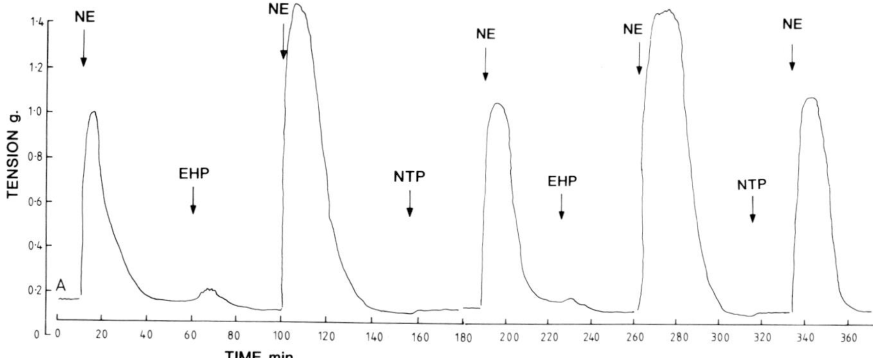

FIGURE 4. Contractile response of a strip of rabbit aorta to 10^{-7} M norepinephrine (NE) during incubation in plasma from two normotensive subjects (NTP) and from two patients with essential hypertension (EHP). The initial contraction (at 10 min) was elicited with the tissue bathed in PSS. (From Blaustein & Hamlyn.[42] With permission from *American Journal of Medicine*.)

cardiotonic steroids,[19] the dose of strophanthidin used here, 5×10^{-5} M, would only be expected to inhibit the Na pump partially in the rat aorta. This modest concentration of strophanthidin, applied for only a brief period, did not increase tonic tension, although we and others have observed such effects when the Na pump in VSM was more completely inhibited for longer times.[14] Like the effects produced by K removal, those produced by strophanthidin were dependent upon external Ca but were not blocked by Ca channel blockers (FIG. 2C).

Cardiotonic steroids are specific blockers of the Na pump.[20] Thus, the likely explanation for the effects observed with both strophanthidin and external K removal is that Ca entry is promoted by these manipulations, presumably as a result of the rise in $[Na^+]_i$ and consequent Na/Ca exchange.[14] Much of the entering Ca is sequestered in the SR. The increase in the amount of stored Ca is demonstrated by the enhanced contractions in response to NE and caffeine—even in the presence of Ca channel blockers. This follows from the evidence that the contractile responses to caffeine, for example, are proportional to the size of the SR Ca store.[21]

These key observations indicate that partial inhibition of the Na pump in VSM may not increase resting tension but may, nevertheless, increase the responsiveness of the tissue to agonists because of the enhanced storage of Ca. Our data are consistent with those of Aalkjaer and Mulvany[22] who found that cardiotonic steroids did not affect the resting tension in small resistance vessels, but did potentiate the "active" response to agonists. The increased vascular reactivity that is observed in normal human subjects treated with digoxin[23] may be comparable to the effects described here, and thus may be attributable tc increased storage of Ca.

EFFECTS OF PLASMA FROM HYPERTENSIVE PATIENTS ON VASCULAR SMOOTH MUSCLE: COMPARISON WITH THE EFFECTS OF STROPHANTHIDIN

The immediate effect of the low dose of strophanthidin was a slight decline in unstimulated tonic tension, followed by the aforementioned increase in responsiveness to NE and caffeine (FIG. 3). The cause of this decline in tonic tension is not known, but might be due to release of endothelial relaxing factor. Nevertheless, virtually identical effects were observed when aortic tissue (in this case, from rabbit) was superfused with plasma from patients with essential hypertension (EHT): there was an initial small decline in tonic tension and an increased response to NE (FIG. 4). In contrast, plasma from normotensive individuals (NTP) had no effect on resting tension and did not increase vascular contractility. These findings seem particularly noteworthy in view of the evidence that plasma from patients with essential hypertension contains a Na pump inhibitor.[10,24]

The increased vascular reactivity in hypertension, like that following digitalization, may be attributed to increased storage of Ca in the SR as a result of partial Na pump inhibition and a slight rise in $[Ca^{2+}]_i$ (albeit, below the contraction threshold). In addition, the presence of chronic vascular tone *in vivo* implies that most VSM's are tonically activated and (partially) contracted, and not fully relaxed, as we usually find *in vitro*. Thus, the increased tone and TPR observed in hypertension may reflect the increased VSM SR storage and release of Ca.

EFFECTS OF REDUCING $[Na^+]_o$ ON CONTRACTION: FURTHER EVIDENCE FOR Na/Ca EXCHANGE

If the effects of Na pump inhibition, produced by strophanthidin or external K removal, are a consequence of Na/Ca exchange promoted by the decline in $\Delta\bar{\mu}_{Na}$, similar effects should be observed when $[Na^+]_o$ is lowered. Indeed, as shown in FIGURE 5, reduction of $[Na^+]_o$ resulted in increased responsiveness to caffeine and, in the presence of caffeine, increased tonic tension. The rate of tonic tension development was inversely related to $[Na^+]_o$ (FIG. 6). Caffeine was used in these experiments to minimize SR Ca sequestration and, thus, to emphasize the plasma membrane systems that transport Ca and help to regulate $[Ca^{2+}]_i$ (see FIG. 1); at the lowest $[Na^+]_o$ rat aorta tonic tension increased even in the absence of caffeine (not shown). N-Methylglucamine (NMG) was used as the (isoosmotic) Na replacement in the experiments illustrated here; however, similar results were obtained when Tris or tetramethylammonium (TMA) was used to replace Na.

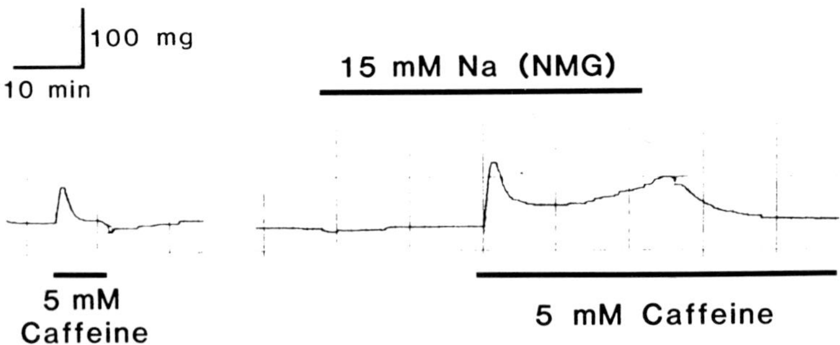

FIGURE 5. Effect of reducing $[Na^+]_o$ on the contractile response of a rat aortic ring to 5 mM caffeine. $[Na^+]_o$ was reduced from 139.2 mM to 15 mM (replaced with N-methyl glucamine, NMG) during the period indicated by the bar. There was a 100 min gap between the left-hand and right-hand records.

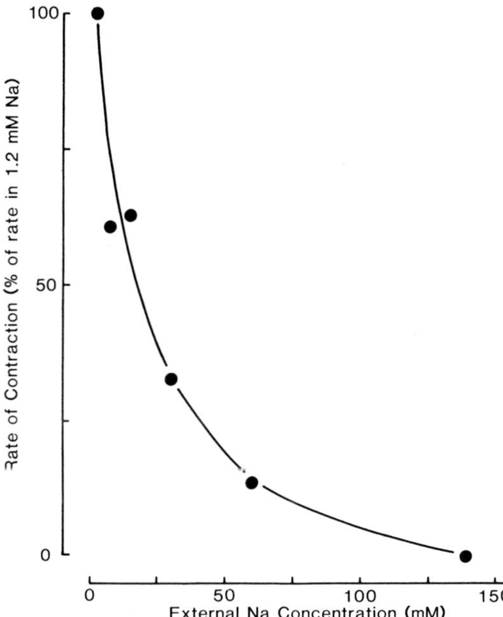

FIGURE 6. Relationship between $[Na^+]_o$ and the relative rate of tension development in rat aorta. The data were obtained in the presence of 10 μM phentolamine, 10 μM verapamil, and 5 mM caffeine. The rate of tension development was calculated from the rising slope of the tension record; the rate of tension development in 1.2 mM Na was taken as 100%. Each symbol represents the mean of data from two rat aortic rings. The absolute rate of tension development in 1.2 mM Na was 10.6 ± 1.6 mg/min ($N = 2$).

The aforementioned observations are all consistent with the view that the rat aorta smooth muscle plasma membrane contains a Na/Ca exchange system, and that lowering $\Delta\bar{\mu}_{Na}$ tends to drive Ca into the cells. This is in accord with tracer flux data on rat aorta and mesenteric artery sarcolemmal vesicles, which provide direct evidence for Na/Ca exchange in these tissues.[25,26]

Although not illustrated here, we have made similar observations on the contractions of bovine tail artery, a more muscular artery, in response to reduction of $[Na^+]_o$. However, in this tissue we could not test the effect of caffeine on the size of the SR Ca store because it produced a profound relaxation. In the absence of caffeine, however, the bovine tail artery seemed to be more sensitive to reductions in $\Delta\bar{\mu}_{Na}$ than the rat aorta. A possible explanation is that the bovine tail artery has a less extensive SR (unpublished electron microscope data), and is therefore more dependent on the plasmalemma Ca transport systems for both long- and short-term regulation of $[Ca^{2+}]_i$.

The changes in $[Ca^{2+}]_i$ that accompany alterations in $\Delta\bar{\mu}_{Na}$ can now be measured directly with Ca-sensitive fluorescent dyes, such as fura 2,[27] and digital imaging microscopy.[28–30] Representative data from a single, enzymatically dissociated, bovine tail artery myocyte loaded with fura 2 are shown in FIGURE 7. As expected from the aforementioned findings, $[Ca^{2+}]_i$ increased and the cell contracted when external K was removed, even in the presence of verapamil; these effects were both augmented

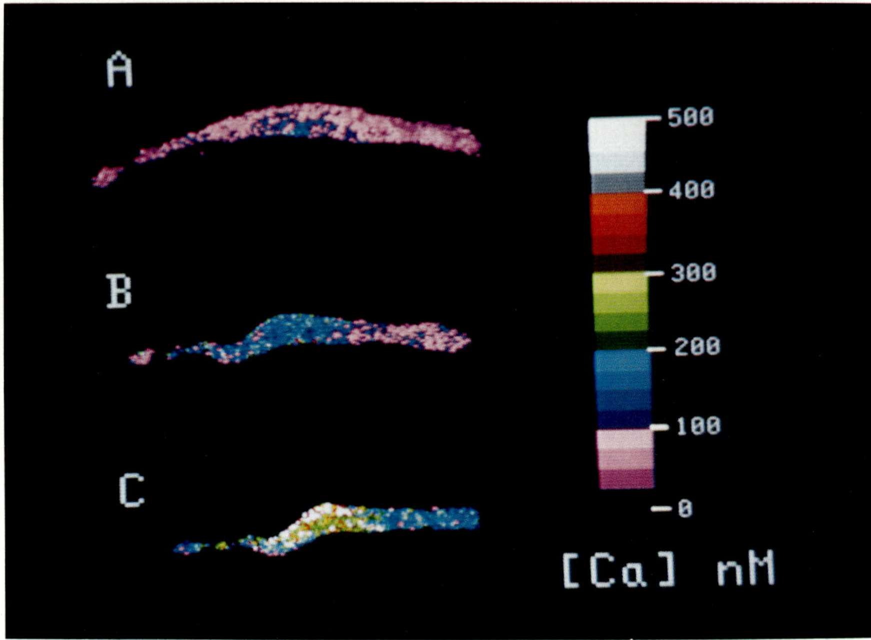

FIGURE 7. "Free calcium images" of a single, isolated, fura 2-loaded bovine tail artery cell. The normal resting free Ca^{2+} (A), and the effects of K-free media (B) and K-free, low-Na media (C) on $[Ca^{2+}]_i$, are illustrated. All solutions contained 10 μM verapamil; NMG replaced most of the Na in the low (1.2 mM) Na solution. The $[Ca^{2+}]_i$ in the quiescent cell (A) was 98 nM. When the cell was superfused with K-free solution, the $[Ca^{2+}]_i$ increased to about 160 nM in the central segment of the cell. Subsequent reduction of $[Na^+]_o$ to 1.2 mM caused $[Ca^{2+}]_i$ to increase to 310 nM in the central segment. The images were constructed by digital analysis of videotaped fura 2 fluorescent images.[28–30]

when most of the external Na was replaced by NMG. This provides direct confirmation that the contraction activated by reduction of $\Delta\bar{\mu}_{Na}$ is associated with a rise in $[Ca^{2+}]_i$.

THE EFFECT OF EXTERNAL Na ON Ca EXTRUSION: THE PHYSIOLOGICAL IMPORTANCE OF Na/Ca EXCHANGE

Some of the experiments described above indicate that a modest reduction in $\Delta\bar{\mu}_{Na}$ may drive Ca into ASM cells, even under conditions in which tonic tension is not altered. The increased cell Ca content could then be demonstrated by examining the augmented contractions induced when Ca was released from the SR. Additional information about the physiological role of Na/Ca exchange may be obtained by examining the effect of external Na on the rate of relaxation following the induction of contractions of modest amplitude. Such relaxation is usually dependent upon removal of Ca from the cytosol either by sequestration in the SR or by extrusion across the sarcolemma via the ATP-driven Ca pump or Na/Ca exchange (see FIG. 1). The influence of SR Ca sequestration in the rat aorta was minimized by carrying out the experiments in the presence of caffeine, but this was not possible in the bovine tail artery because of the aforementioned relaxing effect of caffeine in this tissue.

To determine the effect of Ca extrusion via the ATP-driven Ca pump, we measured the rate of relaxation in Na-depleted media, following removal of external Ca (to avoid the Na/Ca exchange-mediated Ca entry that occurs when $[Na^+]_o$ is reduced; see above). The increment in the rate of relaxation, on adding back external Na, could then be attributed to Na/Ca exchange. The experiments were all carried out in the presence of phentolamine and verapamil to minimize the effects of endogenous catecholamines and Ca movements through Ca-selective channels.

In both rat aorta and bovine tail artery, the relaxation from low-Na induced contractions was much slower in media containing only 1.2–7.5 mM Na, than in media containing much more Na. FIGURES 8 and 9 show data from experiments in which external Na was replaced by equimolar NMG in the low-Na solutions. Similar results were obtained when Na was replaced by Li, K, Tris, or TMA; thus, the nature of the Na substitute did not appear to be important. The Na-dependent relaxation rate was a sigmoid function of $[Na^+]_o$ (FIG. 10); the concentration of Na required to half-maximally increase the rate of relaxation above that in 1.2–7.5 mM Na was 20–25 mM Na for both artery preparations. Normal (139.5 mM) Na produced about a five-fold increase in relaxation rate in rat aorta (see FIG. 8) and about a 10-fold increase in bovine tail artery (FIG. 10), relative to that observed in 1.2–7.5 mM Na. Although not illustrated here, we also observed that the rate of relaxation of rat aorta in media containing 139.5 mM Na was substantially slower following a K-free contraction than following a low-Na contraction. The implication is that a high $[Na^+]_i$, caused by Na pump inhibition, reduces external Na-dependent Ca efflux, perhaps because of competition between internal Na and Ca.

These data demonstrate that, when $[Ca^{2+}]_i$ is raised above the contraction threshold in VSM, the dominant mode of Ca extrusion involves an external Na-dependent pathway, presumably Na/Ca exchange. Our physiological observations on the rate of VSM relaxation may be contrasted with Ca flux and Ca ATPase data from VSM plasma membranes; the latter have led to the suggestion that Na/Ca exchange may be less important than the Ca ATPase for the extrusion of Ca.[16,25] A possible explanation for the difference in conclusions is that Na/Ca exchange apparently requires Ca on both sides of the membranes (see FIG. 1)[31] and is modulated by internal

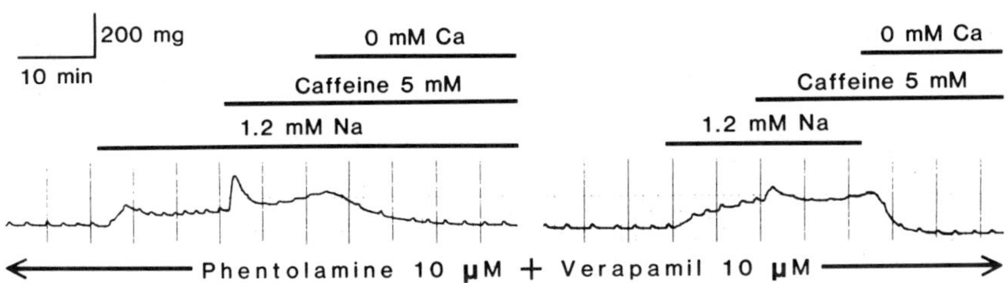

FIGURE 8. Effect of external Na on the rate of relaxation from low-Na contractures upon the removal of external Ca in a ring of rat aorta. The Ca-free solution contained 1.2 mM Na (NMG substituted for Na) in the left-hand record, and 139.2 mM Na in the right-hand record; Ca was replaced by Mg in both Ca-free solutions. There was a 60 min gap between the two records. Caffeine (5 mM) was present where indicated; the superfusion fluid contained 10 μM phentolamine and 10 μM verapamil.

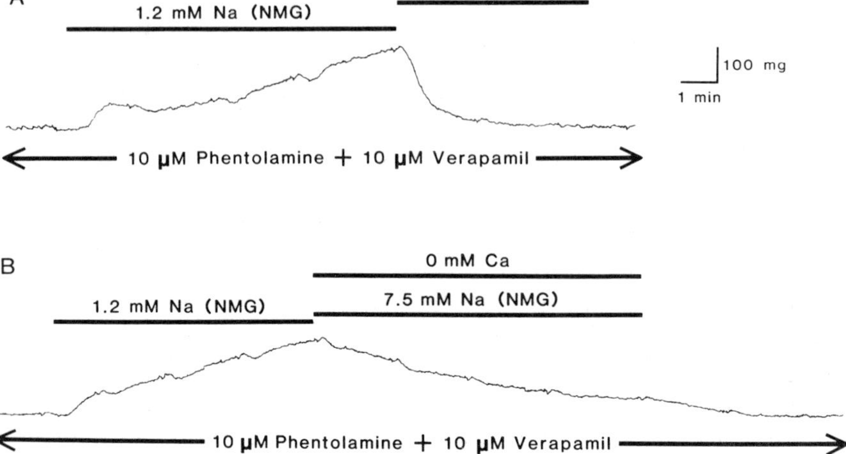

FIGURE 9. Effect of external Na on the rate of relaxation from a low-Na (1.2 mM) induced contraction in a ring of bovine tail artery. The methods used for these experiments were virtually identical to those used for the studies on rat aorta. (A) The Ca-free (relaxing) solution contained 139.2 mM Na. (B) The Ca-free solution contained 7.5 mM Na (remaining Na replaced by NMG). A and B represent data from the same aortic ring; there was an 80 min gap between A and B. All solutions contained 10 μM phentolamine and 10 μM verapamil.

ATP.[32] It is difficult to mimic all the physiological conditions in the isolated vesicle experiments.

Our conclusion also contrasts with that of Petersen and Mulvany,[33] who suggested that "Na-Ca exchange mechanisms (in resistance vessels) . . . only play a role under extreme conditions." These workers found that relaxation of small resistance arteries in the rat, following K contractures, was modestly slowed when $[Na^+]_o$ was reduced to 25 mM.[33] Indeed, our observations on rat aorta and bovine tail artery (FIG. 10) are comparable. But it should be apparent that the "extreme conditions" of nearly complete replacement of external Na are necessary to inhibit Na/Ca exchange-mediated Ca extrusion because so little external Na is required to activate the exchange.[34,35] Under normal conditions (i.e., with $[Na^+]_o$ on the order of 140 mM), the exchanger will be saturated with external Na and the Na/Ca exchange system will be the predominant mode of Ca extrusion.

The role of intracellular Ca also needs to be emphasized. When $[Ca^{2+}]_i$ is very low (i.e., well below contraction threshold), most of the exchanger sites to which transported Ca must bind will be unoccupied. In addition, as illustrated in FIGURE 1, internal Ca is required at a site not involved in translocation, in order to activate Na/Ca exchange-mediated Ca entry.[36–38] In other words, the turnover of the exchanger, and the exchanger-mediated transport of Ca in both directions across the sarcolemma, are activated by internal Ca.

TRANSPORT OF Ca VIA Na/Ca EXCHANGE IS VOLTAGE SENSITIVE

We have mentioned that Na/Ca exchange-mediated Ca transport is influenced by $\Delta\bar{\mu}_{Na}$, but have thus far focused only on the chemical concentration portion of this gradient. Studies in numerous other preparations indicate that the exchange has a stoichiometry of about 3 Na:1 Ca, that it can generate a current, and that it is voltage sensitive.[31] Because activation of VSM may, in some instances, involve depolarization,[39] it is important to know how the membrane potential influences Na/Ca exchange in this tissue. The results of an experiment designed to examine this point are shown in FIGURE 11. When the ring of bovine tail artery was superfused with media containing phentolamine and verapamil, reduction of $[Na^+]_o$ to 43.9 mM, by replacing Na with Tris, produced a very small increase in tonic tension. Then, with $[Na^+]_o$ held constant, Tris was replaced by K to depolarize the tissue; this elicited a much larger increase in tension. All of these effects were reversible. Following treatment with 5×10^{-5} M strophanthidin, the contractions in both the low-Na (Tris) and low-Na (K) solutions were greatly increased, but the effect was most marked when the Tris was replaced by K. This suggests that the contraction in the K-rich solution was internal Na dependent. FIGURE 11 also shows that these effects were external Ca dependent. Since the voltage-gated Ca channels were blocked (by verapamil), the most straightforward

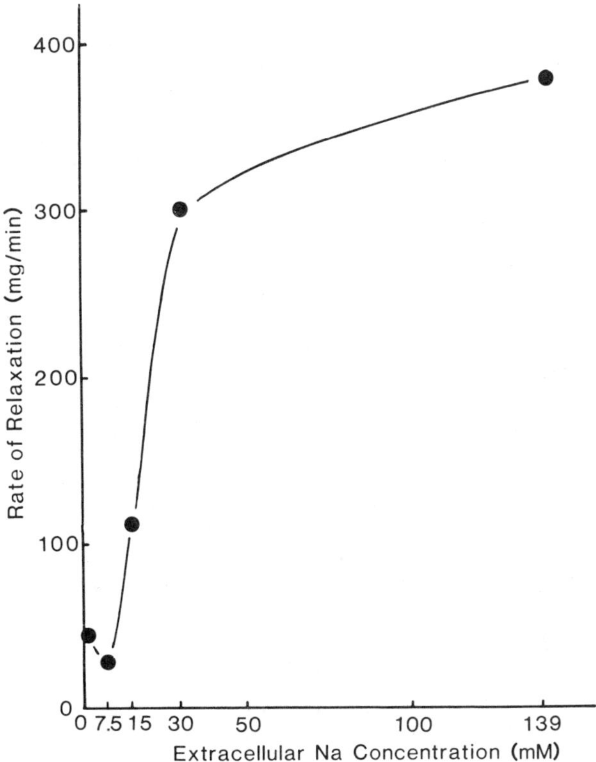

FIGURE 10. Relationship between $[Na^+]_o$ and the rate of relaxation of rings of bovine tail artery. In solutions with reduced $[Na^+]_o$, NMG replaced Na; the sum of the Na and NMG concentration was 139.2 mM. The relaxation rates were calculated from the slopes of the tension curves during relaxation as: (half-maximal amplitude of contraction tension) ÷ (time for tension to decline from peak to half-maximal amplitude). Each data point represents the mean of values from four experiments. All solutions contained 10 μM phentolamine and 10 μM verapamil.

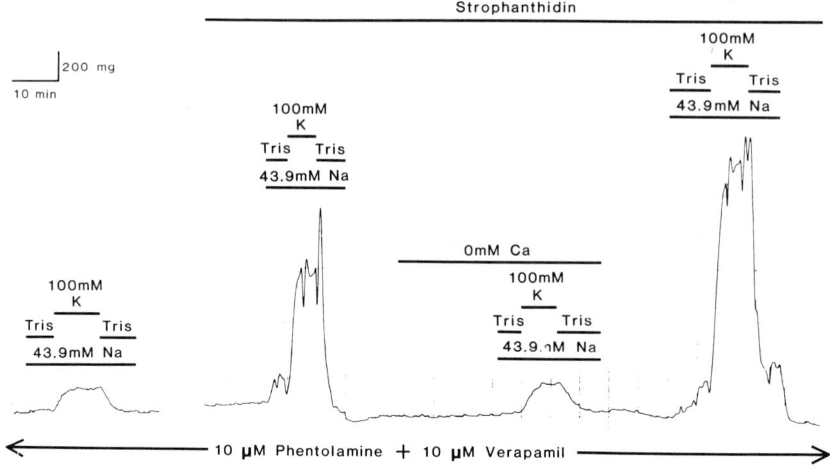

FIGURE 11. Comparison of the effects of partial replacement of external Na by Tris and by K on the development of tension in a ring of bovine tail artery treated with 10 μM phentolamine and 10 μM verapamil. Two short periods of superfusion with 43.9 mM Na (Tris) were used to bracket each period of superfusion with 43.9 mM Na (100 mM K). There was a 150 min gap between the first and second exposures to the low Na media. Strophanthidin (5×10^{-5} M) was added to the superfusion fluid 140 min before the second exposure to the low Na media. External Ca was removed from all superfusion solutions for the period indicated by the 0 mM Ca bar.

explanation is that the K-activated contraction was the result of Na/Ca exchange-mediated Ca entry that was enhanced by depolarization.

In sum, these data indicate that the Na/Ca exchange in VSM is very similar to the exchange systems that has been described in mammalian cardiac muscle and many other types of tissues, and that it has a stoichiometry of more than 2 Na:1 Ca.[31] In relaxed, polarized VSM cells, with $[Ca^{2+}]_i$ below the contraction threshold, the exchange should mediate net Ca movements, but turnover of the exchanger will be slow and the ATP-driven Ca pump may play an important role in the control of the resting $[Ca^{2+}]_i$. Nevertheless, even resting $[Ca^{2+}]_i$ and the size of the SR Ca store will be biased by the Na/Ca exchanger (i.e., by $\Delta\bar{\mu}_{Na}$), as illustrated in FIGURE 3. When the cells are activated and $[Ca^{2+}]_i$ is increased, however, turnover of the exchanger will increase markedly: it will then promote Ca entry when the cells are depolarized (FIG. 11) and Ca exit when the cells are repolarized (see FIGS. 8–10 and Ref. 31). When the cells are tonically activated, and $[Ca^{2+}]_i$ is maintained above the contraction threshold for long periods of time (seconds to minutes or even longer), the level of $[Ca^{2+}]_i$ and, thus, tonic tension, are likely to be modulated by $[Na^+]_i$ (and, thus by $\Delta\bar{\mu}_{Na}$). As discussed elsewhere,[14] control of $[Ca^{2+}]_i$ must reside in the plasma membrane Ca transport mechanisms.

ALTERED Na METABOLISM IN HYPERTENSION: THE ROLE OF A CIRCULATING Na PUMP INHIBITOR (NATRIURETIC HORMONE)

Having identified a cellular mechanism that links Na metabolism to VSM contraction, we now need to elucidate the mechanisms that help to regulate $[Na^+]_i$ and $\Delta\bar{\mu}_{Na}$ that may be altered in hypertensive individuals.

To begin with, changes in $[Na^+]_o$ are not likely to be a substantial factor because plasma $[Na^+]$ is generally well regulated and there is no evidence that it is altered significantly in most hypertensive patients. Furthermore, modest chronic changes in $[Na^+]_o$ may have only a very small effect on $\Delta\bar{\mu}_{Na}$, in part because $[Na^+]_i$ may also then be altered in parallel fashion.

The membrane potential of VSM cells may be modulated by sympathetic nerve activation,[39,40] and sympathetic nerve function may be altered in hypertensive patients.[41] For example, there may be enhanced release and/or decreased re-uptake of catecholamines, both due perhaps to altered $\Delta\bar{\mu}_{Na}$ in the sympathetic neurons.[41,42] In addition, however, there must be a primary alteration in VSM cell Ca metabolism in order to explain the increased vascular reactivity to vasopressors observed in most forms of hypertension.[11] Therefore, we need to consider mechanisms by which $[Na^+]_i$ might be altered in hypertensive individuals.

$[Na^+]_i$ appears to be governed primarily by the balance between the "leak" entry of Na (which may include Na/H exchange, Na, K, Cl cotransport and Na/Ca exchange, for example), and the Na pump-mediated extrusion of Na.[43,44] In VSM cells, steady-state $[Na^+]_i$ could be modified as a consequence of a direct alteration in one of these mechanisms, or as a result of a secondary (perhaps hormonally mediated) modification in one of the transport systems.

In various strains of rats that are genetically predisposed to the development of hypertension, renal transplant experiments demonstrate that the hypertension "goes with the kidneys."[45–47] In human essential hypertension, too, there is evidence that the disease can be cured by transplanting normal kidneys into hypertensive patients.[48] Thus, it seems likely that the genetic defects in Na transport are manifested directly in the kidneys, and that alterations in Na metabolism in the VSM cells, as well as in erythrocytes and leukocytes,[10] are influenced by secondary (perhaps humoral) factors.

Dahl[49] was the first to suggest that a humoral agent with a natriuretic action might be involved in the generation of hypertension. His pioneering studies led to further speculation[14,50–53] as well as experimental testing of the idea that a circulating Na pump inhibitor may be involved in salt-dependent (or volume-dependent) hypertension in

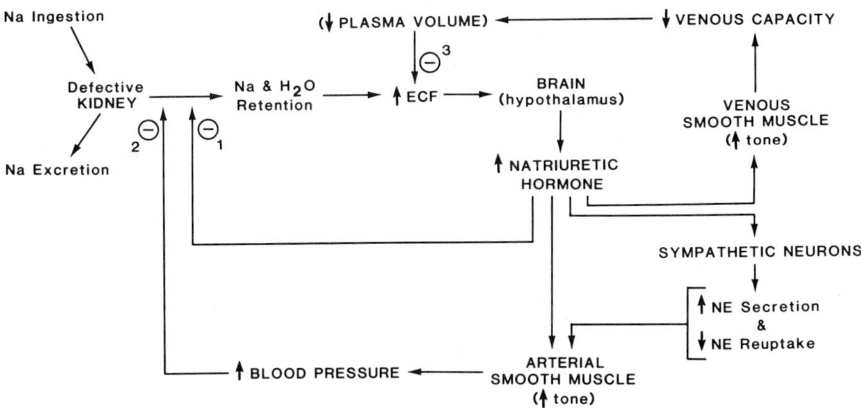

FIGURE 12. Diagram illustrating the three negative feedback loops (1, 2, and 3) that help to prevent extracellular fluid (ECF) volume expansion when excessive Na is ingested relative to the kidney's innate ability to excrete the Na load. (From Blaustein & Hamlyn.[42] With permission from *American Journal of Medicine*.)

rats, and in essential hypertension in man. The physiologically relevant circulating inhibitor of the Na pump has not yet been purified, although numerous laboratories are currently engaged in this effort. Nevertheless, several observations appear to support Dahl's hypothesis[49] about the central role of a natriuretic hormone in hypertension. For example, there is evidence that human plasma contains a Na,K-ATPase inhibitor, and that the level of this inhibitor is correlated with the level of the blood pressure.[24] Rats with mineralocorticoid-induced hypertension and other forms of volume-dependent hypertension also have elevated levels of a Na pump inhibitor in their plasma; the levels of the inhibitor are greatly reduced, and the hypertension is prevented by lesions in the medial, antero-ventral (AV3V) region of the hypothalamus.[54,55] Such lesions also prevent other types of volume-dependent hypertension[56] and reduce the secretion of a circulating inhibitor of the Na pump following expansion of plasma volume.[54,57]

The fact that there are at least two different molecular forms of Na,K-ATPase (Na pump) with different sensitivities to cardiotonic steroids[58] has increased speculation that the Na pump may be modulated, *in vivo*, by endogenous digitalis-like substances analogous to the endorphins. Taken in conjunction with the aforementioned observations, this has led to the aggressive search for such endogenous regulators of the Na pump—i.e., circulating digitalis-like factors that would be expected to have a weak natriuretic action. These substances would be expected to modulate $[Na^+]_i$.[59] They could then contribute, indirectly, to the control of vascular tone and BP as a consequence of the Na/Ca exchange mechanism in VSM described above.

THE ROLE OF SALT RETENTION AND PLASMA VOLUME EXPANSION IN THE ETIOLOGY OF HYPERTENSION: THE NATRIURETIC HORMONE–Na/Ca EXCHANGE–HYPERTENSION HYPOTHESIS

We now have a number of key elements that can be fitted into a comprehensive hypothesis to help explain how salt causes high blood pressure,[14,42,51,53] as diagrammed in FIGURE 12.

The primary problem appears to be a renal defect associated with reduced Na excretion, which may be offset by limiting Na intake. Certain aspects of renal function appear to be abnormal in many, if not all, patients with essential hypertension, and some evidence suggests that altered proximal tubular reabsorption of Na may be at fault.[8,9] This may be one of the important links between salt and hypertension. Indeed, Guyton[60] has stressed the role of the kidneys and extracellular fluid volume regulation in the control of blood pressure.

In the case where dietary Na exceeds the capacity of the kidneys to excrete Na, there will be net retention of Na (and Cl) and water and expansion of plasma volume. The initial effect may be an increase in CO and BP. The blood volume expansion may also lead to secretion of atrial natriuretic peptides (which may be short-acting natriuretic agents) as well as secretion of the natriuretic agent influenced by the hypothalamus. The latter, by virtue of its activity as a Na pump inhibitor, may (1) have a direct renal action as a weak natriuretic, (2) increase secretion of catecholamines and/or prolong their action by inhibiting re-uptake in sympathetic nerve cells,[41] and (3) tend to increase vascular tone directly via the Na/Ca exchange mechanism in VSM described above. The elevated BP will induce a pressure natriuresis that ultimately compensates for the volume expansion. The net effect will be a protection of blood volume at the expense of an elevated blood pressure.[1]

In sum, the diagram in FIGURE 12 illustrates an hypothesis regarding the way in which a tendency to salt retention, Na pump modulation by a circulating Na pump inhibitor (a natriuretic hormone), and the Na/Ca exchange system, all contribute to the long-term regulation of vascular tone, blood pressure, and blood volume. While we have focused on the role of the Na/Ca exchange system in VSM, it is important to note that this transport system may also play a role in catecholamine release from the sympathetic nerve varicosities[41] and in the natriuresis that results directly from Na pump inhibition in the renal tubules (loop −1 in FIG. 12).[61] Na/Ca exchange obviously plays a pivotal role in many physiological processes: one of the more important ones may be in the regulation of BP and control of blood volume.

ACKNOWLEDGMENT

We thank Ms. A. Wilder for preparing the typescript.

REFERENCES

1. HAMLYN, J. M. & M. P. BLAUSTEIN. 1986. Sodium chloride, extracellular fluid volume, and blood pressure regulation. Am. J. Physiol. **251** (Renal Fluid Electrolyte Physiol. **20**): F563–F575.
2. KELLY, M. R., R. C. DAVIDSON, P. L. HOOVER & B. H. SCRIBNER. 1974. Management of resistant hypertension: Case reports of four patients. Prog. Biochem. Pharmacol. **9:** 249–258.
3. BLACKBURN, H. & R. PRINEAS. 1983. Diet and hypertension: Anthropology, epidemiology and public health implications. Prog. Biochem. Pharmacol. **19:** 31–79.
4. FREIS, E. D. 1979. Salt in hypertension and the effects of diuretics. Ann. Rev. Pharmacol. Toxicol. **19:** 13–23.
5. MORGAN, T., S. CARNEY & J. MYERS. 1980. Sodium and hypertension. A review of the role of sodium in pathogenesis and the action of diuretic drugs. Pharmacol. Ther. **9:** 395–418.
6. MACGREGOR, G. A., N. D. MARKANDU, F. E. BEST, D. M. ELDER, J. M. CAM, G. A. SAGNELLA & M. SQUIRES. 1982. Double-blind randomized crossover trail of moderate sodium restriction in essential hypertension. Lancet **i:** 351–355.
7. GRIM, C. E., F. C. LUFT, J. Z. MILLER, P. L. BROWN, M. A. GANNON & M. C. WEINBERGER. 1979. Effects of sodium loading and depletion in normotensive first degree relatives of essential hypertensives. J. Lab. Clin. Med. **94:** 764–771.
8. SKRABAL, F., H. HERHOLZ, M. NEUMAYR, L. HAMBERGER, M. LEDOCHOWSKI, H. SPORER, H. HORTNAGL, S. SCHWARZ & D. SCHONITZER. 1984. Salt sensitivity in humans is linked to enhanced sympathetic responsiveness and to enhanced proximal tubular reabsorption. Hypertension **6:** 152–158.
9. WEDER, A. B. 1986. Red cell lithium-sodium countertransport and renal lithium clearance in hypertension. N. Eng. J. Med. **314:** 198–201.
10. HILTON, P. J. 1986. Cellular sodium transport in essential hypertension. N. Engl. J. Med. **314:** 222–229.
11. FRIEDMAN, S. M. 1983. Vascular reactivity. *In* Hypertension. J. Genest, O. Kuchel, P. Hamet & M. Cantin, Eds. :457–473. McGraw Hill Book Co. New York.
12. BOHR, D. F., C. SEIDEL & J. SOBIESKI. 1969. Possible role of sodium-calcium pumps in tension development of vascular smooth muscle. Microvasc. Res. **1:** 335–343.
13. REUTER, H., M. P. BLAUSTEIN & G. HAUSLER 1973. Na-Ca exchange and tension development in arterial smooth muscle. Phil. Trans. Roy. Soc. Lond. (Ser. B) **265:** 87–94.
14. BLAUSTEIN, M. P. 1977. Sodium ions, calcium ions, blood pressure regulation and hypertension: a reassessment and a hypothesis. Am. J. Physiol. **232** (Cell Physiol. **1**): C165–C173.
15. JOHANSSON, B. & A. P. SOMLYO. 1980. Electrophysiology and excitation-contraction coupling. *In* Handbook of Physiology: Vascular Smooth Muscle. D. F. Bohr, A. P. Somlyo, H. V. Sparks. Eds.: 301–324. American Physiological Society. Bethesda, MD.

16. WUYTACK, F., L. RAEYMAEKERS & R. CASTEELS. 1985. The Ca^{2+}-transport ATPases in smooth muscle. Experientia **41:** 900–905.
17. BRADING, A. F. & T. W. LATEGAN. 1985. Na-Ca exchange in vascular smooth muscle. J. Hypertension **3:** 109–116.
18. MULVANY, M. J. 1985. Changes in sodium pump activity and vascular contraction. J. Hypertension **3:** 429–436.
19. DETWEILER, D. K. 1967. Comparative pharmacology of cardiac glycosides. Fed. Proc. **26:** 1119–1124.
20. GLYNN, I. M. 1964. The action of cardiac glycosides on ion movements. Pharmacol. Rev. **16:** 381–407.
21. HASHIMOTO, T., M. HIRATA, T. ITOH, Y. KANAMURA & H. KURIYAMA. 1986. Inositol 1,4,5-triphosphate activates pharmacomechanical coupling in smooth muscle of the rabbit mesenteric artery. J. Physiol. **379:** 605–618.
22. AALKJAER, C. & M. J. MULVANY. 1985. Effect of ouabain on tone, membrane potential and sodium efflux compared with [^{3}H]ouabain binding in rat resistance vessels. J. Physiol. **362:** 215–231.
23. GUTHRIE T. P. JR. 1984. Effects of digoxin on responsiveness to the pressor actions of angiotensin and norepinephrine. J. Clin. Endocrinol. Metab. **58:** 76–80.
24. HAMLYN, J. M., R. RINGEL, J. SCHAEFFER, P. D. LEVINSON, B. P. HAMILTON, A. A. KOWARSKI & M. P. BLAUSTEIN. 1982. A circulating inhibitor of (Na+K)-ATPase associated with essential hypertension. Nature **300:** 650–652.
25. MOREL, N. & T. GODFRAIND. 1984. Sodium-calcium exchange in smooth-muscle microsomal fractions. Biochem. J. **218:** 421–427.
26. MATLIB, M., A. SCHWARTZ & Y. YAMORI. 1985. A Na^+-Ca^{2+} exchange process in isolated sarcolemmal membranes of mesenteric arteries from WKY and SHR. Am. J. Physiol. **249** (Cell Physiol. **17**): C166–C172.
27. GRYNKIEWICZ, G., M. POENIE & R. Y. TSIEN. 1985. A new generation of Ca^{2+} indicators with greatly improved fluorescence properties. J. Biol. Chem. **260:** 3440–3450.
28. WILLIAMS, D. A., K. E. FOGARTY, R. Y. TSIEN & F. S. FAY. 1985. Calcium gradients in single smooth muscle cells revealed by the digital imaging microscope using fura-2. Nature **318:** 558–561.
29. CANNELL, M. B., W. G. WIER, J. R. BERLIN, E. MARBAN & W. J. LEDERER. 1986. Free intracellular calcium in normal and calcium-overloaded rat heart cells: Digital imaging fluorescent microscopy using fura-2. Biophys. J. **49:** 466a.
30. GOLDMAN, W. F., W. G. WIER & M. P. BLAUSTEIN. 1986. Use of fura-2 and digital imaging methods to determine cytosolic free Ca^{2+} in living single arterial smooth muscle cells. Biophys. J. **49** (No. 2 Part 2): 465a.
31. SHEU, S.-S. & M. P. BLAUSTEIN. 1986. Sodium/calcium exchange and the regulation of cell calcium and contractility in cardiac muscle, with a note about vascular smooth muscle. *In* The Heart and Cardiovascular System. H. A. Fozzard, E. Haber, R. B. Jennings, A. M. Katz & H. E. Morgan. Eds. Raven Press. New York. (In press.)
32. BLAUSTEIN, M. P. 1977. Effects of internal and external cations and ATP on sodium-calcium exchange and calcium-calcium exchange in squid axons. Biophys. J. **20:** 79–111.
33. PETERSEN, T. T. & M. J. MULVANY 1984. Effect of sodium gradient on the rate of relaxation of rat mesenteric small arteries from potassium contractures. Blood Vessels **21:** 279–289.
34. BLAUSTEIN, M. P. 1974. The interrelationship between sodium and calcium fluxes across cell membranes. Rev. Physiol. Biochem. Pharmacol. **70:** 32–82.
35. JUNDT, H., H. PORZIG, H. REUTER & J. W. STUCKI. 1975. The effect of substances releasing intracellular calcium ions on sodium dependent calcium efflux from guinea-pig auricles. J. Physiol. **246:** 229–253.
36. DIPOLO, R. & L. BEAUGÉ. 1984. Interaction of physiological ligands with the Ca pump and Na/Ca exchange in squid axons. J. Gen. Physiol. **84:** 895–914.
37. KIMURA, J., A. NOMA & H. IRISAWA. 1986. Na-Ca exchange current in mammalian heart cells. Nature **319:** 596–597.
38. RASGADO-FLORES, H. & M. P. BLAUSTEIN. 1986. Calcium influx and sodium efflux

mediated by the Na/Ca exchanger in giant barnacle muscle cells are promoted by intracellular Ca^{2+}. Biophys. J. **49** (No. 2, Part 2): 546a.
39. HIRST, G. D. S., G. D. SILVERBERG & D. F. VAN HELDEN. 1986. The action potential and underlying ionic currents in proximal rat middle cerebral arterioles. J. Physiol. **371:** 289–304.
40. FINKEL, A. S., G. D. S. HIRST & D. F. VAN HELDEN. 1984. Some properties of excitatory junction currents recorded from submucosal arterioles of guinea-pig ileum. J. Physiol. **351:** 87–98.
41. BLAUSTEIN, M. P. & J. M. HAMLYN. 1983. Role of natriuretic factor in essential hypertension. Ann. Int. Med. **98** (suppl.): 785–792.
42. BLAUSTEIN, M. P. & J. M. HAMLYN. 1984. Sodium transport inhibition, cell calcium, and hypertension. The natriuretic hormone—Na^+/Ca^{2+} exchange—hypertension hypothesis. Am. J. Med. **77**(4A): 45–59.
43. TOSTESON, D. C. & J. F. HOFFMAN. 1960. Regulation of cell volume by active transport in high and low potassium sheep red cells. J. Gen. Physiol **44:** 169–194.
44. MILANICK, M. A. & J. F. HOFFMAN. 1986. Ion transport and volume regulation in red blood cells. Ann. N.Y. Acad. Sci. This volume.
45. DAHL, L. K. & M. HEINE. 1975. Primary role of renal homografts in setting chronic blood pressure levels in rats. Circ. Res. **36:** 692–696.
46. BIANCHI, G., U. FOX, G. F. DIFRANCESCO, A. M. GIOVANNETTI & D. PAGETTI. 1974. Blood pressure changes produced by kidney cross transplantation between spontaneously hypertensive rats and normotensive rats. Clin. Sci. Mol. Med. **47:** 435–448.
47. KAWABE, K., T. X. WATANABE, K. SHIONO & H. SOKABE. 1978. Influence on blood pressure of renal isografts between spontaneously hypertensive and normotensive rats, utilizing the F_1 hybrids. Jpn. Heart J. **19:** 886–894.
48. CURTIS, J. J., R. G. LUKE, H. P. DUSTAN, M. KASHGARIAN, J. D. WHELCHEL, P. JONES & A. G. DIETHELM. 1983. Remission of essential hypertension after renal transplantation. N. Engl. J. Med **309:** 1009–1015.
49. DAHL, L. K., D. K. D. KNUDSEN & J. IWAI 1969. Humoral transmission of hypertension: evidence from parabiosis. Circ. Res. **24–25** (suppl I): I21–I31.
50. HADDY, F. J. & H. W. OVERBECK. 1976. Role of humoral agents in volume expanded hypertension. Life Sci. **19:** 935–948.
51. BLAUSTEIN, M. P. 1977. The role of Na-Ca exchange in the regulation of tone in vascular smooth muscle. *In* Excitation-Contraction Coupling in Smooth Muscle. R. Casteels, T. Godfrained & J. C. Reugg, Eds. :101–108. Elsevier/North-Holland Biomedical Press B. V.. Amsterdam.
52. DE WARDENER, H. E. & G. A. MACGREGOR. 1980. Dahl's hypothesis that a saluretic substance may be responsible for a sustained rise in arterial pressure: its possible role in essential hypertension. Kidney Int. **18:** 1–9.
53. DE WARDENER, H. E. & G. A. MACGREGOR. 1983. The relation of a circulating sodium transport inhibitor (the natriuretic hormone) to hypertension. Medicine **62:** 310–326.
54. PAMNANI, M. B., J. BUGGY, S. J. HUOT & F. J. HADDY. 1981. Studies on the role of a humoral sodium-transport inhibitor and the anteroventral third ventricle in experimental low renin hypertension. Clin. Sci. **61:** 57s–60s.
55. SONGU-MIZE, E., S. L. BEALER & R. W. CALDWELL. 1982. Effect of AV3V lesions on development of DOCA-salt hypertension and vascular Na-pump activity. Hypertension **4:** 575–580.
56. BUGGY, J., S. HUOT, M. PAMNANI & F. HADDY. 1984. Periventricular forebrain mechanisms for blood pressure regulation. Fed. Proc. **43:** 25–33.
57. BEALER, S. L., J. R. HAYWOOD, K. A. GRUBER, V. M. BUCKALEW JR., G. D. FINK, M. J. BRODY & A. K. JOHNSON. 1983. Preoptic-hypothalamic periventricular lesions reduce natriuresis to volume expansion. Am. J. Physiol. **244:** R51–R57.
58. SWEADNER, K. J. & R. C. GILKESON. 1985. Two isozymes of the Na, K-ATPase have distinct antigenic determinants. J. Biol. Chem. **260:** 9016–9022.
59. BLAUSTEIN, M. P. 1981. What is the link between vascular smooth muscle sodium pumps and hypertension? Clin. Exp. Hypertension **3:** 173–178.
60. GUYTON, A. C., T. G. COLEMAN, A. W. COWLEY, K. W. SCHEEL, R. D. MANNING & R. A.

NORMAN. 1972. Arterial pressure regulation: overriding dominance of the kidneys in long-term regulation and in hypertension. Am. J. Med. **52:** 584–594.
61. BLAUSTEIN, M. P. 1985. The cellular basis of cardiotonic steroid action. Trends Pharmacol. Sci. **6:** 289–292.

DISCUSSION OF THE PAPER

F. R. BÜHLER (*University Hospital of Basel, Basel, Switz.*): Your concept puts great emphasis on volume expansion as an initiating trigger for hypothalamic release of the natriuretic hormone. But, volume expansion of any kind never has been detected in essential or genetic hypertension. Therefore, your hormone may be released in later hypertensive states when volume (e.g. cardiopulmonary) may be increased.

M. BLAUSTEIN (*University of Maryland, Baltimore, MD*): We don't know what the signal is for secretion of natriuretic hormone. Nevertheless, according to the expected blood pressure–blood volume curve, the body, in essential hypertension, behaves as if blood volume is elevated (i.e., there is a virtual hypervolemia). This may be contrasted with hypertension due to excessive pressor secretion (e.g., as occurs with pheochromocytoma), in which case volume is reduced due to the pressure natriuresis. The fact that blood volume is not substantially reduced in chronic essential hypertension implies that the blood volume "appears to be" greater than can be measured. This is discussed in detail in reference 1 of our paper.

Chemical and Clinical Studies of Endogenous Digitalis-Like Factor in Hypertension

J.F. CLOIX, M.A. DEVYNCK, AND P. MEYER

U7 INSERM
Department of Pharmacology
Hopital Necker
75015 Paris, France

Acute and chronic sodium loading induces the release of several natriuretic compounds.[1-5] Atrial natriuretic peptides (ANP) act primarily through a dilatation of renal arteries[5] whereas other compond(s) inhibit the Na^+ pump in renal tubules.[2,3] The origin and chemical structure of the former are well established and those of the latter are not defined with certainty. However, the Na^+-pump inhibitor(s) seem(s) to have digitalis-like properties with inhibitory effects on Na^+,K^+-ATPase and ouabain binding.

The present report summarizes the investigations performed in our laboratory on such endogenous Na^+-pump inhibitors.[6-10] The compound was purified to apparent chemical homogeneity from human urine and appeared to be non-peptidic, possibly steroidal with a molecular weight of around 500. The name of endalin was tentatively given to the inhibitor to evoke its endogenous origin and its digitalis-like properties.

Previous studies have indicated that such a compound is increased in various types of hypertension, in experimental Na^+-dependent hypertension and also in human essential (primary) hypertension. We report here several investigations confirming these results. Chemical, experimental, and clinical data will be presented subsequently.

PURIFICATION OF A Na^+-PUMP INHIBITOR FROM HUMAN URINE

Methods

Urine was collected on sodium azide (0.02% w/vol) and frozen at −20°C until used. Donors of urine were selected according to the capacity of their urine to inhibit Na^+,K^+-ATPase as follows: 10% aliquots of 24 hr urine were analyzed on a small column filled with 2 g of 40 μm octadecyl reversed-phase packing material. After sample loading the column was washed with water followed by 25% and 40% acetonitrile. The mobile phases and the urine were acidified by 0.1% (vol/vol) TFA. The 40% acetonitrile eluates were freeze-dried and assessed for inhibitory activity on Na^+,K^+-ATPase. Such effects were expressed as ouabain equivalent units excreted per 24 hr. One thousand liters of human urine were processed to achieve the present data.

Chromatographic Procedures

The low pressure (1–2 bars of nitrogen) flash chromatography was developed according to previously described methods.[10] In brief, 2–3 liters of pooled urine (containing 0.1% vol/vol TFA) were processed per 100 g of 40 μm octadecyl packing and chromatographed, using a step-wise gradient of acetonitrile in water. The reversed-phase HPLCs (Lichrosorb RP18, wide pore diphenyl and small pore monophenyl) were run using a linear gradient of acetonitrile in water. The mobile phases contained 0.1%, vol/vol TFA. The HPLCs were developed using a chain system from LKB (LKB, Bromna, Sweden). The optical density was monitored at 226 nm. The anion exchange chromatography was performed on a column (140 × 26 mm) of DEAE-Trisacryl (IBF, Villeneuve-La Garenne, France), and run using a linear (10–500 mM) ammonium acetate gradient.[11]

Bioassays

The chromatographic fractions were freeze-dried and the residues reconstituted with 1 mM acetic acid. These fractions were then assayed for their capacity to inhibit both Na^+,K^+-ATPase activity,[11] and [^{3}H]ouabain binding to purified dog kidney Na^+,K^+-ATPase[12] and for their capacity to cross-react with antidigoxin antibodies according to previously described methods[6,7] using a kit from NEN.

Physical Analysis

The purified Na^+-pump inhibitor was analyzed by proton nuclear magnetic resonance (Bruker AM 360) and by fast atomic bombardment (FAB) mass spectrometry (Variant MAT 212).

Results

The 24 hr excretion of a Na^+-pump inhibitor in urine was measured in normotensive control subjects (NT^-) and in normotensive subjects with a known family history of hypertension (NT^+). This value is expressed in nmole of ouabain equivalent units excreted per 24 hr. The NT^- excrete smaller amounts (0.48 ± 0.05 nmole/24 hr, N = 58, $p < 0.05$) of such a compound than NT^+ (1.24 ± 0.11 nmole/24 hr, N = 60). For urine collection, we selected subjects producing the highest amount of the endogenous Na^+-pump inhibitor.

The pooled urine was processed on 40 μm octadecyl packing material (200 g) by flash chromatography. Six liter batches of urine were applied to such a column, which is washed by water after sample loading, and by 25% and 40% acetonitrile. The 40% was then evaporated to dryness using a rotavapor. The residue was capable of inhibiting Na^+,K^+-ATPase activity, ouabain binding, and of cross-reaction with antidigoxin antibodies as previously described.[10]

The active fractions obtained by flash chromatographies were resolved into three peaks inhibiting Na^+,K^+-ATPase activity by anion exchange chromatography. The first peak, eluted off the DEAE-Trisacryl by 250 mM ammonium acetate, was further purified as it was the most important and the most active.

The active materials from the semi-preparative RP18 column were further purified on analytical RP18 followed by a wide pore diphenyl and a small pore monophenyl reversed-phase columns. The active fractions were then rechromatographed on the monophenyl column, giving a sharp peak inhibiting Na^+,K^+-ATPase activity.

The purification factor was calculated by comparing the IC_{50} (expressed in dry weight per assay tube) obtained at each step to that of pooled desalted crude urine

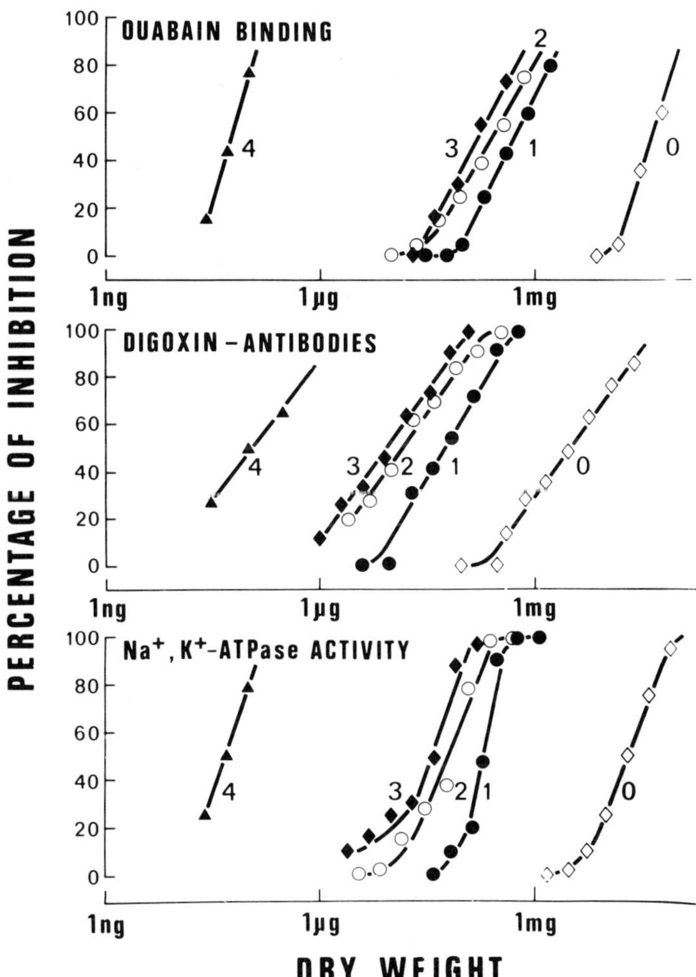

FIGURE 1. Dose-response curves of active material obtained at each step of the purification procedure. *Upper panel:* ouabain-binding inhibition. *Center panel:* cross-reaction with antidigoxin antibodies. *Lower panel:* Na^+,K^+-ATPase inhibition. 0 = desalted crude urine; 1 = flash chromatography; 2 = DEAE-Trisacryl; 3 = analytical RP18; 4 = last phenyl.

(FIG. 1). In terms of Na^+,K^+-ATPase inhibition the purification factor of the final product was estimated to be 360,000. This value was found to be 270,000 and 280,000 for ouabain binding inhibition and cross-reaction with antidigoxin antibodies, respectively.

This purified compound was then analyzed by NMR and by mass spectrometry. This active fraction, which was equal to 2 μg of material, did not contain peptides, amino acids, or fatty acids in sufficient amounts to give the corresponding characteristic NMR signals. The molecular mass was determined by mass spectrometry to be around 500 Da. These data are consistent with the possibility that this Na^+-pump inhibitor extracted from human urine could be a steroid.

The purified inhibitor was identified throughout the whole procedure of purification by its inhibitory effects on ouabain binding, digoxin-antibody reaction, and Na^+,K^+-ATPase activity. Its digitalis-like properties were further substantiated by its natriuretic effect (FIG. 2), by its capacity to inhibit ouabain-sensitive active Na^+ efflux from red blood cells[11] and to inhibit a Na^+-dependent process, such as serotonin uptake in human platelets.[13]

Discussion

Thus, human urine contains a Na^+-pump inhibitor with digitalis-like properties. Mass spectrometry analysis performed on a compound purified to apparent homogeneity and obtained in sufficient amounts for biophysical studies indicated that endalin has a low molecular weight of around 500. This result is in agreement with other studies.[14,15]

The absence of peptide or amino acid residues in the active compound is in disagreement with some previous reports.[14,16] Nevertheless, the previous publications dealt with crude or impure material and the identification of the active compound was hampered accordingly. The same objection applies to the report of Kramer *et al.*[16] who described a Na^+,K^+-ATPase inhibitor extracted from human urine, containing four amino acids. In addition, one cannot exclude the possibility that several endogenous Na^+-pump inhibitors can be found in urine. Wainer *et al.*[10] indeed, reported that urine contains two different compounds inhibiting Na^+,K^+-ATPase and ouabain binding, and cross-reacting with antidigoxin antibodies.

The chemical nature of our purified compound is under investigation. However, we do have sufficient data to forward a hypothesis. The adsorption onto DEAE-Trisacryl depended upon the pH of the fraction obtained after flash chromatography. When the pH of this material was adjusted to 6, our inhibitor was retained by the anion exchange resin. Conversely, when the pH was maintained at 3, the active compound was eluted into the void volume of the column. This means that the pKa of the active substance lies within the range 3–6, indicating the possibility of the presence of a carboxyl acid group. It is thus tempting to propose that the purified Na^+-pump inhibitor with digitalis-like properties could be a steroid containing a carboxyl acid function.

VARIATIONS OF AN ENDOGENOUS DIGITALIS-LIKE COMPOUND IN THE RAT

Variations of an endogenous digitalis-like compound were studied in two circumstances: throughout the course of a high sodium diet maintained during three months and in rats with a reduced renal mass. The overall data suggest that the increased levels of the endogenous digitalis-like factor may contribute to the elevation in blood pressure.

Methods

Male Wistar rats (225–220 g) were used throughout. The control groups received a standard diet (0.2% NaCl) and tap water. High Na^+ regimen was ensured by an

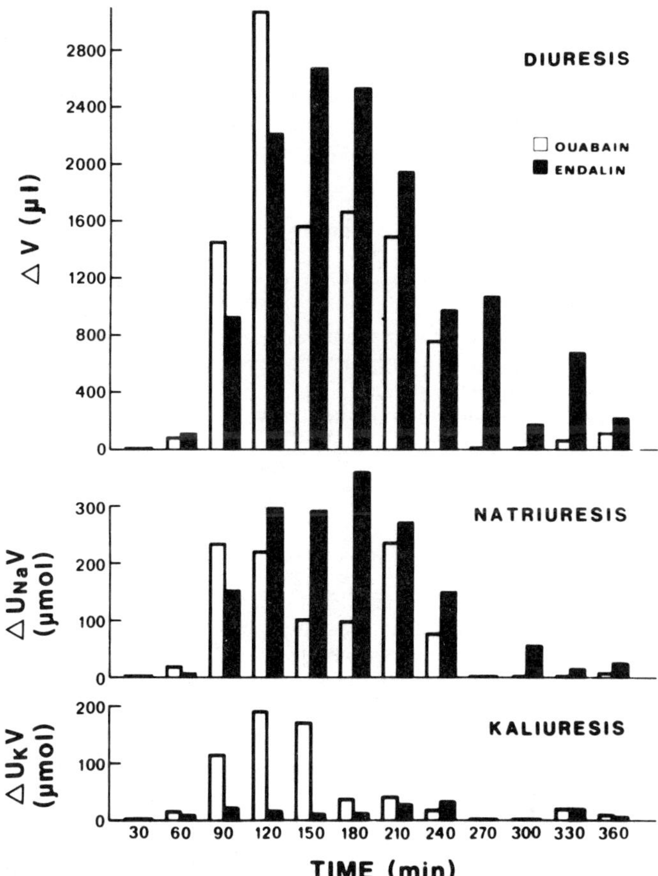

FIGURE 2. *In vivo* effects of ouabain (0.25 mg) and Na^+,K^+-ATPase extracted from human urine (equivalent to four times the amount producing 50% of inhibition of Na^+,K^+-ATPase activity). The diuresis and sodium and potassium excretions were expressed as the difference between either ouabain or urinary Na^+,K^+-ATPase inhibitor effects and the blank, in µl per 30 min and in µmol of Na^+ and K^+ excreted per 30 min, respectively. The values represent the average of four experiments.

artificial chow containing 8% NaCl. The renal sodium excretion averaged 1.2 ± 0.1 and 26.3 ± 1.9 mmoles per day in control and sodium-loaded rats, respectively (mean values of 14–22 determinations).

Reduced renal mass hypertension was produced according to the technique of Huot *et al.*[17] Systolic blood pressure was recorded by the tail-cuff procedure. The ability of rat plasma to inhibit ATPase was investigated on Na^+,K^+-ATPase–enriched membranes from rat kidney, according to another publication.[8]

Plasma digoxin-like immunoreactivity was assayed with the commercially available radioimmunoassay kit RIANEN digoxin [^{125}I] (New England Nuclear). The original technique was modified to obtain a detection limit of 15 pg/ml. Red cell

Na^+,K^+-pump activity was determined as the ouabain-sensitive sodium efflux of sodium-loaded erythrocytes in a magnesium-sucrose medium as previously described.[18] The concentration at which the Na^+, K^+-pump was inhibited by 50% (IC_{50}) was calculated on an Apple II computer from measurement of sodium efflux in the presence of seven concentrations of ouabain ranging from 0 to 10^{-3} M. Intra-erythrocyte sodium was determined according to a previously described method.[18]

Results

High Na^+ Diet

After one week, neither systolic blood pressure nor intra-erythrocytic Na^+ content was modified, but plasma extracts slightly inhibited renal Na^+,K^+-ATPase (70.9 ± 1.7 vs. 76.3 ± 2.1 μmol Pi/mg/hr, $p = 0.05$).

After two weeks, the plasma inhibitory activity, the systolic blood pressure, and the intra-erythrocytic Na^+ content were higher than corresponding values in control animals (65.5 ± 1.6 vs. 79.1 ± 2.8 moles Pi/mg/hr, $p < 0.001$; 132 ± 2 vs. 114 ± 4 mm Hg, $p < 0.001$ and 4.95 ± 0.32 vs. 3.81 ± 0.36 mM/1 cells, $p < 0.05$, respectively).

After three months, the plasma digoxin-like immunoreactivity and its ability to inhibit the Na^+ pump were elevated (68.7 ± 7.9 vs. 48.2 ± 5.4 pg/ml, $p < 0.02$; 57.8 ± 1.8 vs. 72.9 ± 1.8 μmol Pi/mg/hr, $p < 0.001$, respectively) whereas intra-erythrocytic Na^+ content had returned to control level. The results demonstrated that this high salt intake led to simultaneous increases in the systolic blood pressure and in the activity of a digitalis-like compound present in plasma (FIG. 3). The inhibition of the Na^+,K^+-ATPase was correlated with systolic blood pressure and digoxin-like immunoreactivity ($r = 0.569$, $N = 76$, $p < 0.001$ and $r = 0.414$, $N = 34$, $p < 0.02$, respectively).

Reduced Renal Mass

When compared to plasma from control rats, plasma of reduced renal mass (RRM) rats exhibited three weeks after surgery an enhanced digoxin-like immunoreactivity (57.5 ± 5.0 vs. 42.1 ± 3.8 pg/ml, $p < 0.01$) and inhibited both the renal Na^+,K^+-ATPase activity (135 ± 5 vs. 154 ± 5 μmol P_i/mg/hr, $p < 0.01$) and the net Na^+ extrusion from Na^+-loaded erythrocytes of normotensive rats (5.96 ± 0.40 vs. 7.68 ± 0.34 mmol/l cells/hr, $p < 0.01$). In red cells of RRM rats, the Na^+,K^+-pump activity was reduced (7.34 ± 0.29 vs. 10.88 ± 0.41 mmol/l cells/hr, $p < 0.001$) and Na^+ content increased (4.66 ± 0.08 vs. 4.16 ± 0.11 mmol/l cells, $p < 0.01$).

Discussion

Chronic Na^+ loading increases the level of a circulating digitalis-like compound as previously reported by others.[1,2] The present results are interesting in the sense that the salt-induced rise in blood pressure and the plasma capacity to inhibit the Na^+,K^+-ATPase are positively correlated. The concomitant rises in blood pressure and in the activity of the digoxin-like compound indicate parallel changes but do not prove that these two phenomena are causally related. Furthermore, the inhibition of Na^+,K^+-ATPase activity and the digoxin-like immunoreactivity may not be due to the same circulating compound. Some investigators have indeed shown that several compounds having one or the other of these two properties as well as different chromatographic

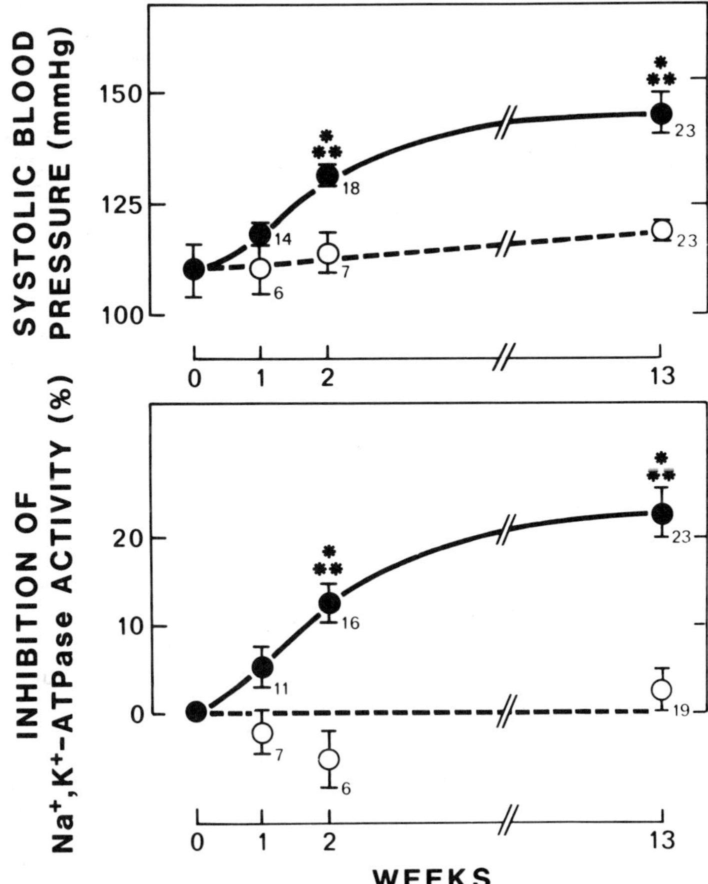

FIGURE 3. Effects of chronic salt load on the systolic blood pressure of Wistar rats and on the activity of a circulating inhibitor of the renal Na^+,K^+-ATPase activity. Values measured in control (○, $N = 6$–23) and Na^+-loaded rats (●, $N = 11$–23) are given as means ± SEM. The inhibition of Na^+,K^+-ATPase activity is expressed as percent of the enzymatic activity measured in parallel in the presence of plasma extracts from control rats. *, **, ***: $p < 0.05$, <0.01, and <0.001 when compared to values obtained with control rats, respectively.

properties are present simultaneously in plasma or serum.[6,19,20] Thus fractions possessing both activities represent only a part of the total digoxin-like immunoreactivity measured. However, a positive correlation between a Na^+,K^+-ATPase inhibitor and the level in blood pressure has also been observed in human primary hypertension.[21]

The presence of high levels of endogenous Na^+,K^+-pump inhibitor in rats with RRM hypertension has been previously reported by Huot *et al.*[17] using both the measurements of ouabain-sensitive rubidium uptake in tail artery and the activity of myocardial Na^+,K^+-ATPase. The present findings confirm these data and the presence of a high digitalis-like immunoreactivity is likely to be linked to increased levels of this factor. The observation that both myocardial Na^+,K^+-ATPase[17] and

Na^+,K^+-pump in red cell are depressed in RRM rats suggests that the factor must be strongly bound because these measurements were obtained from thoroughly washed cells and membranes. The functional activity of this factor was suggested by the observation of an enhanced erythrocyte sodium content.

INVESTIGATION OF AN ENDOGENOUS DIGITALIS-LIKE COMPOUND IN PRIMARY HUMAN HYPERTENSION

Most of the studies performed so far used only one test to quantify endogenous digitalis-like compounds in human plasma. However, the interaction of an endogenous digitalis-like substance with antibodies may have a different pattern than that to ouabain binding sites. In addition, more than one chromatographic fraction obtained from plasma and urine extracts appear to inhibit both digitalin-antibody interaction and ouabain binding.[6,10,19,20] These estimations were therefore run in parallel on plasma samples obtained from essential hypertensives: inhibition of ouabain binding, inhibition of antidigoxin-antibody reaction, and inhibition of Na^+,K^+-ATPase.

The details of the method are published elsewhere.[6,22,23] Na^+,K^+-ATPase was purified from human kidney; the NEN Digoxin kit was used for studying cross-reactivity with digoxin-antibodies; [^{3}H]ouabain binding was studied on human erythrocytes.

Controls and Patients

Normotensive Subjects

This group comprises 60 Caucasian healthy volunteers. All of them were on an unrestricted Na^+ diet. They were free of any medication.

Eighteen males and 9 females aged from 9 to 51, mean 30.2 ± 1.5 had no known family history of hypertension among their first or second degree relatives. Their blood pressure averaged 119 ± 2.3 / 76.4 ± 1.6 mm Hg.

Twenty-one males and 12 females aged 20 to 50, mean 33.3 ± 1.6, reported at least a first or a second degree relative with sustained high blood pressure. Their blood pressure averaged 129.8 ± 2.8 / 78.8 ± 1.5 mm Hg.

Essential Hypertensive Patients

Ninety-five Caucasian patients (58 males and 37 females) aged 19 to 74 mean 45.1 ± 1.8 had sustained and uncomplicated hypertension, which was diagnosed after complete investigation. Their blood pressure averaged 169 ± 2/102 ± 2 mm Hg. None of these patients had been treated for at least 2 weeks before blood sampling. Most of them had remained untreated for more than one month. They were on an unrestricted Na^+ diet.

Results

Results obtained with the three methods were comparable showing significant differences between hypertensives and normotensive controls. One third of the patients had elevated values. There were no significant differences between controls with or without hypertension in their family, except for ouabain binding inhibition, which was more important in the former than in the latter.

The relationship between the three methods used to quantify the digoxin-like properties of plasma was analyzed by rank correlations. The apparent digoxin-like immunoreactivity of whole plasma was positively correlated with the inhibitory effect of plasma extract on both ouabain binding and Na^+,K^+-ATPase activity. The capacity of plasma extracts to decrease the apparent affinity of the Na^+-pump for ouabain was also positively correlated with their ability to inhibit the renal Na^+,K^+-ATPase activity and to cross-react with antidigoxin antibodies. However, analysis of individual values indicated some discrepancies that are reflected by differences in the analysis of blood pressure relationship: inhibitory effect on Na^+,K^+-ATPase and on digoxin-antibody reaction were correlated to blood pressure, but not the inhibitory effect on ouabain-binding.

Discussion

An increased level of endogenous digitalis-like substance(s) may be observed in primary human hypertension. It is obvious that the phenomenon cannot be considered as a common feature and that the Na^+-pump inhibitor cannot be considered as the sole pathogenic factor in contradiction with previous suggestion.[15,24] However, a responsibility in the increase in blood pressure is supported by some positive correlation observed within the present group of hypertensives. The present data do not provide any further insights concerning the factors governing the release of the endogenous Na^+-pump inhibitor; a multiple regression analysis is in progress. The present data cast some doubt on the chemical homogeneity of the circulating inhibitor(s); further improvements in its characterization are necessary and presently studied in our laboratory.

SUMMARY

Endogenous digitalis-like factor (endalin) was investigated by measuring the ability of rat and human plasma and urine to inhibit [^{3}H]ouabain-specific binding, digoxin-antidigoxin antibodies interaction, and renal Na^+,K^+-ATPase activity.

Endalin was detected in plasma (and urine) of one third of 112 patients with sustained and moderate hypertension (Na^+ intake = 110 mmol/l). Endalin tended to be increased in the more pronounced hypertensives. No correlation with any other clinical and biological parameter could be detected.

An activity to inhibit Na^+,K^+-ATPase was also detected in the rat after acute and chronic Na^+ loading, in reduced renal mass-type hypertension and in SHRs as compared to WKY rats.

Comparison of the plasma and urine inhibitory effects in the different tests revealed some chemical heterogeneity. However, a compound possessing the biochemical and pharmacological characteristics of digitaline was extracted from human urine. Chromatographic and spectral analysis of about 1,000 liters revealed a compound with apparent chemical homogeneity, molecular weight around 500, devoid of peptidic bound and of aliphatic structure.

REFERENCES

1. BUCKALEW, V. M. & M. AZER. 1974. Kidney Int. **5:** 12–22.
2. GONICK, H. C., H. J. KRAMER, O. PAUL & E. LU. 1977. Clin. Sci. Mol. Med. **53:** 329–334.
3. DE WARDENER, H. E., G. A. MACGREGOR, E. M. CLARKSON, J. ALAGHBANZADEH, J. BITENSKI & J. CHAYEN. 1981. Lancet **i:** 411–412.
4. GAULT, M. H., S. C. VASDEC, L. L. LONGERICH, P. FERNANDEZ, V. PRABHKARAM, M. DAWE & C. MAILLER. 1983. New Engl. J. Med. **309:** 1459–1460.
5. MAACK, T, M. J. F. CAMARGO, H. D. KLEINERT, J. LARAGH & S. A. ATLAS. 1984. Kidney Int. **27:** 607–615.
6. CRABOS, M, I. W. WAINER & J. F. CLOIX. 1984. FEBS Lett. **176:** 223–228.
7. CLOIX, J. F., M. CRABOS, I. W. WAINER, U. RUEGG, M. SEILER & P. MEYER. 1985. Biochem. Biophys. Res. Commun. **131:** 1234–1240.
8. WAUQUIER, I., M. G. PERNOLLET, P. DELVA & M. A. DEVYNCK. 1986. J. Hypertension. (In press.)
9. DE MENDONCA, M., M. L. GRICHOIS, M. G. PERNOLLET, B. THORMAN, P. MEYER, M. A. DEVYNCK & R. P. GARAY. 1985. J. Hypertension S73–S75.
10. WAINER, I. W., M. CRABOS & J. F. CLOIX. 1985. J. Chromatography **338:** 417–421.
11. CLOIX, J. F., G. DAGHER, M. CRABOS, M. G. PERNOLLET & P. MEYER. 1984. Experientia **40:** 1380–1382.
12. KRACKE, G. 1983. J. Lab. Clin. Med. **101:** 105–113.
13. KAMAL, L. A., J. F. CLOIX, M. A. DEVYNCK & P. MEYER. 1983. Eur. J. Pharmacol. **92:** 167–168.
14. CLARKSON, E. M., S. M. RAW & H. E. DE WARDENER. 1976. Kidney Int. **10:** 387–394.
15. DE WARDENER, H. E. & E. M. CLARKSON. 1985. Physiol. Rev. **65:** 658–759.
16. KRAMER, H. J., M. HEPPE, E. WEILER, A. BACKER, C. LIDDIARD & D. KLINGMULLER. 1985. Renal Physiol. **8:** 80–89.
17. HUOT, S. J., M. B. PAMNANI, D. L. CLOUGH & F. J. HADDY. 1983. Am. J. Nephrol. **3:** 92–99.
18. DE MENDONCA, M., M. L. GRICHOIS, G. DAGHER, R. P. GARAY & P. MEYER. 1983. Pflügers Arch. **398:** 64–68.
19. KOJIMA, I. 1984. Biochem. Biophys. Res. Commun. **122:** 129–136.
20. KRAMER, H. J., E. M. HEPPE, G. PENNING, J. KIPNOWSKI, D. KLINGMULLER, R. DUSSING & F. KRUCK. 1985. Klin. Wochenschr. **63:** S107–S110.
21. HAMLYN, J. M., R. RINGEL, J. SCHAEFFER, P. D. LEVINSON, B. P. HAMILTON, A. A. KOWARSKI & M. P. BLAUSTEIN. 1982. Nature **300:** 650–652.
22. DERAY, G., M. RIEU, M. A. DEVYNCK, M. G. PERNOLLET, P. CHANSON, J. P. LUTON & P. MEYER. 1986. New Engl. J. Med. (submitted).
23. DEVYNCK, M. A., M. G. PERNOLLET, J. B. ROSENFELD & P. MEYER. 1983. Brit. Med. J. **287:** 631–634.
24. DE WARDENER, H. E. & G. A. MACGREGOR. 1983. Medicine **62:** 310–326.

DISCUSSION OF THE PAPER

E. CARAFOLI (*Swiss Federal Institute of Technology, Zurich*): Your factor causes Na^+ increases also in cells that do not have the Na^+/Ca^{2+} exchanger in the plasma membrane; e.g., the erythrocytes. According to Blaustein's hypothesis, the increase in intracellular Na^+ would lead to increased intracellular Ca^{2+} via the Na^+/Ca^{2+} exchanger. If Ca^{2+} increases also in red cells, as it seems to do, this cannot be mediated by the Na^+/Ca^{2+} exchanger.

MEYER: We have not measured the intracellular calcium concentration in erythrocytes. However, free calcium is increased in the platelets of our patients in agreement with the work of F. Bühler.

LUCIANI (*University of Padova, Padova*): Have you tried the positive inotropic effect of endalin and tried to correlate this effect with the purification steps?

MEYER: This investigation of endalin on cardiac contractility is in progress in our laboratory.

Q. AL-AWQATI (*Columbia University, New York, NY*): Is the effect reversible and if so how fast is the reversibility?

MEYER: The endalin effect is reversible and this reversibility is slower than that of ouabain.

L. H. OPIE (*University of Cape Town, Observatory, South Africa*): Can the levels of endaline be correlated with the level of the BP?

MEYER: In the Na-loaded rat, the positive correlation is obvious. In human hypertension, we found it less clear. Possibly in relation to the studied sample of patients, restricted to moderate hypertension.

G. HAUPERT (*Harvard University, Boston, MA*): Have you been able to calculate an affinity constant for the binding of the urinary factor to the Na,K-ATPase?

MEYER: The affinity of endalin to a purified human renal ATPase was close to that of ouabain.

P. GMAJ (*University of Zurich, Zurich*): What do you think about the report published in the *Journal of Biological Chemistry* in early 1986, where the plasma ouabain-like factor in hypertensive patients was identified as oleic acid?

MEYER: There was no oleic acid in our purified extract of endaline and no major lipid change in our hypertensive patients.

Calcium and Sodium in Vascular Smooth Muscle[a]

ANDREW P. SOMLYO, RAYMOND BRODERICK, AND AVRIL V. SOMLYO

Pennsylvania Muscle Institute
Departments of Physiology and Pathology
University of Pennsylvania School of Medicine
Philadelphia, Pennsylvania 19104

INTRODUCTION

Calcium has a unique, central role in regulating smooth and striated muscle contraction, in addition to its multiple effects on cell metabolism, membrane transport, and permeability. Thus, a clear insight into the mechanisms of vascular smooth muscle pathology requires both the structural identification of the sources and sinks of cytoplasmic Ca^{2+} and an understanding of the physiological mechanisms modulating cytoplasmic $[Ca^{2+}]$. The clinically most important abnormality of vascular Ca^{2+} metabolism is its probable role in the pathogenesis of hypertension,[1] as elevation of blood pressure is most likely to be the expression of vascular smooth muscle contraction due to a rise in cytoplasmic $[Ca^{2+}]$. Vascular calcification, such as occurs in coronary artery disease, is another obvious example of the effects of pathological Ca^{2+} metabolism.

The role of abnormal Na^+ metabolism in vascular pathology is not well understood, in spite of its having been widely implicated in the pathogenesis of hypertension,[2,3] largely because the localization of normal and abnormal sites of cellular Ca^{2+} and measurements of cell Na^+ content have been technically difficult: the multicellular composition of blood vessels and their heterogeneous connective tissue matrix complicated the interpretation of bulk ion and radioisotope flux measurements.[2,4] More recently, developments in rapid freezing and electron probe X-ray microanalysis (EPMA) permitted the unambiguous quantitation of Ca^{2+},[5–8] and Na^+,[9,10] and other elements within the cytoplasm and in cell organelles. This communication will summarize our results with rapid freezing and EPMA in identifying the sarcoplasmic reticulum (SR) as the organelle responsible for the physiological regulation of cytoplasmic Ca^{2+} and will review evidence, obtained with other methods, showing the role of inositol trisphosphate in releasing Ca^{2+} from the SR. EPMA has also been highly valuable in demonstrating that normal mitochondrial Ca is low in vascular smooth muscle, as in skeletal[11–14] and cardiac muscle,[15] retinal rods,[16] liver,[17] and neural cells,[18] although under pathological conditions mitochondria can, reversibly, accumulate large amounts of Ca^{2+} *in situ*. Cell Na^+ and its translocations have also been studied with EPMA of rapidly frozen tissues, and lead to the demonstration of a highly active ouabain-resistant, Ca^{2+}-sensitive Na transport system in vascular smooth muscle. The methods used for these studies have been published in detail.[19–22] The

[a]Supported by Grant HL15835 and Training Grant NHLBI 07499 to the Pennsylvania Muscle Institute, and by Post-doctoral Fellowship to R.B. from the Heart Association of Southeastern Pennsylvania.

spatial resolution attainable with EPMA is somewhat better than 10 nm,[23] and the minimum practically measurable Ca concentration in a 100–150 nm thick cryosection is approximately 300 μmol/kg dry wt.[24]

IDENTIFICATION OF THE SARCOPLASMIC RETICULUM AS THE SITE OF PHYSIOLOGICAL CALCIUM STORAGE AND RELEASE

The capacity of the SR (FIGS. 1 and 2) of smooth muscle to accumulate divalent cations *in situ* was initially demonstrated through the localization of strontium within the SR of smooth muscles incubated in solutions in which Sr^{2+} replaced Ca^{2+}, as the SR ATPase transports Sr^{2+} in a manner similar to Ca^{2+}.[25] Subsequently, EPMA of smooth muscles permeabilized with saponin showed that, at levels of low (physiological) [Ca^{2+}], calcium was accumulated by the SR rather than by mitochondria.[6] The quantitation of Ca^{2+} in the SR of normal smooth muscle was more difficult, because the SR membranes cannot be visualized in unstained cryosections and their Ca^{2+} content is lower than that of the terminal cisternae (120 mmol/kg dry wt[12]) of skeletal muscle. However, by exploring, with focused electron probes, the subplasmalemmal region of the cytoplasm that contains the junctional SR[7] or by identifying the central SR through its high phosphorus content,[8] it was possible to estimate the Ca content of the SR in relaxed smooth muscle. The approximately 48 mmol Ca^{2+}/kg dry wt central SR, estimated following correction for partial overlap of the electron probe with adjacent cytoplasm,[8] or the uncorrected value of 28 mmol/kg dry wt in junctional SR in portal vein,[7] are higher than would be expected on the basis of the probable free [Ca^{2+}] in the SR lumen.[26] This, together with the proteinaceous content of the SR lumen found in both tannic acid–treated and deep etched rotary shadowed preparations,[27] suggest the existence of a Ca^{2+} binding protein, similar to calsequestrin, the Ca-binding protein in the SR of striated muscle. The Ca^{2+} content of the SR of relaxed smooth muscle, if released, is sufficient to maximally activate even those smooth muscles in which the SR constitutes only 2% of the cell volume. This conclusion is also supported by the persistence of maximal contractions of guinea pig portal vein in Ca-free solutions.[7] The release of Ca from the junctional [7] as well as from the central SR[8] by norepinephrine was directly demonstrated by measuring, with EPMA, Ca^{2+} in the SR in relaxed and in maximally contracted smooth muscles. Approximately 50% of the Ca content of the SR was released (in main pulmonary artery) during a maximal contraction. Transmitters (e.g., norepinephrine) released Ca^{2+} from the SR of smooth muscles depolarized with Ca-free high K^+ solutions as well as from the SR of normally polarized smooth muscle, in the presence of normal extracellular Ca^{2+}. The release of intracellular Ca^{2+} by norepinephrine has also been demonstrated by ^{45}Ca flux studies.[28,29]

PHARMACOMECHANICAL Ca^{2+} RELEASE FROM THE SARCOPLASMIC RETICULUM: THE ROLE OF INOSITOL 1,4,5-TRISPHOSPHATE

Pharmacomechanical coupling is the term originally coined to include pathway(s) of activation or inactivation (relaxation) of smooth muscle that are independent of the membrane potential, although, under physiological conditions, can operate in parallel with (voltage-dependent) electromechanical coupling.[30,31] Pharmacomechanical coupling occurred in completely depolarized smooth muscles in Ca-free solutions,[7,32,33] indicating unequivocally that the release of intracellular Ca^{2+} is one of its mechanisms.

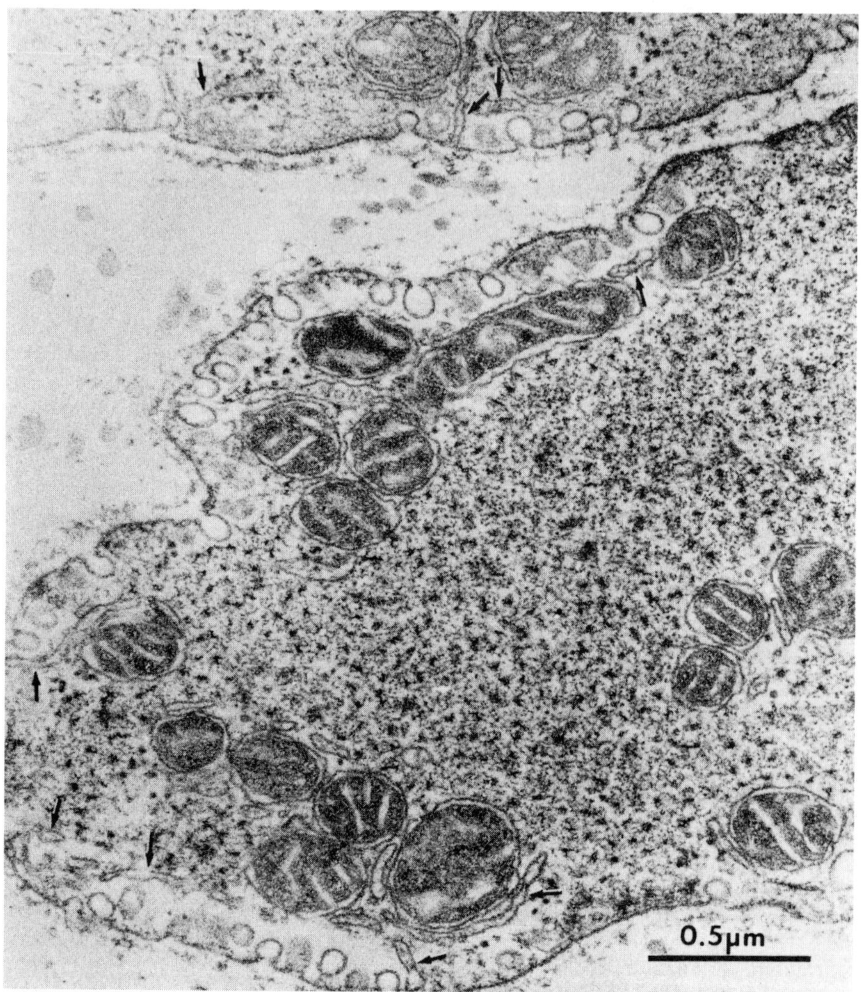

FIGURE 1. Transverse section of smooth muscle cells from rabbit portal anterior mesenteric vein showing typical distribution of tubules of sarcoplasmic reticulum (SR) (arrows) at the cell periphery. This smooth muscle contains a relatively small reticulum, approximately 2% of cell volume, which is nevertheless sufficient for storing enough Ca to activate maximal force development (see text). (From Bond *et al.*[7] With permission from *Journal of Physiology*.)

It now appears likely that inositol 1,4,5-trisphosphate ($InsP_3$) is the second messenger of pharmacomechanical Ca^{2+} release in smooth muscle[34-36] and is analogous to Ca^{2+} release from the endoplasmic reticulum of nonmuscle cells mediated by $InsP_3$ liberated by phospholipase C from phosphatidyl inositol bisphosphate.[37] We have pursued the question of whether $InsP_3$ is a "pharmacomechanical messenger" in rabbit main pulmonary artery (RMPA) smooth muscle, because this preparation has a well-developed SR[33] from which Ca^{2+} is known to be released by a physiological transmit-

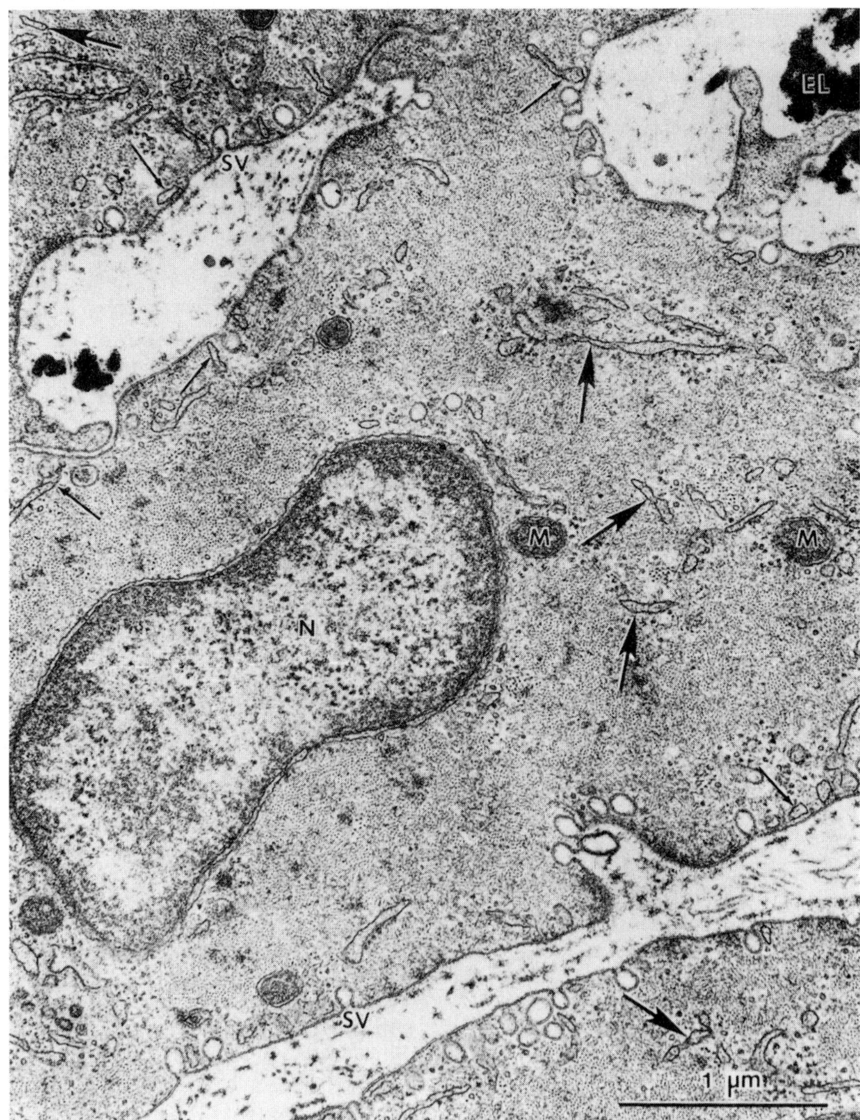

FIGURE 2. Transverse section of smooth muscle cells from rabbit main pulmonary artery, illustrating the location of SR tubules throughout the cytoplasm (large arrows) as well as at the cell periphery (small arrows). Contrast the greater volume of SR (5–7% of cell volume) and its location in this blood vessel with that shown in FIG. 1. Ca content of the central SR is decreased during stimulation with norepinephrine. SV, surface vesicle; EL, elastin; M, mitochondrion. (From Kowarski *et al.*[8] With permission from *Journal of Physiology.*)

ter, norepinephrine.[8] Ca^{2+} electrode measurements have shown that $InsP_3$ also releases Ca^{2+}, in the presence of physiological free [Mg^{2+}], from the SR of RMPA smooth muscle permeabilized with saponin.[36] The amount of Ca^{2+} released by $InsP_3$ is sufficient to cause contractions that, furthermore, are graded according to $InsP_3$ concentration. The latency and rate of contraction induced by $InsP_3$ generated by laser photolysis of "caged $InsP_3$" (a photolyzable precursor of $InsP_3$) are compatible with the physiological delays and rates observed in main pulmonary artery.[38] However, the caged compound currently available is a moderately active mixture of isomers in which the photolabile group is attached in various (i.e., 1, 4, or 5) positions. Therefore, it will be desirable to determine the time course of Ca^{2+} release and contraction, in further experiments, with well characterized and homogeneous isomers of $InsP_3$ "caged" at specific positions.

MITOCHONDRIAL CALCIUM: NORMAL CONTENT AND THE REVERSIBILITY OF PATHOLOGICAL CALCIUM LOADING

It is now well established, through EPMA of rapidly frozen cells, that endogenous mitochondrial Ca^{2+} is normally low (approximately 2 nmol/mg mitochondrial protein) in smooth muscles as in all other cells (TABLE 1). For reasons discussed in detail elsewhere,[39–41] this normal mitochondrial [Ca^{2+}] is too low for the saturation of the mitochondrial Ca^{2+} efflux pathway, indicating that mitochondria do not play a significant role in the physiological regulation of cytoplasmic [Ca^{2+}]. This conclusion is also supported by the absence of a significant increase in mitochondrial Ca^{2+} during

TABLE 1. Mitochondrial Calcium Content *in Situ* (mmole/kg dry Mitochondrion)

Animal	Tissue	Condition		Reference
Rabbit	Portal vein	Relaxed	1.6	1
	Portal vein	Contracted 30′	2.3	1
	Portal vein	0-Ca	0.1[a]	2
	Portal vein	Na-loaded	2.0	3
	Main Pulmonary artery	Relaxed	2.6	4
	Main Pulmonary artery	Contracted	2.0	4
	Heart (papillary)		0.5	5
Guinea pig	Portal vein	Relaxed	0.6	6
	Portal vein	Contracted	1.1	6
Rat	Liver	In animal	0.8	7
	Brain cortex	In animal	1.5	8
Rana pipiens	Retinal rod	Dark adapted	0.0	9
	Retinal rod	Illuminated	0.5	9

[a]Significantly different ($p < .01$) from relaxed rabbit portal vein.

References:

1. BOND, M. *et al.* 1984. J. Physiol. **357:** 185–201.
2. BOND, M. *et al.* 1985. Biophys. J. **47:** 414a.
3. This study.
4. KOWARSKI, D. *et al.* 1985. Biophys. J. **47:** 280a.
5. BRODERICK, R. & A. P. SOMLYO. Unpublished.
6. BOND, M. *et al.* 1984. J. Physiol. **355:** 677–695.
7. SOMLYO, A. P. *et al.* 1985. Nature **314:** 622–625.
8. SOMLYO, A. P. *et al.* 1985. Biochem. Biophys. Res. Commun. **132:** 1071–1078.
9. SOMLYO, A. P. & B. WALZ. 1985. J. Physiol. **358:** 183–195.

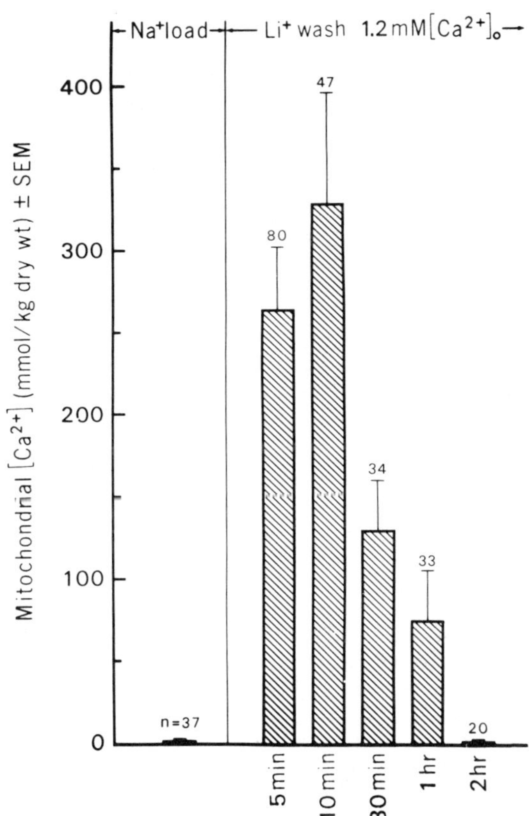

FIGURE 3. Mitochondrial Ca concentration in Na^+-loaded vascular smooth muscle and during Na^+ efflux into Na^+-free, ouabain-containing solution. Note that the massive amount of Ca^{2+} accumulated during the initial 5–10 min of Na^+ efflux is returned to normal values within 2 hr in the Na^{2+}-free, Li^+-solution that contained 1.2 mM Ca^{2+}.

maximal contractions in normal smooth muscle[5,8,42] (TABLE 1). The low *in vivo* mitochondrial Ca observed at normal cytoplasmic Ca^{2+} (50 nM to less than 5 μM) is also consistent with the low affinity of isolated mitochondria for Ca^{2+} in the presence of physiological [Mg^{2+}] (K_m = 10–20 μM[43,44]). Under pathological conditions, mitochondria in smooth muscle, as in other cells, can become heavily Ca^{2+} loaded.[5,6,45–50] Such abnormal Ca^{2+} loading of mitochondria is the consequence of abnormal increases in cytoplasmic free Ca^{2+}, whether due to pathological damage or experimental permeabilization (with saponin) of the cell membrane, or to abnormal increase in inward Ca^{2+} transport (see below).

Massive accumulation of mitochondrial Ca^{2+} occurs during rapid Na^+ efflux, through a ouabain-resistant pathway, from Na^+-loaded smooth muscle (FIG. 3). This large, two orders of magnitude increase in mitochondrial Ca^{2+} was found in rabbit portal vein smooth muscle without any evidence of general cell membrane damage, such as hyperpermeable "light cells,"[9,10] enabling us to determine its reversibility.

Following the rise of mitochondrial Ca^{2+} during the initial 10 minutes of rapid, ouabain-resistant Na^+ efflux, mitochondrial Ca^{2+} content declined and returned to normal values by two hours (FIG. 3). Inasmuch as no abnormal "light" cells were found at this time and the maximal contractility of these strips was also normal, we conclude that even such extensive, pathological increase in mitochondrial Ca^{2+} content is reversible and compatible with cell survival. Furthermore, this extensive decrease in mitochondrial Ca^{2+} occurred in the virtual absence of extracellular Na^+, and while mitochondrial Na was also decreasing.[10] Therefore, while Na^+-induced mitochondrial Ca^{2+} release[45] could occur in smooth muscle, it is clearly not an obligatory mechanism. The changes in mitochondrial monovalent ions during ouabain-resistant Na^+ efflux[10] indicate that K^+ and Na^+ are much more rapidly exchangeable *in situ* than in isolated mitochondria.[5,7,8,51,52]

CELL SODIUM CONTENT, OUABAIN-RESISTANT Na^+ EFFLUX, AND Na^+/Ca^{2+} EXCHANGE

The existence of a ouabain-resistant, relatively temperature-insensitive component of Na^+ efflux in smooth muscle was detected by EPMA of smooth muscles rapidly frozen during Na^+ washout into cold Li^+ solutions.[9] These studies, originally designed to explore whether cellular Na^+ is increased as the result of hypertension, also revealed that cellular Na^+ did not change during vascular smooth muscle hypertrophy,[9] and that massive Na-loading of rabbit portal vein smooth muscles did not cause an increase in cellular or in mitochondrial Ca^{2+}.[9,10]

Na^+ efflux from Na^+-loaded smooth muscles into ouabain-containing, Li^+ solutions at 37°C was rapid (FIG. 4), with cell Na^+ decreasing below normal values, within ten minutes.[10] The results of EPMA are, therefore, consistent with isotope flux studies that have shown that (in *Taenia coli*) at least 50% of cell Na is exchanged within 5 min through a ouabain-resistant pathway.[53,54]

The ouabain-resistant Na^+ efflux in portal vein smooth muscles, as well as the accompanying Cl^- efflux and K^+ influx, are Ca^{2+} sensitive, and markedly reduced in Ca^{2+}-free solutions.[10] More recent $^{22}Na^+$-flux studies have further revealed that the effect of extracellularly added Ca^{2+} on Na^+ efflux is due to accompanying changes in cytoplasmic $[Ca^{2+}]$, revealing the presence of a cytoplasmic $[Ca^{2+}]$-dependent Na^+ channel or transport system in smooth muscle.[55] The existence of $[Ca^{2+}]_i$-sensitive Na^+, Cl^-, and K^+ channels and/or transport[10,55–57] raises the possibility that abnormalities of monovalent ion transport found in hypertensive cells[55,59] could be secondary to changes in $[Ca^{2+}]_i$.

The gain in cell Ca^{2+} [10] and the previously demonstrated[25] increase in cell Sr^{2+} in Na^+-free solutions could reflect increased Ca^{2+} influx due to competition between Ca^{2+} and Na^+ through the influx pathway and/or cellular Ca^{2+} uptake via the Na^+/Ca^{2+} transport.[3,55,59] Studies on several smooth muscles[9,10,53,54,60,61] suggest that Na/Ca^{2+} exchange plays a very small, if any, role in the normal physiological regulation of cytoplasmic $[Ca^{2+}]$, that is controlled primarily by the SR and the plasma membrane Ca^{2+} ATPase. The relatively high cytoplasmic Na^+ and low resting membrane potential of most smooth muscles,[4,62] as well as the low affinity of the Na/Ca^{2+} exchanger for Ca^{2+},[63] also argue against this pathway playing a major physiological role in Ca^{2+} efflux. Finally, the loss, into Na^+-free (Li^+) solutions, of the cellular and mitochondrial Ca^{2+} (FIG. 3) gained during massive Na efflux indicates that Ca^{2+} efflux through the Na^+/Ca^{2+} exchanger does not play an essential role in the regulation of cytoplasmic Ca^{2+} in smooth muscle.

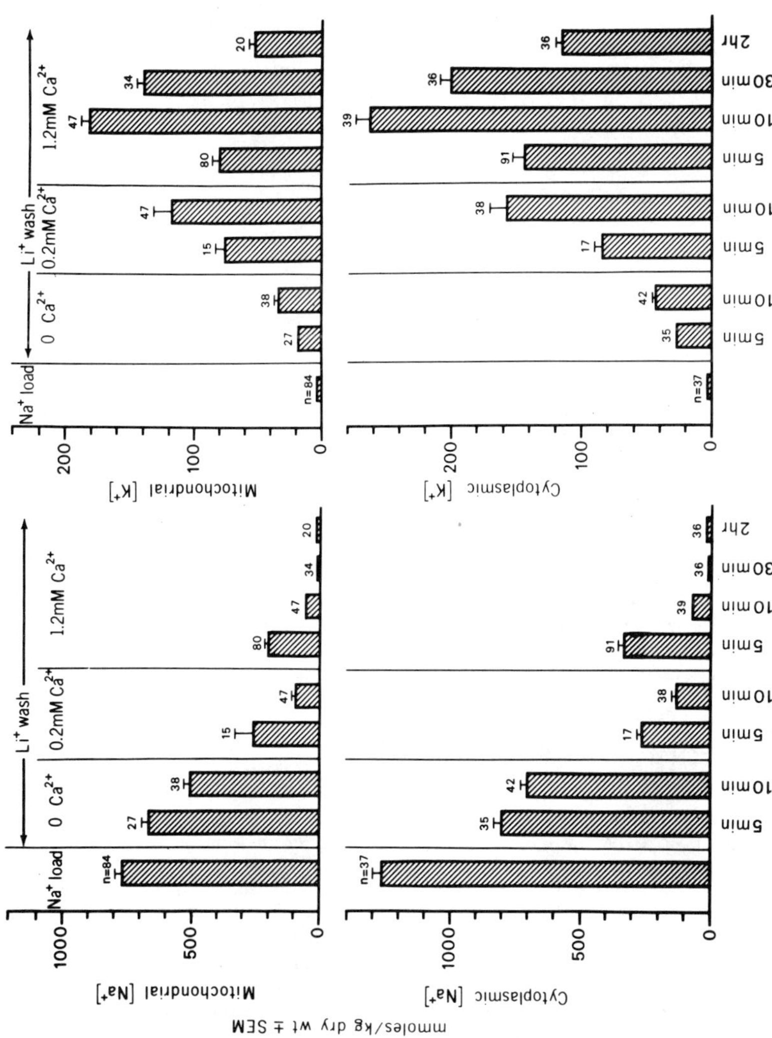

FIGURE 4. Mitochondrial (upper panels) and cytoplasmic (lower panels) [Na^+] and [K^+] in rabbit portal anterior mesenteric vein smooth muscle Na^+-loaded at 37°C, under various experimental conditions; Na^+-loaded, normal (1.2 mM Ca^+) Li^+-washed + ouabain, O Ca^+ + 2 mM EGTA + 1mM ouabain Li^+-washed and 0.2 mM Ca^+ + 1mM ouabain Li^+-washed. Note the marked inhibition of Na^+ efflux and K^+ uptake in Ca^{2+}-free solutions. (From Wasserman *et al.* With permission from *Circulation Research.*)

REFERENCES

1. A. P. Somlyo & A. V. Somlyo. 1983. Calcium, magnesium, and vascular smooth muscle function. *In* International Textbook on Hypertension. J. Genest *et al.*, Eds.: 441–457 McGraw-Hill, Inc. New York.
2. Jones, A. W. 1982. Ionic dysfunction and hypertension. Microcirc. **11:** 134–159.
3. Blaustein, M. P. 1977. Sodium ions, calcium ions, blood pressure regulation and hypertension: A reassessment and a hypothesis. Am. J. Physiol. **232:** C165–C173.
4. Jones, A. W. 1980. Content and fluxes of electrolytes. *In* Handbook of Physiology: The Cardiovascular System. **II:** 253–299. American Physiological Society. Bethesda, MD.
5. Somlyo, A. P., A. V. Somlyo & H. Shuman. 1979. Electron probe analysis of vascular smooth muscle: Composition of mitochondria, nuclei and cytoplasm. J. Cell Biol. **81:** 316–335.
6. Somlyo, A. P., A. V. Somlyo, H. Shuman & M. Endo. 1982. Calcium and monovalent ions in smooth muscle. Fed. Proc. **41:** 2883–2890.
7. Bond, M., T. Kitazawa, A. P. Somlyo & A. V. Somlyo. 1984. Release and recycling of calcium by the sarcoplasmic reticulum in guinea pig portal vein smooth muscle. J. Physiol. **355:** 677–695.
8. Kowarski, D., H. Shuman, A. P. Somlyo & A. V. Somlyo. 1985. Calcium release by norepinephrine from central sarcoplasmic reticulum in rabbit main pulmonary artery smooth muscle. J. Physiol. (London) **366:** 175–184.
9. Junker, J. L., A. J. Wasserman, P. F. Berner & A. P. Somlyo. 1984. Electron probe analysis of sodium and other elements in hypertrophied and Na-loaded smooth muscle. Circ. Res. **54:** 254–266.
10. Wasserman, A. J., G. McClellan & A. P. Somlyo. 1986. Cellular and subcellular transport of sodium, potassium, magnesium and calcium in sodium loaded vascular smooth muscle. Circ. Res. **58:** 790–802.
11. Somlyo, A. V., H. Gonzalez-Serratos, H. Shuman, G. McClellan & A. P. Somlyo. 1981. Calcium release and ionic changes in the sarcoplasmic reticulum of tetanized muscle: An electron probe study. J. Cell Biol. **90:** 577–594.
12. Somlyo, A. V., G. McClellan, H. Gonzalez-Serratos & A. P. Somlyo. 1985. Electron probe X-ray microanalysis of post tetanic Ca and Mg movements across the sarcoplasmic reticulum *in situ*. J. Biol. Chem. **260:** 6801–6807.
13. Kitazawa, T., A. P. Somlyo & A. V. Somlyo. 1984. The effects of valinomycin on ion movements across the sarcoplasmic reticulum in frog muscle. J. Physiol. (London) **350:** 253–268.
14. Yoshioka, T. & A. P. Somlyo. 1984. The calcium and magnesium contents and volume of the terminal cisternae in caffeine-treated skeletal muscle. J. Cell Biol. **99:** 558–568.
15. Chiesi, M., M. M. Ho, G. Inesi, A. V. Somlyo & A. P. Somlyo. 1981. Primary role of sarcoplasmic reticulum in phasic contractile activation of cardiac myocytes with shunted myolemma. J. Cell Biol. **91:** 728–742.
16. Somlyo, A. P. & B. Walz. 1985. Elemental distribution in *Rana pipiens* retinal rods: Quantitative electron probe analysis. J. Physiol. (London) **358:** 183–195.
17. Somlyo, A. P., M. Bond & A. V. Somlyo. 1985. The calcium content of mitochondria and endoplasmic reticulum in liver rapidly frozen *in situ*. Nature **314:** 622–625.
18. Somlyo, A. P., R. Urbanics, G. Vadasz, A. G. B. Kovach & A. V. Somlyo. 1985. Mitochondrial calcium and cellular electrolytes in brain cortex frozen *in situ:* Electron probe analysis. Biochem. Biophys. Res. Commun. **132:** 1071–1078.
19. Shuman, H., A. V. Somlyo & A. P. Somlyo. 1976. Quantitative electron probe microanalysis of biological thin sections: Methods and validity. Ultramicroscopy **1:** 317–339.
20. Kitazawa, T., H. Shuman & A. P. Somlyo. 1983. Quantitative electron probe analysis: Problems and solutions. Ultramicroscopy **11:** 251–262.
21. Karp, R. D., J. C. Silcox & A. V. Somlyo. 1982. Cryoultramicrotomy: Evidence against melting and the use of a low temperature cement for specimen orientation. J. Microsc. (Oxford) **125:** 157–165.
22. Somlyo, A. V. & J. Silcox. 1979. Cryoultramicrotomy for electron probe analysis. *In*

Microbeam Analysis in Biology. C. Lechene & R. Warner, Eds.: 535–555. Academic Press, Inc. New York.

23. SOMLYO, A. P. & H. SHUMAN. 1982. Electron probe and electron energy loss analysis in biology. Ultramicroscopy **8:** 219–234.
24. SOMLYO, A. P. 1985. Calcium measurement with electron probe and electron energy loss analysis. Cell Calcium **6:** 197–212.
25. SOMLYO, A. V. & A. P. SOMLYO. 1971. Strontium accumulation by sarcoplasmic reticulum and mitochondria in vascular smooth muscle. Science **174:** 955–958.
26. WEBER, A. 1971. Regulatory mechanisms of the calcium transport system of fragmented rabbit sarcoplasmic reticulum. I. The effect of accumulated calcium on transport and adenosine triphosphate hydrolysis. J. Gen. Physiol. **57:** 64–70.
27. SOMLYO, A. V. & C. FRANZINI-ARMSTRONG. 1985. New views of smooth muscle structure using freezing, deep-etching and rotary shadowing. Experientia **41:** 841–856.
28. DETH, R. & C. VAN BREEMEN. 1977. Agonist induced release of intracellular Ca^{2+} in the rabbit aorta. J. Membr. Biol. **30:** 363–380.
29. DETH, R. & R. CASTEELS. 1977. A study of releasable Ca fractions in smooth muscle cells of the rabbit aorta. J. Gen. Physiol. **69:** 401–416.
30. SOMLYO, A. V. & A. P. SOMLYO. 1968. Electromechanical and pharmacomechanical coupling in vascular smooth muscle. J. Pharmacol. Exp. Ther. **159:** 129–145.
31. SOMLYO, A. P. 1985. Excitation-contraction coupling and the ultrastructure of smooth muscle. Circ. Res. **57:** 497–507.
32. SOMLYO, A. P., A. V. SOMLYO, C. E. DEVINE & S. R. NORTH. 1971. Sarcoplasmic reticulum and the temperature-dependent contraction of smooth muscle in calcium-free solutions. J. Cell Biol. **51:** 722–741.
33. DEVINE, C. E., A. V. SOMLYO & A. P. SOMLYO. 1972. Sarcoplasmic reticulum and excitation-contraction coupling in mammalian smooth muscle. J. Cell Biol. **52:** 690–718.
34. SUEMATSU, E., M. HIRATA, T. HASHIMOTO & H. KURIYAMA. 1984. Inositol 1,4,5-trisphosphate releases Ca^{2+} from intracellular store sites in skinned single cells of porcine coronary artery. Biochem. Biophys. Res. Commun. **120:** 481–485.
35. BARON, C. B., M. CUNNINGHAM, J. F. STRAUSS & R. F. COBURN. 1984. Pharmacomechanical coupling in smooth muscle may involve phosphatidylinositol metabolism. Proc. Natl. Acad. Sci. USA **81:** 6899–6903.
36. SOMLYO, A. V., M. BOND, A. P. SOMLYO & A. SCARPA. 1985. Inositol trisphosphate-induced calcium release and contraction in vascular smooth muscle. Proc. Natl. Acad. Sci. USA **82:** 5231–5235.
37. STREB, H., R. F. IRVINE, M. J. BERRIDGE & I. SCHULZ. 1983. Release of Ca^{2+} from a nonmitochondrial intracellular store in pancreatic acinar cells by inositol 1,4,5-trisphosphate. Nature (London) **306:** 67–69.
38. GOLDMAN, Y., G. P. REID, A. P. SOMLYO, A. V. SOMLYO, D. R. TRENTHAM & J. W. WALKER. 1986. Activation of skinned vascular smooth muscle by photolysis of 'caged inositol trisphosphate' to inositol 1,4,5-trisphosphate ($InsP_3$) J. Physiol. (London) **376:** 109P (Abs).
39. HANSFORD, R. G. 1985. Relation between mitochondrial calcium transport and control of energy metabolism. Rev. Physiol. Biochem. Pharmacol. **102:** 2–72.
40. DENTON, R. M. & J. G. MCCORMACK. 1980. On the role of the calcium transport cycle in the heart and other mammalian mitochondria. FEBS Lett. **119:** 1–8.
41. SOMLYO, A. P. 1984. Cellular site of calcium regulation. Nature **308:** 516–517.
42. BOND, M., H. SHUMAN, A. P. SOMLYO & A. V. SOMLYO. 1984. Total cytoplasmic calcium in relaxed and maximally contracted rabbit portal vein smooth muscle. J. Physiol. (London) **357:** 185–201.
43. SCARPA, A. & P. GRAZIOTTI. 1973. Mechanisms for intracellular calcium regulation in the heart. J. Gen. Physiol. **62:** 756–772.
44. VALLIERES, J., A. SCARPA & A. P. SOMLYO. 1975. Subcellular fractions of smooth muscle: Isolation, substrate utilization and Ca^{++} transport by main pulmonary artery and mesenteric vein mitochondria. Arch. Biochem. Biophys. **170:** 659–669.
45. CARAFOLI, E. & M. CROMPTON. 1978. The regulation of intracellular calcium by mitochondria. Ann. N.Y. Acad. Sci. **307:** 269–283.

46. SOMLYO, A. V., J. SILCOX & A. P. SOMLYO. 1975. Electron probe analysis and cryoultramicrotomy of cardiac muscle: Mitochondrial granules. *In* Proc. Thirty-Third Annual EMSA Meeting. G. W. Bailey, Ed.: 532. San Francisco Press, Inc. San Francisco, CA.
47. JAMES-KRACKE, M. R., B. F. SLOANE, H. SHUMAN, R. KARP & A. P. SOMLYO. 1980. Electron probe analysis of cultured vascular smooth muscle. J. Cell. Physiol. **103:** 313–322.
48. GUPTA, B. L. & T. A. HALL. 1978. Electron probe X-ray microanalysis of calcium. Ann. N.Y. Acad. Sci. **307:** 28–49.
49. BARNARD, T. 1982. Thin frozen-dried cryosections and biological X-ray microanalysis. J. Microsc. **126:** 317–332.
50. SOMLYO, A. P., A. V. SOMLYO, H. SHUMAN, B. SLOANE & A. SCARPA. 1978. Electron probe analysis of calcium compartments in cryosections of smooth and striated muscles. Ann. N.Y. Acad. Sci. **307:** 523–544.
51. SCARPA, A. 1979. Transport across mitochondrial membranes. *In* Membrane Transport in Biology. G. Giebisch *et al.*, Eds. **II:** 263–355. Springer. New York.
52. SOMLYO, A. P., H. SHUMAN & A. V. SOMLYO. 1978. Mitochondrial and sarcoplasmic reticulum contents *in situ:* Electron probe analysis. *In* Frontiers of Biological Energetics. A. Scarpa, P. L. Dutton & J. Leigh, Eds. **1:** 742–751. Academic Press, Inc. New York.
53. CASTEELS, R., G. DROOGMANS & H. HENDRICKX. 1973. Effect of sodium and sodium-substitutes on the active ion transport and on the membrane potential of smooth muscle cells. J. Physiol. (London) **228:** 733–748.
54. BRADING, A. F. 1975. Sodium/sodium exchange in the smooth muscle of the guinea-pig taenia coli. J. Physiol. (London) **251:** 79–105.
55. KAPLAN, J. H., B. G. KENNEDY & A. P. SOMLYO. Calcium-stimulated sodium efflux from rabbit vascular smooth muscle. J. Physiol. (London) (In press.)
56. HAEUSLER, G. 1983. Contraction, membrane potential and calcium fluxes in rabbit pulmonary arterial muscle. Fed. Proc. **42:** 263–268.
57. SMITH, J. M. & A. W. JONES. 1985. Calcium-dependent fluxes of potassium-42 and chloride-36 during norepinephrine activation of rat aorta. Circ. Res. **56:** 507–516.
58. CANESSA, M. 1984. The polymorphism of red cell Na and K transport in essential hypertension: Findings, controversies and perspectives. *In* Erythrocyte Membranes. 3. Recent Clinical and Experimental Advances. W. C. Kruckeberg *et al.*, Eds.: 293–315. Alan R. Liss, Inc. New York.
59. SITRIN, M. D. & D. F. BOHR. 1971. Ca and Na interaction in vascular smooth muscle contraction. Am. J. Physiol. **220:** 1124–1128.
60. AARONSON, P. & C. VAN BREEMEN. 1981. Effects of sodium gradient manipulation upon cellular calcium, ^{45}Ca fluxes and cellular sodium in the guinea-pig taenia coli. J. Physiol. **319:** 443–461.
61. MULVANY, M. J., C. AALKJAER & T. T. PETERSEN. 1984. Intracellular sodium, membrane potential and contractility of rat mesenteric small arteries. Circ. Res. **54:** 740–749.
62. JOHANSSON, B. & A. P. SOMLYO. 1980. Electrophysiology and excitation-contraction coupling. *In* The Handbook of Physiology: The Cardiovascular System: Vascular Smooth Muscle. D. F. Bohr, A. P. Somlyo & H. V. Sparks, Eds. **II:** 301–324. American Physiological Society. Bethesda, MD.
63. BAKER, P. F. 1978. The regulation of intracellular calcium in giant axons of *loligo* and *myxicola.* Ann. N.Y. Acad. Sci. **307:** 251–268.

DISCUSSION OF THE PAPER

R. GARAY (*Necker Hospital, Paris*): Li^+ is a Na^+ (or K^+) analogue for several ion transport system. In addition it can rapidly enter into the cell and modify phosphatidyl inositol metabolism. Why have you used Li^+ for washing the cells and not choline?

SOMLYO: We also used choline. We originally used Li^+ to verify the Li^+-washout method of cell Na^+ measurements, and found that cell Na^+ was often lost in the cold Li^+ wash. We continued, because of the possible interest of Na^+/Li^+ exchange transport in hypertension.

M. BLAUSTEIN (*University of Maryland, Baltimore, MD*): You have demonstrated that contraction and Ca loading are activated by Na loading and replacement of external Na^+ by Li^+. The vascular smooth muscle relaxes, and Ca is extruded over the next two hours. Have you done the appropriate control of adding back the external Na^+ and comparing the rates of relaxation and Ca extrusion to those in the presence of Li^+ (when only an ATP-driven Ca pump should be operating)? Assuming (as Bohr, 1959; Luciani during this conference; and we, this morning, have observed) that external Na^+ accelerates the relaxation and Ca^{2+} extrusion, how would you explain the result if not by Na^+/Ca^{2+} exchange?

SOMLYO: Ca^{2+} loading of mitochondria during massive Na^+ efflux is less if the Na^+ washout is performed in the absence of ouabain. This further supports the notion that the Na^+/Ca^+ pathway plays a subsidiary role. Reports of the well known effects of Na^+ on relaxation (for review of earlier experiments, see Somlyo and Somlyo, *Pharmacol. Rev.*, 1968) have not been accompanied by measurements of membrane potential or consideration of the effects of changes in intracellular Na^+. Finally, I believe that these effects are observed generally under extremely unphysiological Na^+ deprivation, and phasic smooth muscles relax in high K^+,Na^+-free solutions.

R.M. DENTON (*University of Bristol, Bristol, U.K.*): What is the evidence that the mitochondria are undamaged after loading with calcium to 300 mmol/kg protein?

SOMLYO: We do not find "light cells" (Junker *et al., Circ. Res.*, 1984; Wassermann *et al., Circ. Res.*, 1986) indicative of cell damage and maximal contractility was normal. However, we do not have more refined measures of cell functions.

E. CARAFOLI (*Swiss Federal Institute of Technology, Zurich*): It's interesting that everybody emphatically negates a role of mitochondria in the cellular homeostasis of Ca^{2+} based on their K_m (Ca^{2+}) (about 10 μM), yet no problem seems to exist (and this goes back to the paper by Dr. Blaustein) in accepting a role for the plasma membrane Na^+/Ca^{2+} exchanger, which has about the same K_m (Ca^{2+}). It would be interesting to hear Dr. Somlyo's (or Dr. Blaustein's) comments on this.

SOMLYO: I agree that the low affinity of the Na^+/Ca^{2+} exchanger for Ca^{2+} is an additional argument against a major role of this system in regulating cytoplasmic $[Ca^{2+}]$.

Ca^{2+} Signalling in Exocrine Glands in Comparison to That in Vascular Smooth Muscle Cells

I. SCHULZ, S. SCHNEFEL, H. BANFÍC, AND L. ECKHARDT

*Max-Planck-Institut für Biophysik
6000 Frankfurt (Main 70), Federal Republic of Germany*

INTRODUCTION

Intracellular mechanisms involved in the regulation of cytosolic free Ca^{2+} concentration are very similar in different cell types including exocrine glandular and smooth muscle cells. Since altered calcium metabolism in vascular smooth muscle cells is one important factor involved in the pathogenesis of hypertension, understanding of regulation of cytosolic free Ca^{2+} concentration is necessary for effective treatment of this disease. In the following, main intracellular pathways in receptor-mediated stimulation of intracellular Ca^{2+} release, of Ca^{2+} influx into cells, and of Ca^{2+}-dependent processes in stimulus-response coupling will be discussed. Furthermore, mechanisms involved in the regulation of stimulatory events that lead back to the resting state of the cell will be described. The cell types to be discussed are the acinar cell of the exocrine pancreas and the vascular smooth muscle cell. Although there are basic differences between both types of cells in that smooth muscle cells are excitable, whereas exocrine glandular cells are not, there are also striking similarities in receptor-mediated intracellular events. These pathways, termed "stimulus-secretion coupling" in exocrine glands and "pharmacomechanical coupling" in smooth muscle cells, will be compared in the following.

MAIN PATHWAYS IN STIMULUS-RESPONSE COUPLING

In the transmembranous signal-transduction process hormones and neurotransmitters can activate two important enzymes, located in the plasma membrane, adenylate cyclase, and phospholipase C. Hormone receptors are coupled to adenylate cyclase by GTP-binding proteins (N or G proteins), which stimulate adenylate cyclase when coupled to stimulatory proteins (N_s) and inhibit the enzyme when coupled to inhibitory proteins (N_i).[1] Stimulation of adenylate cyclase and consequent rise in the cellular cAMP level leads to activation of protein kinase A and phosphorylation of a set of proteins involved in stimulus-response coupling. Evidence suggests that in the process of hormone or neurotransmitter-induced activation of phospholipase C, coupling of receptors to this enzyme is also mediated by N proteins.[2,3] This results in consequent hydrolysis of membrane-located phosphatidyl 4,5-bisphosphate (PIP_2) and in a rise in both inositol 1,4,5-trisphosphate (IP_3) and diacylglycerol (DG).[4] Further steps in stimulus-response coupling implicate IP_3-induced Ca^{2+} release from endoplasmic reticulum, hormone-induced Ca^{2+} influx into the cell, and consequent rise in cytosolic free Ca^{2+} concentration. This rise activates protein kinases, which depend on Ca^{2+}-

binding proteins such as calmodulin. Increased production of diacylglycerol in the presence of increased cytosolic free Ca^{2+} concentration will lead to activation of protein kinase C.[5]

Pancreatic Acinar Cells

In acinar cells of the exocrine pancreas, stimulatory peptide hormones such as secretin or vasoactive intestinal polypeptide stimulate adenylate cyclase, which leads to an increase in cellular cyclic adenosine monophosphate (cAMP) level.[6] Secretagogues such as acetylcholine (Ach) or cholecystokinin-pancreozymin (CCK-Pz) activate phospholipase C and increase both IP_3 and DG production.[7] Increase in cytosolic free Ca^{2+} concentration due to IP_3-induced Ca^{2+} release[8] and hormone-stimulated Ca^{2+} influx into the cell[9] do not appear to be sufficient in stimulating enzyme secretion.[10] Diacylglycerol (DG), the other hydrolysis product of PIP_2 breakdown that activates Ca^{2+}-dependent protein kinase C, seems to be more important.[10,11] However, a regulatory role of protein kinase C in stimulation of secretion so far has been implicated only indirectly by use of the phorbolester 12-O-tetradecanoylphorbol 13-acetate (TPA) that mimicks diacylglycerol in its ability to activate protein kinase C.[12] When added together with the Ca^{2+} ionophore A23187 to isolated acini, it stimulates enzyme secretion.[11] In contrast the protein kinase C inhibitor H 7 has been shown to augment CCK-OP, carbachol, or TPA-stimulated amylase secretion from pancreatic acini.[13] The authors concluded from their studies that protein kinase C does not have a stimulatory role in pancreatic stimulus-secretion coupling but may have an inhibitory one. Distal steps in stimulus-secretion coupling are assumed to implicate activation of cAMP-, Ca^{2+}-calmodulin-, and Ca^{2+}-diacylglycerol-dependent protein kinases and phosphorylation of proteins critically important for the regulation of exocytosis.[14–16] However, direct involvement of these proteins in the cascade of events that leads to exocytosis has not yet been established.

Vascular Smooth Muscle Cells

In vascular smooth muscle cells both adenylate cyclase and phospholipase C–activated pathways are also present. Similar to exocrine glands, receptor-mediated stimulation of phospholipase C and production of both inositol 1,4,5-trisphosphate and diacylglycerol is present and could be one way that leads to contraction. Ca^{2+} and calmodulin-dependent protein kinase is the myosin light chain kinase and the substrate is the 20 kD protein myosin light chain that can interact with actin, when it is phosphorylated so that tension can be developed.[17] There is evidence that myosin light chain is also phosphorylated by protein kinase C but at a different site (threonine) than that one which is phosphorylated (serine 19) by myosin light chain kinase.[18] The threonine site is also phosphorylated by myosine light chain kinase but more slowly than the serine 19 site and requires relatively high concentrations of this enzyme. Phosphorylation of the threonine site is accompanied by marked increase in the actin-activated MgATPase activity of myosin.[19] Although stimulation of a smooth muscle contraction with the phorbol ester TPA is suggestive for involvement of protein kinase C,[21] phosphorylating effects of protein kinase C so far have been demonstrated only *in vitro*. Reservation in the interpretation of its role in stimulus-contraction coupling therefore is appropriate as long as more direct evidence for the action of protein kinase C is not available.

BIPHASIC RESPONSE PATTERN IN STIMULATION OF BOTH SECRETION AND CONTRACTION

When vascular smooth muscle cell contracts in response to stimulants, such as angiotensin II or norepinephrine, the cell displays a biphasic pattern of response: an initial rapid phase followed by a slow tonic phase.[20] According to a concept put forward by Rasmussen and Barrett[21] two branches of the Ca^{2+} messenger system have distinct roles in the temporal integration of cellular response (FIG. 1). The Ca^{2+}-calmodulin branch is thought to be largely responsible for the initial cellular response. This pathway is stimulated by rise in cytosolic free Ca^{2+} concentration following hormonal stimulation of phospholipase C, IP_3-induced Ca^{2+} release from endoplasmic reticulum, and concomitant or subsequent increased Ca^{2+} influx into the cell. Activation of response elements in the calmodulin branch of the system includes Ca^{2+}-calmodulin–dependent protein kinases leading to phosphorylation of a set of cellular proteins involved in stimulation of the initial cellular response. However, because Ca^{2+} pumps in plasma membrane and endoplasmic reticulum are activated and much of released Ca^{2+} is taken up by mitochondria, cytosolic free Ca^{2+} concentration falls after a few minutes to low steady-state levels $<5 \times 10^{-7}$ mol/l, so flow of information through the Ca^{2+}-calmodulin branch is incapable of sustaining cellular response. Sustained cellular response is explained by rise in diacylglycerol content of plasma membrane due to activation of phospholipase C. This event along with initial rise in cytosolic free Ca^{2+} concentration brings about activation of protein kinase C branch of the system. Protein kinase C catalyzes phosphorylation of a different set of cellular proteins responsible for sustained phase of cellular response.

This biphasic pattern can be resolved by stimulation with either Ca^{2+} or TPA. In the presence of the Ca^{2+} ionophore A23187 brief contractile responses to each Ca^{2+} addition of 1.5 mM once every 10 min, which did not vary with time, could be observed.[22] However, if a similar vessel was exposed to 100 μM TPA in the presence of

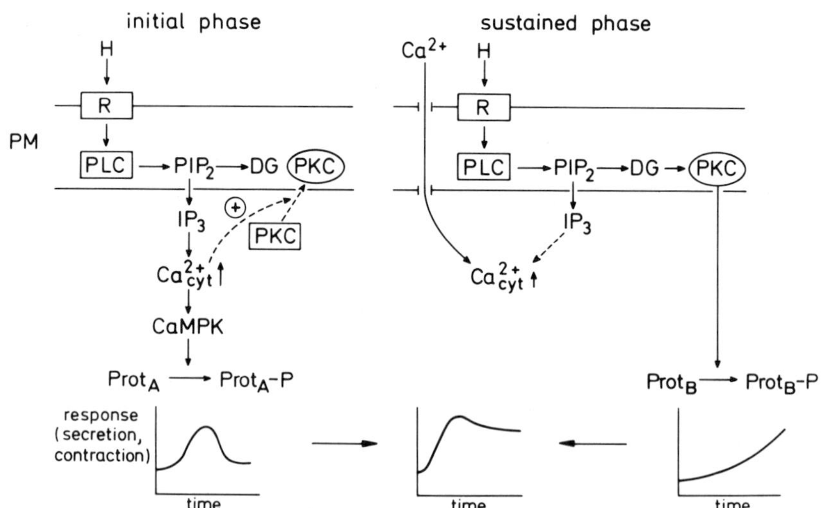

FIGURE 1. Schematic representation of intracellular pathways in stimulus-response coupling. For explanation see text.

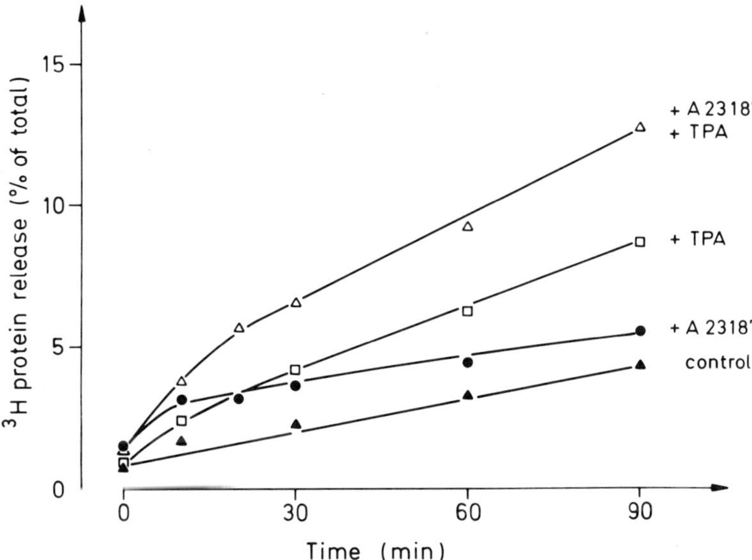

FIGURE 2. Effect of Ca^{2+} ionophore A23187 (10^{-5} mol/l), TPA (10^{-7} mol/l), and A23187 + TPA on enzyme secretion (^{3}H-labeled protein release) in isolated pancreatic acini. Rat pancreatic tissue was prelabeled with [^{3}H] leucine for 30 min and acini were isolated as described previously[10] by collagenase treatment and preincubated in the absence of added Ca^{2+} for 10 min with or without A23187 in a buffer containing in mmol/l: NaCl 145, KCl 4.7, MgCl 1.2, KH_2PO_4 1.2, HEPES 10 adjusted with NaOH to pH 7.4 at 37°C gassed with 100% O_2 in a shaking water bath. At 0 time Ca^{2+} (1 mmol/l) was added to the incubation media and TPA (10^{-7} mol/l), where indicated. Samples were taken immediately after these additions (time 0) and after 10, 30, 60 and 90 min of incubation. Radioactivity was counted and ^{3}H-labeled protein released by acini is expressed as percent of the total trichloroacetic acid-precipitable ^{3}H-labeled protein content present in the cells as described.[10]

A23187 and treated with the same brief pulses of Ca^{2+}, there was a progressive increase in the peak and duration of the contractile response to successive calcium pulses. Similarly in isolated acini from exocrine pancreas, addition of Ca^{2+} in the presence of A23187 caused a short increase in the rate of enzyme secretion that ceased within 10 min. Addition of TPA slowly increased protein release during 90 min of observation, whereas addition of both A23187 and TPA caused rapid and sustained progressive increase in enzyme secretion similar to that observed with secretagogues (FIG. 2).

DIFFERENCES IN cAMP- AND cGMP-MEDIATED CELL RESPONSE IN EXOCRINE AND SMOOTH MUSCLE CELLS

In contrast to exocrine glands, stimulation of the adenylate cyclase–cAMP system leads to inhibition of the smooth muscle cell response. In precontracted smooth muscle cells cAMP-protein kinase A–dependent phosphorylation of myosin light chain kinase inactivates this enzyme and the level of phosphorylated myosin light chain decreases.

Consequently cell tension decreases and the muscle relaxes.[20] Relaxation due to cAMP protein kinase A activation is further promoted by phosphorylation and subsequent stimulation of Ca^{2+} ATPases located in both the sarcoplasmic reticulum and the sarcolemma. This leads to decrease in cytosolic free Ca^{2+} concentration.[20,23]

In exocrine glandular as well as in smooth muscle cells both plasma membrane bound and soluble guanylate cyclases are present. In smooth muscle, membrane-bound guanylate cyclase is coupled to a receptor specific for the atrionatriuretic factor (ANF),[24] whereas for the exocrine pancreas cells similar data are not available. The soluble guanylate cyclase is probably stimulated by degradation products of diacylglycerol such as arachidonic acid, as a consequence of phospholipase C activation and by other fatty acids due to activation of phospholipase A_2 following increase in cytosolic free Ca^{2+} concentration at hormonal stimulation.[25–28] In exocrine glands, secretagogues, which act by stimulating phosphatidylinositol 4,5-bisphosphate breakdown, also give rise to cellular cGMP levels. However, the function of this increase is not yet clear. In smooth muscle cells increase in cellular cGMP concentration leads to relaxation of the cell. Similar to the adenylate cyclase cAMP system, activation of guanylate cyclase and rise of cGMP also activates Ca^{2+} pumps in sarcolemma and leads to decrease in the cytosolic free Ca^{2+} concentration.[29] Such feedback regulation has not yet been established in exocrine glands.

ROLE OF N PROTEINS IN SIGNAL TRANSDUCTION

In the transmembranous signal transduction process several guanine-nucleotide binding proteins (N- or G-proteins) have been identified as membranous signal-transducing components in the adenylate cyclase system (stimulatory, N_s, and inhibitory, N_i, proteins[1]) as well as in retinal cGMP phosphodiesterase activation (transducin). Observations in many cell systems suggest that N-proteins may be also involved in hormonal phospholipase C activation, catalyzing phosphatidylinositol 4,5-bisphosphate hydrolysis.[2,3] For smooth muscle cell, however, such data are not yet available. In mesangial cells, however, which have contractile elements and resemble smooth muscle cells in many respects, angiotensin II stimulated PIP_2 hydrolysis, and production of prostaglandin E_2 was inhibited by pretreatment of cells with pertussis toxin, indicating that an N protein is involved in this effect.[32] Evidence for the role of N-proteins in hormonal activation of phospholipase C in the exocrine pancreas will be given in the following sections.

Effect of GTP and GTP Analogues on IP_3 Production and Ca^{2+} Release

When isolated pancreatic acinar cells are permeabilized by washing in nominally Ca^{2+}-free media[33] and incubated in a buffer with concentrations similar to that of the cytosol (high K^+ HEPES buffer, free $[Ca^{2+}] \sim 2 \times 10^{-6}$ mol/l) in the presence of MgATP (5 mmol/l), cells take up Ca^{2+} until a steady state of about 0.2 μmol/l is reached.[33] Subsequent addition of secretagogues of enzyme secretion, such as cholecystokinin-pancreozymin (CCK-Pz) or its octapeptide (CCK-OP) (FIG. 3A); or of the acetylcholine analogue carbachol (Cch); or of inositol 1,4,5-tris-phosphate (IP_3) causes Ca^{2+} release followed by Ca^{2+} reuptake until the previous steady state is reached.[8]

In order to investigate involvement of N-proteins in hormone-induced Ca^{2+} release, GTP or its weakly hydrolyzable analogues GTPγS and GppNHp were tested. When GTPγS was added together with CCK-OP, increase in Ca^{2+} release by 40% as

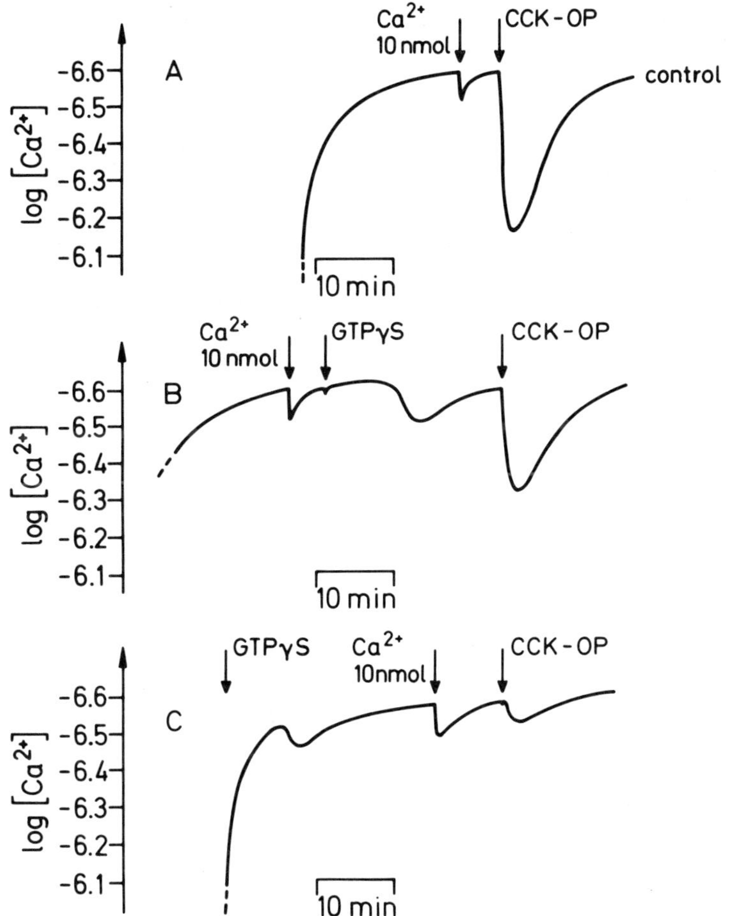

FIGURE 3. Effect of GTPγS and CCK-OP on Ca^{2+} release in isolated permeabilized rat pancreatic acinar cells. Cells were permeabilized by washing them in a buffer without added Ca^{2+} (free Ca^{2+} concentration ~ 2×10^{-6} mol/l), prelabeled with [^{3}H]myoinositol and incubated in a buffer containing in mmol/l: KCl 110, $MgCl_2$ 7, K_2ATP 5, creatine phosphate 10, K^+-succinate 5, K-pyruvate 5, creatine kinase 8 units/ml, HEPES 25, pH 7.4 adjusted with KOH, at 25°C. The medium was stirred and gassed continuously with 100% oxygen. Medium free-Ca^{2+} concentration (ordinate) was measured with a Ca^{2+}-specific macroelectrode as described previously.[30] Where indicated $CaCl_2$ calibration pulse (10 nmol) GTPγS (10^{-5} mol/l) was added at steady state (B) or during initial Ca^{2+} uptake phase (C). CCK-OP (3×10^{-7} mol/l) was added in the absence of GTPγS (A) or subsequently to GTPγS (B and C).

compared to addition of CCK alone was observed (not shown). When GTPγS was added at steady state, Ca^{2+} release occurred with some delay and subsequent CCK-OP effect was diminished (FIG. 3B). This inhibition appeared to be dependent on the period that cells had been preincubated with GTPγS, since even greater inhibition was observed, when GTPγS was added during the Ca^{2+}-uptake phase (FIG. 3C), GTPγS did not appear to alter the time course of Ca^{2+} uptake. However, when steady state was reached, Ca^{2+} release in the presence of GTPγS occurred and Ca^{2+} release to subsequent addition of CCK-OP was nearly abolished (FIG. 3C). Also IP_3-induced Ca^{2+} release was diminished in the presence of GTPγS (FIG. 4) indicating that GTP is involved in the Ca^{2+} release mechanism as had been suggested by Dawson[40] and Ueda *et al.*[41] GTPγS-induced Ca^{2+} release was reflected in increased IP_3 production indicating activation of phospholipase C. Inhibition of CCK-OP effect on both Ca^{2+} release and IP_3 production (FIG. 4) can be explained by decreased affinity of the receptor to its hormone ligand in the presence of GTP.[36,43] Similar results as with GTPγS were also obtained with GTP and GppNHp (data not shown).

Striking similar effects on Ca^{2+} release and IP_3 production were also seen with fluoride. When 10 mmol/l KF was added to the incubation medium, a small but consistent release of Ca^{2+} was seen at steady state (not shown) followed by decreased responsiveness to CCK-OP (TABLE 1). Potassium fluoride also increased IP_3 production, while subsequent CCK-OP effect on IP_3 production was diminished (TABLE 1).

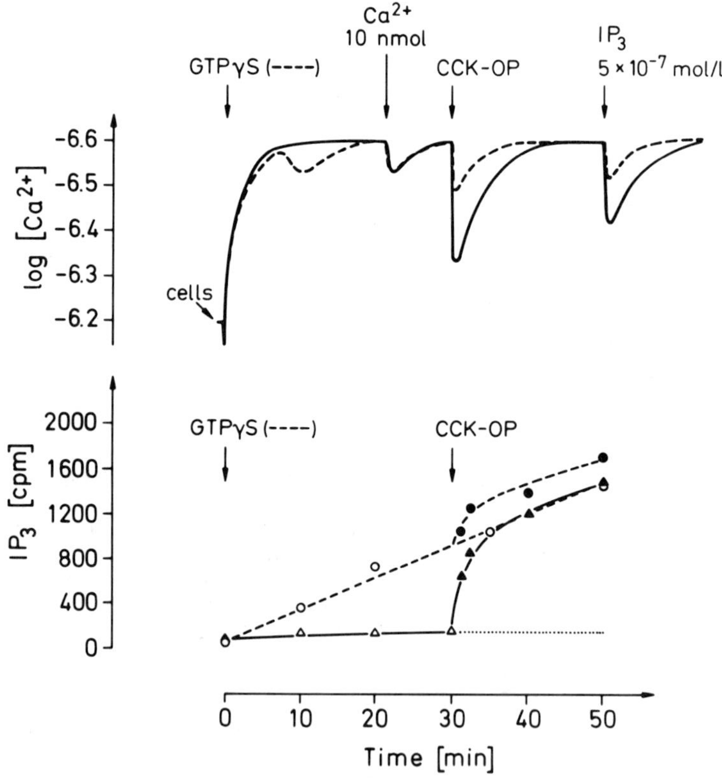

FIGURE 4. Effect of GTPγS (10^{-5} mol/l), CCK-OP (3×10^{-7} mol/l) and IP_3 (5×10^{-7} mol/l) on Ca^{2+} release (upper curve) and of GTPγS and CCK-OP on IP_3 production (lower curve) in isolated permeabilized rat pancreas cells. Incubation condition as described in FIGURE 3. Radioactivity recovered as IP_3 was measured following separation of inositides on Dowex anion exchange columns as described previously.[30]

TABLE 1.

Addition	A IP$_3$ Production after 30 min of Incubation %	B CCK-OP Induced IP$_3$ Production after Further 10 min %	C Maximal CCK-OP Induced Ca^{2+} Release %	*N*
No addition	100 ± 10.3	100 ± 24.6	100 ± 27.7	9
GTPγS 10^{-5} M	477.9 ± 22.8	25.6 ± 6.8	39.8 ± 10.3 (5)	3
Fluoride 10^{-2} M	397.7	64.9	50.5	2
Inactive CT	102.3	104.7	112.9	2
CT 10 μg/ml	127.2	90.0	119.6	1
CT 20 μg/ml	95.8	56.3	61.6	2
CT 40 μg/ml	102.2 ± 7.83	69.2 ± 12.9	56.2 ± 12.1	4
Forskolin 5×10^{-5} M	99.7 ± 27.0	74.0 ± 9.71	76.0 ± 15.8	4
cAMP 5×10^{-3} M	138.6	117.5	163.0	2
8-bromo cAMP 10^{-4} M	111.7	121.0	186.3	1
8-bromo cAMP 10^{-5} M	93.6	116.3	142.7	1

Effects of GTPγS, KF, cholera toxin (CT), heat inactivated CT, forskolin, cAMP, and 8-bromo cAMP on IP$_3$ production (A) on CCK-OP (3×10^{-7} mol/l) induced IP$_3$ production (B) and on CCK-OP induced Ca^{2+} release in permeabilized pancreatic acinar cells. Values are means ± SE and are expressed in percent of the control, which means in column A: IP$_3$ production without any addition (100% is 96.4 ± 9.9 cpm/mg protein). Column B: CCK-OP (3×10^{-7} mol/l) induced IP$_3$ production during 10 min (i.e. from 30 to 40 min of incubation, 100% is 232.7 ± 54.4 cpm/mg protein). Column C: maximal CCK-OP induced Ca^{2+} release, which was obtained 3 min after addition of the hormone (100% is 5.12 ± 1.42 nmol/mg protein).

Addition of IP$_3$, however, showed a higher effect on Ca^{2+} release in the presence of fluoride as compared to the control (data not shown). This effect might be due to the ability of F^- to inhibit IP$_3$ degradation.[34] The similar effect of both KF and GTPγS on IP$_3$ production and on Ca^{2+} release is explained with a model in which trace amounts of Al^{3+} present as contaminant in glassware and chemicals forms an active complex with F^-. The AlF_4^- complex binds close to GDP and mimicks the role of γ-phosphate of GTP.[35]

As should be expected GDPβS, the weakly hydrolyzable analogue of GDP, inhibited both hormone-induced Ca^{2+} release and IP$_3$ production (not shown).

These results on GTPγS and fluoride indicate that a guanine nucleotide binding protein (N_p) is involved in coupling of receptors to phospholipase C in exocrine pancreas cells.

Effect of Bacterial Toxins and Forskolin on Phospholipase C

Both cholera toxin and pertussis toxin have been reported to influence stimulatory (N_s) and inhibitory (N_i) proteins of the adenylate cyclase system, respectively.[1] In the presence of NAD^+ these toxins transfer ADP-ribose to α-subunits of these proteins. While ADP-ribosylation of α_s by cholera toxin results in persistent stimulation of the N_s protein,[36] ADP-ribosylation of N_i by pertussis toxin inhibits exchange of GDP for GTP at α_i and thereby abolishes the activation process of N_i.[36] In some[37,38] but not in all systems[39] pertussis toxin has been shown to suppress the response to agonists that stimulate PIP_2 breakdown. This indicates that pertussis toxin interacts with N_p. In order to obtain more information about the hypothetical N-protein (N_p) that couples receptors to phospholipase C, both pertussis and choleratoxin have been tested on secretagogue-induced IP_3 production and Ca^{2+} release in the exocrine pancreas. In intact or permeabilized cells that had been preincubated for 1 hr in the presence of pertussis toxin (20–2,000 μg/ml) the effect of CCK on both IP_3 production and Ca^{2+} release had not been changed. However, pretreatment of permeabilized cells with dithiothreitol-activated cholera toxin (40 μg/ml) in the presence of NAD^+ for 30 min inhibited subsequent CCK-OP–induced IP_3 production and Ca^{2+} release by about 40% (TABLE 1). A similar effect on IP_3 production and Ca^{2+} release to that with cholera toxin was also seen with forskolin (TABLE 1), which can stimulate adenylate cyclase by direct interaction with its catalytic subunit.[44,45] Evidence suggests that high affinity binding sites for forskolin are associated with the complex of N_s protein and the catalytic unit of the adenylate cyclase system.[44,46] Since both cholera toxin and forskolin elevate cAMP levels in many cell types, including pancreatic acinar cells,[36,42,47] and since inhibitory effects of cAMP on phospholipase C activity have been reported in other systems[5] it could be possible that effects of cholera toxin and forskolin on phospholipase C are mediated by cAMP. We have therefore tested cAMP and its poorly hydrolyzable analogue 8-bromo-cAMP on IP_3 production and Ca^{2+} release in the presence or absence of CCK-OP. However, neither cAMP nor its analogue 8-br-cAMP produced any effect on basal or CCK-OP–induced IP_3 production. In contrast, Ca^{2+} release tended to be elevated when cells were stimulated with CCK-OP (TABLE 1). The data on cholera toxin and forskolin are readily explained by interaction of these substances with N_s protein of the adenylate cyclase system. Dissociation of N_s by activation into its subunits α_s and $\beta\gamma$ increases available $\beta\gamma$ complex. The $\beta\gamma$ complex, similar in all N-proteins, could recombine with α_p of N_p, thereby shifting N_p to its inactive, undissociated form. A similar model had also been proposed to explain inactivation of the adenylate cyclase system by N_i.[31] However, direct inhibition of phospholipase C by α_s could be also possible.

PATHWAYS OF SUSTAINED Ca^{2+} INFLUX DURING STIMULATION

It is likely that N-proteins further mediate stimulatory events by coupling Ca^{2+} channels in the plasma membrane to receptors that mediate hormone-induced Ca^{2+} influx into the cell. There are various mechanisms regulating the influx of Ca^{2+} into cells. Excitable cells possess voltage-gated Ca^{2+} channels, which can be inhibited by channel blockers like verapamil.[48,49] In other cell types, for example exocrine pancreas cells,[50] resting and stimulated Ca^{2+} levels are unaffected by membrane depolarization and Ca^{2+} influx is not susceptible to inhibition by Ca^{2+} channel blockers.[51] There is increasing evidence for the existence of "receptor-operated" Ca^{2+} channels, which are coupled in some way directly to the receptor without intermediation of an intracellular messenger. Thus, in smooth muscle cells it had been shown that α_2-adrenergic stimulation causes Ca^{2+} influx into the cell, which is not accompanied by significant depolarization.[52] Since hormonally induced Ca^{2+} influx was blocked by pretreatment with pertussis toxin this observation indicates that coupling of the receptor to the Ca^{2+} channel is mediated by an N-protein. This could be N_i of the adenylate cyclase system or another pertussis toxin–sensitive N-protein whose nature has not yet been clarified. In exocrine glands hormone-induced Ca^{2+} influx into the cell has been observed by $^{45}Ca^{2+}$ flux measurement.[9] On the other hand, patch-clamp methods have failed to detect hormone-induced opening of Ca^{2+} channels in isolated pancreatic acinar cells. Since Ca^{2+}-dependent non-selective cation channels are present in plasma membranes of pancreatic acinar cells,[53] it is quite possible that Ca^{2+} entry also occurs through these types of channels. Charge transfer by Ca^{2+} ions might be too small, however, to be detected.

A recent patch-clamp study on pancreatic acinar cells[54] demonstrated an external Ca^{2+} requirement for sustained Ca^{2+}-activated opening of K^+ channel evoked by CCK-receptor agonist. The requirement of external Ca^{2+} could be localized at a site separated from the hormone-receptor interaction indicating messenger-mediated rather than receptor-mediated Ca^{2+} uptake. This messenger is not yet known. Because intracellular Ca^{2+} had been chelated by EGTA in these experiments, it is not likely that Ca^{2+}, which had been released by IP_3, is the messenger to open channels through which Ca^{2+} influx might occur. No evidence is yet available that IP_3 acts on the plasma membrane to induce Ca^{2+} influx into the cell. Another candidate for a messenger that induces Ca^{2+} influx into the cell would be diacylglycerol, which activates protein kinase C. Indirect evidence suggests that protein kinase C activation leads to Ca^{2+} influx into mesangial cells.[32,55] The question of whether this enzyme phosphorylates proteins that are involved in opening of Ca^{2+} channels remains to be answered by direct verification using patch-clamp methods.

CONCLUSION

This comparison between stimulus-secretion coupling in exocrine cells and pharmaco-mechanical coupling in smooth muscle cells emphasizes the importance of Ca^{2+} in both types of cells. There are many similarities in the mechanism of regulation of cytosolic free Ca^{2+} concentration, including hormonal activation of phospholipase C, that leads to production of both inositol 1,4,5-trisphosphate (IP_3) and diacylglycerol. In both cell types IP_3 releases Ca^{2+} from endo- and sarcoplasmic reticulum, respectively, which leads to rise in cytosolic free Ca^{2+} concentration. In smooth muscle cells other mechanisms, which are lacking in exocrine cells, such as voltage-dependent and Ca^{2+}-induced Ca^{2+} release are probably of greater importance than IP_3-induced Ca^{2+} release. In both systems the rise of cytosolic free Ca^{2+} concentration triggers further events such as activation of protein kinases, which leads to phosphorylation of proteins involved in secretion or contraction. Whether or not the diacylglycerol-protein kinase C system plays a direct role in both stimulus-secretion and stimulus-contraction coupling is not yet clear. However, with increasing knowledge of signal transduction mechanisms, including the importance of N-proteins in coupling receptors to regulatory enzymes or to ion channels, and of the role of protein kinases in protein phosphorylation-dephosphorylation probably more similarities between stimulus-secretion and stimulus-contraction coupling will emerge and better understanding of each step involved will be obtained.

REFERENCES

1. RODBELL, M. 1980. Nature **284:** 17–22.
2. COCKCROFT, S. & B. D. GOMPERTS. 1985. Nature **314:** 534–536.
3. GOMPERTS, B. D. 1983. Nature **306:** 64–66.
4. BERRIDGE, M. J. & R. F. IRVINE. 1984. Nature **312:** 315–321.
5. NISHIZUKA, Y. 1983. Trends Biochem. Sci. **8:** 13–16.
6. ROBBERECHT, P., T. P. CONLON & J. D. GARDNER. 1976. J. Biol. Chem. **251:** 4635–4639.
7. PUTNEY, J. W., G. M. BURGESS, S. P. HALENDA, J. S. MCKINNEY & R. P. RUBIN. 1983. Biochem. J. **212:** 655–659.
8. STREB, H., R. F. IRVINE, M. J. BERRIDGE & I. SCHULZ. 1983. Nature **306:** 67–69.
9. KONDO, S. & I. SCHULZ. 1976. Biochim. Biophys. Acta **419:** 76–92.
10. KIMURA, T., K. IMAMURA, L. ECKHARDT & I. SCHULZ. 1986. Am. Physiol. Soc. **250:** G. (In press.)

11. De Pont, J. J. H. H. M. & A. M. M. Fleuren-Jakobs. 1984. FEBS Lett. **170:** 64–68.
12. Castagna, M., Y. Takai, K. Kaibuchi, K. Sano, U. Kikkawa & Y. Nishizuka. 1982. J. Biol. Chem. **257:** 7847–7851.
13. Pandol, S. J. & M. S. Schoeffield. 1986. J. Biol. Chem. **261:** 4438–4444.
14. Burnham, D. B. & J. A. Williams. 1984. Am. J. Physiol. **246:** G500–G508.
15. Gorelick, R. S., J. A. Cohn, S. D. Freedman, N. G. Delahunt, J. M. Gershoni & J. D. Jamieson. 1983. J. Cell Biol. **97:** 1294–1298.
16. Burnham, D. B., P. Munowitz, S. R. Hootman & J. A. Williams. 1986. Biochem. J. **235:** 125–131.
17. Adelstein, R. S., M. A. Conti & M. D. Pato. 1980. Ann. N.Y. Acad. Sci. **356:** 142–150.
18. Nishikawa, M., J. R. Sellers, R. S. Adelstein & H. Hidaka. 1984. J. Biol. Chem. **259:** 8808–8814.
19. Ikebe, M. & D. J. Hartshorne. 1985. J. Biol. Chem. **260:** 10027–10031.
20. Bolton, T. B. 1979. Physiol. Rev. **59:** 606–718.
21. Rasmussen, H. & P. Q. Barrett. 1984. Physiol. Rev. **64:** 938–984.
22. Rasmussen, H., J. Forder, I. Kojima & A. Scriabine. 1984. Biochem. Biophys. Res. Commun. **122:** 776–784.
23. Rasmussen, H. 1981. Calcium and cAMP as Synarchic Messengers. Wiley. New York.
24. Waldman, S. A., R. M. Rapoport & F. Murad. 1984. J. Biol. Chem. **259:** 14332–14334.
25. Rubin, R. P. 1982. Calcium and Cellular Secretion. Plenum. New York.
26. Goldberg, N. D. & M. K. Haddox. 1977. Ann. Rev. Biochem. **46:** 823–896.
27. Peach, M. J. 1981. Biochem. Pharmacol. **30:** 2745–2751.
28. Gerzer, R., P. Hamet, A. H. Ross, J. A. Lawson & J. G. Hardman. 1983. J. Pharmcol. Exp. Ther. **226:** 180–186.
29. Suematsu, E., M. Hirata & H. Kuriyama. 1984. Biochim. Biophys. Acta **773:** 83–90.
30. Streb, H., H. Heslop, R. F. Irvine, I. Schulz & M. J. Berridge. 1985. J. Biol. Chem. **260:** 7309–7315.
31. Gillman, A. G. 1984. Cell **36:** 577–579.
32. Pfeilschifter, J. & C. Bauer. 1986. Biochem. J. **236:** 289–294.
33. Streb, H. & I. Schulz. 1983. Am. J. Physiol. **245:** G347–G357.
34. Storey, D. J., S. B. Shears, C. J. Kirk & R. Michel. 1984. Nature **312:** 374–376.
35. Bigay, J., P. Deterre, C. Pfister & M. Chabre. 1985. FEBS Lett. **191:** 181–185.
36. Levitzki, A. 1984. J. Recept. Res. **4:** 399–409.
37. Nakamura, T. & M. Ui. 1985. J. Biol. Chem. **260:** 3584–3593.
38. Bradford, P. G. & R. P. Rubin. 1985. FEBS Lett. **183:** 317–320.
39. Masters, S. B., M. W. Martin, T. K. Harden & J. H. Brown. 1985. Biochem. J. **227:** 933–935.
40. Dawson, A. P. 1985. FEBS Lett. **185:** 147–150.
41. Ueda, T., S.-H. Chueh, M. W. Noel & D. L. Gill. 1986. J. Biol. Chem. **261:** 3184–3192.
42. Svoboda, M., J. Furnelle, F. Eckstein & J. Christophe. 1980. FEBS Lett. **109:** 275–279.
43. Koo, C., R. J. Lefkowitz & R. Syndermann. 1983. J. Clin. Invest. **72:** 748–753.
44. Nelson, C. A. & K. B. Seamon. 1985. FEBS Lett. **183:** 349–352.
45. Metzger, H. & E. Lindner. 1981. IRCS Med. Sci. **9:** 99.
46. Yamashita, A., T. Kurokawa, K. Higashi, T. Dan'ura & S. Ishibashi. 1986. Biochem. Biophys. Res. Commun. **137:** 190–194.
47. Dehaye, J.-P., M. Gillard, P. Poloczek, M. Stievenard, J. Winand & J. Christophe. 1985. J. Cyclic Nucleotide Res. **10:** 269–280.
48. Taylor, W. M., E. van de Pol, D. F. van Helden, I. P. H. Reinhart & F. L. Bygrave. 1985. FEBS Lett. **183:** 70–74.
49. Oettgen, H. C., C. Terhorst, L. C. Cantley & P.M. Rosoff. 1985. Cell **40:** 583–590.
50. Petersen, O. H. 1981. *In* Physiology of the Gastrointestinal Tract. L. R. Johnson, Ed. **1:** 749–772. Raven Press. New York.
51. Doyle, V. M. & U. T. Ruegg. 1985. Biochem. Biophys. Res. Commun. **127:** 161–167.

52. HÄUSLER, G., J.-E. DE PEYER, M. YAJIMA & G. SCHULTZ. 1986. J. Cardiovasc. Pharm. (In press.)
53. MARUYAMA, Y. & O. H. PETERSEN. 1984. J. Membr. Biol. **81:** 83–87.
54. SUZUKI, K., C. C. H. PETERSEN & O. H. PETERSEN. 1985. FEBS Lett. **192:** 307–312.
55. PFEILSCHIFTER, J. A. KURTZ & C. BAUER. 1986. Biochem. J. **234:** 125–130.

DISCUSSION OF THE PAPER

T. POZZAN (*University of Padova, Padova*): Quin 2 acts as an intracellular Ca^{2+} buffer as well as an indicator. Thus, quin 2 alters the $[Ca^{2+}]_i$ transients. If you use lower quin 2, or fura 2, the $[Ca^{2+}]_i$ transients are larger and $[Ca^{2+}]_i$ returns faster to resting.

SCHULZ: I agree with you.

C. SILVIA: It is possible that pancreatic cells have two classes of CCK receptors, one coupled to adenylate cyclase and one to phospholipase C, and that the effects of cholera toxin are on Ns and not on a phospholipase C–coupled N protein?

SCHULZ: It is very likely that there are two CCK-receptors as you said. However, the effect we saw with cholera toxin (CT) cannot be explained by increase in cAMP due to CT-activated adenylate cyclase activity. We have tested the effects of cAMP and 8-bromo-cAMP at different concentrations which it did not decrease CCK-induced IP_3 production nor Ca^{2+} release. A direct inhibitory effect of α_s on phospholipase C is possible, however.

G. BIANCHI (*University of Milan, Milan*): A decrease of cytosolic Ca increases renin secretion, whereas the same change decreases the secretion of most enzymes or hormone. As renin and Ca are relevant to hypertension do you have any explanation of this rather selective effect of Ca on renin secretion?

SCHULZ: I cannot explain it, but I could speculate that in the mechanism of renin release Ca^{2+} inhibits secretion e.g., by Ca^{2+}-dependent phosphorylation or dephosphorylation of proteins involved in the mechanism of renin release and that dephosphorylation activates such hypothetical proteins.

A. SCARPA (*Case Western Reserve University, Cleveland, OH*): There is another system, parathyroid gland, where the intracellular free Ca^{2+} may not be changing during secretion.

Platelet Calcium-Linked Abnormalities in Essential Hypertension[a]

THÉRÈSE J. RESINK, DIMITAR DIMITROV, AINO ZSCHAUER, PAUL ERNE, VSEVOLOD A. TKACHUK, FRITZ R. BÜHLER[b]

Departments of Research and Medicine
University Hospital
Basel, Switzerland

CELLULAR Ca^{2+}: FLUXES AND COMPARTMENTALIZATION

Changes in cytoplasmic Ca^{2+} ($[Ca^{2+}]_i$) provide pivotal links in many fundamental and clinically important (patho-)physiological processes. A simple generalized representation of cellular Ca^{2+} compartmentalization and flux systems is given in FIGURE 1. Extracellular $[Ca^{2+}]$ is approximately 5,000–10,000 times greater than intracellular $[Ca^{2+}]$, but due to the low permeability of plasma membranes for Ca^{2+}, the entry of Ca^{2+} depends on calcium channel openings, voltage dependent and/or receptor operated. Cellular Ca^{2+} storage compartments comprise the plasma membrane, endoplasmic reticulum, mitochondria, and cytosolic Ca^{2+} binding proteins. Although Ca^{2+} influx is important in cell activation, the increase in cytoplasmic free Ca^{2+} is also dependent on release of Ca^{2+} from the endoplasmic reticulum and/or plasma membrane.

There are control systems that maintain Ca^{2+} homeostasis. The Ca^{2+} concentration gradient across the plasma membrane necessitates Ca^{2+} extrusion by energy-dependent processes: the action of a calmodulin-dependent Ca^{2+} ATPase and a Na^+/Ca^{2+} exchange system whereby influx of Na^+ down its concentration gradient drives Ca^{2+} efflux against its concentration gradient. The operation of the latter is dependent on the Na^+/K^+ ATPase pump, which functions to maintain the Na^+ concentration gradient. Ca^{2+} is also sequestered from the cytosol via ATP-dependent uptake into the endoplasmic reticulum and mitochondria, although mitochondria only become significant Ca^{2+} transporters when cytosolic Ca^{2+} rises above 2.0 μM. These systems work in parallel and with different relative affinities in different cells. Their roles are dictated by their specific kinetic properties. For example, the plasma membrane and endoplasmic reticulum are high affinity/low capacity systems, whereas mitochondrial Ca^{2+} uptake and Na^+/Ca^{2+} exchange are low affinity/high capacity systems.[1–6]

Ca^{2+} AS INTRACELLULAR MESSENGER: DUAL INFORMATION FLOW

There are primarily two Ca^{2+}-dependent transduction pathways: the calmodulin (CaM) branch[7] and the kinase C branch,[8] and depending on the nature of the stimulus

[a]Supported by the Swiss National Fund Nr. 3.924.083 and the Swiss Cardiology Foundation.
[b]Address correspondence to: F.R.B., Division of Cardiology, University Hospital, CH-4031 Basel, Switzerland.

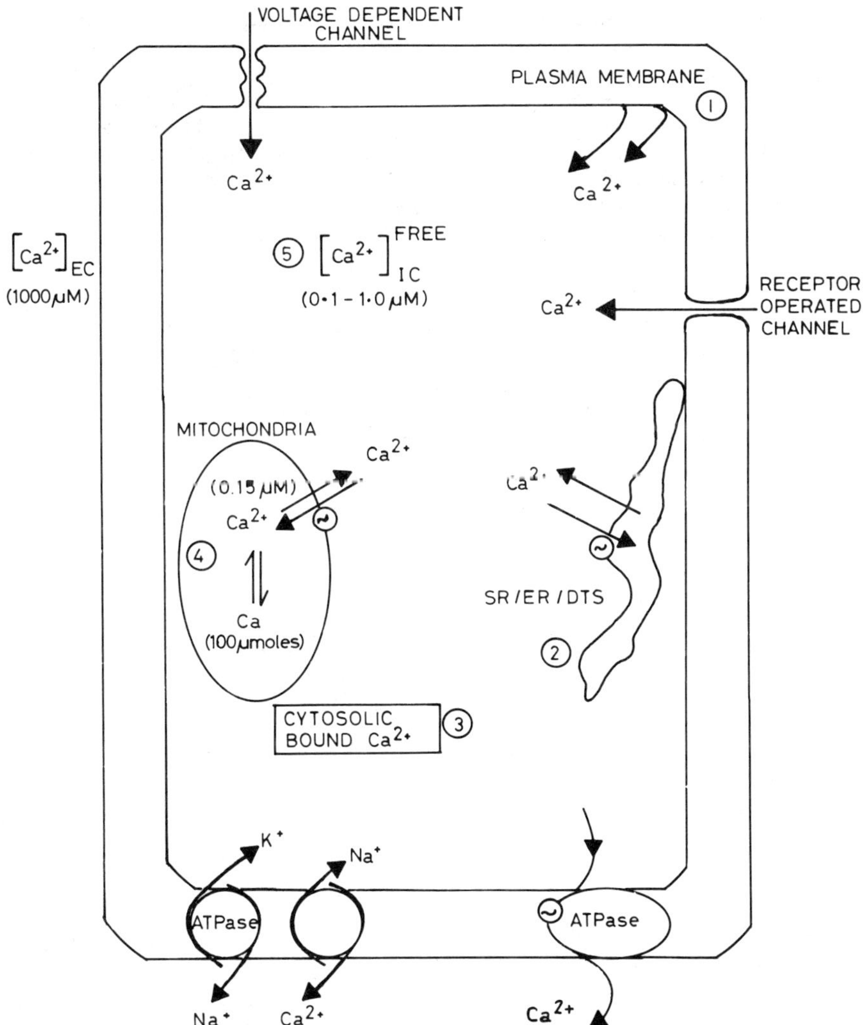

FIGURE 1. Cellular Ca^{2+} pools and autoregulation of cytosolic Ca^{2+} concentrations. $[Ca^{2+}]_{EC}$ and $[Ca^{2+}]_{IC}$ indicate extracellular and intracellular Ca^{2+} concentrations, respectively. Arrows indicate direction of Ca^{2+} flux (or Na^+ and K^+ as specified). ATP requiring Ca^{2+} transport systems are indicated by ~ and encircled numbers indicate intracellular Ca^{2+} stores.

these pathways operate independently or synergistically (FIG. 2). The calmodulin pathway includes events regulated by calmodulin, other Ca^{2+} receptor proteins (e.g. troponin C), and those in which Ca^{2+} directly activates an enzyme. The Ca^{2+} calmodulin complex modulates some enzymes by direct association and others by controlling the activity of an intermediary enzyme.[7,9] In the kinase C pathway regulatory events involve phosphorylation of a specific subset of proteins.[8,10] Although the metabolic consequences of the two pathways are different, events in both are required to obtain a maximum integrated response, each having a distinct temporal role in the cellular response cascade.[11]

Activation of the CaM pathway requires only an increase in cytosolic Ca^{2+}, while activation of the kinase C pathway has a lower Ca^{2+} requirement but is also absolutely dependent on diacylglycerol, a product of receptor-coupled phosphoinositide breakdown.[7-11] These different activation requirements have enabled dissociation experiments using controlled concentrations of Ca^{2+} ionophores and diacylglycerol mimetics. The calmodulin pathway is largely responsible for initiating, whereas the kinase C pathway is largely responsible for maintaining, the cellular response.[8-11]

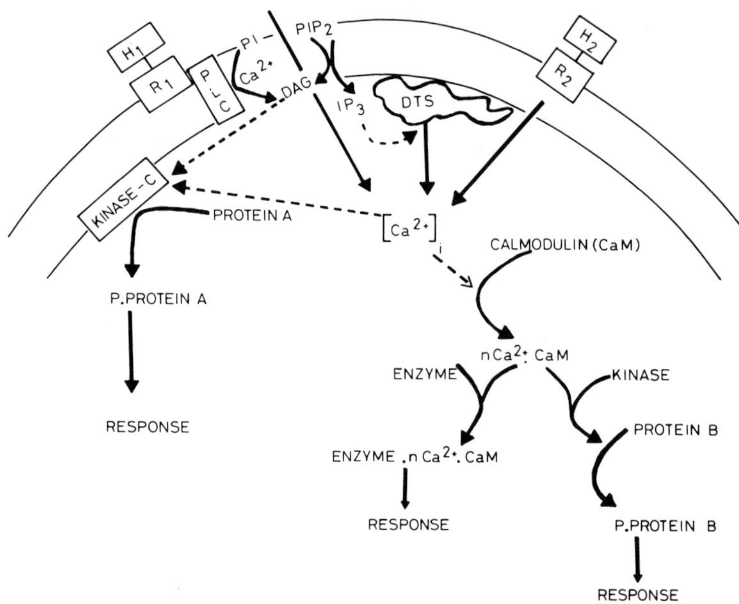

FIGURE 2. Two pathways of the Ca^{2+} messenger system: independence and synergism. H_1 represents hormones that bind to receptors (R_1) coupled to phosphoinositide turnover. Phospholipase C(s) (PLC) that act on monophosphatidylinositol (PI) and phosphatidylinositol 4,5-bisphosphate (PIP_2) produce diacylglycerol (DAG), and in the case of PIP_2, inositol 1,4,5-trisphosphate (IP_3). IP_3 induces Ca^{2+} release from the dense tubular system (DTS), which together with Ca^{2+} influx across the plasma membrane, increases cytosolic free Ca^{2+} concentrations ($[Ca^{2+}]_i$) (After Rasmussent *et al.*[11])

PLATELETS: Ca^{2+} AND THE ACTIVATION CASCADE

In platelets, as with other cells, receptor-agonist interactions modulate the levels of second messengers that govern integrated cellular responses (FIG. 3).[8] Stimulation of platelets is accompanied by a rapid increase in $[Ca^{2+}]_i$ via receptor-operated and/or voltage-sensitive Ca^{2+} channel influx as well as via mobilization of internal plasma membrane and dense tubular Ca^{2+} stores. The relative contributions from influx or release depend on the nature of the stimulus and whether the receptor is coupled to phosphoinositide turnover. In the case of thrombin,[13] vasopressin,[14] and platelet

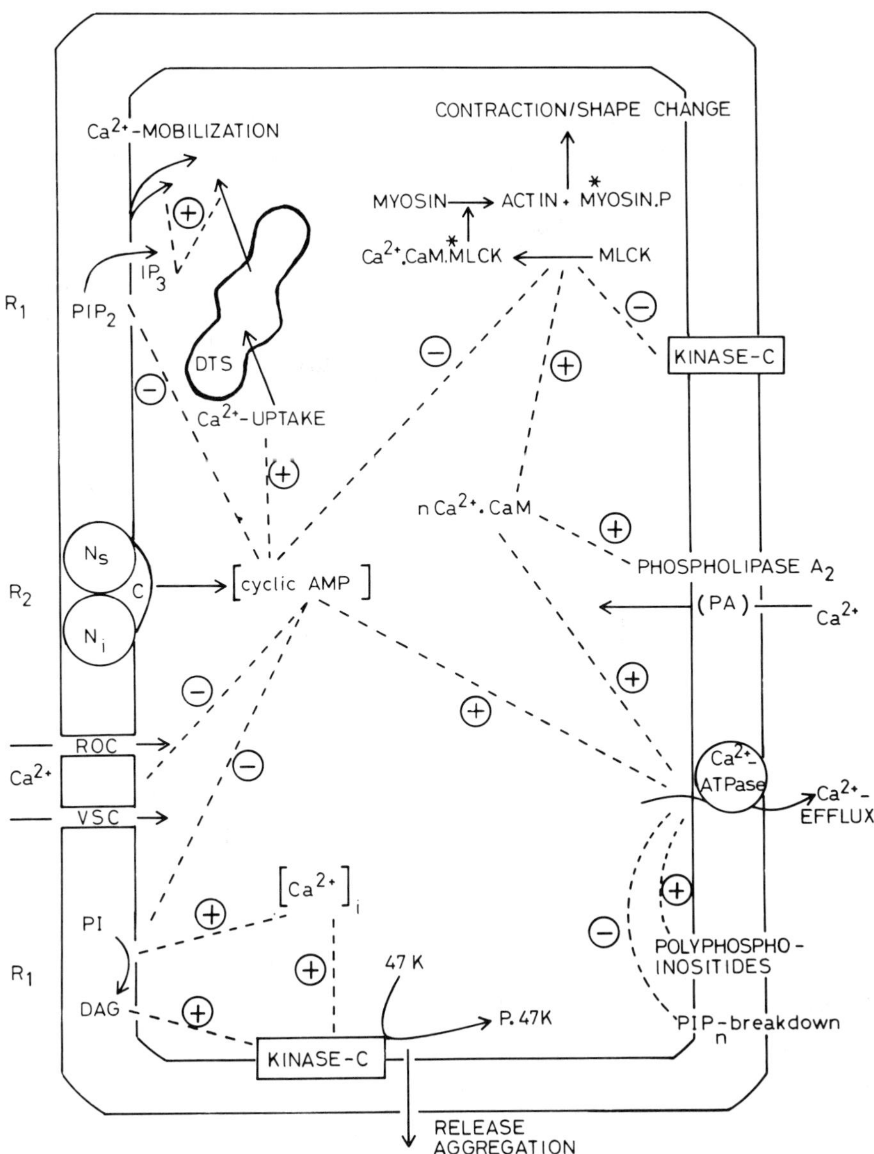

FIGURE 3. cAMP and Ca^{2+} metabolic integration of platelet biochemical and physiological processes. R_1 and R_2 represent receptors coupled to phosphoinositide and cAMP metabolism, respectively. ROC, receptor operated channels; VSC, voltage sensitive channels; N_s, N_i, C, adenylate cyclase complex; CaM, calmodulin; PI, PIP, PIP_2, mono-, di- and triphosphatidylinositol, respectively; PA, phosphatidic acid; IP_3, 1,4,5-inositol triphosphate; MLCK, myosin light chain kinase. DTS, dense tubular system; − and + indicate negative and positive modulation, respectively by Ca^{2+} and/or cAMP.

activating factor[15] the initial stimulated event is hydrolysis of PIP_2, which generates a soluble product IP_3 that elicits release from the dense tubular system[16] and the plasma membrane.[17] Hydrolysis of PIP_2 may also decrease plasma membrane ATP-dependent Ca^{2+} efflux[18,19] thus providing another mechanism whereby $[Ca^{2+}]_i$ can be increased. This apparent disruption of one of the most important Ca^{2+} homeostatic systems is transient as resynthesis of PIP_2 occurs rapidly after platelet stimulation.[13–15]

The resultant increase in $[Ca^{2+}]_i$ activates a number of biochemical events, the first being linked to the CaM pathway of the second messenger system.[11] For example, Ca^{2+}-CaM complexes with and activates myosin light chain kinase.[20] Phosphorylation of myosin light chain activates the contractile system to produce the platelet shape change response. Plasma membrane Ca^{2+} ATPase is also activated by Ca^{2+}-CaM[21] and this promotes Ca^{2+} efflux to restore Ca^{2+} homeostasis. A secondary consequence of increased $[Ca^{2+}]_i$ is activation of the phospholipase C, which hydrolyzes PI to produce the bulk of membrane-associated DAG.[13–15,20] An important difference between hydrolysis of PI and PIP_2 is that the phospholipase C acting on PIP_2 functions at basal $[Ca^{2+}]$ while that acting on PI requires an increase in $[Ca^{2+}]_i$ and is CaM dependent.[11,27] DAG activates kinase C by dramatically increasing its affinity for Ca^{2+}.[8,11] The substrate for kinase C is a M_r 47,000 protein and phosphorylation of this protein is associated with release of granular contents such as ADP and serotonin.[8,20] Subsequent conversion of DAG to PA and lyso PA (with ionophoretic properties) may promote inward Ca^{2+} flux.[23] In addition, Ca^{2+}-CaM-dependent phospholipase A_2 mediates liberation of arachidonic acid from phospholipids and is followed by immediate conversion to thromboxane A_2, itself a potent platelet activator.[8,23] Persistent activation of platelets either by the initial stimulus or by the positive feedback action of secreted ADP and/or thromboxane A_2 facilitates aggregation.[23,24]

PLATELETS AND cAMP METABOLISM

Not all platelet stimuli are directly coupled to the phosphoinositide system and some (e.g. adrenaline) act via inhibition of adenylate cyclase (FIG. 3).[25–29] Adenylate cyclase is a multicomponent enzyme regulated by both N_s-catalytic subunit coupling (stimulatory) or N_i-catalytic subunit coupling (inhibitory).[30] cAMP inhibits platelet activation and its effects are mediated via cAMP-dependent protein kinase. cAMP promotes Ca^{2+} uptake into the DTS and Ca^{2+} efflux and may inhibit voltage-sensitive and receptor-operated Ca^{2+} channel influx. cAMP also prevents Ca^{2+}-CaM activation of myosin light chain kinase and inhibits phospholipase C hydrolysis of PI and PIP_2. The net effect of cAMP is to limit an increase in $[Ca^{2+}]_i$ and to diminish operation of the Ca^{2+} messenger systems. Therefore, hormones that exert their effects via inhibition of adenylate cyclase (with a net reduction in cAMP levels) can promote platelet activation.[1,31,32]

While cAMP modulates $[Ca^{2+}]_i$ and the activity of Ca^{2+}-linked enzymes, $[Ca^{2+}]_i$ also controls cAMP.[33] Platelet adenylate cyclase shows a biphasic response to Ca^{2+} with activation at low $[Ca^{2+}]$ (0.08–0.3 μM) and inhibition at higher $[Ca^{2+}]$ (FIG. 4). The activating phase is CaM dependent and occurs at submaximal concentrations of Ca^{2+} (relative to the maximal Ca^{2+} response of platelets)[12,34] and may promote reversal of platelet activation. Concentrations of Ca^{2+} greater than 0.3 μM override CaM-dependent stimulation of adenylate cyclase, and its subsequent inhibition may facilitate platelet activation by reducing cAMP and thereby permit the unopposed operation of the Ca^{2+} messenger systems. Platelet responsiveness and biochemical reactions governing cellular responses are thus determined by the balance between the two functionally opposing second messengers, Ca^{2+} and cAMP.

PLATELET ABNORMALITIES IN ESSENTIAL HYPERTENSION

In hypertension, aberrant Ca^{2+}-linked cellular processes have been found in vascular smooth muscle, myocardium, adipocytes, erythrocytes, leukocytes, and platelets.[35–39] The platelet possesses many properties common to smooth muscle cells and this together with their clinical accessibility and cellular homogeneity make platelets a suitable model for investigation of hypertension.

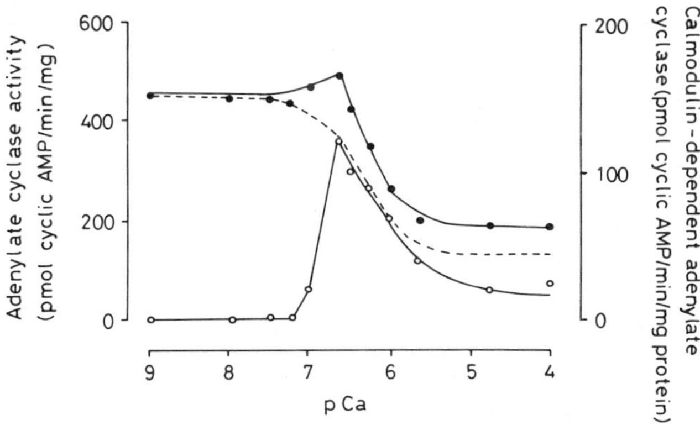

FIGURE 4. Regulation of platelet adenylate cyclase by Ca^{2+} and calmodulin. Calmodulin-depleted membranes were prepared from platelets using hypertonic/EGTA extraction procedures and assayed for adenylate cyclase activity in the absence (---) and presence (●—●) of exogenous calmodulin (1 μM), and in the presence of varying concentrations of free Ca^{2+}.[33] pCa represents the negative logarithm of the molar concentration of free Ca^{2+}. Calmodulin-dependent adenylate cyclase activity (O—O) was obtained by subtracting activity values in the absence of calmodulin from activities in the presence of calmodulin. Assay conditions were as follows: 20 mM HEPES-NaOH (pH 7.4), 2 mM EGTA, 5 mM $MgCl_2$, 1 mM dithiothreitol, membrane protein (0.5 mg/ml), 100 μM cyclic AMP, 100 μM [α-^{32}P]ATP, 3.3 mM phosphocreatine, 0.3 mg/ml creatine phosphokinase, and 37°C.[33]

Cytosolic Free Ca^{2+}

Platelets from patients with essential hypertension (TABLE 1) have an elevated $[Ca^{2+}]_i$.[40–42] In addition both systolic and diastolic blood pressures are directly correlated with $[Ca^{2+}]_i$.[40] Following antihypertensive therapy with Ca^{2+} antagonists $[Ca^{2+}]_i$ becomes normalized (TABLE 1) suggesting that the elevated $[Ca^{2+}]_i$ may be due to enhanced Ca^{2+} influx. Whatever the defect, it is not corrected by antihypertensive therapy since $[Ca^{2+}]_i$ can be stimulated (e.g. by adrenaline) to the same maximum in treated and untreated patients (TABLE 1). Because stimulation of Ca^{2+} influx appears not to be a primary response to adrenaline the elevation of $[Ca^{2+}]_i$ could also be due to a reduction of Ca^{2+} extrusion or sequestration or to an increase in Ca^{2+} release from internal pools.

TABLE 1. Platelet Cytosolic Free Ca^{2+} in Normotensive and Treated or Untreated Hypertensive Subjects

	Normotensive	Hypertensive Untreated	Hypertensive Treated
Blood pressure (mm Hg)	122 ± 2	173 ± 3	134 ± 4
	78 ± 2	106 ± 1	85 ± 1
$[Ca^{2+}]_i$ in unstimulated platelets (nM)	116 ± 4	176 ± 8[a]	112 ± 7
Δ $[Ca^{2+}]_i$ in stimulated platelets (1 μM adrenaline) (nM)	51 ± 7	63 ± 12	102 ± 6[a]
Total $[Ca^{2+}]_i$ (nM)	167	241[a]	214[a]

Platelets were isolated by gel-filtration from 22 normotensive subjects (age 38 ± 6), 22 untreated hypertension (age 43 ± 5), and 8 treated (nitrendipine or betablocker) patients with essential hypertension (age 42 ± 5).[42] Cytosolic free Ca^{2+} concentrations ($[Ca^{2+}]_i$) were measured using the Quin2 fluorescence spectrophotometric method.[40] $[Ca^{2+}]_i$ was measured in Quin2 loaded platelets after 30 min incubation at 37°C in the absence or presence of 10^{-6} M adrenaline. [a]p at least <0.01 and represents significance of difference from normotensive subjects. The data are given as mean ± SEM.

Membrane Potential

Membrane-bound Ca^{2+}, both on inner and outer surfaces of the plasma membrane, can influence membrane potential.[39] For example, increased Ca^{2+} binding causes hyperpolarization.[43] Thus, the decreased Ca^{2+} binding in many cell types in hypertension[35,37,39] implies membrane depolarization. Platelets from patients with essential hypertension are partially depolarized (TABLE 2). In platelets a comparable KCl-induced depolarization (10 mV) causes significant Ca^{2+} influx (TABLE 2), which can be inhibited by Ca^{2+} channel blockers.[44] This together with the partial depolarization suggest activated voltage-sensitive channels in platelets from patients with essential hypertension. The consequent increase in Ca^{2+} influx may contribute towards elevated $[Ca^{2+}]_i$ under basal nonstimulated conditions.

TABLE 2. Membrane Potential Measurements in Platelets from Normotensive Subjects and Patients with Essential Hypertension

	Normotensive	Hypertensive
Membrane potential (mV)	−49.5 ± 4 (N = 9)	−40.8 ± 6[a] (N = 8)
Δ mV induced by KCl (17 mM)	~10	
$^{45}Ca^{2+}$ uptake at 5 mM KCl (pmol/10^8 cells/2 min)	85 ± 20 (N = 14)	
$^{45}Ca^{2+}$ uptake at 17 mM KCl (pmol/10^8 cells/2 min)	116 ± 32 (N = 14)[b]	

Membrane potential measurements were performed using the fluorescent dye 3′3′dipropyl-thiodicarbocyanine.[55] Values are given as mean ± SD and [a]$p < 0.005$ represents significance of difference between subject groups (N = number of individuals).

Δ mV represents the membrane potential difference between experiments performed at physiological KCl (5 mM) and higher KCl (17 mM) concentrations and is an index of depolarization.[44] $^{45}Ca^{2+}$ uptake measurements are presented to indicate that platelet depolarization is accompanied by significant ([b]$p < 0.01$) Ca^{2+} influx.[44]

Ca^{2+} ATPase Activity

In patients with essential hypertension platelet membrane Ca^{2+} ATPase had a higher capacity for Ca^{2+} transport and an apparent blunted degree of CaM stimulation, although the absolute increase in CaM-stimulated Ca^{2+} ATPase activity was unaltered (FIG. 5). These findings[21] may be interpreted to imply that the reduced degree of CaM stimulation is unable to normalize the increase in $[Ca^{2+}]_i$ following

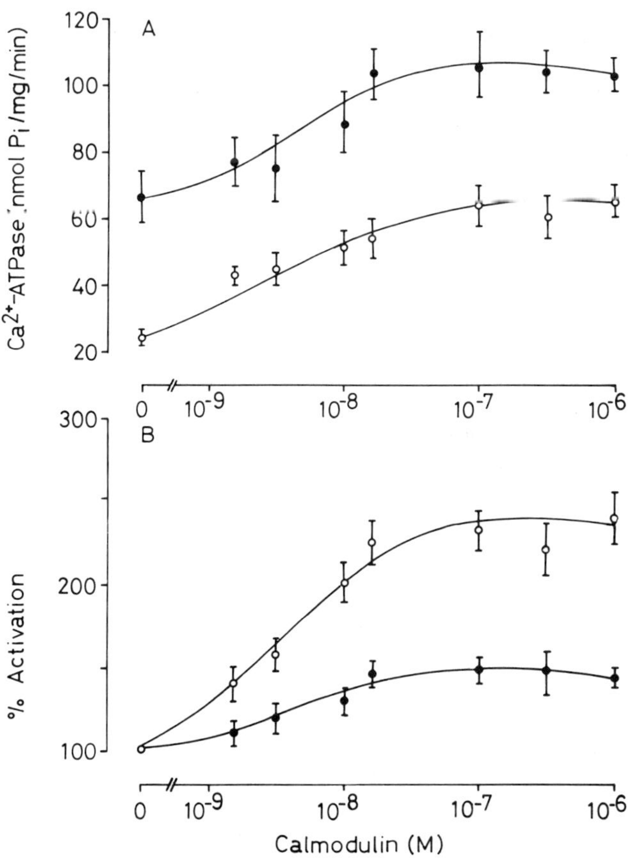

FIGURE 5. Calmodulin stimulation of platelet membrane Ca^{2+} ATPase in normotensive and essential hypertensive patients. Calmodulin-depleted membranes were prepared from platelets and assayed for Ca^{2+}. ATPase activity in the presence of varying concentrations of calmodulin.[21] Assays were performed at 37°C. Reaction medium contained 120 mM KCl, 5 mM $MgCl_2$, 2 mM EGTA, 1 mM ATP, 20 mM TES-NaOH (pH 7.5), 2.5 μM free Ca^{2+}. Stimulation by calmodulin is expressed relative to the activity obtained in the absence of calmodulin (taken as 100%) (B). These values were derived from Ca^{2+} ATPase activity measurements in (A). Data are given as mean ± SEM, 13 normotensive (○) and 14 hypertensive (●) subjects. (From Resink *et al.*[21] With permission from *Hypertension.*)

platelet stimulation. On the other hand, the increased capacity of the Ca^{2+} transport system may protect the cell against Ca^{2+} overload.

Hormone-Receptor-Stimulus Coupling

Platelets from untreated patients with essential hypertension have enhanced sensitivity to a variety of activators.[42,45,46] This is reflected in a number of Ca^{2+}-linked events (TABLE 3) including the $[Ca^{2+}]_i$ response, the shape-change response, and phosphorylation of M_r 20,000 by Ca^{2+} calmodulin-dependent myosin light chain kinase and that of the M_r 47,000 substrate for kinase C. From above data obtained in untreated patients, it is not clear whether an intrinsically enhanced hormone receptor coupling is responsible for the increased $[Ca^{2+}]_i$ response and hence sensitized Ca^{2+}-dependent biochemical processes or whether an elevated $[Ca^{2+}]_i$ per se potentiates hormone responsiveness. Treated patients have a normalized platelet $[Ca^{2+}]_i$ and no longer display hormone supersensitivity (at least in terms of the $[Ca^{2+}]_i$ response). Thus, it appears that $[Ca^{2+}]_i$ influences hormone-receptor coupling and elevated $[Ca^{2+}]_i$ facilitates hormone stimulation of platelets.

cAMP Metabolism

In view of the integrated bi-directional control of platelet function by Ca^{2+} and cAMP (FIG. 3) it is possible that increased $[Ca^{2+}]_i$ may be associated with anomalies in platelet cAMP metabolism.[32,47] cAMP accumulation in platelets from patients with essential hypertension was more sensitive both to stimulation by PGE_1 (which acts via N_s) and to inhibition by adrenaline (which acts via N_i) (TABLE 4). The enhanced cAMP accumulation response to PGE_1 probably results from increased adenylate

TABLE 3. Evidence for Enhanced Ca^{2+}-Linked Receptor-Stimulus Coupling in Essential Hypertension

		Hypertensive	
	Normotensive	Untreated	Treated
$[Ca^{2+}]_i$ (EC_{50})			
Thrombin (U/ml)	0.027 ± 0.009	0.009 ± 0.002[a]	0.036 ± 0.009
Adrenaline (μM)	1.038 ± 0.288	0.149 ± 0.020[a]	0.944 ± 0.190
Shape change (EC_{50})			
Adrenaline (μM)	0.64 ± 0.10	0.32 ± 0.08[a]	
Serotonin (μM)	0.92 ± 0.08	0.65 ± 0.09[a]	
Phosphorylation (EC_{50} for adrenaline, μM)			
M_r 20,000	0.60 ± 0.08	0.20 ± 0.03[a]	
M_r 47,000	0.70 ± 0.15	0.15 ± 0.10[a]	

Hormone responsiveness in platelets from normotensive and treated or untreated patients with essential hypertension was examined in three independent studies.[40,42] $[Ca^{2+}]_i$ was measured in gel-filtered Quin2 loaded platelets,[40,42] shape change measurements were performed on washed platelets using turbidimetric procedures,[42] phosphorylation determinations were made on ^{32}P-prelabeled washed platelets.[42] EC_{50} indicates the concentration of hormone required to elicit a half-maximal response. Data are given as mean ± SEM, N = 5–10. [a]p at least <0.05 and indicates significance of difference from normotensive subjects.

TABLE 4. Evidence for Enhanced cAMP-Linked Receptor-Stimulus Coupling in Essential Hypertension

	Normotensive	Hypertensive
cAMP production (D_{50})		
PGE_1 (μM) stimulation	0.73 ± 0.16	0.37 ± 0.05[a]
Adrenaline (μM) inhibition of PGE_1 stimulatory effects	0.17 ± 0.08	0.04 ± 0.01[a]
Adenylate cyclase activity ratio (stimulated/basal)		
PGE_1 (1 μM)	7.87 ± 0.43	9.30 ± 0.37[a]
Forskolin (1 μM)	12.51 ± 0.53	15.83 ± 0.66[a]

Hormone-regulated cAMP production and adenylate cyclase activity ratios were determined in two independent studies.[46] cAMP levels were measured by radioimmunoassay following hormone treatment of intact platelets.[46] Adenylate cyclase activity was determined in membrane preparations by measuring production of [^{32}P]cAMP from [α^{32}P]ATP.[46] Data are given as mean ± SEM (number of individuals = 9–11). D_{50} indicates the concentration of hormone required to elicit half-maximal alterations in cAMP levels. [a]p at least <0.05 and indicates significance of difference from normotensive subjects.

cyclase activation, since forskolin, which bypasses receptor N_s coupling and acts directly on the catalytic subunit, also activated adenylate cyclase to a greater extent in patients with essential hypertension (TABLE 4). Conceivably, increased catalytic turnover could protect against Ca^{2+} overload. Such an adaptive mechanism is seemingly contradicted by the parallel enhanced sensitivity to adrenaline because adrenaline inhibits adenylate cyclase. Nevertheless, since a small increase in $[Ca^{2+}]_i$ can activate adenylate cyclase (FIG. 4), stimulation of platelets by circulating subthreshold concentrations of adrenaline could dampen the platelet activation response (as for PGE_1).

Phosphoinositide Metabolism

The phosphoinositides PI, PIP, and PIP_2, although quantitatively minor components of membranes, are particularly reactive membrane phospholipids. Their metabolism is involved in the Ca^{2+} gating system of the plasma membrane.[8,11,19] The polyphosphoinositides also participate in membrane fluidity,[48] Ca^{2+} ATPase activity,[6,49–51] and Ca^{2+} binding.[51,52] DAG has been implicated in the regulation of adenylate cyclase.[53]

In patients with essential hypertension, no significant change in PA was observed, while ^{32}P incorporation into PIP and PIP_2 was greater in hypertensive patients at the apparent expense of PI (TABLE 5). The data suggest an alteration in the interconversion equilibrium of the phosphoinositides, which is directed towards polyphosphoinositide formation. This could result from increased phosphoinositide kinase(s) or decreased phosphoinositide phosphomonoesterase(s) activities. Phosphodiesteratic cleavage, by phospholipase C(s), and the resynthesis of PI appear not to be involved since ^{32}P incorporation into PA was not different between subject groups. The nature of alterations in phosphoinositide metabolism in the platelets of hypertensive patients could explain both an elevated Ca^{2+} ATPase activity and an increased $[Ca^{2+}]_i$. An increase in membrane content of polyphosphoinositides is correlated with an increase in Ca^{2+} ATPase activity and ATP-sensitive Ca^{2+} binding.[51,52,54] In addition, acidic phospholipids (e.g. PIP_2) activate purified erythrocyte Ca^{2+} ATPase.[6,49] The elevated level of PIP_2 could provide a greater source for IP_3, and thus magnify the IP_3-induced release of Ca^{2+} from intracellular stores.

CONCLUSIONS

The plasticity of cellular Ca^{2+} control and the large number of events regulated by $[Ca^{2+}]_i$ and cAMP make it difficult to assign causative or consequential roles to deranged platelet Ca^{2+}-linked processes in the pathophysiology of essential hypertension. Our studies support an underlying membrane pathology as being causative since all the above-described derangements (potential, Ca^{2+} ATPase, hormone responsiveness, adenylate cyclase, and even cytosolic $[Ca^{2+}]_i$ concentrations) are membrane-associated systems. Modification of phosphoinositide metabolism may be a key factor

TABLE 5. Platelet Phosphoinositide Metabolism in Essential Hypertension

	^{32}P Incorporation into Phosphoinositides (%)			
	PA	PI	PIP	PIP_2
Normotensive subjects ($N = 5$)	9.3 ± 3.5	31.9 ± 14.9	33.0 ± 6.5	26.6 ± 8.5
Hypertensive patients ($N = 9$)	4.0 ± 1.1	5.9 ± 1.2[a]	46.2 ± 2.4[a]	43.3 ± 2.7[a]

The ^{32}P incorporation into phosphatidic acid (PA), monophosphatidylinositol (PI), phosphatidylinositol 4-phosphate (PIP), and phosphatidylinositol 4,5-bisphosphate was measured in resting (60–90 min, 37°C) ^{32}P-prelabeled and washed platelets. Lipids were extracted from platelets by addition of 3.75 vol. of cold chloroform/methanol/HCl (20:40:1) and the resultant single phase partitioned by addition of 1.25 vol. each of chloroform and H_2O. The upper aqueous phase and protein interface were removed and the lower chloroform phase evaporated to dryness under N_2. Lipids were re-extracted into chloroform and dried again before dissolution in chloroform. Lipid extracts were spotted onto oxalate-pretreated silica gel plates that were developed in chloroform/acetone/methanol/acetic acid/H_2O (40:15:13:12:8) ^{32}P labeled phospholipids were visualized after radioautography, scraped out, and radioactivity determined by scintillation counting.[56] Results (mean ± SEM) are expressed as percentage of total ^{32}P incorporated (PA + PI + PIP + PIP_2 = 100%). [a]$p < 0.05$ and indicates significance of difference from normotensive subjects.

accounting for the multi-faceted membrane abnormalities associated with essential hypertension. Whether the modification in phosphoinositide metabolism is a membrane abnormality, also found in smooth muscle cells, has yet to be determined.

REFERENCES

1. RASMUSSEN, H. & D. B. P. GOODMAN. 1977. Physiol. Rev. **57:** 421–509.
2. BYGRAVE, F. L. 1978. Biol. Rev. Camb. Philos. Soc. **53:** 43–79.
3. CARAFOLI, E. & M. CROMPTON. 1978. Curr. Top. Membr. Transp. **10:** 151–216.
4. KRETSINGER, R. H. 1976. Int. Rev. Cytol. **46:** 323–393.
5. BLAUSTEIN, M. P. & M. I. NELSON. 1982. *In* Membrane Transport of Calcium. E. Carafoli, Ed.: 217–236. Academic Press. London.
6. CARAFOLI, E. & M. ZURINI. 1982. Biochim. Biophys. Acta **683:** 279–301.
7. CHEUNG, W. Y. 1980. Science **207:** 19–27.
8. TAKAI, Y., U. KIKKAWA, K. KAIBUCHI & Y. NISHIZUKA. 1984. Adv. Cycl. Nucl. Res. **18:** 119–158.

9. KLEE, C. B., T. H. CROUCH & P. G. RICHMAN. 1980. Ann. Rev. Biochem. **49:** 489–515.
10. ASHENDEL, C. L. 1985. Biochim. Biophys. Acta **822:** 219–242.
11. RASMUSSEN, H., I. KOJIMA, K. KOJIMA, W. ZWALICH & W. APFELDORF. 1984. Adv. Cycl. Nucl. Prot. Phos. **18:** 159–193.
12. MACINTYRE, D. E., A. M. SHAW, M. BUSHFIELD, L. J. MACMILLAN, A. MCNICOL & W. K. POLLOCK. 1981. Nouv. Res. Fr. Hematol. **27:** 285–292.
13. SEISS, W., M. STIFEL, H. BINDER & P. L. WEBER. 1986. Biochem. J. **233:** 83–91.
14. MACINTYRE, D. E. & W. K. POLLOCK. 1983. Biochem. J. **212:** 433–437.
15. BILLAH, M. M. & E. G. LAPETINA. 1982. J. Biol. Chem. **257:** 12705–12708.
16. O'ROURKE, F. A., S. P. HALENDA, G. B. ZAVOICO & M. B. FEINSTEIN. 1985. J. Biol. Chem. **260:** 956–962.
17. BROEKMAN, M. J. 1984. Biochem. Biophys. Res. Commun. **120:** 226–231.
18. PENNISTON, J. 1982. Ann. N.Y. Acad. Sci. **402:** 296–303.
19. COCKROFT, S. 1984. Biochem. Soc. Trans. **12:** 966–968.
20. GERRARD, J. M., S. J. ISRAELS & L. L. FRIESEN. 1985. Nouv. Rev. Fr. Hematol. **27:** 267–273.
21. RESINK, T. J., V. A. TKACHUK & F. R. BÜHLER. 1986. Hypertension **8:** 159–166.
22. MAJERUS, P. W., D. B. WILSON, T. CONNELY, T. E. BROSS & E. J. NEUFIELD. 1985. Trends Biochem. Sci. **10:** 168–171.
23. LAPETINA, E. G. & S. P. WATSON. 1985. Nouv. Rev. Fr. Hematol. **27:** 235–238.
24. KRISHNAMURTHI, S., J. WESTWICK & V. V. KAKKAV. 1984. Biochem. Pharmacol. **33:** 3025–3035.
25. JAKOBS, K. H., W. SAUR & G. SCHULTZ. 1978. FEBS Lett. **85:** 167–170.
26. MELLWIG, K. P. & K. H. JAKOBS. 1980. Thromb. Res. **18.** 7–17.
27. VANDERWEL, M., D. S. LUM, & R. J. HASLAM. 1983. FEBS Lett. **164:** 340–344.
28. WILLIAMS, K. A. & R. A. HASLAM. 1984. Biochim. Biophys. Acta **770:** 216–223.
29. AKTORIES, K. & K. H. JAKOBS. 1984. Eur. J. Biochem. **145:** 333–338.
30. ROSS, E. M. & A. G. GILMAN. 1980. Ann. Rev. Biochem. **49:** 533–564.
31. HASLAM, R. J., M. M. L. DAVIDSON, T. DAVIES, J. A. LYNHAM & M. D. MCCLENAGHAN. 1978. Adv. Cycl. Nucl. Res. **9:** 533–552.
32. MILLS, D. C. B. 1987. Handb. Exp. Pharmacol. **58:** 723–761.
33. RESINK, T. J., S. STUCKI, G. Y. GRIGORIAN, A. ZSCHAUER & F. R. BÜHLER. 1986. Eur. J. Biochem. **154:** 451–456.
34. HALLAM, T. J., A. SANCHEZ & T. J. RINK. 1984. Biochem. J. **218:** 819–827.
35. POSTNOV, Y. V. & S. N. ORLOV. 1985. Physiol. Rev. **65:** 904–945.
36. KWAN, C. Y. 1985. Can. J. Physiol. Pharmacol. **63:** 366–374.
37. POSTNOV, Y. V. & S. N. ORLOV. 1984. J. Hypertension **2:** 1–6.
38. FRIEDMAN, S. M. 1983. J. Hypertension **1:** 109–114.
39. ROBINSON, B. F. 1984. J. Hypertension **2:** 453–460.
40. ERNE, P., F. R. BÜHLER, H. AFFOLTER & E. BÜRGISSER. 1984. N. Engl. J. Med. **310:** 1084–1088.
41. LE QUAN SANG, K. H., T. MONTENAY-GARESTIER & M. A. DEVYNCK. 1985. Nouv. Rev. Fr. Hematol. **27:** 279–283.
42. ERNE, P., T. J. RESINK, P. BOLLI, A. HEFTI, R. RITZ & F. R. BÜHLER. J. Hypertension **2** (Suppl. 3): 159–161.
43. ISENBERG, G. 1977. Pfluger's Arch. **371:** 71–72.
44. ZSCHAUER, A., T. J. RESINK & F. R. BÜHLER. 1986. (unpublished observations)
45. ERNE, P., T. J. RESINK, E. BÜRGISSER & F. R. BÜHLER. 1985. J. Cardiovasc. Pharmacol. **7** (Suppl. 6): S103–S108.
46. RESINK, T. J., E. BÜRGISSER & F. R. BÜHLER. 1986. Hypertension. (In press.)
47. HAMET, P., A. J. FRANKS, S. ADNOT & J. F. COQUIL. 1980. Adv. Cycl. Nucl. Res. **12:** 11–23.
48. ALLAN, D. 1982. Cell Calcium **3:** 451–465.
49. NIGGLI, V., E. S. ADUNYAH & E. CARAFOLI. 1981. J. Biol. Chem. **256:** 8588–8592.
50. DOWNES, C. P. & R. H. MICHELL. 1982. Biochem J. **202:** 53–58.
51. KAWAGUCHI, T. & K. KONISHI. 1980. Biochem. Biophys. Acta **597:** 577–586.
52. BUCKLEY, J. T. & J. N. HAWTHORNE. 1972. J. Biol. Chem. **247:** 7218–7223.
53. JAKOBS, K. H., S. BAUER & Y. WATANABE. 1985. Eur. J. Biochem. **151:** 425–430.

54. Varsanyi, M., H. G. Tolle, L. M. G. Heilmayer, R. M. C. Dawson & R. F. Irvine. 1983. EMBO J. **2:** 1543–1548.
55. Greenberg-Sepersky, S. M. & E. R. Simons. 1984. J. Biol. Chem. **259:** 1502–1508.
56. Tysne, O. B., G. M. Aarbakke, A. J. M. Verhoeven & H. Holmsen. 1985. Thromb. Res. **40:** 329–338.

DISCUSSION OF THE PAPER

A. Scarpa (*Case Western Reserve University, Cleveland, OH*): (1) In platelets of hypertensive patients, the cytosolic Ca^{2+} as measured by Quin2 appears elevated. However, there is a large scattering in your data and problems with Quin2 measurements. In view of those potential problems a more direct measurement by atomic absorbance of the total Ca^{2+} will indicate if this is increased. Although the action is at the level of free Ca^{2+}, this pool is in equilibrium with the total.

(2) A possible and challenging explanation for some of the data of norepinephrine effect in platelets is the effect of extracellular ATP, which in other systems, including smooth muscle, stimulates intracellular Ca^{2+} release via IP_3. The reason is that any time you release one norepinephrine from chromaffin granules you also release one ATP stored within the granules.

Bühler: So far we have not looked at the total platelet calcium and at an ATP amplifying system, which most likely is enhanced in platelets of hypertensive patients similar to other release reactions e.g. serotonin.

G. Bianchi (*University of Milan, Milan*): If any type of antihypertensive therapy is able to lower platelet Ca then two obvious questions may be asked: (1) What is the rationale for using Ca antagonists in the therapy of hypertension? (2) What is the role of platelet Ca in the understanding of the primary mechanisms leading to hypertension?

Bühler: (1) Calcium antagonists have proven most effective and they were well tolerated. Slow channel calcium influx inhibition, however, is only part of the antihypertensive mode of action. Consequent to vasodilation counter regulatory mechanisms—if intact as in a young patient often with a high renin—may prevent a full fall in pressure and cytosolic calcium. Conversely, older patients with blunted counter regulatory reflexes exhibit greater falls in blood pressure.

(2) It is not more or less than a corollary for the vascular smooth muscle cell, which so far we have not been able to study in man. But all we learn from platelet research in hypertensives will help to better understand their increased thromboembolic complications.

L. H. Opie (*University of Capetown, Capetown, South Africa*): Please comment on the steps between increased circulating catecholamines in some hypertensives, decreased acidic AMP in platelets, and increased platelet calcium.

Bühler: In human platelets, adrenaline leads via α_2-adrenoceptors to N_i coupling of adenylate cyclase with a decrease in cAMP and enhanced slow channel calcium influx.

S. Corvera (*University of Massachusetts, Worcester, MA*): You have shown an increased responsiveness of platelets from hypertensive subjects to adrenaline and serotonin. Are the numbers of the receptors for these hormones increased in this condition?

Bühler: We only know it in the case of alpha$_2$adrenocytes and they are not different. The defect appears to be at the α-subunit of the N-protein coupling complex.

L. CANTLEY (*Tufts University, Boston, MA*): John Penniston has shown that when the red cell Ca^{2+}-ATPase is reconstituted in high concentrations of PIP_2 the enzyme is maximally activated in the absence of calmodulin. Could the increased PIP_2 level you see in platelet membranes of hypertensive patients explain the elevated Ca ATPase activity and loss of calmodulin sensitivity?

BÜHLER: This is a plausible interpretation that we have not tested yet. It may well be that the shift in phosphatidylinositol metabolism towards more PIP_2 production is a Ca platelet membrane event providing a link for other observed defects as well.

Membrane Abnormalities in Essential Hypertension:

Physiologic and Genetic Links

GIUSEPPE BIANCHI,[a] PATRIZIA FERRARI, DANIELE CUSI, BARRY R. BARBER, SERGIO SALARDI, LUCIA TORIELLI, MARIA GRAZIA TRIPODI, ENRICO NIUTTA, GIUSEPPE VEZZOLI, AND CRISTINA BARLASSINA

Istituto di Scienze Mediche
Università di Milano
Farmitalia Carlo Erba Research Center
Nerviano (Milan) Italy

INTRODUCTION

Essential hypertension develops from the interaction between genetic and environmental factors. The genetic component of this disease is rather complex because polygenes of both minor and major effect are likely to be involved.[1] This complex heritability is not surprising in view of the multiplicity of factors involved in blood pressure regulation and the network of the feedback systems that determine the set point of blood pressure.[2] In fact, each of these factors or systems could potentially be affected by one or more genes, thus a polygenic influence can reasonably be assumed as a background. On the other hand it is also reasonable to look for a major gene that affects the modification of biochemical functions such as those involved in transmembrane ion transport. In fact these transport systems have a key role in the regulation of the function of those organs, such as kidney, CNS etc., which are particularly important in determining the set point of blood pressure. For all these reasons a combined approach, involving techniques of physiology, biochemistry, protein chemistry, molecular biology, and population genetics should be applied in order to elucidate the sequence of events from the abnormal major gene to the abnormal cell and organ functions responsible for hypertension. To be fruitful, such an approach should first be applied to a genetic animal model of hypertension to allow the application of the appropriate experimental maneuvers necessary to fulfill the criteria for establishing a cause-effect relationship between a given gene abnormality and the subsequent development of hypertension.[3] For obvious reasons, the cell membranes of those organs involved in blood pressure regulation in humans are not available for these studies. Red blood cell membranes from readily available erythrocytes have therefore been chosen as the material for study in both humans and animal models.

When studying the abnormalities in blood cell membranes in essential hypertension we have to face two main questions: Are these abnormalities related in some way to hypertension or are they epiphenomena with no relation to the disease? If the

[a]Address correspondence to: Prof. Giuseppe Bianchi, Istituto Scienze Mediche, Via F. Sforza, 35, 20122 Milano, Italy.

relation exists, then what is its nature? In fact, to demonstrate that the abnormalities we are describing are caused by the same genetic mechanism(s) responsible for the development of the disease, we have to consider at least three alternative possibilities. First, the blood cell membrane abnormality may be caused by other physiological changes, including the real primary ones. Second, there may be a genetic linkage between blood cell membrane alterations and the membrane alterations of the cells of the organs more directly involved in blood pressure regulation. In both of these possibilities the relationship is present, but the blood cell abnormalities are not the direct expression of the genetic molecular mechanism causing hypertension. Third, there may be a pleiotropic effect of the same gene(s) both on blood cell function and on the function of those cells (renal, vascular, or nervous) more directly involved in blood pressure regulation. Only if this possibility is true may the blood cell abnormalities be considered as useful tools for approaching the genetic molecular mechanisms of hypertension.

The aim of this paper is to review our most recent findings on this subject in the genetic hypertension of the Milan hypertensive strain of rats (MHS) studied in comparison to its appropriate normotensive control (MNS), where a precise answer to these questions has been sought with the suitable experimental design. Moreover, we shall try to verify how much of the information obtained in the rat model is useful for understanding the role of blood cell membrane abnormalities in human essential hypertension.

RAT STUDIES

The most relevant alterations in the red blood cells of the MHS are the following: the cell volume and sodium content are reduced, the outward Na-K cotransport is increased, and the Ca-ATPase at V_{max} is reduced (FIG. 1) when compared to the MNS.[4–6] In order to look for possible biochemical-physiological links between the erythrocyte alteration and those of the organ responsible for the development of hypertension we will summarize the main abnormalities in MHS present before the development of hypertension and that are relevant to its pathogenesis.

Prehypertensive MHS rats have faster glomerular filtration rate, both *in vivo*[4,7] and *in vitro,*[8] when the kidney is perfused with an artificial medium; lower plasma renin activity;[9] lower urinary kallikrein excretion;[10] greater 24 hr urinary output; and they retain more sodium and water than the MNS.[9] Kidney cross-transplantation experiments clearly indicate that the MHS kidney is responsible for the rise in blood pressure[11,12] during and after the development of hypertension, in fact the kidney coming from a MNS is able to reduce the blood pressure of the MHS recipient, whereas the MHS kidney increases the blood pressure in the MNS recipient. Moreover retention of water and Na takes place during the development of hypertension. A primitive increase in tubular Na reabsorption may explain these findings. In fact, by increasing the body fluid volumes it will reduce the plasma renin activity and increase the glomerular filtration rate through a depression of the tubulo-glomerular feedback. The results of micropuncture studies,[13–15] where single nephron filtration rate, tubulo-glomerular feed-back, proximal tubular reabsorption, micropressures in the tubuli and interstitium, and interstitial oncotic pressure were measured, demonstrate that the absolute proximal tubular reabsorption is similar in the two strains, in spite of a greater net hydrostatic pressure in the renal interstitium in the MHS. This suggests that the intrinsic ability of the MHS proximal tubular epithelium to reabsorb solute and water is greater in MHS during the prehypertensive stage. These data are confirmed *in vitro* by an increased rheogenic sodium flux in isolated brush border vesicles.[16,17] Moreover

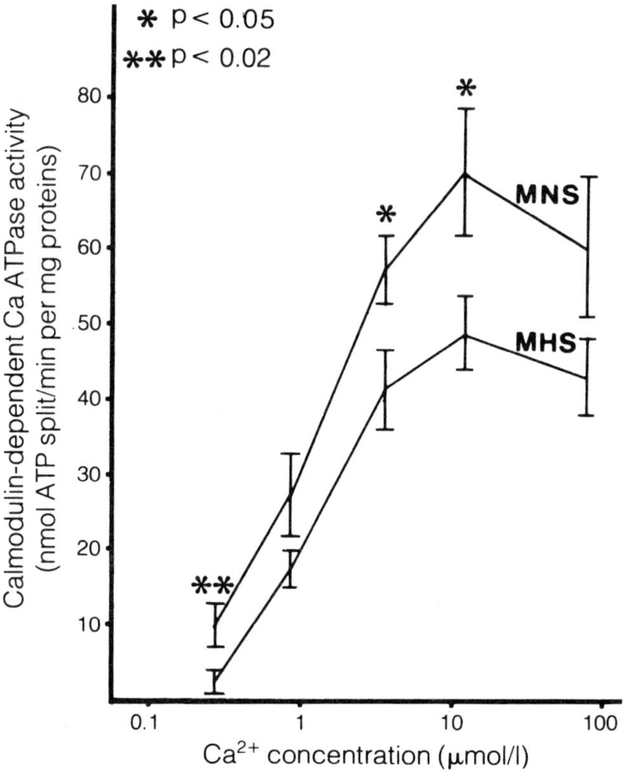

FIGURE 1. Calmodulin-dependent Ca ATPase activity at increasing Ca^{2+} concentrations in erythrocyte ghosts of 16 MHS and 15 MNS rats (mean values ± s.e.m.). MHS, genetically hypertensive rats of the Milan strain; MNS, controls. (From Vezzoli *et al.*[5] *Journal of Hypertension.*)

the active calcium transport in vesicles prepared by basolateral membranes from tubular cells is lower.[18] The intracellular sodium concentration and the volume of the proximal tubular cells are lower.[19]

In order to prove that a primary and genetically determined cell abnormality is responsible for the MHS tubular cell biochemical changes, we need another type of cell that possesses the same functional abnormalities as the tubular cell but, conversely, is more suitable for applying the appropriate experimental manipulations. Such a cell could be the erythrocyte. In fact there are many biochemical and physiological similarities between the MHS erythrocytes and MHS proximal tubular cell abnormalities (TABLE 1). When compared to MNS, MHS have smaller volume of proximal tubular and red blood cells, both with lesser intracellular sodium, while at least one pathway of Na transport across the plasma membrane is faster (Na-K cotransport for the erythrocytes and rheogenic Na transport in isolated brush border vesicles of proximal tubular cells). Moreover the Ca pump activity at V_{max} is lower in both types of cells. These alterations seem to be peculiar and rather selective for erythrocytes and proximal tubular cells. In fact the volume and the sodium content of distal tubular cells and of liver cells are similar in MHS and in MNS.[19,20] At present the reason for this selectivity is unknown. We have shown[6] that the increase in Na-K cotransport and the decrease in erythrocyte volume is genetically associated with hypertension with a back-cross experiment. In the F2 hybrids obtained by crossing the F1 (MHS × MNS) there is a significant correlation of both to blood pressure; i.e. the rats that attain the highest blood pressures have faster Na-K cotransport and smaller erythrocyte volume (FIG. 2). These data were recently confirmed after several crossings of the hybrids, thus ruling out the possibility of genetic linkage as the cause of genetic correlation. Moreover, with bone marrow transplantation from MHS or MNS to irradiated F1 (MHS × MNS) hybrids, we demonstrated that the characteristics of MHS or MNS erythrocytes are found in the recipients of their bone marrow, indicating that the stem cells of MHS or MNS contain the genetic information responsible for the peculiar function of the mature erythrocyte (FIG. 3).[6]

The results discussed so far clearly demonstrate that a genetically determined primary cell abnormality may cause hypertension in MHS. But how can we distinguish, at the cellular level, the genetic molecular alteration that eventually causes the secondary cellular biochemical changes? We are approaching this question in many ways. (1) By different manipulations of the erythrocytes we try to define the relationships among the different cellular or membrane abnormalities. The experiments shown in Ferrari *et al.*[21] are along this line and showed that the difference in cell volume between MHS and MNS persists in resealed ghosts also when they are incubated in different solutions ranging from 300 to 100 mOsm. As the amount of cholesterol, taken as an index of the lipid bilayer of the cell membrane, per 10^6 cells is similar in the two strains it is very likely that the difference in erythrocyte volume between MHS and MNS is due to an intrinsic property of the membrane cytoskeleton. Data from Ferrari *et al.*[21] also show that the difference in outward Na-K cotransport is lost when this system is measured in inside-out vesicles (which are deprived of

TABLE 1. Comparison of Prehypertensive Humans with Rats of the MHS Strain

	Humans Essential Hypertension	Rats MHS
Pressor effect of the kidney		
after transplantation	↑ (*)	↑
Total	↑	= (x)
Renal blood flow		
% CO	↑	↑ (x)
GFR	↑ (+)	↑ (o)
Na excretion after load	↑	↑
24 hr urinary output	↑	↑
Plasma renin	↓	↓
Urine kallikrein	↓	↓
Plasma aldosterone	=	=
Plasma Na and K	=	=
CO	=	= (x)
Erythrocytes Na concentration	↓	↓
Net erythrocyte cell membrane Na transport	↑	↑

↑, ↓, = higher, lower or equal in the prehypertensive humans or rats compared to the appropriate controls. (x) Measured in anesthetized rats, (o) expressed per unit of kidney weight, (+) expressed per unit of body surface, and (*) Guidi *et al.*[26]

cytoskeleton) prepared from MHS and MNS erythrocytes. Taken together, these observations support the suggestion that the difference in volume and Na-K cotransport may be due to a primary abnormality of a cytoskeleton protein. This suggestion is also supported by recent findings showing that the membrane cytoskeleton may be involved in the regulation of Na-K cotransport.[22–24] (2) Using immunochemical techniques we tried to evaluate possible differences in structure between the cell membrane proteins of MHS and MNS. As shown in Ferrandi *et al.*[25] this approach has demonstrated that a cytoskeletal 105 KDa protein may have a structural difference in MHS and MNS. The isolation of the gene coding for this protein is in progress (Sidoli

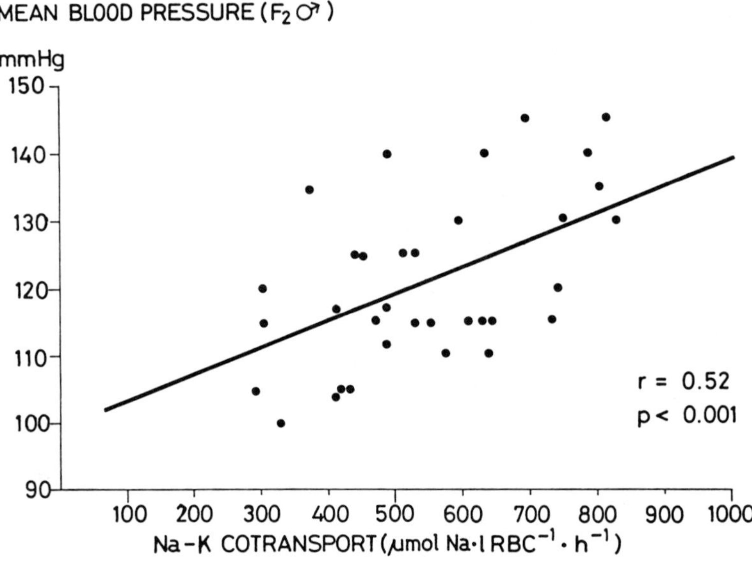

FIGURE 2. Correlation between RBC Na-K cotransport and mean blood pressure measured simultaneously in 60-day-old male rats of the F2 generation (N = 34) obtained by crossing the (MHS × MNS) F1 hybrids. $y = 0.04\,x + 98.7$. Mean blood pressure was measured on the carotid artery in the conscious rats through a chronic catheter connected to a pressure transducer (Bell & Howell, Pasadena, CA) and recorded on a Beckman R511 Polygraph. Na-K cotransport was measured in fresh RBC suspended in a Na^+-free choline medium and was calculated as ouabain resistant-bumetanide sensitive Na^+ efflux. (From Bianchi *et al.*[6] With permission from *Hypertension.*)

et al., unpublished data). (3) Using an intercross-backcross system between MHS and MNS we are isolating different lines of rats carrying different proportions of the hypertensive and normotensive genomes. We are also isolating a cogenic line of normotensive rats that, at present, has reached a 0.93 coefficient of relationship with the MHS rats (that is 93% of the genome is identical in the two lines). Using these different lines of rats, and in particular the cogenic line, it should be possible to establish the relationship between abnormalities of gene, gene products, cell membrane, and organ leading to hypertension.

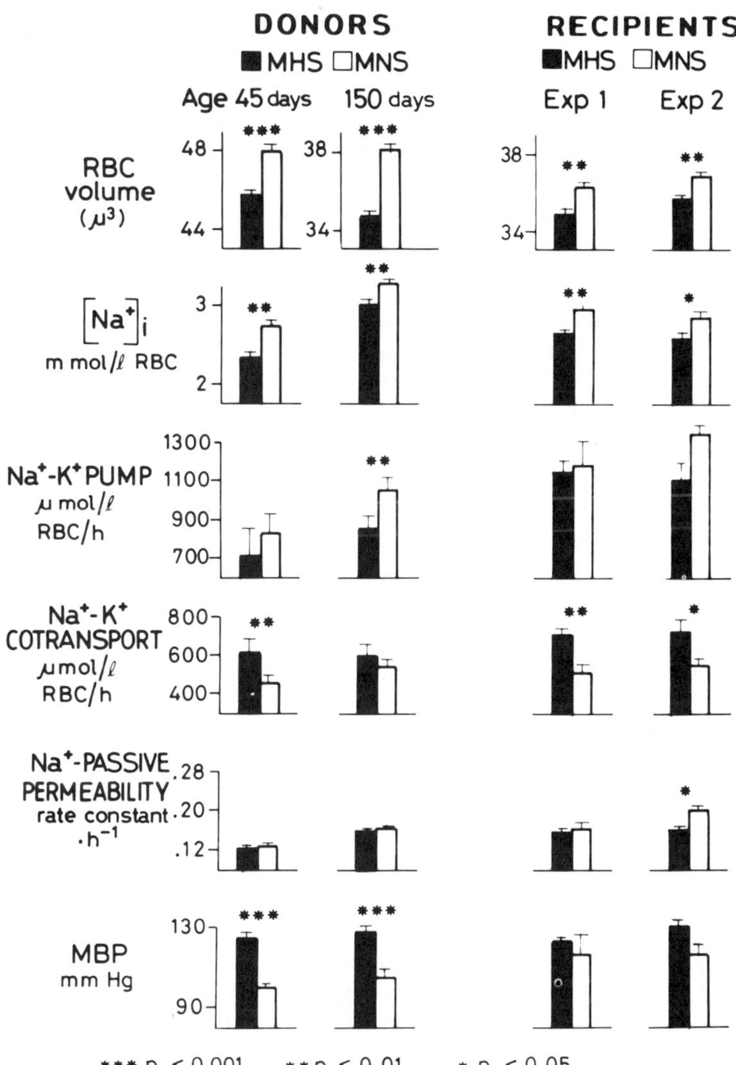

FIGURE 3. Bone marrow transplantation from MHS and MNS parental donor strains respectively into two groups of previously irradiated F1 hybrids, as recipients. Reported parameters are: RBC volume, RBC intracellular Na^+ concentration ($[Na]_i$), Na-K pump, Na-K cotransport, Na passive permeability, and mean blood pressure (MBP) of the donor strains (MHS and MNS), at 45 and 150 days of age, and of the F1 hybrids, three months after they received bone marrow from MHS and MNS. Two complete, separate sets of experiments (Exp. 1 and 2) were carried out with the same protocol using 45-day-old rats as donors. RBC volume was measured by Coulter Counter+Channalyzer C 1000 (Coulter Electronics, U.K.). $[Na]_i$ was measured in fresh RBC washed at 37°C in Na-free choline medium and then incubated in the same solution for Na flux measurements. Na-K pump was calculated as ouabain-sensitive Na efflux, Na-K cotransport as ouabain-resistant, bumetanide-sensitive Na efflux and Na passive permeability as ouabain and bumetanide-resistant Na efflux. (From Bianchi *et al.*[6] With permission from *Hypertension.*)

HUMAN STUDIES

Many alterations of blood cell membrane function in patients with essential hypertension have been reported by different investigators. However a wide overlap exists between normotensive and hypertensive patients. The influence of the different environmental factors on the blood cell membrane function is not clear. For the reasons given above we are looking for possible correlations between erythrocyte and renal function alterations to see if the results obtained in the rats are applicable also to humans, especially due to the similarities between MHS and humans shown in TABLE 2.[4,26,28,29] That the kidney may be the culprit for hypertension was recently shown also in humans by a kidney transplantation study. For obvious reasons we could perform only a retrospective study on the influence of the kidney donor's familial hypertension on blood pressure and the requirement of antihypertensive therapy of the recipient.[26] In fact, the actual blood pressure of the donor and of the recipient could not be taken into account—the donor was in a deep coma and the recipient was under chronic dialysis treatment for uremia. Therefore we considered as "hypertensive" those kidneys from a donor with at least one hypertensive parent. Kidneys were assumed to be "normotensive" if both parents were normotensive. Taking into account similar plasma creatinine levels, numbers of rejection crises, and steroid therapy, "normotensive" recipients of "hypertensive" kidneys required significantly more antihypertensive therapy than recipients of "normotensive" kidneys (FIG. 4). This difference in therapy requirement was not seen in "hypertensive" recipients. The phenomenon of genetic coadaptation may be involved in this peculiar interaction between the familial predisposition hypertension of the donor and of the recipient.[27] Whatever the explanation for this is, the renal transplantation and function data suggest that in MHS and in at least some forms of human essential hypertension a genetic abnormality of kidney function has a key role in the pathogenesis of the disease. So far the nature of this renal alteration in humans is unknown. However, because of the similarities between MHS and humans (TABLE 2), an increased Na and water reabsorption through the proximal tubular cells may be proposed. In keeping with this hypothesis are some recent findings[30] showing an increase of proximal tubular reabsorption of lithium in patients with essential hypertension and a negative correlation between erythrocyte Na-Li countertransport and renal lithium clearance in these patients. We have also found similar results measuring erythrocyte function and proximal tubular cell function in humans: the lithium fractional reabsorption of urate and phosphate reabsorption (which can be taken as an index of proximal tubular cell function) are positively correlated to erythrocyte Na-Li countertransport, thus suggesting that the subjects with greater Na-Li countertransport have faster proximal tubular reabsorption (Niutta *et al.* unpublished results).

TABLE 2. Characteristics of Proximal Tubular Cells and Erythrocytes of MHS Rats Compared with the Same Cells of the Control MNS Rats

	MHS and MNS Kidney (Proximal tubular cells)	Erythrocytes
Cell volume	↓	↓
Intracellular Na concentration	↓	↓
Cell membrane Na transport	↑	↑
Ca ATPase (V_{max})	↓	↓

↑, ↓ = higher or lower in MHS.

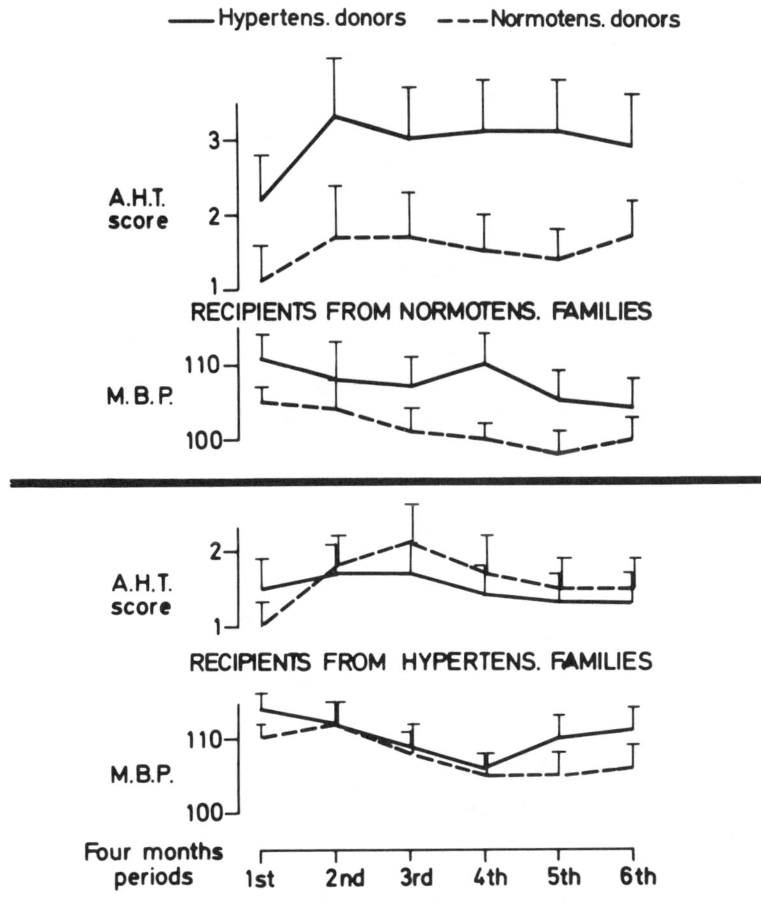

FIGURE 4. Antihypertensive therapy (AHT), expressed in arbitrary scores, and mean blood pressure (MBP) of 50 patients after kidney transplantation. The recipients were subdivided into four groups according to the absence (upper part) or presence (lower part) of hypertension in their own families and in the families of the donors. AHT scores of normotensive recipients of a kidney from a donor with hypertension in his family are statistically different from those of normotensive recipients of a kidney from a donor without hypertension in his family ($p < 0.05$ during the first year and $p < 0.01$ during the second year of follow up). (From Guidi *et al.*[6] With permission from *Nephron.*)

In conclusion, the results so far obtained are consistent with the hypothesis that a genetically determined alteration in cell membrane function may be a cause of genetic hypertension in rats or "essential" hypertension in humans.

REFERENCES

1. McManus, I. C. 1983. Statistics Med. **2:** 253–258
2. Guyton, A. C., T. G. Coleman & H. J. Granger. 1972. Ann. Rev. Physiol. **34:** 13–46.

3. RAPP, J. P. 1983. Hypertension **5**(1): I/198–I/203.
4. FERRARI, P., D. CUSI, B. R. BARBER, C. BARLASSINA, G. VEZZOLI, L. DUZZI, L. MINOTTI & G. BIANCHI. 1982. Clin. Sci. **63:** 61s–64s
5. VEZZOLI, G., A. ELLI, G. TRIPODI, G. BIANCHI & E. CARAFOLI. 1985. J. Hypertension **3:** 645–648.
6. BIANCHI, G., P. FERRARI, D. TRIZIO, M. FERRANDI, L. TORIELLI, B. R. BARBER & E. POLLI. 1985. Hypertension **7:** 319–325.
7. BIANCHI, G. 1986. *In* Handbook of Hypertension (Physiology and Pathophysiology of Hypertension). Elsevier Science Publ. Amsterdam.
8. SALVATI, P., G. P. PINCIROLI & G. BIANCHI. 1984. J. Hypertension **2**(suppl. 3): 351–353.
9. BIANCHI, G., B. R. BAER, U. FOX, L. DUZZI, D. PAGETTI & A. M. GIOVANNETTI. 1975. Circ. Res. 36 & 37 (suppl. I): I-153–I-161.
10. PORCELLI, G., G. BIANCHI & H. R. CROXATTO. 1975. Medicine **149:**: 983–986.
11. BIANCHI, G., U. FOX, G. F. DI FRANCESCO, A. M. GIOVANNETTI & D. PAGETTI. 1974. Clin. Sci. Mol. Med. **47:** 435–448.
12. FOX, U. & G. BIANCHI. 1976. Clin. Exper. Pharm. Physiol. **Suppl. 3:** 71–74.
13. PERSSON, A. E., G. BIANCHI & U. BOBERG. 1984. Acta Physiol. Scand. **122:** 217–219.
14. PERSSON, A. E., G. BIANCHI & U. BOBERG. 1985. Acta Physiol. Scand. **123:** 139–146.
15. BOBERG, U. & A. E. PERSSON. 1986. Am. J. Physiol. **250:** F967–F974.
16. PARENTI, P., G. HANOZET & G. BIANCHI. 1986. Hypertension **8:** 932–939.
17. HANOZET, G., P. PARENTI & P. SALVATI. 1985. Biochim. Biophys. Acta **819:** 179–186.
18. GMAJ, J., H. MURER & G. BIANCHI. (In preparation.)
19. BECK, F., G. BIANCHI, A. DORGE, R. RICK, M. SCHRAMM & K. THURAU. 1983. J. Hypertension **1**(suppl. 2): 38–39
20. FERRARI, P., G. NUSSDORFER, L. TORIELLI, P. SALVATI, M. G. TRIPODI, E. NIUTTA & G. BIANCHI. 1986. Proceedings of Congress on Early Pathogenesis of Biliary Hypertension. Rotterdam. Elsevier Publ. Co.
21. FERRARI, P., L. TORIELLI, M. FERRANDI & G. BIANCHI. 1986 Ann. N.Y. Acad. Sci. (This volume.)
22. BURNHAM, C., S. J. D. KARLISH & P. L. JORGENSEN. Biochim. Biophys. Acta **821:** 461–469.
23. FOSKETT, J. K. & K. R. SPRING. 1985. Am. J. Physiol. **248:** C27–C36.
24. BEALL, P. T., K. J. KARNAKY, L. T. GARRETSON, Y. J. KUO & L. L. SHANBOUR. 1983. Biochim. Biophys. Acta **763:** 19–26.
25. FERRANDI, M., S. SALARDI, R. MODICA & G. BIANCHI. 1986 Ann. N.Y. Acad. Sci. (This volume.)
26. GUIDI, E., G. BIANCHI, E. RIVOLTA, C. PONTICELLI, F. QUARTO DI PALO, E. MINOTTI & E. POLLI. 1985. Nephron **41:** 14–21.
27. PARKIN, D. T. 1979. *In* Introduction to Evolutionary Genetics. Edward Arnold. London.
28. BIANCHI, G., D. CUSI, C. BARLASSINA, G. P. LUPI, P. FERRARI, G. B. PICOTTI, M. GATTI & E. POLLI. 1983. Kidney Int. **23:** 870–875.
29. BIANCHI, G. & C. BARLASSINA. 1983. *In* Hypertension: Physiopathology and Treatment. 2nd edit. J. Genest, O. Kuchel, P. Hamet & M. Cantin, Eds. **4:** 54–73. McGraw Hill Book Company.
30. WEBER, A. B. 1986. N. Engl. J. Med. **314:** 198–201.

DISCUSSION OF THE PAPER

R. GARAY (*Necker Hospital, Paris*): Did you measure ANF in MSH rats?

BIANCHI: No. But we measured ouabain-like factor in plasma (using the technique of Dr. Wardener *et al.*) and we found that this factor is higher in plasma of MSH.

P. MEYER (*Necker Hospital, Paris*): The genetic association of a given characteristic to hypertension does not necessarily imply that the abnormal characteristic is a cause of the disease. Is it thus necessary to study the gene of that abnormality?

BIANCHI: For this reason we are applying a combined approach: physiological (linking the organ dysfunction responsible for hypertension to the protein abnormality) and genetic (linking the protein abnormality to the gene abnormality that is associated to the hypertension).

P. COLEMAN (*New York University, New York, NY*): In addition to differences in membrane protein composition that you have uncovered between cells from hypertensive and normotensive animals, are there correlative differences in the general (or specific) lipid of such cells, that may also correlate with the hypertensive syndrome?

BIANCHI: There are minor differences in lipid composition: PI is slightly lower in MHS and so are some other fatty acids.

Pathophysiology of the Na Exchange and Na-K-Cl Cotransport in Essential Hypertension:

New Findings and Hypotheses

MITZY L. CANESSA

Endocrine-Hypertension Unit
Brigham and Women's Hospital and
Department of Medicine
Harvard Medical School
Boston, Massachusetts 02115

INTRODUCTION

In order to provide a tool for the study of genetic-environmental interactions in the etiology of essential hypertension, we have investigated two gradient-driven transporters in human red cells, the Na exchanger and Na-K-Cl cotransport. At the time we initiated our studies, we knew very little about the kinetic and equilibrium properties of these transporters and about their physiological functions in target cells. The effects of alterations of these two transporters on the pathophysiology of essential hypertension were not understood. However, recent studies on the cellular function of these transporters in the kidney and in vascular and red cells promise to illuminate the problem of the regulation of Na transport by hormonal and vasoactive substances. The present paper aims to discuss recent findings from my laboratory and others regarding the physiology of Na countertransport and Na-K-Cl cotransport in essential hypertension in the context of Blaustein's hypothesis.

GENETIC FACTORS AND Na TRANSPORT

Blaustein's hypothesis proposes that an undefined renal defect in Na excretion initiates enhanced production of a natriuretic hormone that inhibits the Na pump in the kidney and vascular smooth muscle. The hypothesis also considers Na intake the only triggering environmental factor. However, extensive studies by the Framingham group have established that fat intake is an important risk factor in hypertension. The association of elevated Li/Na exchange and hyperlipidermia indicates that lipid metabolism can be a critical pathogenic factor in the cellular functions of membrane Na transporters.

Studies of twins[1] and families[2] have provided information that indicates that the V_{max} of the red cell Na exchanger and the Na-K-Cl cotransport are largely determined by genetic factors. Both transporters play important cellular functions in the kidney and in vascular cells. It is, therefore, possible that a defective Na-K-Cl cotransport, Na/H exchange, or an abnormal number of sites may be determined by genetic factors and influence cellular functions, such as tubular Na reabsorption. From work carried out by Garay *et al.*,[3] there is evidence that a high K_m for internal Na, which activates outward reaction of the Na-K-Cl cotransport, may render it defective in its cellular function. Our studies of the equilibrium properties of this transporter[4] indicate that the system can perform net Na extrusion in cells containing high chloride (such as the vascular smooth muscle) and net ion gain in cells containing low Cl. However, in kidney tubular cells, in which net ion entry is promoted, a high K_m for cell Na may favor Na reabsorption. It has been more difficult to answer the question of whether or not the elevation of the V_{max} of the Li/Na exchange represents an abnormal pathway or an enhanced Na transport, because its cellular function was unknown.

CELLULAR FUNCTIONS OF THE Li/Na EXCHANGE

The elevation of the V_{max} of red cell Li/Na exchange in essential hypertension and pregnancy and its high heritability values[1,2] have raised the question of its cellular function. Several investigators have hypothesized that the Li/Na exchange is a mode of operation of the Na/H exchange. At the time of these speculations there was no evidence that a Na/H exchanger might be present in human red cells. Red cell physiologists hold that cell pH is solely regulated by the Cl/HCO_3 exchanger. Because we knew that the Na-K-Cl cotransport was present, we thought at early stages of our research that the Na/Na exchange might be a mode of functioning of the furosemide-sensitive transporter. However, from our studies[5,6] it became apparent that both systems have very different properties; moreover, a negligible Na/Na exchange was found in high potassium cells.[6]

In experiments designed to test the effect of a rise in cytosolic calcium on the Na/Na exchange, we found that it markedly stimulated Na influx, which could be inhibited by amiloride.[7] Further studies on the effects of proton gradients led us to provide evidence for the presence of a Na_o/H_i exchanger in human red cells[8] that could also transport Li in the inward direction.[9] FIGURE 1 shows different modes of operation of a Na/H exchange. One can imagine that the same transporter can perform Li_o/H_i exchange. It has been shown that the Na/Na exchange can perform Li_i/Na_o or Na_i/Li_o exchange.[5] The sensitivity to amiloride of the Li_o/H_i exchange is ten times lower than that of Na_o/H_i exchange, a property that can be explained by the higher affinity for Li than for Na, as it is for the Na/Na exchanger.[10]

We have also found that both transporters behave highly asymmetric for H^+, Li, and Na ions.[10] Cytosolic protons and Li have stimulatory effects on Na influx by Na_o/H_i and Li_i/Na_o exchange and inhibitory properties on the external side[10]; cell Na inhibits Na/H exchange and stimulates Na/Na exchange. The Li/Na exchange appears to be sensitive to pH gradients; in the absence of pH gradients ($pH_i = pH_o =$ 6), Li/Na and Na/H are inhibited; and outward H^+ gradient ($pH_i = 6$, $pH_o = 8$) stimulates Na_o/H_i and Li_i/Na_o exchange.[9] These experimental data support the conclusion that the Li/Na exchange might be a mode of operation of the Na/H. Further studies on the experimental conditions to study the kinetic properties of the Na/H exchanger may determine whether this transporter is operating abnormally or normally in red cells of patients with essential hypertension.

THE RELATIONSHIP BETWEEN HORMONAL FACTORS, Na-K-Cl COTRANSPORT, AND Li/Na EXCHANGE

Blaustein's hypothesis proposes that the same hormone that produces natriuresis inhibits the Na pump in the kidney to increase Na excretion and, in smooth muscle cells, to increase sensitivity to vasoconstrictive hormones. The intensive work on Na pump inhibitors of the last years indicates that there are several chemical substances acting on this enzyme, such as those circulating in plasma, eliminated in the urine, or present in amniotic fluid. However, all these "ouabain-like" materials inhibit the Na pump equally in red cell, kidney, brain, or vascular cells. There is no evidence yet for tissue specificity of the "natriuretic" action on the kidney, nor of specific vasoconstriction. It is not yet clear if it is a consequence of hypertension or an already functionally abnormal kidney.

The mechanism proposed to increase vasoconstriction is the elevation of cytosolic

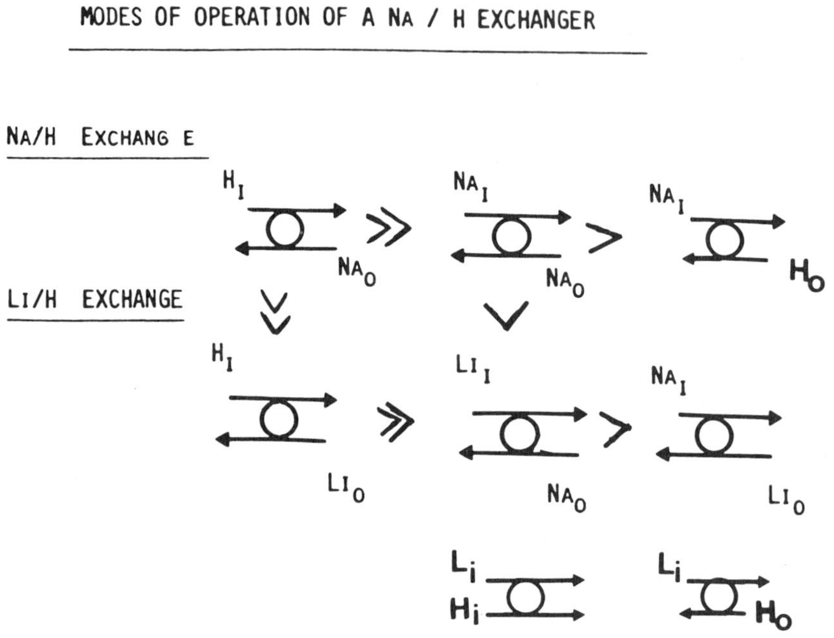

MODES OF OPERATION OF A NA / H EXCHANGER

NA/H EXCHANGE

LI/H EXCHANGE

FIGURE 1. The diagrams explain that a Na_o/H_i exchanger may perform Na_o/Na_i exchange at a lower rate and also Na_i/H_o exchange; this last pathway appears to be significantly lower than Na_o/H_i exchange, because H^+ have asymetric effects on the transporter (activator inside, inhibiting outside). The Na_o/H_i exchange can also transport Li_o in the inward direction but at lower rate than Na_o. The Na_i/Na_o and Na_o/H_i exchangers can also transport Li in the inward or outward direction, but at very lower rates. Li/Na_o exchanger has higher V_{max} than Na_i/Li_o exchange. Alternative routes such as Li_i-H_i cotransport and Li_i/H_o exchange are also considered. It is worthwhile to remember that the kinetic properties of the Na/Na exchangers and the effect of protons in its activity have not yet been studied.

calcium caused by reduced activity of the Na/Ca exchange secondary to reduced Na gradient generated by Na pump inhibition. Na pump inhibition by "ouabain-like" factors will also influence the Na gradient available to the Na/H exchanger to regulate cell pH, a parameter that can also affect contractibility.

The explosion of research on cardiac peptides has shown that the heart releases a potent natriuretic and vasodilatory peptide (ANF) that does not operate by inhibiting the Na pump. Recent work by O'Donnell *et al.*[11] has provided evidence that ANF and its second messenger cGMP are powerful stimulators of the Na-K-Cl cotransport in vascular smooth muscle cells (VSM). We have also reported[12] that another vasodilator peptide, such as bradykinin, also stimulates Na-K-Cl cotransport in vascular endothelial cells. These findings raise the question of whether there are alterations of circulating hormone involved in the regulation of Na transport and/or the transporter activated by this hormone. It would be important to investigate whether or not the abnormal kinetic properties of outward Na-K-Cl cotransport may hamper the physiological action of ANF.

Recent findings in my laboratory also indicate that there might be a pathophysiological link between the activity of the Li/Na exchanger and essential hypertension. We have recently found that Na/H is extremely active in VSM cells.[13] External Na removal produces cytosolic acidification and turns on Na/H exchange. This transporter also appears to be modulated by a vasoactive hormone that transiently increases cytosolic calcium. Na/H is activated by angiotensin II. This peptide mobilizes calcium and stimulates breakdown of phosphatidyl inositol and generates IP_3 diacylglycerol. The AII action takes place for a long period (20 minutes) after cytosolic calcium is already returning to normal values. It would be important to clarify the role of cell pH changes in contractibility. Owen[14] has also shown that the VSM Na/H exchanger is modulated by PDGF, another vasoconstrictor.

On the basis of these observations it is, therefore, possible to postulate that the elevation of the Li/Na exchanger in hypertensive subjects and their adolescent offspring might be an indicator of enhanced Na/H exchanger that is activated by cytosolic calcium signals; a defective Na/H exchanger performing more Na/Na than Na/H exchange; or a Na/H exchange activated by PDGF or other circulating growth factor released by early vascular damage. Since three different studies have shown high correlation of Li/Na exchange with hyperlipidemias, further studies are required to determine the role of abnormal lipids on the cellular functions of Na/H exchange. A distinction should be drawn between the changes of Li/Na exchange in pregnancy and those in essential hypertension.

THE RELATIONSHIP BETWEEN Ca AND Na IONS IN ESSENTIAL HYPERTENSION

Studies indicate that the link between Ca and Na ions is not solely determined by the Na/Ca exchanger as proposed in Blaustein's hypothesis. Calcium ions are powerful modulators of Na gradient transporters of vascular and kidney cells, such as Na/H exchange and Na-K-Cl cotransport. At low values, cytosolic calcium is kept very well buffered by the Ca pump and, at higher levels, by the Na/Ca exchange. We propose that the second messenger action of cytosolic calcium (or others such as cAMP or cGMP) on the regulation of Na transport, vasoactive hormone, and growth factors in the kidney and vascular membrane is likely to play an important pathogenic role in essential hypertension. The human red cell appears to be recording in its gradient-driven transporters these second messenger signals in hypertensive subjects and their offspring. It is at this level that interactions between genetic factors (determining kinetic properties of gradient-driven transporters) and hormonal and vasoactive peptides may occur to regulate blood pressure and renal Na excretion at a given level of Na or fat intake. Genetic factors might also be involved in determining the kinetic parameters (K_m and V_{max}) of the transporters or in controlling the enzymes involved in the modulation of the activity of the transporters by vasoactive substances in VSM.

To complete our picture that gradient-driven transporters are modulated by second messengers action of vasoactive peptides and growth factors, it would be important to address the question of the modulation of Na/Ca exchanger in vascular smooth muscle as it has been done in the heart.

REFERENCES

1. LEWITTER, F. & M. CANESSA. 1984. Am. J. Human Gen. **36:** 172s.
2. DADONE, M., S. J. HASSTEDT, S. C. HUNT, J. B. SMITH, K. O. ASH & R. R. WILLIAMS. 1984. Am. J. Med. Gen. **17:** 565.
3. GARAY, R. P., C. NAZARET, P. HANNAERT & M. PRICE. 1983. Eur. J. Clin. Inv. **13:** 311–320.
4. BRUGNARA, C., M. CANESSA, D. CUSI & D. C. TOSTESON. 1986. J. Gen. Physiol. **87:** 91–112.
5. CANESSA, M., I. BIZE, N. ADRAGNA & D. C. TOSTESON. 1982. J. Gen. Physiol. **80:** 149–168.
6. CANESSA, M., C. BRUGNARA, D. CUSI & D. C. TOSTESON. 1986. J. Gen. Physiol. **87:** 113–142.
7. ESCOBALES, N. & M. CANESSA. 1985. J. Biol. Chem. **260:** 11914–11923.
8. ESCOBALES, N. & M. CANESSA. 1986. J. Membr. Biol. **89:** 21–28.
9. CANESSA, M., A. SPALVINS & N. ESCOBALES. 1986. Biophys. J. **69:** 141a.
10. CANESSA, M. & A. SPALVINS. 1987. Biophys. J. (in press.)
11. O'DONNELL, C. & N. E. OWEN. 1986. Fed. Proc. **45:** 653.
12. BROCK, T., C. BRUGNARA, M. CANESSA & M. GIMBRONE. 1986. Am. J. Physiol. **250:** C888–895.
13. VALLEGA, G., M. CANESSA, B. C. BERK, T. A. BROCK, M. A. GIMBRONE & R. W. ALEXANDER. 1986. J. Gen. Physiol. **88:** 60a.
14. OWEN, N. E. 1984. Am. J. Physiol. **247:** C-501–505.

DISCUSSION OF THE PAPER

M. BLAUSTEIN (*University of Maryland, Baltimore, MD*): Like vascular smooth muscle, cardiac muscle has parallel Ca-ATPase, Ca channels, Na/H exchange, and Na/Ca exchange. Why should removal of external Na, which produces the same effects on contraction in cardiac and vascular smooth muscle, be attributed to different mechanisms in the two tissues?

CANESSA: Heart muscle has many differences in the activity ratio of the Na^+-Ca^+ exchange, Na^+/H^+ and Na pump than VSM. I wonder whether the Na removal has an effect not only on cytosolic Ca but on cellular pH and hence in contractility.

S. LUCIANI (*University of Padova, Padova*): I notice that you have been using amiloride at millimolar concentrations for inhibiting Na^+/H^+ exchange in erythrocytes. The normal concentration of amiloride to inhibit the Na^+/H^+ exchange is in the micromolar range.

CANESSA: We have titrated with amiloride Na^+ influx in vascular smooth muscle cell lines (rat). Its IC_{50} is 10–20 μM amiloride. In human red cells, the Ca^{2+}-activated Na^+ influx also has an IC_{50} (100 mM Na^+) of 20–30 μM amiloride; however, when the Na^+/H^+ exchange is stimulated by an outward H^+ gradient the IC_{50} for amiloride (100 mM Na) is 0.4 mM. Amiloride is a competitive inhibitor of Na^+ ions these data indicate that there are various Na^+/H^+ exchangers with different K_m for Na^+ ions.

Quantitative Histochemical Approaches Are Essential for Investigating Fundamental Questions of Diabetes Research

FRANZ M. MATSCHINSKY, FRANCISCO BEDOYA,[b]
LESLIE MACGREGOR, TAKAO SHIMIZU,
AND JEANNE WILSON

Diabetes Center
Department of Biochemistry and Biophysics
School of Medicine
University of Pennsylvania
Philadelphia, Pennsylvania 19104

[b]*Departamento de Bioquimica*
Facultad de Medicina
Universidad de Sevilla
Avda. Sanchez Pizjuan
4 41009 Sevilla, Spain

INTRODUCTION

For many years we have used quantitative histochemical approaches to study important diabetes-related research questions. Quantitative histochemistry has proven such a powerful tool because it provides a unique way to take into account the cellular heterogeneity of tissues and organs involved in the pathogenesis of diabetes mellitus. In this paper we briefly describe a few selected examples of this work to illustrate the capacity of quantitative histochemistry. The method is based on sampling of microscopic specimens from freeze-dried cryostat sections of quick frozen tissues followed by weighing of the sample and ultrasensitive, specific chemical analysis.[1] The method has been applied for studying the hormone-producing pancreatic islet cells (e.g. pure B cells), the tissues that are the target of the islet cell hormones (e.g. the liver acinus) and tissues that are affected by the pathologies of diabetic complications (e.g. the layers of the retina).

QUANTITATIVE HISTOCHEMISTRY OF GLUCOSE PHOSPHORYLATION IN PANCREATIC ISLET CELLS

Quantitative histochemistry methods have been particularly useful in studies of the biochemical design features of pancreatic islet cells. Only highlights of these studies are discussed here and the work from our laboratory is emphasized for reasons of brevity.

We proposed as early as 1968 that the enzyme glucokinase (ATP-1-glucose 6-phosphotransferase, Ec 2.7.1.1) might serve as glucose sensor of the pancreatic B-cells.[2] Measurements with quantitative histochemical methods led to this proposal.

These early observations were later confirmed and expanded by a series of studies using large batches (i.e. mg amounts) of isolated pancreatic islets obtained by collagenase digestion of the pancreas and by using gram quantities of islet cell tumor tissue.[3] It was found that islet cell glucokinase is kinetically and chromatographically indistinguishable from liver glucokinase. The two preferred substrates for the enzyme are glucose and mannose. The K_ms for glucose and mannose are approximately 8 mM and 16 mM, respectively. It was discovered that glucokinase prefers the α-anomer of the two hexoses over the β-anomer. We believe that the ability of glucokinase to discriminate between α and β anomers of glucose and mannose explains the greater potencies of the α-anomers of glucose and mannose to stimulate insulin release. Our latest studies focused on measuring glucokinase in human pancreatic islet tissue to explore whether the glucokinase glucose sensor idea might have general applicability.[4] To accomplish this we have designed a new radiometric oil well assay for glucokinase.[5] The assay is sufficiently sensitive to detect as little as 1 pg glucokinase and allowed us to reliably measure important kinetic constants of glucokinase (i.e. the K_m and Hill

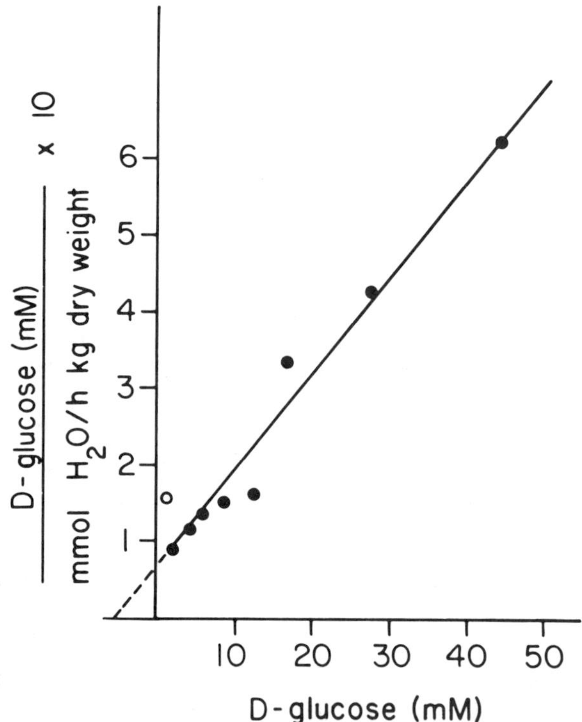

FIGURE 1. Hanes-Woolf plot of the glucose dependency of glucokinase in microdissected pieces from human pancreatic islets. Freeze-dried pieces of human islets weighing 0.05–0.150 μg were used. The assays at each glucose concentration were performed in quintuplicate and the results were corrected for temperature (37 °C) assuming a Q_{10} of 2.1. Glucokinase activity was corrected for the hexokinase component. The velocity at 0.9 mM glucose (O) was not considered when computing kinetic parameters for glucokinase because it was considered indicative of deviation from Michaelis-Menten behavior. (From Bedoya *et al.*[4] With permission from *Diabetes*.)

TABLE 1. Differential Regulation of Glucokinase Activity in Pancreatic Islets and Liver of the Rat

Experimental Conditions	Blood Plasma Chemistry glucose (mg%)	insulin (ng/ml)	Islet Glucokinase (mmol/kg dry tissue × hr 37°C)	Liver Glucokinase (μmol/g wet tissue × min 30°C)
Control	139 ± 15 (3)	2.0 ± 0.3 (3)	135.5 ± 12.7 (5)	1.52 ± 0.25 (5)
Insulinoma bearing rat	20 ± 2 (9)	9.9 ± 0.9 (9)	38.1 ± 4.1 (7)	5.61 ± 0.31 (4)
1 day after resection of insulinoma	340 ± 33 (7)	0.4 ± 0.1 (6)	120.0 ± 12.1 (6)	3.11 ± 0.41 (4)
6 days after resection of insulinoma	125 ± 2 (6)	1.5 ± 0.1 (6)	160.3 ± 25.0 (3)	1.50 ± 0.42 (4)
Insulinoma bearing rat after 24 hr of glucose infusion	238 ± 39 (3)	12.1 ± 1.8 (3)	72.6 ± 6.3 (3)	4.8 ± 0.73 (4)

number) in microscopic samples of tissue. We found that glucokinase is a regular constituent of human islet tissue. The measured capacity of human islet glucokinase to phosphorylate glucose (i.e. 63.5 ± 0.6 mmol/kg dry tissue × hour at 37 °C) was equal to the rate of glucose use measured independently with intact human islets.[6] The apparent K_m for glucose was about 5.0 mM, both in the phosphorylation assay with freeze-dried islet samples (FIG. 1) and in the glucose use studies with intact islets. The radiometric oil well method is sufficiently sensitive and specific to be of potential benefit in future studies of glucokinase in pancreatic islet tissue of normal and diabetic subjects.

The micromethod was also useful in investigations of islet tissue from hypoglycemic insulinoma-bearing rats and from streptozotocin-treated diabetic rats.[7] These studies revealed that the measurable activities of islet glucokinase and liver glucokinase are regulated in a differential manner (TABLE 1). Islet glucokinase is decreased by a combination of hypoglycemia and hyperinsulinemia and is induced by hyperglycemia combined with hypoinsulinemia. Liver glucokinase is influenced in an opposite manner. These findings raise important questions about the molecular biology of glucokinase in liver and islet cells.

The literature on islet glucokinase implies that the enzyme is preferentially located in the β cells.[2–5] We have used quantitative histochemistry approaches and the radiometric oil well method to explore whether this is true, using for this purpose β-cell–depleted or β-cell–enriched pancreatic islet.[8] It was found (TABLES 2 and 3) that β-cell–depleted islets from diabetic rats contain glucokinase activities comparable to glucokinase activities in whole islet samples or trimmed islet cores (i.e. β-cell–enriched islets). The data presented in TABLE 3 clearly show that A and/or D cells contain glucokinase. Yet it remains unsettled whether the enzyme is a constituent of normal A and/or D cells or was induced by hyperglycemia. These questions are amenable to study with the present methodology.

This brief review shows that quantitative histochemistry has facilitated crucial research of pancreatic islet glucose metabolism and that it might be the method of choice for important future investigation.

TABLE 2. Glucokinase Activity and Hormone Content of Cores and Mantle Trimmings of Freeze-Dried Islet Tissue from Normal Rats

	Glucokinase Activity V_{max} (37°C) (mmol/kg dry weight/hr)	Insulin (ng/μg dry weight)	Glucagon
A. Whole islet sections	134.2 ± 4.3 (5)[a,b]	94.7 ± 4.7 (3)	2.5 ± 0.3 (3)
B. Trimmed islet cores	125.6 ± 10.7 (4)	130.0 ± 10.8 (5)	0.4 ± 0.2 (4)
C. Islet mantle trimmings	59.1 ± 11.6 (4)	39.4 ± 3.6 (5)	3.6 ± 0.7 (5)
Statistical treatment of data			
A versus B	N.S.[c]	0.005	0.001
A versus C	0.001	0.005	N.S.
B versus C	0.005	0.001	0.001

[a]All values represent the means ± S.E. with the number of animals in parentheses.

[b]Hexokinase activities measured with 0.5 mM glucose present but resistant to inhibition by 10 mM G6P were (in terms of mmol/Kg dry weight/hr, 37°C) 36.1 ± 4.8 (5) for the group A, 43.3 ± 2.8 (4) for the group B, and 35.3 ± 5.3 (4) for the group C.

[c]N.S., not significant.

TABLE 3. Glucokinase Activity and Hormone Content of Microdissected Freeze-Dried Islet Samples from the Pancreas of Normal as well as Treated and Untreated Diabetic Animals

Condition	Glucokinase Activity V_{max} (mmol/kg dry weight/hr, 37°C)	Insulin	Glucagon (ng/μg dry weight)	Somatostatin
A. Controls	134.1 ± 7.3 (5)[a]	113.8 ± 11.8 (5)	2.4 ± 0.1 (5)	0.25 ± 0.06 (5)
B. Untreated Diabetics	143.1 ± 13.6 (5)	3.8 ± 1.5 (5)	12.5 ± 1.1 (5)	3.28 ± 1.03 (5)
C. Treated Diabetics	124.4 ± 10.7 (5)	1.9 ± 0.5 (5)	16.1 ± 3.7 (5)	5.10 ± 0.70 (5)
Statistical treatment of data				
A versus B	N.S.[b]	0.001	0.001	0.001
A versus C	N.S.	0.001	0.001	0.001
B versus C	N.S.	N.S.	N.S.	0.05

[a]All values represent the means ± S.E. with the number of animals in parentheses.
[b]N.S., not significant.

QUANTITATIVE HISTOCHEMISTRY OF GLUCOSE PHOSPHORYLATION IN THE HEPATIC ACINUS

The liver plays an active role in glucose homeostasis and hepatic glucose metabolism.[9] An important aspect of investigating the biochemical basis of normal and abnormal hepatic metabolism is the microscopic diversity of the organ, in particular the liver's organization into functional and anatomical lobular units (FIGURE 2).[9] This lobular architecture of the liver constitutes a largely unmet challenge to the biochemist.[11]

We have used quantitative histochemistry sampling procedures and oil well methods for glucose phosphorylation enzymes and for glucose-6-phosphatase to explore the interrelationship between hepatic microarchitecture and glucose metabolism. We discovered that the high K_m enzyme glucokinase has an intralobular distribution quite distinct from the distribution of the low K_m glucose phosphorylation enzymes.[12] Using the Rappaport model of the hepatic acinus as a guide[10] we observed a glucokinase gradient with activities rising from zone 1 to zone 3 of the acinus and a hexokinase gradient in the opposite direction. We are currently developing a radiometric oil well method for measuring glucose-6-phosphatase in lobular structures to complement our studies of glucose phosphorylation. In that new assay ^{14}C-labeled glucose-6-P serves as substrate. After completion of the reaction, substrate and product are separated on an ion exchange column and the amount of free ^{14}C-glucose is

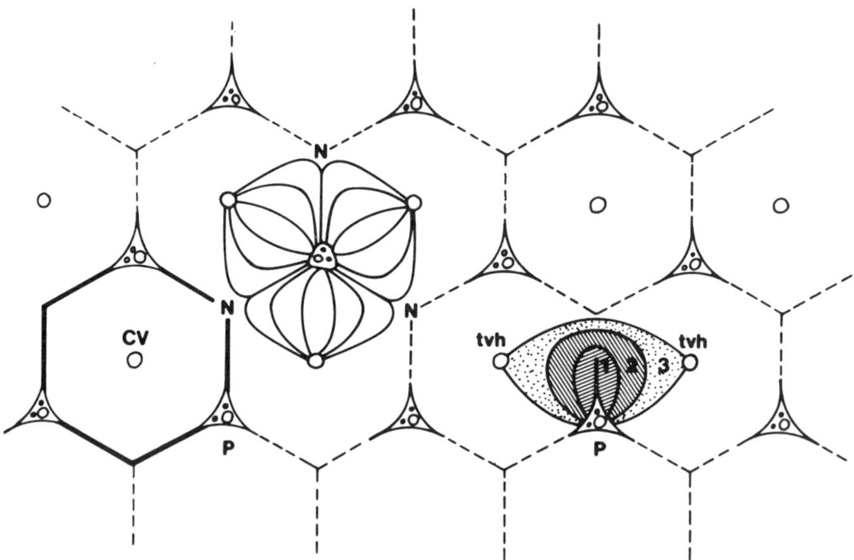

FIGURE 2. Schematic drawing of the three main liver units. *Left*: Hexagonal liver lobule, surrounded by Glisson's portal triads (P) and drained by the central vein (CV). *Middle:* Portal unit, the center of which is formed by the triad. The lines point to the more or less bent sinusoids, the periphery of the portal unit is marked by central veins or by nodal points (N). *Right:* Liver acinus, the center of which is formed by the terminal afferent vessels and the periphery of which is drained by terminal afferent venules (tvh). There are three different acinar zones: the periportal zone (1), the intermediate zone (2), and the perivenous zone (3).

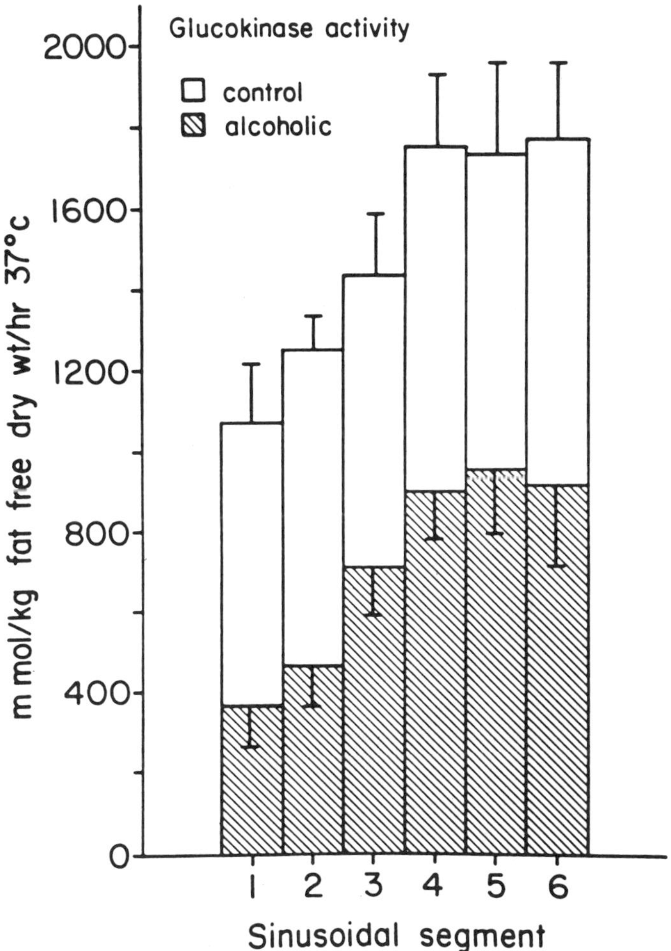

FIGURE 3. Acinar distribution of glucokinase in the liver of normal and alcohol fed rats. Segment 1 stems from the acinar zone 1 and segment 6 from the acinar zone 3.

measured by liquid scintillation counting. This new procedure promises to be more convenient than the fluorimetric assay now employed by quantitative histochemists.[13]

We have begun to use this approach to study the hepatic glucose homeostasis in rats fed an artificial alcohol-enriched diet (FIGURE 3). We found that chronic ethanol intake decreased total hepatic glucokinase and that the enzyme gradient was steepened, i.e. the periportal hepatocytes showed a relatively larger decline of the enzyme than the perivenous hepatocytes. With the limited data now available, it is difficult to fathom the possible functional significance of the finding. A comprehensive picture of lobular patterns of enzymes involved in glycolysis and gluconeogenesis needs to be composed before we comprehend what chronic ethanol ingestion might do to carbohydrate metabolism of the liver. It is however clear from the illustrative example,

TABLE 4. Comparison of Pathologic Changes in Nerve and Retina of Diabetic Animals

Nerve	Retina
↓ Axonal transport velocity[a]	—
↑ Glucose concentration	↑ Glucose concentration
↑ Sorbitol concentration	↑ Sorbitol concentration
↓ *myo*-inositol concentration[a]	↓ myoinositol concentration
↑ Na^+-K^+-ATPase activity[a]	↑ Na^+-K^+-ATPase activity
↑ Nerve Na^+ (indirectly measured)	↑ Retinal Na^+
↓ Conduction velocity[a]	↓ Electroretinogram c-wave[a]

[a]Changes that are normalized by treatment with aldose reductose inhibitor or with myoinositol supplementation.

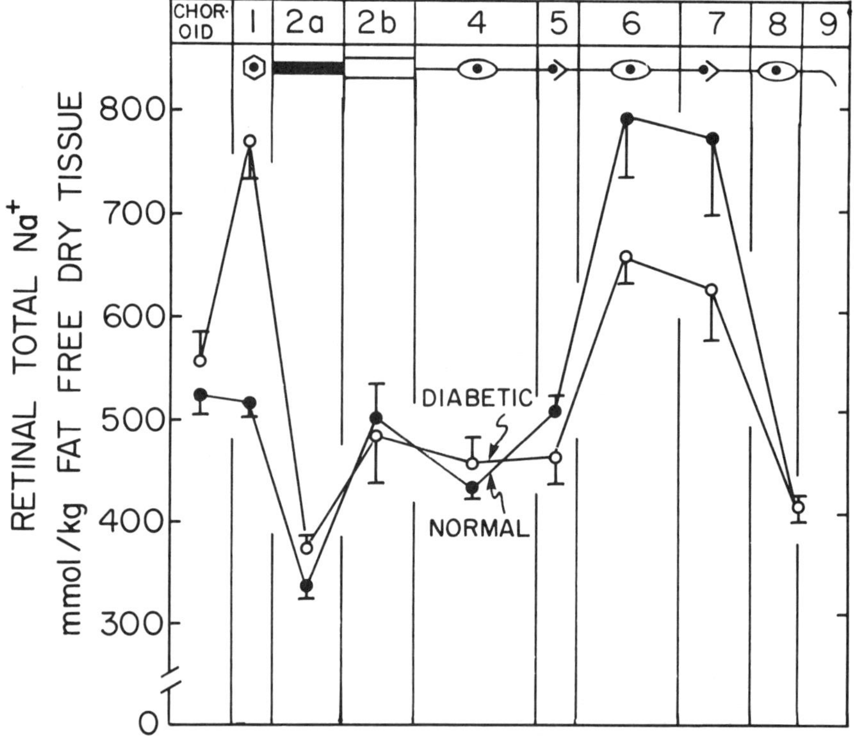

FIGURE 4. Total sodium of retinas from normal and from diabetic rabbits. Samples were microdissected from lyophilized sections of eyes from normal and from alloxan-diabetic rabbits. For each retinal layer, nine samples were analyzed. Nine normal eyes and eight eyes from diabetics were incorporated. Sodium was measured using atomic absorption with a carbon rod atomizer. R.P.E., retinal pigment epithelium, O.P., outer plexiform layer, I.P., inner plexiform layer. (From MacGregor & Matschinsky.[22] With permission from *Journal of Clinical Investigation.*)

that quantitative histochemistry provides an invaluable tool to study the complexities of hepatic carbohydrate metabolism in metabolic diseases as for instance diabetes or alcoholism.

QUANTITATIVE HISTOCHEMISTRY APPROACHES IN STUDYING THE HISTOCHEMICAL BASIS OF DIABETIC COMPLICATIONS

Quantitative histochemistry has also been an extremely useful approach in exploring the biochemical derangements in tissues affected by diabetic complications.

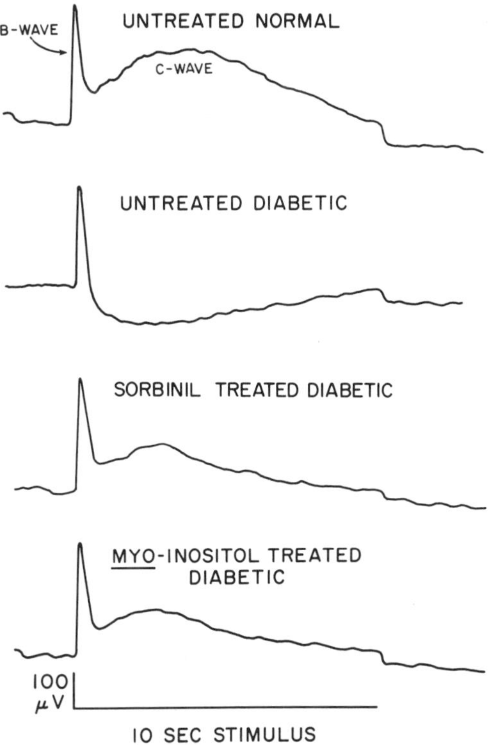

FIGURE 5. Representative ERGs of rats 6 wk after induction of diabetes. The ERG patterns of all nondiabetic rats were similar, regardless of treatment. (From MacGregor & Matschinsky.[22] With permission from *Journal of Clinical Investigation.*)

The heuristically most influential hypotheses explaining the pathogenesis of diabetic complications are undoubtedly the polyol accumulation hypothesis developed in the laboratories of van Heyningen[14] and Kinoshita[15] and the myoinositol depletion hypothesis developed by Winegrad and his associates.[16] Recently a unifying concept is evolving that combines these two hypotheses.[17] This evolving hypothesis implies that the pathologies of diabetic complications might develop in cells and tissues characterized by insulin-independent high capacity glucose transport systems and by the presence of an active polyol pathway. High levels of intracellular glucose resulting from diabetes enhance flux through the polyol pathway usually manifested by an accumulation of sorbitol and fructose. In ways not understood, this enhanced polyol pathway activity is thought to deplete cellular myoinositol. Myoinositol depletion is believed to compromise phospholipid metabolism and transmembranous ion gradients. It remains to be explored in detail how the outlined metabolic abnormalities produce the pathologic lesions in eyes, kidneys, and peripheral nerves.

One of the difficulties confronting an investigator exploring the biochemical processes that lead to diabetic complications is the high degree of cellular complexity of the tissues involved.[18]

We have developed quantitative histochemistry methodology to study several crucial parameters of the metabolic pathways involved in the pathogenesis of complications.[19–21] As a result of this development microassays are now available for glucose, sucrose, sorbitol, myoinositol, Na^+, K^+, and Na^+K^+ ATPase. These assay have been applied to a layer-by-layer analysis of the retina from normal and diabetic rabbits, paying particular attention to the retinal pigment epithelium. The retinal pigment epithelium is known to be affected by diabetes as indicated by altered morphological and electrophysiological parameters.[22] The studies provide the first set of biochemical data from a homogeneous cell population affected by diabetic hyperglycemia.

A summary of the results in TABLE 4 shows that the alterations and the cellular monolayer of the retinal pigment epithelium are comparable to the changes observed in the composite structure of the sciatic nerve. The biochemical data for the retinal pigment epithelium are unambiguously confined to one cell population whereas it is not clear which cell types are affected in the nerve. Probably, the most impressive result is apparent in the transretinal sodium profile, which shows a circumscribed accumulation of Na^+ in the retinal pigment epithelium (FIGURE 4).

It is not unreasonable to assume that the reported disappearance of the C-wave of the ERG is a manifestation of the Na^+ accumulation in the retinal pigment epithelium.[22] It was observed that myoinositol feeding or sorbinil therapy arrest the deterioration of the electroretinogram of diabetic rats (FIGURE 5).[22]

Quantitative histochemistry is the method of choice to test whether myoinositol feeding and/or sorbinil treatment have the predicted effect on the sorbitol, myoinositol, and Na^+ contents of the retinal pigment epithelium monolayer of diabetic animals.

CONCLUSION

It should be apparent from the examples selected here that quantitative histochemistry provides unique opportunities to vigorously test several attractive concepts that strongly influence present day diabetes research. Such concepts are often developed with the help of simplifying *in vitro* system raising the concern whether extrapolation to the complex *in vivo* state is warranted. With the help of quantitative histochemistry the investigative cycle can often be closed as illustrated here by studies with human pancreatic islet tissue and with the retinal pigment epithelium.

REFERENCES

1. LOWRY O. H. & J. V. PASSONNEAU. 1972. A Flexible System of Enzymatic Analysis. Academic Press. New York.

2. MATSCHINSKY, F. M. & J. E. ELLERMAN. 1968. Metabolism of glucose in the islets of Langerhans. J. Biol. Chem. **243:** 2730–2736.
3. MEGLASSON, M. D. & F. M. MATSCHINSKY. 1984. New perspectives on pancreatic islet glucokinase. Am. J. Physiol. **246:** E1–E13.
4. BEDOYA, F. J., J. M. WILSON, A. K. GHOSH, D. FINEGOLD & F. M. MATSCHINSKY. 1986. The glucokinase glucose sensor in human pancreatic islet tissue. Diabetes **35:** 61–67.
5. BEDOYA, F. J., M. D. MEGLASSON, J. M. WISLON & F. M. MATSCHINSKY. 1985. Radiometric oil well assay for glucokinase in microscopic structures. Anal. Biochem. **144:** 504–513.
6. HARRISON, D. E., M. R. CHRISTIE & D. W. R. GRAY. 1985. Properties of isolated human islets of Langerhans: insulin secretion, glucose oxidation and protein phosphorylation. Diabetes **28:** 99–103.
7. BEDOYA, F. J., F. M. MATSCHINSKY, T. SHIMIZU, J. J. O'NEIL & M. C. APPEL. 1986. Differential regulation of glucokinase activity in pancreatic islets and liver of rat. J. Biol. Chem. **261:** 10760–10764.
8. BEDOYA, F. J., J. C. OBERHOLTZER & F. M. MATSCHINSKY. 1986. Glucokinase is found in B-cell depleted islets of Langerhans. J. Histochem. Cytochem. (In press.)
9. JUNGERMANN, K. & N. KATZ. 1986. Metabolism of carbohydrates. *In* Regulation of Hepatic Metabolism. R. G. Thurman, F. C. Kauffman & K. Jungermann, Eds.: 211–235. Plenum Press. New York.
10. SASSE, D. 1986. Liver structure and innervation. *In* Regulation of Hepatic Metabolism. R. G. Thurman, F. C. Kauffman & K. Jungermann, Eds.:3–25. Plenum Press. New York.
11. KAUFFMAN, F. C. & F. M. MATSCHINSKY. 1986. Quantitative histochemical measurements within sublobular zones of the liver lobules. *In* Regulation of Hepatic Metabolism. R. G. Thurman, F. C. Kauffman & K. Jungermann, Eds.: 119–136. Plenum Press. New York.
12. TRUS, M., M. ZAWALICH, D. GAYNOR & F. M. MATSCHINSKY. 1980. Hexokinase and glucokinase distribution in the liver lobules. J. Histochem. Cytochem. **28:** 579–581.
13. TEUTSCH, H. F. 1978. Quantitative determination of G-6-Pase activity in histochemical defined zones of the liver acinus. Histochemistry **58:** 281–288.
14. VAN HEYNINGEN R. 1959. Formation of polyols by the lens of the rat with sugar cataract. Nature **184:** 194–196.
15. KINOSHITA, J. H., S. FUTTERMAN, K. SATOH, *et al.* 1963. Factors affecting the formation of sugar alcohols in the ocular lens. Biochim. Biophys. Acta **74:** 340–350.
16. GREENE, D. A., P. V. DEJESUS, JR. & A. I. WINEGRAD. 1975. Effects of insulin and dietary myoinositol on impaired peripheral motor nerve conduction velocity in acute streptopotoein diabetes. J. Clin. Invest. **55:** 1326–1336.
17. GREENE, D. A. 1986. A sodium-pump defect in diabeteic peripheral nerve corrected by sorbinil administration: relationship to myoinositol metabolism and nerve conduction slowing. Metabolism **35**(suppl. 1): 60–65.
18. MACGREGOR, L. C. & F. M. MATSCHINSKY. 1986. Experimental diabetes mellitus impairs the function of the retinal pigment epithelium. Metabolism **35**(suppl. 1): 28–34.
19. MACGREGOR, L. C. & F. M. MATSCHINSKY. 1984. An enzymatic fluorimetric assay for myoinositol. Anal. Biochem. **141:** 382–389.
20. MACGREGOR, L. C., L. R. ROSECAN, A. M. LATIES & F. M. MATSCHINSKY. 1986. Altered retinal metabolism in diabetes. I. Microanalysis of lipid, glucose, sorbitol and myoinositol in the choroid and in the individual layers of the rabbit retina. J. Biol. Chem. **261:** 4046–4051.
21. MACGREGOR, L. C. & F. M. MATSCHINSKY. 1986. Altered retinal metabolism in diabetes. II. Measurement of sodium potassium ATPase and total sodium and potassium in individual retinal layers. J. Biol. Chem. **261:** 4052–4058.
22. MACGREGOR, L. C. & F. M. MATSCHINSKY. 1985. Treatment with aldose reductose inhibitor or with myoinositol arrests deterioration of the electroretinogram of diabetic rats. J. Clin. Invest. **76:** 887–889.

The Morphology of Proinsulin Processing[a]

L. ORCI

Institute of Histology and Embryology
University of Geneva Medical School
1211 Geneva 4, Switzerland

ULTRASTRUCTURAL AND BIOCHEMICAL DETECTION OF (PRO)INSULIN POLYPEPTIDES

Insulin starts its intracellular journey when synthesized as preproinsulin on the rough endoplasmic reticulum (RER) of the pancreatic B cell.[1] The co-translational removal of the pre(signal) sequence yields proinsulin in the cavities of the RER. Proinsulin is next transferred to the Golgi apparatus via vesicles issued from transitional elements of the RER. These vesicles fuse with the RER-related or *cis* Golgi pole while secretory granules are released at the *trans* pole (FIG. 1).[2–4] The Golgi complex has been implicated in the initiation of the proteolytic maturation of proinsulin to the definitive hormone.[2,5–7] By the integration of biochemical and morphological techniques a precise picture of proinsulin processing is now emerging.

Electron microscope autoradiography allows the ultrastructural localization of radioactively labeled (pro)insulin polypeptides in the B cell at various time points of a chase after a 5-min pulse of [^{3}H]leucine. At the end of pulse, the autoradiographic reaction is localized predominantly over the cisternae of the RER. Ten min later, the radioactive front reaches the cisternae of the Golgi apparatus (the Golgi stack), and after 25 min of chase, a specific population of secretory granules in the Golgi area becomes preferentially labeled. The limiting membrane of these granules is decorated by protein bristles (a coat) on its outer aspect and they originate from the pinching off of the dilated extremities of cisternae at the *trans* Golgi pole, which are similarly coated (FIG. 2). From fifty-five min post-pulse on and until the end of the observation period, the radioactive front predominates over secretory granules that are devoid of coats (FIG. 2). The quantitative analysis of the autoradiographic reaction during the pulse-chase experiment is shown in FIGURE 3. The pattern of autoradiographic labeling of coated and non-coated granules appears superposable to the temporal profile of radioactivity in proinsulin and insulin immunoreactive fractions and shows inversion of the proinsulin/insulin ratio (FIG. 4). This pattern suggests a critical role for the coated granules in conversion. By immunocytochemistry, the coat on coated granules and coated *trans* Golgi cisternae is shown to contain clathrin (FIG. 5).

Attempts towards a more precise demarcation of the subcellular site of proinsulin conversion were carried out by two strategies: one consisted of perturbing the function (and structure) of the Golgi with the ionophore monensin[8]; another was to render the proinsulin molecule resistant to proteolytic cleavage by substituting the amino acids arginine and lysine with their respective analogues canavanine and thialysine.[9,10] Arginine in position 31–32 and arginine and lysine in position 64–65 of the proinsulin molecule are recognized and removed by the converting enzyme(s) to release insulin and C-peptide. After the usual pulse-chase labeling of isolated islets with [^{3}H]leucine, the subcellular localization of ^{3}H-labeled (pro)insulin polypeptides was determined by

[a]Supported by the Swiss National Science Foundation grant 3.404.86.

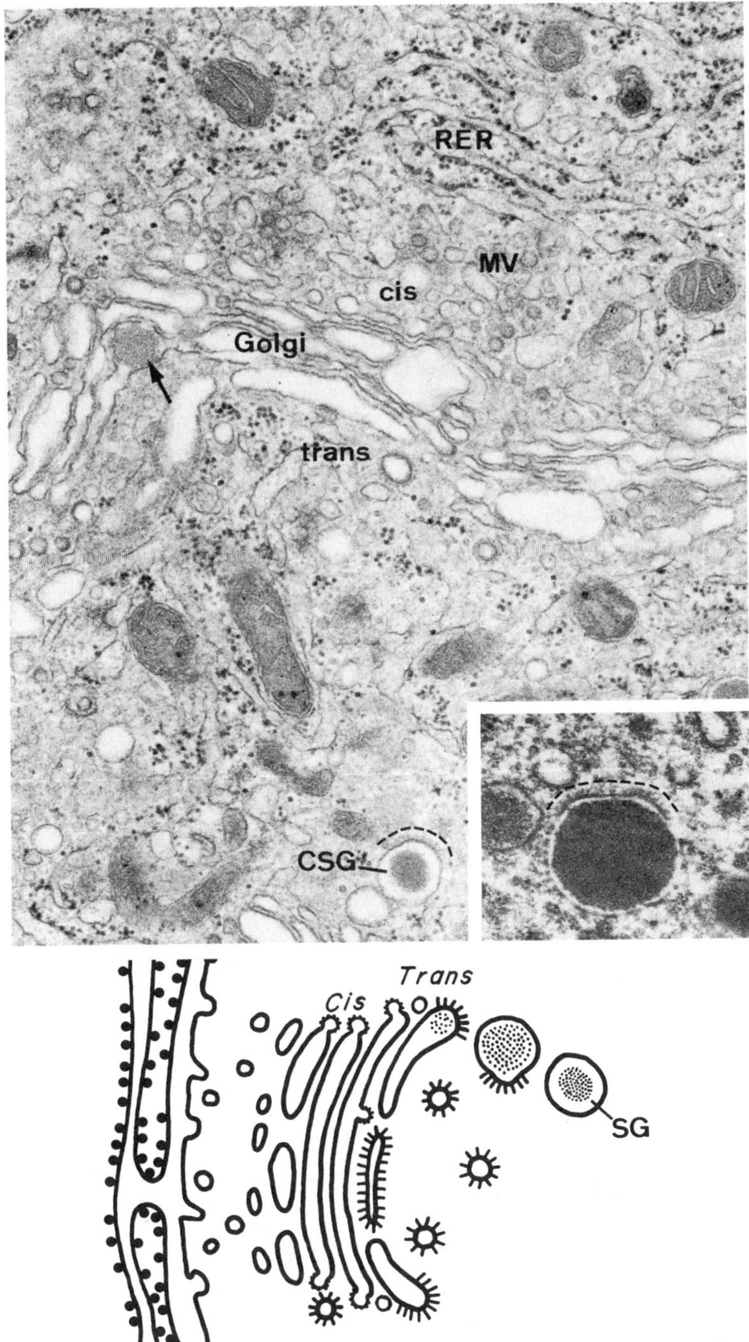

FIGURE 1. Thin section of B cell cytoplasm showing an overview of the compartments involved in (pro)insulin processing. RER, rough endoplasmic reticulum; MV, transfer microvesicles; Golgi, Golgi stack with one cisterna at the *trans* pole showing condensing secretory material (black arrow); CSG, coated secretory granule freshly released from a *trans* cisterna. A coated granule is shown at higher magnification in the inset. The coat is indicated by the dotted line. Non-coated mature secretory granules are not visible on this field but are indicated (SG) in the schematic drawing of the compartments (see also FIG. 2B). × 30,000 and inset, × 36,000.

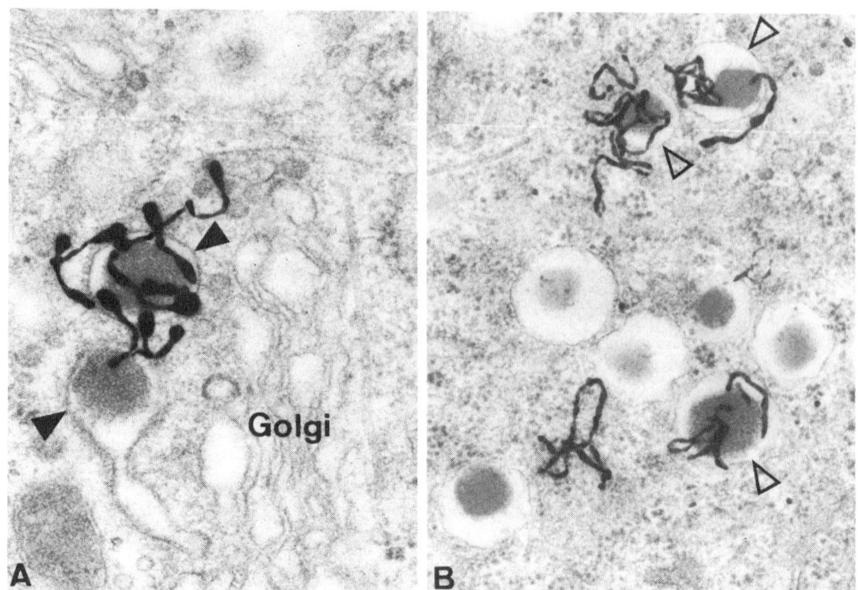

FIGURE 2. (A) and (B) Autoradiographs of B cells during a pulse-chase experiment with [^{3}H]leucine. (A) Example of autoradiographic labeling of the expanded end of a Golgi cisterna with condensing secreting material and of a neighboring secretory granule (black arrowheads). The labeled cisterna and secretory granule show a coat on part of their limiting membranes (coated cisternae and coated granule). (B) Example of autoradiographic labeling of dense-core mature secretory granules outside the Golgi area (white arrowheads). The labeled granules are devoid of coat (non-coated granules). See FIGURE 3 for the quantitative evaluation of the autoradiographic labeling during the pulse-chase experiment. (A) × 45,000 and (B) × 28,000.

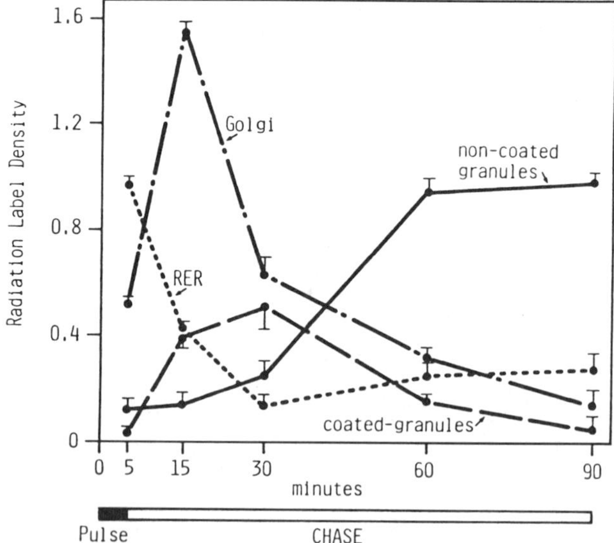

FIGURE 3. Pattern of autoradiographic reaction in B cell compartments involved in (pro)insulin processing during a pulse-chase labeling with [^{3}H]leucine. The radiation label density expresses the number of autoradiographic grains in each compartment per unit cell volume. RER = rough endoplasmic reticulum.

autoradiography and their cellular and extracellular (secreted) levels assessed by gel chromatography and immunoassay.

In the presence of monensin (50 nM), we found that the normal pattern of transfer of radioactivity from the RER to the non-coated secretory granules was perturbed: radioactivity accumulated during the chase period in the dilated extremities of clathrin-coated *trans* Golgi cisternae and in dilated clathrin-coated secretory granules. At the same time, non-coated mature secretory granules showed a low level of radioactivity as compared to controls. At the end of the 85 min chase period under monensin treatment, only 35% of the immunoprecipitable radioactive products were recovered in the form of insulin as compared to 82% in controls.[11]

When amino acid analogues replaced arginine and lysine in the newly synthesized proinsulin molecule, the morphology of the Golgi stack and clathrin-coated granules did not appear modified; however, as compared to controls, clathrin-coated granules remained strongly labeled at the end of the 85 min chase period and non-coated granules showed only a very weak autoradiographic reaction. At this time point,

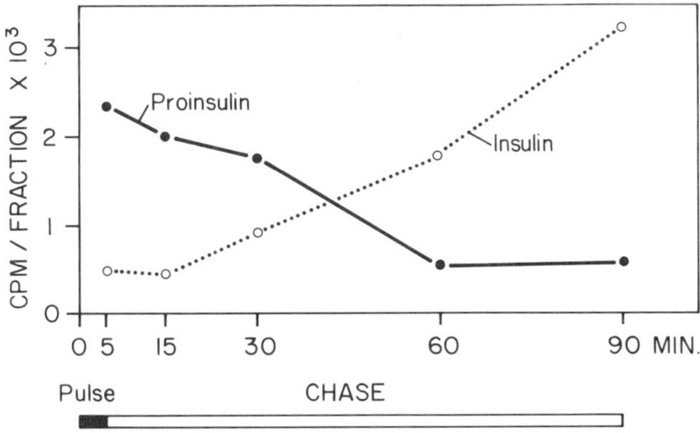

FIGURE 4. Radioactivity in proinsulin and insulin fractions during a pulse-chase labeling of B cells with [^{3}H]leucine. See FIGURE 3 for the autoradiographic labeling of B cell compartments during a similar pulse-chase experiment.

analogue-treated islets contained only 11% of the labeled immunoprecipitable products in the form of insulin, as compared to 78% in controls.[12] Thus, using two different experimental approaches, one that interferes with the intracellular transfer (and release) of (pro)insulin polypeptides, the other that renders the proinsulin molecule insensitive to proper proteolytic cleavage, the same result was achieved, i.e. impairment of proinsulin conversion and accumulation of radioactive secretory products in a clathrin-coated, Golgi-related compartment of the B cell. Taken together, these data represent compelling evidence that conversion of proinsulin is linked to its unperturbed passage through this clathrin-coated compartment and to the subsequent maturation of the latter into non-coated secretory granules. According to this interpretation, the coated granules would represent the proinsulin-rich compartment in which the bulk of conversion takes place.

The validity of this view could be demonstrated by the immunocytochemical detection of proinsulin in B cells during various states of cell functioning.

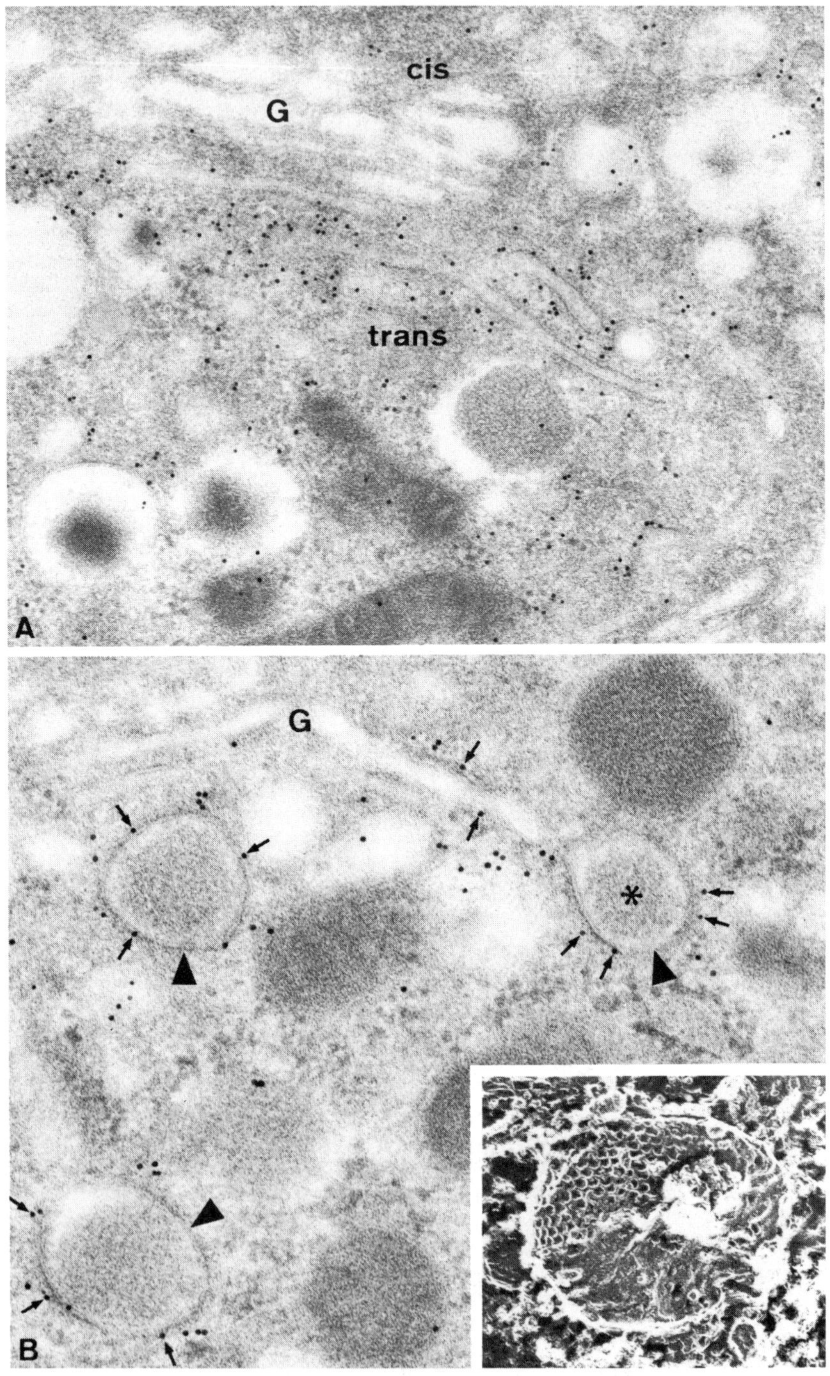
cis
G
trans
A
G
*
B

SPECIFIC IMMUNOLOCALIZATION OF PROINSULIN IN HUMAN AND RAT ISLETS

A series of monoclonal antibodies (Mabs) were recently raised against recombinant proinsulin,[13] with one group showing unique specificity towards an epitope spanning the two arginine at the B-C chain junction. We found this epitope, present in human and rat proinsulins, but not in insulin nor in C-peptide, to be preserved in aldehyde-fixed, plastic-embedded tissues, thus allowing the specific localization of the immunoreactive sites on thick and thin sections.[14]

By immunofluorescence after application of the GS-9A8 Mab, an intense, crescent-shaped proinsulin-specific staining was evident in the perinuclear region of B cells (FIG. 6); positive dots were also detectable in the vicinity of the large fluorescent areas. This pattern of staining differed from that obtained with an anti-insulin serum (cross-reacting with proinsulin), which elicited a spotty fluorescence throughout most of the B-cell cytoplasm (FIG. 6). On thin sections for electron microscopy, a marked proinsulin-specific labeling (revealed by the protein A–gold technique) was observed in Golgi stacks and in a population of secretory granules located mainly in the Golgi area (FIG. 7). These granules were characterized by a tightly fitting core. Serial sections immunostained with proinsulin and clathrin antibodies showed that virtually all proinsulin-rich granules presented clathrin immunoreactivity on their limiting membrane (FIG. 8). By contrast, non-coated secretory granules had a very low level of proinsulin labeling (FIG. 7), while they were intensely marked with the anti-insulin serum. The quantitation of the immunoreactive sites in each compartment (evaluated as number of gold particles/μm^2) showed that the bulk of proinsulin-specific immunoreactivity was situated over coated secretory granules.[14] The intensity of proinsulin immunolabeling in Golgi stacks and coated granules was dependent on the concentration of glucose in the incubation medium while that of non-coated granules remained invariably very low (FIG. 9). We studied the time course of this modulation.

When islets of Langerhans were preincubated in the presence of 1.67 mM glucose (below the threshold for stimulation of insulin biosynthesis),[15,16] then transferred to a medium containing 16.7 mM glucose, the low proinsulin-specific fluorescence visible at the end of the low glucose incubation increased progressively to reach a maximum after 60 min of incubation in high glucose (FIG. 10). On the other hand, islets preexposed to a high glucose concentration and presenting maximal proinsulin immunofluorescence showed a progressive decrease of fluorescence intensity upon transfer to a low glucose-containing medium (FIG. 11). These changes of proinsulin immunolabeling were quantitated at the ultrastructural level by measuring the number of gold particles per μm^2 of Golgi stacks and coated granules. The number of gold particles/μm^2 decreased over time after transfer of islets from high glucose to low glucose-containing medium, and increased in the opposite condition (low then high

---

FIGURE 5. (A) and (B) Golgi areas of B-cells immunostained with anti-clathrin antibody and revealed by the protein A–gold technique. (A) Asymetric labeling of the Golgi stack (G): most clathrin immunoreactive sites are concentrated over cisternal elements at the *trans* pole (see Orci *et al.*[30] for the quantitative evaluation of the labeling). (B) Clathrin immunostaining of the membrane of a *trans* Golgi cisterna with an expanded extremity (black arrowhead) containing condensing secretory material (asterisk); the limiting membrane of two neighboring secretory granules (black arrowheads) is also labeled by clathrin antibody. Individual gold particles revealing immunoreactive sites are indicated by arrows. (Inset). The membrane of a clathrin-coated granule as revealed by the quick freeze–deep etching method. A basketwork pattern characteristic of clathrin coats is exposed. (A) × 45,000, (B) × 61,000, and (Inset) × 48,000.

glucose medium) (FIG. 12). The decrease of the total number of proinsulin immunoreactive sites following transfer from high to low glucose occurred with a time course fully superposable to that obtained with an immunoassay of radiolabeled, newly synthesized proinsulin during a pulse-chase experiment (cf. FIG. 4). The inhibition of protein synthesis by cycloheximide[17,18] during continuous incubation in high glucose-containing medium had an effect comparable to that of lowering glucose concentra-

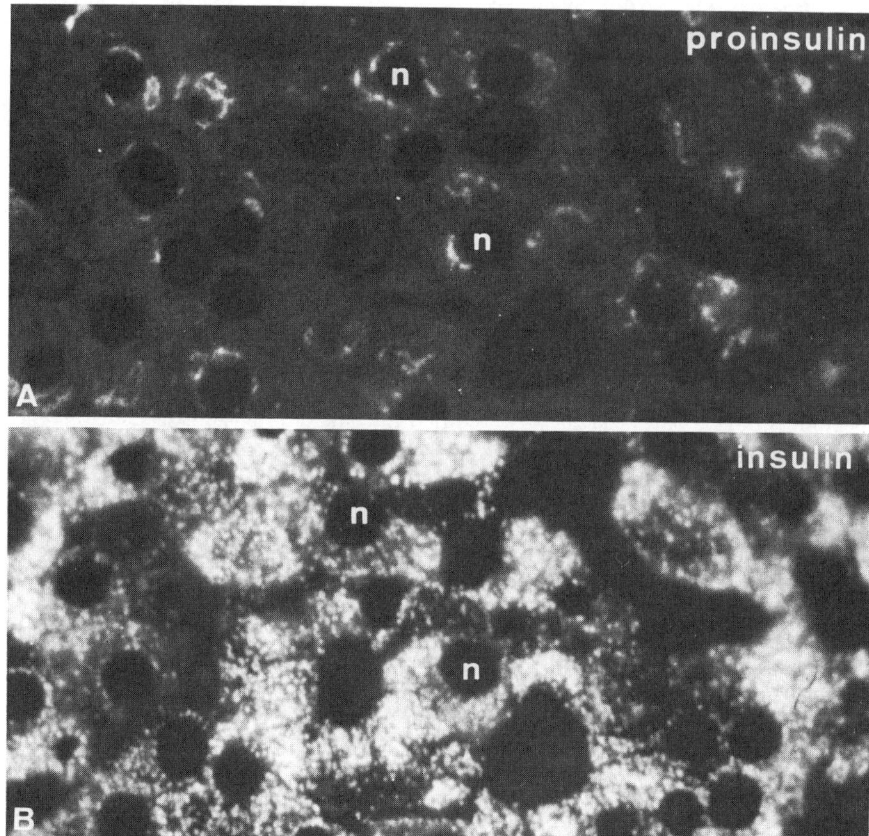

FIGURE 6. Consecutive thick serial sections showing immunofluorescent staining of B cells with proinsulin monoclonal antibody (A) and insulin polyclonal antibody (B). Proinsulin antibody determines a perinuclear, crescent-shaped staining in most cells in the field. Insulin antiserum elicits a spotty fluorescence throughout the cytoplasm. Negative nuclei (n) are evident. $\times$ 1,300.

tion: proinsulin-specific fluorescence in the Golgi areas progressively disappeared. The quantitation of proinsulin immunoreactive sites showed a rapid and marked decrease of the density of immunolabeling in the Golgi stacks and a progressive reduction of labeling in coated granules (FIG. 13). Non-coated granules, as in controls, had a very low constant degree of labeling. Both low glucose and cycloheximide experiments thus confirmed that the intracellular transit of proinsulin and its conversion (as evidenced

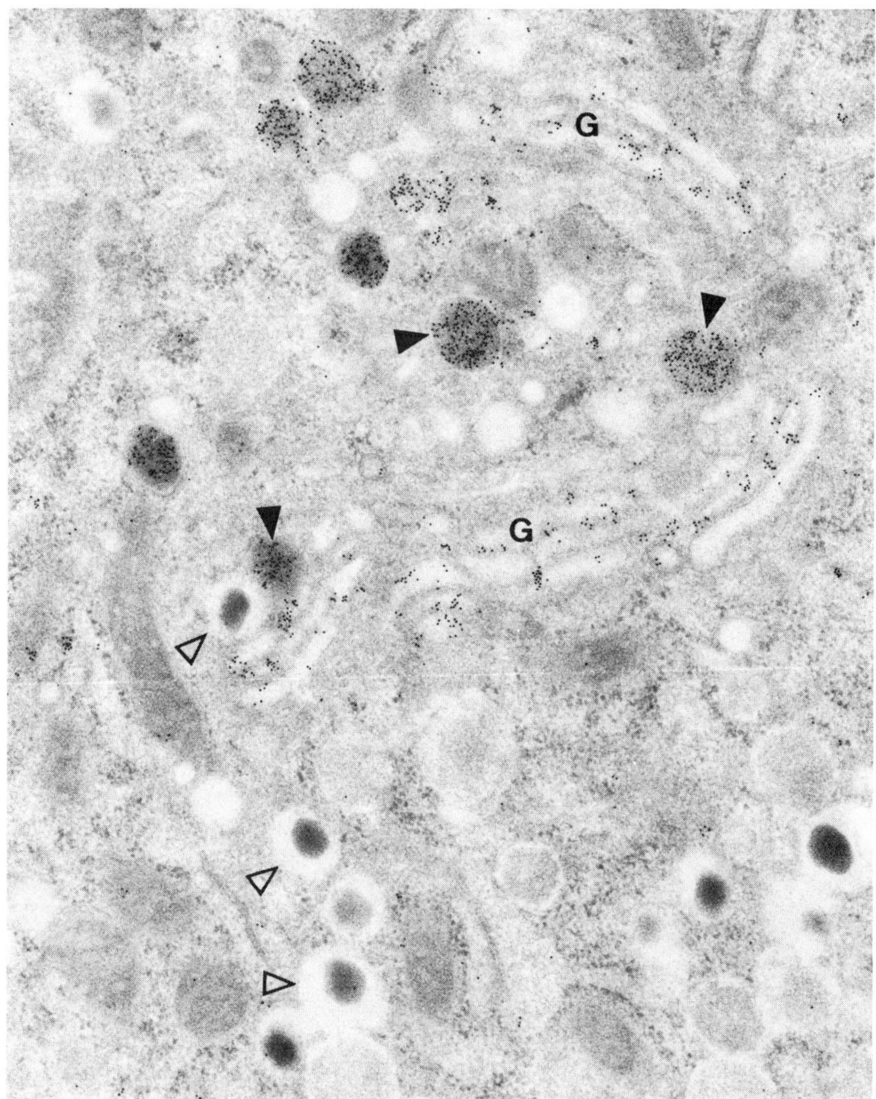

FIGURE 7. Thin section of a B cell immunostained with proinsulin specific monoclonal antibody revealed by the protein A–gold technique. Immunoreactive sites appear numerous on the stacked Golgi cisternae (G) and on a population of secretory granules with a tightly fitting core (black arrowheads). These correspond to newly formed clathrin-coated granules (see FIG. 5B and FIG. 8). By contrast, secretory granules with a dense core surrounded by a wide halo (white arrowheads) appear virtually unlabeled. These correspond to non-coated mature granules. × 31,000.

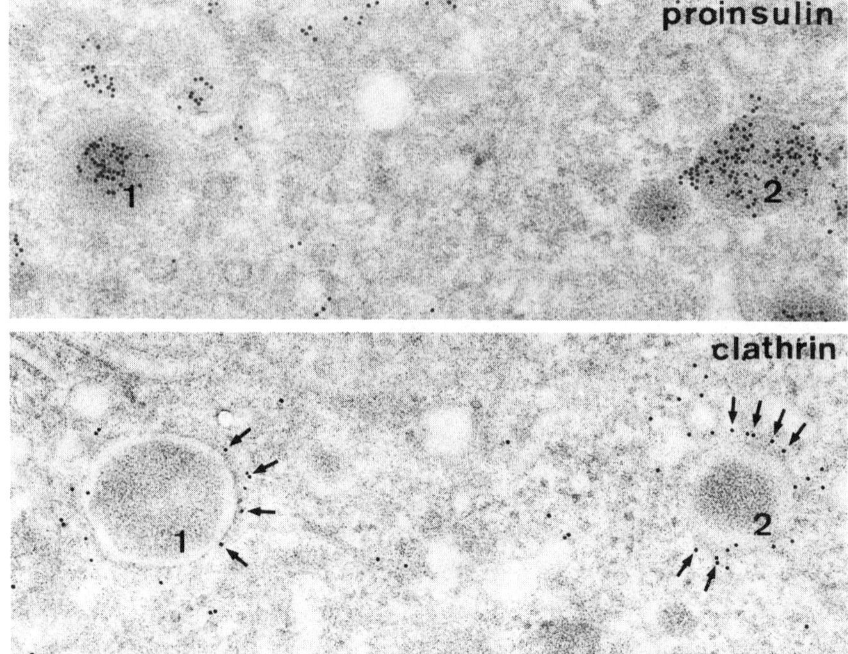

FIGURE 8. Consecutive thin serial sections showing proinsulin immunolabeling of the core and clathrin immunolabeling of the limiting membrane of the same granules (1 and 2). Individual gold particles revealing clathrin antibody on the outer aspect of the secretory granule membrane are indicated by arrows. × 48,000.

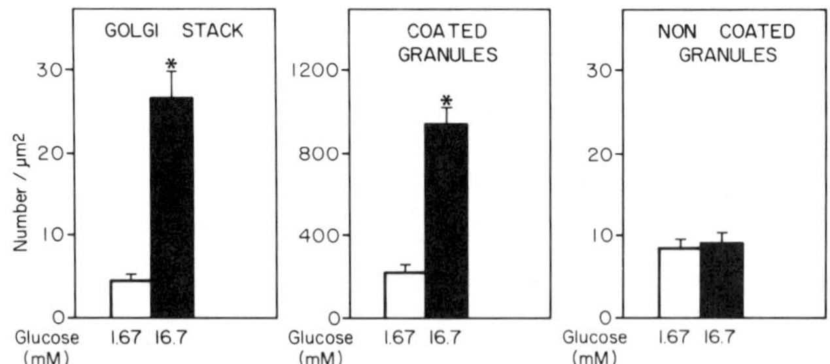

FIGURE 9. Quantitation of proinsulin immunolabeling of Golgi stack, coated (proinsulin-rich) granules and non-coated secretory granules of B cells following a 120-min incubation in presence of low (1.67 mM) or high (16.7 mM) glucose concentration. The asterisk indicates a significant difference between the two columns ($p < 0.001$). The intensity of immunoreaction was evaluated as the number of gold particles per μm^2 of the compartment. (From Orci *et al.*[14] With permission from *Cell.*)

by the loss of reactivity towards the proinsulin-specific Mab) are independent of ongoing protein synthesis[5–7]; they also showed that extensive conversion occurs in clathrin-coated granules. Whether the Golgi stack itself was involved in conversion, as previously assumed,[5–7,19] could not be determined by this experiment. Indeed proinsulin

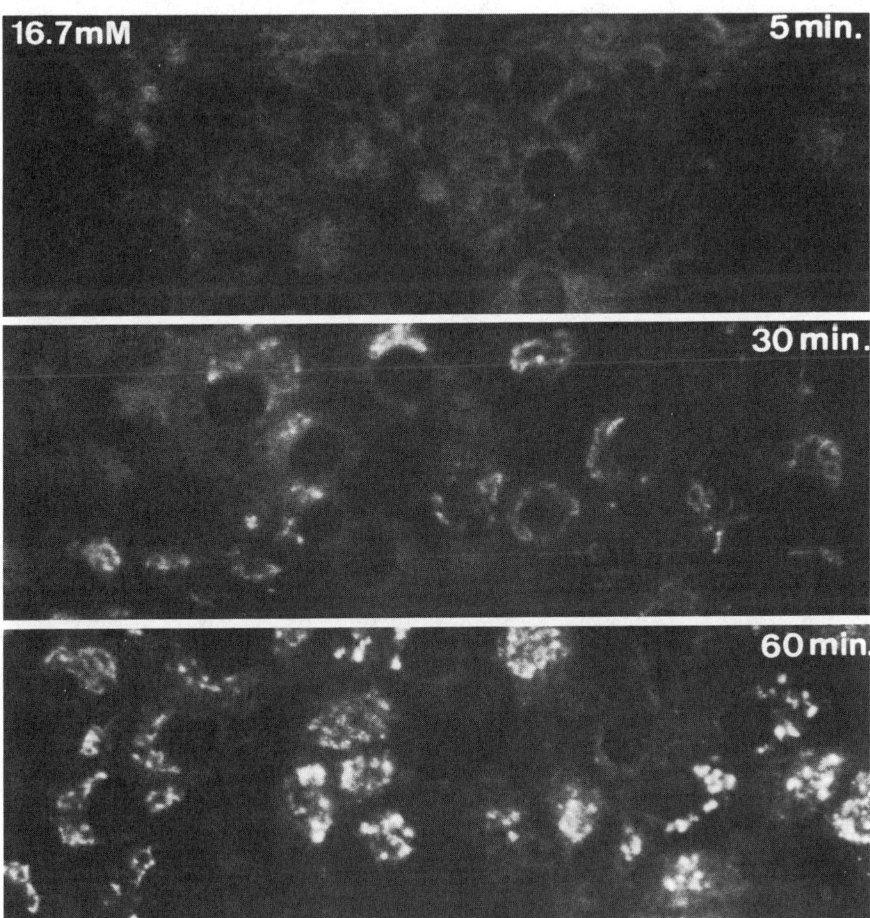

FIGURE 10. Proinsulin immunofluorescence in thick sections of isolated islets preincubated in low (1.67 mM) glucose, then transferred for 5, 30, and 60 min in a high glucose (16.7 mM)-containing medium. Note the progressive build up of the immunofluorescent reaction. × 1,200.

disappearance from the Golgi stack may be due not only to conversion but also to transfer and exit of proinsulin into coated granules.

To distinguish between these two possibilities, we used the respiration blocker antimycin A.[20] In the B cell, it is known from previous studies that antimycin A does not prevent conversion if added after the hormone has reached the converting

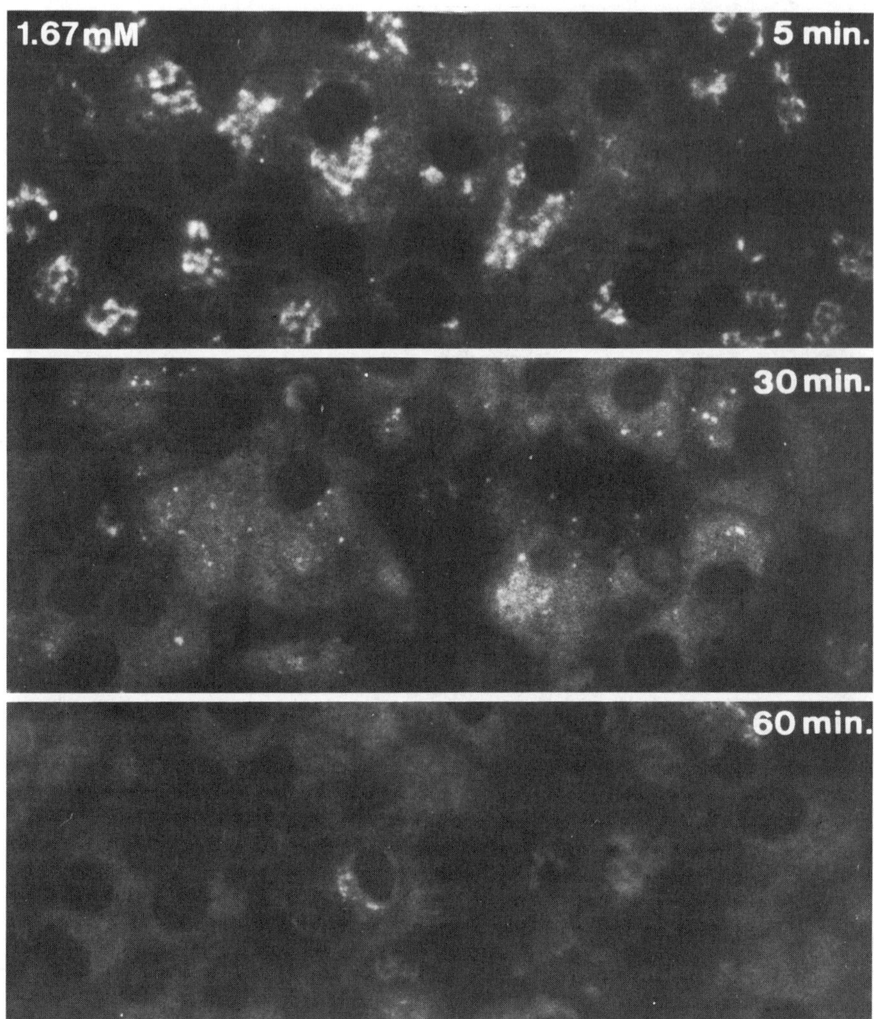

FIGURE 11. Proinsulin immunofluorescence in thick sections of isolated islets preincubated in high (16.7 mM) glucose, then transferred for 5, 30, and 60 min in a low (1.67 mM) glucose-containing medium. The crescent-shaped perinuclear fluorescence (see FIG. 6) progressively disappears. × 1,200.

compartment.[5-7] In the present experiments, isolated islets were preincubated in high glucose for 60 min then transferred to a low glucose-containing medium with antimycin, or without (control) for a further 60 min. In presence of antimycin, B cells failed to show the decrease of Golgi-associated immunoreactivity as observed in controls (FIGS. 14 and 15). In marked contrast, the number of proinsulin-rich granules preexisting to the antimycin exposure, and their degree of proinsulin labeling, were reduced as in untreated cells. Non-coated granules maintained their very low constant immunoreactivity throughout. The quantitation of these changes is shown in FIG. 16.

How can these data be interpreted? First, they indicate that proinsulin, which has reached the coated secretory granules before antimycin is added, is converted. Second, they show that in addition to the RER-Golgi transfer,[21,22] there is another hitherto unrecognized energy-requiring step to allow transport through the Golgi stack and the formation of coated granules at the *trans* pole; it is tempting to suggest that this effect of the drug is related to the absolute requirement for ATP in the transport of proteins between successive Golgi compartments, most probably through microvesicular carriers.[23,24] Third, the Golgi stack itself does not appear involved in conversion. Our results thus indicate that clathrin-coated granules are the major, and probably only, subcellular site of proinsulin conversion to mature insulin. Very recently, we were able to confirm this with a monoclonal antibody against insulin showing no more than 1% cross-reaction with proinsulin. With this probe (FIG. 17), the Golgi stack appeared

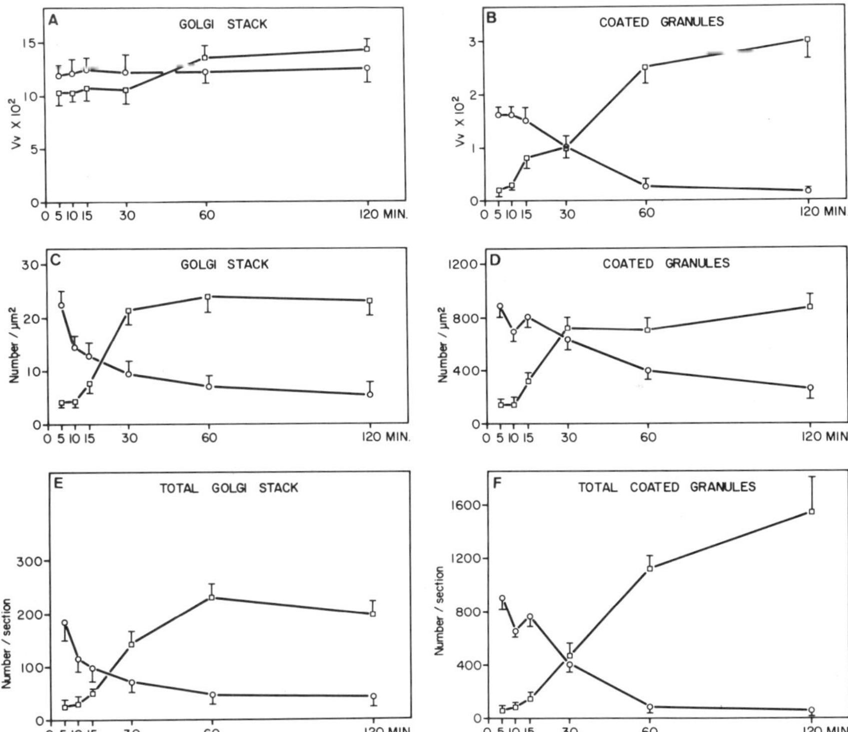

FIGURE 12. Quantitative evaluation of the size and density of proinsulin immunolabeling of Golgi stacks and coated granules. Open squares: B cells of islets preincubated (60 min) in low (1.67 mM) glucose and transferred in high glucose (16.7 mM)-containing medium for increasing periods of time. Open circles: islets preincubated (60 min) in high (16.7 mM) glucose and transferred in low glucose (1.67 mM) for increasing periods of time. (A) and (B) Size (volume density, Vv $\times$ 10^2) of the compartments. (C) and (D) Density (number of gold particles/μm^2) of proinsulin immunolabeling in the compartments. (E) and (F) Total number of immunoreactive sites per compartment. (From Orci *et al.*[14] With permission from *Cell.*)

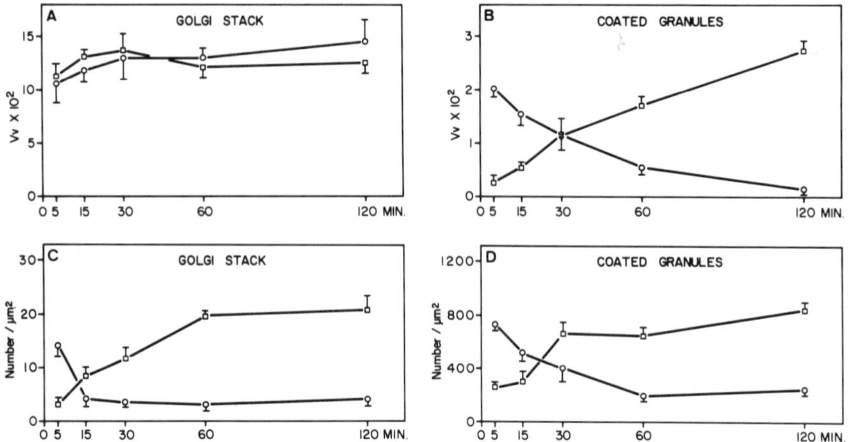

FIGURE 13. Quantitative evaluation of the size and density of proinsulin immunolabeling of Golgi stacks and coated granules during cycloheximide treatment. Open circles represent islets preincubated (60 min) in 16.7 mM glucose and transferred to the same medium containing 5 μg/ml cycloheximide for increasing periods of time. Open squares represent islets preincubated (60 min) in high glucose in presence of cycloheximide and transferred to a high glucose-containing medium without cycloheximide for increasing periods of time (reversibility experiment). (A) and (B) Size (volume density, Vv $\times$ 10^2) of the compartments. (C) and (D) Density (number of gold particles/μm^2) of proinsulin immunolabeling in the compartments. (From Orci *et al.*[14] With permission from *Cell.*)

virtually free of insulin-specific immunoreactivity, the coated granules showed a few sites indicating the onset of conversion, and a large amount of labeling was present in mature non-coated granules (L. Orci and M.J. Storch, unpublished data).

REGULATION OF PROINSULIN EXIT FROM THE GOLGI STACK

We have shown above that the disappearance of the proinsulin immunoreactivity from the Golgi stack under a condition of reduced protein synthesis was due to proinsulin exit from this compartment. We thus explored whether the transit (and exit) of proinsulin across the Golgi stack could be modulated. It is well established that an elevation of the glucose concentration in the islet environment causes an increase in insulin secretion by triggering exocytosis of storage granules, as well as by stimulating synthesis of the hormone. The possibility that glucose may also control intermediate steps in the intracellular transport and processing of insulin polypeptides has, however, received only little attention. We have analyzed the rate of transit of proinsulin through the Golgi apparatus as a function of the concentration of glucose in the incubation medium. Proinsulin immunoreactivity in the Golgi stack fell off very rapidly when synthesis of new hormone was arrested by cycloheximide and the islets kept in the presence of a high concentration of glucose (FIG. 18). By contrast, if the islets were transferred to low glucose-containing medium at the time of inhibition of protein synthesis, the decrease in Golgi proinsulin immunoreactivity was strikingly reduced (FIG. 18). Since all the available evidence indicates that proinsulin to insulin conversion is initiated only after the prohormone has left the Golgi stack, the observed effect cannot be ascribed to a decrease in processing of the prohormone, but rather to a marked slowdown of its transport out of the Golgi stack. Too little is known at present about the mechanism of protein transport through and out of the Golgi to allow formulation of a hypothesis regarding the mode of action of glucose in accelerating this transport. However, our result provides firm evidence for an unexpected effect of this secretagogue on the intracellular transit of insulin polypeptides.

THE CLATHRIN COAT MAY BE INVOLVED IN THE FORMATION AND MATURATION OF SECRETORY GRANULES

We have seen previously that in analogue-treated B cells, the formation of clathrin-coated secretory granules, which contain non-convertible proinsulin, occurs as in control B cells[12]; this indicates that the sorting of proinsulin from other RER-derived proteins (e.g. lysosomal enzymes) into coated granules can take place in the absence of

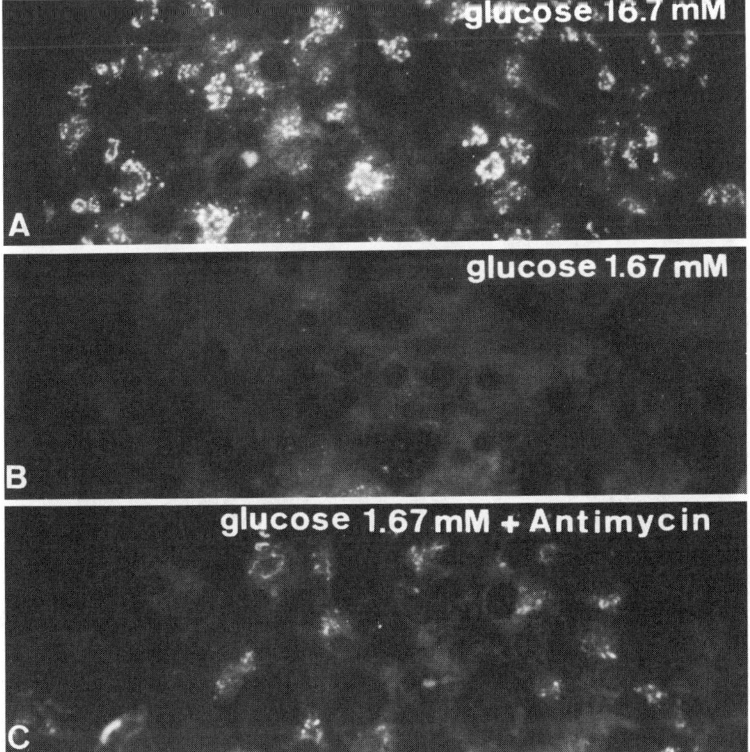

FIGURE 14. Proinsulin immunofluorescence in isolated islets preincubated for 60 min in high (16.7 mM) glucose and further incubated for 60 min in high glucose (A), low (1.67 mM) glucose without antimycin (B), or low glucose with 10 μM antimycin (C). Antimycin treatment decreases the loss of fluorescence induced by the low glucose incubation alone (see FIGS. 15 and 16 for the ultrastructural appearance and quantitation of these changes). $\times$ 800.

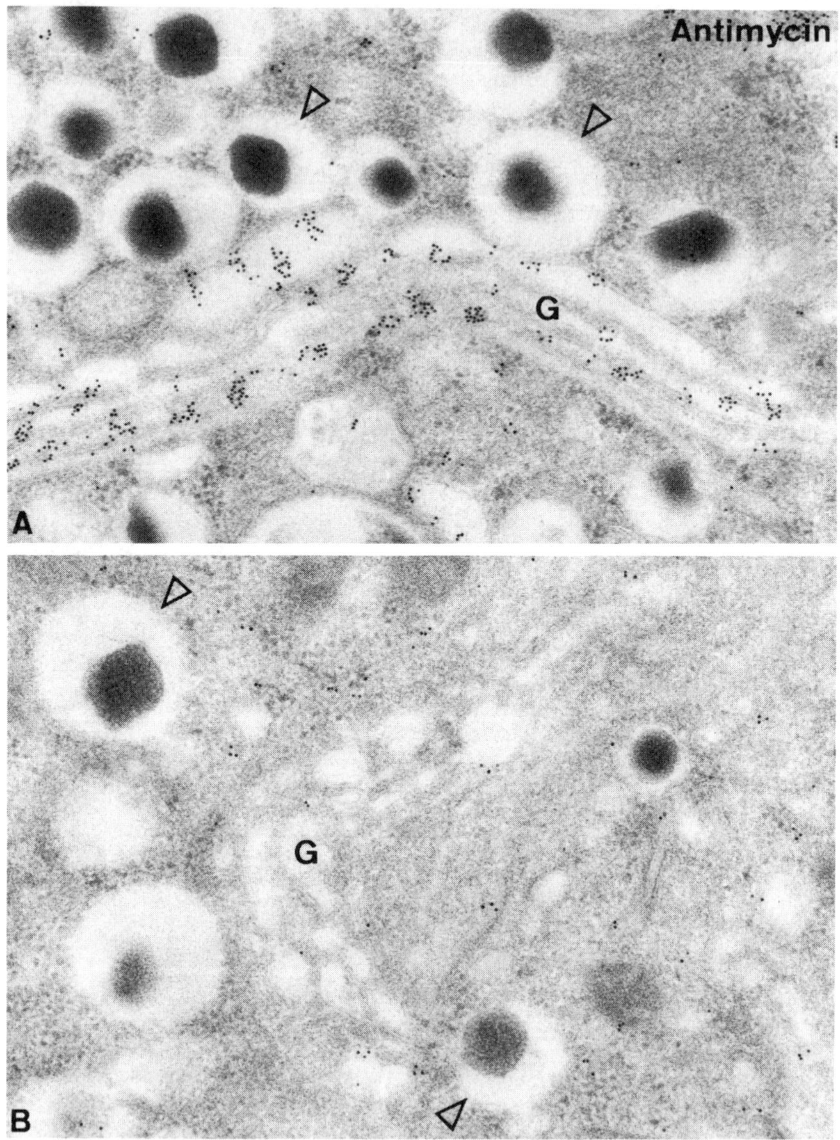

FIGURE 15. Golgi areas of B cells immunostained with the proinsulin antibody revealed by the protein A–gold method. (A) Islets preincubated (60 min) in presence of high (16.7 mM) glucose and further incubated (60 min) in low (1.67 mM) glucose with addition of 10 μM antimycin. Note the persistence of the Golgi stack immunolabeling. (B) Same experimental protocol as in (A) but without antimycin added in the second incubation period. The Golgi stack shows an extremely low level of proinsulin immunoreactivity. In both (A) and (B) non-coated secretory granules (white arrowheads) are virtually free of immunolabeling. (A) × 40,000 and (B) × 40,000.

proteolytic cleavage. If the sorting process occurs before conversion, i.e. proximally to the clathrin-coated compartment, it is expected to occur in the Golgi stack. It is well known that the process of lysosomal enzyme sorting is based on the presence of specific receptors in Golgi membranes.[25–27] In B cells, we were struck by the fact that (pro)insulin immunoreactivity (revealed with an anti-insulin serum cross-reacting with proinsulin) is associated with the inner aspect of Golgi membranes while in the secretory granules, the immunoreactive sites are localized over the granule's core.[28] This was recently confirmed with the proinsulin-specific antibody (FIG. 19). This pattern suggests the presence of specific proinsulin binding sites at the Golgi level. An intracellular sorting mechanism for proinsulin involving a specific binding process

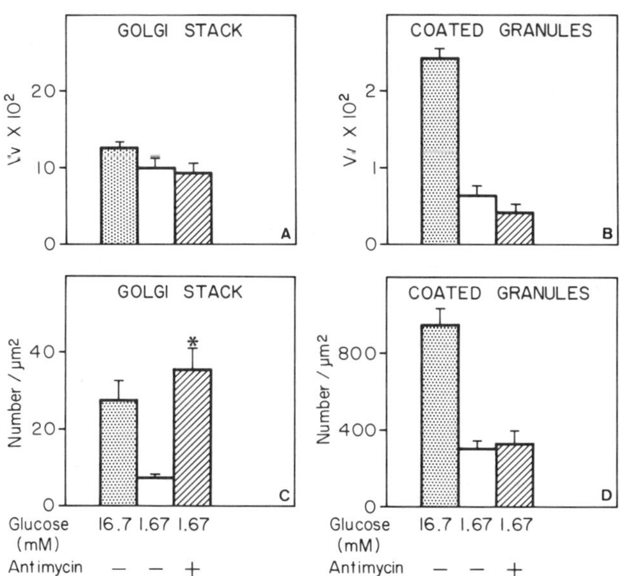

FIGURE 16. Quantitation of the size and density of immunolabeling of Golgi stacks and coated granules in antimycin experiments. (A) and (B) Size (volume density, $Vv \times 10^2$) of the compartments. (C) and (D) Density (number of gold particules/μm^2) of proinsulin immunolabeling in the compartments. (C) The high proinsulin labeling of the Golgi stack in presence of low glucose and antimycin (striped column) as compared to low glucose incubation alone (white column). The asterisk indicates a significant difference ($p < 0.05$) between these two values. (From Orci *et al.*[14] With permission from *Cell.*)

seems compatible with the ultrastructural localization of clathrin in the Golgi stack. Using an affinity-purified anti-clathrin antibody,[29] we observed that most clathrin immunoreactive sites were observed at the *trans* pole of the Golgi[30] (FIG. 5). At the plasma membrane level, receptor-mediated endocytosis is characterized by the association of a clathrin coat with the cytoplasmic leaflet of the membrane[31] and clathrin is believed to be involved in the internalization of membrane segments with clustered receptors.[32] If, by analogy with the plasma membrane, we take the typical clathrin coats on *trans* Golgi membranes as morphological markers of "receptor-mediated" intracellular transport, our data would therefore support the working hypothesis of a

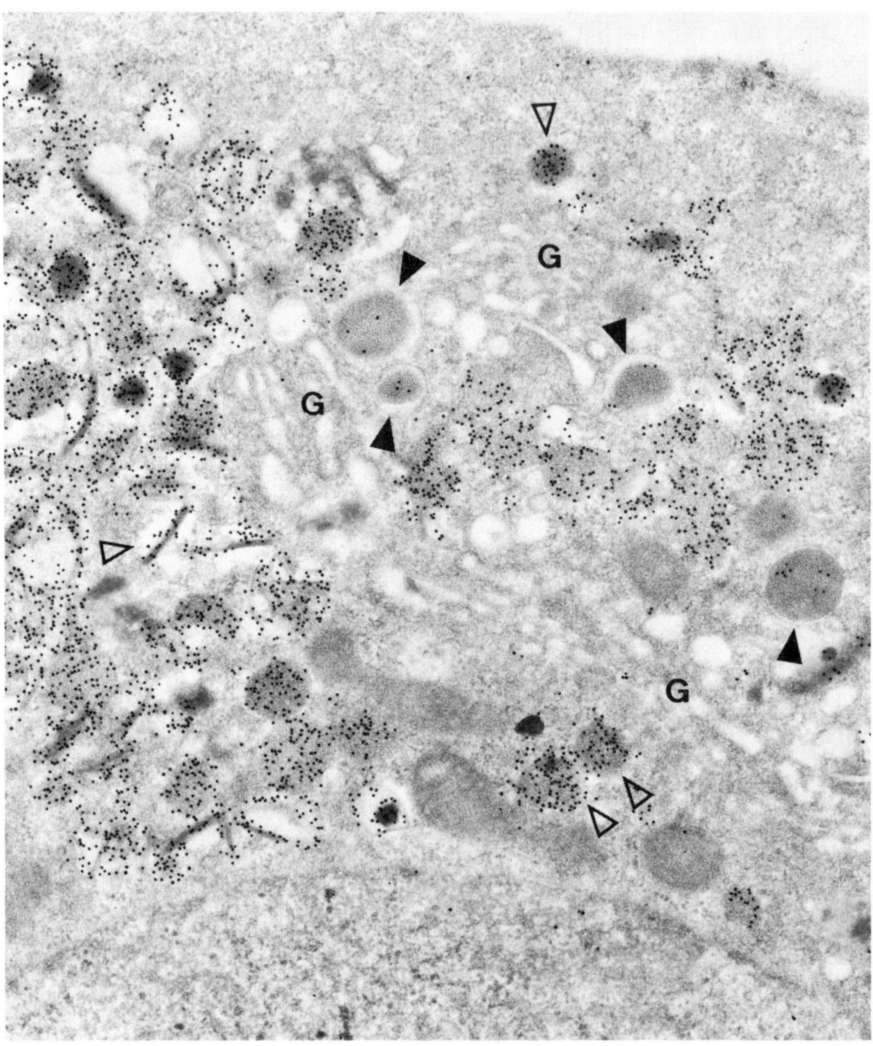

FIGURE 17. Thin section of a B cell immunostained with an insulin-specific monoclonal antibody[41] revealed by the protein A–gold method. The Golgi stack (G) appears virtually free of insulin immunoreactive sites. A population of secretory granules in the Golgi area (black arrowheads) shows a few sites, while another population of granules (some with a typical crystalloid core) is heavily labeled with insulin immunoreactivity (white arrowheads). The two granule populations correspond to coated and non-coated granules, respectively. Compare with FIGURE 7 showing the distribution of proinsulin immunoreactivity in these compartments. × 26,000.

proinsulin receptor mechanism involved in the detachment of coated secretory granules from coated cisternae, and thereby, in the sorting of this polypeptide to secretory granules.

THE INTRAGRANULAR pH

Clathrin-coated vesicles have been shown to undergo acidification *in vitro*[33,34] and isolated Golgi membranes also contain an electrogenic H^+ pump.[35] It is thus pertinent to consider the relationship between the presence of clathrin on the limiting membrane

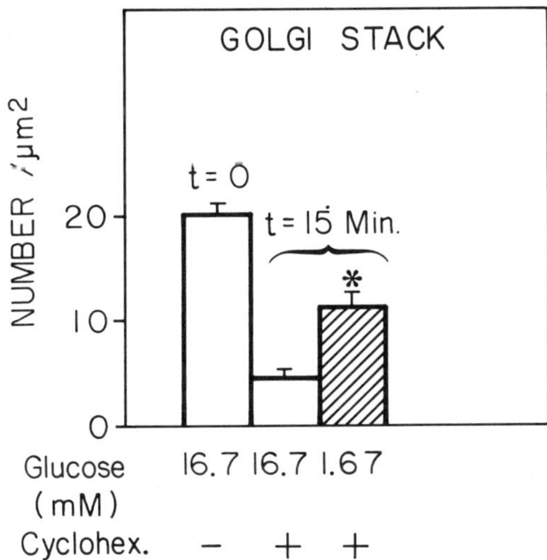

FIGURE 18. Quantitative evaluation of proinsulin labeling in Golgi stacks of B cells preincubated (60 min) in high glucose (t = 0) and transferred (15 min) to a medium containing cycloheximide in a high (16.7 mM) or low (1.67 mM) glucose concentration (t = 15 min). Proinsulin immunoreactivity (number of gold particles/μm^2) decreases rapidly when protein synthesis is inhibited during incubation in high glucose (white columns). The decrease is strikingly reduced by low glucose incubation (hatched column). The asterisk indicates a significant difference between the two columns ($p < 0.005$).

of the Golgi-derived, coated compartment where conversion of proinsulin takes place, and the putative role of a proinsulin-converting thiol-protease with acidic pH optimum.[36] The intragranular pH has been estimated to be between 5 and 6.[37] Recently, Anderson *et al.*[38,39] showed that intracellular compartments with low pH can be visualized with a basic congener of dinitrophenol, 3-(2,4-dinitroanilino)-3′amino-N-methyldipropylamine (DAMP). DAMP accumulates in acidic compartments and can be detected immunocytochemically with a monoclonal anti-dinitrophenol antibody revealed by immunofluorescence (FIG. 20) or by the protein A–gold technique (FIG. 21). Using this probe we found that the Golgi stack is practically devoid of DNP immunoreactivity; moderate numbers of immunoreactive sites appear in clathrin-

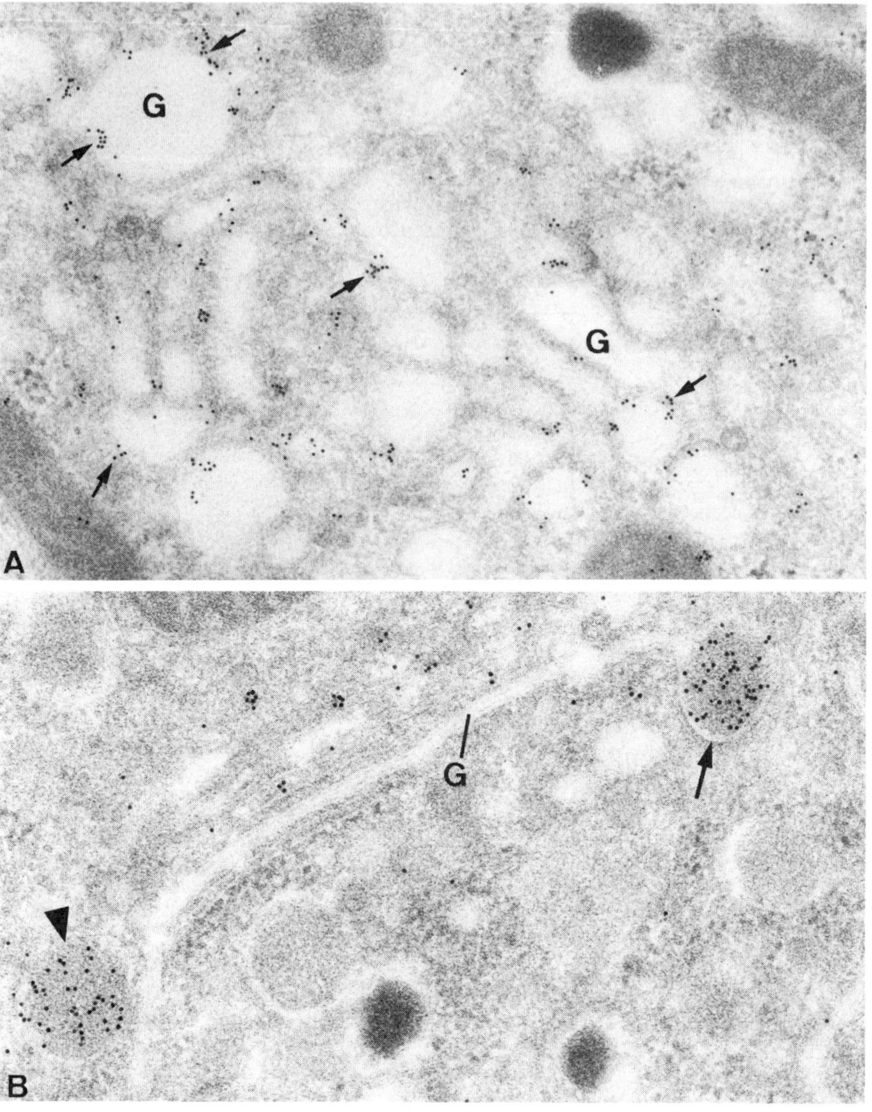

FIGURE 19. Thin sections of Golgi stacks immunostained with proinsulin-specific monoclonal antibody revealed by the protein A-gold technique. Islets were incubated in high glucose (16.7 mM) before fixation. (A) Several Golgi elements (G) are visible and the gold particles revealing proinsulin antigenic sites appear associated with the periphery of the cisternal space (arrows). (B) *Trans* cisterna with its limiting membrane lacking proinsulin immunoreactivity. The latter has been sorted and concentrated to the condensing secretory material at the expanded cisternal extremity (arrow). Proinsulin labeling also appears concentrated on the core of a proinsulin-rich coated granule neighboring the cisterna (arrowhead). (A) × 44,000 and (B) × 52,000.

coated (proinsulin-rich) granules and large amounts are seen in non-coated, mature (insulin-rich) granules (FIGS. 21 and 22). Although it remains to be determined whether the partitioning properties of DAMP between different subcellular compartments are exclusively a function of local pH, our current working hypothesis is that proinsulin exits from the Golgi stack in a non acidic pH compartment, and that this compartment gradually becomes acidic. Coordinate with acidification is the conver-

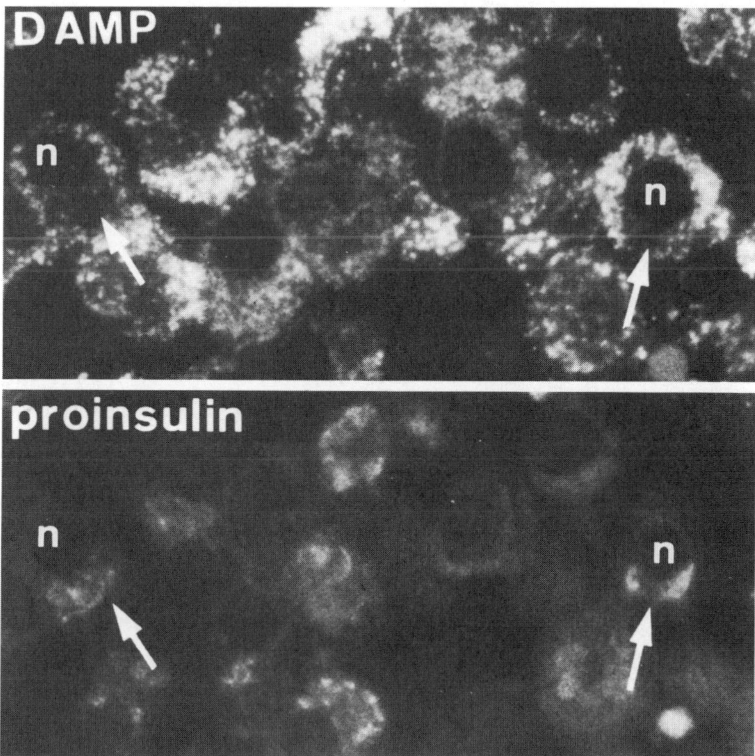

FIGURE 20. Consecutive thick serial sections immunostained with anti-proinsulin or anti-dinitrophenol (to reveal DAMP) antibody. DAMP staining appears as a spotty fluorescence throughout the B-cell cytoplasm. Crescent-shaped areas of lesser intensity are seen in the perinuclear region (arrows). The latter correspond to the regions preferentially stained in the same cells by the proinsulin antibody (arrows) (cf. FIG. 6A). n = nuclei. × 1,400.

sion of proinsulin to insulin and the shedding of the coat.[40] The regulation of intragranular pH may be critical in providing an optimal milieu for the processing of insulin precursors, as well as for terminating the activity of the relevant proteolytic enzyme(s). It is easy to envision how physiological or pathological differences in the timing or extent of such pH changes could affect the intragranular postsynthetic modifications of proteins, including prohormones and zymogens.

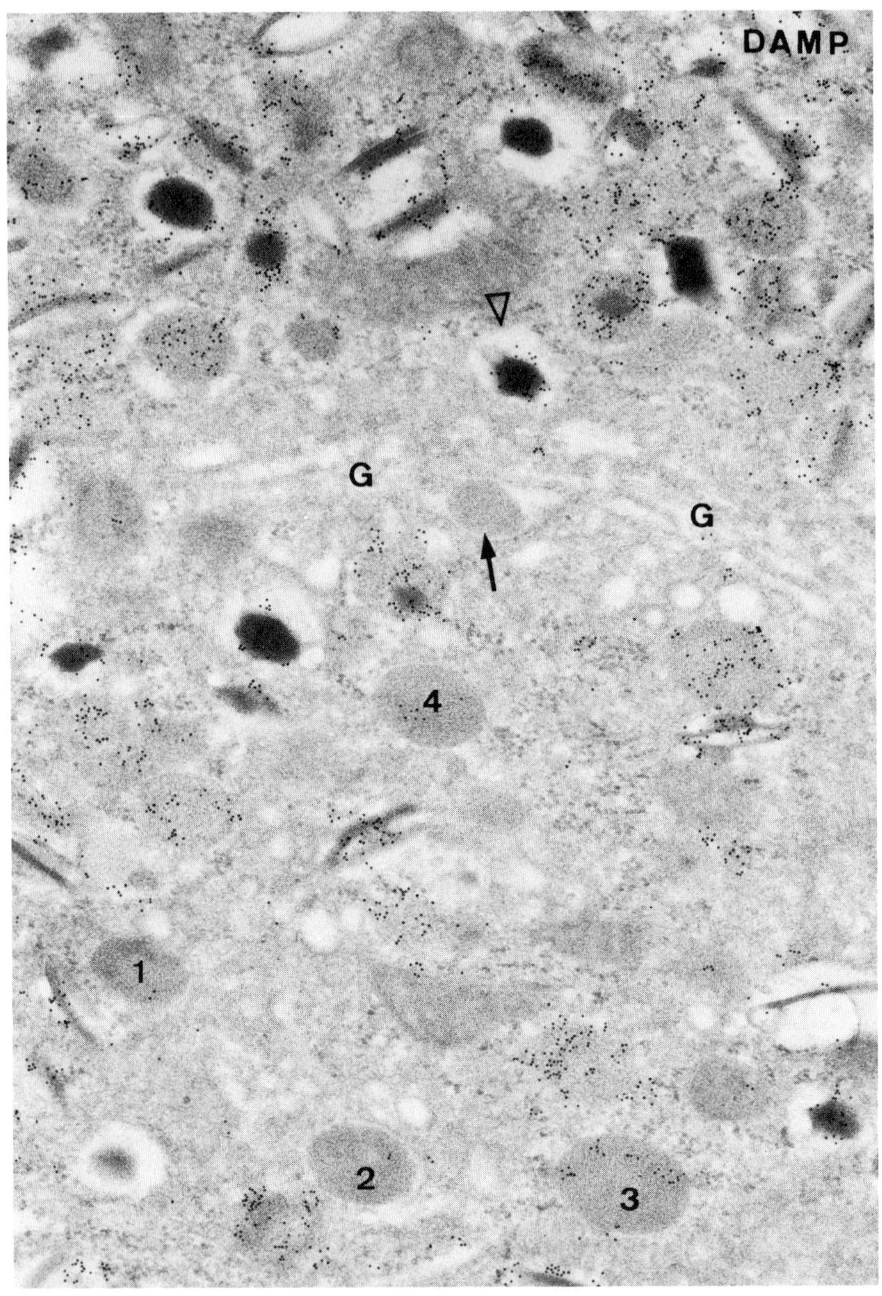

FIGURES 21 AND 22. Consecutive thin serial sections stained with anti-dinitrophenol (to reveal DAMP) or anti-proinsulin antibody revealed by the protein A–gold technique. DAMP and proinsulin immunoreactive sites are distributed with opposite intensities over the Golgi stack (G), coated secretory granules (black arrowheads), and non-coated mature granules (white arrowheads) of the same cell. Condensing secretory material (proinsulin-rich, DAMP-poor) in a Golgi

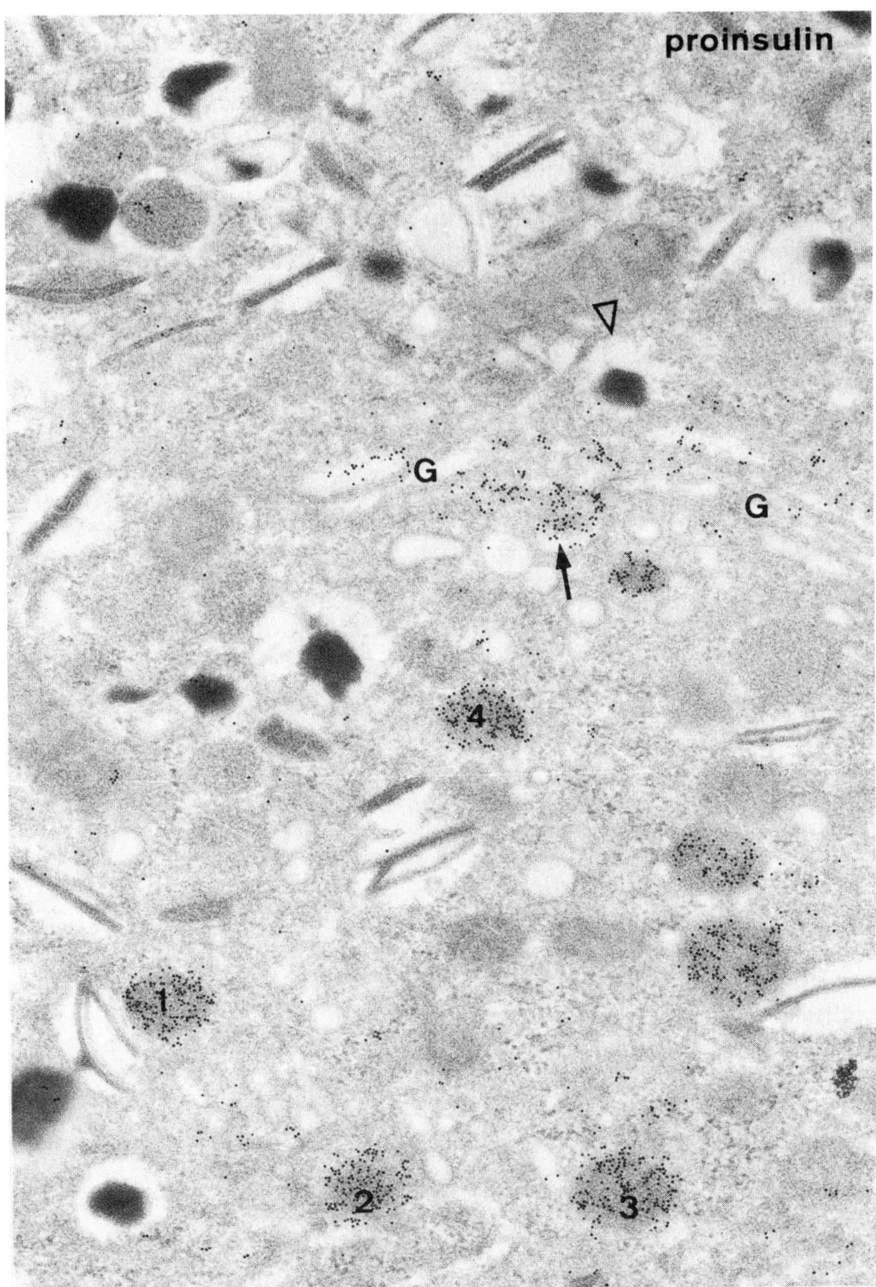

cisterna is indicated by the black arrow. Four coated secretory granules that appear rich in proinsulin and poor in DAMP immunoreactivity are identified by the same numbers (1–4) in the two figures. A non-coated mature secretory granule showing the reverse pattern (proinsulin-poor, DAMP-rich) is indicated by the white arrowhead. Both figures, × 36,000.

CONCLUDING REMARKS

The data summarized above have shown progress towards a better integration between morphology and function in the understanding of proinsulin processing. Further advances towards a more complete picture of the B cell will be made by answering the following questions. (1) At which step of transport/sorting out of proinsulin is the converting enzyme(s) added? Is the enzyme(s) in active or inactive (proenzyme) form at the time of addition? Is there an optimal pH decrease for initiating or terminating conversion, respectively? Is the enzyme(s) used in several cycles of conversion? (2) What is the signal for uncoating proinsulin-rich, clathrin-coated secretory granules during their maturation into non-coated granules?

ACKNOWLEDGMENTS

I thank Mylène Amherdt, Roberto Montesano, Alain Perrelet, Mariella Ravazzola, and Jean-Dominique Vassalli for their invaluable contribution in the various aspects of this work. I am grateful to Anne-Marie Lucini, Gorana Perrelet, Gyslaine Tripet, and Ariane Widmer for technical assistance, to Gérard Negro and Pierre-Alain Rüttimann for photographic work, and to Isabelle Bernard for typing the manuscript.

REFERENCES

1. PATZELT, C., A. D. LABRECQUE, J. R. DUGUID, R. J. CARROLL, P. S. KEIM, R. L. HEINRIKSON & D. F. STEINER. 1978. Detection and kinetic behavior of preproinsulin in pancreatic islets. Proc. Natl. Acad. Sci. USA **75:** 1260–1264.
2. ORCI, L. 1974. A portrait of the pancreatic B-cell. Diabetologia **10:** 163–187.
3. ORCI, L. 1982. Macro- and micro-domains in the endocrine pancreas. Diabetes **31:** 538–565.
4. ORCI, L. 1985. The insulin factory: a tour of the plant surroundings and a visit to the assembly line. Diabetologia **28:** 528–546.
5. STEINER, D. F., J. L. CLARK, C. NOLAN, A. H. RUBENSTEIN, E. MARGOLIASH, F. MELANI & P. E. OYER. 1970. The biosynthesis of insulin and some speculations regarding the pathogenesis of human diabetes. *In* The Pathogenesis of Diabetes Mellitus. E. Cerasi & R. Luft, Eds.: 123–132. Almquist and Wiksell. Stockholm.
6. STEINER, D. F., W. KEMMLER, H. S. TAGER & J. D. PETERSON. 1974. Proteolytic processing in the biosynthesis of insulin and other proteins. Fed. Proc. **33:** 2105–2115.
7. STEINER, D. F., S. TERRIS, S. J. CHAN & A. H. RUBENSTEIN. 1976. Chemical and biological aspects of insulin and proinsulin. *In* Insulin. R. Luft, Ed.: 53–103. Nordisk Insulin Laboratorium. Gentofte, Denmark.
8. TARTAKOFF, A. M. 1983. Perturbation of vesicular traffic with the carboxylic ionophore monensin. Cell **32:** 1026–1028.
9. NOE, B. D. 1981. Inhibition of islet prohormone to hormone conversion by incorporation of arginine and lysine analogs. J. Biol. Chem. **256:** 4940–4946.
10. HALBAN, P. A. 1982. Inhibition of proinsulin to insulin conversion in rat islets using arginine and lysine analogs. Lack of effect on rate of release of modified products. J. Biol. Chem. **257:** 13177–13180.
11. ORCI, L., P. HALBAN, M. AMHERDT, M. RAVAZZOLA, J.-D. VASSALLI & A. PERRELET. 1984. A clathrin-coated, Golgi-related compartment of the insulin secreting cell accumulates proinsulin in the presence of monensin. Cell **39:** 39–47.
12. ORCI, L., P. HALBAN, M. AMHERDT, M. RAVAZZOLA, J.-D. VASSALLI & A. PERRELET. 1984. Nonconverted, amino acid analog-modified proinsulin stays in a Golgi-derived clathrin-coated membrane compartment. J. Cell Biol. **99:** 2187–2192.

13. MADSEN, O. D., B. H. FRANK & D. F. STEINER. 1984. Human proinsulin-specific antigenic determinants identified by monoclonal antibodies. Diabetes **33:** 1012–1016.
14. ORCI, L., M. RAVAZZOLA, M. AMHERDT, O. MADSEN, J. -D. VASSALLI & A. PERRELET. 1985. Direct identification of prohormone conversion site in insulin-secreting cells. Cell **42:** 671–681.
15. KAELIN, D., A. E. RENOLD & G. W. G. SHARP. 1978. Glucose-stimulated proinsulin biosynthesis. Rates of turn off after cessation of the stimulus. Diabetologia **14:** 329–335.
16. MORRIS, G. E. & A. KORNER. 1970. The effect of glucose on insulin biosynthesis by isolated islets of Langerhans of the rat. Biochim. Biophys. Acta **208:** 404–413.
17. ENNIS, H. L. & M. LUBIN. 1964. Cycloheximide: Aspects of inhibition of protein synthesis in mammalian cells. Science **146:** 1474–1476.
18. JAMIESON, J. D. & G. E. PALADE. 1968. Intracellular transport of secretory proteins in the pancreatic exocrine cell. III. Dissociation of intracellular transport from protein synthesis. J. Cell Biol. **39:** 580–588.
19. HOWELL, S. L. 1972. Role of ATP in the intracellular translocation of proinsulin and insulin in the rat pancreatic B-cell. Nature **235:** 85–86.
20. CHANCE, B. & G. R. WILLIAMS. 1956. The respiratory chain and oxidative phosphorylation. Adv. Enzymol. **17:** 65–134.
21. JAMIESON, J. D. & G. E. PALADE. 1968. Intracellular transport of secretory proteins in the pancreatic exocrine cell. IV. Metabolic requirements. J. Cell Biol. **39:** 589–603.
22. JAMIESON, J. D. & G. E. PALADE. 1971. Condensing vacuole conversion and zymogen granule discharge in pancreatic exocrine cells: metabolic studies. J. Cell Biol. **48:** 503–522.
23. BALCH, W. E., B. S. GLICK & J. E. ROTHMAN. 1984. Sequential intermediates in the pathway of intercompartmental transport in a cell-free system. Cell **39:** 525–536.
24. BALCH, W. E., W. G. DUNPHY, W. A. BRAELL & J. E. ROTHMAN. 1984. Reconstitution of the transport of protein between successive compartments of the Golgi measured by the coupled incorporation of N-acetylglucosamine. Cell **39:** 405–416.
25. SLY, W. S., & H. D. FISCHER. 1982. The phosphomannosyl recognition system for intracellular and intercellular transport of lysosomal enzymes. J. Cell Biochem. **18:** 67–85.
26. BROWN, W. J. & M. G. FARQUHAR. 1984. The mannose-6-phosphate receptor for lysosomal enzymes is concentrated in *cis* Golgi cisternae. Cell **36:** 295–307.
27. GEUZE, H. J., J. W. SLOT, G. J. A. M. STROUS, A. HASILIK & K. VON FIGURA. 1984. Ultrastructural localization of the mannose 6-phosphate receptor in rat liver. J. Cell Biol. **98:** 2047–2054.
28. ORCI, L., M. RAVAZZOLA & A. PERRELET. 1984. (Pro)insulin associates with Golgi membranes of pancreatic B-cells. Proc. Natl. Acad. Sci. USA **81:** 6743–6746.
29. LOUVARD, D., C. MORRIS, G. WARREN, K. STANLEY, F. WINKLER & H. REGGIO. 1983. A monoclonal antibody to the heavy chain of clathrin. EMBO J. **2:** 1655–1664.
30. ORCI, L., M. RAVAZZOLA, M. AMHERDT, D. LOUVARD & A. PERRELET. 1985. Clathrin immunoreactive sites in the Golgi apparatus are concentrated at the *trans* pole in polypeptide hormone-secreting cells. Proc. Natl. Acad. Sci. USA **82:** 5385–5389.
31. GOLDSTEIN, J. L., R. G. W. ANDERSON & M. S. BROWN. 1979. Coated pits, coated vesicles, and receptor-mediated endocytosis. Nature **279:** 679–685.
32. PEARSE, B. M. F., & M. S. BRETSCHER. 1981. Membrane recycling by coated vesicles. Ann. Rev. Biochem. **50:** 85–101.
33. FORGAC, M., L. CANTLEY, B. WIEDENMANN, L. ALTSTIEL & D. BRANTON. 1983. Clathrin-coated vesicles contain an ATP-dependent proton pump. Proc. Natl. Acad. Sci. USA **80:** 1300–1303.
34. STONE, D. K., X. S. XIE & E. RACKER. 1983. An ATP-driven proton pump in clathrin-coated vesicles. J. Biol. Chem. **258:** 4059–4062.
35. GLICKMAN, J., K. CROEN, S. KELLY & Q. AL-AWQATI. 1983. Golgi membranes contain an electrogenic H^+ pump in parallel to a chloride conductance. J. Cell Biol. **97:** 1303–1308.
36. DOCHERTY, K., R. J. CARROLL & D. F. STEINER. 1982. Conversion of proinsulin to insulin: Involvement of a 31,500 molecular weight thiol protease. Proc. Natl. Acad. Sci. USA **79:** 4613–4617.

37. Hutton, J. C. 1982. The internal pH and membrane potential of the insulin-secretory granules. Biochem. J. **204:** 171–178.
38. Anderson, R. G. W., J. R. Falck, J. L. Goldstein & M. S. Brown. 1984. Visualization of acidic organelles in intact cells by electron microscopy. Proc. Natl. Acad. Sci. USA **81:** 4838–4842.
39. Anderson, R. G. W. & R. K. Pathak. 1985. Vesicles and cisternae in the trans Golgi apparatus of human fibroblasts are acidic compartments. Cell **40:** 635–643.
40. Orci, L., M. Ravazzola, M. Amherdt, O. Madsen, A. Perrelet, J.-D. Vassalli & R. G. W. Anderson. 1986. Conversion of proinsulin to insulin occurs coordinately with acidification of maturing secretory vesicles. J. Cell Biol. (In press.)
41. Storch M.-J., K.-G. Petersen, T. Licht & L. Kerp. 1985. Recognition of human insulin and proinsulin by monoclonal antibodies. Diabetes **34:** 808–811.

Signal Transduction in Insulin Secretion:

Comparison between Fuel Stimuli and Receptor Agonists

CLAES B. WOLLHEIM AND TREVOR J. BIDEN

Institut de Biochimie Clinique
Centre Médical Universitaire
University of Geneva
1211 Geneva 4, Switzerland

The secretion of insulin from the B cells in the pancreatic islets is the single most important factor governing normal blood glucose homeostasis. It is therefore not surprising that insulin secretion is subject to multifactorial control by circulating fuels and hormones as well as by neural input via autonomic nerve fibers ending inside the pancreatic islets.[1-3] Like other hormones, neurotransmitters, and enzymes, insulin is secreted by exocytosis during which the hormone contained within secretory granules is extruded from the cell. The study of secretion in the B cell therefore offers the possibility of comparing hormones and neurotransmitters acting on surface receptors to that of fuels, which are generally only active following their metabolic degradation.

It has been known for many years that Ca^{2+} plays an important regulatory role in insulin secretion.[3,4] Thus, in the absence of extracellular Ca^{2+}, glucose and other secretagogues are incapable of eliciting insulin release. Moreover, increased Ca^{2+} uptake occurs in response to glucose.[3] However, fuller understanding of the role of Ca^{2+} has been made possible with two recently developed approaches: direct measurement of cytosolic Ca^{2+} concentrations ($[Ca^{2+}]_i$) in small cells with fluorescent indicators and assessment of Ca^{2+} transport by intracellular organelles.[4] Since neither approach lends itself easily to use with normal B cells, we have routinely used the clonal insulin-secreting cell line, RINm5F, or tissue from rat insulinomas. The cell line exhibits altered glucose metabolism and no longer recognizes the sugar as a secretagogue.[5] However, it does respond to the triose glyceraldehyde, which in RINm5F cells mimicks most of the action of glucose in pancreatic islets.[6,7] In the present paper we report on the manner in which nutrient secretagogues, on the one hand, and muscarinic agonists on the other, alter Ca^{2+} handling in cells during the initiation of insulin release.

FUEL STIMULI DEPOLARIZE THE MEMBRANE AND RAISE $[Ca^{2+}]_i$

It has been known for many years that glucose and glyceraldehyde depolarize the B-cell membrane potential and evoke rhythmical electrical activity.[8] The underlying mechanism, as demonstrated in both normal B cells[9] and RINm5F cells,[7] is the closure of a specific K^+ channel called the "inward rectifier." Since ATP also closes the channel, this nucleotide has been proposed as a link between carbohydrate metabolism and alterations in membrane potential.[10,11] However, total cellular ATP levels are undoubtedly too high to be consistent with such a role, and change little over the glucose range stimulatory for insulin release.[12,13] It remains to be seen whether localized changes in ATP, consistent with such a hypothesis, can be demonstrated in insulin-secreting cells.

Changes in the average membrane potential of a cell suspension can be measured qualitatively with the fluorescent probe bisoxonol.[6] Thus, as shown in FIGURE 1A, glyceraldehyde after a lag of 15 sec, depolarizes RINm5F cells: patch clamping studies with the same clone indicate that the membrane potential is lowered from −70 to −20 mV by glyceraldehyde.[7] The triose also promotes a doubling of the resting $[Ca^{2+}]_i$ of 100 nM, as measured with the fluorescent indicator quin 2 (FIG. 1B). This effect is partially blocked by diltiazem and verapamil, which block voltage-dependent Ca^{2+} channels.[4,6] Depolarization-induced Ca^{2+} currents have been directly demonstrated in islet [14] and RINm5F cells.[15]

As shown in FIGURE 1C, the rate of glyceraldehyde-stimulated insulin secretion was increased after a lag of around 1 min compared to the peak in $[Ca^{2+}]_i$. The rate of insulin release remained elevated longer than $[Ca^{2+}]_i$, suggesting that $[Ca^{2+}]_i$ does not act as a moment-to-moment regulator of insulin release.[16]

To date the parallel analysis of insulin secretion and $[Ca^{2+}]_i$ in the same preparation has only been achieved in the RINm5F cells. Although using quin 2 glucose has also been shown to raise $[Ca^{2+}]_i$ in normal islet cells,[17] the cell suspensions do not exhibit completely normal insulin release in response to glucose.[4] Therefore, in our preliminary experiments we used monolayer cultures of adult rat islets, a preparation retaining glucose sensitivity.[4] When these monolayers were loaded with the new fluorescent Ca^{2+} indicator, fura-2, it was possible to measure changes in

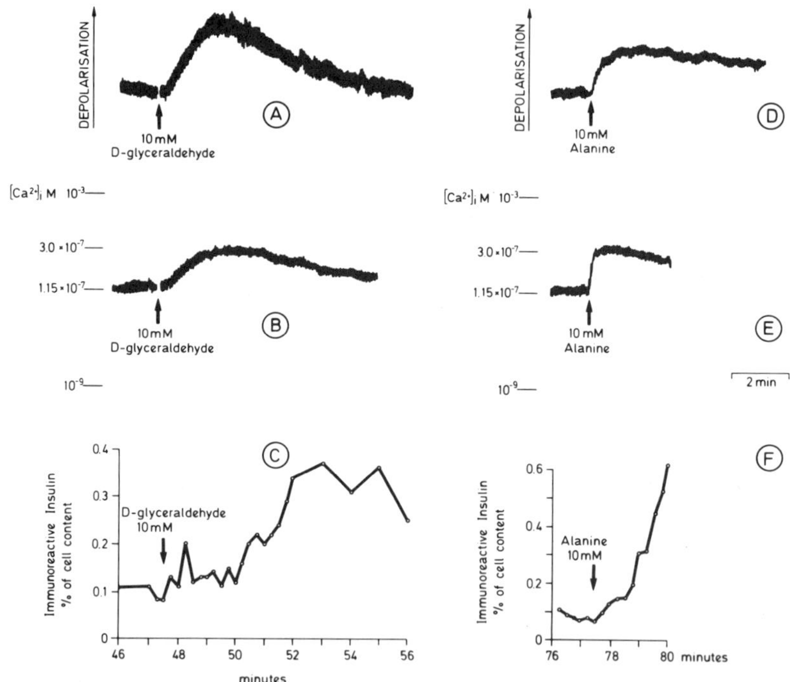

FIGURE 1. Effect of glyceraldehyde and alanine on membrane potential (A and D) $[Ca^{2+}]_i$ (B and E) and insulin secretion (C and F) in RINm5F cells. Membrane potential and $[Ca^{2+}]_i$ were measured with the fluorescent probe bisoxonol and with quin 2, respectively.[6] Immunoreactive insulin release was measured during perifusion of quin 2–loaded cells as described in Wollheim *et al.*[16]

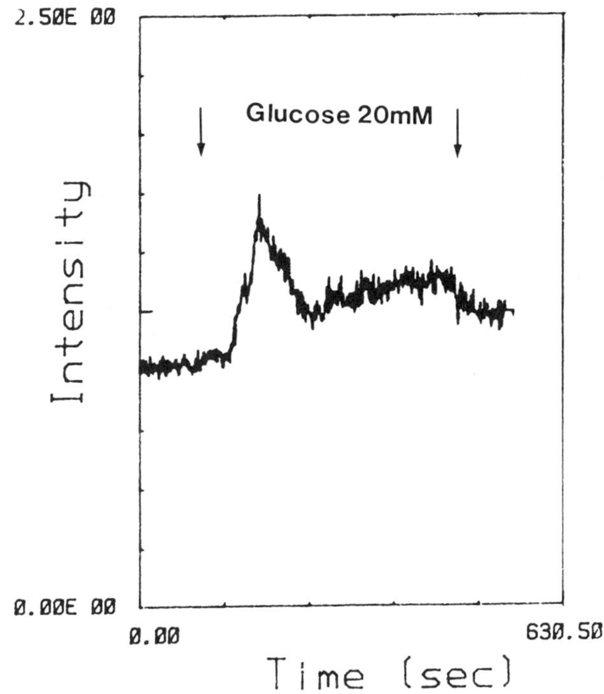

FIGURE 2. Effect of glucose on $[Ca^{2+}]_i$ in a single islet cell. Monolayer cultures of adult dispersed rat islets were prepared as outlined in Prentki & Wollheim.[4] On the third day of culture the cells were loaded with the fluorescent Ca^{2+} indicator fura-2 (1 μM of its acetoxymethyl ester for 30 min). A single cell was monitored in a fluorescence microscope attached to a double beam fluorometer (Fluorolog 2, Spex International). Excitation wavelengths were 350 and 380 nm, emission wavelength 505 nm. The ordinate gives the ratio between the fluorescence recorded at the two excitation wavelengths. For details see Tsien *et al.*[18] The cells were maintained at 37°C and superfused with the indicated glucose concentrations.

$[Ca^{2+}]_i$ in single cells by fluorescence microscopy.[18] A representative experiment is depicted in FIGURE 2. A "square-wave" increase of the glucose concentration from 2.8 mM to 20 mM in the superfusion medium raised $[Ca^{2+}]_i$ in a biphasic manner. After a lag of about 40 sec, $[Ca^{2+}]_i$ rose to a peak 90 sec after the beginning of the stimulation. In perifused, cultured rat islets the rate of insulin release is stimulated by glucose after a 1 min lag and reaches peak rates after 2–3 min.[3] Thus, both the onset and maximal increase of $[Ca^{2+}]_i$ precede the changes in insulin release both in glucose-stimulated islets and glyceraldehyde-challenged RINm5F cells.

Various amino acids are capable of releasing insulin from the pancreatic B cell. The most extensively studied, leucine, can act both as a fuel and an activator of glutamate dehydrogenase.[2,19] Leucine elicits membrane potential changes similar to those of glucose.[20] While normal B-cell preparations are only marginally sensitive to alanine, RINm5F cells respond well to this amino acid.[4,6] It can be seen in FIGURE 1D that alanine depolarizes the RINm5F cells. Alanine caused a rapid rise in $[Ca^{2+}]_i$ (FIG. 1E) and stimulation of insulin release (FIG. 1F). These effects were due to Ca^{2+} influx through voltage-gated Ca^{2+} channels, since verapamil abolished both the rise in

$[Ca^{2+}]_i$[4,6] and insulin release (TABLE 1).[4,6] This conclusion is further supported by the finding that alanine failed to raise $[Ca^{2+}]_i$ in the absence of extracellular Ca^{2+}.[21] We are thus unable to confirm an earlier suggestion that alanine elicits Ca^{2+} mobilization secondary to the Na^+ influx[22] necessary for co-transport of the amino acid.[23,24] Accordingly, when Na^+ was replaced by choline, alanine effects on $[Ca^{2+}]_i$ were markedly inhibited (not shown). In the presence of Na^+, alanine (10 mM) caused an incremental increase of insulin release of 17.8 ± 4.8 ng/10^6 cells/10 min and in choline medium 4.6 ± 1.3 ng/10^6 cells/10 min (mean ± S.E.M, $N = 6$). Although glucose-induced secretion is also inhibited under some circumstances in the absence of Na^+[25] this is not due to inhibition of glucose transport, but to a derangement of cytoplasmic pH buffering secondary to an inhibition of Na^+/H^+ exchange.[26]

The mode of action of alanine appears therefore to involve a Na^+-dependent depolarization of the membrane potential and gating of voltage-dependent Ca^{2+} channels. In pancreatic B cells, alanine causes a small depolarization but no electrical activity.[20] Moreover, alanine metabolism by pancreatic islets is low in the absence of glucose, but is enhanced by glucose and pyruvate.[27] As RINm5F cells display exaggerated glycolysis relative to islets,[5] it is possible that they metabolize alanine more avidly than the normal B cell. Another amino acid, glutamine, does not elicit insulin release from pancreatic islets when added alone but markedly potentiates release evoked by leucine or its non-metabolizable analogue BCH.[28] In RINm5F cells glutamine depolarized the membrane potential and raised $[Ca^{2+}]_i$ (FIG. 3). Glutamine also stimulated insulin release, which was abolished by verapamil (TABLE 1). It appears that glutamine like alanine elicits insulin release by gating of Ca^{2+} channels. The effect of glutamine on the membrane potential of B cells has not been reported, but it can be assumed to be marginal as the reduction of $^{86}Rb^+$ outflow (a K^+ substitute) was not as marked as that seen with glucose.[29] Although glutamine is well metabolized by islets, it was concluded that the amino acid, because of its sparing effect on the utilization of endogenous substrates, does not enhance islet cell respiration sufficiently to induce insulin release.[29]

Whereas glucose and glyceraldehyde, on the one hand, and alanine and glutamine, on the other, raise $[Ca^{2+}]_i$ by opening voltage-gated Ca^{2+} channels, there is a clear difference between the carbohydrates and the amino acids. The actions of the amino acids both on $[Ca^{2+}]_i$ and insulin release are abolished by verapamil, a situation also seen with depolarizing concentrations of K^+.[6] In contrast, glyceraldehyde-induced rises in $[Ca^{2+}]_i$ are attenuated but not abolished by verapamil or diltiazem.[4,6] This suggests that the carbohydrate stimulants in addition to increasing Ca^{2+} influx may also raise $[Ca^{2+}]_i$ by another mechanism. As previously proposed the carbohydrates

TABLE 1. Effect of Verapamil on Alanine and Glutamine-Stimulated Insulin Release from RINm5F Cells

Amino Acid	Immunoreactive Insulin Release (ng/10^6 cells/10 min) Control	Verapamil (20 μM)
None	14.2 ± 0.8	12.4 ± 1.2
Alanine (10 mM)	26.8 ± 1.9	15.8 ± 1.2
Glutamine (6 mM)	22.2 ± 1.6	14.8 ± 0.9

Cells were incubated in buffer containing glucose 2.8 mM as in Wollheim & Pozzan.[6] Results are given as mean ± SEM, $N = 12$.

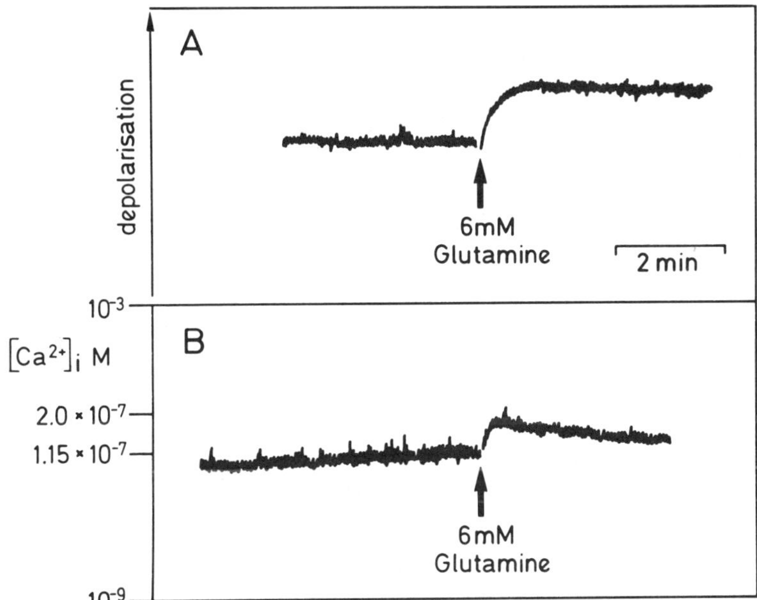

FIGURE 3. Effect of glutamine on membrane potential (A) and $[Ca^{2+}]_i$ (B) in RINm5F cells. Cells were loaded with quin 2 for the measurement of $[Ca^{2+}]_i$ as described in Wollheim & Pozzan.[6] Membrane potential was monitored with bisoxonol.[6]

could mobilize Ca^{2+}.[3,4] However, to date it has not been possible to directly demonstrate such an effect by using quin 2 or fura-2. A possible explanation for this is discussed below.

CARBACHOL RAISES $[Ca^{2+}]_i$ FOLLOWING THE GENERATION OF INOSITOL 1,4,5-TRISPHOSPHATE

Acetylcholine and carbachol have been shown to stimulate insulin release and enhance $^{45}Ca^{2+}$ fluxes in pancreatic islets.[30–32] To investigate the mode of action of muscarinic agonists we measured in parallel $[Ca^{2+}]_i$ and insulin release in RINm5F cells.[33] As can be seen in FIGURE 4 carbachol caused an almost three-fold increase in $[Ca^{2+}]_i$ (TABLE 2). The initial kinetics of the $[Ca^{2+}]_i$ rise were analyzed in fura-2 loaded RINm5F cells, since this indicator has less buffering capacity and is more sensitive to $[Ca^{2+}]_i$ changes than quin 2.[18] It was found that the rise in $[Ca^{2+}]_i$ occurred after a 2 sec lag and was maximal at 13 sec. Carbachol raised $[Ca^{2+}]_i$ in the absence of extracellular Ca^{2+}, clearly demonstrating mobilization by the agonist.[33] This conclusion was also borne out by the finding that carbachol did not depolarize the cells and that the rise in $[Ca^{2+}]_i$ was seen in the presence of verapamil.[33]

It was first shown in pancreatic acinar cells that the Ca^{2+} mobilizing action of carbachol is mediated by the generation of inositol 1,4,5-trisphosphate (Ins 1,4,5-P_3).[34,35] This compound is formed concurrently with diacylglycerol, also a second messenger,[36] upon hydrolysis of a minor lipid component of the plasma membrane,

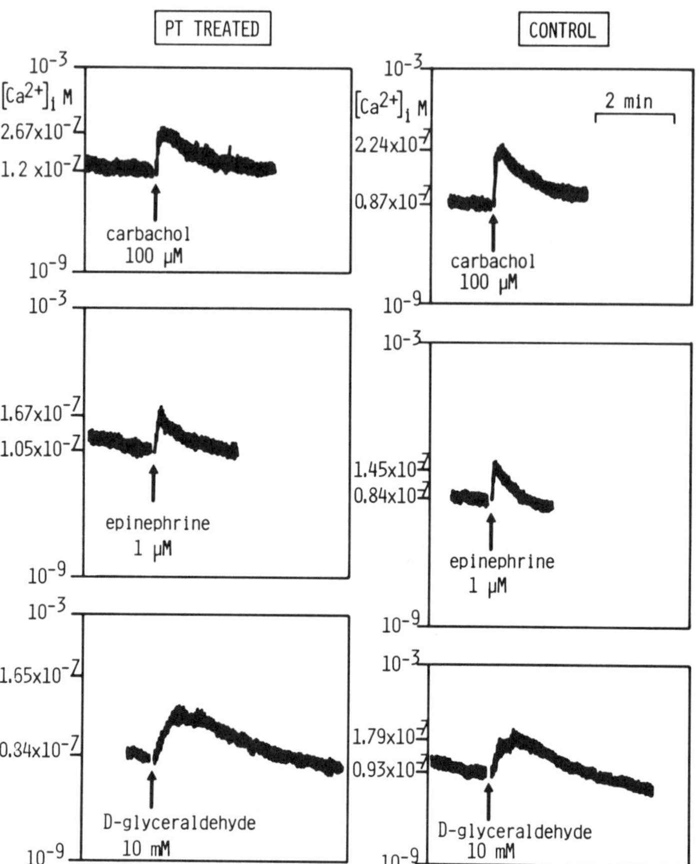

FIGURE 4. Changes in $[Ca^{2+}]_i$ in response to carbachol, epinephrine, and glyceraldehyde in pertussis toxin pretreated RINm5F cells. Cells were treated with 100 ng/ml pertussis toxin (PT) or vehicle (control) during a 3 hr spinner culture period preceding the loading with quin 2. For details on cell culture and the quin 2 method see Wollheim & Pozzan.[6]

phosphatidyl inositol 4,5-bisphosphate (Ptd Ins 4,5-P_2).[37] Carbachol-stimulated Ptd Ins 4,5-P_2 hydrolysis had been previously reported in islets incubated for 1 min.[32] In a study carried out in RINm5F cells, carbachol raised Ins 1,4,5-P_3 maximally at 2–5 sec which then returned towards pre-stimulatory levels.[33] Thus, the peak in Ins 1,4,5-P_3 clearly preceded that of $[Ca^{2+}]_i$. With regard to carbachol-stimulated insulin release from perifused, quin 2–loaded RINm5F cells, it was observed that the release rate only increased after 1 min, and was maximal when $[Ca^{2+}]_i$ had returned towards basal levels.[33] It appears, therefore, that for carbachol-induced insulin release, as is the case for carbohydrates, $[Ca^{2+}]_i$ is not a moment-to-moment regulator of insulin secretion.

It is now more than 10 years since the discovery by Ui and his co-workers that injection of rats with pertussis vaccine results in drastic alterations of insulin secretion from the pancreas.[38] The secretory responses to several stimuli including glucose were exaggerated and the alpha adrenergic inhibition of secretion by epinephrine was converted into a beta adrenergic stimulation.[38] The active principle, pertussis toxin, was therefore called islet activating protein and found to be active also when added directly to isolated islets in tissue culture.[39] Further studies revealed that pertussis toxin ADP ribosylates the alpha subunit of the inhibitory guanine nucleotide binding protein of the adenylate cyclase system, N_i (also referred to as G_i) in a variety of cell types.[40] Pertussis toxin thus blocks the action of receptor agonists, which exert inhibitory actions on adenylate cyclase. It was subsequently found that pretreatment of certain cell types like mast cells,[41] neutrophils,[42] and HL-60 cells[42] interferes with subsequent agonist-induced hydrolysis of Ptd Ins 4,5-P_2 and generation of Ins-P_3. This also resulted in an impairment of agonist-induced rises in $[Ca^{2+}]_i$.[42] As the hydrolysis of Ptd Ins 4,5-P_2 was also demonstrated in plasma membrane fractions of neutrophils[43] and hepatocytes by the addition of GTP or its non-hydrolyzable analogues, it was concluded that the stimulation of phospholipase C by Ca^{2+} mobilizing agonists is mediated by a GTP-binding protein also termed Np.[43,44]

We examined the possible involvement of a pertussis toxin substrate (Np) in the rise of $[Ca^{2+}]_i$ evoked by the two agonists carbachol (acting on muscarinic receptors) and epinephrine, which can activate $alpha_1$ adrenoceptors.[21] As can be seen in FIGURE 4, the effects of the two receptor agonists, as well as those of glyceraldehyde, were not affected by pretreatment of the cells with pertussis toxin. The results of several experiments are given in TABLE 2. It could be argued that the toxin treatment was ineffective. This does not seem to be the case, as indicated by measurements of insulin secretion from pertussis toxin–pretreated RINm5F cells (TABLE 3). Both glyceraldehyde and alanine approximately doubled the rate of insulin secretion. In the control cells this stimulation was abolished by epinephrine. In contrast, epinephrine was completely ineffective in inhibiting insulin release from toxin pretreated cells. It can thus be concluded that, as in pancreatic islets, pertussis toxin abolishes the $alpha_2$ adrenergic inhibitory action of epinephrine on insulin secretion.[39] However, as in several other cell types,[45,46] the toxin does not affect the Ca^{2+} generating signal. However, this cannot be equated with the lack of involvement of an N protein in the effect of the Ca^{2+} mobilizing actions of carbachol or $alpha_1$ agonists. We have recently observed that non-hydrolyzable analogues of GTP are capable of generating Ins P_3 in electrically permeabilized RINm5F cells.[47] It appears therefore that the N protein,

TABLE 2. Effects of Carbachol, Epinephrine, and Glyceraldehyde on $[Ca^{2+}]_i$ in RINm5F Cells Pretreated with Pertussis Toxin

	Control		Pertussis Toxin	
	$[Ca^{2+}]_i$ (nM)			
	Basal	Peak	Basal	Peak
Carbachol (100 μM)	97 ± 10	193 ± 12 (7)	115 ± 6	258 ± 31 (7)
Epinephrine (1 μM)	102 ± 9	166 ± 19 (6]	101 ± 12	163 ± 19 (6)
Glyceraldehyde (10 mM)	82 ± 7	156 ± 12 (3)	83 ± 1	149 ± 8 (3)

Cells were treated with pertussis toxin (100 ng/ml) or vehicle for 3 hr during spinner culture. They were then loaded with quin 2 acetoxymethylester and further handled as described in Wollheim & Pozzan.[6] The mean basal $[Ca^{2+}]_i$ of the controls was 96 ± 6 nM and that of pertussis-treated cells 104 ± 6 nM, $N = 16$. Results are given as mean ± SEM and the number of observations is indicated in parentheses.

TABLE 3. Pertussis Toxin Pretreatment of RINm5F Cells Abolishes Epinephrine-Induced Inhibition of Insulin Release

	Immunoreactive Insulin Release (ng/10^6 cells/10 min)			
	Control Cells		Pertussis Toxin–Treated Cells	
	−epi	+epi	−epi	+epi
Glucose 2.8 mM	12.5 ± 1.0	10.1 ± 1.5	14.8 ± 0.9	16.5 ± 1.4
Glucose 2.8 mM + glyceraldehyde 10 mM	23.7 ± 2.9	13.6 ± 0.8	31.1 ± 3.6	28.9 ± 1.4
Glucose 2.8 mM + alanine 10 mM	25.0 ± 3.1	12.1 ± 2.2	29.6 ± 3.9	27.2 ± 3.1

Cells were pretreated with pertussis toxin (100 ng/ml) or vehicle during 3 hr in spinner culture. They were then incubated as described in Wollheim & Pozzan.[6] When present epinephrine (epi) was used at a final concentration of 1 μM. Results are the mean ± SEM of 8–10 observations.

which activates phospholipase C in the insulin-secreting cells, is not a substrate for pertussis toxin.

INTRACELLULAR Ca^{2+} HOMEOSTASIS IN INSULIN-SECRETING CELLS

Although the flux across the plasma membrane must ultimately determine the amount of Ca^{2+} inside the cell, there can be little doubt that intracellular organelles must also play an important role in the regulation of $[Ca^{2+}]_i$. This is true for all cell types. Although, for many years it was thought that the mitochondria played the predominant role, recent evidence involving studies with both intact[48,49] and broken cells,[50,51] strongly suggests that the endoplasmic reticulum (ER) is the organelle chiefly responsible for intracellular Ca^{2+} regulation, at least under resting conditions. A similar conclusion can be reached by considering studies done with insulin-secreting cells. Perhaps surprisingly, the granules in these cells, although containing large amounts of Ca^{2+}, seem to play no role in the short-term regulation of $[Ca^{2+}]_i$.[4,52] On the other hand, the ER, as studied in subcellular fractions of pancreatic islets, exhibits active Ca^{2+} uptake that is dependent on MgATP and has a high affinity for Ca^{2+}.[53] The functional interrelationships between the ER and mitochondria have been well characterized by the use of Ca^{2+}-specific minielectrodes to monitor, under different conditions, the ambient free Ca^{2+} concentration ($[Ca^{2+}]_a$) maintained by subcellular fractions[52,54,55] or permeabilized cells.[56] The basic approach is exemplified in FIGURE 5. In this instance electrically permeabilized cells were added to an incubation medium the ionic composition of which approximates that normally found in cytosol.[50] In the presence of MgATP the cells immediately lowered the initial $[Ca^{2+}]_a$ of the medium, thereby indicating net Ca^{2+} uptake into an intracellular pool (since the plasma membranes of these cells have been rendered freely permeable to low molecular weight substances including Ca^{2+}). The organelle maintaining this pool displayed a high affinity for Ca^{2+} since it established and then maintained a new Ca^{2+} steady state of approximately 200 nM, close to the resting $[Ca^{2+}]_i$ found in intact cells (TABLE 2). This most probably represents Ca^{2+} transport by the ER, in view of its high affinity, dependence on MgATP, sensitivity to vanadate,[56] and its functional similarity to Ca^{2+} uptake by isolated vesicles enriched in ER.[52]

As shown previously,[56,57] Ins 1,4,5-P_3 releases Ca^{2+} from the ER pool of insulin secreting cells (FIG. 5). The increase in $[Ca^{2+}]_a$ is transitory since Ins 1,4,5-P_3 is metabolized in these preparations.[56] When Ins 1,4,5-P_3 is continuously infused, $[Ca^{2+}]_a$ is maintained at a new elevated steady state.[58] This latter observation rules out the possibility, previously entertained[56,57] that the ER might become "densensitized" to Ins 1,4,5-P_3. It should also be noted that the responsiveness of the ER is highly selective, since neither inositol 1,4-bisphosphate (Ins 1,4-P_2),[56,57] nor the newly discovered inositol 1,3,4,5-tetrakisphosphate (Ins 1,3,4,5-P_4)[59,60] had any effect at concentrations that are maximal for Ins 1,4,5-P_3 (FIG. 5).

Under conditions in which MgATP is omitted from the incubation medium, but respiratory substrate is included, permeabilized cells maintain a $[Ca^{2+}]_a$ steady state of about 800 nM.[56] This is due to mitochondrial Ca^{2+} transport since it is abolished by the respiratory poison antimycin,[56] and displays characteristics similar to those seen with preparations of isolated mitochondria.[54] This transport system shows a lower affinity but higher capacity for Ca^{2+} than does that of the ER.[55,56] Nevertheless the capacity of the latter pool in RINm5F cells is relatively large (around 14 nmol/mg cell protein),[58] so it can be concluded that under non-stimulatory conditions of Ca^{2+} loading, the ER would make a larger contribution to the regulation of $[Ca^{2+}]_i$ than mitochondria. Thus, it has been clearly shown in studies with permeabilized cells,[56] or mixtures of ER vesicles and mitochondria,[55] that the ER maintains $[Ca^{2+}]_a$ at 100–200 nM, independent of functional mitochondrial transport. However, the existence of two intracellular Ca^{2+} pools raises the possibility that mitochondria may compete with the ER during the reuptake phase after Ca^{2+} mobilization by Ins 1,4,5-P_3. We have shown that this actually occurs using the clonal pituitary GH_3 cell line,[61] and if also seen in other systems it might account for the fact that RINm5F cells are slightly refractory to sequential additions of Ins 1,4,5-P_3.[56]

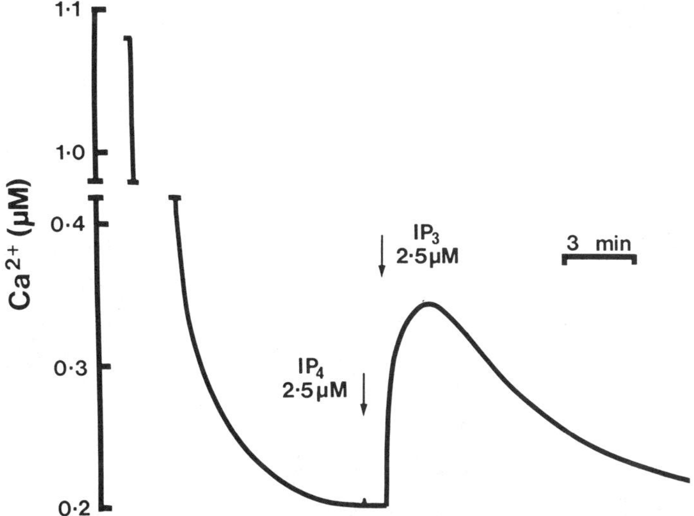

FIGURE 5. Mobilization of Ca^{2+} from the ER of permeabilized RINm5F cells by Ins 1,4,5-P_3 (IP_3) but not Ins 1,3,4,5,-P_4 (IP_4). Cells were permeabilized by 10 × 100 μs electrical discharges (3 kV/cm) and resuspended in an intracellular medium containing MgATP, antimycin A, and oligomycin.[56] Changes in $[Ca^{2+}]_a$ were monitored with a Ca^{2+} selective electrode.[54]

EFFECTS OF GLUCOSE ON INTRACELLULAR Ca^{2+} HANDLING

As well as exerting their major effect on gating plasma membrane Ca^{2+} channels, there is considerable evidence that carbohydrate nutrients such as glucose also mobilize Ca^{2+} from intracellular stores.[3,4] However, it has been alternatively proposed that glucose promotes net uptake of Ca^{2+} into an intracellular site.[62] This is undoubtedly true during prolonged stimulation[63] and by analogy with other systems the most likely site of accumulation would be the mitochondria.[61] However, according to the hypotheses in question, glucose promotes an early net uptake of Ca^{2+} into an intracellular store now identified as the ER,[64] thereby causing a decrease in $[Ca^{2+}]_i$.[17] All of the key experiments in support of this hypothesis[17,62,64,65] have been performed with cells preincubated in the absence of nutrient substrate and therefore depleted in ATP.[12] In many instances intracellular Ca^{2+} stores were also depleted.[17,64,65] Thus in a recent study with RINm5F cells glucose caused a net uptake of Ca^{2+} into a non-mitochondrial store that could be partially emptied after addition of carbachol.[64] Net uptake under these conditions is hardly surprising in view of the well documented dependence of ER Ca^{2+} transport on ATP; indeed the fact that glucose did not stimulate Ca^{2+} uptake, under conditions where ATP synthesis was inhibited with antimycin A, clearly points to a fuel effect. That sequestration by non mitochondrial stores depends on a mitochondrial function unrelated to ATP production is unlikely in view of the fact that isolated ER vesicles,[55] or permeabilized RINm5F cells,[56] can maintain $[Ca^{2+}]_a$ at 100–200 nM in the presence of respiratory poisons, as long as exogenous ATP is provided. However, since normal ATP levels can be maintained in intact cells by concentrations of glucose that are substimulatory for insulin release,[12] the physiological significance of the glucose effect discussed above is unclear. Therefore the onus remains on the proponents of this hypothesis to show that glucose can also stimulate net uptake of Ca^{2+} into the ER of cells that contain normal ATP levels and non-depleted Ca^{2+} stores.

PUTATIVE MECHANISMS BY WHICH GLUCOSE MOBILIZES Ca^{2+}

Changes In Phosphorylation Potential

Alterations in the ATP/ADP ratio have also been proposed as a mechanism by which glucose might mobilize Ca^{2+} from the ER of insulin-secreting cells.[52] Thus in a microsomal preparation from rat insulinoma, it has been shown that a rise in ADP relative to ATP increased the $[Ca^{2+}]_a$ steady state.[52] Since it is possible that ATP levels fall during the initial seconds following glucose challenge, due to increased glucose phosphorylation, these results are not without interest. However, at present the proposal that glucose mobilizes Ca^{2+} through alterations in the phosphorylation potential is somewhat speculative.

Arachidonic Acid

An alternative proposal that has recently enjoyed some attention is that arachidonic acid, which is liberated upon stimulation of B cells, may act as a second messenger to release Ca^{2+} from intracellular stores.[66] This proposal is based on experiments performed with the "digitonin-permeabilized pancreatic islet model".[66] However, it is obvious that a model based upon detergent treatment of a piece of tissue, with its inherent problems of a large extracellular space and non-uniform permeabilization, is hardly ideal for this type of study. Indeed, despite the extremely well documented sensitivity of insulin secretory cells to Ins 1,4,5-P_3,[56,67] the permeabilized islet model was barely responsive.[68] Thus the K_m for Ins 1,4,5-P_3 induced Ca^{2+} release was at least an order of magnitude higher than previously reported,[56,67] and the kinetics of release much slower (maximal release occurring after 2–3 min rather than within seconds).[56,67,68] Moreover, submaximal concentrations of Ins 1,4,5-P_3 were able to promote significant release, only when vanadate was added to inhibit the reuptake of Ca^{2+}.[68] When arachidonic acid was used in this model it promoted a Ca^{2+} release that was very similar to that seen with Ins 1,4,5-P_3—with two important exceptions.[66] Firstly, arachidonic acid released Ca^{2+} from mitochondria as well as ER. Secondly, little molecular specificity was shown since other non-saturated fatty acids such as oleic acid were also effective.[66] These observations are already strong arguments against the hypothesis that arachidonic acid plays any specific role in intracellular Ca^{2+} homeostasis. Nevertheless, we have reinvestigated its effects using a more sensitive experimental system than previously employed. It can be readily seen from FIGURE 6, that arachidonic acid does indeed release Ca^{2+} from the ER but that this effect is in marked contrast to that induced by Ins 1,4,5-P_3, the release is much slower and is irreversible. Although irreversibility may be due to the fact that arachidonic acid is not necessarily degraded by this preparation, it should be stressed that under conditions where Ins 1,4,5-P_3 is continually present, there is no irreversible depletion of Ca^{2+} from the ER, but a controlled release that results in the maintenance of a new $[Ca^{2+}]_a$ steady state.[58] It must be concluded, therefore, that arachidonic acid acts *in vitro* in a completely non-specific fashion to disrupt membrane permeability, thereby provoking uncontrolled leakage of Ca^{2+} down its concentration gradient. It remains extremely dubious whether such a potentially deleterious phenomenon plays any physiological role *in vivo*.

Generation of Ins 1,4,5-P_3

Although it has been known for many years that nutrient secretagogues stimulate phosphoinositide turnover in islets,[69–73] it is only recently that concomitant increases in inositol phosphates including Ins-P_3 have been demonstrated.[74,75] In TABLE 4 we have compared the effects of glucose and carbachol on inositol phosphate generation in pancreatic islets. Consistent with experiments carried out with RINm5F cells (see above), carbachol was a potent stimulus in islets, even when extracellular Ca^{2+} was depleted with EGTA. Glucose, on the other hand, produced smaller effects that may be secondary to the gating of voltage-dependent Ca^{2+} channels, since they were inhibited by removal of extracellular Ca^{2+}. Finally, 2-deoxyglucose, which is not metabolized by islets and does not act as a secretagogue, had no effect on inositol phosphate levels.

We have now confirmed, by HPLC analysis, that glucose also raises Ins 1,4,5-P_3 (not just total Ins-P_3 as measured above) in a manner consistent with its proposed role in glucose-induced Ca^{2+}-mobilization.[76] The demonstration that this rise was largely (though perhaps not entirely) dependent on extracellular Ca^{2+},[76] is therefore consistent with the experiments measuring $[Ca^{2+}]_i$,[4,6] in which evidence for Ca^{2+} mobilization with carbohydrate secretagogues was only obtained when Ca^{2+} was present in the incubation medium. The findings are also in keeping with the repeated demonstrations that the effects of glucose on phosphoinositide turnover,[72,77] Ptd Ins 4,5-P_2 hydrolysis[73,74,78] and inositol phosphate generation[74] are Ca^{2+} dependent. It is therefore likely that most of the Ins-P_3 generated in response to glucose is due to a Ca^{2+}-mediated hydrolysis of Ptd Ins 4,5-P_2, secondary to the gating of voltage-dependent Ca^{2+} channels. Such a mechanism has been previously proposed on the basis of experiments

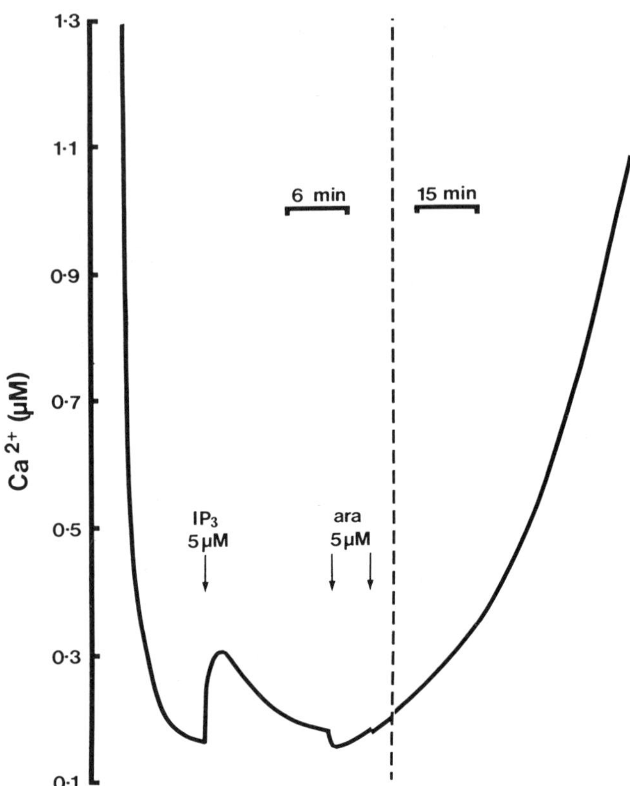

FIGURE 6. Release of Ca^{2+} from the ER of permeabilized RINm5F cells by Ins 1,4,5-P_3 (IP_3) and arachidonic acid (ara). Cells were permeabilized and incubated as described in the legend to FIGURE 5. The downward deflection after addition of arachidonic acid is an artefact due to the solvent used. Solvent alone did not release Ca^{2+} from intracellular stores, nor did arachidonic acid alter the potential in the absence of cells. Note the change in time scale.

with a Ca^{2+} ionophore and pancreatic islets.[73] We have subsequently characterized the Ca^{2+} mediated generation of Ins 1,4,5-P_3 in considerable detail, using K^+ depolarized islets and RINm5F cells.[76] However, it should be stressed that a transitory, Ca^{2+}-independent rise in Ins 1,4,5-P_3 may also be involved in the process whereby nutrient secretagogues stimulate insulin release.[76] Possible mechanisms by which this rise might occur are discussed below.

INOSITOL PHOSPHATE METABOLISM IN INSULIN SECRETORY CELLS

Knowledge of inositol phosphate metabolism in other tissues suggests that stimulation of Ptd Ins 4,5-P_2 hydrolysis is not the only mechanism by which glucose might raise Ins 1,4,5-P_3. It is known for example that glucose C-atoms can be directly incorporated into myoinositol in many tissues[79]; the rate-limiting step of this metabolic sequence, conversion of glucose-6-P to L-inositol-1-phosphate (L-Ins-P_1) is catalyzed by Ins-P_1 synthase (FIG. 7.)[79] The possibility therefore arises that should free inositol be in short supply in islet cells, glucose might activate phosphoinositide turnover through the activity of Ins-P_1 synthase. However, we have demonstrated recently that inositol is concentrated in islet cells via an active transport mechanism, energized by co-transport with Na^+.[80] It is therefore unlikely that metabolic conversion makes a substantial contribution to the myoinositol pool in these cells. However, L-Ins-P_1 derived from glucose-6-P might still account for a significant proportion of the total Ins-P_1 pool. Since both D and L forms of Ins-P_1 compete for the same phosphomonoesterase enzyme,[81] any rise in L-Ins-P_1 could lead to a build-up of Ins-P_2 and Ins-P_3. Glucose, through the activity of Ins-P_1 synthase, might thus exert effects on inositol phosphate metabolism similar to those seen with Li^+, which also inhibits breakdown of Ins-P_1 (FIG. 7).

Alternatively, glucose metabolites may exert even more direct effects in a manner similar to that in which 2,3-bisphosphoglycerate inhibits Ins-P_3 phosphomonoesterase.[82] Interestingly, a recent study[83] lends support to this hypothesis with the demonstration of inhibitory effects of phosphoenolpyruvate and fructose 1,6-P_2 at various points during the stepwise dephosphorylation of Ins-P_3 (FIG. 7).

However, dephosphorylation is not the only metabolic fate to befall this compound (FIG. 7). It can also be converted by a two-step isomerization process to Ins 1,3,4-P_3[60] via the newly discovered intermediate Ins 1,3,4,5-P_4.[59,60] The initial reaction in this sequence, catalyzed by a specific Ins 1,4,5-P_3-kinase, may thus be as important in removing Ins 1,4,5-P_3 as the dephosphorylation reaction. In fact, in a recent study with RINm5F cells, we have shown that the kinase step may be quantitatively more important, at least in the initial seconds after muscarinic stimulation.[84] Moreover, the demonstration that the maximal activity of this enzyme could be stimulated by an increase of Ca^{2+} over the physiological range suggests that the rise in $[Ca^{2+}]_i$ due to the action of Ins 1,4,5-P_3, may in turn stimulate its own catabolism.[84] In this manner Ins 1,4,5-P_3-kinase may counterregulate increases in $[Ca^{2+}]_i$.

TABLE 4. Effects of Various Additions on Inositol Phosphate Accumulation in Isolated Pancreatic Islets

Additions	Inositol Phosphates		
	Ins-P_1	Ins-P_2	Ins-P_3
		% control	
2.8 mM glucose	100 ± 2 (31)	100 ± 4 (31)	99 ± 2 (31)
16.7 mM glucose	114 ± 4 (30)[a]	115 ± 6 (30)[a]	133 ± 6 (30)[a]
16.7 mM glucose + 2 mM EGTA	106 ± 5 (21)	97 ± 6 (21)	108 ± 3 (21)
500 μM carbachol	181 ± 11 (12)[a]	194 ± 7 (12)[a]	266 ± 6 (12)[a]
500 μM carbachol + 2 mM EGTA	157 ± 7 (13)[a]	174 ± 10 (13)[a]	208 ± 8 (13)[a]
16.7 mM 2-deoxyglucose	116 ± 9 (9)	100 ± 8 (9)	100 ± 6 (9)

Islets were cultured in the presence of myo-[2^3H]inositol for 2–3 days. Groups of 50 islets were incubated for 20 min in Krebs-Ringer bicarbonate buffer and then stimulated for 1 min with the additions as described. Unless otherwise specified 2.8 mM glucose was present during each incubation. Inositol phosphates in the aqueous cell extracts were separated by anion exchange chromatography using Dowex 1 × 8 columns and a stepwise gradient of 0–1.0 M ammonium formate. Radioactivity in the eluate was determined by liquid scintillation spectrometry. Results are expressed as a percentage of the 2.8 mM glucose response at the head of each column, which averaged 628 ± 34, 52 ± 2, and 217 ± 9 c.p.m. for Ins-P_1, Ins-P_2, and Ins-P_3 respectively. Each value is the mean ± S.E.M. with the number of observations in parentheses.

[a]Significantly different from control ($p < 0.05$).

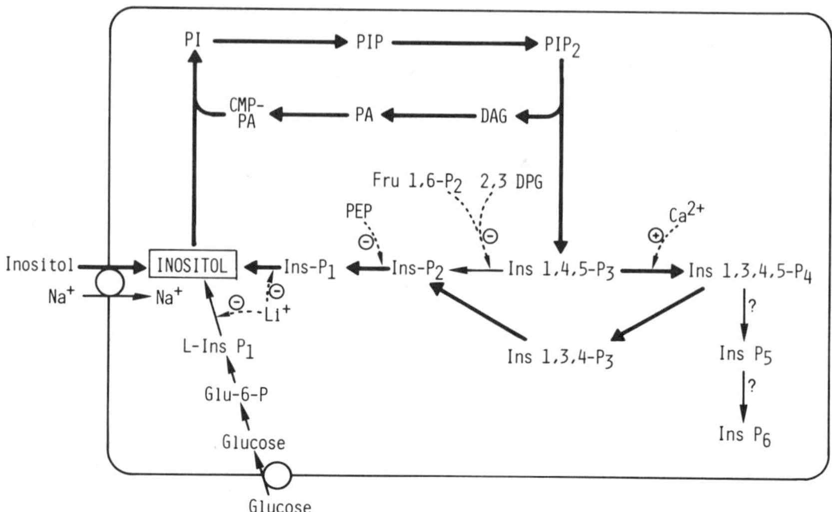

FIGURE 7. Inositol metabolism in insulin-secreting cells. Ptd Ins (PI), Ptd Ins 4-P (PIP), Ptd Ins 4,5-P_2 (PIP_2), diacylglycerol (DAG), and phosphatidic acid (PA) are hydrophobic compounds associated with the plasma membrane. Ptd Ins 4,5-P_2 is cleaved by phospholipase C activity to yield DAG and the water-soluble Ins 1,4,5-P_3. The latter is metabolized either by a series of phosphomonoesterase enzymes to yield eventually free inositol, or by Ins 1,4,5-P_3-3-kinase to form Ins 1,3,4,5-P_4. Possible sites of action of phosphoenolpyruvate (PEP), fructose-1,6-P_2 (fru 1,6-P_2), and 2,3-bisphosphoglycerate (2,3 DPG) are also shown.

Finally, it is of interest to note that the levels of Ins 1,3,4,5-P_4 and Ins 1,3,4-P_3 are also increased in pancreatic islets upon stimulation with glucose.[76] This raises the possibility that these compounds may also exert some function in the activation of the secretory process. Such a proposal is especially interesting with regard to Ins 1,3,4,5-P_4, which is significantly increased within 2 sec of glucose addition.[76] However, at present no precise function for this metabolite has been demonstrated. It would not, for example, appear to play a role in the mobilization of intracellular Ca^{2+} stores (FIG. 5).

SUMMARY

The initial events in signal transduction in insulin-secreting cells are summarized in FIGURE 8. Both nutrient stimuli, such as glucose and amino acids and the muscarinic agonist carbachol (carbamylcholine) raise $[Ca^{2+}]_i$. Although the rise in $[Ca^{2+}]_i$ precedes the stimulation of insulin release, it is not a moment-to-moment regulator of release. The metabolizable fuel stimuli cause Ca^{2+} influx through voltage-dependent Ca^{2+} channels following depolarization of the membrane potential. In contrast, carbachol, which does not depolarize, elicits Ptd Ins 4,5-P_2 hydrolysis, a reaction catalyzed by phospholipase C. The generation of Ins 1,4,5-P_3 in this instance is Ca^{2+} independent, but appears to involve a GTP-binding protein. However, this protein is not a substrate for pertussis toxin. The levels of Ins 1,4,5-P_3, which releases Ca^{2+} from an ATP-dependent Ca^{2+} pool of the endoplasmic reticulum, are increased prior to the

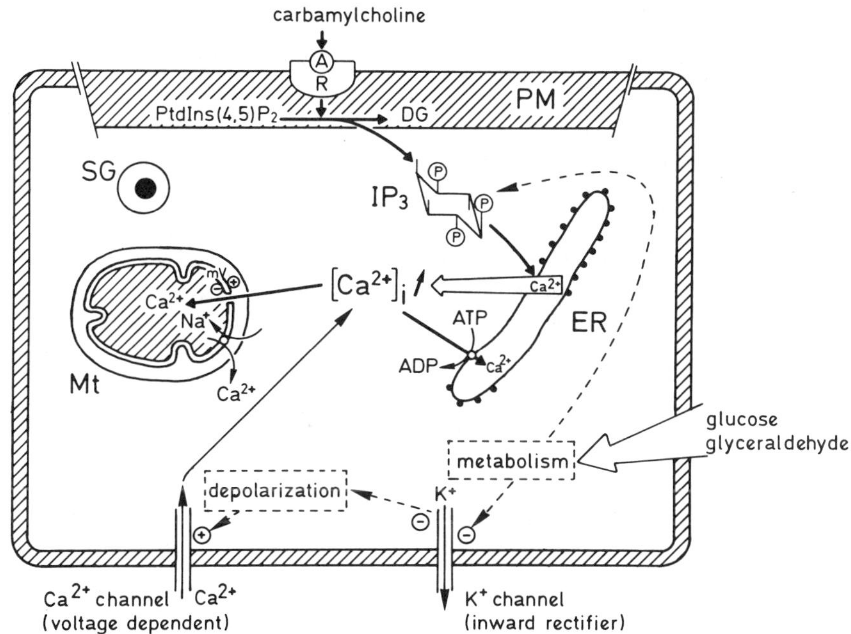

FIGURE 8. Ca^{2+} handling in insulin-secreting cells stimulated with fuels (glucose, glyceraldehyde, and amino acids) or the receptor agonist carbachol (carbamylcholine). Phospholipase C catalyzes the hydrolysis of Ptd Ins 4,5-P_2 generating diacylglycerol (DAG) and Ins 1,4,5-P_3. The former remains in the membrane while the latter mobilizes Ca^{2+} from the endoplasmic reticulum (ER). M, mitochondria; SG, insulin containing secretory granules; and PM, plasma membrane. For further explanations, see text.

rise in $[Ca^{2+}]_i$. The mitochondria may take up Ca^{2+} after large increases in $[Ca^{2+}]_i$. A previously proposed second messenger, arachidonic acid, is much less selective than Ins 1,4,5-P_3 in that it releases Ca^{2+} from mitochondria as well as from the endoplasmic reticulum in a slow and irreversible manner. As Ins 1,4,5-P_3 is also generated during glucose stimulation of islets, albeit in a Ca^{2+}-dependent manner, this metabolite could mediate not only the action of carbachol but also contribute to amplifying the $[Ca^{2+}]_i$ rise in response to glucose.

REFERENCES

1. MILLER, R. E. 1981. Endocrine Rev. **2:** 471–494.
2. SENER, A. & W. J. MALAISSE. 1984. Experientia **40:** 1026–1035.
3. WOLLHEIM, C. B. & G. W. G. SHARP. 1981. Physiol. Rev. **61:** 914–973.
4. PRENTKI, M. & C. B. WOLLHEIM. 1984. Experientia **40:** 1052–1060.
5. HALBAN, P. A., G. PRAZ & C. B. WOLLHEIM. 1983. Biochem. J. **212:** 439–443.
6. WOLLHEIM, C. B. & T. POZZAN. 1984. J. Biol. Chem. **259:** 2262–2267.
7. DUNNE, M. J., I. FINDLAY, O.H. PETERSEN & C. B. WOLLHEIM. 1986. J. Membr. Biol. **93:** 271–279.
8. DEAN, P. M., E. K. MATTHEWS & Y. SAKAMOTO. 1975. J. Physiol. (London) **246:** 459–478.

9. Ashcroft, F. M., D. E. Harrison & S. J. H. Ashcroft. 1984. Nature **312:** 446–448.
10. Cook, D. L. & C. N. Hales. 1984. Nature **311:** 271–273.
11. Findlay, I., M. J. Dunne & O. H. Petersen. 1985. J. Membr. Biol. **88:** 165–172.
12. Ashcroft, S. J. H., L. C. C. Weerasinghe & P. J. Randle. 1973. Biochem. J. **132:** 223–231.
13. Malaisse, W. J., J. C. Hutton, S. Kawazu, A. Herchuelz, I. Valverde & A. Sener. 1979. Diabetologia **16:** 331–341.
14. Satin, L. S. & D. L. Cook. 1985. Pflügers Arch. **400:** 385–387.
15. Findlay, I. & M. J. Dunne. 1986. FEBS Lett. **189:** 281–285.
16. Wollheim, C. B., S. Ullrich & T. Pozzan. 1984. FEBS Lett. **177:** 17–22.
17. Rorsman, P., H. Abrahamsson, E. Gylfe & B. Hellman. 1984. FEBS Lett. **170:** 176–200.
18. Tsien, R. Y., T. J. Rink & M. Poeni. 1985. Cell Calcium **6:** 145–157.
19. Sener, A. & W. J. Malaisse. 1980. Nature **288:** 187–189.
20. Henquin, J.-C. & H. P. Meissner. 1981. Am. J. Physiol. **240:** E245–E252.
21. Ullrich, S. & C. B. Wollheim. 1985. Mol. Pharmacol. **28:** 100–106.
22. Charles, S. & J.-C. Henquin. 1983. Biochem. J. **214:** 899–907.
23. Hellman, B., J. Sehlin & I.-B. Täljedal. 1971. Endocrinology **89:** 1432–1439.
24. Prentki, M. & A. E. Renold. 1983. J. Biol. Chem. **258:** 14239–14244.
25. Hales, C. N. & R. D. G. Milner. 1968. J. Physiol. (London) **194:** 725–743.
26. Biden, T. J., D. Janjic & C. B. Wollheim. 1986. Am. J. Physiol. **250:** C207–C213.
27. Hellman, B., J. Sehlin & I.-B. Täljedal. 1971. Biochem. J. **123:** 513–521.
28. Malaisse-Lagae, F., A. Sener, P. Garcia-Morales, I. Valverde & W. J. Malaisse. 1982. J. Biol. Chem. **257:** 3754–3758.
29. Malaisse, W. J., A. Sener, A. R. Carpinelli, K. Anjaneyulu, P. Lebrun, A. Herchuelz & J. Christophe. 1980. Mol. Cell. Endocrinol. **20:** 171–189.
30. Wollheim, C. B., E. G. Siegel & G. W. G. Sharp. 1980. Endocrinology **107:** 924–929.
31. Nenquin, M., P. Awouters, F. Mathot & J. C. Henquin. 1984. FEBS Lett. **176:** 457–461.
32. Morgan, N. G., G. M. Rumford & W. Montague. 1985. Biochem. J. **228:** 713–718.
33. Wollheim, C. B. & T. J. Biden. 1986. J. Biol. Chem. **261:** 8314–8319.
34. Streb, H., R. F. Irvine, M. J. Berridge & I. Schulz. 1983. Nature **306:** 67–69.
35. Streb, H., H. P. Heslop, R. F. Irvine, I. Schulz & M. J. Berridge. 1985. J. Biol. Chem. **260:** 7309–7315.
36. Nishizuka, Y. 1984. Nature **308:** 693–698.
37. Berridge, M. J. & R. F. Irvine. 1984. Nature **312:** 315–321.
38. Katada, T. & M. Ui. 1977. Endocrinology **101:** 1247–1255.
39. Katada, T. & M. Ui. 1979. J. Biol. Chem. **254:** 469–479.
40. Ui, M. 1984. TIPS **5:** 277–279.
41. Nakamura, T. & M. Ui. 1985. J. Biol. Chem. **260:** 3584–3593.
42. Krause, K.-H., W. Schlegel, C. B. Wollheim, T. Andersson, F. A. Waldvogel & P. D. Lew. 1985. J. Clin. Invest. **76:** 1342–1354.
43. Cockcroft, S. & B. D. Gomperts. 1985. Nature **315:** 534–536.
44. Wallace, M. A. & J. N. Fain. 1985. J. Biol. Chem. **260:** 9527–9530.
45. Masters, S. B., M. W. Martin, T. K. Harden & J. H. Brown. 1985. Biochem. J. **227:** 933–937.
46. Schlegel, W., F. Wuarin, C. Zbaren, C. B. Wollheim & G. R. Zahnd. 1985. FEBS Lett. **189:** 27–32.
47. Vallar, L., T. J. Biden & C. B. Wollheim. 1986. Diabetologia **29:** 603A.
48. Somlyo, A. P., M. Bond & A. V. Somlyo. 1985. Nature **314:** 622–625.
49. Eisen, A. & G. T. Reynolds. 1985. J. Cell Biol. **100:** 1522–1527.
50. Becker, G. L., G. Fiskum & A. L. Lehninger. 1980. J. Biol. Chem. **255:** 9009–9012.
51. Streb, H. & I. Schulz. 1983. Am. J. Physiol. **245:** G347–G357.
52. Prentki, M., D. Janjic, T. J. Biden, B. Blondel & C. B. Wollheim. 1984. J. Biol. Chem. **259:** 10118–10123.
53. Colca, J. R., J. M. McDonald, N. Kotagal, C. Patke, C. J. Fink, M. H. Greider, P. E. Lacy & M. L. McDaniel. 1982. J. Biol. Chem. **257:** 7223–7228.

54. PRENTKI, M., D. JANJIC & C. B. WOLLHEIM. 1983. J. Biol. Chem. **258:** 7597–7602.
55. PRENTKI, M., D. JANJIC & C. B. WOLLHEIM. 1984. J. Biol Chem. **259:** 14054–14058.
56. BIDEN, T. J., M. PRENTKI, R. F. IRVINE, M. J. BERRIDGE & C. B. WOLLHEIM. 1984. Biochem. J. **223:** 467–473.
57. PRENTKI, M., T. J. BIDEN, D. JANIC, R. F. IRVINE, M. J. BERRIDGE & C. B. WOLLHEIM. 1984. Nature **309:** 562–564.
58. PRENTKI, M., B. E. CORKEY & F. M. MATSCHINSKY. 1985. J. Biol. Chem. **260:** 9185–9190.
59. BATTY, I. R., S. R. NAHORSKI & R. F. IRVINE. 1985. Biochem. J. **232:** 211–215.
60. IRVINE, R. F., A. J. LETCHER, J. P. HESLOP & M. J. BERRIDGE. 1986. Nature **320:** 631–634.
61. BIDEN, T. J., C. B. WOLLHEIM & W. SCHLEGEL. 1986. J. Biol. Chem. **261:** 7223–7229.
62. HELLMAN, B. 1985. Diabetologia **28:** 494–501.
63. SIEGEL, E. G., C. B. WOLLHEIM, D. JANJIC, G. RIBES & G. W. G. SHARP. 1983. Diabetes **32:** 993–1000.
64. GYLFE, E. & B. HELLMAN. 1986. Biochem. J. **223:** 865–870.
65. GYLFE, E., T. ANDERSSON, P. RORSMAN, H. ABRAHAMSSON, P. ARKHAMMAR, P. HELLMAN, B. HELLMAN, H. K. OIE & A. F. GAZDAR. 1983. Biosci. Rep. **3:** 939–946.
66. WOLF, B. A., J. TURK, W. R. SHERMAN & M. L. MCDANIEL. 1986. J. Biol. Chem. **261:** 3501–3511.
67. JOSEPH, S. K., J. R. WILLIAMS, B. E. CORKEY, F. M. MATSCHINSKY & J. R. WILLIAMSON. 1984. J. Biol. Chem. **259:** 12952–12955.
68. WOLF, B. A., P. G. COMENS, K. E. ACKERMAN, W. R. SHERMAN & M. L. MCDANIEL. 1985. Biochem. J. **227:** 965–969.
69. FEX, G. & A. LERNMARK. 1972. FEBS Lett. **25:** 287–291.
70. FREINKEL, N., C. EL YOUNSI & R. M. C. DAWSON. 1975. Eur. J. Biochem. **59:** 245–252.
71. CLEMENTS, R. S. Jr. & W. B. RHOTEN. 1976. J. Clin. Invest. **57:** 684–691.
72. CLEMENTS, R. S. Jr., M. H. EVANS & C. S. PACE. 1981. Biochim. Biophys. Acta **674:** 1–9.
73. LAYCHOCK, S. G. 1983. Biochem. J. **216:** 101–106.
74. BEST, L. & W. J. MALAISSE. 1983. Biochem. Biophys. Res. Commun. **116:** 9–16.
75. MONTAGUE, W., N. G. MORGAN, R. M. RUMFORD & C. A. PRINCE. 1985. Biochem. J. **227:** 483–489.
76. BIDEN, T. J., B. PETER-RIESCH, W. SCHLEGEL & C. B. WOLLHEIM. 1987. J. Biol. Chem. **262.** (In press.)
77. RANA, R. S., R. J. MERTZ, A. KOWLURU, J. F. DIXON, L. E. HOKIN & M. J. MCDONALD. 1985. J. Biol. Chem. **260:** 7861–7867.
78. DUNLOP, M. E. & R. G. LARKINS. 1984. J. Biol. Chem. **259:** 8407–8411.
79. EISENBERG, F., JR. 1967. J. Biol. Chem. **242:** 1375–1382.
80. BIDEN, T. J. & C. B. WOLLHEIM. 1986. Biochem. J. **236:** 889–893.
81. HALLCHER, L. M. & W. R. SHERMAN. 1980. J. Biol. Chem. **255:** 10896–10901.
82. DOWNES, C. P., M. C. MUSSAT & R. H. MICHELL. 1982. Biochem. J. **203:** 169–177.
83. RANA, R. S., M. C. SEKAR, L. E. HOKIN & M. J. MACDONALD. 1986. J. Biol. Chem. **261:** 5237–5240.
84. BIDEN, T. J. & C. B. WOLLHEIM. 1986. J. Biol. Chem. **261:** 11931–11934.

Altered Sorbitol and *myo*-Inositol Metabolism as the Basis for Defective Protein Kinase C and (Na,K)-ATPase Regulation in Diabetic Neuropathy[a]

DOUGLAS A. GREENE AND SARAH A. LATTIMER

Diabetes Research and Training Center
Department of Medicine
Ann Arbor, Michigan 48109

INTRODUCTION

Diabetic neuropathy is probably the most common and one of the most troubling late complications of diabetes mellitus. It is responsible for a major portion of the pain and disability associated with diabetes and is the leading cause of amputations in diabetic patients. Until recently, little was known about its pathogenesis and effective therapy was unavailable. Recent studies performed *in vitro* in animal models, and now in humans, suggest that perturbations in membrane function, specifically phosphoinositide, protein kinase C, and (Na,K)-ATPase regulation, constitute an important permissive link between elevations in blood glucose and the functional and structural defects in peripheral nerve that characterize diabetic neuropathy.[1,2] *myo*-Inositol depletion is induced in tissues susceptible to diabetic complications (retina, renal glomerulus, arterial wall, and peripheral nerve) as a result of an increase in intracellular glucose and its subsequent metabolism to sorbitol by the enzyme aldose reductase. This conclusion is based on the ability of aldose reductase inhibitors of varying molecular structures to block *myo*-inositol depletion when administered *in vivo,*[1,3–5] although competitive inhibition of *myo*-inositol uptake by glucose cannot be excluded as a contributing element in *myo*-inositol depletion (FIGURE 1).[6]

myo-Inositol depletion, by interfering with membrane phosphoinositide metabolism,[5,7] links cytosolic carbohydrate and polyol metabolism with membrane function. Altered membrane phosphoinositide metabolism impairs (Na,K)-ATPase activity[1,4,5,7–10] presumably by a protein kinase C–mediated mechanism (FIGURE 1).[2] Reduced activation of protein kinase C by phosphoinositide-derived diacylglycerol was initially implicated in the impaired (Na,K)-ATPase activity of diabetic nerve by experiments in which phorbol esters or exogenous diacylglycerol acutely restored normal ouabain-inhibitable oxygen consumption, a measure of (Na,K)-ATPase activity, in diabetic peripheral nerve *in vitro* (FIGURE 2).[2] Since diminished enzymatically measured (Na,K)-ATPase can be normalized by exposure of broken cell preparations or isolated membrane fractions derived from diabetic rat or rabbit nerve to phorbol esters or diacylglycerol, the functional association between the phorbol ester receptor (presumably protein kinase C) and the (Na,K)-ATPase must be preserved in the diabetic membrane.[11,12] The ability of protein kinase C agonists to reactivate the (Na,K)-ATPase in broken cell preparations or in the absence of added calcium

[a]Supported in part by U.S. Public Health Service research grant RO-1 AM29892.

effectively excludes a major role for protein kinase C–mediated activation of (Na,H)-exchange in this process.[11,12]

Impaired (Na,K)-ATPase activity is responsible for the acutely reversible slowing of nerve conduction in diabetic animals[13,14] and probably humans.[15] The associated rise in intra-axonal [Na^+] decreases the axolemmal Na^+ gradient by 50%, diminishing nodal excitability and producing a selective conduction block of large myelinated nerve fibers and/or a nodal conduction delay.[16] Furthermore, *myo*-inositol depletion and the accompanying reduction in (Na,K)-ATPase activity result in acutely reversible axonal swelling in the region of the node of Ranvier.[17] This physical distortion of the paranodal apparatus is followed by a characteristic loss of strategic junctional complexes between

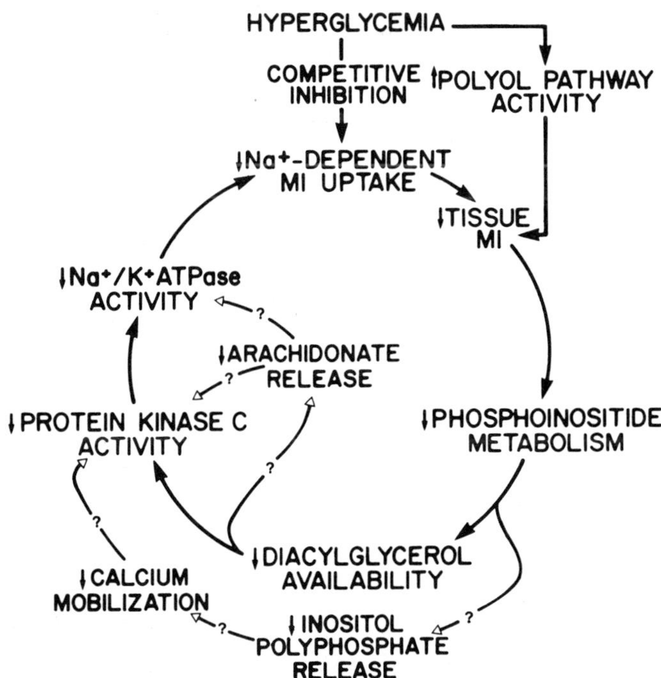

FIGURE 1. Proposed self-reinforcing metabolic defect involving polyol (sorbitol) pathway activity, phosphoinositide metabolism, protein kinase C, and the (Na,K)-ATPase. See Greene *et al. N. Engl. J. Med.* (Submitted for publication.) for further details.

terminal loops of myelin and the paranodal axolemma, that is temporally associated with a poorly reversible reduction in nodal Na^+ permeability.[16,18] It has been proposed that loss of these paranodal axo-glial junctions ("axo-glial dysjunction") permits lateral diffusion of highly concentrated membrane-bound Na^+ channels from the nodal to the internodal region of the axolemma, thereby diminishing nodal Na^+ permeability. Morphometric studies have identified a marked increase in axo-glial dysjunction in sural nerve biopsies of patients with insulin-dependent diabetes and diabetic neuropathy, which appeared to represent an early stage of paranodal demyelination.[19] Finally, 12 months of administration of the aldose reductase inhibitor sorbinil to patients with overt diabetic polyneuropathy resulted in marked resolution of axo-glial dysjunction, paranodal swelling, and paranodal demyelination, and a striking appearance of new myelinated nerve fibers that increased myelinated fiber number by approximately 50%.[20] Taken together with the earlier observation that sorbinil administration acutely increased (but did not normalize) nerve conduction velocity in diabetic patients without neuropathy,[13] these observations support the relevance of increased levels of nerve glucose and sorbitol, depletion of *myo*-inositol, impaired phosphoinositide metabolism, decreased activation of protein kinase C, and diminished (Na,K)-ATPase activity in the pathogenesis of human diabetic neuropathy. Despite much speculation[1,5,7,21–23] a clearly defined mechanism linking *myo*-inositol metabolism

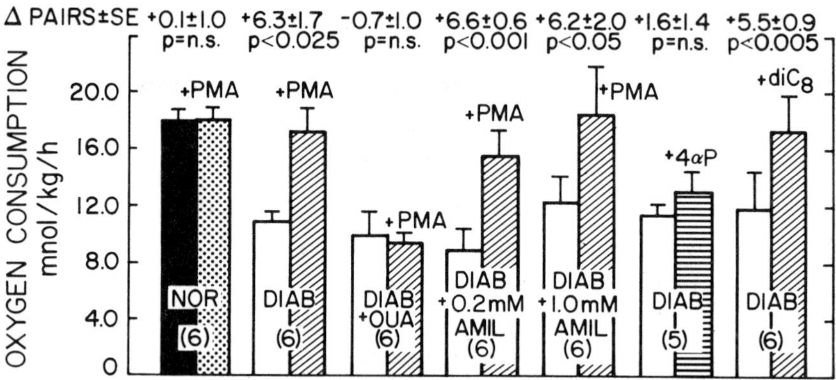

FIGURE 2. Effect of experimental diabetes and protein kinase C agonists on ouabain-inhibitable respiration in peripheral nerve. Oxygen consumption, expressed as mmol/kg/hr, was measured in pairs of endoneurial preparations derived from normal (NOR, solid bars) and 2-wk alloxan diabetic (DIAB, open bars) rabbits. One member of each pair of endoneurial preparations was studied after *in vitro* exposure to active protein kinase C agonists (phorbol myristate acetate [+PMA] or *sn*-1,2-dioctanoylglycerol [+diC$_8$], diagonally hatched bars) or an inactive phorbol analogue (4-alpha phorbol [+4aP], horizontally hatched bar). The other member of each pair was exposed to dimethylsulfoxide, the vehicle for phorbol and diacylglycerol additions. In one set of experiments, 2 mM ouabain (OUA) was present in the incubation medium for 10 min before and during the measurement of oxygen consumption to inhibit the (Na,K)-ATPase. In two sets of experiments, amiloride (AMIL) in 0.2 or 1.0 mM concentrations was present for 30 min before and during the measurement of oxygen consumption in order to inhibit (Na^+,H^+)-exchange. Numbers of paired samples are shown in parentheses. Error bars = SEM. Mean differences of paired samples are given numerically above each bar, and were used to calculate p values by t-tests for paired comparisons. (From Greene & Lattimer.[2] With permission from *Diabetes*.)

and (Na,K)-ATPase regulation in nerve or other tissues has not emerged until recently.[2] A role for *myo*-inositol in the modulation of (Na,K)-ATPase was first described by Charalampous in inositol-requiring mutant KB cells over 15 years ago.[21] More recent studies have suggested[22] but not proven[23] that phosphatidylinositol may be a specific endogenous activator of (Na,K)-ATPase. Effects of *myo*-inositol on cell regulation are currently viewed as occurring primarily through its incorporation into phosphoinositide,[24] which may be rate-limiting for basal phosphoinositide metabolism in some tissues.[5,7] Stimulated phosphoinositide turnover gives rise to two classes of catabolites with regulatory properties: inositol phosphates[24–26] and arachidonate-enriched diacylglycerol.[24,27] Inositol-1,4,5-triphosphate specifically mobilizes intracellular calcium sequestered in non-mitochondrial stores[24] while the role(s) of other inositol-poly phosphates or cyclic phosphates remains unknown.[28,29] Arachidonate-enriched diacylglycerol specifically binds to and activates protein kinase C by lowering its calcium requirement to within the physiological intracellular concentration range[24,27] and also releases arachidonate for cyclooxygenase- and lipooxygenase-mediated metabolism.[27] It is noteworthy that most evidence implicating *myo*-inositol metabolism in (Na,K)-ATPase regulation has emerged under circumstances of *myo*-inositol depletion either related[1,8–12] or unrelated[5,7,21] to diabetes. Thus, *myo*-inositol metabolism would appear normally to exert a tonic rather than a phasic effect on (Na,K)-ATPase activity. Initial observations suggesting that (Na,K)-ATPase activity was stimulated by protein kinase mechanisms[30] was subsequently called into serious question,[31] but recently Ling and Cantley[32] documented phosphorylation of the alpha subunit of the (Na,K)-ATPase at a site distinct from the catalytic one, that was phosphorylated by an unidentified membrane-bound cyclic-AMP-independent protein kinase. Hence, a putative regulatory phosphorylation of the (Na,K)-ATPase could be linked to phosphoinositide metabolism by at least two potential mechanisms, protein kinase C activation by phosphoinositide-derived diacylglycerol or calcium-calmodulin–dependent protein kinase activation secondary to intracellular mobilization of Ca^{2+} by inositol-trisphosphate (FIG. 1). Although the studies cited above demonstrate selective and complete reactivation of (Na,K)-ATPase activity in diabetic nerve with protein kinase C, and agonists would definitely seem to favor the former mechanism for regulatory phosphoprotein formation; other mechanisms, including some independent of a regulatory phosphorylation, cannot be excluded entirely. For instance, phorbol esters and diacylglycerols are lipophilic compounds that intercalate within and modify the biophysical characteristics of lipid bilayers[33,34] thus possibly interacting directly with the hydrophobic domain of the (Na,K)-ATPase and thereby compensating for a putative alteration in membrane phosphoinositide composition.[2,35] Arachidonic acid, a diacylglycerol-derived metabolite, partly compensates for *myo*-inositol depletion in (Na,K)-ATPase regulation in normal[7] but not diabetic[10] peripheral nerve possibly via its ability to directly activate protein kinase C independent of diacylglycerol (FIG. 1).[36] Inositol-1,4,5-triphosphate or other inositol poly- or cyclic-phosphates may be important mediators of *myo*-inositol depletion on nerve function, particularly with regard to calcium-dependent processes such as axoplasmic transport.[1,37] On the other hand, the available data are most compatible with the hypothesis that changes in protein kinase C activation, reflecting the limiting availability of phosphoinositide-derived diacylglycerol in diabetic nerve, regulate nerve (Na,K)-ATPase activity in the presence of *myo*-inositol depletion. Studies presently underway are attempting to characterize a putative protein kinase C-mediated regulatory phosphorylation of the (Na,K)-ATPase.

If electrically stimulated[38] and/or receptor agonist–stimulated phosphoinositide turnover and subsequent activation of protein kinase C is seriously diminished in diabetic peripheral nerve, then a whole range of cellular processes in addition to (Na,K)-ATPase regulation may be defective due to the broad substrate specificity of protein kinase C.[35] Furthermore, since elements of the phosphoinositide cascade are thought to function synergistically,[24,35] and since release of other phosphoinositide metabolites including inositol phosphates and arachidonate might be correspondingly reduced in diabetic nerve, the *myo*-inositol–related (Na,K)-ATPase defect in diabetic nerve may constitute a marker for far more widespread defects stemming from multiple abnormalities in the phosphoinositide cascade. For instance, protein kinase C has been implicated in the neuronal response to various growth or trophic factors including insulin and insulin-like growth factor II (IGF-II).[39] Therefore, peripheral nerve neurons might be resistant to the effect of factors that promote compensatory nerve regeneration in other degenerative neuropathies. Such resistance to agonists that normally stimulate nerve fiber regeneration might explain the relatively rapid appearance of immature nerve fibers in sural nerve biopsies of diabetic patients treated with aldose reductase inhibitors for 12 months.[20] Thus, defects in nerve phosphoinositide metabolism in diabetes might not only promote nerve fiber dysfunction and degeneration via abnormalities in (Na,K)-ATPase activity, but also interfere with compensatory fiber regeneration by blocking the response to receptor-mediated trophic agents as well. Diabetic neuropathy thus may provide an interesting model in which to explore various membrane and nuclear effects of phosphoinositide metabolism in the nervous system, and aldose reductase inhibitors, by blocking *myo*-inositol depletion, may become important therapeutic agents in the treatment of diabetic complications including neuropathy.

REFERENCES

1. Greene, D. A., S. Lattimer, J. Ulbrecht & P. Carroll. 1985. Glucose-induced alterations in nerve metabolism: current perspective on the pathogenesis of diabetic neuropathy and future directions for research and therapy. Diabetes Care **8:** 290–299.
2. Greene, D. A. & S. A. Lattimer. 1986. Protein kinase C agonists acutely normalize decreased ouabain-inhibitable respiration in diabetic rabbit nerve: implications for (Na,K)-ATPase regulation and diabetic complications. Diabetes **35:** 242–245.
3. MacGregor, L. C., L. R. Rosecan, A. M. Laties & F. M. Matchinsky. 1986. Altered retinal metabolism in diabetes. I. Microanalysis of lipid, glucose, sorbitol and *myo*-inositol in the choroid and in the individual layers of the rabbit retina. J. Biol. Chem. (In press.)
4. Cohen, M. P. 1985. Reduced glomerular sodium-potassium adenosine triphosphatase activity in acute streptozotocin diabetes an its prevention by oral sorbinil. Diabetes **34:** 1071–1074.
5. Simmons, D. A., E. F. O. Kern, A. I. Winegrad & D. B. Martin. 1986. Basal phosphatidylinositol turnover controls aortic Na^+/K^+-ATPase activity. J. Clin. Invest. **77:** 503–513.
6. Greene, D. A. & S. A. Lattimer. 1982. Sodium- and energy-dependent uptake of *myo*-inositol by rabbit peripheral nerve: effect. J. Clin. Invest. **70:** 1009–1018.
7. Simmons, D. A., A. I. Winegrad & D. B. Martin. 1982. Significance of tissue *myo*-inositol concentrations in metabolic regulation in nerve. Science **217:** 848–851.
8. Greene, D. A. & S. A. Lattimer. 1983. Impaired rat sciatic nerve sodium-potassium adenosine triphosphatase in acute streptozocin diabetes and its correction by dietary *myo*-inositol supplementation. J. Clin Invest. **72:** 1058–1063.
9. MacGregor, L. C. & F. M. Matchinsky. 1986. Altered retinal metabolism in diabetes. II. Measurement of sodium-potassium ATPase and total sodium and potassium in individual retinal layer. (Submitted for publication.)
10. Greene, D. A. & S. A. Lattimer. 1984. Impaired energy utilization and sodium-potassium ATPase in diabetic peripheral nerve. Am. J. Physiol. **246:** E311–318.
11. Greene, D. A. & S. A. Lattimer. Manuscript in preparation.
12. Kim, J., H. Kyriazi & D. A. Greene. 1987. Manuscript in preparation.
13. Brismar, T. & A. A. F. Sima. 1981. Changes in nodal function in nerve fibers of the spontaneously diabetic BB-Wistar rat. Potential clamp analysis. Acta Physiol. Scand. **113:** 499–506.
14. Greene, D. A., S. Yagihashi, S. A. Lattimer & A. A. F. Sima. 1984. Nerve Na^+-K^+-ATPase, conduction and *myo*-inositol in the insulin deficient BB rat. Am. J. Physiol. **247:** E534–539.
15. Judzewitsch, R. G., J. B. Jaspan, K. S. Polonsky, C. R. Weinberg, J. B. Halter, E. Halar, M. A. Pfeifer, C. Vukadinovic, L. Bernstein, M. Schneider, K.-Y. Liang, K. H. Gabbay, A. H. Rubenstein & D. Porte Jr. 1983. Aldose reductase inhibition improves nerve conduction in diabetic patients. N. Engl. J. Med. **308:** 119–125.

16. BRISMAR, T., A. A. F. SIMA & D. A. GREENE. 1987. Reversible and irreversible nodal dysfunction in diabetic neuropathy. (Submitted for publication.)
17. GREENE, D. A., S. A. LATTIMER & A. A. F. SIMA. 1986. Acute paranodal nerve fiber swelling and conduction slowing in the insulin-deficient BB rat reflects *myo*-inositol depletion and Na/K-ATPase deficiency rather than sorbitol accumulation. Clin. Res. **34:** 683A.
18. SIMA, A. A. F., S. A. LATTIMER, S. YAGIHASHI & D. A. GREENE. 1986. Axo-glial dysjunction: a novel structural lesion that accounts for poorly-reversible conduction slowing in the spontaneously-diabetic Bio-breeding rat. J. Clin. Invest. **77:** 474–485.11.
19. SIMA, A. A. F., V. BRIL & D. A. GREENE. 1986. A new characteristic ultrastructural abnormality, and morphological evidence for pathogenetic heterogeneity in human diabetic neuropathy. Clin. Res. **34:** 688A.
20. SIMA, A. A. F., V. BRIL & D. A. GREENE. 1987. Manuscript in preparation.
21. CHARALAMPOUS, F. C. 1971. Metabolic functions of myoinositol VIII. Role of inositol in the Na^+-K^+ transport and in Na^+- and K^+-activated adenosine triphosphatase of KB cells. J. Biol. Chem. **246:** 455–460.
22. MANDERSLOOT, J. G., B. ROELOFSEN & J. DE GIER. 1978. Phosphatidylinositol as the endogenous activator of the (Na^+ + K^+)-ATPase in microsomes of rabbit kidney. Biochim. Biophys. Acta **508:** 478–485.
23. ROELOFSEN, B. 1981. The (non)specificity in the lipid requirement of the calcium and (sodium plus potassium)-transporting adenosine triphosphatases. Life Sci. **29:** 2235–2247.
24. BERRIDGE, M. J. 1984. Inositol triphosphate and diacylglycerol as second messengers. Biochem. J. **220:** 345–360.
25. MAJERUS, P. W., D. B. WILSON, T. M. CONNOLLY, T. E. BROSS & E. J. NEUFELD. 1985. Phosphoinositide turnover provides a link in stimulus-response coupling. Trends Biochem. Sci. **10:** 168–171.
26. EXTON, J. H. 1985. Role of calcium and phosphoinositides in the actions of certain hormones and neurotransmitters. J. Clin. Invest. **75:** 1753–1757.
27. NISHIZUKA, Y. 1986. Studies and perspectives of protein kinase C. Science **233:** 305–311.
28. DIXON, F. J. & L. E. HOKIN. 1985. The formation of inositol-1,2-cyclic phosphate on agonist stimulation of phosphoinositide breakdown in mouse pancreatic minilobules. Evidence for direct phosphodiesteratic cleavage of phosphatidylinositol. J. Biol. Chem. **260:** 16068–16071.
29. BATTY, I. R., R. NAHORSKI & R. F. IRVINE. 1985. Rapid formation of inositol-1,3,4,5-tetrakisphosphate following muscarinic receptor stimulation of rat cerebral cortical slices. Biochem J. **232:** 211–215.
30. SPECTOR, M. S., S. O'NEAL & E. RACKER. 1981. Regulation of phosphorylation of the beta-subunit or the Ehrlich ascites tumor Na^+K^+-ATPase by a protein kinase cascade. J. Biol. Chem. **26:** 4219–4227.
31. RACKER, E. 1981. Warburg effect revisited (letter). Science **213:** 131.
32. LING, L. & L. CANTLEY. 1984. The (Na,K)-ATPase of Friend erythroleukemia cells is phosphorylated near the ATP hydrolysis site by an endogenous membrane-bound protein kinase. J. Biol. Chem. **259:** 4089–4095.
33. BACKARD, B. S., M. J. SAXTON, M. J. BISSELL & M. P. KLEIN. 1984. Plasma membrane reorganized induced by tumor promoters in an epithelial cell line. Proc. Natl. Acad. Sci. USA **81:** 449–452.
34. DAS, S., & R. P. RAND. 1985. Diacylglycerol causes structural transitions in phospholipid bilayer membranes. Biochem. Biophys. Res. Commun. **124:** 491–496.
35. KIKKAWA, U. & Y. NISHIZUKA. 1986. Protein kinase C. *In* The Enzymes. E. Krebs, Ed. Academic Press. New York. (In press.)
36. MURAKAMI, K. & A. ROUTTENBERG. 1985. Direct activation of purified protein kinase C by unsaturated fatty acids (oleate and arachidonate) in the absence of phospholipid and Ca^{2+}. Fed. Eur. Biochem. Soc. Lett. **192:** 189–193.
37. MAYER, J. H. & D. R. TOMLINSON. 1983. Prevention of defects of axonal transport and nerve conduction velocity by oral administration of *myo*-inositol or an aldose reductase inhibitor in streptozotocin-diabetic rats. Diabetologia **25:** 433–438.

38. HAWTHORNE, J. N., M. R. PICKARD & H. D. GRIFFIN. 1978. Phosphatidylinositol, triphosphoinositide and synaptic transmission. *In* Cyclitols and Phosphoinositides. W. W. Wells & F. J. Eisenberg, Eds.: 145–151. Academic Press. New York.
39. ISHII, D. N., E. RECIO-PINTO, W. SPINELLI, J. F. MILL & K. H. SONNENFELD. 1985. heurite formation modulated by nerve growth factor, insulin and tumor promoter receptors. Int. J. Neurosci. **26:** 109–127.

DISCUSSION OF THE PAPER

MELDOLESI (*University of Milan, Milan*): You showed that oxygen consumption of your nerve preparation obtained from diabetic animals was not decreased by ouabain. This suggests that the (Na^+,K^+)ATPase is severely inhibited and, as a consequence, that these animals should be extremely susceptible to ouabain.

D. GREENE (*University of Pittsburgh, Pittsburgh, PA*): The fact that ouabain does not inhibit resting oxygen consumption in diabetic nerve preparations simply means that (Na^+,K^+)ATPase activity is no longer rate-limiting for oxygen consumption under basal (non-stimulated) conditions. Since enzymatically assayable (Na^+,K^+) ATPase activity is only partially decreased (about 40%) in diabetic nerve, it is presumed that diabetes would only partially decrease ouabain-inhibitable energy utilization in the stimulated state, more relevant to the *in situ* condition. With regard to *in situ* sensitivity to ouabain, the conditions of ouabain used during *in vitro* incubation (2 mM) are far higher than those attained *in vivo* with systemic ouabain administration.

L. CANTLEY (*Tufts University, Boston, MA*): Does addition of EGTA block the PMA activation of (Na^+,K^+)ATPase in isolated membranes?

GREENE: This is an extremely important question. The two alternative interpretations of our data are: (1) that PMA stimulates (Na^+,K^+)ATPase in diabetic membranes by stimulating protein kinase C thereby phosphorylating the (Na^+,K^+)ATPase or a closely associated regulatory protein, or (2) that PMA directly interacts with a hydrophobic domain of the Na^+/K^+-ATPase directly stimulating its activity independent of protein kinase C.

One way to distinguish between these two possibilities would be to remove protein kinase C from the membrane preparation by procedures such as exposure to EGTA. We are now studying whether EGTA or other compounds known to inhibit protein kinase C influence the ability of PMA and other protein kinase C agonists to stimulate (Na^+,K^+)ATPase in diabetic membranes.

Transmembrane Ion Distribution and Insulin Secretion[a]

PETER RONNER[b]

Joslin Diabetes Center
Research Division
and
Harvard Medical School
Department of Medicine
Boston, Massachusetts 02146

INTRODUCTION

How B cells recognize a glucose stimulus and translate this information into insulin release is an important question in diabetes research that has remained largely unanswered. Glucose metabolism is required for insulin release, but the step generating the signal for insulin release has not been identified.[1] However, it is clear that glucose rapidly equilibrates across the B-cell plasma membrane and that glucokinase exerts control over the rate of glycolysis within physiological concentrations of glucose.[2–5] Electrophysiological experiments with microelectrode-impaled mouse B cells show that the cell membrane depolarizes upon glucose stimulation.[6,7] This depolarization occurs after a lag phase of about one minute and it is due to the inhibition of a potassium channel.[8,9] The specific intracellular signal leading to inhibition of the potassium channel is not known, but several hypotheses about its nature exist. According to one hypothesis, a stimulatory concentration of glucose leads to an increase of the cytosolic concentration of ATP, and ATP then inhibits the potassium channel.[10–12] In fact, in B cells such a potassium channel was half-maximally inhibited by 15 μM ATP and it also appeared to determine the resting potential of excised membrane patches.[10,12] It is questionable, though, whether the high sensitivity of this channel for ATP allows it to participate in stimulus-secretion coupling of glucose-induced insulin release, since islets contain an overall averaged concentration of ATP in the mM range.[13]

According to another hypothesis, a fall in intracellular pH due to glucose metabolism regulates the permeability of yet another type of potassium channel.[14] This particular channel was found to be inhibited at a low pH but also activated by an elevated concentration of calcium.[14] This channel is probably not responsible for the initial depolarization; experimental manipulations designed to decrease the intracellular pH of B cells do not induce insulin release, and furthermore, pH measurements with permeable acids indicate that glucose leaves unaffected or even increases the intracellular pH.[15–21]

Once the potassium channel has been inhibited and the plasma membrane has depolarized, the low membrane potential is believed to trigger the opening of

[a]Supported by Grant AM-35577 and Grant AM-35449 to G.C. Weir from the National Institutes of Health.

[b]Present address: Department of Biochemistry, School of Medicine and Dentistry, University of Rochester, Rochester, NY 14642.

voltage-dependent calcium channels.[22–25] As these channels open, extracellular calcium flows into the cell thereby increasing the cytosolic concentration of calcium and activating the exocytotic machinery that moves granules containing insulin to the plasma membrane surface. This concept is supported by the findings that insulin release is inhibited in the absence of extracellular calcium or in the presence of a phenylalkylamine calcium channel blocker.[26,27] At present, it is not clear whether calcium released from the endoplasmic reticulum by the action of inositol phosphates also plays a significant role in glucose-induced insulin release.[28,29]

About one to two minutes after the initial depolarization and an ensuing series of calcium channel gatings, the membrane potential starts assuming a rhythmic pattern consisting of a depolarization with a series of calcium channel gatings followed by a transient repolarization and inactivity of the calcium channels.[6,7] Possibly, a calcium-dependent potassium channel similar to the one described above is responsible for these intermittent repolarizations and silent phases.[8,12,14,30]

The repolarization phase at the end of a stimulus is not well understood. It may involve reactivation of the potassium channel that initiated the depolarization, as well as activation of a voltage- and calcium-dependent potassium channel. The intracellular calcium concentration is probably lowered by a combination of Na:Ca countertransport and active pumping by a calmodulin-dependent Ca,Mg-ATPase, perhaps also by calcium uptake into the endoplasmic reticulum.

SODIUM DEPENDENCE OF INSULIN RELEASE

Since the experiments of Hales and Milner in 1968 with pieces of rabbit pancreas, insulin release is known to be inhibited in the absence of extracellular sodium.[31] However, sodium plays no major role in the scheme outlined above. Furthermore, at the outset of the studies described here, no explanation previously offered for the sodium dependence of hormone release appeared entirely satisfactory. Clearly, a voltage-dependent sodium channel is not necessary for glucose-induced insulin release since the highly specific inhibitor tetrodotoxin has no effect on it.[32–34] It also appears unlikely that glucose uptake into B cells is driven by extracellular sodium, since glucose uptake is quite independent of the concentration of sodium.[35]

In order to study the role of sodium in insulin release we chose as a model the *in vitro* perfused principal pancreatic islet of channel catfish, which provides an unusually large aggregate of readily accessible endocrine tissue and contains very little exocrine tissue.[36–39] The principle pancreatic islet is also called Brockmann body and it is located in the peritoneal cavity adjacent to duodenum, spleen, and gall bladder. In the fish used for the present study it typically weighed about 5 mg. The *in vitro* perfusion is similar to the perfusion of a rat pancreas. The surgery is performed under a microscope at ten-fold magnification using thin silk thread for ligating the blood vessels. The surgical procedure prior to perfusion takes 2–3 hours. The principal islet is then perfused *in vitro* from the celiacomesenteric artery to the hepatic portal vein at a flow rate of 29 μl/min and at a temperature of 23°C. Compared to the perfused rat pancreas the fish pancreas can be perfused for a longer time period, at least 7 hours. This makes it possible to obtain control and experimental data from the same preparation.

The *in vitro* perfused catfish principal islet responds with insulin release to stimulation with glucose, arginine, or potassium chloride (FIG. 1). It also releases glucagon in response to arginine or potassium chloride. These responses are compara-

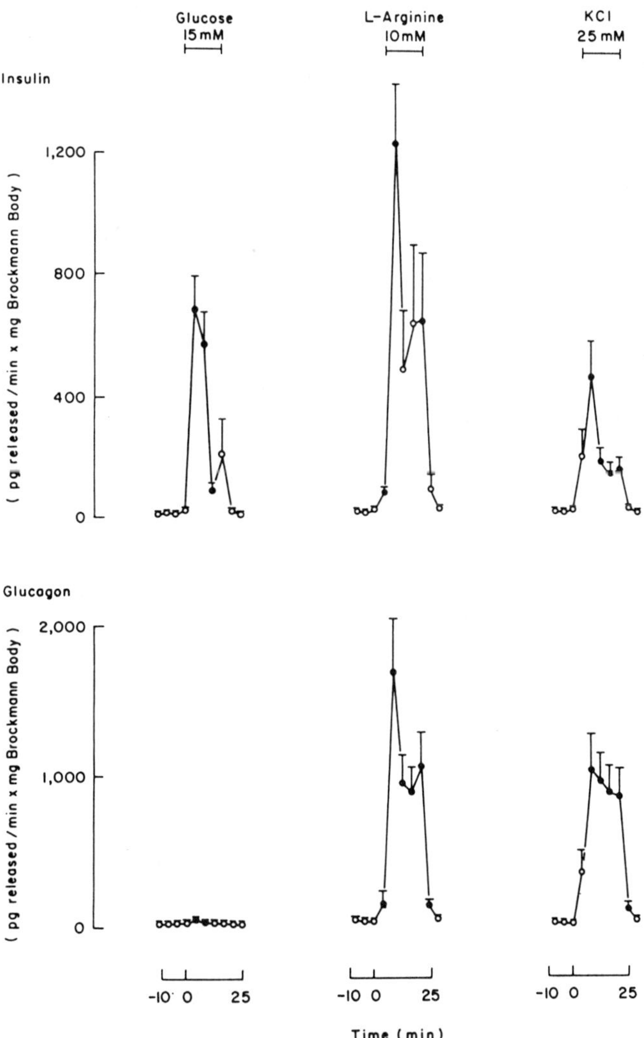

FIGURE 1. Effect of glucose, arginine, and KCl on hormone release from the *in vitro* perfused principal pancreatic islet (Brockmann body) of channel catfish (*Ictalurus punctatus*). Perfusion medium consisted of (in mM): NaCl, 130; KCl, 3; $MgSO_4$, 0.5; $CaCl_2$, 1; NaH_2PO_4, 0.1; glucose, 2; HEPES/NaOH, 10; 40 g/l dextran T-40, and 1 g/l bovine serum albumin; pH 7.6. Stimuli consisted of 15 mM glucose, 10 mM arginine in the presence of 2 mM glucose, or 25 mM KCl in the presence of 2 mM glucose. When test substances were added, concentration of NaCl was reduced to maintain isotonicity. Flow rate: 29 μl/min. Temperature: 23°C. Weight of Brockmann bodies (w), number of preparations (N), and number of test phases (t) examined: for glucose, w:3.6 ± 0.4 mg, $N = 11$, t = 26; for arginine, w: 5.6 ± 1.5 mg, $N = 5$, t = 16; for KCl, w: 5.8 ± 1.1 mg/ $N = 8$, t = 23. Hormone release was measured by radioimmunoassays using catfish hormones as standards. Means and standard errors are shown. Hormone release significantly different (at the $p \leq 0.05$ level) from average of first three baseline measurements is indicated by filled circles.

ble to those of other pancreas perfusion systems. Furthermore, catfish and mammalian B cells have a similar glucose sensitivity.[40]

With this background in mind, this study aimed to determine whether glucose-, arginine-, and potassium-induced hormone release from the fish principal islet were sodium dependent and if true, why this was the case.

Thus, the Brockmann body was perfused at the normal concentration of sodium (130 mM) in the perfusion medium and stimulated with glucose (30 mM), or sodium chloride in the perfusion medium was isotonically replaced with choline chloride before and during a stimulation with glucose. As shown in FIGURE 2, clearcut insulin release was reproducibly obtained in the presence of sodium. When all of the sodium was replaced with choline, after a 20-min equilibration period, glucose failed to stimulate insulin release. Upon simultaneous reintroduction of sodium and removal of the glucose stimulus a very small transient release of insulin occurred. After a further 20-min perfusion period in the presence of sodium, glucose again induced the same pronounced insulin release as in the beginning of the perfusion, thus indicating that the inhibition was fully reversible. In separate experiments (not shown) when sodium was removed while glucose was maintained at a stimulatory concentration, the onset of inhibition was rapid, terminating secretion as rapidly as simple removal of glucose. On the other hand, when sodium was reintroduced still in the presence of stimulatory

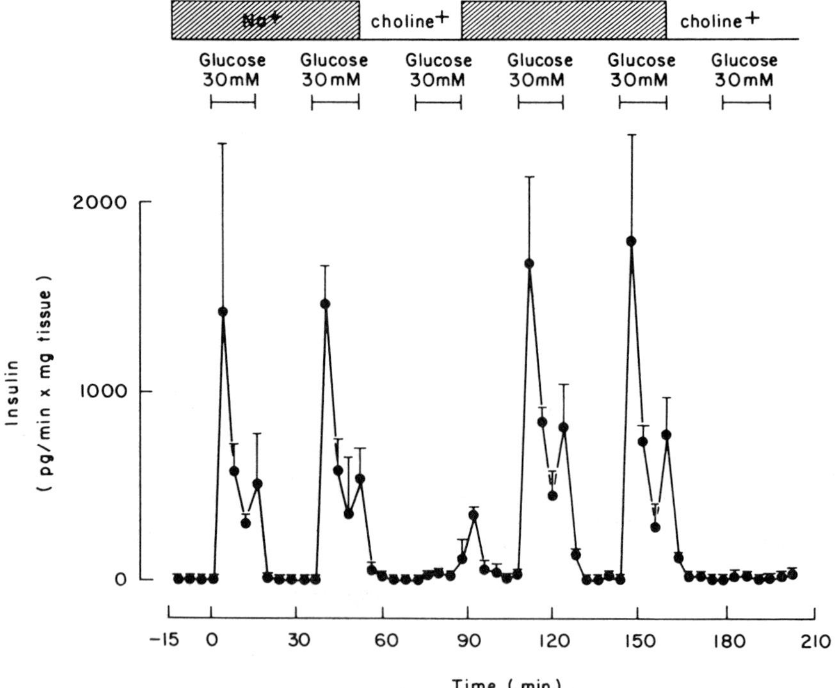

FIGURE 2. Effect of absence of extracellular sodium on glucose-induced insulin release from perfused catfish Brockmann body. Perfusion conditions as in FIGURE 1, except that NaCl was isotonically replaced with choline chloride during the periods indicated. Average with standard errors of two perfusion preparations. Weight of Brockmann bodies: 5.9 and 7.2 mg.

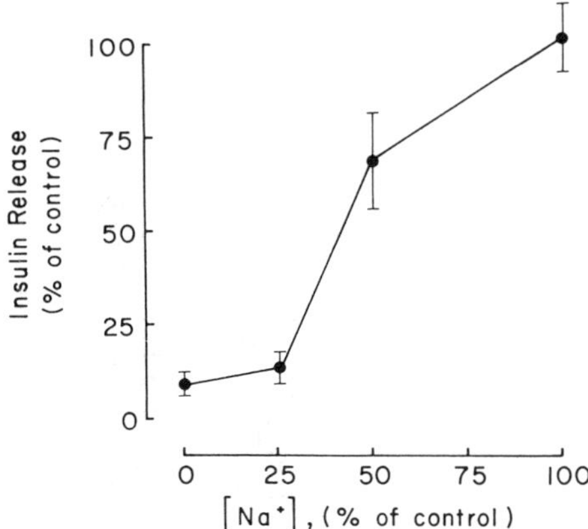

FIGURE 3. Effect of concentration of extracellular sodium on cumulative glucose-induced insulin release. Perfusion conditions as in FIGURE 2. Brockmann bodies were perfused at normal concentration of sodium before and in between periods of altered sodium concentration. First, Brockmann bodies were equilibrated for 20 min at the concentration of sodium shown, then stimulated with 15 mM glucose for 16 min. Total hormone release for stimulatory period is shown as percentage of same at normal concentration of sodium. Values are not corrected for baseline release. Weight of Brockmann bodies 5.5 ± 0.7 mg (mean ± S.E.M.; $N = 3$). Five stimulations at 0%, 2 at 25%, 6 at 50%, 11 at 100% sodium.

glucose, recovery of insulin release from inhibition was slow and gradual, requiring about 10 min.

Additional experiments were performed to determine the exact relationship between the concentration of sodium in the perfusion medium and the degree of inhibition of glucose-induced insulin release. Thus, principal islets were perfused at various levels of extracellular sodium; first for 20 min at 2 mM glucose, then for 16 min at 15 mM glucose. Insulin release during the 16 min stimulatory period was then plotted against the concentration of sodium in the perfusion medium. As shown in FIGURE 3, replacing sodium with choline progressively inhibited glucose-induced insulin release.

Similar experiments using 10 mM arginine as the stimulus in the presence of 2 mM glucose throughout the perfusion experiment revealed only a minor inhibition of insulin release in the absence of sodium (FIG. 4). Under these same conditions arginine-induced glucagon release was stimulated threefold in the absence of sodium. Interestingly, with potassium chloride as the stimulus, again in the presence of 2 mM glucose throughout the experiment, insulin release was affected only to a minor degree in the absence of sodium (FIG. 5). Potassium-induced glucagon release, on the other hand, was stimulated about twofold in the absence of sodium.

In summary, in the absence of sodium, glucose-induced insulin release was completely inhibited, whereas arginine-induced insulin release was only somewhat decreased and potassium-induced insulin release was not significantly affected. Since in the perfused rat pancreas arginine-induced insulin release is known to be glucose-dependent, our results raise the question whether this is also true for arginine-induced insulin release from the Brockmann body.[41,42] Thus, selective inhibition of glucose recognition in the absence of sodium might account for the complete inhibition of glucose-induced and the only partial inhibition of arginine-induced insulin release.

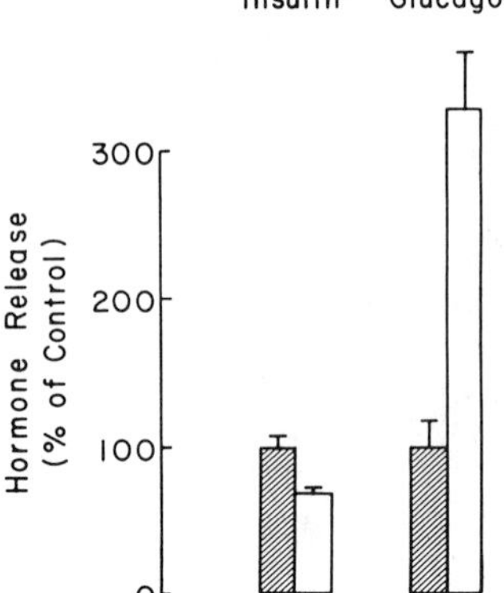

FIGURE 4. Effect of concentration of extracellular sodium on arginine-induced insulin and glucagon release from the perfused Brockmann body. Perfusion conditions as in FIGURE 3, except that stimulus consisted of 10 mM arginine in the continued presence of 2 mM glucose. Hatched bars: control stimuli ($N = 9$) at normal sodium; open bars: test stimuli ($N = 3$) in the absence of sodium. Two preparations, weight of Brockmann bodies: 5.4 and 6.8 mg.

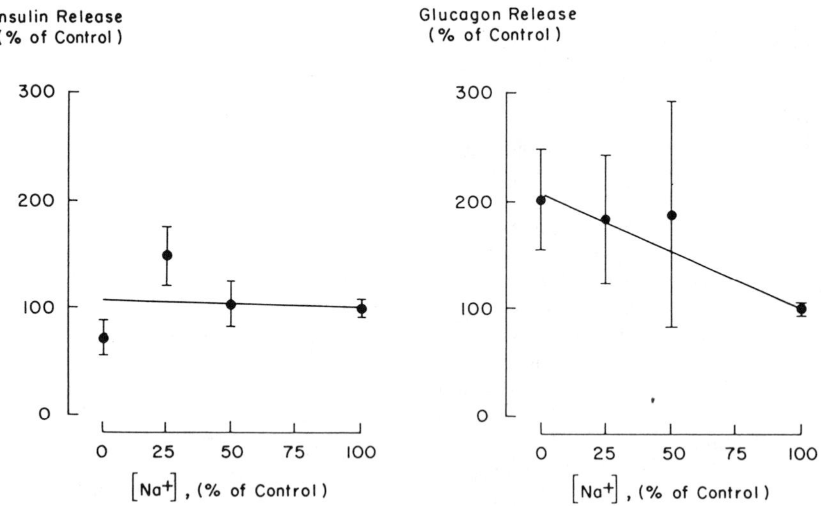

FIGURE 5. Effect of concentration of extracellular sodium on potassium-induced insulin and glucagon release from the perfused Brockmann body. Perfusion conditions as in FIGURE 3 except that stimulus consisted of 25 mM KCl in the continued presence of 2 mM glucose. Data from four perfusions and four stimuli at 0% Na^+, 4 at 25%, 2 at 50%, and 17 at 100% Na^+. Weight of Brockmann bodies: 5.4 ± 1.6 mg (mean ± S.E., $N = 4$).

Likewise, selective inhibition of glucose recognition in the A cells in the absence of sodium might account for the observed stimulation of arginine-induced glucagon release. Principal islets were perfused at 2 mM glucose, then stimulated with 10 mM arginine, either in the presence of 2 mM glucose or in its absence (FIG. 6). Arginine-induced insulin release was significantly inhibited in the absence of glucose, supporting the concept that glucose recognition is selectively inhibited in the absence of sodium. Arginine-induced glucagon release was only insignificantly affected by the absence of glucose, and it may be that the short period of exposure to low glucose is responsible for the lack of stimulation.

It is concluded that in the absence of sodium, glucose recognition is impaired selectively, affecting both the role of glucose as a direct stimulus of insulin release and its secondary role in modulating arginine-induced hormone release. In both rat islets and the catfish principal islet, glucose-induced insulin release is inhibited in the absence of sodium, but in the rat the inhibition does not appear to be as selective, since in the absence of sodium potassium-induced insulin release is also inhibited.[31,43–45] It is of great interest to determine the specific step in glucose sensing, which is affected by the absence of sodium. In the catfish principal islet this is complicated by the fact that mannose is the only other carbohydrate besides glucose that enters glycolysis and stimulates insulin release.[46] As expected, mannose-induced insulin release also was sodium dependent. Glyceraldehyde, which enters glycolysis at the triose phosphate step and is a potent secretagogue in rat islets, does not elicit any hormone release from the catfish islet.[46–48] However, very recently Biden *et al.* could demonstrate that glyceraldehyde-induced insulin release from rat islets is also significantly inhibited in the absence of extracellular sodium.[43] Our finding of a selective inhibition of glucose recognition in the absence of sodium is well complemented by the observations in rat islets, which showed a 30% decrease in the rate of glucose utilization, a 60% increase in glucose-6-phosphate, and a 140% increase in the content of fructose diphosphate in the absence of extracellular sodium.[35,49]

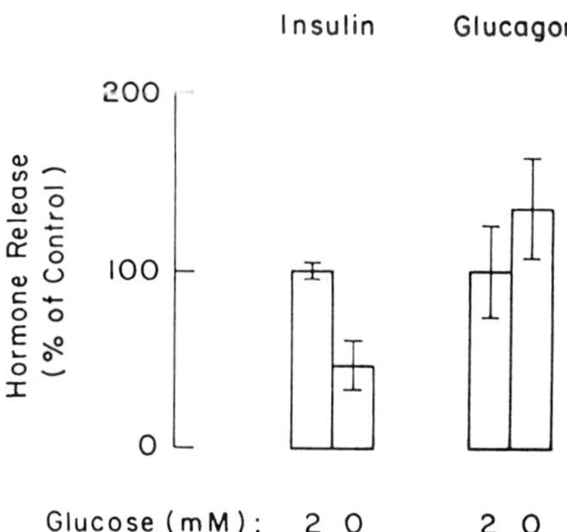

FIGURE 6. Effect of glucose on arginine-induced insulin and glucagon release. Perfusion conditions as in FIGURE 1. Brockmann bodies were repeatedly perfused at 2 mM glucose for 20 min, then stimulated with 10 mM arginine in the presence of 2 mM glucose or in its absence for 16 min. Two preparations, four stimuli at 2 mM glucose, three without glucose. Weight of Brockmann bodies: 2.3 and 2.8 mg.

ROLE OF THE Na:H ANTIPORT

It was next hypothesized that the sodium dependence of glucose-induced insulin release was due to the Na:H antiport. This antiport found in virtually every cell type is involved in pH homeostasis by pumping protons out of the cell thereby utilizing the inwardly directed sodium gradient. It can be inhibited by the absence of a sufficient sodium gradient, by a low pH outside the cell, or with amiloride.[50] As a consequence intracellular proton liberation will gradually acidify the cytosol. The hypothesis further postulates that such a decrease in cytosolic pH inhibits glycolysis and hence the glucose-sensing mechanism. In another mode, the Na:H antiport can also use an inwardly directed gradient of lithium to remove protons from the cytosol. Thus, if the hypothesis is correct, lithium should be able to replace sodium without inhibiting glucose-induced insulin release. Finally, in a third mode, the Na:H antiport may use an outwardly directed sodium gradient, such as in the absence of extracellular sodium, to pump protons into the cell, leading to acidification of the cytosol. This process may lead to more rapid acidification of the cytosol than liberation of protons from metabolic reactions alone. Here, the influence of extracellular pH, amiloride, and lithium on

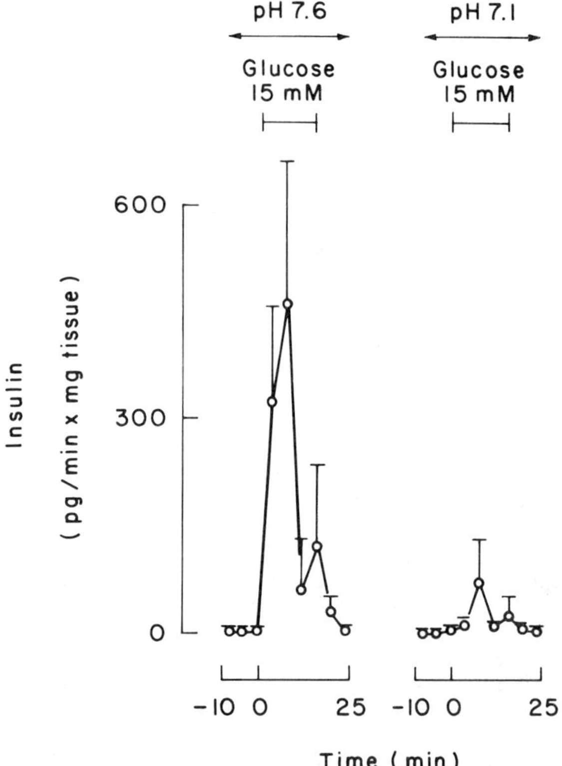

FIGURE 7. Effect of low extracellular pH on glucose-induced insulin release from the perfused Brockmann body. Perfusion conditions as in FIGURE 1, but pH changed as indicated. Two preparations; weight of Brockmann bodies: 4.1 and 6.5 mg.

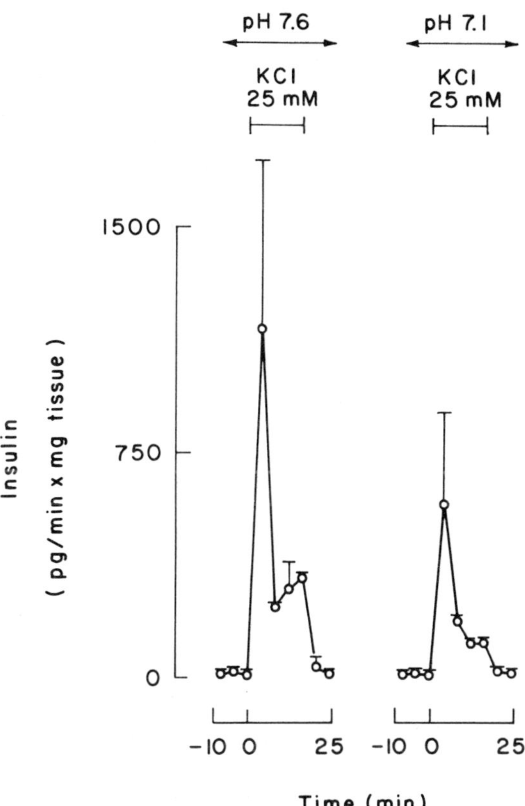

FIGURE 8. Effect of low extracellular pH on KCl-induced insulin and glucagon release from the perfused Brockmann body. Conditions as in FIGURE 7. Two preparations, weight of Brockmann bodies; 2.3 and 6.2 mg.

glucose-induced insulin release was studied. First, the effect of inhibiting the Na:H antiport with a low extracellular pH was investigated (FIG. 7). When the pH of the perfusion medium was lowered from 7.6 to 7.1, glucose-induced insulin release was almost completely inhibited even in the presence of sodium. When the same experiment was performed with potassium chloride as the stimulus instead of glucose there was only a modest inhibition at the low pH (FIG. 8). Thus, in agreement with the hypothesis, a low extracellular pH had a similar effect on glucose-induced insulin release as the absence of sodium.

In a second set of experiments the Na:H antiport was inhibited with amiloride, while sodium was present throughout the perfusion. FIGURE 9 shows that glucose once again reproducibly elicited insulin release. Amiloride (500 μM) was then introduced into the perfusion medium, first in the absence, then in the presence of stimulatory glucose. Contrary to the expectations, insulin release was not inhibited in the presence of amiloride. The same result was obtained when amiloride was used at a higher concentration of 1 mM. In a separate experiment, sodium was temporarily removed then reintroduced during continuous glucose stimulation, either in the continuous

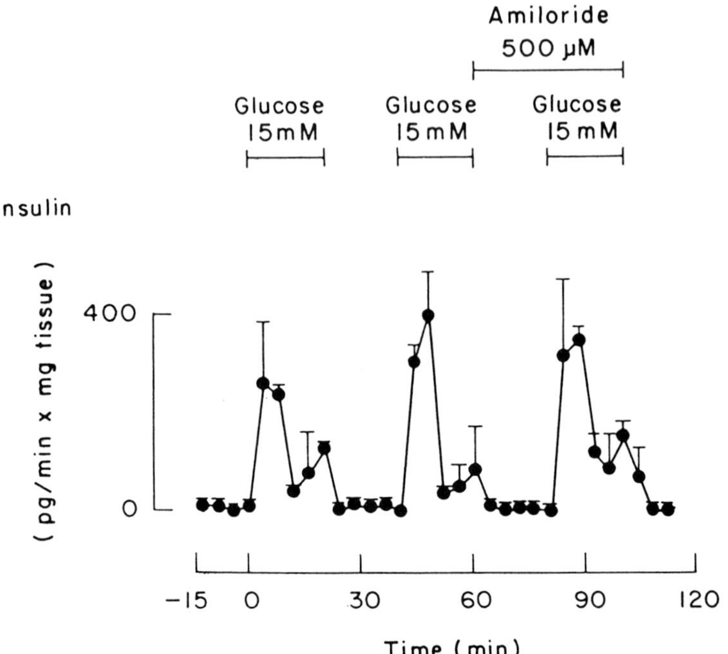

FIGURE 9. Effect of amiloride on glucose-induced insulin release from the perfused Brockmann body. Perfusion medium with normal concentration of sodium as in FIGURE 1. Means and S.E.s of two perfusions; weight of Brockmann bodies: 3.3 and 6.9 mg.

presence of 1 mM amiloride or in its absence. Recovery from inhibition of insulin release in the absence of sodium was markedly faster during amiloride treatment. These findings suggest that amiloride did not effectively inhibit the Na:H antiport in the presence of normal sodium, that the Na:H antiport reversed its pumping direction in the absence of sodium, and that under these conditions 1 mM of amiloride effectively inhibited the reversal of the Na:H antiport. Clearly, further experiments concerning the effects of amiloride or of amiloride analogs are needed before firm conclusions can be reached.

Finally, sodium in the perfusion medium was replaced with lithium in an attempt to maintain sufficient proton pumping activity to permit some residual glucose-induced insulin release (FIG. 10). Once again, glucose (15 mM) elicited reproducible insulin release in the presence of sodium. Lithium was then introduced in exchange for sodium during a 20-min perfusion period at baseline glucose, followed by a 16-min stimulatory phase at 15 mM glucose. Unexpectedly, glucose induced only a small amount of insulin release when sodium was replaced with lithium and a rapid transient release when sodium was reintroduced and glucose removed simultaneously. Whether the initial release is significantly higher than the release seen with choline substitution remains to be demonstrated in a number of further experiments. However, the Na:H antiport does not function as efficiently with lithium as with sodium, and may thus never provide adequate transport rates necessary for normal glucose stimulation of insulin release.[50] In addition, in isolated rat islets lithium does not relieve the inhibition of insulin release caused by the absence of sodium and may cause a poorly reversible inhibition of insulin release even at a concentration as low as 20 mM.[44,51]

In summary, it is shown that glucose-induced insulin release is inhibited in the absence of extracellular sodium or at a low pH of the perfusion medium. These findings are consistent with the hypothesis that the sodium dependency of glucose-induced insulin release is due to inhibition or reversal of the Na:H antiport. However, the findings are marred by the fact that amiloride did not inhibit glucose-induced insulin release and that substitution of sodium with lithium could not restore glucose-induced insulin release.

The hypothesis gains support, though, from a recent study by Biden *et al.* on collagenase-isolated rat islets.[43] They reported that glucose-induced insulin release is inhibited considerably in either the absence of extracellular sodium or the presence of amiloride, and that the inhibition by the absence of sodium can be overcome by an elevated concentration of bicarbonate. In addition, their results and ours cast doubt on the hypothesis that a decrease in cytosolic pH triggers insulin release. It is expected that an elucidation of the step in the glucose-sensing mechanism that is impaired in the absence of extracellular sodium or at low pH will help our understanding of stimulus secretion coupling in glucose-induced insulin release.

It is well worth mentioning that both glucose-induced insulin release and glucose modulation of arginine-induced insulin release are preferentially impaired in type II and pre-type I diabetic patients.[52,53] The present results suggest that pH homeostasis in pathologically altered B cells should be studied.

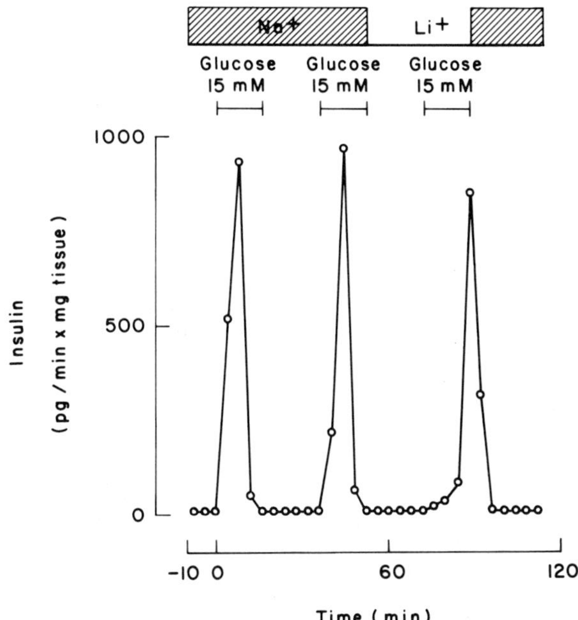

FIGURE 10. Effect of replacing extracellular sodium with lithium on glucose-induced insulin release from the perfused Brockmann body. Perfusion conditions as in FIGURE 1. Where indicated, sodium in perfusion medium was isotonically replaced with lithium. Single experiment, weight of Brockmann body: 8.5 mg.

SUMMARY

The sodium dependence of insulin release was studied with the *in vitro* perfused principal islet of channel catfish. When extracellular sodium was replaced with choline, glucose-induced insulin release and glucose enhancement of arginine-induced insulin release were selectively abolished. Lowering the extracellular pH from 7.6 to 7.1 had a similar selective effect on glucose-induced insulin release. It is suggested that the Na:H antiport is responsible for the sodium dependence of insulin release and that a lowered cytosolic pH preferentially inhibits the glucose sensing mechanism.

ACKNOWLEDGMENTS

I am most grateful for the help of the following persons: Dr. M. A. Permutt, Washington University, St. Louis, MO for providing an antiserum against catfish insulin; Dr. P. C. Andrews, Purdue University, West Lafayette, IN, for generous gifts of catfish insulin and catfish glucagon; Ms. Dzidra Rumba and Mr. Benjamin Burg for excellent technical assistance; Dr. G. C. Weir for financial support; Dr. R. Maron for help in producing an antibody to catfish insulin; Drs. Biden and C. A. Wollheim for sharing some of their results in the early stages of both of our work; Drs. A. Scarpa, A. Epple, F. M. Matschinsky, and G. C. Weir for helpful criticism and encouragement; and Ms. D. Poczatek and Ms. A. Rosenstein for expert typing.

REFERENCES

1. MALAISSE, W. J., A. SENER, A. HERCHUELZ & J. C. HUTTON. 1979. Insulin release: the fuel hypothesis. Metabolism **28:** 373–386.
2. HELLMAN, B., J. SEHLIN & I. -B. TÄLJEDAL. 1971. Evidence for mediated transport of glucose in mammalian pancreatic β-cells. Biochim. Biophys. Acta **241:** 147–154.
3. MATSCHINSKY, F. M. & J. E. ELLERMAN. 1968. Metabolism of glucose in the islets of Langerhans. J. Biol. Chem. **243:** 2730–2736.
4. GARFINKEL, D., L. GARFINKEL, M. D. MEGLASSON & F. M. MATSCHINSKY. 1984. Computer modeling identifies glucokinase as glucose sensor of pancreatic β-cells. Am. J. Physiol. **247:** R527–R536.
5. MEGLASSON, M. & F. M. MATSCHINSKY. 1984. New perspectives on pancreatic islet glucokinase. Am. J. Physiol. **246:** E1–E13.
6. DEAN, P. M. & E. K. MATTHEWS. 1970. Glucose-induced electrical activity in pancreatic islet cells. J. Physiol. (Lond.) **210:** 255–264.
7. MEISSNER, H. P. 1976. Electrical characteristics of the beta-cells in pancreatic islets. J. Physiol. (Paris) **72:** 757–767.
8. RIBALET, B. & P. M. BEIGELMAN. 1979. Cyclic variation of K^+ conductance in pancreatic β-cells: Ca^{2+} and voltage dependence. Am. J. Physiol. **237:** C137–C146.
9. SEHLIN, J. & N. FREINKEL. 1983. Biphasic modulation of K^+ permeability in pancreatic islets during acute stimulation with glucose. Diabetes **32:** 820–824.
10. COOK, D. L. & C. N. HALES. 1984. Intracellular ATP directly blocks K^+ channels in pancreatic B-cells. Nature **311:** 271–273.
11. ASHCROFT, F. M., D. E. HARRISON & S. J. H. ASHCROFT. 1984. Glucose induces closure of single potassium channels in isolated rat pancreatic β-cells. Nature **312:** 446–448.
12. FINDLAY, I., M. J. DUNNE & O. H. PETERSEN. 1985. ATP-sensitive inward rectifier and voltage- and calcium-activated K^+ channels in cultured pancreatic islet cells. J. Membr. Biol. **88:** 165–172.
13. TRUS, M., H. WARNER & F. M. MATSCHINSKY. 1980. Effects of glucose on insulin release

and on intermediary metabolism of isolated perifused pancreatic islets from fed and fasted rats. Diabetes **29:** 1–14.

14. Cook, D. L., M. Ikeuchi & W. Y. Fujimoto. 1984. Lowering of pH_i inhibits Ca^{2+}-activated K^+ channels in pancreatic B-cells. Nature **311:** 269–271.
15. Smith, J. S. & C. S. Pace. 1983. Modification of glucose-induced insulin release by alteration of pH. Diabetes **32:** 61–66.
16. Pace, C. S., J. T. Tarvin & J. S. Smith. 1983. Stimulus-secretion coupling in β-cells: Modulation by pH. Am. J. Physiol. **244:** E3–E18.
17. Eddlestone, G. T. & P. M. Beigelman. 1983. Pancreatic β-cell electrical activity: The role of anions and the control of pH. Am. J. Physiol. **244:** C188–C197.
18. Hellman B., J. Sehlin & I. -B. Täljedal. 1972. The intracellular pH of mammalian pancreatic B-cells. Endocrinology **90:** 335–337.
19. Deleers, M., P. Lebrun & W. J. Malaisse. 1983. Increase in CO_3H-influx and cellular pH in glucose-stimulated pancreatic islets. FEBS Lett. **154:** 97–100.
20. Lindström, P. & J. Sehlin. 1984. Effect of glucose on the intracellular pH of pancreatic islet cells. Biochem. J. **218:** 887–892.
21. Deleers, M., P. Lebrun & W. J. Malaisse. 1985. Nutrient-induced changes in the pH of pancreatic islet cells. Horm. Metabol. Res. **17:** 391–395.
22. Dean, P. M. & E. K. Matthews. 1970. Electrical activity in pancreatic islet cells: Effect of ions. J. Physiol. (Lond) **210:** 265–275.
23. Matthews, E. K. & Y. Sakamoto. 1975. Electrical characteristics of pancreatic islet cells. J. Physiol. **246:** 421–437.
24. Meissner, H. P. & M. Preissler. 1980. Ionic mechanisms of the glucose-induced membrane potential changes in B-cells. Horm. Metabol. Res. Suppl. **10:** 91–99.
25. Ribalet, B. & P. M. Beigelman. 1981. Effects of divalent cations on β-cell electrical activity. Am. J. Physiol. **241:** C59–C67.
26. Wollheim, C. B. & G. W. G. Sharp. 1981. Regulation of insulin release by calcium. Physiol. Rev. **61:** 914–973.
27. Henquin, J. C., S. Charles, M. Nenquin, F. Mathot & T. Tamagawa. 1982. Diazoxide and D600 inhibition of insulin release; distinct mechanisms explain the specificity for different stimuli. Diabetes **31:** 776–783.
28. Biden, T. J., M. Prentki, R. F. Irvine, M. J. Berridge & C. B. Wollheim. 1984. Inositol 1,4,5-triphosphate mobilizes intracellular Ca^{2+} from permeabilized insulin-secreting cells. Biochem. J. **223:** 467–473.
29. Prentki, M., B. E. Corkey & F. M. Matschinsky. 1985. Inositol-1,4,5-triphosphate and the endoplasmic reticulum Ca^{2+} cycle of a rat insulinoma cell line. J. Biol. Chem. **260:** 9185–9190.
30. Findlay, I., M. J. Dunne & O. H. Petersen. 1985. High conductance K^+ channel in pancreatic islet cells can be activated and inactivated by internal calcium. J. Membr. Biol. **83:** 169–175.
31. Hales, C. N. & R. D. G. Milner. 1968. The role of sodium and potassium in insulin secretion from rabbit pancreas. J. Physiol. (Lond) **194:** 725–743.
32. Meissner, H. P. & H. Schmelz. 1974. Membrane potential of β-cells in pancreatic islets. Pflueger's Arch. **351:** 195–206.
33. Pace, C. S. 1979. Activation of Na channels in islet cells: Metabolic and secretory effects. Am. J. Physiol. **237:** E130–E135.
34. Ronner, P., unpublished observation.
35. Hellman, B., L. -A. Idahl, A. Lernmark, J. Sehlin & I. -B. Täljedal. 1974. The pancreatic β-cell recognition of insulin secretagogues; effects of calcium and sodium on glucose metabolism and insulin release. Biochem. J. **138:** 33–45.
36. Ronner, P. & A. Scarpa. 1982. Isolated perfused Brockmann body as a model for studying pancreatic endocrine secretion. Am. J. Physiol. **243:** E352–E359.
37. Epple, A. 1969. The endocrine pancreas. *In* Fish Physiology. W. S. Hoar & D. J. Randall, Eds. **2:** 275–319. Academic Press. New York.
38. Falkmer, S. & Y. Östberg. 1977. Comparative morphology of pancreatic islets in animals. *In* The Diabetic Pancreas. B. W. Volk & K. F. Wellmann, Eds.: 15–59. Plenum Press. New York.

39. EPPLE, A., J. E. BRINN & J. B. YOUNG. 1980. Evolution of pancreatic islet functions. *In* Evolution of Vertebrate Endocrine Systems. P. K. T. Pang & A. Epple, Eds.: 269–321. Texas Technical Press. Lubbock, TX.
40. RONNER, P. & A. SCARPA. 1984. Difference in glucose dependency of insulin and somatostatin release. Am. J. Physiol. **246:** E506–E509.
41. PAGLIARA, A. S., S. N. STILLING, B. HOVER, D. M. MARTIN & F. M. MATSCHINSKY. 1974. Glucose modulation of amino acid-induced glucagon and insulin release in the isolated perfused rat pancreas. J. Clin. Invest. **54:** 819–832.
42. GERICH, J. E., M. A. CHARLES & G. M. GRODSKY. 1974. Characterization of the effects of arginine and glucose on glucagon and insulin release from the perfused rat pancreas. J. Clin. Invest. **54:** 833–841.
43. BIDEN, T. J., D. JANJIC & C. B. WOLLHEIM. 1986. Sodium requirement for insulin release: putative role in regulation of intracellular pH. Am. J. Physiol. **250:** C207–C213.
44. LAMBERT, A. E., J.-C. HENQUIN & P. MALVAUX. 1974. Cationic environment and dynamics of insulin secretion. I. Effect of low concentrations of sodium. Endocrinology **95:** 1069–1077.
45. HENQUIN, J.-C. & A. E. LAMBERT. 1974. Cationic environment and dynamics of insulin secretion. II. Effect of a high concentration of potassium. Diabetes **23:** 933–942.
46. RONNER, P. & A. SCARPA. 1987. Secretagogues for pancreatic hormone release in the channel catfish (*Ictalurus punctatus*). Gen. Comp. Endocrinol. (In press.)
47. HELLMAN, B., L.-A. IDAHL, A. LERNMARK, J. SEHLIN & I. -B. TÄLJEDAL. 1974. The pancreatic β-cell recognition of insulin secretagogues; comparison of glucose with glyceraldehyde isomers and dihydroxyacetone. Arch. Biochem. Biophys. **162:** 448–457.
48. ZAWALICH, W. S., E. S. DYE, R. ROGNSTAD & F. M. MATSCHINSKY. 1978. On the biochemical nature of triose- and hexose-stimulated insulin secretion. Endocrinology **103:** 2027–2034.
49. ASHCROFT, S. J. H., J. M. BASSETT & P. J. RANDLE. 1972. Insulin secretion mechanisms and glucose metabolism in isolated islets. Diabetes **21**(Suppl. 2): 538–545.
50. ARONSON, P. S. 1985. Properties of the renal Na^+-H^+ exchanger. Ann. NY. Acad. Sci. **456:** 220–228.
51. ANDERSON, J. H. & W. G. BLACKARD. 1978. Effect of lithium on pancreatic islet insulin release. Endocrinology **102:** 291–295.
52. PFEIFER, M. A., J. B. HALTER & D. PORTE. 1981. Insulin secretion in diabetes mellitus. Am. J. Med. **70:** 579–588.
53. WARD, K., J. C. BEARD, J. B. HALTER, M. A. PFEIFER & D. PORTE. 1984. Pathophysiology of insulin secretion in non-insulin-dependent diabetes mellitus. Diabetes Care **7:** 491–502.

DISCUSSION OF THE PAPER

M. BLAUSTEIN (*University of Maryland, Baltimore, MD*): You showed, in one of your slides, that Na^+/Ca^{2+} exchange plays a role in Ca extrusion in B cells. Most data (from other types of cells) indicate that the exchange stoichiometry is about $3Na^+:1Ca^{2+}$, and that the exchange is voltage sensitive. Have you considered how the effects of an altered Na^+ electrochemical gradient on insulin release might be complicated by alterations of Ca^{2+} movements mediated by Na^+/Ca^{2+} exchange?

RONNER: I doubt that the Na^+/Ca^{2+} exchanger is involved in the phenomena presented here. On the one hand, I do not expect that the Na^+/Ca^{2+} exchanger is involved in raising the cytosolic concentration of calcium during initiation of secretion.

At any rate, arginine- and KCl-induced insulin release function also in the absence of extracellular sodium, whereas glucose-induced insulin release does not; I would certainly expect that the Na^+/Ca^{2+} exchanger behaved similarly with all three stimuli. On the other hand, with respect to removal of cytosolic calcium by the Na^+/Ca^{2+} exchanger at the end of a stimulus, the exchanger may have a physiological function in the catfish islets; but the experimental results also suggest that there is certainly no absolute need for the Na^+/Ca^{2+} exchanger to remove cytosolic calcium, since removal of sodium in the presence of a glucose stimulus leads to immediate inhibition of insulin release and I would presume that this coincides with a corresponding decrease of the cytosolic concentration of calcium.

I. SCHULZ (*Max-Planck-Institute for Biophysics, Frankfurt*): Can glucose-stimulated insulin release be inhibited by phlorizin?

RONNER: I have not tested the effect of phlorizin on glucose-induced insulin release from the perfused catfish principal islet. But phlorizin is known to inhibit glucose-induced insulin release from isolated rat islets if applied at a relatively high concentration of 10–20 mM.

I. W. ARIAS (*Tufts University, Boston, MA*): Is there a difference in ionic control of insulin secretion in fresh water fish, such as catfish, and salt water fish that also have separated islet tissue, such as *Opsanus tau* (toadfish)?

RONNER: I do not know, simply because the preparation described here is the only *in vitro* perfusion preparation of a Brockmann body developed thus far.

Cell Biology of Insulin's Stimulatory Action on Glucose Transport and Its Perturbation in Altered Metabolic States

BARBARA B. KAHN AND SAMUEL W. CUSHMAN[a]

Experimental Diabetes, Metabolism and Nutrition Section
Molecular, Cellular and Nutritional Endocrinology Branch
National Institute of Diabetes
and Digestive and Kidney Diseases
National Institutes of Health
Bethesda, Maryland 20892

INTRODUCTION

Glucose metabolism is essential to the survival of animals and humans. A necessary precursor to metabolism is the entry of glucose into cells, which takes place in most tissues by facilitated diffusion via specific glucose-carrier proteins. Insulin is known to stimulate the transport of glucose in a variety of cell types, most importantly muscle[1,2] and adipose cells.[3,4] In insulin-responsive tissues, the majority of glucose-transporter proteins have been shown to reside in an intracellular pool from which they are recruited to the plasma membrane with insulin stimulation.[5] The size of the intracellular pool appears to be an important determinant of the magnitude of the response to insulin. Both the size of this pool and the number of glucose transporters recruited to the plasma membrane in response to insulin can be dramatically changed by various nutritionally altered or disease states,[6] resulting in a marked difference in the ability of cells to transport glucose in response to insulin. Recent data further indicate that *in vivo* metabolic alterations may modify an additional factor, glucose transporter intrinsic activity.[7–9] This review will describe current concepts of the cellular mechanism for insulin's stimulatory effect on glucose transport and its perturbation in altered metabolic states.

MECHANISM FOR INSULIN'S STIMULATORY ACTION ON GLUCOSE TRANSPORT

In adipose cells from metabolically normal, lean, growing, 170–230 g male rats, the basal rate of glucose transport is relatively low and is enhanced 20–30 fold by insulin, reaching maximal stimulation within 7–10 min with a half-time of 3–5 min (FIG. 1).[10] This effect of insulin is concentration dependent and reversible as seen with the addition of a 300-fold excess of anti-insulin antibody. Kinetic experiments show that insulin stimulation results from a change in the maximum transport velocity (V_{max}) and not in the apparent affinity (K_m) of the glucose transporter for glucose.[3,4] Two potential mechanisms could account for this: an increase in glucose transporter intrinsic activity,

[a]Address correspondence to: S.W. Cushman, EDMNS/MCNEB/NIDDK, Bldg. 10, Room 5N-102, National Institutes of Health, Bethesda, MD 20892.

i.e. turnover number, and/or an increase in the number of functional glucose transporters present in the plasma membrane of the cell.

Two groups of investigators who tackled this problem were able to quantify glucose transporters using independent techniques. Both carried out studies in rat adipose cells incubated in the absence (basal) or presence (insulin-stimulated) of insulin. Cells were homogenized and fractionated either by differential ultracentrifugation[5,10,11] or by sucrose density-gradient centrifugation[12,13] into plasma membranes, high-density microsomes enriched in endoplasmic reticulum, and low-density microsomes enriched in components of the Golgi apparatus. The number of glucose transporters in these subcellular fractions was then quantitated using different methodology.

One group, Wardzala *et al.*,[5,14] developed a specific binding assay using cytochalasin B, a potent inhibitor of glucose transport that binds competitively to the glucose transporter. Specificity is conferred by performing the assay in the absence or presence

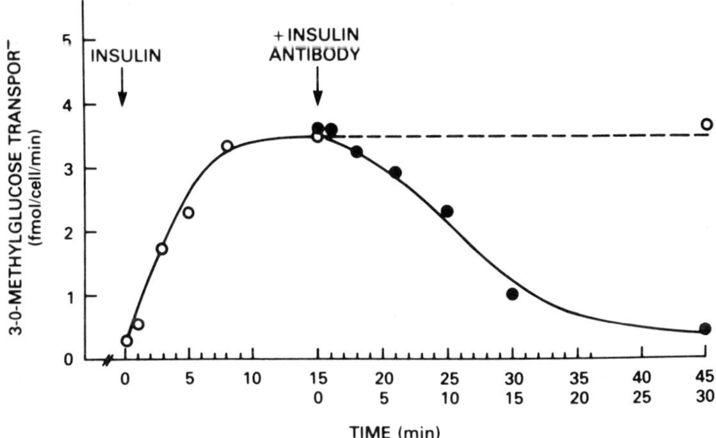

FIGURE 1. Stimulation of glucose transport activity by insulin at 37°C in the isolated rat adipose cell: time course (○) in response to 6.7 nM (1000 μU/ml) insulin and reversal (●) using a 300-fold excess of the IgG fraction of an anti-insulin antiserum. (From Karnieli *et al.*[10])

of a saturating concentration of D-glucose. Non-D-glucose–inhibitable binding is subtracted from total binding to yield specific binding. FIGURE 2 shows the steady-state distribution of D-glucose-inhibitable cytochalasin B binding sites, representing glucose transporters, among the three subcellular membrane fractions.[11] In the basal state, the highest concentration of the glucose transporters is observed in the low-density microsomes, the intracellular pool, and few are detected in the plasma membranes or high-density microsomes. With insulin stimulation, ~60% of the glucose transporters are lost from the low-density microsomes, concurrent with an ~five-fold increase in their concentration in the plasma membranes and an ~two-fold increase in the high-density microsomes. Specific marker enzyme assays reveal that most of the glucose transporters present in the high-density microsomes are due to contamination from the other two membrane fractions.[11]

A separate group, Kono and colleagues,[12,13] reconstituted rat adipose cell plasma membranes and an intracellular membrane fraction enriched in enzyme markers of the

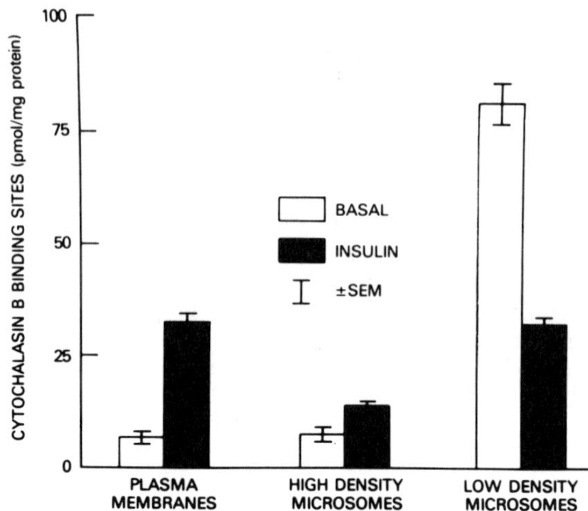

FIGURE 2. Steady-state subcellular distribution of glucose transporters in the isolated rat adipose cell incubated at 37°C in the absence (basal) or presence of 6.7 nM (1000 μU/ml) insulin. (From Simpson *et al.*[11])

Golgi apparatus into artificial liposomes in which glucose transport activity was measured directly. Basal and insulin-stimulated glucose transport into these reconstituted vesicles demonstrated changes paralleling those observed using the cytochalasin B binding assay. Thus, both groups concluded independently that the primary mechanism for insulin's stimulatory effect on glucose transport appeared to be a translocation of glucose transporters from a large intracellular pool to the plasma membrane. This process is rapid, with maximal stimulation in 6–8 min, and fully reversible.[10] Furthermore, the translocation is stoichiometric as determined by estimating a total number of transporters per cell using protein recoveries and specific marker enzyme activitites.[5,11] Subsequently, studies utilizing rat diaphragm[15,16] and cardiac muscle,[17] and guinea pig[18] and human[19,20] adipose cells have indicated that this mechanism operates in other tissues and species.

Since these initial experiments documenting the redistribution of glucose transporters with insulin stimulation, two additional techniques have been developed that confirm the translocation hypothesis and provide insights into the structure of the glucose transporter. Both Western blotting, using a rabbit antiserum prepared against the purified human erythrocyte glucose transporter,[21,22] and covalent cross-linking of [^{3}H]cytochalasin B to the glucose transporter by either direct photolysis or use of the bifunctional reagent hydroxysuccinimidyl-4-azidobenzoate[26] demonstrate a 45–55 kDa protein whose subcellular distribution is altered in response to insulin consistent with translocation.

ACUTE MODULATION OF INSULIN'S STIMULATORY EFFECT

Recent experiments acutely exposing adipose cells to adenylate cyclase stimulators and inhibitors *in vitro*[27–29] and chronically exposing them to certain altered metabolic states *in vivo*[6] have provided evidence for alterations in the intrinsic activity of the glucose transporter. Lipolytic agents that stimulate adenylate cyclase, such as catecholamines, ACTH, and glucagon, acutely reduce the V_{max} for insulin-stimulated glucose transport activity and this inhibition is reversible by antilipolytic agents that inhibit adenylate cyclase, such as adenosine, nicotinic acid, and prostaglandins.[27,30,31] The interchangeability of ligands known to operate through specific stimulatory and inhibitory receptors suggests that these effects are mediated by GTP-binding proteins, N_s and N_i.[27] Of particular interest, the alterations in glucose transport activity are not accompanied by changes in the number or subcellular distribution of glucose transporters as illustrated in FIGURE 3. *In vitro* experiments have been carried out in the presence of adenosine deaminase in order to prevent the accumulation of adenosine in the incubation medium due to cell lysis[32] and/or secretion.[33] Although saturating concentrations of isoproterenol in combination with adenosine deaminase induce a 60% inhibition of insulin-stimulated glucose transport in the intact cell, there is no change in the number of glucose transporters in the plasma membranes or the low-density microsomes. Subsequent addition of N^6-phenylisopropyladenosine, a nonmetabolizable adenosine analogue, restores the transport activity to 75% of the initial rate with again no change in the numbers of glucose transporters in the subcellular membrane fractions. Thus, these variations in cellular glucose transport activity appear to occur through alterations in the intrinsic activity of those glucose transporters already present in the plasma membrane and not in their translocation.

Further experiments by Joost *et al.*[29] show that the inhibitory effect of isoproterenol on glucose transport is lost after cooling cells but can be preserved by KCN treatment prior to cell cooling. Glucose transport activity in isolated plasma membranes from isoproterenol-treated cells that are exposed to KCN prior to cooling correlates with glucose transport activity in the intact cell but not with glucose transporter number, thus confirming modification of glucose transporter intrinsic activity.

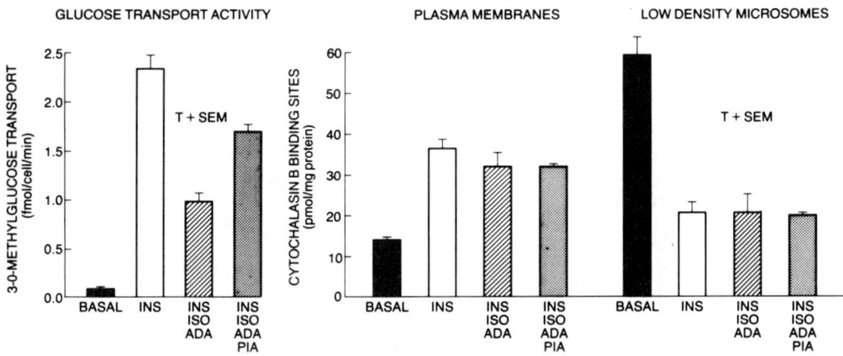

FIGURE 3. Counterregulation of insulin-stimulated glucose transport activity and the subcellular distribution of glucose transporters by adenylate cyclase stimulators and inhibitors at steady state at 37°C in isolated rat adipose cells: basal, the response to 6.7 nM (1,000 μU/ml) insulin (INS), the subsequent response to 200 nM isoproterenol (ISO) and simultaneous adenosine removal using 1 U/ml adenosine deaminase (ADA), and the further subsequent response to 1 μM phenylisopropyladenosine (PIA).

EFFECTS OF *IN VIVO* METABOLIC ALTERATIONS ON *IN VITRO* INSULIN RESPONSIVENESS

In vivo metabolic alterations in the rat are proving to have rapid, progressive, and dramatic effects on insulin's ability to stimulate glucose transport *in vitro*.[6] Insulin resistance at the cellular level associated with widely divergent metabolic states ranging from fasting[7] to high fat feeding[34] or obesity[35] appears to result from a depleted intracellular pool of glucose transporters in the basal state resulting in a diminished translocation of glucose transporters to the plasma membrane with insulin stimulation. In contrast, the augmented glucose transport response to insulin seen in hyperresponsive conditions appears to involve at least two distinct subcellular mechanisms. Chronic exposure of normal rats to insulin[36] or to physical training[37] results in an increase in the net synthesis of glucose transporters whereas reversal of a catabolic state by refeeding after fasting[7] or insulin treatment of diabetes[8,9] appears to increase glucose transporter intrinsic activity as well.

Streptozotocin Diabetes: An Example of Insulin Resistance

Although the major lesion in streptozotocin diabetes is insulin deficiency, insulin resistance is also present at the cellular level.[38,39] FIGURE 4 shows that basal 3-O-methylglucose transport decreases slightly per cell in diabetic animals but remains constant per unit surface area when the smaller cell size is taken into account (data not shown). However, insulin-stimulated glucose transport activity is decreased by 67% per cell (FIG. 5) or 55% per unit surface area.[39] Correspondingly, glucose-transporter concentration is decreased 53% in the plasma membranes after insulin stimulation with no change in the basal state. Concomitantly, low-density microsomal transporters are decreased 45% in the basal state. Thus, the insulin resistance appears to result from a reduced translocation of glucose transporters to the plasma membrane in response to insulin due to a depleted intracellular pool.

A similar cellular mechanism for resistance to insulin's stimulatory effect on glucose transport is seen in adipose cells from several other rat models including the marked obesity associated with aging in the male rat,[35] high fat feeding,[34] and total fasting.[7] These states appear divergent in terms of nutritional status. In addition, ambient insulin levels vary from markedly reduced in streptozotocin diabetes to

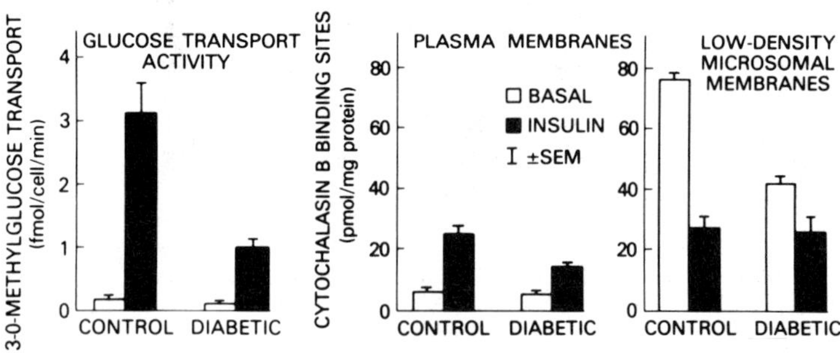

FIGURE 4. Effects of streptozotocin diabetes on glucose transport activity and the subcellular distribution of glucose transporters in basal and insulin-stimulated rat adipose cells. (From Karnieli *et al.*[39])

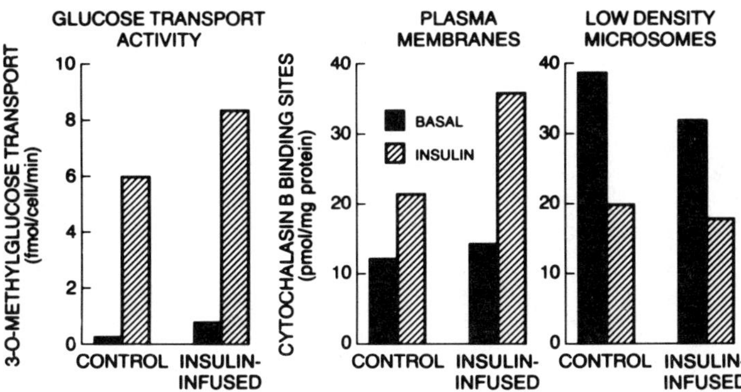

FIGURE 5. Effects of experimentally induced chronic hyperinsulinemia on glucose transport activity and the subcellular distribution of glucose transporters in basal and insulin-stimulated rat adipose cells.

significantly elevated in obesity, raising the question of the role of chronic exposure to insulin in modulating cellular responsiveness to insulin.

Chronic Hyperinsulinemia: One Form of Insulin Hyperresponsiveness

To address the above question directly, two models of chronic hyperinsulinemia have been studied: normal rats treated by injection[40–43] or infusion of insulin[44,45] and young, genetically obese Zucker rats.[46] Initially, studies in normal rats using daily insulin injections showed that hyperinsulinemia was associated with reduced sensitivity to insulin and either no change[42] or an increase[40,41] in basal and maximally insulin-stimulated glucose transport activity. More recently, Wardzala *et al.*[45] infused male, 175 g rats with insulin via subcutaneously implanted Alzet osmotic mini-pumps and found a 1.6–2.0-fold increase in adipose cell insulin-stimulated glucose transport activity with no change in the sensitivity to insulin. Surprisingly, after two weeks of treatment insulin-infused animals had identical mean body weights and epididymal adipose cell sizes as controls (TABLE 1). Non-fasting plasma glucose was reduced 55% and plasma insulin was increased 9.5-fold. No change was found in plasma free fatty acid levels while plasma triglycerides were decreased in the insulin-infused rats.

Kahn *et al.*[36] investigated the mechanism for the enhanced *in vitro* adipose cell insulin responsiveness seen with chronic hyperinsulinemia in normal rats. The results of a representative experiment are illustrated in FIGURE 5. While the basal rate of 3-O-methylglucose transport appears to be increased in the cells from the hyperinsulinemic rats compared to the controls, the results of several additional experiments suggest that these basal rates are essentially the same in both preparations of cells. However, a significant increase of 44% is observed in the glucose transport activity of the insulin-stimulated cells from the hyperinsulinemic rats. Cytochalasin B binding in subcellular membrane fractions reveals a corresponding increase in the concentration of glucose transporters in the plasma membranes in the insulin-stimulated state and not in the basal state, and a marginal decrease in the low-density microsomes in both the basal and insulin-stimulated states. However, the cells from the hyperinsulinemic rats have much more intracellular protein. After taking this into account, more glucose

TABLE 1. Clinical Characteristics of Control and Chronically Insulin-Infused Rats[a]

	Control ($N = 22$)[b]	Insulin-Infused ($N = 23$)[b]
	(mean ± SEM)	
Rat weight (g)	264 ± 5	257 ± 5
Adipose cell size (μg of lipid/cell)	0.15 ± 0.02	0.14 ± 0.01
Plasma glucose (mg/dl)	157 ± 4	71 ± 16[c]
Plasma insulin (μU/ml)	62 ± 10	589 ± 90[c]
Plasma free fatty acids (μmol/l)	451 ± 24	429 ± 42
Plasma triglyceride (mg/dl)	217 ± 9	109 ± 9[c]

[a]After Wardzala *et al.*[45]
[b]Number of rats.
[c]Statistically significant difference.

transporters are apparent in the intracellular pool. Thus, the increase in insulin-stimulated glucose transport can be explained by a greater number of glucose transporters translocated to the plasma membrane from an enlarged intracellular pool. The additional transporters appear to be the result of a generalized increase in the net synthesis of intracellular protein. A similar mechanism has been proposed to explain insulin hyperresponsive glucose transport in adipose cells from naturally hyperinsulinemic, young, genetically obese Zucker rats[46] and from physically trained rats.[37]

These phenomena do not clearly answer the question of the role of ambient insulin concentrations in modulating cellular responsiveness to insulin. Hyperinsulinemia in normal[36] and young Zucker rats[46] appears to result in an increase in the number of glucose transporters in adipose cells and therefore in the magnitude of insulin's stimulatory effect on glucose transport. This same mechamism may be seen in the absence of hyperinsulinemia in physically trained rats.[37] Furthermore, hyperinsulinemia in aging/obese rats[35] or in adult obese Zucker rats[47] is associated with an attenuated response to insulin and, in the former case, a depletion of intracellular glucose transporters.

These discrepancies remain unresolved but may, in fact, reflect the modification of insulin's effect at the cellular level by the ambient glucose level, alone or in combination with other hormones released in response to experimental or pathophysiological alterations. For example, catecholamines have been documented to be elevated in at least one model of experimental hyperinsulinemia,[44] presumably in response to hypoglycemia. Such elevations could result in decreased beta receptor sensitivity[48,49] thereby reducing the inhibitory effects of catecholamines, resulting in enhanced stimulatory effects of insulin.[50] In addition, the glycemic state of the animal may directly affect glucose transporter number and distribution. A precedent for this can be seen in cultured cells where glucose starvation has been reported to markedly increase glucose transport because of more transporters present in the plasma membrane.[51] This is not accompanied by changes in the level of *in vitro* translatable glucose transporter mRNA or in the rate of glucose transporter polypeptide synthesis[52] and therefore suggests decreased turnover of the glucose transporter.

Reversal of a Catabolic State: Cellular Recovery and Overshoot

Further complexity is introduced by studying the cellular events associated with insulin treatment of diabetic animals.[8,9] In studies by Kahn and Cushman,[8] diabetes was induced by intraperitoneal injection of 85 mg/kg of streptozotocin. After seven days of hyperglycemia, subcutaneous insulin infusion was begun via Alzet osmotic minipumps. Complete reversal of the state of cellular insulin resistance associated with untreated diabetes[39] is seen. In fact, a time course of insulin therapy shows a sequential recovery of insulin responsiveness with restoration to control levels at four days and a rapid increase to a maximum overshoot of 3.5-fold at 7–8 days.[8] The insulin-stimulated glucose transport rate then returns very gradually toward control levels but remains slightly elevated even at five weeks. This effect may bear some relationship to the "honeymoon" phenomenon of transiently reduced insulin requirements seen in some newly diagnosed Type I human diabetics shortly after initiating insulin therapy.

Preliminary investigations[8] of the insulin hyperresponsive glucose transport during insulin therapy of these diabetic rats (not illustrated) show an increase in the number of glucose transporters in the low-density microsomes from basal cells with only a small increase in the number translocated to the plasma membrane in response to insulin. There is a marked discrepancy between the dramatic overshoot in insulin-stimulated glucose transport rate and the modest increase in plasma membrane transporters. Kinetic experiments have confirmed an increase in glucose transport V_{max} with no change in K_m. Glucose transporters assessed by Western blotting also show only slightly increased antibody binding to plasma membranes from the cells of diabetic/insulin-treated animals compared to control.

FIGURE 6 illustrates a comparison of glucose transport activities and the concentrations of plasma membrane glucose transporters in maximally insulin-stimulated adipose cells from normal rats, nondiabetic rats made chronically hyperinsulinemic, diabetic rats, and insulin-treated diabetic rats. In the nondiabetic control rats, glucose transport activity and the concentration of plasma membrane glucose transporters have been normalized. Chronic insulin infusion of nondiabetic rats is then associated with a 56% increase in insulin-stimulated glucose transport activity and a corresponding increase in plasma membrane glucose transporters. The diabetic state, on the other hand, is accompanied by a 67% reduction in adipose cell glucose transport activity

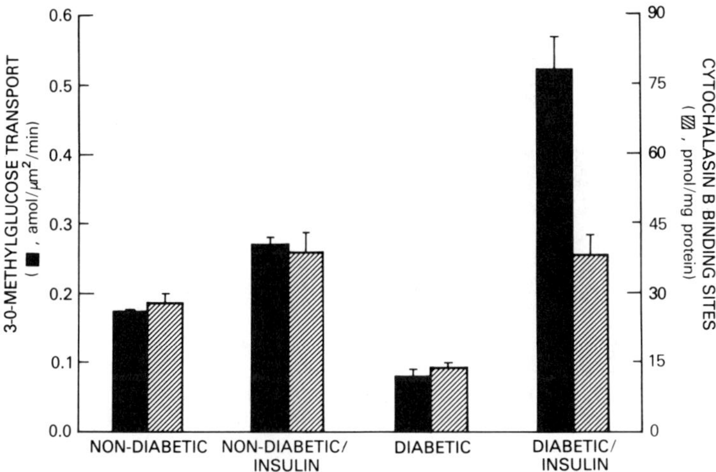

FIGURE 6. Comparison between glucose transport activity in intact cells and glucose transporters in plasma membranes from insulin-stimulated rat adipose cells obtained from nondiabetic and diabetic animals with and without insulin infusion.

compared to nondiabetic controls and a similar decrease in plasma membrane glucose transporters. In marked contrast, however, insulin treatment of streptozotocin diabetic rats results in an ~3.5-fold increase in insulin-stimulated glucose transport activity but only a 36% increase in plasma membrane glucose transporters. The latter increase is equivalent to that seen with chronic insulin infusion in normal rats and reflects the increased protein synthesis effect of insulin. The further overshoot in glucose transport activity with insulin treatment of diabetic rats suggests an additional striking increase in glucose transporter intrinsic activity.

A similar reversal of cellular insulin resistance resulting in an overshoot in insulin-stimulated glucose transport activity is seen when rats are refed after fasting.[7] While smaller than that observed with insulin treatment of diabetes, this hyperresponsiveness can be only partially explained by increased numbers of glucose transporters and therefore, increased glucose transporter intrinsic activity must also be invoked. These states have in common nutritional repletion after cellular starvation, which may have a unique effect on intracellular mechanisms to augment substrate entry into the cell.

SUMMARY: AN UPDATED MODEL FOR INSULIN'S STIMULATORY ACTION ON GLUCOSE TRANSPORT

The original model for insulin's stimulatory action on glucose transport in adipose and muscle cells, proposed by Karnieli *et al.*[10] and Kono and colleagues,[12,13] can now be updated (FIG. 7). Under normal metabolic conditions, insulin's action is thought to be initiated by binding to its receptor in the plasma membrane (step 1), then generating a signal (step 2) the nature of which is unkown. This results in the exocytic-like movement of membrane vesicles containing glucose transporters from an intracellular pool to the plasma membrane (step 3) where they are thought to first bind (step 4) and subsequently fuse (step 5), exposing glucose transporters to the extracellular medium and increasing the glucose transport rate (step 6). When the insulin dissociates from its receptor due to either physiological events or experimental treatment of cells with anti-insulin antibody (step 7), the process is reversed. Glucose transporters which are present in the plasma membrane are then reinternalized by an endocytic-like process and translocated back to the intracellular pool (step 8).

Recent findings now indicate that the magnitude of insulin's stimulatory effect depends on both the number and intrinsic activity (light compared to heavy arrows) of the glucose transporters present in the plasma membrane. The number of transporters may be altered by (1) changing the size of the intracellular pool without altering the proportion of transporters translocated as seen with *in vivo* nutritional alterations,[7,34] aging/obesity,[35] hypoinsulinemia,[39] and hyperinsulinemia[36] or exercise training,[37] or (2) altering the proportion of transporters translocated without changing the total pool size as has been demonstrated with the sulfonylurea, glyburide.[53] The data presented here show that glucose transporter intrinsic activity may be modified by both acute exposure to hormones and other agents that stimulate or inhibit adenylate cyclase and chronic exposure to certain *in vivo* metabolic alterations. In addition, insulin itself may alter both glucose transporter number and intrinsic activity as suggested by estimations of the turnover number of the glucose transporter[11] and measurements with fluorescein isothiocyanate,[54] the membrane-impermeant, amino-group modifying agent. Other investigators have shown that dexamethasone treatment of thymocytes[55] and adipose cells[56] may alter the rate of synthesis of another protein that affects the activity of the glucose transporter. Finally, recent experiments performed under

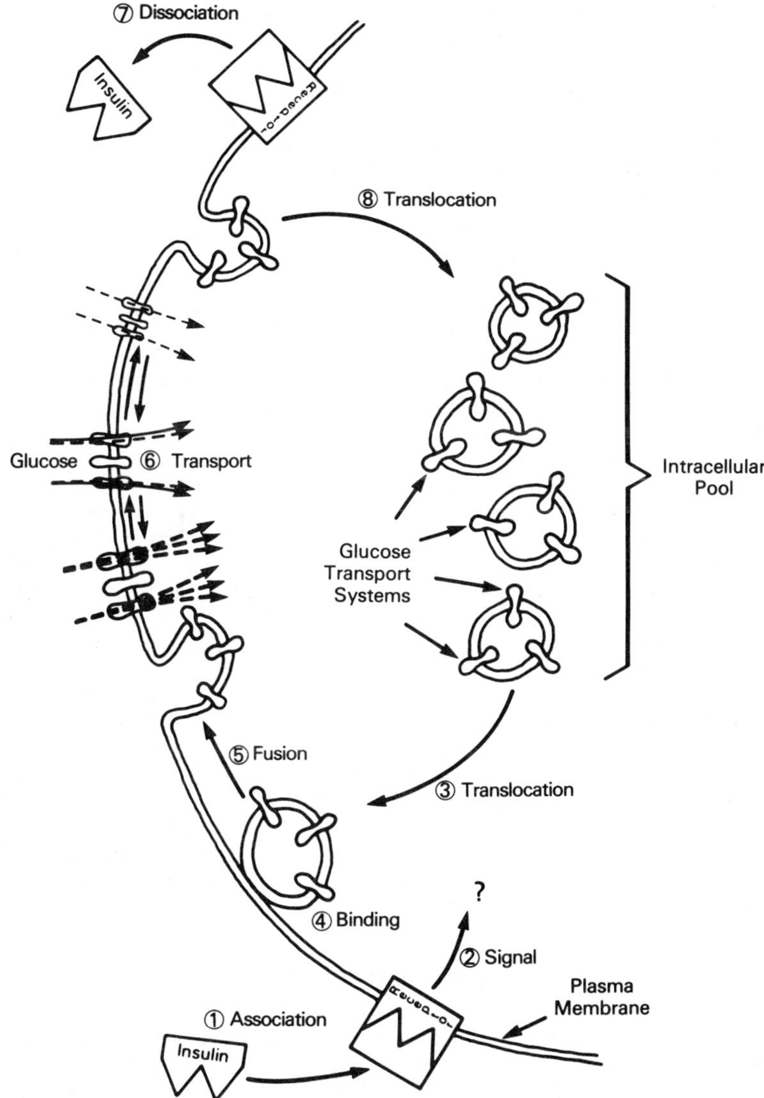

FIGURE 7. Hypothetical model of insulin's stimulatory action on glucose transport. (Adapted from Karnieli *et al.*[10])

significantly different incubation conditions have suggested that regulation may also occur by changes in K_m.[57]

Thus, the dramatic alterations in glucose transport activity seen in various pathophysiological states may involve any of these mechanisms. Interestingly, insulin resistance associated with divergent metabolic conditions appears to result from one common mechanism: a depletion of intracellular glucose transporters with fewer translocated to the plasma membrane in response to insulin. In contrast, insulin hyperresponsiveness has been shown to result from at least two distinct mechanisms: (1) an enlarged intracellular pool of glucose transporters due to a generalized increase in net intracellular protein synthesis, which may be due to the growth promoting effects of insulin, and (2) an alteration in glucose transporter intrinsic activity due to cellular events associated with nutritional repletion after depletion. In particular, the changes in glucose transport seen with experimental diabetes and its treatment involve all three of these mechanisms.

FUTURE DIRECTIONS

Experiments to date suggest that the translocation mechanism is operable in human adipose cells[19,20] and that changes in *in vitro* glucose transport activity in these cells correlate with total glucose disposal *in vivo*.[6] Further investigations need to be carried out to document the applicability to human disease states and the relative importance in various tissues involved in glucose homeostasis.

With the recent cloning of the glucose transporter from a human hepatoma cell line[58] and rat brain,[59] the probes are now available to investigate the molecular regulation of glucose transporter number and intrinsic activity as well as structure/function relationships. Currently, little is known about the biochemical features of the glucose transporter that are critical for translocation, binding and fusion with the plasma membrane, or for insulin responsiveness. The nature of the insulin-activated signaling mechanism resulting in translocation remains undefined and study of possible separate proteins involved in modifying glucose transporter activity[55,56] is in its infancy.

Evidence has recently been reported for two forms of the transporter in the rat adipose cell[60] only one of which appears to be translocated. In addition, recent data suggest that the non-insulin-responsive glucose transporter in liver is structurally different from the glucose transporter present in most insulin-responsive tissues.[59,61] Studies are in progress to better define these differences and uncover the structural features that are of key importance for insulin action on glucose transport. These investigations promise to enhance our understanding of the regulation of altered membrane phenomena associated with disease states and may aid in differentiating the extent to which these changes are adaptive or causative.

ACKNOWLEDGMENTS

The authors wish to thank their many colleagues, both former and current, for their indispensable contributions to the concepts and experimental results described here. These investigators include: Paul J. Hissin, Rupert C. Honnor, Richard Horuk, Hans G. Joost, Eddy Karnieli, Masao Kuroda, Constantine Londos, Lester B. Salans, Ian A. Simpson, Lawrence J. Wardzala, Thomas J. Wheeler, Dena R. Yver, and Mary Jane Zarnowski. The authors also wish to thank Dr. Simpson for his critical comments regarding this report.

REFERENCES

1. LEVINE R. & M. GOLDSTEIN. 1955. Rec. Prog. Horm. Res. **11:** 343–380.
2. PARK, C. R., D. REINWEIN, M. J. HENDERSEN, E. CADENAS & H. E. MORGAN. 1959. Am. J. Med. **26:** 674–684.

3. CROFFORD, O. B. & A. E. RENOLD. 1965. J. Biol. Chem. **240:** 14–21.
4. CROFFORD, O. B. & A. E. RENOLD. 1965. J. Biol. Chem. **240:** 3237–3244.
5. CUSHMAN, S. W. & L. J. WARDZALA. 1980. J. Biol. Chem. **255:** 4758–4762.
6. KAHN, B. B. & S. W. CUSHMAN. 1985. Diabetes Metab. Rev. **1:** 203–227.
7. KAHN, B. B. & S. W. CUSHMAN. 1984. Diabetes **33** (Suppl. 1): 71A (abstr.).
8. KAHN, B. B. & S. W. CUSHMAN. 1987. J. Biol. Chem. **262:** In press.
9. KARNIELI, E., M. ARMONI, P. COHEN, Y. KANTER & R. RAFAELOFF. 1987. Diabetes. **36**: In press.
10. KARNIELI E., M. J. ZARNOWSKI, P. J. HISSIN, I. A. SIMPSON, L. B. SALANS & S. W. CUSHMAN. 1981. J. Biol. Chem. **256:** 4772–4777.
11. SIMPSON, I. A., D. R. YVER, P. J. HISSIN, L. J. WARDZALA, E. KARNIELI, L. B. SALANS & S. W. CUSHMAN. 1983. Biochim. Biophys. Acta **763:** 393–407.
12. SUZUKI, K. & T. KONO. 1980. Proc. Natl. Acad. Sci. USA **77:** 2542–2545.
13. KONO, T., K. SUZUKI, L. E. DANSEY, R. W. ROBINSON & T. L. BLEVINS. 1981. J. Biol. Chem. **256:** 6400–6407.
14. WARDZALA, L. J., S. W. CUSHMAN & L. B. SALANS. 1978. J. Biol. Chem. **253:** 8002–8005.
15. WARDZALA, L. J. & B. JEANRENAUD. 1981. J. Biol. Chem. **256:** 7090–7093.
16. WARDZALA, L. J. & B. JEANRENAUD. 1983. Biochim. Biophys. Acta **730:** 4956.
17. WATANABE, T., M. M. SMITH, F. W. ROBINSON & T. KONO. 1984. J. Biol. Chem. **259:** 13117–13122.
18. HORUK, R., M. RODBELL, S. W. CUSHMAN & L. J. WARDZALA. 1983. J. Biol. Chem. **258:** 7425–7429.
19. KARNIELI, E., B. CHERNOW, P. J. HISSIN, I. A. SIMPSON & J. E. FOLEY. 1986. Herm. Metabol. Res. **18:** 860–861.
20. KARNIELI, E., A. BARZILAI, R. RAFAELOFF & M. ARMONI. 1986. J. Clin. Invest. **78:** 1051–1055.
21. WHEELER, T. J., I. A. SIMPSON, D. C. SOGIN, P. C. HINKLE & S. W. CUSHMAN. 1982. Biochem. Biophys. Res. Commun. **105:** 89–95.
22. LIENHARD, G. E., H. K. KIN, K. J. RANSOME & J. C. GORGA. 1982. Biochem. Biophys. Res. Commun. **105:** 1150–1156.
23. SHANAHAN, M. F., S. A. OLSON, M. J. WEBER, G. E. LIENHARD & J. C. GORGA. 1982. Biochem. Biophys. Res. Commun. **107:** 38–43.
24. CARTER-SU, C., J. E. PESSIN, E. MORA, W. GITOMER & M. P. CZECH. 1982. J. Biol. Chem. **257:** 5419–5425.
25. OKA, Y. & M. P. CZECH. 1984. J. Biol. Chem. **259:** 8125–8133.
26. HORUK, R., M. RODBELL, S. W. CUSHMAN & I. A. SIMPSON. 1983. FEBS Lett. **164:** 261–266.
27. SMITH U., M. KURODA & I. A. SIMPSON. 1984. J. Biol. Chem. **259:** 8758–8763.
28. KURODA, M., R. C. HONNOR, S. W. CUSHMAN, C. LONDOS & I. A. SIMPSON. 1987. J. Biol. Chem. **262:** In press.
29. JOOST, H. G., T. M. WEBER, S. W. CUSHMAN & I. A. SIMPSON. 1986. J. Biol. Chem. **261:** 10033–10036.
30. TAYLOR, W. M. & M. L. HALPERIN. 1979. Biochem. J. **178:** 381–389.
31. KASHIWAGA, A., T. P. HEUCKSTEADT & J. E. FOLEY. 1983. J. Biol. Chem. **258:** 13685–13692.
32. HONNOR, R. C., G. S. DHILLON & C. LONDOS. 1985. J. Biol. Chem. **260:** 15122–15129.
33. SCHWABE, U., R. EBERT & H. C. ERBLER. 1973. Arch. Pharmacol. **273:** 133–148.
34. HISSIN, P. J., E. KARNIELI, I. A. SIMPSON, L. B. SALANS & S. W. CUSHMAN. 1982. Diabetes **31:** 589–592.
35. HISSIN, P. J., J. E. FOLEY, L. J. WARDZALA, E. KARNIELI, I. A. SIMPSON, L. B. SALANS & S. W. CUSHMAN. 1982. J. Clin. Invest. **70:** 780–790.
36. KAHN, B. B., E. S. HORTON & S. W. CUSHMAN. 1987. J. Clin. Invest. **79:** in press.
37. VINTEN, J., L. N. PETERSEN, B. SONNE & H. GALBO. 1985. Biochim. Biophys. Acta **841:** 223–227.
38. KOBAYASHI, M. & J. M. OLEFSKY. 1979. Diabetes **28:** 87–95.
39. KARNIELI, E., P. J. HISSIN, I. A. SIMPSON L. B. SALANS & S. W. CUSHMAN. 1981. J. Clin. Invest. **68:** 811–814.

40. KOBAYASHI, M. & J. M. OLEFSKY. 1978. Am. J. Physiol. **253:** E53–E62.
41. KOBAYASHI, M. & J. M. OLEFSKY. 1978. J. Clin. Invest. **62:** 73–81.
42. WHITTAKER, J., K. G. M. M. ALBERTI, D. A. YORK & J. SINGH. 1979. Biochem. Soc. Trans. **7:** 1055–1066.
43. MARTIN, C., K. S. DESAI & G. STEINER. 1983. Can. J. Physiol. Pharmacol. **61:** 802–807.
44. TRIMBLE, E. R., G. C. WEIR, A. GJINOVCI, F. ASSIMACOPOULOS-JEANNET, R. BENZI & A. E. RENOLD. 1984. Diabetes **33:** 444–449.
45. WARDZALA, L. J., M. HIRSHMAN, E. POFCHER, E. D. HORTON, P. M. MEAN, S. W. CUSHMAN & E. S. HORTON. 1985. J. Clin. Invest. **76:** 460–469.
46. GUERRE-MILLO, M., M. LAVAU, J. S. HORNE & L. J. WARDZALA. 1985. J. Biol. Chem. **260:** 2197–2201.
47. CUSHMAN, S. W., M. J. ZARNOWSKI, A. J. FRANZUSOFF & L. B. SALANS. 1978. Metabolism. **27** (suppl. 2): 1930–1940.
48. SAHA, J., R. LOPEZ-MONDRAGON & H. T. NARAHARA. 1968. J. Biol. Chem. **243:** 521–527.
49. BIHLER, L., P. C. SAWH & I. G. SLOAN. 1978. Biochim. Biophys. Acta **510:** 349–360.
50. SCHEIDEGGER, K., D. C. ROBBINS & E. DANFORD, JR. 1984. Diabetes **33:** 1144–1149.
51. YAMADA, K., L. G. TILLOTSON & K. J. ISSELBACHER. 1983. J. Biol. Chem. **258:** 9786–9792.
52. HASPEL, H. C., E. W. WILK, M. T. BIRNBAUM, S. W. CUSHMAN & O. M. ROSEN. 1986. J. Biol. Chem. **261:** 6778–6789.
53. JACOBS, D. B. & C. Y. JUNG. 1985. J. Biol. Chem. **260:** 25693–2596.
54. HYSLOP, P. A., C. E. KUHN & R. D. SAUERHEBER. 1985. Biochem. J. **232:** 245–254.
55. MOSHER, K. M., D. A. YOUNG & A. MUNCK. 1971. J. Biol. Chem. **246:** 654–659.
56. CARTER-SU, C. & K. OKAMOTO. 1985. Am. J. Physiol. **248:** E215–E223.
57. WHITESELL, R. R. & N. A. ABUMRAD. 1985. J. Biol. Chem. **260:** 2894–2899.
58. MUECKLER, M., C. CARUSO, S. A. BALDWIN, M. PANICO, I. BLENCH *et al.* 1985. Science **229:** 941–945.
59. BIRNBAUM, M. J., H. C. HASPEL & O. M. ROSEN. 1986. Proc. Natl. Acad. Sci. USA **83:** 5784–5788.
60. HORUK, E., S. MATTHAEI, J. M. OLEFSKY, D. L. BALY, S. W. CUSHMAN & I. A. SIMPSON. 1986. J. Biol. Chem. **261:** 1823–1828.
61. FLIER, J. S., M. MUECKLER, A. L. MCCALL & H. F. LODISH. 1987. J. Clin. Invest. **79:** in press.

DISCUSSION OF THE PAPER

R. M. DENTON (*University of Bristol, Bristol, U.K.*): Many of insulin's effects on cells are brought about by changes in phosphorylation. To what extent can reversible phosphorylation be implicated in the translocation of the glucose transporters and/or in the changes in specific activity you have described?

KAHN: We are currently carrying out studies examining phosphorylation of the glucose transporter. We have encountered more technical difficulties than anticipated and the data are too preliminary to report. However, we have looked extensively at A-kinase ratios, reflecting cyclic AMP activity, in relation to changes in the intrinsic activity of the glucose transporter seen with lipolytic and antilipolytic agents. These changes appear to be independent of cyclic AMP.

G. ESPOSITO (*University of Milan, Milan*): Is there any evidence that hyperglycemia per se in normal animals increases sugar transport and the number of sugar transporters? This is true, for instance, for the enterocyte.

KAHN: Hyperglycemia *in vivo* appears to result in an impaired glucose transport response to insulin in isolated adipocytes from diabetic animals. This is accompanied by a decrease in glucose transporters. We have not made normal animals hyperglycemic but normal humans who are infused with glucose during euglycemic clamp studies show increased glucose disposal with increased rate and duration of glucose infusion.

S. CORVERA (*University of Massachusetts, Worcester, MA*): Can you comment further on the fact that in cells from normal rats the stimulation of 3-O-methylglucose uptake by insulin is 20–30-fold, while the increase in transporter numbers is 3–5-fold? Especially in light of your new data illustrating a direct stimulation by insulin on the transporter activity.

KAHN: Measurement of specific marker enzymes for the subcellular membrane fractions indicates that there is substantial contamination of the plasma membranes from cells in the basal state with low-density microsomes. Because of the relative enrichment of the low-density microsomes from basal cells with glucose transporters (80 pmol/mg protein) compared to the plasma membranes (7 pmol/mg protein) a small amount of contamination could significantly raise the value in the basal plasma membranes and so diminish the fold stimulation observed after insulin treatment of the cells. If the basal value in the plasma membranes were closer to zero, as it would be if the contaminating glucose transporters were subtracted, we would see a fold stimulation of transporters that would approach the fold stimulation of transport activity in the intact cell. I do not think that this discrepancy reflects activation of the glucose transporter by insulin in adipose cells from metabolically normal animals, although this remains a possibility.

J. M. MCDONALD (*Washington University, St. Louis, MO*): Don't you think that the consistent 20–30-fold stimulation of glucose transport induced by insulin is more compatible with activation of glucose transport activity as the major mechanism rather than the twofold increase in translocation? This is supported by the studies of Whitesell who demonstrated a major change in the K_m for glucose after insulin treatment and only a small change in V_{max}.

KAHN: As discussed in response to Dr. Corvera's question, I think the small fold increase in glucose transporters in the plasma membranes from cells after insulin stimulation is due to contamination of the plasma membranes with low density microsomes, which are relatively enriched in glucose transporters. Activation of the glucose transporter by insulin may be part of its stimulatory effect. However, the stoichiometry of translocation implies that it is a major mechanism for the stimulatory effect of insulin under normal metabolic conditions. Whitesell's studies were done under significantly different incubation conditions, which probably account for his findings.

Mechanisms Whereby Insulin and Other Hormones Binding to Cell Surface Receptors Influence Metabolic Pathways within the Inner Membrane of Mitochondria[a]

RICHARD M. DENTON, JAMES G. McCORMACK,[b]
AND ANDREW P. THOMAS[c]

Department of Biochemistry
University of Bristol Medical School
Bristol BS8 1TD, United Kingdom

INTRODUCTION

A wide array of intracellular changes occurs within a few minutes of the binding of insulin to receptors on the cell surface of its target tissues such as muscle, fat, and liver.[1-4] These changes often involve changes in phosphorylation and hence activity of many key enzymes and other proteins within target cells. Some proteins, such as ribosomal protein S_6, ATP-citrate lyase, acetyl-CoA carboxylase, and the insulin receptor itself, exhibit increases in phosphorylation, while other proteins are dephosphorylated. Examples of the latter group of proteins are glycogen synthase, triglyceride lipase, liver pyruvate kinase, and the enzyme complex most relevant to the current topic—pyruvate dehydrogenase. This enzyme complex is located exclusively within mitochondria. In the early 1970s, it was shown by Jungas,[5] Denton *et al.*,[6] and Weiss *et al.*[7] that exposure of rat epididymal fat cells to insulin leads to a marked increase in pyruvate dehydrogenase activity within 3–5 minutes. Subsequent studies by Hughes *et al.*[8] showed directly that the activation was the result of dephosphorylation. Insulin has also been shown to be capable of increasing pyruvate dehydrogenase activity in liver,[9] mammary tissue[10] and brown adipose tissue.[11] The effect is important in the stimulation of fatty acid synthesis from precursors of pyruvate such as intracellular glucose, glycogen, and lactate.

The regulation of intramitochondrial events is achieved by other hormones that have plasma membrane receptors (TABLE 1). This regulation is a topic of particular fascination because the mechanisms involved must allow signal transmission across the plasma membrane and the inner membrane of the mitochondria. The inner membrane is essentially impermeable to all charged molecules unless a specific carrier or transport system is present in the membrane. Of the known second messengers of

[a]Supported by grants from the Medical Research Council, British Diabetic Association, and Percival Waite Salmond Bequest.

[b]Present address: Department of Biochemistry, University of Leeds, Leeds, LS2 9JT, U.K.

[c]Present address: Department of Pathology, Thomas Jefferson University, Philadelphia, PA 19107.

hormone action only Ca^{2+} may be transferred into mitochondria by a specific transport system.

Because of our interest in the activation of pyruvate dehydrogenase, we investigated the kinetic properties of pyruvate dehydrogenase phosphatase and showed that the enzyme was activated by Ca^{2+}.[12] This raised the possibility that insulin might activate pyruvate dehydrogenase by increasing intracellular Ca^{2+}. However, subsequent studies, which are summarized within this article, showed that insulin action was independent of changes in intramitochondrial Ca^{2+}. Nevertheless, changes in intramitochondrial Ca^{2+} are very probably involved in the actions of the many other hormones and extracellular stimuli that increase cytoplasmic Ca^{2+} in mammalian cells.[13,14] Increases in cytoplasmic Ca^{2+} may be relayed into mitochondria and lead to stimulated rates of pyruvate oxidation and citrate cycle flux. In this way, ATP supply is increased when there is an enhanced demand.[13,14] It is this aspect of our studies which we will summarize first.

TABLE 1. Regulation of Intramitochondrial Metabolism by Hormones with Plasma Membrane Receptors

Hormone	Tissue	Main Cytoplasmic Signal	Increases in Mitochondrial Metabolism
Insulin	Fat, mammary, liver	(?)	Pyruvate dehydrogenase (PDH)
Many	Many	IP_3, Ca^{2+}	Oxidative metabolism (PDH, NAD-ICDH, OGDH)
Noradrenaline	Brown fat	cAMP	Uncoupling and oxidative metabolism
Glucagon	Liver	cAMP	Respiratory chain and general metabolism
Trophic (eg. ACTH, LH)	Target	(cAMP?)	Cholesterol metabolism
PTH	Kidney	(cAMP?)	25-OHD_3, 1,25-diOHD_3

REGULATORY PROPERTIES OF PYRUVATE DEHYDROGENASE COMPLEX, NAD-ISOCITRATE DEHYDROGENASE, AND OXOGLUTARATE DEHYDROGENASE

The activity of the pyruvate dehydrogenase complex from mammalian sources may be regulated both by end-product inhibition (through increases in concentration ratios of both $NADH/NAD^+$ and acetyl-CoA/CoA) and by phosphorylation.[15,16] Interconversion of the active unphosphorylated form (PDH_a) and the inactive phosphorylated form (PDHP) is catalyzed by an ATP-requiring kinase and a specific phosphatase. Studies on the properties of these interconverting enzymes have led to the recognition of a number of effectors of potential physiological importance (FIG. 1). It appears that the major regulators of the kinase are the ATP/ADP ratio plus substrates and products of the pyruvate dehydrogenase reaction. In contrast, the major regulators of the phosphatase appear to be the divalent metal ions Ca^{2+} and Mg^{2+}.[6,12,16,17] Purified preparations have an absolute requirement for Mg^{2+} (K_a about 1 mM) and Ca^{2+} can increase activity a further 5–10 fold with a K_a of about 1 μM. The effects of Ca^{2+} are

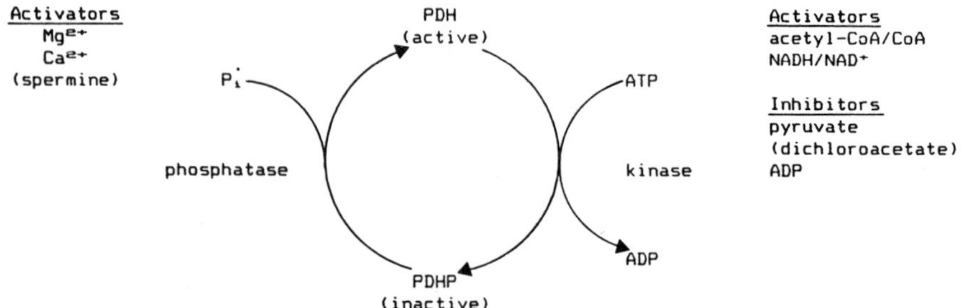

FIGURE 1. Regulation of the mammalian pyruvate dehydrogenase system by reversible phosphorylation.

both to diminish the apparent K_m of the phosphatase[16,18] for PDHP and also the K_a for Mg^{2+}.[6]

Further studies in Bristol showed that NAD-isocitrate dehydrogenase[19] and oxoglutarate dehydrogenase[20] from mammalian tissues are also activated by Ca^{2+}—again with a K_a of 1 μM or less. In these cases, the effect of Ca^{2+} is directly on the dehydrogenases and is characterized by marked decreases in the K_m value for their respective substrates, isocitrate and oxoglutarate.

Like pyruvate dehydrogenase, NAD-isocitrate dehydrogenase and oxoglutarate dehydrogenase may in addition be regulated by decreases in ADP/ATP and NAD^+/NADH ratios.[19,20] We have suggested that these activations should be viewed as examples of "intrinsic" control, which ensures that under all circumstances changes in ATP and NADH concentrations may directly initiate appropriate changes in the provision of reducing power for the respiratory chain.[13,14] On the other hand, the regulation by Ca^{2+} could then be an example of "extrinsic" control since it is potentially a means whereby hormones and other extracellular stimuli may override the local "intrinsic" regulation. The advantage for the cell is that oxidative metabolism and ATP synthesis may be stimulated without the need to disturb the rather important ADP/ATP and NAD^+/NADH ratios.

The three dehydrogenases from non-vertebrate sources such as insects and plants appear not to be activated by Ca^{2+} but still to exhibit regulation by the nucleotide ratios.[21] Since only mitochondria from vertebrate sources contain the ruthenium red–inhibited transporter, which allows Ca^{2+} to be taken up into mitochondria under physiological conditions (see next section), this is also powerful support for the notion that there is a close functional link between the transfer of Ca^{2+} into the mitochondria and the activation of the dehydrogenases by Ca^{2+}.

STUDIES ON THE Ca^{2+}-SENSITIVE DEHYDROGENASES USING INTACT ISOLATED MITOCHONDRIA

The calcium transport system in mammalian mitochondria has separate uptake and efflux components.[22,23] Uptake of calcium occurs as Ca^{2+} via an electrophoretic uniporter mechanism that is driven by the membrane potential across the inner membrane. The uniporter is inhibited by the dye ruthenium red and also by Mg^{2+} ions at concentrations likely to be present in the cytoplasm of mammalian cells. The transfer of calcium out of mitochondria occurs mainly via an electroneutral exchange, or antiport, of Ca^{2+} for 2 Na^+. There is also a sodium-independent efflux pathway that may be particularly active in liver and kidney mitochondria. The 2 Na^+/Ca^{2+} antiport can be inhibited by several compounds that are perhaps better known as plasma membrane Ca^{2+} channel blockers, the most potent being diltiazem. It is generally agreed that there is continuous cycling of Ca^{2+} ions across the inner membrane through the operation of the separate uptake and efflux pathways. The distribution of Ca^{2+} across the membrane will thus depend on the relative activities of the two pathways.[14,22,23]

TABLE 2 summarizes some observations on the sensitivity of pyruvate dehydrogenase and oxoglutarate dehydrogenase within mitochondria from rat heart, liver, and epididymal adipose tissue. In uncoupled mitochondria it is evident that the dehydrogenases are activated with a half-maximal effect being observed at about one micromolar.[24-28] This is similar to values obtained with the isolated enzymes, as might be expected as the gradient of Ca^{2+} across the inner membrane under these circumstances will be very low. In the case of pyruvate dehydrogenase, it has recently become evident that the major effect of Ca^{2+} exerted within mitochondria is probably to lower the $K_{0.5}$ for Mg^{2+}. This is illustrated in FIGURE 2, using uncoupled rat epididymal adipose tissue mitochondria incubated with the divalent metal ion ionophore A23187 to ensure near equilibration of intramitochondrial Ca^{2+} and Mg^{2+} with the extramitochondrial concentrations of these divalent metal ions. In direct contrast to studies with isolated pyruvate dehydrogenase phosphatase there is little or no effect of Ca^{2+} at saturating concentration of Mg^{2+}. The likely explanation for this discrepancy is that Ca^{2+} can increase the apparent K_m for PDHP but this is not significant in intact mitochondria because the concentration of PDHP is probably considerably greater than the apparent K_m, whereas studies with the isolated phosphatase are carried out at non-saturating concentrations of PDHP. Since the pyruvate dehydrogenase system is clearly sensitive to changes in Ca^{2+} in both uncoupled and coupled mitochondria (TABLE 2), it follows that the concentration of free Mg^{2+} in these

TABLE 2. Sensitivity of Pyruvate Dehydrogenase and Oxoglutarate Dehydrogenase within Uncoupled and Coupled Mitochondria to Changes in Extramitochondrial Concentrations of Ca^{2+}

		Ca^{2+} Concentration (nM) Giving 50% Activation	
Mitochondria	Condition	Pyruvate Dehydrogenase	Oxoglutarate Dehydrogenase
Rat heart	Uncoupled	980	940
	Coupled (minus Na^+/Mg^{2+})	39	21
	Coupled + NaCl + $MgCl_2$	468	328
Rat liver	Uncoupled	1059	—
	Coupled (minus Na^+/Mg^{2+})	98	121
	Coupled + NaCl + $MgCl_2$	484	560
Rat epididymal adipose tissue	Uncoupled	940	—
	Coupled (minus Na^+/Mg^{2+})	33	13
	Coupled + NaCl + $MgCl_2$	419	—

Results taken from Denton *et al.*[24] (rat heart), McCormack (rat liver),[25] and Marshall *et al.*[26] (adipose tissue) from which full details can be obtained. In all cases, mitochondria were prepared from control tissue. Activities in the presence of saturating extramitochondrial Ca^{2+} were at least three times that observed in the absence of Ca^{2+}, when added NaCl and $MgCl_2$ were present in the mitochondrial incubation medium at 10 mM and 1 mM, respectively.

mitochondria is non-saturating for PDHP-phosphatase and is probably less than 1 mM. This is in agreement with recent estimates for rat liver mitochondria of 0.4 mM.[42]

In coupled mitochondria the range of extramitochondrial Ca^{2+} concentrations to which the dehydrogenases respond is influenced by the presence of Mg^{2+} (which inhibits the uptake of Ca^{2+} into mitochondria) and Na^{+} (which facilitates the efflux of Ca^{2+}). In the absence of both, the gradient of Ca^{2+} across the mitochondria inner membrane is high and half-maximal activation is observed at about 10–100 nM. However, under the conditions likely to occur in cells, namely about 1 mM Mg^{2+} and 10 mM Na^{+}, half-maximal activation is observed at 300–600 nM (TABLE 2). Thus, the calcium-sensitive intramitochondrial dehydrogenases would be expected to be activated by increases in the concentration of Ca^{2+} in the physiological range of 0.1 to 1 μM.

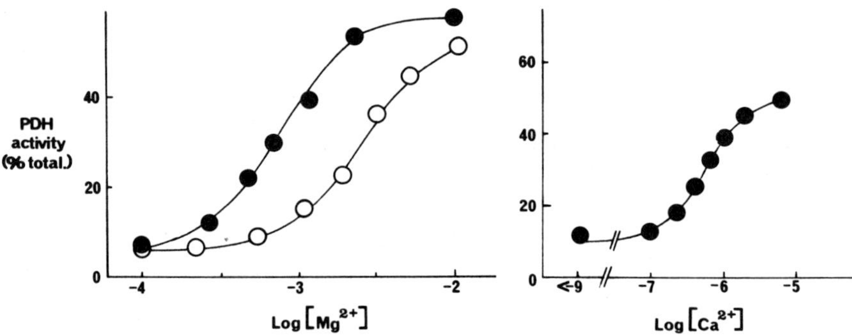

FIGURE 2. Effects of Mg^{2+} and Ca^{2+} on the activity of pyruvate dehydrogenase within rat epididymal adipose tissue mitochondria. Mitochondria were prepared from rat epididymal adipose tissue that had been pre-incubated in the absence of added hormones for 30 min. The mitochondria were then incubated in KCl-based medium containing the ionophore A23187, the uncoupler FCCP, oligomycin, ATP, and various concentrations of free Mg^{2+} and Ca^{2+} using EGTA and EGTA/HEDTA buffers. In the left-hand panel, the Ca^{2+} concentration was either less than 1 nM (○) or about 100 μM (●). The calculated $K_{0.5}$ for Mg^{2+} under these conditions were 2.7 mM and 1.1 mM, respectively. In the right-hand panel, the calculated Mg^{2+} concentration were in the range 0.39–0.45 mM. The calculated $K_{0.5}$ for Ca^{2+} was 0.5 mM. (Results taken from Thomas *et al.*[27] and McCormack & Denton.[20])

DIRECT EVIDENCE THAT HORMONES MAY REGULATE INTRAMITOCHONDRIAL OXIDATIVE METABOLISM THROUGH ALTERATIONS IN INTRAMITOCHONDRIAL Ca^{2+} IN HEART AND LIVER

In the rat heart, the concentration of Ca^{2+} in the cytoplasm is increased by adrenaline (acting through β-adrenergic receptors), glucagon, and other positive inotropic agents, whereas in the liver, it is increased by α-adrenergic agonists, vasopressin, angiotenin II, and to a less marked extent by glucagon. In all these circumstances, the hormones increase oxygen uptake, flux through the citrate cycle, and the proportion of pyruvate dehydrogenase in its active, non-phosphorylated form without detectable sustained decreases in either ATP/ADP or $NADH/NAD^{+}$ ratios.[30–32]

TABLE 3. Effects of Hormones and Other Treatments in the Activity of Pyruvate Dehydrogenase and Total Calcium Associated with Rapidly Prepared Mitochondrial Fractions in the Perfused Rat Heart, Perfused Rat Liver, and Incubated Rat Epididymal Adipose Tissue

Preparation	Treatment	Pyruvate Dehydrogenase Activity (% total)	Estimated Total Ca Associated with Mitochondria (nmol/mg protein)
Rat heart	Control	10	1.8
	Adrenaline	41	4.2
	High calcium	42	—
	Control + ruthenium red	11	—
	Adrenaline + ruthenium red	12	—
	High calcium	13	—
Rat liver	Control	5	1.2
	Vasopressin	27	2.1
	Glucagon + vasopressin	45	4.9
Rat epididymal adipose tissue	Control	22	—
	Insulin	58	—
	High calcium	39	—
	Control + ruthenium red	24	—
	Insulin + ruthenium red	62	—
	High calcium + ruthenium red	29	—

Data taken from McCormack and England[33] and Crompton *et al.*[34] (rat heart), Assimacopoulos-Jeannet *et al.*,[29] (rat liver) and Marshall *et al.*[26] (rat epididymal adipose tissue) from which full details can be obtained.

TABLE 4. Extent of Persistence of the Activations of Pyruvate Dehydrogenase in Mitochondria Prepared from Intact Rat Heart and Liver Previously Exposed to Adrenaline and Rat Epididymal Adipose Tissue Previously Exposed to Insulin

Source of Mitochondria	Pyruvate Dehydrogenase Activity (as % total) in Mitochondria Incubated with Medium Containing			
	No Additions	NaCl	NaCl + Diltiazem	Ca^{2+}
Rat heart perfused with medium containing				
no hormone	8	8	8	45
adrenaline	20	7	20	47
Liver from rats injected with				
no hormone	12	13	13	49
adrenaline	20	14	23	51
Rat epididymal adipose tissue incubated with				
no hormone	11	11	—	39
insulin	25	25	—	59

Data taken from McCormack & Denton[36] (rat heart), McCormack[32] (rat liver), and Marshall *et al.*[26] (rat epididymal adipose tissue) from where full details can be obtained. In brief, after exposure to appropriate concentrations of hormone, tissue was rapidly homogenized and mitochondria prepared. The mitochondria were then incubated at 30°C for 5 min in KCl-based medium containing respiratory substrates, EGTA, and where indicated additions of NaCl (10 mM), diltiazem (300 μM), and Ca^{2+} (to give a final concentration that maximally activates pyruvate dehydrogenase).

TABLES 3 and 4 summarize some observations that lend considerable support to the view that the activation of pyruvate dehydrogenase observed under these conditions may involve increases in the intramitochondrial concentration of Ca^{2+}. For example, in the perfused rat heart the effects of adrenaline can be mimicked by increasing the concentration of Ca^{2+} in the perfusing medium and can be blocked by ruthenium red, which inhibits the transfer of Ca^{2+} into mitochondria. With both heart and liver, the total amount of calcium associated with mitochondria is found to be increased[29,34] by hormones that increase the activity of pyruvate dehydrogenase.[29,34] In such studies, it is essential that precautions are taken to inhibit translocation of calcium during the rapid preparations of mitochondrial fractions free of appreciable contamination by other intracellular organelles. The resting levels found in these studies are similar to those found directly by X-ray probe analysis[25] and much less that those obtained in many earlier studies where adequate precautions were not taken.

The most convincing evidence that hormones do act on intramitochondrial oxidative metabolism via increases in the intramitochondrial concentration of Ca^{2+} has been obtained in studies on mitochondria rapidly prepared by similar procedures from rat heart or liver previously exposed to an appropriate stimulating hormone. Under such conditions, activations of pyruvate dehydrogenase initiated by prior hormone treatment of the tissues persist not only during the preparation of the mitochondria but also during their subsequent incubation at 30°C in sodium-free medium containing respiratory substrates and EGTA (TABLE 4). However, these increases are lost if Na^+ is added to the mitochondrial incubation medium. Since these effects of Na^+ are blocked by diltiazem, which inhibits the sodium-dependent efflux pathway, it can be concluded that the activations of pyruvate dehydrogenase are the result of increases in Ca^{2+} concentration in the mitochondria from the hormone-treated tissues. Moreover, incubation of the mitochondria with sufficient Ca^{2+} to elicit a maximum stimulation also results in the disappearance of the differences in mitochondria from control and hormone-treated tissue (TABLE 4). Very similar observations have been made when the activity of oxoglutarate dehydrogenase was followed rather than pyruvate dehydrogenase.[32,36]

Taken together, there is now excellent evidence in rat heart and liver that increases in cytoplasmic Ca^{2+} are relayed into mitochondria and result in activation of pyruvate dehydrogenase and oxoglutarate dehydrogenase and most probably also NAD-isocitrate dehydrogenase.

STUDIES ON THE MECHANISMS INVOLVED IN THE ACTIVATION OF ADIPOSE TISSUE PYRUVATE DEHYDROGENASE BY INSULIN

The insulin activation of pyruvate dehydrogenase persists during the preparation and subsequent incubation of mitochondria from white and brown adipose tissue.[37] The intramitochondrial concentrations of kinase regulators ATP, ADP, NADH, NAD, acetyl CoA, and CoA have all been determined but no changes found that might result in inhibition of kinase activity.[37] This suggested that insulin may cause activation of the phosphatase and further evidence for this view was obtained from the study of the rate of incorporation of ^{32}P from intramitochondrial [γ-^{32}P]-ATP into pyruvate dehydrogenase phosphate.[38] The rate was found to be increased in mitochondria from insulin-treated adipose tissue.[38] The simplest explanation of this observation is that insulin increases the rate of dephosphorylation of pyruvate dehydrogenase and that this in turn leads to the increased turnover of the pyruvate dehaydrogenase phosphorylation-dephosphorylation cycle under the steady-state conditions studied. Until recently, the only known regulators of pyruvate dehydrogenase phosphatase were Ca^{2+}, Mg^{2+}, and perhaps NADH (FIG. 2). Since no increases in either mitochondrial total Mg^{2+} or $[NAD^+]/[NADH]$ were found,[37] the possibility that insulin might act through increases in mitochondrial Ca^{2+} was explored.[26] It was realized that if insulin were to act in this way it would be very unlikely that is was secondary to an increase in the cytoplasmic concentration of Ca^{2+}. However, one mechanism that could be envisaged is that insulin might increase the cytoplasmic concentration of spermine since spermine has recently been shown to activate calcium transport into mitochondria.[29]

Similar approaches were therefore employed to those described in the previous section that had established that hormones that increase the cytoplasmic Ca^{2+} concentration in heart and liver probably activate pyruvate dehydrogenase and the other Ca^{2+}-sensitive intramitochondrial dehydrogenases through increases in the intramitochondria concentration of Ca^{2+} (TABLES 3 and 4).[26] Using these approaches it became quite clear that insulin does not in fact act through an increase in intramitochondrial Ca^{2+}. The main evidence was as follows: (1) addition of ruthenium red does not block the effect of insulin (TABLE 3); (2) activation of pyruvate dehydrogenase in mitochondria from insulin-treated tissue persists not only when Na^+ is added to the incubation medium but also when mitochondria were incubated with a saturating concentration of Ca^{2+} (TABLE 4); (3) there was no evidence that activation of NAD-isocitrate dehydrogenase or oxoglutarate dehydrogenase accompanied the increased activity of pyruvate dehydrogenase in mitochondria from insulin-treated adipose tissue.[26]

We have been unable to detect any changes in phosphatase activity in extracts of mitochondria from insulin-treated tissue[17,26] and therefore turned our attention to the use of permeabilized mitochondria as a means of exploring in greater detail the basis of the increased activation of pyruvate dehydrogenase in adipose tissue. The mitochondria were rendered either specifically permeable to Mg^{2+} and Ca^{2+} by incubation with A23187 in the presence of an uncoupler (as in FIG. 2) or permeable to all substances up to a molecular weight of 1,000–2,000 by treatment with toluene (FIG. 3). With the latter preparation the activity of pyruvate dehydrogenase can be followed continuously by assaying the production of acetyl-CoA spectrophotometrically (FIG. 3). The entire pyruvate dehydrogenase system complete with phosphatase and kinase remains located within the mitochondria so changes in the activity of the phosphatase and kinase can, for the first time, be followed either independently or together while the complex is catalyzing the conversion of pyruvate to acetyl-CoA as it would in the intact cell. In the traces shown in FIGURE 3, the expected increased pyruvate dehydrogenase activity is evident in the mitochondria from insulin-treated tissue in the absence of both kinase and phosphatase activity. What is of particular interest is the apparent increase in the activity of the phosphatase in the mitochondria from insulin-treated tissue but it is difficult to make precise conclusions from this type of experiment because of the initial differences in the mitochondrial content of the phosphorylated form of pyruvate dehydrogenase. However, this difference can be overcome by prior treatment of the mitochondria with MgATP (FIG. 4). It is now quite evident that phosphatase activity is greater in the mitochondria from insulin-treated tissue by observing the time course of either acetyl-CoA production (main figure) or NADH formation (inset). This figure also illustrates the use of sodium fluoride to block phosphatase activity after a fixed time as a further means of convincingly demonstrating the activation of phosphatase activity. Using these techniques it was found that the effects of insulin were evident at all Ca^{2+} concentrations but were lost at high Mg^{2+} concentrations. Indeed, there appeared to be a decrease in the apparent K_m for Mg^{2+}. This was confirmed in separate experiments in both A23187- and toluene-permeabilized mitochondria in which

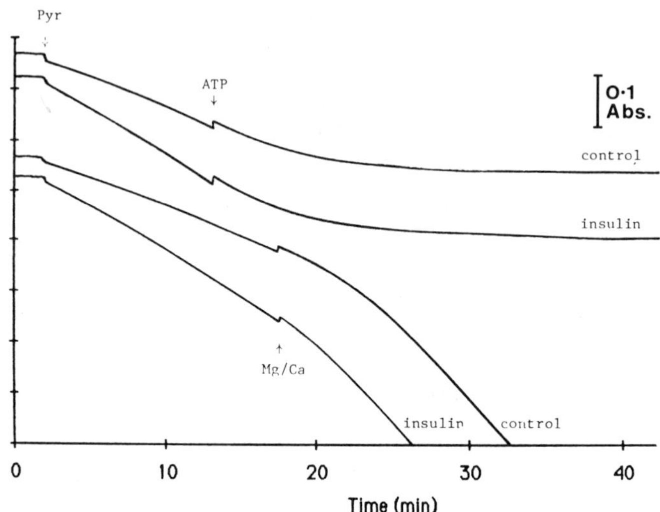

FIGURE 3. Use of toluene-permeabilized mitochondria to study the effects of pretreatment of intact epididymal adipose tissue with insulin on the activities of pyruvate dehydrogenase phosphatase and kinase. Adipose tissue mitochondria from control or insulin-treated tissue were treated for a brief period with 0.2% toluene in the presence of 8.5% polyethylene glycol as described by Thomas & Denton.[40] Pyruvate dehydrogenase activity was followed continuously by measuring acetyl-CoA production as the decrease in absorption at 460 nm resulting from the acetylation of the dye *p*-(*p*-aminoplenylazo)-benzene sulphonate in the presence of arylamine acetyl-transferase. In the experiments shown, incubations were carried out in medium containing initially NAD, thiamin pyrophosphate, CoA, a low concentration of Mg^{2+} (7 μM free), and the calcium chelators EGTA and HEDTA. The pyruvate dehydrogenase reaction was then initiated by adding 1 mM pyruvate. It should be noted that at this stage the proportion of pyruvate dehydrogenase in the active form is greater in the mitochondria from insulin-treated tissue. In the upper pair of traces, pyruvate dehydrogenase kinase activity was initiated at 12 min by adding 0.2 mM MgATP. In the lower pair of traces, pyruvate dehydrogenase phosphatase activity was initiated at 16 min by increasing the free concentrations of Mg^{2+} and Ca^{2+} to about 0.8 mM and 100 μM, respectively.

changes in the steady-state activity of pyruvate dehydrogenase were investigated under conditions where kinase activity was kept constant but activity of the phosphatase was altered by adding various concentrations of Ca^{2+} and Mg^{2+} (FIG. 5).[27,40]

A most intriguing aspect of the finding that insulin acts on pyruvate dehydrogenase phosphatase by decreasing the apparent K_m for Mg^{2+} is that the polyamine spermine has a very similar effect on purified preparations of the phosphatase (FIG. 5).[27,41] However, no effects of spermine are found when the polyamine is added directly to toluene-permeabilized mitochondria although it seems almost certain that the inner membrane would be permeable to this small but highly charged molecule under these conditions.

GENERAL CONCLUSIONS

The level of phosphorylation and hence activity of pyruvate dehydrogenase may be regulated by a number of different hormones with plasma membrane receptors. Most of these hormones act by increasing the cytoplasmic and hence the intramitochondrial concentration of Ca^{2+} but insulin acts by a calcium-independent mechanism. In both situations the hormones act by decreasing the apparent K_m of pyruvate dehydrogenase phosphatase for Mg^{2+}, albeit by different independent mechanisms. The studies of Corkey *et al.*[42] and the sensitivity of the phosphatase to Ca^{2+} within intact mitochon-

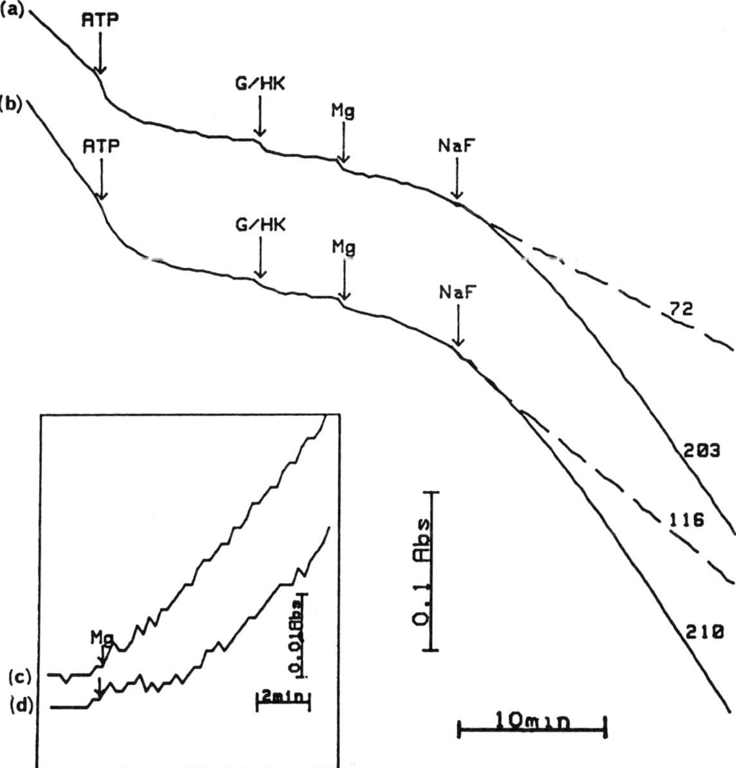

FIGURE 4. Different activities of pyruvate dehydrogenase phosphatase in toluene-permeabilized mitochondria from control and insulin-treated adipose tissue. Activity of pyruvate dehydrogenase was either followed using arylamine acetyl-transferase (a and b) as described in legend to FIGURE 2 or by direct measurement of NADH production at 340 nM (c and d). Permeabilized mitochondria were prepared from either control (a and c) or insulin-treated (b and d) adipose tissue. In traces a and b, pyruvate dehydrogenase was initially fully phosphorylated and inactivated by adding 0.2 mM ATP plus 50 μM $MgCl_2$; subsequently, 10 mM glucose plus 1.4 U/ml hexokinase was added to remove the ATP and the phosphatase reaction was initiated by adding $MgCl_2$ to give 0.3 mM free Mg^{2+}. A further parallel pair of cuvettes were treated exactly as for traces a and b but 6 min after $MgCl_2$ addition the phosphatase reaction was stopped by adding 25 mM NaF (resultant traces are shown as dashed lines). (Inset) Permeabilized mitochondria from control (c) and insulin-treated (d) tissue were treated as for traces a and b except that arylamine acetyl-transferase and the dye *p*-(*p*-aminophenylazo)-benzene sulphonate was not added and pyruvate dehydrogenase substrates were not added until 2 min before initiating the phosphatase reaction with 0.3 mM Mg^{2+}. (Results taken from Thomas & Denton.[40])

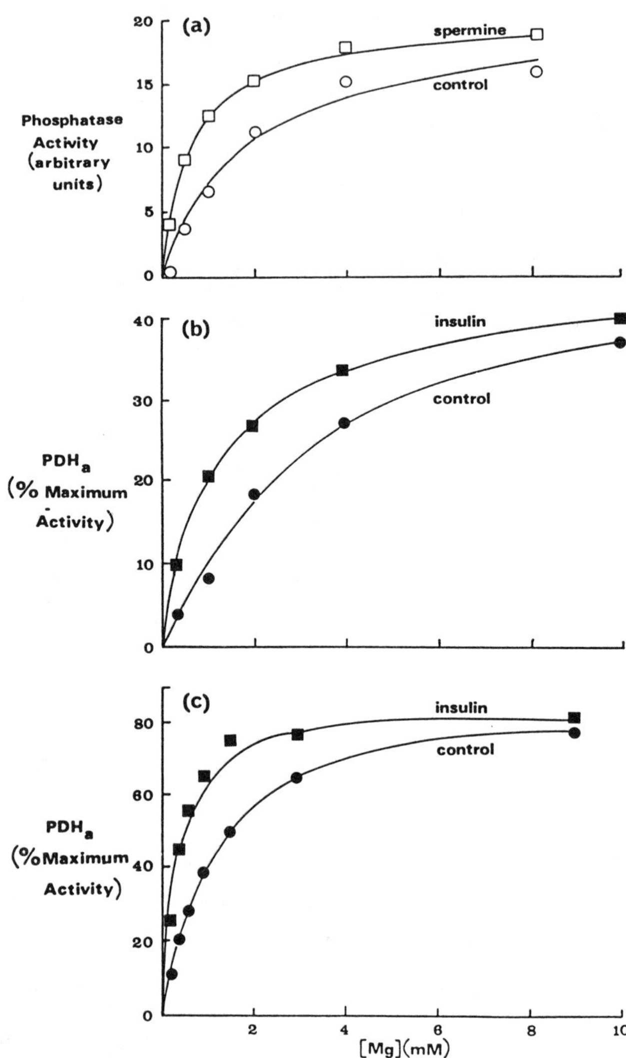

FIGURE 5. Effects of spermine and insulin on the sensitivity of pyruvate dehydrogenase phosphatase to Mg^{2+}. (*a*) Partially purified pyruvate dehydrogenase phosphatase assayed with about 100 μM Ca^{2+} in the presence and absence of 0.5 mM spermine. (*b*) Mitochondria were prepared from control and insulin-treated adipose tissue and incubated in KCl-based medium containing A23187, the uncoupler FCCP, rotenone, oligomycin, ATP, and about 100 μM Ca^{2+} plus varying Mg^{2+} concentrations, rapidly harvested by centrifugation and steady-state pyruvate dehydrogenase activity determined. (*c*) Toluene-permeabilized mitochondria from control and insulin-treated tissue were incubated as in FIGURE 4 and steady-state values of pyruvate dehydrogenase activity obtained in the presence of 0.2 mM ATP, 100 μM Ca^{2+}, and various Mg^{2+} concentrations. (Results are replotted from Thomas *et al.*[27] and Thomas & Denton.[40])

dria clearly demonstrate that the intramitochondrial Mg^{2+} concentration is sufficiently low (i.e. less than 1 mM) for this type of regulation to operate within intact cells.

Activation of pyruvate dehydrogenase and the other Ca^{2+}-sensitive dehydrogenases in mitochondria under conditions of increased cytoplasmic concentrations of Ca^{2+} should be seen as part of the role of Ca^{2+} in stimulus-response–metabolism coupling (FIG. 6). Thus, increases in cytoplasmic Ca^{2+} not only initiate a range of energy-requiring responses within cells such as muscle contraction, secretion, and ion-pumping, but also stimulate key steps in metabolism so that there is an increased supply of ATP to meet the increased requirements of these stimulated processes. The activation of phosphorylase kinase has been viewed in this way for some time. The particular importance of activation of the intramitochondrial dehydrogenase is that it

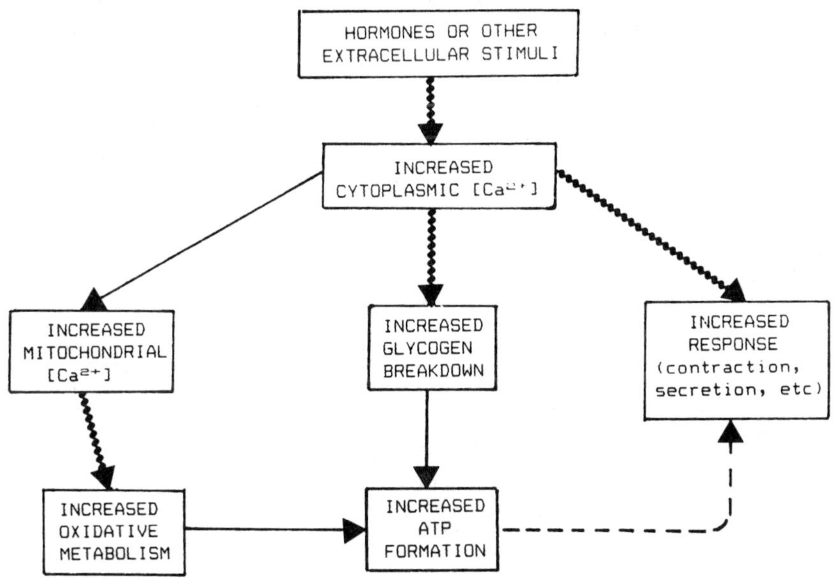

FIGURE 6. Role of Ca^{2+} in stimulus-response–metabolism coupling.

may allow increased ATP synthesis without any decreases in the critical ATP/ADP and $NADH/NAD^+$ ratios. Indeed, the latter ratio probably increases this stimulating oxidative phosphorylation. If this line of reasoning is correct then two long-standing assumptions are brought into question. Firstly, the calcium transport system in mitochondria is not primarily involved in the regulation of cytoplasmic Ca^{2+} but is rather a means of ensuring that changes in the mitochondrial concentration of Ca^{2+} follow those occurring in the cytoplasm. Secondly, the rate of oxidative phosphorylation is not necessarily determined by increases in the concentration of ADP but may be increased by increases in the intramitochondrial Ca^{2+} and hence $NADH/NAD^+$ ratio under conditions where there is no increase (and may in fact be a decrease) in the concentration of ADP.

The molecular basis of the activation of pyruvate dehydrogenase phosphatase in fat cells exposed to insulin remains unknown. Our studies have shown that the change can be mimicked with the isolated phosphatase by the polyamine spermine. Spermine has also been shown to activate the cytoplasmic phosphatases 1 and 2A especially when these phosphatases are acting on the sites in glycogen synthase, which are dephosphorylated specifically in tissues exposed to insulin.[43] Thus the intriguing situation has arisen that spermine can be linked with action of insulin not only on glycogen synthase but on pyruvate dehydrogenase. In addition, early studies showed that spermine had insulin-like actions on isolated adipocytes. However, it would seem unwise to conclude that insulin acts by the simple means of increasing the concentration of spermine in the appropriate cell compartments. Firstly, the cell content of spermine is high and insulin appears to have little or no short-term effect on its concentration. Secondly, the insulin-like effects on fat cells have been shown to be due to formation of hydrogen peroxide.[44]

Although the simplest explanation of our recent work on pyruvate dehydrogenase would appear to be that insulin causes a change in the concentration of a spermine-like compound in mitochondria, it is surprising that this activation should persist in toluene-permeabilized mitochondria if the compound is small and water-soluble. It is also difficult to reconcile our findings with the putative insulin mediator proposed by Jarett, Seals, and others[45] especially as the mediator is reported to be negatively charged and thus must have fundamentally different properties to spermine. Further studies are in progress to establish the means whereby the insulin activation of pyruvate dehydrogenase persists in toluene-permabilized mitochondria but is lost on complete disruption of the mitochondria. An attractive hypothesis would be that insulin leads to changes in the interactions of the pyruvate dehydrogenase system with the inner mitochondrial membrane.

REFERENCES

1. Denton, R. M., R. W. Brownsey & G. J. Belsham. 1981. Diabetologia **21:** 347–363.
2. Kahn, C. R., K. L. Baird, J. S. Flier, C. Grunfeld, J. T. Harmon, L. C. Harrison, F. A. Karlsson, M. Kasuga, G. L. King, U. C. Lang, J. M. Podskalny & E. Van Obberghen. 1981. Rec. Prog. Hormone Res. **37:** 477–538.
3. Houslay, M. D. 1985. *In* Molecular Mechanisms of Transmembrane Signalling. P. Cohen & M. D. Houslay, Eds.:279–333. Elsevier Biomedical Press. Amsterdam.
4. Denton, R. M. 1986. Adv. Cyclic Nucl. Prot. Phos. Res. (In press.)
5. Jungas. R. L. 1971. Metabolism **20:** 43–53.
6. Denton, R. M., H. G. Coore, B. R. Martin & P. J. Randle. 1971. Nature **231:** 113–116.
7. Weiss, L., G. Loffler, A. Shirmann & O. H. Wieland. 1971. FEBS Lett. **15:** 229–231.
8. Hughes, W. A., R. W. Brownsey & R. M. Denton. 1981. Biochem. J. **192:** 469–481.
9. Assimacopoulos-Jeannet, F., J. G. McCormack, J. Prentki, B. Jeanrenaud & R. M. Denton. 1982. Biochim. Biophys. Acta **717:** 86–90.
10. Baxter, M. A., A. Goheer & H. G. Coore. 1979. FEBS Lett. **97:** 27–31.
11. McCormack, J. G. & R. M. Denton. 1977. Biochem. J. **166:** 627–630.
12. Denton, R. M., P. J. Randle & B. R. Martin. 1972. Biochem. J. **128:** 161–163.
13. Denton, R. M. & J. G. McCormack. 1980. FEBS Lett. **119:** 1–8.
14. Denton, R. M. & J. G. McCormack. 1985. Am. J. Physiol. **249:** E543–E554.
15. Wieland, O. H. 1983. Rev. Physiol. Biochem. Pharmacol. **96:** 123–170.
16. Reed, L. J. & S. J. Yeaman. 1986. *In* The Enzymes. E. G. Krebs & P. D. Boyer, Eds. **18:** In press.
17. Randle, P. J., R. M. Denton, H. T. Pask & D. L. Severson. 1974. Biochem. Soc. Symp. **39:** 75–87.
18. Pettit, F. H., T. E. Roche & L. J. Reed. 1972. Biochem. Biophys. Res. Commun. **49:** 563–571.
19. Denton, R. M., D. A. Richards & J. B. Chin. 1978. Biochem. J. **176:** 899–906.
20. McCormack, J. G. & R. M. Denton. 1980. Biochem J. **190:** 95–105.

21. McCormack, J. G. & R. M. Denton. 1981. Biochem. J. **196:** 619–624.
22. Akerman, K. E. O. & D. G. Nicholls. 1983. Rev. Physiol. Biochem. Pharmacol. **9:** 149–201.
23. Carafoli, E. & G. Sottocasa. 1984. New Comprehensive Biochemistry. A. Neuberger & L. L. H. Van Reeman, Eds. **9:** 269–289. Elsevier. Amsterdam.
24. Denton, R. M., J. G. McCormack & N. J. Edgell. 1980. Biochem. J. **190:** 107–117.
25. McCormack, J. G. 1985. Biochem. J. **231:** 581–595.
26. Marshall, S. E., J. G. McCormack & R. M. Denton. 1984. Biochem. J. **218:** 249–260.
27. Thomas, A. P., T. A. Diggle & R. M. Denton. 1986. Biochem. J. **238:** 83–91.
28. McCormack, J. G. & R. M. Denton. 1980. Biochem. J. **190:** 95–105.
29. Assimacopoulos-Jeannet, F. D., J. G. McCormack & B. Jeanrenaud. 1986. J. Biol. Chem. **261:** 8799–8804.
30. McCormack, J. G. & R. M. Denton. 1981. Biochem. J. **194:** 639–643.
31. Hems, D. A., J. G. McCormack & R. M. Denton. 1978. Biochem. J. **176:** 627–629.
32. McCormack, J. G. 1985. Biochem. J. **231:** 597–608.
33. McCormack, J. G. & P. J. England. 1983. Biochem. J. **214:** 581–585.
34. Crompton, M., P. Kessar & I. Al-Nassar. 1983. Biochem. J. **216:** 333–342.
35. Somlyo, A. P., M. Bond & A. V. Somlyo. 1985. Nature **314:** 622–625.
36. McCormack, J. G. & R. M. Denton. 1984. Biochem. J. **218:** 235–247.
37. Denton, R. M., J. G. McCormack & S. E. Marshall. 1984. Biochem. J. **217:** 441–452.
38. Hughes, W. A. & R. M. Denton. 1976. Nature **264:** 471–473.
39. Nichitta, C. V. & J. R. Williamson. 1984. J. Biol. Chem. **259:** 12978–12983.
40. Thomas, A. P. & R. M. Denton. 1986. Biochem. J. **238:** 93–101.
41. Damuni, Z., J. S. Humphreys & L. J. Reed. 1984. Biochem. Biophys. Res. Commun. **124:** 95–99.
42. Corkey, B. E., I. J. Duszynski, T. L. Rich, B. Matschinsky & J. R. Williamson. 1986. J. Biol. Chem. **261:** 2567–2574.
43. Tung, H. Y. L., S. Pelech, M. J. Fisher, C. I. Pogson & P. Cohen. 1985. Eur. J. Biochem. **149:** 305–313.
44. Livingston, J. N., P. A. Gurney & D. H. Lockwood. 1977. J. Biol. Chem. **252:** 560–562.
45. Jarett, L. & F. J. Kiechle. 1984. Vitam. Hormones **41:** 51–78.

DISCUSSION OF THE PAPER

E. Carafoli (*Swiss Federal Institute of Technology, Zurich*): (1) I have noticed that you have rather evident Na^+ effects also in liver mitochondria, which are normally considered to be poorly sensitive to Na^+. Could you comment?

(2) Adrenalin induces increases in the Ca^{2+} concentration in your mitochondria to levels that are higher than those which Dr. Somlyo insists are present *in vivo*. Do you think his maximal *in vivo* figures of about 2 nmol per mg of protein may be exceeded under the conditions you have studied?

Denton: (1) At levels of calcium loading sufficient to cause near maximal activation of the dehydrogenases (i.e., less than 5 nmol/mg of protein), Jim McCormack has found marked effects of Na^+ on Ca^{2+} efflux from liver mitochondria. I suspect that the lack of sodium sensitivity found in other studies was because the mitochondria used contained much higher (and in our view, non-physiological) levels of calcium.

(2) There is reasonable agreement on levels of total mitochondrial calcium under the basal conditions found by Somlyo with X-ray probe analysis and those found by Crompton *et al.*, and Assimacopoulos-Jeannet & McCormack using atomic absorption

on appropriately purified mitochondrial fractions. Concerning values after hormone stimulation, it will be necessary to compare values under identical conditions and I do not believe that this is possible on the basis of results published to date. In any case, I would emphasize that the relationship between total calcium associated with mitochondria and free Ca^{2+} within the mitochondrial matrix is likely to be complex. It is very unlikely that the relationship can be described by a single "activity" or "binding" constant.

Characterization of Mediators of Insulin Action

W. K. GOTTSCHALK,[a] S. L. MACAULAY,[b]
J. O. MACAULAY,[b] K. KELLY,[a] J. A. SMITH,[a]
AND L. JARETT[a,c]

[a]*Department of Pathology and Laboratory Medicine*
School of Medicine
University of Pennsylvania
Philadelphia, Pennsylvania 19104-4283

[b]*Department of Medicine*
Royal Melbourne Hospital
University of Melbourne
Victoria 3050, Australia

INTRODUCTION

In the nearly 60 years since the discovery of insulin a number of theories have been proposed to account for the actions of this anabolic hormone on metabolism and growth. None of these theories has withstood the test of time, however. Indeed, because insulin exerts pleiotropic effects, it is unlikely that any unitarian model can satisfactorily explain the mechanism of insulin action. For instance, the effects of insulin on glycogen synthase could be dissociated from the hormone's effects of glucose transport,[1,2] which had long been viewed as the initial mandatory event in insulin action. The dissociation of the effects of insulin on membrane transport systems from its intracellular actions on metabolic pathways has now been documented in a number of laboratories.[3–5]

Recently evidence has been accumulating that a family of low molecular weight substances are generated from the plasma membrane by the interaction of insulin with its receptor and that these substances may mediate a number of intracellular effects of insulin. Larner's laboratory was the first to describe material extracted from rabbit skeletal muscle with the properties consistent with a mediator of insulin action.[6] In an independent series of experiments, Jarett's laboratory was the first to show that the interaction of insulin with its receptor on the cell surface resulted in the generation/release of a low molecular weight material that stimulated pyruvate dehydrogenase in isolated mitochondria.[7] Since these initial reports, insulin mediators have been extracted from a number of membrane, intact cell, and tissue sources following insulin treatment, including: membranes from adipocytes, liver, and human placenta; rat adipocytes and hepatocytes, H4-II-EC3′ and H-35 hepatoma cell, IM-9 lymphocytes, and human monocytes; rabbit and rat skeletal muscle, rat liver, and rat heart. These mediators have been found to affect the following insulin-sensitive enzymes in the same manner as insulin: glycogen synthase phosphatase, pyruvate dehydrogenase, cAMP-dependent protein kinase, low-K_m cAMP phosphodiesterase, adenylate cyclase, acetyl-CoA carboxylase, glucose-6-phosphase dehydrogenase, and phospholipid meth-

[c]To whom correspondence should be addressed.

yltransferase. In addition, mediators have been found to alter cAMP levels and insulin-sensitive metabolic pathways, including lipogenesis and lipolysis, in intact hepatocytes and adipocytes. A recent comprehensive review[8] summarized the sources of insulin mediators and their targets of action.

This report describes new biological and physicochemical characteristics of these mediators. They have been found to mimic insulin action on several insulin-sensitive enzymes in the intact adipocyte. The mediators had no effect on glucose transport in the same cell system, suggesting that these mediators cannot account for all of the pleiotropic effects of insulin. A number of physicochemical properties of the insulin mediator have been discovered during attempts at purification that demonstrate that not only is there a family of mediators, but that the same mediator activity can have different properties from cell to cell. In addition, the divalent cations Ca^{2+} and Mg^{2+} have been found to contaminate mediator preparations and thereby confound purification schemes. Recent efforts have distinguished the effects of these ions from those of the PDH mediator. The ability to separate these ions from the mediator should allow further progress to be made in purification and chemical identification of the insulin mediator.

RESULTS

Biological Properties

Insulin mediators can mimic many of the effects of insulin on intact cells, and this property has led to the development of new bioassays for these compounds. Caro *et al.* demonstrated that mediator generated from a liver particulate fraction stimulated lipogenesis and promoted the down-regulation of the insulin receptors when added to intact hepatocytes.[9] Likewise, Zhang *et al.* found that mediator prepared from adipocyte plasma membranes stimulated lipogenesis, lowered intracellular cAMP levels, and blocked lipolysis in intact adipocytes.[10] These results showing the insulin-like effects of mediators on metabolic systems suggested that mediators could mimic insulin's effects on individual enzymes in intact cells.

Jarett *et al.* explored this possibility by examining the effects of insulin mediator on pyruvate dehydrogenase in fat cells.[11,12] FIGURE 1 shows the effects of insulin mediator extracted from skeletal muscle on PDH activity in intact cells. A physiological concentration of insulin (100 μU/mL) stimulated the enzyme about twofold, which is consistent with results reported from other laboratories using similar systems. Mediator from insulin-treated rats promoted a fourfold stimulation of PDH activity, about twofold greater than the stimulation elicited by insulin itself. Mediator from control rats, which had been injected with saline in place of insulin, also promoted a nearly twofold increase in enzyme activity, and this stimulation probably reflects mediator that was generated by the endogenous circulating insulin present in the control animals at the time they were sacrificed. The immediate conclusion from these experiments is that insulin mediator can modulate the activities of individual enzymes in physiologically appropriate ways in intact cells. In this particular case, it also substantiates the conclusion drawn by Parker and Jarett[13] that insulin mediator can stimulate the activity of PDH in intact, respiring mitochondria. In addition, if it is assumed that the only component in the crude tissue extract that contributed to enzyme activation was the mediator substance, then these results suggest that the stimulation of PDH activity in intact cells is limited by the amounts of mediator made available. Thus, the production of mediator is probably the limiting step in the insulin-elicited activation of pyruvate dehydrogenase.

FIGURE 2 shows that the stimulatory effect of mediator was transient, reaching its maximum at ten minutes and then rapidly declining. Since PDH activity could be restimulated by further additions of mediator (data not shown), this transient effect was most probably due to the rapid inactivation of mediator by the intact fat cells. In contrast to the effects of mediator, the stimulation of PDH activity by insulin persisted for as long as the hormone was present (FIG. 2). This sustained response, coupled with the fact that the insulin mediator is labile in this assay system, suggests that insulin must direct the continuous generation of mediator. Interestingly, the stimulatory effects of the mediator were additive with those of insulin when both were added simultaneously to a suspension of adipocytes (data not shown).[12] This result suggests that the exogenously added mediator fortified the relatively limited amounts produced *in vivo* by insulin, which then led to greater PDH activation than detected in the presence of either insulin or mediator alone.

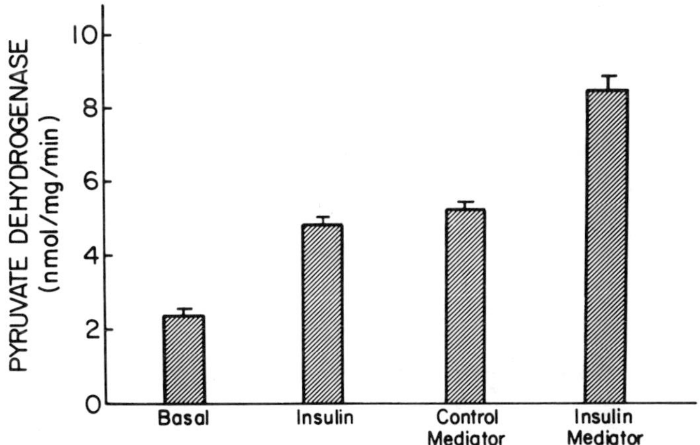

FIGURE 1. Effect of insulin and insulin mediator on pyruvate dehydrogenase activity in intact adipocytes. Adipocytes were incubated 10 minutes at 37°C with insulin (100 μU/ml) or mediators (1/20 final dilutions) from skeletal muscle of control or insulin-treated rats, extracted, and the extracts assayed for PDH activity. The data are means ± SEM of triplicate incubations from the same experiment. (From Jarett *et al.*[12] With permission from *Science.*)

The intact cell assay methodology provided a means to determine whether insulin mediator affected other insulin-sensitive enzymes or pathways, including those that cannot readily be studied in cell homogenates. Of special interest in this regard are enzymes and pathways involved in glucose metabolism. FIGURE 3 shows that insulin mediator from rat skeletal muscle stimulated glycogen synthase by twofold, about the same as by a physiological concentration of insulin (100 μU/mL). Mediator from control muscle was virtually without effect on this enzyme. However, neither glucose transport nor glucose oxidation were affected by insulin mediator. For instance, although insulin stimulated glucose oxidation nearly ninefold, mediator did not stimulate above basal levels even at a dilution that was maximally effective in stimulating PDH (FIG. 4). Thus, this series of experiments showed that insulin mediator could mimic some of insulin's actions on individual enzymes in intact cells, but it could not account for all of the hormone's pleiotropic effects.

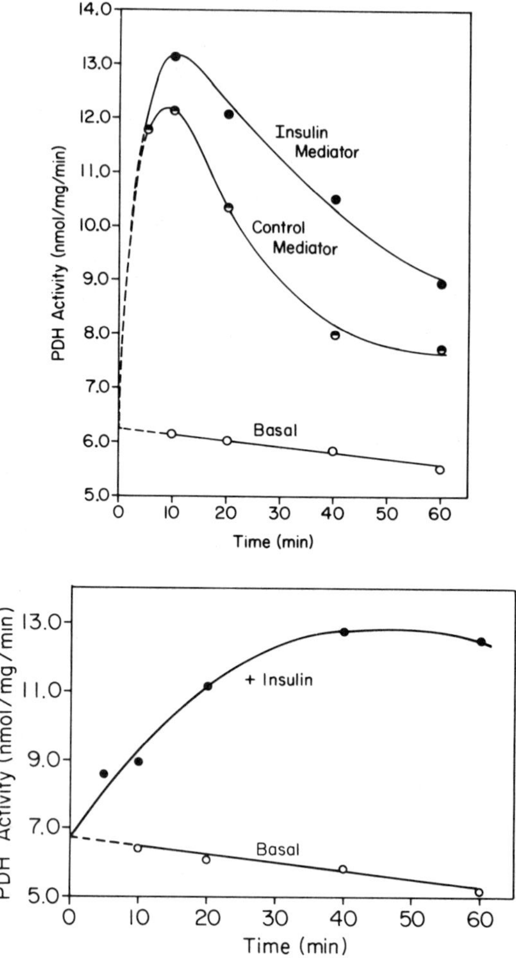

FIGURE 2. Time-course of PDH activity in intact adipocytes. (Upper) Effects of mediators from insulin-treated and control tissue extracts. Adipocytes were incubated for the indicated times in the absence (O) or presence of mediators prepared from insulin-treated (●) or control (◓) rat skeletal muscle before extraction and assay. (Lower) Effects of insulin. Adipocytes were incubated in the absence (O) or presence (●) of 20 μU/ml insulin and extracted for PDH assay. The data in both panels were the means of quadruplicate measurements. (From Jarett *et al.*[11] With permission from *Endocrinology*.)

Kelly *et al.* have recently been investigating the hormonal regulation of phospholipid methyltransferase (PLMT) in rat adipocytes.[14] It has been proposed that PLMT, which catalyzes the three-step conversion of phosphatidylethanolamine into phosphatidylcholine, plays an important role in hormone action, perhaps by transducing receptor-mediated signals.[15] Hormones and other agents that elevate intracellular cAMP levels promoted the activation of PLMT in hepatocytes and Leydig cells.[16]

Mato *et al.* reported that PLMT was phosphorylated *in vivo* by glucagon,[17] and they showed that partially purified PLMT from a rat liver particulate fraction could be phosphorylated and simultaneously activated by exogenously added protein kinase A (cAMP-dependent protein kinase).[18] These workers have proposed that PLMT may be hormonally regulated by phosphorylation/dephosphorylation cycles.[16,19]

TABLE 1 shows that in adipocytes, agents that are known to elevate the intracellular levels of cAMP increased PLMT activity. Insulin inhibited both basal and hormone-stimulated enzyme activity. These results are consistent with the proposal that adipocyte PLMT is hormonally regulated by phosphorylation/dephosphorylation. One reported effect of insulin was to lower hormonally elevated intracellular cAMP levels.[20,21] In addition, insulin blocks the activation of protein kinase A by cAMP,[22,23] and stimulates phosphoprotein phosphatases.[24] Thus it is possible that insulin inhibits PLMT activity by either suppressing the phosphorylation of the enzyme or by promoting its dephosphorylation, or both.

Since insulin mediator mimicks the effects of insulin on other enzymes known to be hormonally regulated via the phosphorylation/dephosphorylation cycle (such as, GS and PDH), it was of interest to examine the effects of mediator on PLMT. These experiments were performed using both a subcellular system and an *in vivo* assay. Mediator blocked PLMT activity in both liver and adipocyte plasma membranes, and mediator from insulin-treated tissue blocked activity to a greater extent than control mediator (data not shown). The effects of mediator on PLMT activity in intact cells are shown in FIGURE 5. Insulin suppressed isoproterenol-stimulated activity by about 30%. Control mediator was less effective than insulin in suppressing the stimulation, but mediator extracted from insulin-treated liver was almost twice as effective as insulin. These results suggest that, in intact cells, mediator can account for insulin's effects on PLMT.

The studies reported in this section show that insulin mediators can modulate the

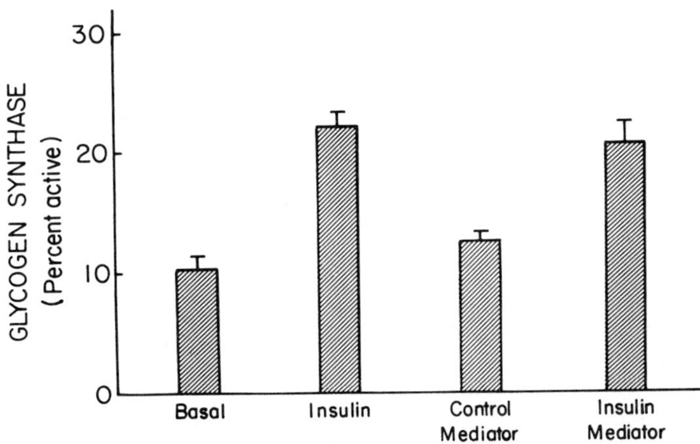

FIGURE 3. Effect of insulin and insulin mediator on glycogen synthase activity in intact adipocytes. After preincubating adipocytes 20 minutes at 37°C, insulin (100 μU/ml) or mediator preparations (1/20 final dilutions) were added and the incubations continued for 15 minutes before the cells were extracted for assay of GS activity. The data are means of triplicate measurements made within the same experiment. (From Jarett *et al.*[12] With permission from *Science.*)

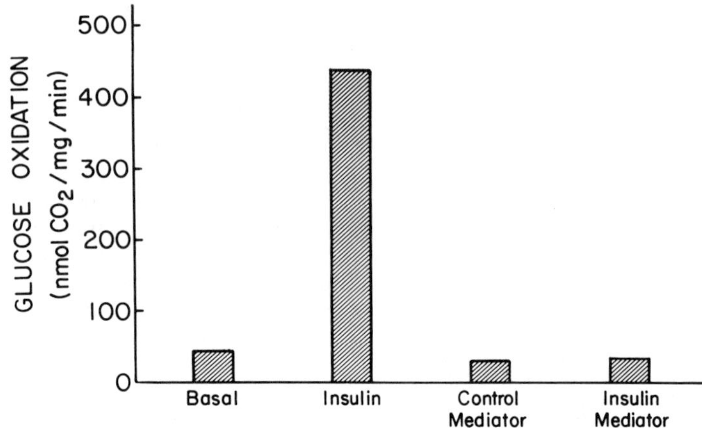

FIGURE 4. Effect of insulin and insulin mediator on glucose oxidation in intact adipocytes. Adipocytes were preincubated 15 minutes at 37°C before being dispensed into solutions containing insulin (100 μU/ml) or mediator preparations (1/20 final dilutions). Incubations were continued for 60 minutes at 37°C and the released ^{14}C was determined. The data are means of triplicate incubations within the same experiment. (From Jarett *et al.*[12] With permission from *Science.*)

activities of a number of insulin-sensitive enzymes in intact cells, including pyruvate dehydrogenase, glycogen synthase, and phospholipid methyltransferase. However, not all insulin-sensitive processes are affected by mediators. Phospholipid methyltransferase represents the basis for a new assay system for insulin mediator activity.

Purification and Physical-Chemical Properties

Although insulin mediator can be obtained in subcellular systems by treating plasma membranes from insulin-sensitive cells with insulin, the miniscule yields obtained by this method are insufficient for thorough biological or chemical characterizations. This laboratory has therefore explored the use of extracts made from tissues of insulin-treated animals as an alternative, potentially richer, source of insulin mediator.

TABLE 1. The Effects of Insulin on Phospholipid Methyltransferase Activity

Agent (concentration)	PLMT Activity (% of Control) Insulin (100 μU/ml) –	+
None	100	40
ACTH (2 mU/ml)	191	110
Isoproterenol (100 mM)	203	107
Forskolin (10 μM)	184	117

Adipocytes (1×10^6 cells/ml) were incubated in the absence or presence of insulin and the maximum stimulatory concentration of the agents indicated for 10 minutes at 37°C. Results are expressed as % of control PLMT activity, which was 6.0 ± 0.8 pmoles/mg.

FIGURE 6 illustrates the ultrafiltration characteristic of PDH-stimulatory mediator extracted from rat liver. Extracts prepared as described previously[11] were applied successively to ultrafilters with nominal cutoffs of 2,000 and 500 daltons, respectively. No detectable PDH-stimulatory activity was retained by the 2,000 MW filter. However, when the flow-through material from this step was ultrafiltered over the 500 MW filter, a significant amount of PDH-stimulatory activity was detected in the retained volume. Similar results have been obtained using extracts from rat skeletal muscle and heart. Thus, insulin mediator can be extracted from intact tissue, and its apparent molecular weight is greater than 500.

Insulin mediators extracted from adipocyte plasma membranes and from a liver particulate fraction had biphasic effects on PDH activity.[25] The PDH-stimulatory activity from a liver particulate fraction could be separated from the inhibitory activity by extraction into ethanol.[26] This result has obvious implications for designing an effective purification scheme for mediator. To determine whether mediator extracted

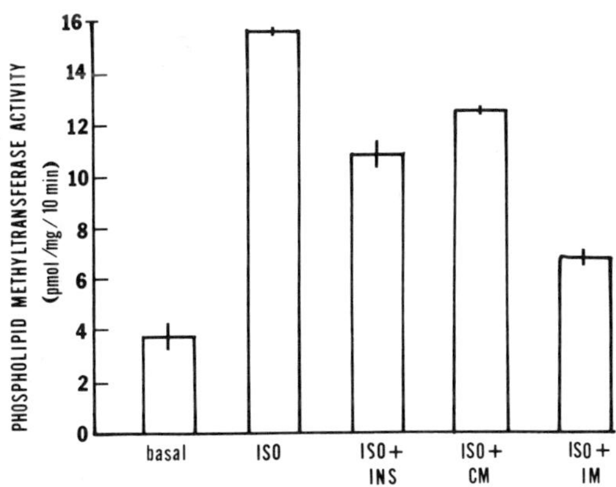

FIGURE 5. Effect of insulin and insulin mediator on phospholipid methyltransferase activity in intact adipocytes. Adipocytes were preincubated 30 minutes at 37°C, isoproterenol (100 nM), insulin (100 μU/mL), or mediator preparations (1/20 final dilutions) were added, and the incubation was continued for 10 minutes. Cells were extracted and assayed for PLMT activity as previously described.[14] The data are means of triplicate measurements from the same experiment.

from tissues also exhibited this behavior, the ethanol solubilities of PDH-stimulatory activity from liver and from the H4-II-EC3′ hepatoma cell line were compared. FIGURE 7 shows that mediator from the H4 cells was soluble in 100% ethanol, while the mediator from liver was found only in the insoluble pellet. Mediator from muscle extract was also found only in the ethanol-insoluble fraction (data not shown). The implications of these results concerning the possible tissue-to-tissue structural variability of mediator molecules that affect the same target enzyme are discussed below.

Ion-exchange chromatography is another potentially useful step for the purification of insulin mediator. FIGURES 8 and 9 illustrate the behavior of mediators from three different sources during anion-exchange chromatography. The PDH-stimulatory activity from H4 hepatoma cells was retarded by the anion exchange column and was eluted with 0.4 M salt. The mediators from liver (shown) and muscle (not shown) were completely excluded by the column. The PDH-stimulatory activity from heart was

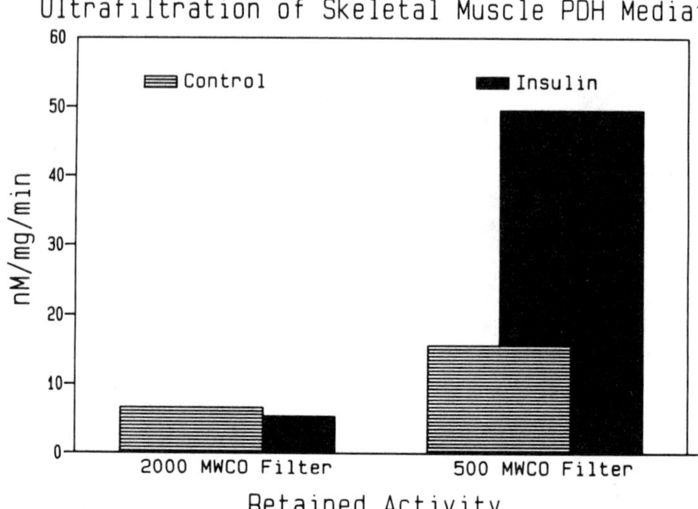

FIGURE 6. Ultrafiltration of insulin mediators. Mediators were extracted as described previously,[11] applied to 2,000 MW ultrafilters, concentrated under N_2, and the retentate washed four times with 50 mM formic acid. The original eluate and the four washings were pooled and applied to 500 MW ultrafilters, which were washed as above. The retentates from each filter were lyophilized and resuspended in 1 mM formic acid to 1 ml per gram of starting tissue before assay. The data are means ± SEM of triplicate determinations from a representative experiment.

unique in that it was partially excluded and partially retained; the retained portion was eluted with a 0.5 M NaCl wash. As might be expected from their behavior on anion exchange columns, the stimulatory activities from liver and skeletal muscle were retained by cation exchangers, including Dowex-50, Chelex-100, and phosphocellulose (FIG. 13).

Although the reasons are not clear for the different apparent physicochemical properties of mediators extracted from different sources, two possibilities must be entertained. First, the tissue extracts are certainly more complex than the extracts from the H4 hepatoma cells or the membrane supernatants used by Saltiel,[25,26] Seals,[7] and others. This implies both a greater number and a greater complexity of interactions between the mediators and other components in these extracts. Such interactions could modify or even mask important properties of the mediator molecules. In this regard, it should be noted that the apparent physicochemical properties of a number of molecules of biological interest were reported to change during purification as interferring substances were removed (for example, compare the ion-exchange behavior of thyrotropin-releasing factor in References 27 and 28). Second, it is possible that the mediators extracted from different sources have uniquely different molecular structures, reflecting, possibly, tissue-to-tissue differences in the enzymes whose actions insulin mediators might modulate. Only the purification and complete chemical characterization of the mediator molecules will resolve this issue.

The PDH complex is dependent upon Mg^{2+} for activity, and is stimulated by micromolar concentrations of Ca^{2+}.[29] These facts, coupled with the observation that PDH-stimulating activity extracted from liver, skeletal muscle, and heart binds to cation resins, including Chelex-100, raised the possibility that the PDH-stimulatory activity detected in these extracts was due to divalent cations. This possibility has been examined in a number of ways, and it has been found that the mediator activity generated by insulin in intact tissue resides in a molecular species that is distinct from calcium and magnesium and that is at least partially organic.

First, extracts from liver, skeletal muscle, and heart that were prepared through the ultrafiltration step on the 500 MW filters have been examined for divalent cations. FIGURE 10 shows there were significant amounts of both divalent cations in muscle extracts, and similar results have been obtained for liver and heart. It is important to note that, although the averages presented here show that there were slightly higher levels of both cations in the tissue extracts from insulin-treated animals than in control extracts, there was, in fact, no consistent insulin effect within experiments. Cadmium, zinc, and manganese have also been found in tissue extracts, but since these ions do not fractionate with PDH-stimulatory activity, they will not be considered further here.

To examine whether the divalent cations present in the tissue extracts could account for the observed PDH-stimulatory activity, artificial cocktails were made up

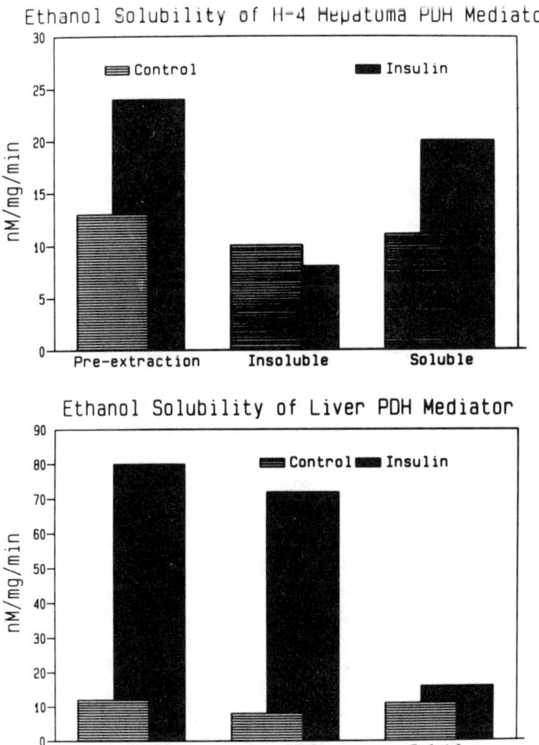

FIGURE 7. Ethanol solubilities of insulin mediators. Mediator preparations from H4-II-EC3′ hepatoma cells (upper) and rat liver (lower) were lyophilized and resuspended in 100% ethanol by sonication. The soluble and insoluble fractions were separated by centrifugation, dried under a stream of N_2, resuspended to their respective starting volumes with 1 mM formic acid, and assayed. The data are means ± SEM of triplicate determinations. The data for the hepatoma cell extract from Parker *et al.*[62]

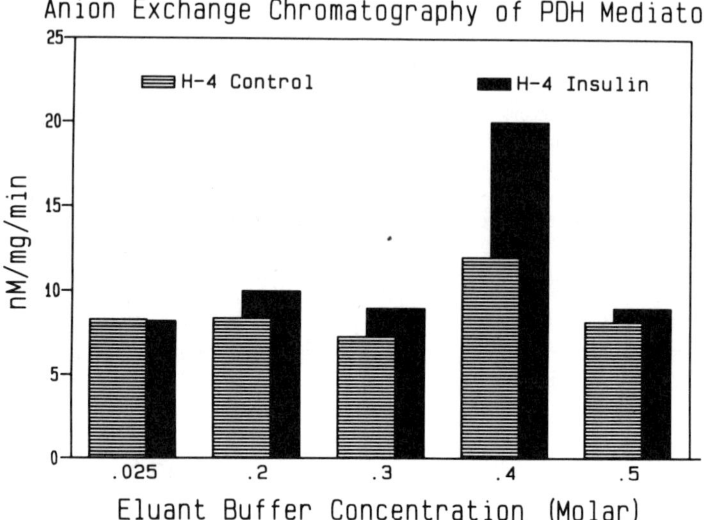

FIGURE 8. Anion exchange chromatography of insulin mediators from hepatoma H4-II-EC3′ cells. Cell extract was applied to a Dowex-1 column that had been equilibrated with 25 mM pyridine-HCl buffer, pH 6.0. The column was washed with successive one-column volume aliquots of buffer containing 0.025, 0.2, 0.3, and 0.4 M pyridine, and the fractions were lyophilized and assayed for PDH activity. The data are means of twelve measurements made in four separate experiments. For additional details, see Parker *et al.*[62]

to contain the same concentrations of Ca^{2+} and Mg^{2+} as were found in individual extracts, and both the cocktails and the original extracts were assayed, side-by-side, using the standard, broken mitochondria assay. The results of one such comparison experiment are presented in FIGURE 11. Clearly, there was an insulin-sensitive PDH-stimulatory activity present in the extract that was not accounted for by calcium and magnesium alone.

The effects of added calcium and magnesium on the *in vivo* PDH assay were also examined. Ion cocktails that were constructed to reflect the dilutions of tissue extracts that are standardly tested in this assay had no detectable effects (FIG. 12, compare with FIG. 1); the addition of extract levels of phosphate to the ion cocktails did not alter the results (not shown). Added ions were similarly without effect in either the subcellular or the *in vivo* PLMT assays (K. Kelly and A. Abler, personal communication, data not shown). Mediator activity detected using any of these assays must therefore be distinct from calcium and magnesium.

The preceding experiments in this section have demonstrated that an insulin-sensitive PDH-stimulatory activity, which cannot be accounted for solely in terms of divalent cations, can be detected in tissue extracts. FIGURE 13 illustrates the physical separation of this mediator activity from calcium and magnesium by ion exchange chromatography using Dowex-50. Liver extracts from control and insulin-treated rats were prepared in the standard way though ultrafiltration on the 2,000 MW filters. The filter eluates were applied to columns of Dowex-50 (volume of applied eluate equaled the bed volumes of the columns) and the columns were then washed with five column volumes of each of the indicated eluants. Both calcium and magnesium eluted in the

0.5 M–0.75 M HCl washes, and these fractions did exhibit marked levels of PDH-stimulatory activity. The most potent PDH-stimulating activity did not elute until the third 2 M HCl wash, in a region that was devoid of detectable magnesium and contained little calcium. An artificial cocktail containing the same calcium concentration as the third 2 M HCl wash of the insulin column did not detectably stimulate PDH activity (data not shown).

Evidence that further substantiates the organic nature of the insulin mediators was provided by the stability studies presented in FIGURE 14. PDH-stimulating activity extracted from skeletal muscle of insulin-treated animals was destroyed, or otherwise inactivated, by digestion with 2.5 N NaOH for one hour at room temperature, and also by a 72-hour digestion with 6 N HCl at 100°C. These data suggest that an intact

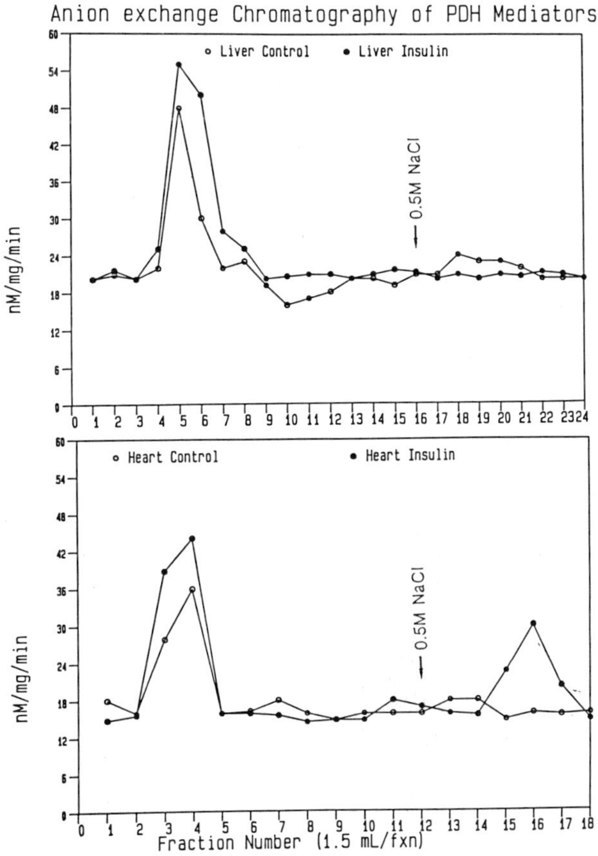

FIGURE 9. Anion exchange chromatography of insulin mediators from rat liver and heart. Mediators were extracted from rat liver (upper) or heart (lower),[56] applied to Sephadex QAE columns that had been equilibrated with 50 mM Tris-HCl, pH 8.5, and the columns were washed with the same buffer until the time when the elutriant was adjusted to contain 0.5 M NaCl. 1.5 ml fractions were collected and assayed for PDH activity. The data are means of triplicate measurements made in the same experiment.

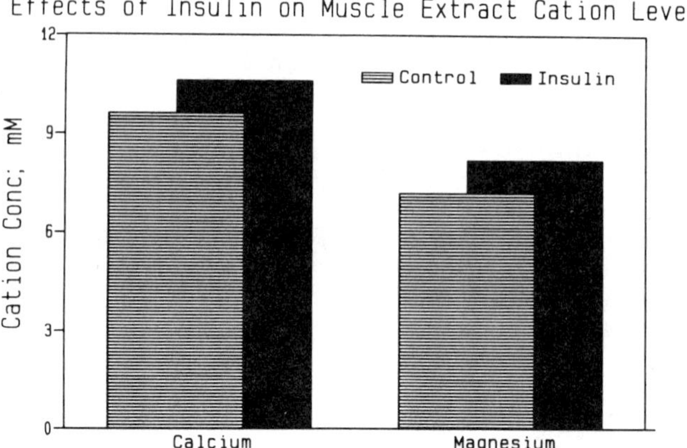

FIGURE 10. The effects of insulin on the levels of calcium and magnesium in rat skeletal muscle extracts. Skeletal muscles from control and insulin-treated rats were extracted as described in Jarett *et al.*[11] and aliquots were diluted into a $LaCl_3$-HNO_3 mixture for atomic absorbance spectroscopy. The data are the means of measurements made on 21 sets of paired extracts.

organic moiety is essential for the biochemical effects of the insulin mediator on PDH activity. Together with the data summarized above (FIGURES 9–12), they show that the molecule(s) mediating insulin's effects on PDH is (are) not calcium or magnesium.

To test whether this substance possesses peptide bonds, as first suggested by Seals and Czech[30] and later by others (see Ref. 8 for references), mediator preparations from

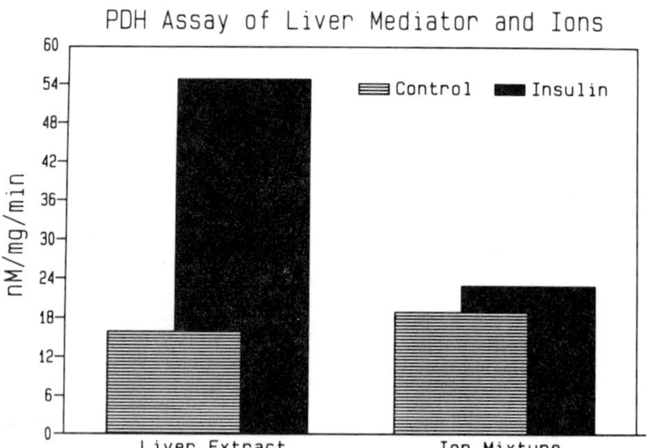

FIGURE 11. The effects of mediators and calcium/magnesium cocktails on pyruvate dehydrogenase activity. Livers from control and insulin-treated rats were extracted as described in Jarett *et al.*,[11] and ion levels were measured as described in FIGURE 10. Ion cocktails were prepared by dissolving $CaCl_2$ and $MgCl_2$ in 1 mM formic acid, and both the extracts and the cocktails were assayed at 1.4 final dilutions. The data are means of triplicate determinations made in the same experiment.

all three rat tissues were treated with proteolytic enzymes. Trypsin was without effect, while chymotrypsin and subtilisin caused partial inactivations (S.L. Macaulay, data not shown). This inactivation was prevented by conducting the incubation in the cold (4°C) or by heat-inactivating the proteases prior to incubating them with the mediators (data not shown). Since, of all the protease inhibitors tested, only TLCK prevented this inactivation, it is possible the detected inactivation was due to some contaminating activity in the enzyme preparations used and was not the result of proteolysis.

An interesting conjecture, first promulgated by Seals and Czech,[30] is that the mediators may, in fact, be proteolytic fragments of the insulin receptor itself. The gene sequence of the insulin receptor has recently been deciphered,[31,32] making it possible to test this interesting proposal. Fragments corresponding to likely proteolytic cleavage

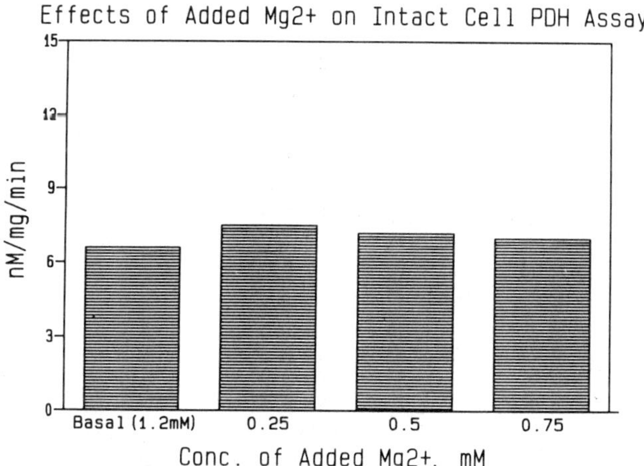

FIGURE 12. The effects of added calcium and magnesium on pyruvate dehydrogenase activity in intact adipocytes. Adipocytes were incubated in Krebs-Ringer bicarbonate buffer adjusted to contain 0.25 mM added calcium and the indicated concentration of added magnesium and incubated at 37°C. After 30 minutes the cells were extracted and assayed for PDH activity. The data are means of triplicate determinations made within the same experiment.

sites on the cytoplasmic portion of the B-subunit, synthesized and provided by R. Herrera and O. Rosen, were examined for PDH-stimulatory activity in the broken mitochondria assay. Four fragments, corresponding to the tyrosine 960 domain, the ATP-binding site, the tyrosine 1150 domain, and the carboxyl-terminus, were examined over the concentration range of 10 μM to 1 fM. Only the ATP-binding site peptide showed any effect on PDH activity; at 10 μM, the only concentration that elicited a response, it stimulated PDH activity only about 1.5-fold (unpublished results). Of course, this result suggests only that none of the tested fragments is the insulin mediator for PDH activation. A better and more conclusive test of this proposal will involve exposing liposomes that contain insulin receptors that have been incorporated into a phospholipid bilayer with trypsin, and assaying the degradation products in a variety of mediator assay systems. In any event, the true identity of the insulin mediator(s) awaits its (their) purification from biologically relevant sources in

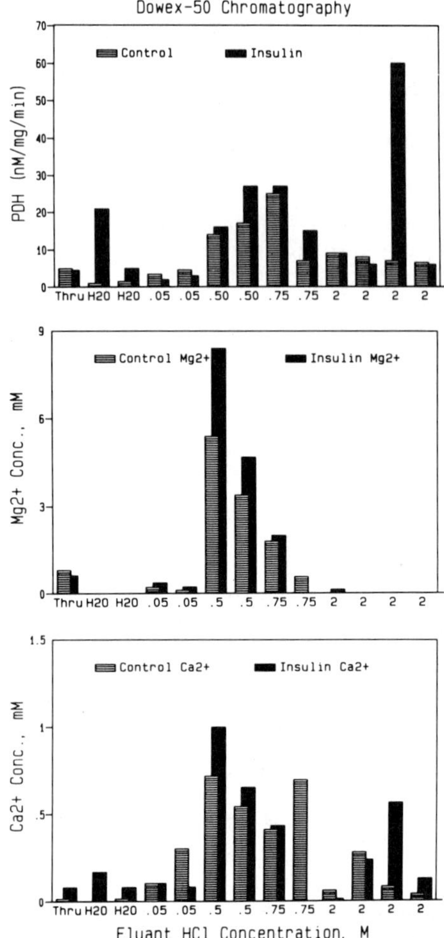

FIGURE 13. Cation-exchange chromatography of insulin mediators from rat liver. Livers from control and insulin-treated rats were extracted as described in Jarett *et al.*[11] except that ultrafiltration on the 500 MW filters was omitted, and applied to Dowex-50 columns that had been pre-equilibrated with 1 mM formic acid. Two successive washes, each equivalent to five column volumes, of each eluant were applied and assayed for calcium and magnesium, and for PDH-stimulatory activity. The PDH data are means of triplicate determinations from the same experiment, which has been repeated three times with similar results.

amounts that will allow its (their) complete and unambiguous chemical characterization.

DISCUSSION

In the fourteen years since Larner first proposed that cytoplasmic factors might mediate many of insulin's intracellular effects,[33] a number of laboratories have provided direct experimental proof that the interaction of insulin with its receptor results in the generation of a family of insulin mediator substances.[8] However, the exact number of such molecules, their chemical identities, and the molecular details of their synthesis and modes of action are questions that remain frustratingly unanswered. There are three main obstacles that must be overcome before the answers for these questions can be found.

First, insulin mediators are extremely labile substances, often not even surviving storage overnight at $-70°C$ (unpublished observations; cf. Ref. 25). To circumvent this problem, extraction and purification procedures must be at once gentle and rapid.

Second, mediators directed against the same target enzyme but extracted from different sources can exhibit strikingly different physicochemical properties. The practical effect of this finding is that methodologies developed for the extraction and

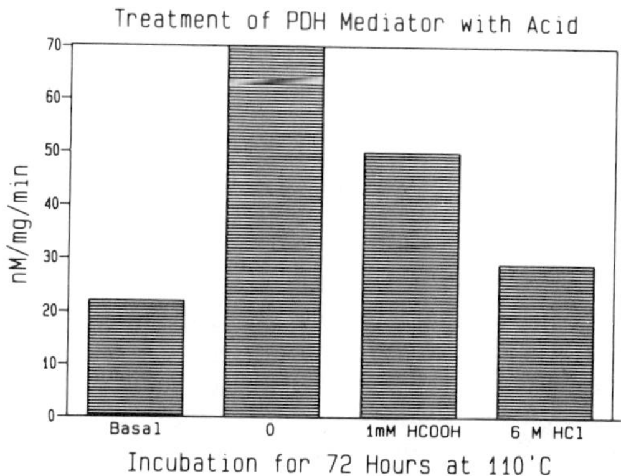

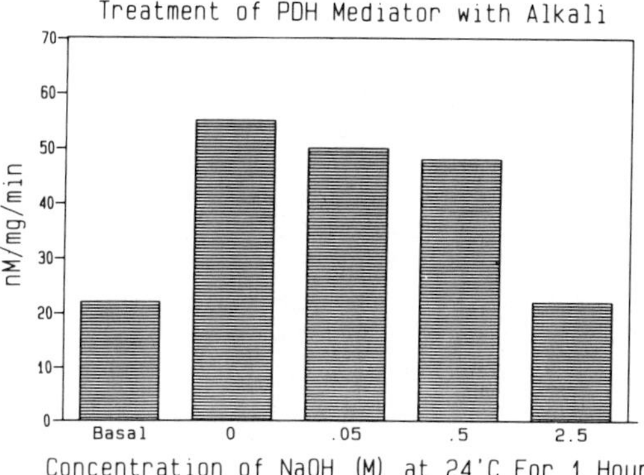

FIGURE 14. Stability of insulin mediator to treatment with acid or base. (Upper) Stability in acid. Mediator was extracted from skeletal muscle as described in Jarett *et al.*,[11] lyophilized, resuspended in the indicated agent, incubated at 100°C, and after 72 hours neutralized for assay. (Lower) Stability in base. Mediator was treated as described above, except it was resuspended in the indicated concentration of NaOH and incubation was at room temperature for 1 hour. The data are means of triplicate determinations made within the same experiment.

purification of mediator from one source may not work when applied to another source.

However, potentially the most serious difficulty is that the assays that have been most routinely used to follow mediator activities during purification are extremely sensitive to contaminants present in the extract that often obscure the presence and/or the biological effects of the bona fide mediators. Since these are not quantitative chemical assays but are instead bioassays that are used essentially as qualitative tests for the presence or absence of mediator activity, contaminants are often difficult to detect. To minimize the effects of this difficulty, it is necessary to examine mediator preparations for possible contamination with known substances that can modulate the activity of the test enzyme and to compare the effects of such substances, at the concentrations found in the extracts/isolation fractions, with those of the mediator preparation itself.

This paper details the evidence that the mediator of insulin's effects on pyruvate dehydrogenase is neither calcium, magnesium, nor a combination of the two. Both cations are necessary for PDH activity,[29] and both are found in high concentrations in extracts of rat skeletal muscle, liver, and heart. However, neither is the putative mediator of insulin action since (1) the tissue levels of these cations were not reproducibly elevated (or depressed) by treatment of the animals with insulin before sacrifice, but mediator activity was increased by insulin treatment; (2) careful comparison of the dose-response effects on PDH of the extracts with those of cocktails made up to contain the same concentrations of calcium and magnesium as found in the extracts showed that there was mediator activity in the extract that could not be accounted for by these cations; (3) the *in vivo* PDH assay used by Jarett *et al.* to study the effects of mediator on intact adipocytes[11,12] was insensitive to the extracellular presence of these ions; (4) the contaminating ions could be physically separated from the insulin-sensitive mediator by cation-exchange chromatography on Dowex-50; and (5) the acid and base instability of insulin mediator suggests that an intact organic moiety is required for biological activity.

Although the chemical nature of insulin mediators is unknown, several minimum characteristics can be assumed. Since mediators can be generated/released from isolated plasma membranes following insulin treatment, it follows that they are probably break-down products of some component(s) of the membrane. Membranes can be depleted of their capacity to generate/release mediator by repeated exposure to mediator-releasing agents, such as Tris,[34] and this implies that the size of the precursor pool of mediator in the membrane is limited. Finally, the mediators must be capable of rapid turnover to enable the cell to exercise precise temporal control over insulin's intracellular signalling mechanism.

Several workers have proposed that insulin mediators possess peptide bonds. This conclusion has been based on three kinds of evidence. First, trypsin elicits many of the same intracellular responses as does insulin, including the generation of mediators.[30,35] In their most literal interpretation these results mean only that perturbing the cell with trypsin caused intracellular changes; the experiments did not address the specific question of whether insulin mediator was the direct product of trypsin action on a surface molecule. In fact, mediator generation may have been temporally and spatially distinct from the point of trypsin action. Second, it has been reported that PDH-stimulating mediator was destroyed by treatment with proteases.[8,30,36,37] These results are difficult to interpret because important controls, such as treating mediator with inactivated protease or conducting the digestion in the presence of excess alternate substrate for the protease, were either not performed or not reported. In our hands, in experiments in which such controls were performed, either no or incomplete inactivation was observed (see above, results). Third, Larner's laboratory has reported that amino acids are detected after acid hydrolysis of preparations of purified mediators.[38] The relationship of these amino acids to mediator activity is obscure because it has not been shown that mediator activity is related in an obligatory manner to the structural integrity of any of these acids. In summary, the available evidence has not conclusively proven that insulin mediators possess peptide bonds.

Recently, a great deal of attention has focused on the changes in phospholipid metabolism that accompanies the interaction of a number of polypeptide hormones with their target tissues.[39] Currently the most interesting of these changes involve phosphatidylinositides, which form a small fraction of the total membrane-associated phospholipid pool and are directly involved with intracellular signalling mechanisms. The hormones presumably activate phospholipase C, which in turn leads to a rapid drop in phosphatidylinositol levels and concomitant increases in the levels of phosphatidylglycerol, phosphatidic acid, diacylglycerol, and inositol phosphates, especially inositol trisphosphate. Many of these products may function as second messengers: phosphatidic acid as a putative Ca^{2+} ionophore, diacylglycerol as an activator of protein kinase C, and inositol trisphosphate as a modulator of intracellular Ca^{2+} levels.

Similar, but not identical, events may be involved in insulin action. Insulin treatment of fat cells altered the metabolism of a number of phospholipids[40,41]; specifically, insulin rapidly (within seconds) stimulated the *de novo* synthesis of inositol-containing phospholipids[42–44] and phosphatidylserine,[45] although this latter result has not been confirmed (B. Corkey, personal communication). Interestingly, Jarett's laboratory found that phosphatidylserine stimulated both pyruvate dehydrogenase and low K_m-cAMP phosphodiesterase, and phosphatidylinositol-4,5-bisphosphate inhibited pyruvate dehydrogenase.[46,47] It has long been recognized that phospholipase C mimicked many of insulin's effects on intact adipocytes, including the stimulation of pyruvate dehydrogenase.[48,49] Koepfer-Hobelsberger and Wieland have reported that insulin activated endogenous phospholipase C in adipocytes with kinetics and dose-response effects that paralleled insulin's effects on pyruvate dehydrogenase.[50] Wieland's lab also reported that the insulin receptor could phosphorylate phosphatidylinositides *in vitro.*[51] Together these data suggest that inositol-containing phospholipids are involved in mediating insulin's actions, perhaps by subserving the role of precursor for the insulin mediators. Indeed, inositol has been identified in partially purified preparations of mediators from hepatocytes.[52] Koepfer-Hobelsberger and Wieland reported that exogenously added inositol trisphosphate stimulated pyruvate dehydrogenase, and suggested that this molecule might be the insulin mediator.[53] However, this is not likely to be the case since insulin did not affect the levels of inositol phosphates in either cultured myocytes[43] or adipocytes.[43,44] Furthermore, IP3 stimulated PDH activity only when it was added to permeabilized cells, and not when added directly to isolated mitochondria.[53] This is in direct contrast to results with mediators extracted from rat tissue, which act on PDH in both intact cell and subcellular systems.[8] Nevertheless, it remains an intriguing possibility that insulin mediators might be related to hitherto undiscovered derivatives of inositol phosphate.

In vivo assays for insulin mediators, first reported by Caro *et al.*[9] and by Zhang *et al.*,[10] have provided a means to examine the effects of mediator substances on a number of insulin-sensitive processes, including their effects on intact metabolic systems that are not readily amenable to examination in cell homogenates. In addition, their use has yielded important new insights into the biology of the insulin/mediator systems. Utilizing this methodology, Jarett *et al.* showed that insulin mediators can modulate, in the physiologically appropriate manner, the *in vivo* activities of individual enzymes as well as intact metabolic systems.[11,12] Exogenously added insulin mediators activated both pyruvate dehydrogenase and glycogen synthase in intact rat epididymal adipocytes. However, not all of insulin's pleiotropic effects on intact cells were mimicked by mediator, since a mediator-containing skeletal muscle extract had no detectable effects on either the transport or the oxidation of glucose. Use of the *in vivo* assay for PDH activity also revealed two biologically important facets of the insulin-mediator system. (1) The effects of exogenously added mediators on intact cells were transitory, in contrast to the sustained effects of insulin. Since cells that had been exposed to mediator could be restimulated by subsequent additions, it is likely that intact cells degraded or otherwise inactivated the mediator substance. (2) Added mediator could stimulate PDH to a greater extent than a physiological concentration of insulin. This suggested that the stimulation of the enzyme by insulin was limited by the amounts of mediator made available by the hormone. Together, these conclusions suggest that insulin elicits the continuous production of mediator but only small amounts are produced per unit time interval.

Data presented in this paper show that phospholipid methyltransferase provides a new assay system for insulin mediators. Mediator preparations inhibited this enzyme in both whole cell and in subcellular system assays. Kelley *et al.* have recently shown that PLMT in adipocytes was insulin sensitive.[14] Insulin inhibited both basal PLMT activity and PLMT activity that had been stimulated by hormones acting through cAMP. The biological steps that lead to the inhibition of PLMT following the interaction of insulin with its receptor remain to be elucidated, but one attractive possibility is that the effects of the hormone are brought about by dephosphorylation of the enzyme—by either inhibition of cAMP-dependent protein kinase, activation of a phosphatase, or both.

Insulin mediators promise to play an important role in post-receptor, insulin-resistant states such as obesity and type II diabetes mellitus. Amatruda and Chang[54] and Trowbridge *et al.*[55] have shown that liver membranes from fasting rats did not generate mediator after insulin treatment, but membranes from refed normal animals did. Macaulay *et al.*[56] confirmed this observation with intact animals, showing that insulin treatment of fasted rats did not increase mediator production in heart tissue. Amatruda and Chang have also reported that insulin failed to elicit the production of mediator from membranes from liver of streptozotocin-treated diabetic rats, but did elicit mediator generation from the membranes of diabetic rats that had been treated with insulin.[54] The ability to use human monocytes[37] as a source of insulin mediator makes clinical studies feasible, especially when coupled with better assay systems.

It is possible that insulin mediators, which probably represent a new family of second messengers, may play important roles in signal transduction for other hormones and membrane perturbing agents. For instance, preliminary evidence has shown that prolactin elicited the production, from plasma membranes of lactating mammary glands, of a substance that stimulated PDH activity and that had properties similar to that from rat liver and skeletal muscle.[57]

An important avenue of future work will be the mechanism of action of the insulin mediators. One means of metabolic regulation is through covalent modification, especially by modifying the extent of phosphorylation of key regulatory enzymes. Insulin promotes the phosphorylation and dephosphorylation of proteins in its target tissues, including the dephosphorylation of glycogen synthase[58] and pyruvate dehydrogenase.[59] However, other insulin-sensitive enzymes, such as adenylate cyclase, low K_m cAMP phosphodiesterase, protein kinase A, and pyruvate dehydrogenase phosphatase are exceptions to this rule. This latter class of enzymes includes those directly involved in cAMP metabolism, as well as kinases and phosphatases interacting directly with other key regulatory enzymes. All are involved with the making or breaking of phosphate bonds, and therefore some commonality of mechanism may be involved. It had been shown previously that PDH mediator enhances PDH activity by stimulating

PDH phosphatase,[60] although the details of this activation remain to be elucidated. Conceivably, then, the mechanism of action of insulin mediators involves their direct interaction with these enzymes, which then leads to changes in both the intracellular cAMP levels and extent of phosphorylation of other key regulatory enzymes.

The complete chemical characterization of insulin mediators awaits the isolation and purification of enough material so that powerful physical techniques such as NMR and mass spectroscopy can be employed. Efforts are continuing in this laboratory to accomplish this goal and to determine the molecular mechanisms of mediator production and action and their role in clinical conditions.

[NOTE ADDED IN PROOF: At the time of this meeting it was learned that Saltiel and Cuatracasas have reported the identification of a molecule that fulfills the role of an insulin second messenger for the stimulation of low K_m cAMP phosphodiesterase from adipocytes.[61,63] This substance appears to be generated by the phosphodiesteratic cleavage, by a phospholipase C, of a novel inositol-containing glycolipid precursor found in liver membranes. Insulin or phospholipase C treatment of the membranes released identical material that appeared to possess, in addition to inositol, glucosamine and phosphate. Mato, Kelly, and Jarett have confirmed and extended these observations. They purified an insulin-sensitive glycophospholipid and partially characterized the polar head group, which is a phospho-oligosaccharide.[64] This latter compound mimicked the inhibitory effect of insulin on phospholipid methyltransferase both in a subcellular system and in intact adipocytes.[65] These exciting observations open new avenues for future investigations concerning the final identification of insulin mediators, the mechanism by which insulin promotes their generation, and their mechanisms of action on various insulin-sensitive pathways.]

REFERENCES

1. Villar-Palasi, C. & J. Larner. 1961. Arch. Biochem. Biophys. **94:** 436–442.
2. Villar-Palasi, C. & J. Larner. 1960. Biochim. Biophys. Acta **39:** 171–173.
3. Gelehiter, T. D., P. D. Shreve & V. M. Dilworth. Diabetes **33:** 428–434.
4. Plehwe, W. E., P. F. Williams, I. D. Caterson, L. C. Harrison & J. R. Turtle. 1983. Biochem. J. **214:** 361–366.
5. Simpson, I. A. & J. A. Hedo. 1984. Science **223:** 1301–1304.
6. Larner, J., Y. Takeda, H. B. Brewer, G. Brooker & F. Murad. 1976. *In* Metabolic Interconversions of Enzymes. S. Shatiel, Ed. Springer-Verlag, New York.
7. Seals, J. R. & L. Jarett. 1980. Proc. Natl. Acad. Sci. USA **77:** 77–81.
8. Gottschalk, W. K. & L. Jarett. 1985. Diabetes/Metab. Rev. **1:** 229–259.
9. Caro, J. R., F. Folli, F. Cacchin & M. K. Sinha. 1983. Biochim. Biophys. Res. Commun. **115:** 375–382.
10. Zhang, S. -R., G. -H. Shi & R. Ho. 1983. J. Biol. Chem. **258:** 6471–6476.
11. Jarett, L., E. H. A. Wong, J. A. Smith & S. L. Macaulay. 1985. Endocrinology **116:** 1011–1016.
12. Jarett, L., E. H. A. Wong & J. A. Smith. 1985. Science **227:** 533–539.
13. Parker, J. C. & L. Jarett. 1985. Diabetes **34:** 92–97.
14. Kelly, K. L., E. H. A. Wong & L. Jarett. 1985. J. Biol. Chem. **260:** 3640–3644.
15. Hirata, F. & J. Axelrod. 1980. Science **209:** 1082–1090.
16. Mato, J. M. & S. Alemany. 1983. Biochem. J. **213:** 1–10.
17. Castano, J. G., S. Alemany, A. Nieto & J. M. Mato. 1980. J. Biol. Chem. **255:** 9041–9043.
18. Valera, I., M. Isabel, M. Pajares, M. Villalba & J. M. Mato. 1984. Biochem. Biophys. Res. Commun. **122:** 1065–1070.
19. Varela, I. M., M. Pajares, I. Merida, M. Villalbo, C. Cabrero, J. Traver & J. M. Mato. 1986. Regulation of phospholipid methyltransferase by reversible phosphoryla-

tion. *In* Biological Methylation and Drug Design. R. T. Borchardt, C. R. Creveling & P. M. Ueland, Eds.: 879–899. Humana Press. Clifton, NJ.
20. Butcher, R. W., J. G. T. Syned, C. R. Park & E. W. Sutherland. 1966. J. Biol. Chem. **241:** 1651–1653.
21. Kono, T. & F. W. Barham. 1983. J. Biol. Chem. **248:** 7417–7426.
22. Walkenbach, R. J., R. Hazen & J. Larner. 1978. Mol. Cell. Biochem. **19:** 31–41.
23. Walkenbach, R. J., R. Hazen & J. Larner. 1980. Biochim. Biophys. Acta **629:** 421–430.
24. Cohen, P. 1985. Eur. J. Biochem. **151:** 439–448.
25. Saltiel, A. R., S. Jacobs, M. Siegel & P. Cuatrecasas. 1981. Biochem. Biophys. Res. Commun. **102:** 1041–1047.
26. Saltiel, A. R., S. Jacobs, M. Siegel & P. Cuatrecasas. 1982. Proc. Natl. Acad. Sci. USA **79:** 3513–3517.
27. Guillemin, R., E. Sakiz & D. N. Ward. 1965. Proc. Natl. Soc. Exp. Biol. Med. **118:** 1132–1137.
28. Schally, A. V., T. W. Redding, C. Y. Bowers & J. F. Barrett. 1969. J. Biol. Chem. **244:** 4077–4088.
29. Hucho, F., D. D. Randall, T. E. Roche, M. W. Burgett, J. W. Kelly & L. J. Reed. 1972. Arch. Biochem. Biophys. **151:** 328–240.
30. Seals, J. R. & M. Czech. 1980. J. Biol. Chem. **255:** 6529–6531.
31. Ullrich, A., J. R. Bell, E. Y. Chen, R. Herrera, L. M. Petruzzelli, T. J. Dull, A. Gray, L. Coussens, Y. -C. Liao, M. Tsubokawa, A. Mason, P. H. Seeburg, C. Grunfield, O. M. Rosen & J. Ramachandran. 1985. Nature **313:** 756–761.
32. Ebina, Y., L. Ellis, K. Jaruagin, M. Edesy, L. Graf, E. Clauser, J. -H. Ou, R. Masiarz, Y. W. Kan, I. D. Goldfine, R. A. Roth & W. J. Rutter. 1985. Cell **40:** 747–758.
33. Larner, J. 1972. Diabetes **22** (Suppl. 2): 428–438.
34. Kiechle, F. L., L. Jarett, N. Kotogal & D. A. Popp. 1981. J. Biol. Chem. **256:** 2945–2951.
35. Kituchi, K. C., C. Schwartz, S. Creary & J. Larner. 1981. Mol. Cell. Biochem. **37:** 125–130.
36. Suzuki, S., T. Toyota, J. Suzuki & Y. Goto. Arch. Biochem. Biophys. **235:** 418–426.
37. Larner, J., K. Cheng, C. Swartz, K. Kituchi, S. Tamura, S. Creacy, R. Bubler, G. Glasko, C. Pullin & M. Katz. 1982. Fed. Proc. **41:** 2724–2729.
38. Cheng, K., M. Thompson, C. Swartz, C. Malchoff, S. Tamura, J. Craig, E. Locher & J. Larner. 1985. *In* Molecular Basis of Insulin Action. M. Czech, Ed. Plenum Press. New York.
39. Hokin, L. E. 1985. Ann. Rev. Biochem. **54:** 205–235.
40. DeTorrontegui, G. & J. Berthet. 1966. Biochim. Biophys. Acta **116:** 477–481.
41. Stein, J. M. & C. M. Hales. 1974. Biochim. Biophys. Acta **337:** 41–49.
42. Farese, R. V., R. E. Larson & M. A. Sabir. 1982. J. Biol. Cell. **257:** 4042–4045.
43. Farese, R. V., J. S. Davis, D. E. Barnes, M. L. Standaert, J. S. Babischkin, R. Hock, N. K. Rosic & R. J. Pollet. 1985. Biochem. J. **231:** 269–278.
44. Pennington, S. R. & B. R. Martin. 1985. J. Biol. Chem. **250:** 11039–11045.
45. Farese, R. V., M. A. Sabir, R. E. Larson & W. L. Trudeau, III. 1983. Biochim. Biophys. Acta **750:** 200–202.
46. Macaulay, S. L., F. L. Kiechle & L. Jarett. 1983. Biochim. Biophys. Acta **760:** 293–299.
47. Kiechle, F. L. & L. Jarett. 1983. Mol. Cell. Endo. **55:** 99–105.
48. Rodbell, M. 1966. J. Biol. Chem. **241:** 130–139.
49. Honeyman, T. W., W. Strohsnitter, C. R. Scheid & R. J. Schimmel. 1983. Biochem. J. **212:** 489–498.
50. Kopefer-Hobelsberger, B. & O. H. Wieland. 1984. Mol. Cell. Endo. **32:** 123–129.
51. Machicao, E. & O. H. Wieland. 1984. FEBS Lett. **175:** 113–116.
52. Wasner, H. K. 1981. FEBS Lett. **133:** 260–264.
53. Koepfer-Hobelsberger, B. & O. H. Wieland. 1984. FEBS Lett. **176:** 411–413.
54. Amatruda, J. M. & C. L. Chang. 1983. Biochim. Biophys. Acta **112:** 35–41.

55. TROWBRIDGE, M., A. SUSSMAN, L. FERGUSON, B. DRAZNIN, N. NEUFELD, N. BEGUM, H. M. TEPPERMAN & J. TEPPERMAN. 1984. Mol. Cell. Biochem. **62:** 25–36.
56. MACAULAY, S. L., J. O. MACAULAY & L. JARETT. 1985. Arch. Biochem. Biophys. **241:** 432–437.
57. JARETT, L., S. L. MACAULAY, J. O. MACAULAY & L. M. HOUDEBINE. 1984. *In* 7th Int. Congress of Endo.: Elsevier Publishing Co. New York. 777.
58. ROACH, P. J. 1981. Curr. Topics Cell. Reg. **20:** 45–105.
59. REED, L. J. 1981. Curr. Topics Cell. Reg. **20:** 95–106.
60. POPP, D. A., F. L. KIECHLE, N. KOTAGAL & L. JARETT. 1980. J. Biol. Chem. **255:** 7540–7543.
61. SALTIEL, A. & P. CUATRACASAS. 1986. Proc. Natl. Acad. Sci. USA. In press.
62. PARKER, J. C., F. L. KIECHLE & L. JARETT. 1982. Arch. Biochem. Biophys. **215:** 339–344.
63. SALTIEL, A. R., J. A. FOX, P. SHERLINE & P. CUATRACASAS. 1986. Science **233:** 967–972.
64. MATO, J. M., K. L. KELLY & L. JARETT. 1987. J. Biol. Chem. (In press.)
65. KELLY, K. L., J. M. MATO & L. JARETT. 1986. FEBS Lett. (In press.)

DISCUSSION OF THE PAPER

E. CARAFOLI (*Swiss Federal Institute of Technology, Zurich*): It appears that the effect of insulin is maintained in organelles isolated from insulin-treated tissues, as Dr. Denton has told us for the case of PDH-P phosphatase in mitochondria. Is it because a small M_r mediator is taken up by the organelles and stays there or is it because it induces a permanent modification of enzymes in the organelles?

JARETT: It is not known which is the correct mechanism. It is more likely that covalent modification of the phosphatase is not involved since the insulin mediator can activate the phosphatase without ATP or other compounds present other than low Mg^{2+} and Ca^{2+}. I believe the mediator remains attached to the phosphatase. The correct answer will be found only with further studies using purified mediator and isolated phosphatase.

The Role of Calcium and Calmodulin in Insulin Receptor Function in the Adipocyte[a]

JAY M. McDONALD

Departments of Pathology and Medicine
Washington University School of Medicine
Barnes Hospital
St. Louis, Missouri 63110

HARRIHAR A. PERSHADSINGH

Department of Laboratory Medicine
University of California, San Francisco
San Francisco, California 94143

JERRY COLCA

The Upjohn Company
7241-126-3
Kalamazoo, Michigan 49001

The mechanism(s) whereby the binding of insulin to its plasma membrane receptor results in the multiplicity of plasma membrane and intracellular effects remains elusive. Despite intensive research efforts, a large variety of mediator substances continue to be implicated in this signal-transduction process. Historically, candidates for mediating the cellular effects of insulin have included cAMP, cGMP, H_2O_2, H^+, Mg^{2+}, membrane hyperpolarization, phospholipids, Ca^{2+}, small molecular weight compounds, and even glucose itself.[1] To date no single candidate fulfills all the criteria for a universal messenger. However, substantial evidence exists supporting the majority of these as playing some role in the receptor-effector system for insulin. The general concept has emerged, therefore, that no single mediator is responsible for mediating the pleiotropic intracellular effects of insulin. Rather, the mechanism is likely to involve several potential mediator candidates acting as a complex network to produce a finely orchestrated insulin-induced metabolic cascade.

The role of one of these candidates, Ca^{2+}, albeit still controversial, has acquired renewed emphasis in light of substantial evidence supporting the concept that Ca^{2+} plays a fundamental role in some aspects of the cellular mechanism of insulin action.[2–5] The original independent proposals made by Clausen,[6] Kissebah,[7] and their associates, that Ca^{2+} was central to insulin action in the adipocyte were largely based on indirect studies. Our general approach to understanding the role of Ca^{2+} has been to evaluate directly the effect of insulin on Ca^{2+} homeostasis in the rat adipocyte. This was accomplished by characterizing the mechanisms responsible for Ca^{2+} homeostasis at the subcellular organelle level and to determine the effects of insulin on these

[a]Supported by U. S. Public Health Service grant AM25897 and a grant from the Juvenile Diabetes Foundation International.

processes.[8] These studies provided substantial direct evidence to support the concept that Ca^{2+} and calmodulin, acting at the plasma membrane domain of the cell, are important regulatory components of the insulin receptor signal-transduction process. This review will emphasize the role of Ca^{2+} and calmodulin in insulin receptor function in the rat adipocyte.

A variety of observations support the concept that Ca^{2+} homeostasis at the plasma membrane plays an important role subsequent to the interaction between insulin and its receptor. Insulin pretreatment of adipocytes increases Ca^{2+} binding to adipocyte plasma membranes[9] and inhibits the high affinity (Ca^{2+} + Mg^{2+})-ATPase responsible for extruding Ca^{2+} from the cell.[10] Insulin, added directly to adipocyte plasma membranes, inhibits the calmodulin-dependent (Ca^{2+} + Mg^{2+})-ATPase,[11] stimulates calmodulin binding,[12] and phosphorylates a 40 kDa protein that is also a substrate for the Ca^{2+}-stimulated, phospholipid-dependent protein kinase, protein kinase C.[13] Furthermore, the insulin receptor has both Ca^{2+} [14] and calmodulin[15] binding sites. Finally, the recent observations that Ca^{2+} and calmodulin enhance stimulation of the insulin receptor tyrosine kinase activity by insulin in a subcellular adipocyte preparation,[16] and that insulin phosphorylates calmodulin both in the subcellular preparation[16] and the intact cell (*vide infra*) will be reviewed here. Before embarking on this task, however, recent data obtained with intact adipocytes, supporting the concept that intracellular Ca^{2+} is an essential component of the insulin-effector system, will be presented.

INHIBITION OF THE INSULIN SIGNAL BY QUIN2-AM

If intracellular Ca^{2+} is important in the insulin signal-transduction process, then chelation of intracellular Ca^{2+} should prevent insulin-dependent modulation of metabolism. This is effectively accomplished by loading adipocytes with quin2 by incubating them with quin2-AM (FIG. 1). Being membrane permeable, quin2-AM (ester form) rapidly enters the cell where it is hydrolyzed by intracellular esterases to yield the trapped Ca^{2+} chelator quin2 (carboxylate form), which is able to bind and inactivate intracellular Ca^{2+} (FIG. 1).[17] Insulin-stimulated glucose transport (FIG. 2A) and glucose oxidation (FIG. 2B) were inhibited in a concentration-dependent fashion by preincubation of adipocytes with quin2-AM. No effect was observed on basal (insulin-independent) activities. The effect of quin2-AM on insulin-stimulated glucose transport was partial (approximately 55% of insulin-independent transport was suppressed at maximum quin2-AM concentration) whereas its effect on insulin-stimulated glucose oxidation was virtually complete. Quin2 loading also blocked the antilipolytic effect of insulin without affecting basal or cAMP-stimulated lipolysis[39] (data not shown). Under the experimental conditions employed in these studies, quin2 loading had no effect on intracellular ATP concentration or ^{125}I-labeled insulin binding to the cells.[39] Preincubation of cells with another membrane-permeable Ca^{2+} chelator, chlortetracycline,[18] had smaller but similar effects on insulin-stimulated glucose transport (FIG. 2A) and glucose oxidation (FIG. 2B). Significantly, chlortetracyline, like quin2-AM, had no effect on basal (hormone-independent) activities.

Rosic *et al.*[19] demonstrated a similar effect of quin2 loading on insulin-stimulated glucose transport in BCH3-1 myoblasts. In contrast, Klip *et al.*[20] found no effect of quin2-AM (20 μM) on insulin-stimulated glucose transport in L_6 muscle cells. Reports of such conflicting findings tend to propagate the controversy surrounding the role of Ca^{2+} in insulin action but eschew explanation of the disparity. For instance, wide variations in tissue responsiveness to stimulation by insulin may account, in part, for

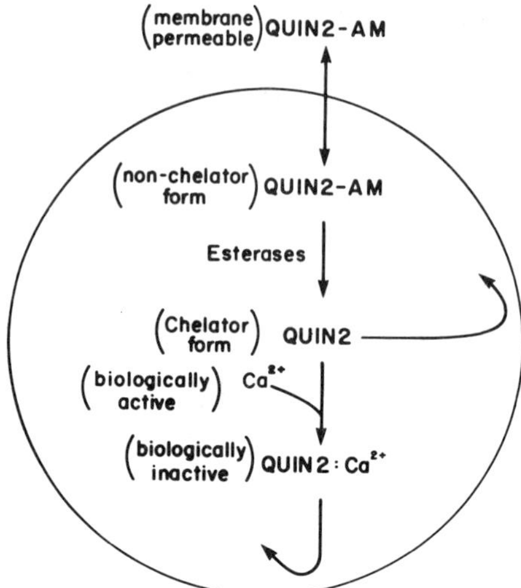

FIGURE 1. Schematic representation of chelation of intracellular Ca^{2+} by quin2 loading. Cells were incubated with quin2-AM (non-chelating ester form), which is membrane permeable. Intracellularly accumulated quin2-AM is hydrolyzed by esterases and trapped within the cell as the Ca^{2+}-chelating carboxylate form, which chelates intracellular Ca^{2+}.

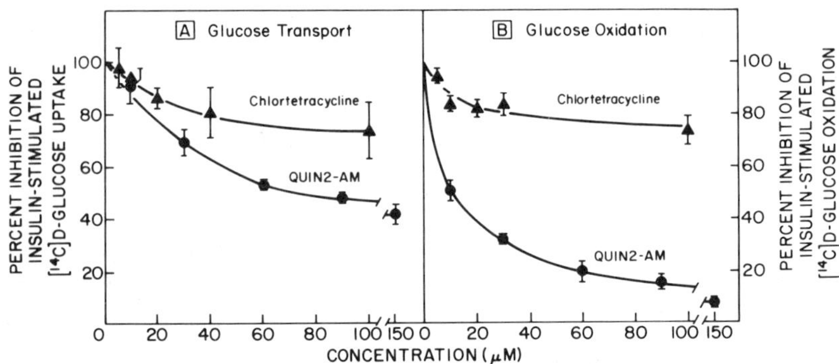

FIGURE 2. (A) Effect of quin2-AM and chlortetracycline on insulin-stimulated glucose transport. Isolated adipocytes were incubated with quin2-AM (10–150 μM, ●) or chloretetracycline (5–100 μM, ▲) for 15 min. [^{14}C]D-glucose uptake is monitored over the next 15-min interval (after addition of isotope) in a medium containing (per liter): 128 mmol NaCl, 5.2 mmol KCl, 1.4 mmol $MgSO_4$, 30 g bovine serum albumin, 10 mmol Na pyruvate, and 10 mmol Na phosphate (pH 7.4) at 37°C in the absence and presence of 1.25 nM insulin. The data are presented as the percent of insulin-stimulated D-glucose uptake in the absence of quin2-AM or chlortetracyline, and is the mean ± SE of three experiments with quin2-AM and two experiments with chlortetracycline. (B) Effect of quin2-AM and chlortetracycline on insulin-stimulated glucose oxidation. Isolated adipocytes were incubated with quin2-AM (10–150 μM, ●) and chlortetracycline (5–150 μM, ▲) as described in (A) except that oxidation was initiated by addition of cells directly to buffer containing quin2-AM or chlortetracycline and the reaction allowed to continue for 60 min at 37°C. Oxidation was monitored as $^{14}CO_2$ released from [^{14}C]D-glucose. The data are presented as the percent of insulin-stimulated glucose oxidation in the absence of quin2-AM or chlortetracycline, and is the mean ± SE of three experiments with quin2-AM and two experiments with chlortetracycline.

the conflicting results. Caution must therefore be employed when attempting to generalize limited data obtained from a single cell type.

It is also worthwhile emphasizing the concept of agonist-induced Ca^{2+} heterogeneities at cytoplasmic microdomains. There is now substantial evidence to support the concept that many primary signals mediated by Ca^{2+} will not result in uniform Ca^{2+} concentration changes in the overall cytoplasmic waterspace. Rather, localized changes in cytoplasmic Ca^{2+} are likely to occur in the vicinity of the site of signal initiation with rapid quenching and/or redistribution at other cytoplasmic domains by a combination of the Ca^{2+} homeostatic processes at the subcellular sites of Ca^{2+} sequestration and the buffering action of cytosolic proteins.[21–24] This concept may be especially important with regard to involvement of Ca^{2+} in insulin action. It is possible that the Ca^{2+} components of the insulin signal are small, transient, and localized at domains related to the plasma membrane. Therefore current techniques for measuring intracellular Ca^{2+} using batch fluorescence of cell suspensions are not likely to have the sensitivity required to reveal these changes. Most encouraging are the recent advances in the technique of individual cell fluorescence microscopy using digital imaging for data acquisition[25] and the discovery of newer fluorescent Ca^{2+} probes with greater quantum efficiency,[26] which may ultimately yield enough sensitivity and resolution to directly evaluate the role of Ca^{2+} in insulin action. In the meanwhile one may employ intracellular chelators to test the hypothesis that increases in intracellular Ca^{2+} are essential for the insulin-induced metabolic cascade, a hypothesis supported by the experimental data presented here.

Ca^{2+} AND CALMODULIN ENHANCE INSULIN-STIMULATED PHOSPHORYLATION OF THE β SUBUNIT OF THE INSULIN RECEPTOR

The interaction of insulin with its receptor results in phosphorylation of the β subunit of the receptor.[1,27] Insulin-stimulated phosphorylation of the β subunit of the insulin receptor is enhanced by Ca^{2+} and calmodulin.[16] Using partially purified preparations of the insulin receptor, obtained by wheat germ agarose chromatography from Triton (1%)-solubilized adipocyte plasma membranes, we demonstrated that calmodulin increased both the rate and maximum amount of insulin-stimulated phosphorylation of the receptor β subunit in a Ca^{2+}-dependent manner. As shown in FIGURE 3, calmodulin, in the presence of Ca^{2+}, enhanced insulin-stimulated phosphorylation of the β subunit (compare lane 4 to lane 2) by approximately twofold. Furthermore, in the absence of insulin (lane 3), calmodulin had no effect on phosphorylation (compare with lane 1).

The effect of calmodulin on insulin-stimulated phosphorylation of the β subunit was further characterized with respect to time (FIG. 4). Calmodulin increased phosphorylation up to fourfold at 0.5 min. Other characteristics of this effect of calmodulin are outlined in TABLE 1. The effect was saturable with a $K_{0.5}$ for calmodulin of 0.4 μM, was concentration-dependent with respect to both insulin and Ca^{2+}, and was inhibited by the calmodulin inhibitor, calmidazolium. Trypsin digestion of the phosphorylated β subunit followed by acid hydrolysis and high pressure liquid chromatography indicated that 75% of the increased phosphorylation that was induced by calmodulin was identified as phosphotyrosine residues.[16] The remainder (25%) of the calmodulin-enhanced component was evenly distributed between phosphoserine and phosphothreonine residues. These features clearly indicate that the effect of calmodulin was characterized predominantly by phosphate incorporation into tyrosine residues of the β subunit of the receptor.

CALMODULIN ENHANCES INSULIN-STIMULATED TYROSINE KINASE ACTIVITY AND CALMODULIN IS PHOSPHORYLATED BY INSULIN

To confirm that Ca^{2+} and calmodulin enhanced activity of the insulin receptor tyrosine kinase, their effects on the phosphorylation of the exogenous substrate, histone

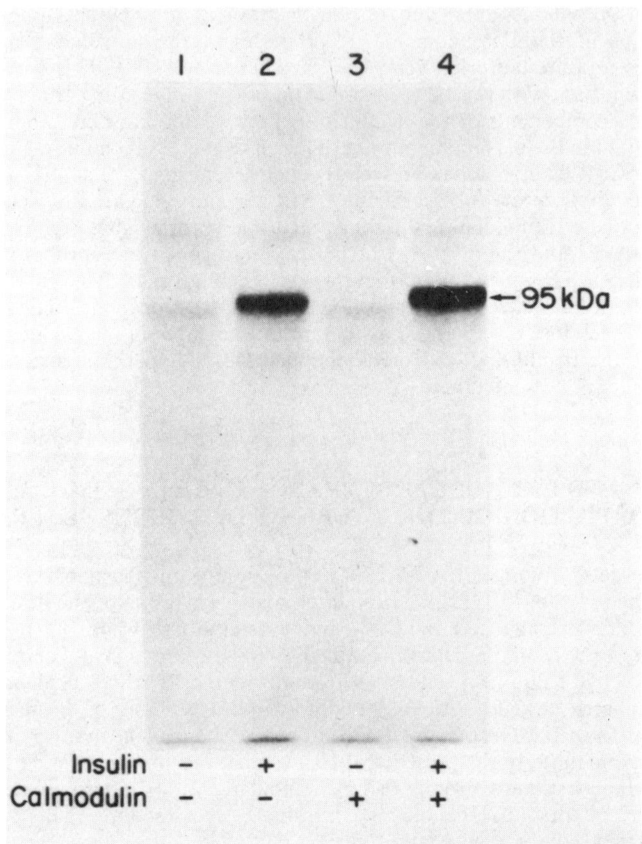

FIGURE 3. Stimulation by calmodulin of insulin-mediated phosphorylation of the insulin receptor. Lectin-purified preparations of adipocyte plasma membranes were phosphorylated as outlined in Graves *et al.*[16] Shown are the autoradiographs following SDS-PAGE. Assays were conducted in the absence (−) and presence (+) of 0.67 nM insulin and/or 2.3 μM calmodulin. The 95 kDa β subunit of the insulin receptor is designated. (From Graves *et al.*[16] With permission from *Journal of Biological Chemistry*.)

H2b, were determined. As shown in FIGURE 5, calmodulin enhanced insulin-stimulated phosphorylation of histone H2b approximately twofold. Similar to the effect of calmodulin on phosphorylation of the β subunit, calmodulin alone (lane 3) had no effect on histone phosphorylation in the absence of insulin. Most interestingly, calmodulin itself was phosphorylated in this system (band marked CaM, FIG. 5). The

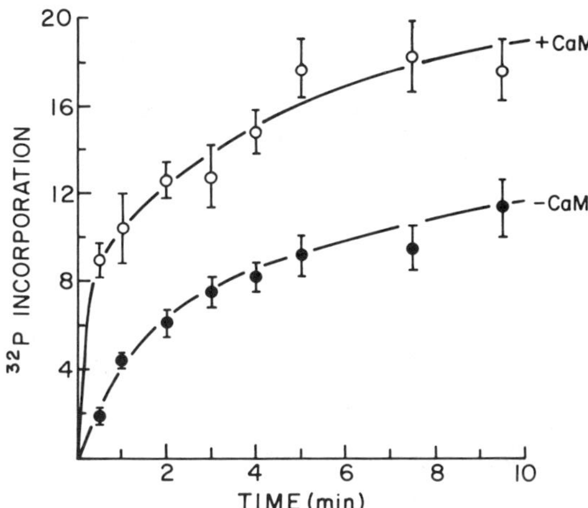

FIGURE 4. Time dependence of the effect of calmodulin on insulin-stimulated phosphorylation of the β subunit of the insulin receptor. The data were quantified from three separate experiments performed as in FIGURE 3. Incorporation of ^{32}P into the β subunit (ordinate) at various incubation times (abscissa) is shown. Quantitation of phosphorylation of the β subunit was accomplished by laser densitometry following SDS-PAGE and autoradiography. Assays were performed in the absence (●) or presence (○) of 2.3 μM calmodulin (CaM). (From Graves *et al.*[16] With permission from *Journal of Biological Chemistry*.)

identity of this band as calmodulin was confirmed not only by electrophoretic comigration with native calmodulin by both one- and two-dimension gel electrophoresis but also by immunospecific adsorption using anticalmodulin antibody bound to protein A Sepharose.

The time courses for the effect of calmodulin on insulin-stimulated phosphorylation of histone H2b and for insulin-stimulated phosphorylation of calmodulin are

TABLE 1. Characteristics of the Effect of Calmodulin on Insulin-Stimulated Phosphorylation of the β Subunit of the Insulin Receptor

Calmodulin dependency	Saturable $K_{0.5} = 0.4\ \mu M$ V_{max} obtained at 2.0 μM
Insulin dependency	Saturable $K_{0.5} = 100\ \mu U/ml^a$
Calcium dependency	Biphasic $K_{0.5} = 0.6\ \mu M$ V_{max} obtained at 3.5 μM Effect abolished at 300 μM
Effect of calmodulin antagonist	Calmidazolium inhibits: $IC_{50} = 1\ \mu M$
Phosphoamino acid analysis	75% of the calmodulin-enhanced phosphorylation in tyrosine residues

[a] 100 μU/ml insulin = 0.67 nM.
Adapted from Graves *et al.*[16]

shown in FIGURE 6. Note that at almost all time points calmodulin causes approximately a doubling of the effect of insulin on phosphorylation of histone H2b (panel A). Phosphorylation of calmodulin increased rapidly after an apparent lag phase of approximately 2 min (FIG. 6, Panel B). Other characteristics of insulin-stimulated phosphorylation of histone H2b and calmodulin are summarized in TABLE 2.

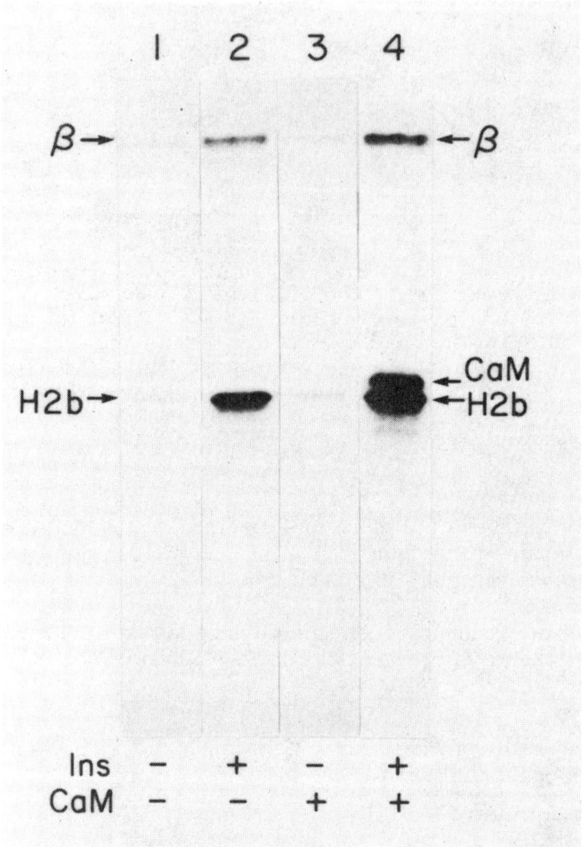

FIGURE 5. Insulin-stimulated phosphorylation of histone H2b and calmodulin. Phosphorylation assays were carried out as outlined in FIGURE 3 except that histone H2b was present in all reactions. Insulin (INS) and calmodulin (CaM) were absent (−) or present (+) as indicated. (From Graves *et al.*[16] With permission from *Journal of Biological Chemistry*.)

The dependence of all three processes (effects of calmodulin on insulin-stimulated β subunit and histone H2b phosphorylation, and insulin-stimulated phosphorylation of calmodulin) on free Ca^{2+} concentration were similar (TABLES 1 and 2). All processes were biphasic with respect to Ca^{2+} concentration with $K_{0.5}$'s for activation in the submicromolar range (0.1 to 0.6 μM); maximal activities were obtained between 3.0 and 10.0 μM Ca^{2+}; and progressive decreases were observed at supramicromolar

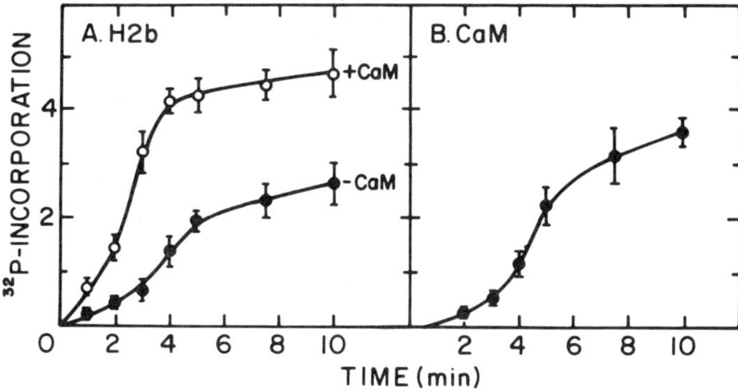

FIGURE 6. Time dependence of insulin-stimulated phosphorylation of histone H2b and calmodulin. Quantitation of ^{32}P incorporated into H2b (panel A) and calmodulin (panel B) at various times of incubation. In panel A the time course is shown in the absence (●) and presence (○) of calmodulin (CaM). All assays contained insulin. (From Graves *et al.*[16] With permission from *Journal of Biological Chemistry*.)

(>30 μM) concentrations of Ca^{2+}. Such biphasic responses to Ca^{2+} are similar to other intracellular Ca^{2+}/calmodulin-dependent enzymes such as adenylate cyclase[28] and the plasma membrane (Ca^{2+} + Mg^{2+})-ATPase.[29] There is one extremely important methodological consideration that must be emphasized when studying the effects of Ca^{2+} on insulin-stimulated phosphorylation *in vitro*. The dependence on free Ca^{2+} concentration must be determined using Ca^{2+}/EGTA buffers containing Mg^{2+} instead of Mn^{2+}. Under the conditions employed in the assay, it is not possible to buffer Ca^{2+} in the submicromolar range with EGTA in the presence of Mn^{2+} because EGTA has a higher affinity for Mn^{2+} than it has for Ca^{2+}, whereas the affinity for Mg^{2+} is several orders of magnitude less than that for Ca^{2+}.[30]

Most intriguing is the observation that calmodulin is only phosphorylated by insulin when histone H2b is present (FIG. 1). Histone is known to bind calmodulin[31,32] and this binding may be a prerequisite for calmodulin to be phosphorylated by the

TABLE 2. Characteristics of the Effect of Calmodulin on Insulin-Stimulated Phosphorylation of Histone H2b and Insulin-Stimulated Phosphorylation of Calmodulin

	Insulin-stimulated Phosphorylation of H2B	Insulin-stimulated Phosphorylation of Calmodulin
Calmodulin dependency	Saturable $K_{0.5}$ = 3.0 μM V_{max} obtained above 4.6 μM	Non-saturable
Insulin dependency	Saturable $K_{0.5}$ = 200 μU/ml	Saturable $K_{0.5}$ = 350 μU/ml
Calcium dependency	Biphasic $K_{0.5}$ = 0.2 μM V_{max} obtained at 10 μM Effect abolished at 100 μM	Biphasic $K_{0.5}$ = 0.1 μM V_{max} obtained at 10 μM Effect abolished at 100 μM

Adapted from Graves *et al.*[16]

insulin receptor kinase. Two observations by Haring *et al.*[33] are particularly relevant. First, using a similar insulin receptor preparation from hepatoma cells, they observed that calmodulin was phosphorylated predominantly on tyrosine residues by insulin. Second, they observed insulin-stimulated phosphorylation of calmodulin only when another calmodulin binding protein, purified calmodulin-dependent kinase, was present in the assay. Since calmodulin was phosphorylated predominantly on tyrosine residues, they concluded that it was phosphorylated by the insulin receptor tyrosine kinase rather than by activation of the calmodulin-dependent kinase. Therefore, calmodulin binding proteins appear to be necessary for insulin to stimulate phosphorylation of calmodulin. This hypothesis is currently under investigation in our *in vitro* system.

Furthermore, it is interesting to speculate about the potential functional significance of insulin-stimulated phosphorylation of calmodulin. Although we have yet to determine the amino acid residues that are phosphorylated on calmodulin by insulin, Haring *et al.*[33] showed that phosphotyrosine was the predominant species. Vertebrate calmodulins have only two tyrosine residues, both of which reside in two of the four Ca^{2+} binding domains.[34] It is therefore reasonable to speculate that phosphorylation at these sites might alter the Ca^{2+} binding properties of calmodulin and therefore alter its biological activity. If, for example, insulin-stimulated phosphorylation of calmodulin resulted in a less biologically active form of calmodulin, then the activity of known Ca^{2+}/calmodulin sensitive enzymes would be altered. Consequently, such enzymes as adenylate cyclase, the plasma membrane (Ca^{2+} + Mg^{2+})-ATPase, and phosphorylase kinase would become less active. Interestingly, calmodulin has been reported to be phosphorylated *in vitro* by a phosphorylase kinase preparation from skeletal muscle.[35] These suggestions are purely speculative at this time and considerable experimentation is necessary to test the hypothesis. We are presently attempting to isolate phosphorylated calmodulin to determine its biological activity.

INSULIN-STIMULATED PHOSPHORYLATION OF CALMODULIN IN INTACT ADIPOCYTES

Finally, one essential criterion for determining the biological relevance of the phosphorylation of calmodulin by insulin, is to determine if it occurs in the intact cell. We have recently employed a newly developed, high resolution two-dimensional gel electrophoresis system[36] to study the effect of insulin on protein phosphorylation in intact adipocytes.[38] In these studies intact rat adipocytes were prelabeled with $^{32}P_i$ and then stimulated with various concentrations of insulin. The cells were rapidly removed from the medium, frozen, and sonicated in a buffer with phosphatase inhibitors. The sonicates were separated by reverse-phase chromatography and then analyzed by two-dimensional polyacrylamide electrophoresis and autoradiography. These studies revealed that insulin stimulated the phosphorylation of at least three 17 kDa proteins with isoelectric points of approximately 3, 4, and 4.3. The protein that was most intensely phosphorylated (pI = 4 in this system) migrated with a pattern identical to that of bovine testicular calmodulin. It is not known whether all of the 17 kDa phosphoproteins represent different forms of phosphocalmodulin; however only one of these proteins (pI = 4) was detectable by Western blot[37] using antibody to bovine brain calmodulin. The maximum increase in the phosphorylation of the 17 kDa proteins that occurred in the intact cells ranged from 600–1,300% (not shown).[38]

The stimulation by insulin of endogenous calmodulin phosphorylation had an insulin concentration dependence similar to that for insulin activation of glucose uptake in adipocytes (FIG. 7). Thus, half-maximal concentration for activation of both glucose transport and calmodulin phosphorylation occurred at 25 μU/ml insulin. However, unlike the stimulation of glucose uptake, the effect of insulin on calmodulin phosphorylation was biphasic with the degree of phosphorylation decreasing at higher physiological concentrations of insulin (FIG. 7). The biphasic nature of this dose-response curve is reminiscent of the effect of insulin on calmodulin binding to the plasma membrane.[12]

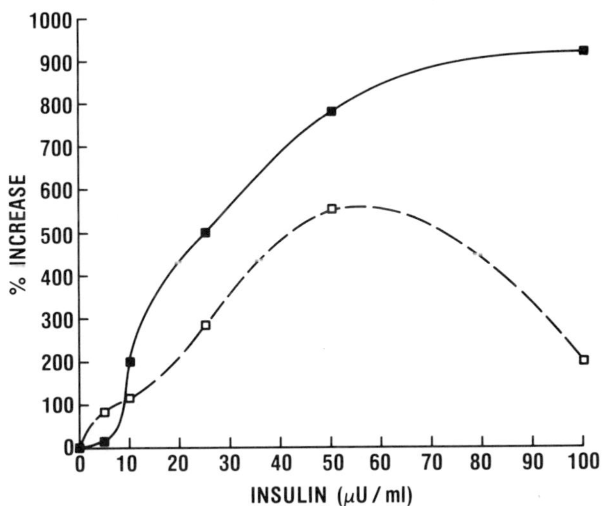

FIGURE 7. Comparison of the dose-dependence of insulin-stimulated phosphorylation of calmodulin and glucose uptake in isolated rat adipocytes. Isolated adipocytes were incubated in Krebs-Ringer bicarbonate buffer with or without 2 mCi $^{32}P_i$ for 2 hr at 37°C. Cells preloaded with $^{32}P_i$ were then stimulated with the indicated concentrations of insulin for 10 min. The cells were then quick frozen and processed for electrophoresis and autoradiography as described in the text. In a separate parallel experiment, the adipocytes were incubated in the same buffer containing [^{14}C]D-glucose (1.1 mM) and the indicated concentrations of insulin for 15 min. The extent of phosphorylation (□) of calmodulin (pI = 4; identified by Western blots) was determined by scanning the two-dimensional gels with a DU-8 spectrophotometer. [^{14}C]D-glucose incorporation (■) was determined by liquid scintillation. Data are the percent increase over basal for the respective parameters; similar results were obtained in three experiments.

CONCLUSIONS

Collectively the observations summarized here support the hypothesis that Ca^{2+} and calmodulin are involved in insulin receptor function. The results indicate that calmodulin may play an important role in modulating the sensitivity of receptor coupling to kinase activity. They do not, however, indicate how they are involved and what relationship they have to the mechanism of insulin action. We recently hypothesized the plasma membrane (Ca^{2+} + Mg^{2+})-ATPase may be functionally related to the insulin receptor[4] and that regulation of calmodulin binding to the insulin receptor may play a key role.[5] For instance, by increasing the affinity of the insulin receptor for calmodulin, insulin might actually catalyze the removal of calmodulin from the calmodulin-stimulated ATPase. This would decrease the activity of the (Ca^{2+} + Mg^{2+})-ATPase, a documented effect of insulin in adipocytes. Now, with the discovery of insulin-stimulated phosphorylation of calmodulin both in subcellular and intact cell preparations, it is tempting to speculate that this is, in fact, a major regulatory event. If phosphocalmodulin has less biological activity than dephosphocalmodulin, one consequence would be the inhibition of the (Ca^{2+} + Mg^{2+})-ATPase. Another might be a negative feedback signal to the insulin receptor kinase, thereby damping the primary signal to the cellular machinery. However, not all calmodulin may be phosphorylated. In other words, data thus far indicate that calmodulin must be bound to a separate protein in order to be a substrate for the receptor kinase. Perhaps binding uncovers a cryptic domain that contains the phosphorylation site(s) (e.g. tryosine containing Ca^{2+} binding pockets). Since phosphorylation of calmodulin did not occur without having another calmodulin binding protein present, calmodulin alone bound to the receptor may not be accessible to the kinase.

These observations clearly support the need to investigate the precise role of Ca^{2+} and calmodulin in insulin action. They also suggest a new and fascinating avenue for Ca^{2+}-mediated regulation of cellular metabolism, i.e. cell regulation via the phospho/dephosphocalmodulin couple.

REFERENCES

1. CZECH, M. P., Ed. 1985. Molecular Basis of Insulin Action. Plenum Press. New York.
2. CLAUSEN, T. 1978. The role of calcium in insulin action. *In* Proceedings of the 11th FEBS Meeting. **45:** 229–238. P. Nichols, *et al.*, Eds. Pergamon. Oxford.
3. MCDONALD, J. M., C. B. GRAVES & R. L. CHRISTENSEN. 1985. *In* Calcium and Cell Function. W. -Y. Cheung, Ed. **5:** 209–277. Academic Press. New York.
4. PERSHADSINGH, H. A. & J. M. MCDONALD. 1984. Cell Calcium **5:** 111–130.
5. MCDONALD, J. M. & H. A. PERSHADSINGH. 1985. *In* Molecular Basis of Insulin Action. M. P. Czech, Ed.: 103–117. Plenum Press. New York.
6. CLAUSEN, T., J. ELBRINK & B. R. MARTIN. 1974. Acta Endocrinol. (Copenh.) Suppl. **191:** 137–143.
7. KISSEBAH, A. H., B. R. TULLOCH, H. HOPE-GILL, P. V. CLARKE, N. VYDELINGUN & T. R. FRASER. 1975. Lancet **1**(7899): 144–147.
8. MCDONALD, J. M., D. E. BRUNS & L. JARETT. 1976. Proc. Natl. Acad. Sci. USA **73:** 1542–1546.
9. SANDRA, A. & D. J. FYLER. 1982. Endocr. Res. Commun. **9:** 107–120.
10. PERSHADSINGH, H. A. & J. M. MCDONALD. 1981. Biochem. Int. **2:** 243–248.
11. PERSHADSINGH, H. A. & J. M. MCDONALD. 1979. Nature (London). **281:** 495–497.
12. GOEWERT, R. R., N. B. KLAVEN & J. M. MCDONALD. 1983. J. Biol. Chem. **258:** 9995–9999.
13. GRAVES, C. B. & J. M. MCDONALD. 1985. J. Biol. Chem. **260:** 11286–11292.
14. WILLIAMS, P. F. & J. R. TURTLE. 1984. Diabetes **33:** 1106–1111.
15. GRAVES, C. B., R. R. GOEWERT & J. M. MCDONALD. 1985. Science **230:** 827–829.
16. GRAVES, C. B., R. D. GALE, J. P. LAURINO & J. M. MCDONALD. 1986. J. Biol. Chem. **261:** 10429–10438.
17. TSIEN, R. Y. 1983. Ann. Rev. Biophys. Bioeng. **12:** 91–116.
18. CASWELL, A. H. 1979. Ann. Rev. Cytol. **56:** 145–181.
19. ROSIC, N., L. MOJSILOVIC, M. STANDAERT & R. POLLET. 1985. Diabetes **34** (Suppl 1): 78A.
20. KLIP, A., G. LI & W. J. LOGAN. 1984. Am. J. Physiol. **247:** E297–E304.
21. ROSE, B. & W. R. LOWENSTEIN. 1975. Science **190:** 1204–1206.
22. MATTHEWS, E. K. 1980. *In* Secretory Mechanisms. C. R. Hopkins & C. J. Duncan, Eds.: 225–249. Cambridge University Press. London.
23. HARARY, H. H. & J. E. BROWN. 1984. Science **224:** 292–294.

24. Keith, C. H., R. Ratan, F. R. Maxfield, A. Bajer & M. L. Shelanski. 1985. Nature (London) **316:** 848–850.
25. Arndt-Jovin, D. J., M. Robert-Nicoud, S. J. Kaufman & T. M. Jovin. 1985. Science **230:** 247–256.
26. Grynkiewicz, G. M. Poenie & R. Y. Tsien. 1985. J. Biol. Chem. **260:** 3440–3450.
27. Gammeltoft, S. & E. Van Obberghen. 1986. Biochem. J. **235:** 1–11.
28. Brostrom, C. O., M. A. Brostrom & D. J. Wolff. 1977. J. Biol. Chem. **252:** 5677–5685.
29. Scharff, O. & B. Foder. 1982. Biochim. Biophys. Acta **691:** 133–143.
30. Bulas, B. A. & B. Sackter. 1979. Anal. Biochem. **95:** 62–72.
31. Grand, R. J. A. & S. V. Perry. 1979. Biochem. J. **183:** 285–295.
32. Malencik, D. A. & S. R. Anderson. 1982. Biochemistry **21:** 3480–3486.
33. Häring, H. -U ., M. F. White, C. R. Kahn, Z. Ahmad, A. A. DePaoli-Roach & P. J. Roach. 1985. J. Cellular Biochem. **28:** 171–182.
34. Dedman, J. R., R. L. Jackson, W. E. Schreiber & A. R. Means. 1978. J. Biol. Chem. **253:** 343–346.
35. Plancke, Y. E. & E. Lazarides. 1983. Molec. Cell. Biol. **3:** 1412–1420.
36. Dewald, D. B., L. A. Adams & J. D. Pearson. 1986. Anal. Biochem. **154:** 502–508.
37. Van Eldik, L. J. & S. R. Worchok. 1984. Biochem. Biophys. Res. Commun. **124:** 752–759.
38. Colca, J. R., D. B. Dewald, J. D. Pearson, B. J. Gilchrist, C. B. Graves & J. M. McDonald. 1987. Submitted for publication.
39. Pershadsingh, H. A., D. L. Shade, D. M. Delfert & J. M. McDonald. 1987. Proc. Natl. Acad. Sci. USA (In press.)

DISCUSSION OF THE PAPER

G. Pozza (*University of Milan, Milan*): (1) What control do you have for the effect of quin2-AM not being due to generation of formaldehyde, a byproduct of acetoxymethylester hydrolysis?

(2) Furthermore, quin2 can bind other divalent cations, for example Zn^{2+}, Mn^{2+}, etc., and the effect of quin2 could be due to an effect of these cations.

McDonald: (1) Our only control was to measure intracellular ATP levels. Under our assay conditions, ATP levels were unaltered. However, if we didn't have pyruvate in the incubation medium, ATP levels fell.

(2) Our best evidence is the ability of low concentration of A23187 in the presence of Ca^{2+} to reverse the inhibitory effect of quin2 on insulin-stimulated glucose transport.

L. Cantley (*Tufts University, Boston, MA*): Since phospholipase C enzyme, which has been investigated, requires a certain basal level of Ca^{2+} in the cytosol for activity, it is possible that the ability of high quin2 concentrations to block insulin stimulation is due to inhibitions of this enzyme. This need not be interpreted as an indication that Ca^{2+} is acting as a signal.

McDonald: This is true for the quin2 loading studies that I showed. Clearly these investigations provide indirect evidence, at best, for some role of calcium. In contrast, the studies I showed on phosphorylation of the receptor B subunit, histone and calmodulin did not employ quin2 and are thus not open to this interpretation.

E. Carafoli (*Swiss Federal Institute of Technology, Zurich*): What is the efficiency of the phosphorylation of calmodulin you have described, and, secondly, is phosphorylated calmodulin more or less active on targets, say PDE, than non-phosphorylated calmodulin?

McDonald: In our *in vitro* system we maximally incorporate about 0.2–0.5 moles of phosphate/mole of calmodulin. We do not know the biological activity of phosphorylated calmodulin as yet. This is a vitally important question. If, as in the case of the work of Haring *et al.*, calmodulin is phosphorylated on tyrosine residues we might anticipate a decreased biological activity. This is because calmodulin has only two tyrosine residues and those are located in two of the four calcium binding domains of the molecule. However, this remains only a speculation at this point.

Mechanism of Receptor Kinase Action on Membrane Protein Recycling[a]

S. CORVERA,[b] R. J. DAVIS,[b] P. J. ROACH,[c]
A. DEPAOLI-ROACH,[c] AND M. P. CZECH[b]

[b]*Department of Biochemistry*
University of Massachusetts Medical Center
Worcester, Massachusetts 01605

[c]*Department of Biochemistry*
Indiana University School of Medicine
Indianapolis, Indiana 46223

Polypeptide growth factors such as insulin, IGF-I, EGF, and PDGF have profound effects on cellular metabolism. The initial step in the action of these hormones is their binding to specific high affinity receptors on the surface of the cell. These receptors have been found to possess a tyrosine kinase enzymatic activity that is stimulated upon binding of their respective ligand.[1-4] This enzymatic activity causes the phosphorylation of tyrosine residues on the receptor, a process called autophosphorylation, and the phosphorylation of exogenous protein substrates. The activation of this tyrosine kinase activity may be the initial event whereby the hormone signal is transduced to the interior of the cell. However, the biochemical mechanisms that link the activation of the tyrosine kinase to the final cellular effects of the hormones have not been elucidated.

The purpose of the present work was to gain insight into the mechanism involved in the production of one of the well known cellular effects of insulin. The effect we have focused on is the rapid redistribution of several membrane components between the plasma membrane and the internal membranes of the cell. The first component shown to be redistributed in this way in response to insulin was the glucose transporter. Using cell fractionation techniques, Cushman and Wardzala[5] and Suzuki and Kono[6] demonstrated that upon addition of insulin, there was a marked increase in the concentration of glucose transporters in a fraction corresponding to the plasma membrane of the cell, and a simultaneous decrease in the concentration of this component in an intracellular low density microsomal fraction. It was subsequently found that another membrane component, the high affinity receptor for IGF-II,[7,8] was also redistributed to the cell surface from intracellular membranes in response to insulin. We have very recently found a third membrane component that responds to insulin in this manner, the transferrin receptor.[9] FIGURE 1 is a composite that illustrates the effect of insulin on the distribution of these three membrane components between the plasma membrane and the low density microsomes of isolated adipocytes. In each case Scatchard analysis of the binding of ligands to these membrane proteins is linear, indicating non-cooperative binding sites on the proteins.

The increase in the number of receptors on the plasma membrane could also be demonstrated by measuring the binding of ^{125}I-labeled ligands or specific anti-receptor

[a]Supported by National Institutes of Health Grants AM 30898, AM 30648, AM 27221, AM 01089, and a Program Project Grant CA 39240.

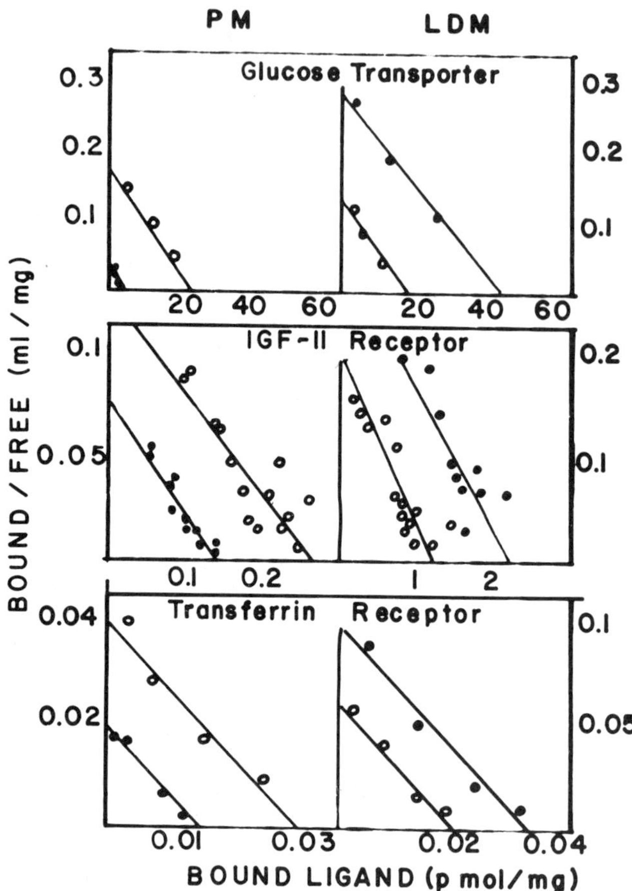

FIGURE 1. Scatchard analysis of equilibrium binding of [^{3}H]cytochalasin B, [^{125}I]IGF-II and [^{125}I]diferric transferrin to isolated adipocyte plasma membrane and low density microsomal fractions. Plasma membranes (PM) and low density microsomes (LDM) were prepared from control (●—●) or insulin-treated adipocytes (○—○). These membranes have been extensively characterized.[5,7] The plasma membrane fraction is enriched in 5′ nucleotidase activity and the microsomal fraction is enriched in UDP-galactose: N-acetylglucosamine galactosyltransferase. The concentration of glucose transporters was measured by specific [^{3}H]cytochalasin B binding (Data from Cushman & Wardzala.[5]). The number of IGF-II receptors was measured by [^{125}I]IGF-II binding (data from Oppenheimer *et al.*[7]), and the number of transferrin receptors was measured by [^{125}I]diferric transferrin binding (data from Davis *et al.*[9]).

antibodies to intact cells. In the experiments shown in FIGURE 2, fat cells were exposed to insulin and then treated with KCN. The concentration of IGF-II and transferrin receptors present on the surface of the cells could be measured because both endocytotic and exocytotic processes were blocked by the cyanide.[8]

There are at least two mechanisms by which increasing the number of membrane proteins on the cell surface might have profound physiological effects. In the case of glucose transporters, increasing the number on the cell surface would be expected to increase the flux of glucose across the plasma membrane and into the cell. Similarly, the cellular responsiveness to hormones could increase when a greater number of receptors to such hormones are exposed on the cell surface. A second type of effect may be to alter the rates of recycling of these proteins in such a way that the internalization rate of a ligand is changed. To investigate the physiological consequences of the increased number of cell surface transferrin receptors caused by insulin the rate of uptake of ^{59}Fe in intact adipocytes was measured. FIGURE 3 illustrates that the rate of uptake of ^{59}Fe is markedly increased in insulin-treated cells compared to controls. This increase in the rate of iron uptake occurs rapidly after the addition of the hormone. These data illustrate that the effect of insulin to cause a redistribution of membrane components has important physiological consequences, two of which are increases in the uptake of glucose and the uptake of iron by the cell.

The mechanism whereby insulin can cause the redistribution of these membrane components is unknown. Other well known targets of insulin action are enzymes that are involved in the intermediary metabolism of carbohydrates and lipids. These enzymes are phosphoproteins, whose activity can be regulated by their state of phosphorylation. The activity of the one subset of these enzymes is inhibited by phosphorylation. In these cases the effect of insulin to stimulate these enzymes correlates with a decrease in their phosphorylation state.[10–12] However, not all effects of

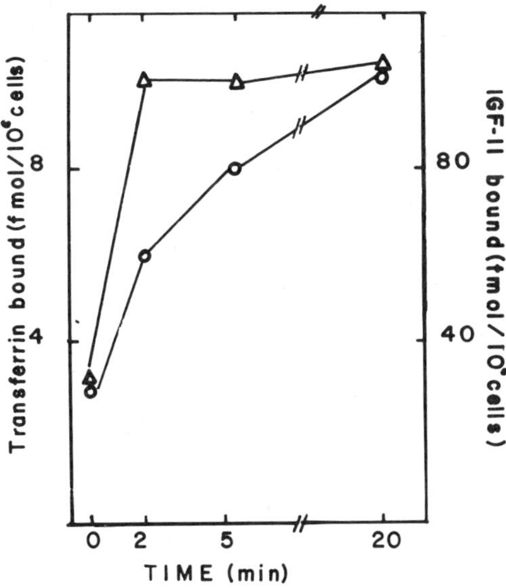

FIGURE 2. Time course of the effect of insulin on the number of transferrin and IGF-II receptors on the cell surface. Isolated rat adipocytes were incubated in Krebs-Ringer HEPES buffered medium supplemented with 3% bovine serum albumin (pH. 7.4) at 22 °C. The cells were exposed to 10^{-8} M insulin for the times indicated in the abcissa, following which KCN was added to a final concentration of 2 mM. 5 nM [^{125}I]diferric transferrin (Δ—Δ) or 5 nM [^{125}I]IGF-II (O—O) were then added to investigate cell surface binding. After 30 min of incubation at 22°C, the adipocytes were separated from the unbound ligand by oil flotation. The radioactivity bound to the cells was assessed by gamma counting. Non-specific binding was determined in incubations containing a 200-fold excess of diferric transferrin or IGF-II, and was typically 20% of the total binding.

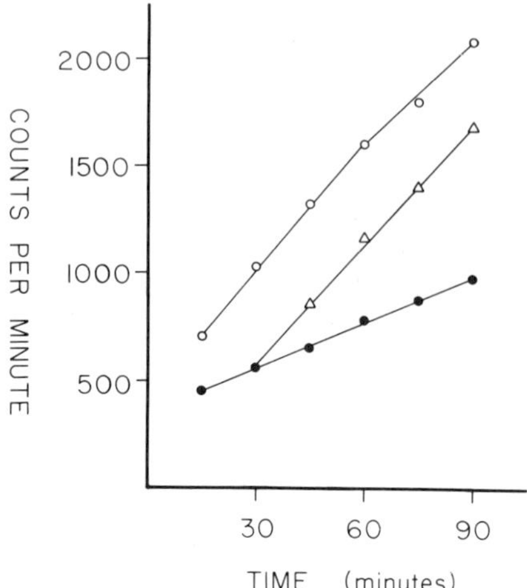

FIGURE 3. Effect of insulin on the accumulation of [^{59}Fe]diferric transferrin by isolated adipocytes. Adipocytes were incubated as described in FIGURE 2. [^{59}Fe]diferric transferrin was added and at indicated times the cells were separated from the incubation medium by oil flotation. The radioactivity associated with the cells was measured by scintillation counting (control (●—●)); 10^{-8} M insulin added 5 min before the [^{59}Fe]transferrin (○—○); 10^{-8} M insulin added 15 min after the [^{59}Fe]transferrin (△—△).

insulin are mediated through dephosphorylation reactions. It has been shown that the action of insulin also leads to an increase in the phosphorylation of some proteins such as the ribosomal protein S-6,[13] ATP citrate lyase,[14] and acetyl CoA carboxylase.[15]

Thus, several effects of insulin on carbohydrate and lipid metabolism are related to phosphorylation or dephosphorylation of target proteins. We therefore considered the possibility that the effect of insulin on the distribution of membrane components might also be mediated through phosphorylation or dephosphorylation reactions. Although the mechanisms that govern the basal distribution of proteins among different cellular membranes are largely unknown, it has been shown that other membrane components such as the epidermal growth factor receptor and the transferrin receptor, are phosphorylated in response to phorbol esters, which activate protein kinase C.[16–18] This phosphorylation correlates with the internalization of these receptors under some experimental conditions.[17,18] These data suggest a model shown in FIGURE 4 in which phosphorylation of membrane components at the cell surface may promote the movement of the receptor into an endocytic pathway. Once inside the cell, the phosphate group can be removed, and the dephosphorylated protein may then recycle back to the cell surface. Thus, the phosphorylation state of membrane components might determine their basal distribution among cellular membrane compartments.

We have investigated the possibility that the effect of insulin to cause a redistribution of IGF-II receptors is accompanied by changes in the phosphorylation state of this molecule.[19] To determine an accurate stoichiometry of receptor phosphorylation we employed immunoblotting as a method to quantitate the number of receptors that had been extracted and immunoprecipitated from ^{32}P-labeled fat cell membranes. Following electrophoresis the receptor was electroeluted onto nitrocellulose paper and incubated with anti IGF-II receptor antibodies and ^{125}I-protein A. In these experiments, we were able to determine the amount of IGF-II receptor by gamma counting of the ^{125}I-protein A and the amount of ^{32}P phosphate covalently attached to the receptor by Cerenkov counting. The ratio of ^{32}P-to-^{125}I counts provided a precise estimate of the stoichiometry of IGF-II receptor phosphorylation.

FIGURE 5 shows the results obtained when intact cells were incubated with ^{32}P phosphate for different periods of time. It can be seen that ^{32}P became rapidly incorporated into the IGF-II receptor, reaching constant specific activity after approximately 120 min of incubation. It can also be observed that the stoichiometry of phosphorylation, reflected by the ratio of ^{32}P-to-^{125}I, is at least two times higher in the receptors obtained from the plasma membrane compared to those obtained from low density microsomes. These data are consistent with the notion that phosphorylation of this receptor occurs in the plasma membrane and may be an important factor in determining its internalization. Using the same methodology, an experiment was performed in which cells were incubated with ^{32}P for two hours to obtain constant specific activity and then exposed to insulin. The results shown in FIGURE 6 show that the change in the distribution of the IGF-II receptor was closely paralleled by a change in the phosphorylation state of the receptor. In this figure, the open symbols are a plot of the amount of [^{125}I]protein A, which reflects the number of IGF-II receptors in each membrane fraction at different times after insulin addition. It can be seen in the left panel that insulin causes a rapid two to threefold increase in the number of IGF-II receptors on the plasma membrane.

The closed circles are a plot of the ratio of ^{32}P-to-^{125}I, which reflects the stoichiometry of receptor phosphorylation. Here it can be seen that insulin causes a very rapid and pronounced decrease in the phosphorylation state of the IGF-II receptor in the plasma membrane. In contrast, it can be seen in the right panel that insulin causes a decrease in the number of IGF-II receptors in the low density microsomes, but

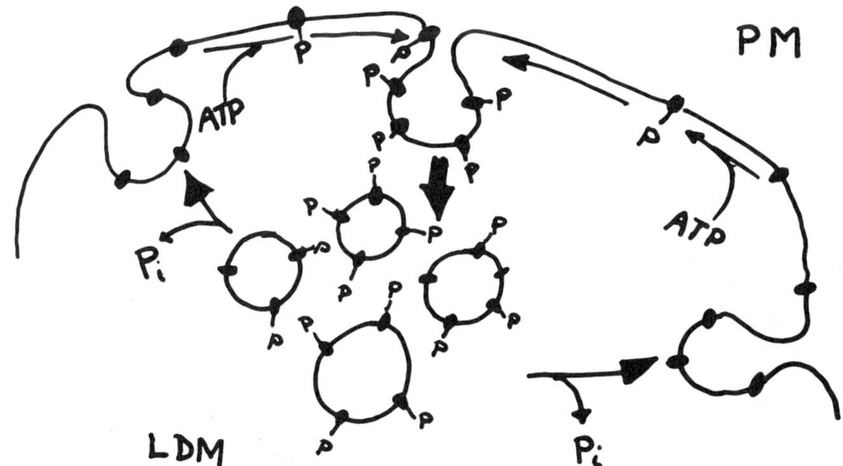

FIGURE 4. Schematic representation of a hypothetical role of phosphorylation in the recycling of membrane proteins. In this model, receptors are phosphorylated in the plasma membrane (PM) as a requisite for receptor internalization. Phosphorylated receptors are able to link into the endocytotic process and dephosphorylated receptors are not. Dephosphorylation in the intracellular compartment (LDM) allows the receptors to return to the PM in the non-phosphorylated state. Kinetics of phosphorylation and dephosphorylation determine the steady-state rate of recycling.

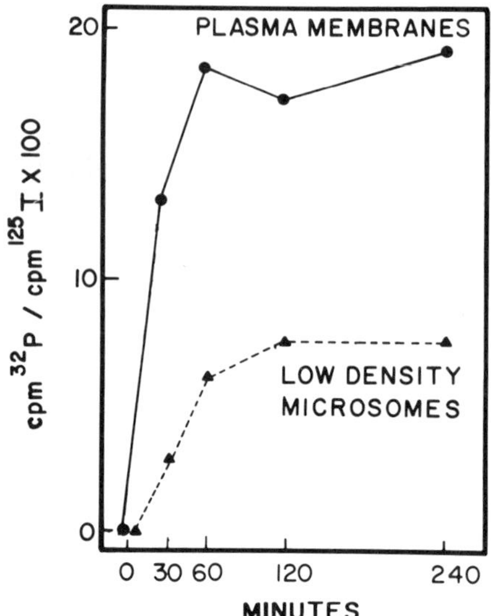

FIGURE 5. Time course for the incorporation of ^{32}P phosphate into the IGF-II receptor in adipocyte membrane fractions. Cells were incubated as described in FIGURE 2, in the presence of 1 mCi/ml [^{32}P]phosphate. At the indicated times, cells were homogenized and the plasma membranes (●——●) and low density microsomes (▲——▲) were prepared. The IGF-II receptor was then immunoprecipitated, electrophoresed, and immunoblotted. The bands were visualized by autoradiography of the immunoblots, cut out, and the amount of ^{125}I and ^{32}P associated with the receptor were measured by gamma and Cerenkov counting, respectively. Plotted is the $^{32}P/^{125}I$ ratio, which provides an estimate of the amount of phosphate covalently associated with the receptor. (Data taken from Corvera & Czech.[19]).

does not appear to cause any change in the overall stoichiometry of phosphorylation of the receptors that are obtained from these membranes.

These data indicate that the effect of insulin to cause a redistribution of the IGF-II receptor is indeed accompanied by a change in the phosphorylation state of this target molecule. This change consists of a dephosphorylation, which appears to occur specificially on the receptors that reside in the plasma membrane.

How could a dephosphorylation event lead to an increase in the number of cell surface receptors? In the previous model (FIG. 4), the possibility was raised that phosphorylation of membrane components in the plasma membrane might be one aspect of the driving force for receptor endocytosis. Therefore, the inhibition of this phosphorylation would be expected to increase the half life of the receptor on the cell surface. Because the receptor is continually recycling from the low density microsomes into the plasma membrane,[20] a delay in the rate of internalization would rapidly lead to an accumulation of receptors on the cell surface. An attractive feature of this model is that it would allow the receptors or other proteins to become redistributed selectively without any changes in the bulk rate of membrane endocytosis or exocytosis.

We have proceeded further to characterize the phosphorylation of the IGF-II receptor and have investigated the effect of insulin to cause dephosphorylation of the receptor in other insulin-sensitive cell types. FIGURE 7 shows the results from experiments done in H-35 rat hepatoma cells. These cells were labeled for 24 hours with [^{32}P]phosphate, and then treated with insulin. The cells were then solubilized and the total cellular IGF-II receptor pool was immunoprecipitated. In these experiments it was observed that insulin caused a 15–20% decrease in the amount of ^{32}P in the IGF-II receptor. This is the magnitude that would be expected if, as was shown in the fat cell, the effect of insulin was specific for the receptors found in the plasma membrane, which make up only 10–20% of the total receptor pool in these cell.

Thin layer electrophoresis of a partial acid hydrolysate of the receptor reveals that the IGF-II receptor is phosphorylated on serine and threonine residues. Interestingly, insulin caused a striking decrease in the phosphothreonine residues in the IGF-II receptor together with a smaller decrease in serine phosphorylation.

These results do not allow us to determine whether the effect of insulin occurs by the inhibition of a kinase or an activation of a phosphatase. It is clear, however, that the enzymes that are involved in the phosphorylation/dephosphorylation of the IGF-II receptor may be important targets for insulin action.

We have therefore proceeded to try to identify and characterize such enzymes. One approach that has been employed has been to test if the IGF-II receptor, purified on an immunoaffinity resin is a substrate for any known protein kinase. These experiments have shown that under these conditions the IGF-II receptor is a substrate for casein kinase 2. It is in contrast a poor substrate for cAMP-dependent protein kinase, Ca^{2+}-phospholipid–dependent protein kinase C and phosphorylase kinase under the conditions of these experiments. FIGURE 8 shows the phosphorylation of the IGF-II receptor immunoaffinity-purified from H-35 hepatoma cells by this enzyme. Thin

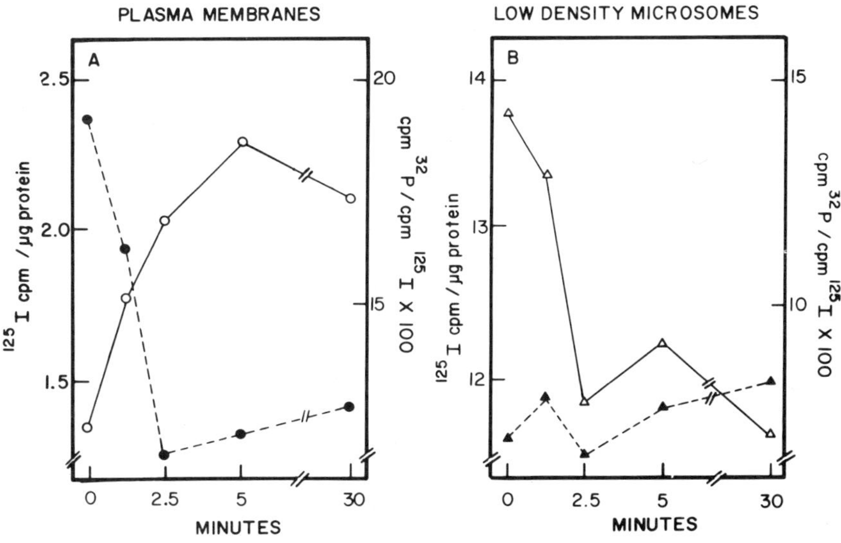

FIGURE 6. Effect of insulin on the distribution and phosphorylation state of the IGF-II receptor. Cells were incubated for two hours with [^{32}P]phosphate, after which 10^{-8} M insulin was added for the times indicated. The number of receptors, measured as the ^{125}I associated with the receptor band on the immunoblot, (empty symbols) and the ^{32}P/^{125}I (filled symbols) ratios were quantitated as described in FIGURE 5. The ^{125}I radioactivity is expressed as a function of the amount of protein initially used for immunoprecipitation. (Data taken from Corvera & Czech.[19])

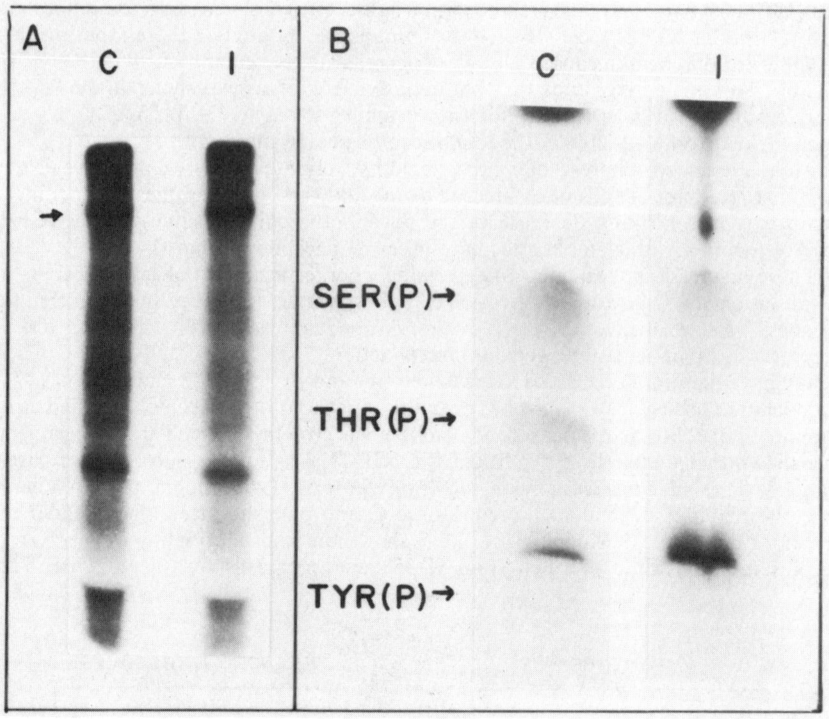

FIGURE 7. Effect of insulin on the phosphorylation of IGF-II receptors from H-35 cells. H-35 hepatoma cells were seeded in 100 mm culture plastic dishes in Dulbecco's modified Eagles medium containing 5% fetal calf serum. When cells were still subconfluent, the medium was replaced by DMEM buffered with HEPES. After 24 hours this medium was supplemented with 0.5 mCi/ml [^{32}P]phosphate. Cells were further incubated for 24 hours, after which insulin was added, where indicated, at a concentration of 10^{-8} M. After 10 min of incubation at 37°C, the medium was removed and cells were solubilized in ionic and non-ionic detergents.[19] The IGF-II receptor was immunoprecipitated and electrophoresed on a 6% polyacrylamide-SDS slab gel. The position of the receptor (panel A) is indicated by the arrow. Panel B shows an autoradiogram of a thin layer electrophoresis of a partial acid hydrolysate of the IGF-II receptor obtained from control (c) or insulin-treated (I) ^{32}P-labeled H-35 cells. The positions of the phosphoamino acids (phosphoserine (SERP)), phosphothreonine (THR (P)) and phosphotyrosine (TYR (P)) are indicated. The amount of the contaminant that migrated above the phosphotyrosine standard varied between different experiments.

layer electrophoresis of a partial acid hydrolysate of the casein kinase 2 phosphorylated receptor shows that the principal phosphoamino acid is phosphothreonine.

Phosphoamino acid analysis of the IGF-II receptor obtained from ^{32}P-labeled cells had shown that *in vivo* the receptor is phosphorylated on threonine and serine residues, and that the former sites are particularly susceptible to dephosphorylation in response to insulin (FIG. 7). It was of interest to investigate the possibility that the insulin-sensitive phosphothreonine sites were also the sites phosphorylated by casein kinase 2. To address this question, experiments were performed in which the extent of phosphorylation by casein kinase 2 of IGF-II receptors obtained from control or insulin-treated H-35 cells was compared. It was observed that the extent of IGF-II receptor phosphorylation by casein kinase 2 was 15–20% higher in receptors obtained from insulin-treated H-35 cells compared to controls (FIG. 7 panel A). This increase compares in magnitude to the decrease in the amount of phosphate in IGF-II receptors obtained from insulin-treated cells labeled *in vivo* with [^{32}P]phosphate. Thus, these data suggest that *in vivo*, the IGF-II receptor may be a substrate for casein kinase 2 or a similar kinase. Furthermore, they raise the possibility that insulin may cause an inhibition of this enzyme or an activation of a phosphatase with specificity towards casein kinase 2 phosphorylation sites.

In conclusion, our data show that (1) in isolated adipocytes, insulin causes the redistribution of at least three membrane components, the glucose transporter, the

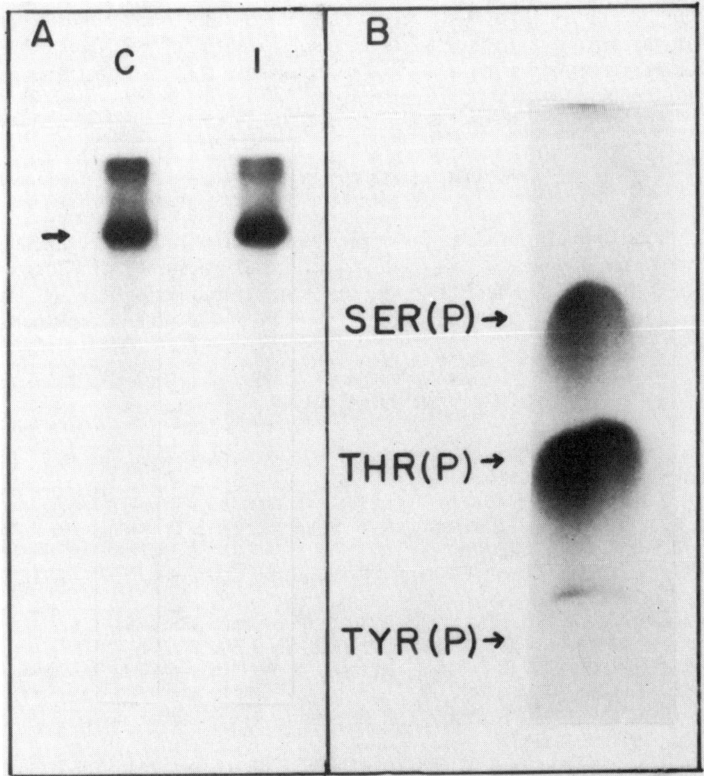

FIGURE 8. Phosphorylation of the IGF-II receptor by purified casein-kinase 2. IGF-II receptors were immunoprecipitated from control (c) or insulin-treated (I) H-35 hepatoma cells as described in FIGURE 7, but omitting the radiolabeled phosphate. The immunoprecipitated receptor (approx.20 fmol) was incubated in 50 μl of a buffer containing 50 mM Tris-HCl; 0.1% mercaptoethanol 6 mM $MgCl_2$, 0.120 M NaCl, 20 μM [^{32}P]ATP, and approximately 50 fmol of purified casein kinase 2 from rabbit liver. After 20 min at 22°C, the reactions were stopped and the receptor was electrophoresed on a 6% polyacrylamide SDS gel. Panel B shows the phosphoamino acid composition of the casein kinase 2 phosphorylated receptor (from lane I, panel A). Abbreviations are defined in FIGURE 7.

IGF-II receptor, and the transferrin receptor. The functional consequences of the insulin-mediated redistribution of two of these components are the increase in glucose and iron uptake. The physiological role of IGF-II receptor redistribution is presently unknown. (2) One of these components, the IGF-II receptor, is a phosphoprotein. The stoichiometry of phosphorylation of this receptor varies depending on the membrane compartment in which it resides, being higher in the plasma membrane and lower in the low density microsomes. (3) The effect of insulin to cause a redistribution of this receptor correlates with an acute decrease in the phosphorylation of threonine residues in the receptors and a smaller decrease in serine phosphorylation. These decreases appear to occur on receptors localized to the plasma membrane of the cell. These data have led to the hypothesis that this decreased phosphorylation may cause a delay in the rate of receptor internalization and lead to the steady-state increase in the number of cell surface receptors observed in response to insulin. (4) One of the enzymes that can phosphorylate the IGF-II receptor on threonine residues is casein kinase 2. This enzyme can phosphorylate IGF-II receptors from insulin-treated cells to a greater extent than those from control cells. Thus, insulin treatment may cause the dephosphorylation *in vivo* of one or more IGF-II receptor sites that are substrates for casein kinase 2 or an enzyme with similar specificity.

ACKNOWLEDGMENTS

The authors thank John Cruz for expert assistance with cell culture, and Mary Halley and Karen Donahue for excellent secretarial assistance. S. Corvera is the recipient of a Juvenile Diabetes Foundation Postdoctoral Fellowship. R. Davis is a recipient of a fellowship from the Damon Runyon-Walter Winchell Cancer Fund.

REFERENCES

1. Rosen, O. M. R. Herrera, Y. Obwe, L. M. Petruzzelli & M. H. Cobb. 1983. Proc. Natl. Acad. Sci. USA **80:** 3232–3240.
2. Rubin, J. B., M. A. Shia & P. F. Pilch. 1983. Nature **302:** 438–441.
3. Pike, L. J., B. Gallis, J. E. Casnellie, P. Bornstein & E. G. Krebs. 1982. Proc. Natl. Acad. Sci. USA **79:** 1443–1447.
4. Pike, L. J., D. F. Bowen-Pope, R. Ross & E. G. Krebs. 1983. J. Biol. Chem. **258:** 9383–9390.
5. Cushman, S. W. & L. J. Wardzala. 1980. J. Biol. Chem. **225:** 4758–4762.
6. Suzuki, Y. & T. Kono. 1980. Proc. Natl. Acad. Sci. USA **77:** 2542–2545.
7. Oppenheimer, C. L., J. E. Pessin, J. Massague, W. Gitomer & M. P. Czech. 1983. J. Biol. Chem. **258:** 4824–4830.
8. Wardzala, L. S., I. A. Simpson, M. M. Rechler & S. W. Cushman. 1984. J. Biol. Chem. **259:** 8378–8383.
9. Davis, R. J., S. Corvera & M. P. Czech. 1986. J. Biol. Chem. **261:** 8708–8711.
10. Parker, P. J., F. B. Cauldwell & P. Cohen 1983. Eur. J. Biochem. **130:** 227–234.
11. Lawrence, J. C., J. F. Hiken, A. A. DePaoli-Roach & P. J. Roach. 1983. J. Biol. Chem. **258:** 10710–10719.
12. Denton, R. M., W. A. Hughes, B. J. Bridges, R. W. Brownsey, S. G. McCormick & D. Stansbie. 1978. Horm. Cell. Regul. **2:** 191–208.
13. Persic, O. & J. A. Traugh. 1983. J. Biol. Chem. **258:** 9589–9592.
14. Pierce, M. W., J. L. Palmer, H. T. Keutmann, T. A. Hall & J. Avruch. 1982. J. Biol. Chem. **257:** 10681–10686.
15. Brownsey, R. W. & R. M. Denton. 1982. Biochem. J. **202:** 77–86.
16. Davis, R. J. & M. P. Czech. 1985. Proc. Natl. Acad. Sci. USA **82:** 1974–1978.

17. LIN, C. R., W. S. CHEN, C. S. LAZAR, C. D. CARPENTER, G. N. GILL, R. M. EVANS & M. G. ROSENFELD. 1986. Cell **44:** 839–848.
18. MAY W. S., S. JACOBS & P. CUATRECASAS. 1984. Proc. Natl. Acad. Sci. USA **81:** 2016–2020.
19. CORVERA, S. & M. P. CZECH. 1985. Proc. Natl. Acad. Sci. USA **82:** 7314–7318.
20. OKA, Y. & M. P. CZECH. 1986. J. Biol. Chem. **261:** 9090–9093.

DISCUSSION OF THE PAPER

E. CARAFOLI (*Swiss Federal Institute of Technology, Zurich*): Perhaps you could speculate on the reason(s) why phosphorylation convinces the receptor to leave the plasma membrane?

CORVERA: The mechanisms that govern the movement and sorting of proteins among cellular membranes are unknown. Internalization of plasma membrane components has been shown to occur through endocytosis, a process that is probably mediated by cytoskeletal membrane components. An allosteric modification such as phosphorylation might anchor surface receptors to the cytoskeletal structures involved in the formation of endocytotic vesicles. Another possibility would be that phosphorylation of a cluster of receptors might be a signal that could trigger the formation of an endocytic vesicle, through which the phosphorylated receptors would become internalized.

L. ROSENBERG (*Yale University, New Haven, CT*): Since many kinds of receptors are internalized by "coated pit" vesicles, doesn't your model require that dephosphorylated IGF receptors are being selected from many other like-internalized receptors at the cell surface?

CORVERA: I don't think the model calls for a selective entry of dephosphorylated receptors into the cellular machinery involved in endocytosis. Dephosphorylated receptors might be entering and leaving coated pits, or other endocytic structures randomly, but phosphorylation might anchor the receptors to such structures, where they could become concentrated and internalized.

L. JARETT (*University of Pennsylvania, Philadelphia, PA*): You stated that the regulation of key metabolic regulatory enzymes by insulin is through phosphorylation and dephosphorylation. This is not true for the low K_m cyclic AMP phosphodiesterase on the plasma membrane or ER of adipocytes. Also, your slide implies that insulin works on these enzymes through tyrosine kinase. There is no evidence to substantiate this that I am aware. Certainly the putative insulin mediators do not involve the insulin receptor tyrosine kinase for their generation.

CORVERA: The regulation of the activity of a significant number of regulatory enzymes such as glycogen synthetase, pyruvic dehydrogenase, ATP citrate lyase, etc. correlates with phosphorylation/dephosphorylation reactions, and I believe that the role of these processes on the regulation of the low K_m phosphodiesterase is controversial. This does not imply that all the cellular effects of insulin are caused through phosphorylation/dephosphorylation reactions. Whether or not the importance of the tyrosine kinase activity of the insulin receptor resides in its ability to directly phosphorylate exogenous cellular substrates, has as far as I know, not been determined yet.

Biological Membranes and Malignancy:

An Overview of Pharmacological Opportunities

RUSSELL G. GREIG[a] AND GEORGE POSTE

Department of Cell Biology
Smith Kline & French Laboratories
Philadelphia, Pennsylvania 19101

INTRODUCTION

The plasma membrane properties of malignant tumor cells have been the center of intensive research for many years. The objectives of these studies have been essentially threefold. First, to define what role, if any, alterations in surface properties play in the pathogenesis of malignant neoplasms; second, to define the specific molecular events underlying these alterations; and third, to use this information in the discovery of novel and effective antineoplastic drugs. A vast amount of information has been published on these topics.[1-8] For focus, and to reflect our own interests, this article will address only the last objective, namely, the current status and future prospects of developing anticancer agents that exploit molecular differences in membrane structure between normal and neoplastic cells.

THE PLASMA MEMBRANE AND MALIGNANCY

Although the plasma membrane is only one of many membranous structures within the cell (others include membranes of the nuclear envelope, mitochondria, endoplasmic reticulum, Golgi apparatus, endosomes, lysosomes, and peroxisomes), it has been a logical focus for cancer research because of the role of cell surface properties in growth, positional control, and the various factors that influence tumor-host interactions. In common with other aspects of cell biology, the analysis of plasma membrane organizations has been determined by technical advances. Four phases of activities are discernible: descriptive (1880–1950); the phenomenological (1950–1960); the biochemical (1960–1970); and the final phase dominated by the advances in molecular biology and the powerful technologies of molecular genetics. In 1889 Paget, based on autopsies of breast carcinoma patients, first described the organ-specific spread of malignant tumor cells and advanced what he termed the "seed and soil" hypothesis.[9] While Paget's original observations have been supported by numerous clinical investigations,[10] biochemical studies to elucidate the precise mechanisms of organ-specific spread were, until recently, technically impossible. Nevertheless, it was clear even from the early microscopic examination of invasive primary tumors and their metastases that neoplastic progression was accompanied by an apparent failure of tumor cells to respond to the various regulatory signals that determine the orderly behavior of normal cell populations. The cell surface and, in particular, the plasma membrane, thus

[a]Address correspondence to: Dr. Russell G. Greig, Department of Cell Biology, Smith Kline & French Laboratories (L-109), 709 Swedeland Road, Swedeland, PA 19479.

emerged as an obvious candidate for detailed biological characterization.[11] With the advent of tissue culture methodologies, experiments to analyze surface properties began. These pioneering studies concentrated on the cohesive properties of tumor cell populations and arrived at the general conclusion that neoplastic cells were less adhesive than their normal counterparts,[12] an observation that seemed to be consistent with the invasive and metastatic behavior of malignant tumors. Rapid improvement in the ability to propagate animal tumor cells in culture and to observe their behavior under controlled conditions *in vitro* initiated the phenomenological era during which many fundamental concepts of cancer biology were established. These included the description of behavioral properties like contact inhibition of locomotion, density-dependent inhibition of growth, and anchorage-independent growth.[13–19] The common theme to emerge from these studies was that tumor cells in culture (as well as *in vivo*) appeared to disobey the rules that governed orderly and territorial cell proliferation. Detailed morphological examination again suggested a central role for the plasma membrane in the expression of these properties.

During the 1960s and 1970s a major objective was to employ state-of-the-art biochemical techniques in the molecular dissection of tumor cell membranes. The goal was to understand the biochemical mechanisms underlying what appeared to be *in vitro* correlates of tumorigenicity, such as altered contact inhibition and anchorage-independent growth. During a ten-year period a massive amount of work was reported on the general topic of cell surface biochemistry. The range of plasma membrane properties examined reflected the full spectrum of evolving technologies and included: membrane transport of ions, sugars, and amino acids (availability of radiolabeled precursors, expanded tissue culture methodologies)[20,21]; characterization of membrane glycoproteins, glycolipids, and proteoglycans[4,22–25] (surface radiolabeling, non-ionic detergents, metabolic labeling, gel electrophoresis, polyclonal antibodies)[21]; surface enzymes such as glycosyltransferases[27] (microanalytical techniques); cell agglutination[28] (lectins); expression of "tumor-specific" antigens[29] (analytical and affinity purification, polyclonal antibodies); analysis of the extracellular matrix[5] (separation technologies); lipid bilayer structure and membrane fluidity[30–35] (spectroscopy, freeze fracture); the dynamic nature of membrane receptors and ligand-induced topographic redistribution of surface receptors (lectins, electron microscopy, photobleaching); membrane linkage[36,37] (electron microscopy, monoclonal antibodies); and signal transduction[38–40] (radioimmunoassays). For each phenotype, there exists at least one report claiming that tumor cells differ from their normal counterparts. The application of molecular genetic techniques to membrane biology began in the late 1970s and already has had a major impact. This is particularly true in growth regulation. Membrane growth factor receptors have been purified, sequenced, and cloned.[41,42]

PROBLEMS IN EXPERIMENTAL INTERPRETATION AND THE DEFINITION OF CORRELATES OF TUMORIGENICITY AND METASTASIS

While the above investigations generated enormous amounts of information, their interpretation within a clinical context has been difficult and frustrating for three main reasons. First, despite the long list of apparent differences between normal and tumor cell populations described from the 1960s onwards, it remains unclear whether these are a cause or consequence of neoplastic conversion. Tumor cells display both genotypic and phenotypic instability *in vivo* and *in vitro*[43,44] and thus any peculiarity in their biochemistry need not necessarily by related to the carcinogenic process. This issue has yet to be resolved, and determining causality between a specific molecular event (e.g. oncogene activation) and acquisition of tumorigenic and metastatic properties continues to be a controversial topic. Second, a number of studies conducted over the last 30 years were performed on poorly characterized cell reagents. Cell populations presumed to be normal displayed tumorigenic properties when injected into experimental animals. Conversely, cultures considered to be tumorigenic and/or malignant either failed to grow *in vivo* and/or did not display metastatic properties.[45–47] Indeed, a sizeable fraction of reported studies never undertook an analysis of the *in vivo* behavior of the cells being subjected to detailed analysis *in vitro*. The third reason is the lack of reproducible and routine methodologies to propagate human carcinoma cells *in vivo*. In the absence of such technology, investigators have been forced to employ mammalian (mainly rodent) or avian (chicken) cell lines as models for human neoplastic disease. Furthermore, the majority of these are of mesenchymal origin and studies using epithelial populations that have more fastidious cultivation conditions have been less popular. This carries with it two significant problems. The first is that experimental data have to be extrapolated across distant species barriers. This is troublesome because most animal systems are poor models for the pathogenesis and clinical course of human cancer. The second issue reflects the choice of cell type. The majority of human neoplasms are carcinomas in that they afflict cells of epithelial origin. However, most investigators selected rodent or avian fibroblasts as their *in vitro* models because these cells are readily propagated in culture. The confidence with which observations made of these systems can be applied to human epithelial membranes is clearly debatable. Better techniques for culturing human tumors are needed badly.

THE PLASMA MEMBRANE AS A PHARMACOLOGICAL TARGET: CURRENT STATUS

The problems described in the preceding section are not insignificant. They partly explain why much of the work on cell surfaces over the last 30 years, while contributing significantly to our basic understanding of membrane structure and function, has failed to provide pharmacological opportunities to disrupt tumor cell growth via drug design directed against well characterized "target" molecules or regulatory pathways. A realistic, if unfortunate, conclusion is that over a quarter of a century of research on normal and cancer cell membranes has had little impact on the techniques and therapies currently employed to diagnose and treat patients with malignant tumors. In diagnostics the once hoped for goal of identifying "tumor-specific" antigens remains unfulfilled.[29] In therapeutics, no biochemical lesion displayed preferentially or uniquely by human tumor cell membranes has been identified that could serve as a target for pharmacological intervention. Antineoplastic drugs that act primarily at the cell surface have yet to be discovered. This lack of progress is not limited to studies on the plasma membrane. A diverse panel of cellular and subcellular phenotypes have been studied to identify potentially exploitable "tumor-specific" changes, but have met with similarly unrewarding answers. Consequently, it is not surprising that our success in treating cancer with new, more effective drugs has been halting. The five-year survival for cancer patients in the USA increased by only 1% between 1973–1975 and 1976–1981.[48]

At this juncture, it is instructive to look at the antineoplastic drugs most commonly employed to treat cancer patients and how they were discovered. In the United States there are approximately forty anticancer drugs approved for clinical use by the FDA. Of the top ten, which together account for 85% of the total market, none is reported to have a membrane locus of action (TABLE 1). Although their precise mechanism of action is uncertain, and in some cases controversial,[49] the top ten therapeutic agents can be classified into four categories according to their postulated pharmacological target, namely, DNA-reactive, hormonal agonists/antagonists, antimetabolites, and microtubule poisons (TABLE 1). Most of these agents were identified by their ability to kill dividing neoplastic cells *in vitro* and in animal tumor models. Our lack of understanding of biochemical differences between normal and neoplastic cells,

TABLE 1. FDA-Approved Antineoplastic Agents: U.S.A. (1983)

Product	% Total Market	Proposed Mechanism of Action	Introductory Year
Adriamycin (doxorubicin)	17	DNA-reactive	1974
Platinol (cisplatin)	17	DNA-reactive	1978
Nolvadex (tamoxifen)	12	Anti-estrogen antagonist	1978
Oncovin (vincristine)	8	Microtubule poison	1963
Cytoxan (cyclophosphamide)	7	DNA-reactive	1959
Blenoxane (bleomycin)	6	DNA-reactive	1973
Methotrexate	5	Antimetabolite	1955
Mutamycin (mitomycin)	5	DNA-reactive	1974
Megace (megestrol)	4	Progestin agonist	1972
Cytosar-U (cytosine arabinoside)	2	DNA polymerase inhibitor	1969

These are the top ten antineoplastic drugs used in the USA, rated according to market share (1983).[63]

whether at the plasma membrane or any other cellular locus, continues to frustrate more efficient attempts at designing better antineoplastic compounds. The top ten chemotherapeutic agents have made a significant impact in the treatment of certain, albeit less frequent, tumors such as certain leukemias, lymphomas, testicular carcinoma, and choriocarcinomas. These achievements are important, yet unfortunately these "responsive" tumors account for only a small percentage of human tumors. The utility of these drugs against the major disseminated solid tumors (lung, colon, and breast), is much less impressive [50] and there is wide agreement that new classes of antineoplastic agents are sorely needed.

FUTURE TRENDS

It is not unreasonable to speculate that over the coming years the plasma membrane will emerge as a major focus for the diagnosis and therapy of malignant tumors. We consider that progress in plasma membrane research will likely advance on five broad fronts. First, increased emphasis will be given to the important need to study human tumor cells and improved techniques will emerge to meet this goal. This will be achieved through an appreciation of the growth requirements (hormones, growth factors, nutrients, matrices)[51] and by the genetic manipulation of primary tumor cultures.[52] The importance of achieving this objective cannot be overstated. Second, the analytical power of molecular genetics will permit the structural characterization of the various membrane (glyco)proteins that are implicated in regulating cell proliferation.[41,42] The availability of sufficient quantities of growth factor receptor to perform X-ray crystallography for the purpose of understanding receptor structure and the design of novel drugs that can interact selectively with the relevant region of the receptor is but one exciting example of the dividends this approach may yield. Third, genetic manipulation will be employed to study membranes structure and function. The ability to insert, modify, and delete genes encoding for membrane molecules will provide valuable insight into the function of individual membrane (glyco)proteins and how they interact with other membrane components to confer particular cellular phenotypes. Fourth, similar strategies will be used to examine the molecular biology of signal transduction. The mechanism by which extracellular signals are conveyed across the plasma membrane to the cytoplasm and the nucleus remains relatively obscure, and the molecules responsible for mediating this information flow are only beginning to be identified.[53] Signal transduction is a crucial membrane property and advances in this area are likely to have substantial impact on our understanding of cell biology. Fifth, increasing attention will be afforded the role of lipid metabolism in the expression of membrane properties. The participation of arachidonic acid metabolites in regulating signal transduction, second messenger levels, and gene expression will likely emerge as a critical focus for understanding how membrane biochemistry can influence overall cell metabolism.

Prediction of exactly which efforts will lead to new opportunities in cancer diagnosis and treatment is difficult, but one or two general areas appear more attractive than others. In cancer diagnostics, the identification of human monoclonal antibodies that detect (*in vitro* and *in vivo*) cell surface antigens preferentially associated with tumor membranes will likely lead to significant improvements in both the early detection and treatment of malignant tumors. Although useful, murine polyclonal and monoclonal antibodies may fail to detect certain human tumor antigens and react against determinants that are expressed by human cell lineages without adequate discrimination of rare antigens associated with the neoplastic phenotype. If the problems frequently found with the production of human monoclonal antibodies (such as instability and low levels of immunoglobulin production)[54,55] can be solved, a new, more productive era in isolating tumor-"specific" antigens might emerge.

Advances in cancer therapy may also be expected from an increased understanding of the regulation of cell proliferation and differentiation by soluble mediators. With the identification of specific growth factors, (acting through autocrine, paracrine, or endocrine mechanisms) whose aberrant expression can be implicated in neoplastic proliferation[56,57] therapeutic modulation of these activitites will offer a new strategy for tumor treatment. This goal could be achieved in several ways. One approach is to design structural antagonists of endogenous mediators either by site-specific mutagenesis of the parent molecule or by synthesizing low molecular weight molecules that mimic the mediator. Another is to administer monoclonal antibodies that recognize and eliminate the unwanted growth regulator. An additional strategy is to interfere with the synthesis, processing, and secretion of the growth factor. Many growth regulators are synthesized only in response to specific stimuli, are produced as large polyprotein precursors that are proteolytically trimmed to yield active pharmacophore, and then are released through specialized secretory processes.[58–62] Each step in this complex process is a possible target for pharmacological intervention. The growth factor receptors themselves may also be targets for manipulation. In many instances it may be difficult to block or antagonize growth factor production and action because of the risk of iatrogenic damage to normal cell populations. Attempts could also be made to usurp the availability and/or the function of the corresponding membrane receptor. In this situation, the purpose is to design compounds that could inhibit growth factor receptor synthesis and/or alter receptor internalization and recycling. Agents that promote receptor desensitization would also be worthy of evaluation. The second messenger systems that convey mitogenic signals from the plasma membrane to the nucleus are still poorly understood. Nevertheless as our insight into these processes grows, we might expect that compounds able to disrupt the activity of tyrosine kinases, protein kinase C, ion channels, and other transduction mechanisms will prove useful tools in examining growth regulation and some may hold the prospect of clinical utility.

While many of these developments are likely to occur, their translation into clincially useful drugs is still a long way off. Indeed, it is unlikely that any of these avenues will produce compounds ready for clinical testing before the mid-to-late 1990s. At the same time, however, current research continues to emphasize the crucial role of the tumor cell plasma membrane as a focus for the development of new cancer diagnostic and therapeutic agents.

REFERENCES

1. Bretscher, M. S. & M. C. Raff. 1975. Mammalian plasma membranes. Nature **258:** 43–49.
2. Harrison, R. & G. G. Lunt. 1980. Biological Membranes. Blackie. Glasgow.
3. Singer, S. J. & G. L. Nicholson. 1972. The fluid mosaic model of the structure of cell membranes. Science **175:** 720–731.
4. Hynes, R. O., Ed. 1979. Surfaces of Normal and Malignant Cells. J. Wiley & Sons. New York.
5. Hynes, R. O. 1976. Cell surface proteins and malignant transformation. Biochim. Biophys. Acta **458:** 73–107.
6. Martz, E. & M. S. Steinberg. 1973. Contact inhibition of what? An analytical review. J. Cell Physiol. **81:** 25–38.
7. Cuatrecacas, P. 1974. Membrane receptors. Ann. Rev. Biochem. **43:** 169–214.
8. Weiss, L. 1969. The cell periphery. Int. Rev. Cytol. **26:** 63–105.
9. Paget, S. 1889. The distribution of secondary growths in cancer of the breast. Lancet **i:** 571–573.
10. Sugarbaker, E. V. 1981. Patterns of metastasis in human malignancies. Cancer Biol. Rev. **2:** 235–278.
11. Takahashi, M. 1915. An experimental study of metastasis. J. Pathol. Bacteriol. **20:** 1–13.
12. Coman, D. R. 1944. Decreased mutual adhesiveness: A property of cells from squamous cell carcinomas. Cancer Res **4:** 625–629.
13. Abercrombie, M. & J. E. M. Heaysman. 1953. Observations on the social behavior of cells in tissue culture. I. Speed of movement of chick heart fibroblasts in relation to their mutual contacts. Exp. Cell Res. **5:** 111–131.
14. Abercrombie, M. & J. E. M. Heaysman. 1954. Observations on the social behavior of cells in tissue culture. II. Monolayering of fibroblasts. Exp. Cell Res. **6:** 293–306.
15. Abercrombie, M., J. E. M. Heaysman & H. M. Karthauser. 1957. Observations on the social behavior of cells in tissue culture. III. Mutual influence of sarcoma cells and fibroblasts. Exp. Cell Res. **13:** 276–291.
16. Gey, G. O. 1955. Some aspects of the constitution and behavior of normal and malignant cells maintained in continuous culture. *In* Harvey Lecture Series. L. pp. 154–338. Academic Press. New York.
17. MacPherson, I. & L. Montagnier. 1964. Agar suspension culture for the selective assay of cells transformed by polyoma virus. Virology **23:** 231–294.
18. Freedman, V. H. & S. Shin. 1974. Cellular tumorigenicity in *nude* mice: correlation with cell growth in semi-solid medium. Cell **3:** 355–359.
19. Gail, M. H. & C. W. Boone. 1971. Density inhibition of motility in 3T3 fibroblasts and their SV40 transformants. Exp. Cell Res. **64:** 156–162.
20. Holley, R. W. 1975. Control of growth of malignant cells in cell culture. Nature **258:** 487–490.
21. Pardee, A. B. 1975. The cell surface and fibroblast proliferation: Some current research trends. Biochim. Biophys. Acta **417:** 153–172.
22. Brady, R. O. & P. H. Fishman. 1974. Biosynthesis of glycolipids in virus-transformed cells. Biochim. Biophys. Acta **355:** 121–148.
23. Critchley, D. R. & M. G. Vicker. 1977. Glycolipids as membrane receptors important in

growth regulation and cell-cell interactions. *In* Dynamic Aspects of Cell Surface Organization. G. Poste & G. L. Nicolson, Eds.: 307–370. North-Holland. Amsterdam.

24. GAHMBERG, C. G. 1977. Cell surface proteins: Changes during cell growth and malignant transformation. *In* Dynamic Aspects of Cell Surface Organization. G. Poste & G. L. Nicolson, Eds.: 371–421. North-Holland. Amsterdam.
25. HAKOMORI, S. I. 1973. Glycolipids of tumor cell membrane. Adv. Cancer Res. **18:** 265–315.
26. ROBLIN, R., I.-N. CHOU & P. H. BLACK. 1975. Proteolytic enzymes and viral transformation. Adv. Cancer Res. **22:** 203–260.
27. ROSEMAN, S. 1970. The synthesis of complex carbohydrates by multiglycosyltransferase systems and their potential function in intercellular adhesion. Chem. Phys. Lipids. **5:** 270–282.
28. NICOLSON, G. L. 1974. The interactions of lectins with animal cell surfaces. Int. Rev. Cytol. **39:** 89–190.
29. SULITZEANU, D. 1985. Human cancer-associated antigens: Present status and implications for immunodiagnosis. Adv. Cancer Res. **44:** 1–42.
30. EDIDIN, M. 1974. Rotational and translational diffusion in membranes. Annu. Rev. Biophys. Bioeng. **3:** 179–201.
31. CHAPMAN, D. 1975. Lipid dynamics in cell membranes. *In* Cell Membranes: Biochemistry, Cell Biology and Pathology. G. Weissman & R. Claiborne, Eds.: 13–22. Hospital Practice. New York.
32. QUINN, A. J. & D. CHAPMAN. 1980. The dynamics of membrane structure. CRC Crit. Rev. Biochem. **8:** 1–117.
33. KIMELBERG, H. K. 1977. The influence of membrane fluidity on the activity of membrane-bound enzymes. *In* Dynamic Aspects of Cell Surface Organization. Cell Surface Reviews. G. Poste & G. L. Nicolson, Eds. **3:** 95–126. Elsevier. Amsterdam.
34. BRANTON, D. 1966. Fracture faces of frozen membranes. Proc. Natl. Acad. Sci. USA **55:** 1048–1056.
35. MCNUTT, N. S. 1977. Freeze-fracture techniques and applications to the structural analysis of the mammalian plasma membrane. *In* Dynamic Aspects of Cell Surface Organization. Cell Surface Reviews. G. Poste & G. L. Nicolson, Eds. **3:** 95–126. Elsevier. Amsterdam.
36. NICOLSON, G. L. 1976. Trans-membrane control of the receptors on normal and tumor cells. II. Surface changes associated with transformation and malignancy. Biochim. Biophys. Acta **458:** 1–71.
37. POSTE, G., D. PAPAHADJOPOULOS & G. L. NICOLSON. 1975. Local anesthetics affect transmembrane cytoskeletal control of mobility and distribution of cell surface receptors. Proc. Natl. Acad. Sci. USA **72:** 4430–4434.
38. JOHNSON, G. S. & I. PASTAN. 1972. Cyclic AMP increases adhesion of fibroblasts to substratum. Nature **236:** 247–249.
39. THOMOPOULOS, P., J. ROTH, E. LOVELACE & I. PASTAN. 1976. Insulin receptors in normal and transformed fibroblasts: relationship to growth and transformation. Cell **8:** 417–423.
40. ABELL, C. W. & T. M. MONAHAN. 1973. The role of adenosine 3′,5′-cyclic monophosphate in the regulation of mammalian cell division. J. Cell Biol. **59:** 549–558.
41. ULLRICH, A., J. R. BELL, E. Y. CHEN, R. HERRERA, L. M. PETRUZZELLI, T. J. DULL, A. GRAY, L. COUSSENS, Y.-C. LIAO, M. TSUBOKAWA, A. MASON, P. H. SEEBURG, C. GRUNFELD, O. M. ROSEN & J. RAMACHANDRAN. 1985. Human insulin receptor and its relationship to the tyrosine kinase family of oncogenes. Nature **313:** 756–761.
42. ULLRICH, A., L. COUSSENS, J. S. HAYFLICK, T. DULL, A. GRAY, A. W. TAM, J. LEE, Y. YARDEN, T. A. LIBERMAN, J. SCHLESSINGER, J. DOWNWARD, E. L. V. MAYES, N. WHITTLE, M. D. WATERFIELD & P. H. SEEBURG. 1984. Human epidermal growth factor receptor cDNA sequence and aberrant expression of the amplified gene in A431 epidermoid carcinoma cells. Nature **309:** 418–425.
43. CIFONE, M. & I. J. FIDLER. 1981. Increasing metastatic potential is associated with increasing genetic instability of clones isolated from murine neoplasms. Proc. Natl. Acad. Sci. USA **78:** 6949–6952.
44. HOWELL, P. 1976. The clonal evolution of tumor cell populations. Acquired genetic lability permits stepwise selection of variant sublines and underlies tumor progression. Science **194:** 23–28.

45. JARRETT, O., & I. MACPHERSON. 1968. The basis of the tumorigenicity of BHK21 cells. Int. J. Cancer **3:** 654–662.
46. DEFENDI, V., J. LEHMAN & P. KRAEMER. 1963. "Morphologically" normal hamster cells with malignant properties. Virology **19:** 592–598.
47. POSTE, G. 1977. The cell surface and metastasis. *In* Cancer Invasion and Metastasis: Biological Metastasis and Therapy. S. B. Daly *et al.,* Eds.: 19–47. Raven Press. New York.
48. HORM, J. M., A. J. ASIRE, J. L. YOUNG, *et. al.,* Eds. 1984. SEER Program: Cancer incidence and mortality in the United States 1973–81. Dept. Health and Human Services. Bethesda, MD.
49. DORR, R. T. & W. L. FRITZ. 1980. Cancer Chemotherapy Handbook. Elsevier. New York.
50. BAILAR, J. C. & E. M. SMITH. 1986. Progress against cancer? New Engl. J. Med. **314:** 1226–1232.
51. HAMMOND, S. L., R. G. HAM & M. R. STAMPFER. 1984. Serum-free growth of human mammary epithelial cells: Rapid clonal growth in defined medium and extended serial passage with pituitary extract. Proc. Natl. Acad. Sci. USA **81:** 5435–39.
52. MOYER, M. P., J. B. AUST. 1984. Human colon cells: culture and *in vitro* transformation. Science **224:** 1445–1447.
53. HORWITZ, A., K. DUGGAN, C. BUCK, M. C. BECKERLE & K. BURRIDGE. 1986. Interaction of plasma membrane fibronectin receptor with talin—a transmembrane linkage. Nature **320:** 531–533.
54. GAFFAR, S. A., I. ROYSTON & M. C. GLASSY. 1986. Strategies for the design and use of tumor-reactive human monoclonal antibodies. BioEssays **4:** 119–123.
55. COLE, S. P. C., B. G. CAMPLING, T. ATLAW, D. KOZBOR & J. C. RODER. 1984. Human monoclonal antibodies. Mol. Cellular Biochem. **62:** 109–120.
56. SPORN, M. B. & G. J. TODARO. 1980. Autocrine secretion and malignant transformation of cells. New Engl. J. Med. **303:** 878–880.
57. GOUSTIN, A. J., E. B. NEOF, G. D. SHIPLEY & H. L. MOSES. 1986. Growth factors and cancers. Cancer Res. **46:** 1015–1029.
58. LAWRENCE, D. A., R. PIRCHER & P. JULLIEN. 1985. Conversion of high molecular weight latent β-TGF from chicken embryo fibroblasts into a lower molecular weight active β-TGF under acidic conditions. Biochem. Biophys. Res. Commun. **133:** 1026–1034.
59. DERYNCK, R., J. A. JARRETT, E. Y. CHEN & D. V. GOEDDEL. 1986. The murine transforming growth factor precursor. J. Biol. Chem. **261:** 4377–4379.
60. SCOTT, J., M. URDEA, M. QUIROGA, R. SANCHEZ-PRESCADOR, N. FONG, M. SELBY, W. J. RUTTER & G. I. BELL. 1983. Structure of a mouse submaxillary messenger RNA encoding epidermal growth factor and seven proteins. Science **221:** 236–240.
61. FREY, P., R. FORAND, T. MACIAG & E. M. SHOOTER. 1979. The biosynthetic precursor of epidermal growth factor and the mechanism of its processing. Proc. Natl. Acad. Sci. USA **76:** 6294–6298.
62. GIRI, J. G., P. T. LOMEDICO & S. B. MIZEL. 1985. Studies of the synthesis and secretion of interleukin 1. I. A 33,000 molecular weight precursor for interleukin 1. J. Immunol. **134:** 343–349.
63. Smith Kline and French Laboratories Marketing Report. 1983.

The Role of Mitochondrial Hexokinase Binding in the Abnormal Energy Metabolism of Tumor Cell Lines[a]

RICHARD A. NAKASHIMA, LAURA J. SCOTT, AND PETER L. PEDERSEN[b]

Department of Biological Chemistry
School of Medicine
The Johns Hopkins University
Baltimore, Maryland 21205

It is commonly observed that cancer cells possess an abnormal pattern of energy metabolism when compared to normal cells from the same tissue of origin.[1–5,8] When examined under artificial conditions of cell culture, tumor cells have been found to possess sufficient metabolic flexibility to satisfy their energy requirements by either oxidative or anaerobic metabolism.[5–7] However, when placed under *in vivo* conditions in tumor-bearing animals many cancer cell lines show a marked preferential utilization of glycolytic metabolism to supply their energy needs.[3–5,8] Rapidly growing, highly malignant tumor cells can obtain up to sixty percent of their total ATP production from glycolysis[3,5,9,10] in contrast to normal cells, which obtain the bulk of their ATP production (80% or more) from mitochondrial oxidative phosphorylation. The increased rates of glycolytic activity seen in tumor cells are a specific effect of the transformation process and do not simply result from increased cell growth rates.[10–12]

Elevated rates of glucose utilization and lactic acid production have been shown to be consistent phenotypic markers in human and other mammalian tumor cells under both *in vivo* and *in vitro* conditions.[1–5,8–12] Significantly, this property has been the basis of recent advances in tumor imaging by the technique of positron emission tomography.[13,14] It would thus appear that elevated rates of glucose utilization and anaerobic energy production must confer some selective advantages for tumor cell growth and survival in comparison with surrounding normal cells. There are several possible advantages that could derive from elevated rates of glucose utilization. (1) It is known that elevated glycolysis in tumor cells results in an increase in the intracellular concentration of glucose-6-phosphate,[12] a key precursor in the *de novo* synthesis of nucleic acids, phospholipids, and other macromolecules that are required for cell growth. (2) Elevated glycolysis also provides an alternate source of ATP production that would supplement that available from mitochondrial respiration. This would be of particular importance in low oxygen environments such as in poorly vascularized areas of solid tumors. (3) The increase in lactate production from anaerobic glucose catabolism produces a lowered extracellular pH in the vicinity of the tumor. It has been suggested that low pH and anaerobic energy production would tend to interfere with normal immunologic reactions to transformed cells,[10,15] thus providing a partial escape from immune surveillance for highly glycolytic tumors.

A number of changes in cell metabolism are correlated with increased glycolysis in

[a]Supported by National Institutes of Health Grant CA 32742.
[b]To whom correspondence should be addressed.

tumors. Among the more significant ones, there is an increase in plasma membrane glucose transporters,[16] increased levels of several key glycolytic enzymes (including hexokinase, phosphofructokinase, and pyruvate kinase),[4,10,11,17,18,24] a shift in isozyme composition towards less regulated fetal type enzymes,[20] and a change in subcellular localization of hexokinase. This initial enzyme of glycolysis occurs in a predominantly mitochondrially bound form in highly glycolytic tumors.[4,9,10,21–24] Mitochondrially bound hexokinase has been reported to be less sensitive to feedback inhibition by glucose-6-phosphate,[9,11,23,25] a potent inhibitory regulator of hexokinase activity in normal cells. In addition, the bound enzyme may obtain preferential access to mitochondrially generated ATP.[9,26,27] A direct correlation has been observed between glycolytic activity and mitochondrially bound hexokinase in a variety of tumor cell lines.[4,8]

Given the apparent importance of bound hexokinase in tumor cell glycolysis and its potential suitability as a target for chemotherapeutic intervention in cancer cell metabolism, our laboratory has focused on the enzyme hexokinase [EC 2.7.1.1] and its interaction with the outer mitochondrial membrane. We report here some of our most

TABLE 1. Specific Binding of Glycolytic Enzymes to Hepatoma Mitochondria

Enzyme	Specific Activity Mitochondrial Fraction / Specific Activity Cytosolic Fraction
Hexokinase	3.5
Phosphoglucose isomerase	0.03
Phosphofructokinase	0.006
Aldolase	0.2
Enolase	0.03
Pyruvate kinase	0.03
Lactate dehydrogenase	0.08

Mitochondrial and cytosolic fractions were purified from AS-30D rat hepatoma cells by homogenization followed by differential centrifugation as described previously.[4] Enzyme activities were measured by standard spectrophotometric assays.[48,49] Protein concentrations were determined by the Lowry assay using BSA as standard.[50] The relative concentrations of glycolytic enzymes in the mitochondrial fraction were determined by dividing the specific enzyme activities present in the mitochondrial fraction by the specific activities present in the cytosolic fraction.

recent studies.[5,34,35] The interested reader may also wish to refer to some of our earlier work on the subject.[4,10,21,23,41]

SPECIFICITY OF BINDING OF TUMOR HEXOKINASE AND OTHER GLYCOLYTIC ENZYMES TO THE OUTER MITOCHONDRIAL MEMBANE

Of all the glycolytic enzymes examined, only hexokinase appears to be specifically and preferentially associated with an outer mitochondrial membrane receptor site.[21] Cells of the highly glycolytic AS-30D rat hepatoma line were fractionated by sonication and homogenization followed by differential centrifugation according to Bustamante *et al.*[4] The mitochondrial and cytosolic compartments were purified and the specific activities for a number of glycolytic enzymes were determined in each fraction. As can be seen from the data in TABLE 1, the specific activity of hexokinase in the mitochondrial fraction is fourfold higher than in the cytosolic fraction. In sharp contrast, all other glycolytic enzymes examined show at least a fivefold higher specific activity in the cytosolic than in the mitochondrial fraction. We should emphasize that although significant amounts of several glycolytic enzymes can be found in the purified mitochondrial fraction, only hexokinase is present in higher specific activity in the mitochondrial fraction than in the cytosol. In confirmation of previous reports,[22,28] we find that mitochondrial binding of hexokinase is also specific for the type of hexokinase used. In the presence of 10 mM magnesium ion almost 40% of added hepatoma hexokinase is recovered in the pelleted mitochondrial fraction (110 mU out of 300 mU added) while less than 1% of added yeast hexokinase is associated with the mitochondrial pellet (2 mU out of 300 mU added). Similarly, low levels of hexokinase binding have been observed with rat liver hexokinase.[22,28] Thus, it appears that the outer membrane binding site is not only highly selective for hexokinase compared to other glycolytic enzymes but can also distinguish between hexokinases from different sources.

IDENTIFICATION OF THE OUTER MITOCHONDRIAL MEMBRANE RECEPTOR FOR TUMOR HEXOKINASE

An outer mitochondrial membrane receptor protein for hexokinase binding was partially purified from rat liver mitochondria by Felgner *et al.*[29] This integral membrane protein of approximately 31,000 M_r has been reported by two laboratories[30,31] to be identical to the outer membrane pore-forming protein of rat liver mitochondria (also known as VDAC or mitochondrial "porin"), a channel-forming protein of apparently ubiquitous distribution in mitochondria. For several reasons, we considered it desirable to isolate and characterize the hexokinase receptor from tumor mitochondria. (1) Although the outer membrane pore protein would appear to be widely distributed in mitochondria,[32] mitochondrial binding of hexokinase is both tissue and species specific.[22,33] (2) The evidence for identity of mitochondrial "porin" and the hexokinase receptor protein is based largely upon peptide mapping data for the two proteins,[30] along with the demonstration of pore-forming activity in the partially purified hexokinase binding protein fraction.[30,31] Reports of hexokinase binding to purified and reconstituted "porin"[31] show several orders of magnitude lower specific binding activity than we find with the partially purified hexokinase receptor of hepatoma mitochondria. Fiek *et al.*[31] report hexokinase binding to reconstituted rat liver mitochondrial "porin" of 121 mU/mg of protein in the presence of 10 mM magnesium. Under similar conditions we have observed binding of as much as 13,000 mU of hexokinase/mg protein to the partially purified hexokinase receptor of AS-30D hepatoma mitochondria prepared by the repetitive octylglucoside solubilization technique of Felgner *et al.*[29] Freshly isolated AS-30D mitochondria contain between 1,000 and 2,000 mU of bound hexokinase per mg of mitochondrial protein. Therefore, the reported values[31] for specific hexokinase binding to reconstituted mitochondrial "porin" (consisting of a single band on SDS-polyacrylamide gel electrophoresis) were an order of magnitude lower than we observe with whole tumor mitochondria and two orders of magnitude lower than we find with the partially purified hexokinase receptor of tumor mitochondria. (3) The saturable level of hexokinase receptor sites appears to be approximately an order of magnitude higher in hepatoma mitochondria than in normal rat liver mitochondria.[22,34] It was therefore not clear whether the same receptor was present in normal and tumor mitochondria or if the hexokinase receptor (pore-forming protein) differed in some way in transformed cells.

We have purified the outer membrane pore-forming protein from mitochondria of the highly glycolytic AS-30D rat hepatoma cell line. This represents the first isolation of this protein from a tumor cell line. AS-30D mitochondria were extracted with the nonionic detergent Genapol X-80 (from Bio-Rad) and the extract was sequentially chromatographed on DEAE- and CM-Sepharose as previously described.[35] The final eluate contained an apparently homogeneous protein of 34,500 daltons that showed voltage-dependent, channel-forming activity by a standard lipid bilayer conductance assay.[32] FIGURE 1 shows a typical trace of channel insertions into an asolectin bilayer in

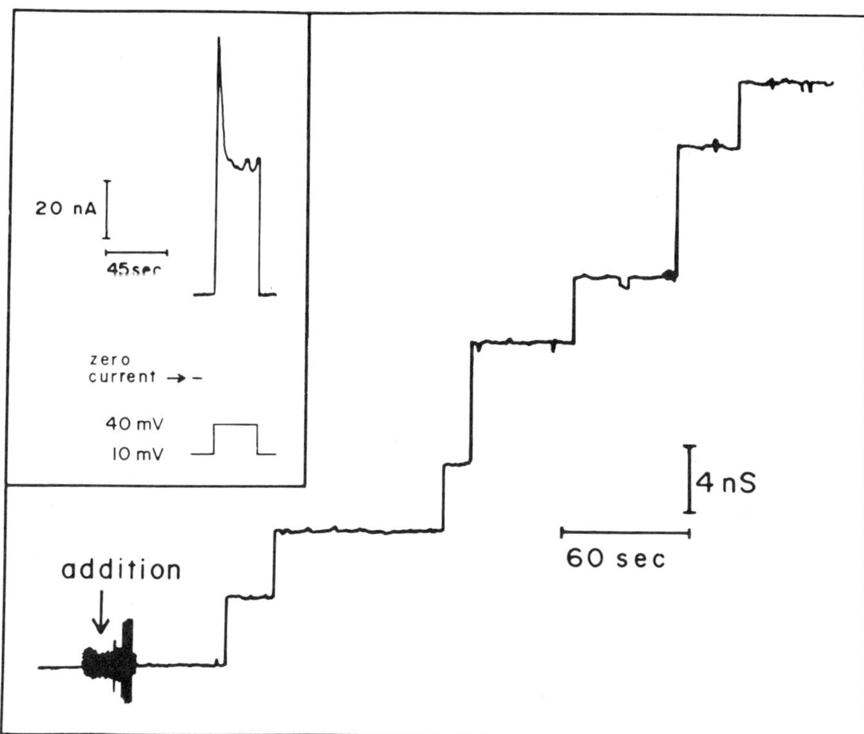

FIGURE 1. Channel forming activity of the purified AS-30D mitochondrial pore protein. The outer membrane pore-forming protein was purified from mitochondria of the highly glycolytic AS-30D rat hepatoma cell line as previously described.[35] At the point indicated by the arrow, 10 μl of a diluted solution containing detergent-solubilized AS-30D protein were added to the aqueous phase bathing a planar phospholipid membrane (made according to Colombini[36]). Both sides of the membrane were in contact with 1.0 M KCl and 5 mM $CaCl_2$ solutions. After a short time, the conductance of the membrane (in nanosiemens) increased in a stepwise fashion characteristic of channel insertions. The inset shows the voltage dependence of these channels for a membrane containing about 200 channels. The assay was performed by Dr. Patrick Mangan.[35] (From Nakashima *et al.*[35] With permission from the American Chemical Society.)

a voltage-clamping experiment. These channels are voltage gated and exhibited partial closure at a membrane potential of 40 mV (see inset to FIG. 1). Return of the transmembrane potential to zero results in a reopening of the channels.

We have compared the functional and structural characteristics of purified pore-forming proteins from AS-30D hepatoma and normal rat liver mitochondria.[35]

Assays of channel-forming activity were conducted in the laboratory of Dr. Marco Colombini at College Park, Maryland by Dr. Patrick Mangan. In ion selectivity and electrical properties the two proteins are almost identical.[35,36] Both proteins show a preference for anions over cations, with a twofold selectivity for chloride ion over potassium ion.[35,36] The conductance of individual channels is approximately 4 nano-

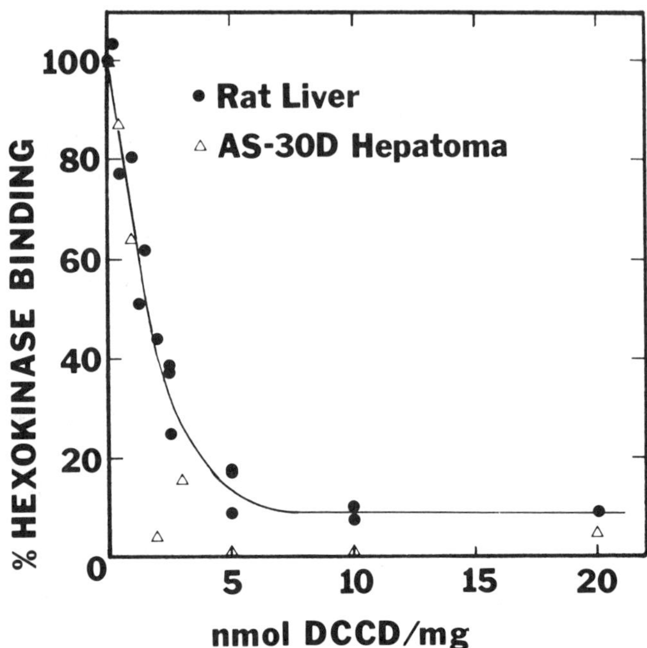

FIGURE 2. Effect of DCCD pretreatment on mitochondrial binding of hexokinase. Whole mitochondria suspended in H medium (210 mM D-mannitol, 70 mM sucrose, 5 mM HEPES, pH 7.4) were treated for 15 min at room temperature with the amounts of DCCD indicated. After being washed with BSA-containing medium, the mitochondria were resuspended in fresh H medium and their ability to bind solubilized AS-30D hexokinase was assayed as previously described.[35] Control mitochondria treated with solvent (methanol) alone were used to determine the 100% level of hexokinase binding. The data are plotted as relative percent hexokinase binding versus the dosage of DCCD and are based upon three experiments with rat liver mitochondria (●) and one experiment with AS-30D mitochondria (△). Under the conditions of these experiments, the 100% level of binding corresponded to 110, 170, and 112 milliunits of hexokinase/mg mitochondrial protein for rat liver and 232 milliunits/mg for AS-30D mitochondria. (From Nakashima *et al.*[35] With permission from American Chemical Society.)

siemens for both the hepatoma and rat liver proteins.[35,36] The voltage dependence of channel conductance and the kinetics of voltage-dependent channel opening and closure are also very similar for the hepatoma and liver proteins.[35,36] The specific channel-forming activity of the purified hepatoma protein (136 channels/min/μg of protein) is in the same range as reported values for the rat liver VDAC protein.[36] It

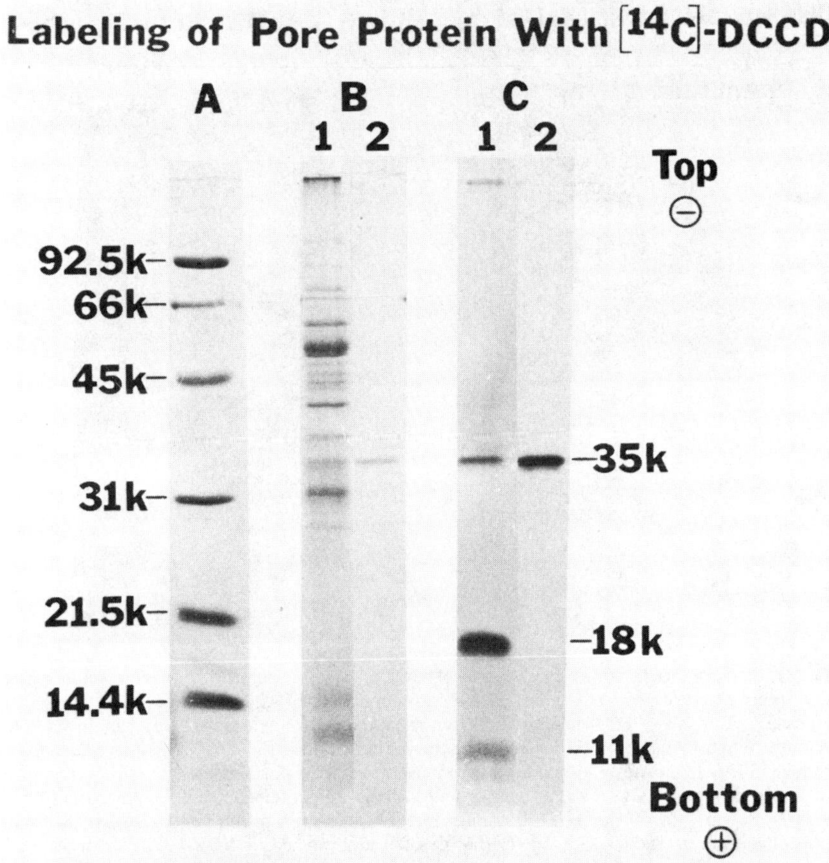

FIGURE 3. [^{14}C]DCCD labeling of AS-30D mitochondria. Freshly isolated AS-30D hepatoma mitochondria were incubated with [^{14}C]DCCD at 2 nmol/mg of protein for 15 min at room temperature. After washing with H medium containing 1 mg/ml of bovine serum albumin (BSA) the [^{14}C]DCCD-labeled mitochondria were sequentially extracted with 2% Triton X-100 and 2.5% Genapol X-80. The outer membrane pore-forming protein was purified from the Genapol X-80 extract as previously described.[35] Aliquots containing 30 μg of Triton X-100 extract (B and C, lanes 1) and 1.5 μg of purified pore protein (B and C, lanes 2) were run on a 14% polyacrylamide gel in the presence of sodium dodecyl sulfate (SDS). (A) Molecular weight markers (from Bio-Rad). (B) The gel containing the Triton X-100 extract (lane 1) and purified AS-30D pore protein (lane 2) was stained with Coomassie brilliant blue R-250 (from Bio-Rad) to visualize the peptide composition. (C) The Coomassie-stained gel was impregnated with En^{3}Hance (New England Nuclear) and dried. The [^{14}C]DCCD-labeled bands were localized by fluorography with Kodak XAR-5 X-ray film. Aside from material that did not enter the gel (lane 1, top) only three DCCD-labeled peptides were present in the Triton X-100 mitochondrial extract (C, lane 1) of apparent M_r 35000,18000, and 11,000. Purification of the outer membrane pore-forming protein from [^{14}C]DCCD-labeled mitochondria resulted in the substantial enrichment of the 35,000 dalton DCCD binding protein (C, lane 2). (After Nakashima *et al.*[35])

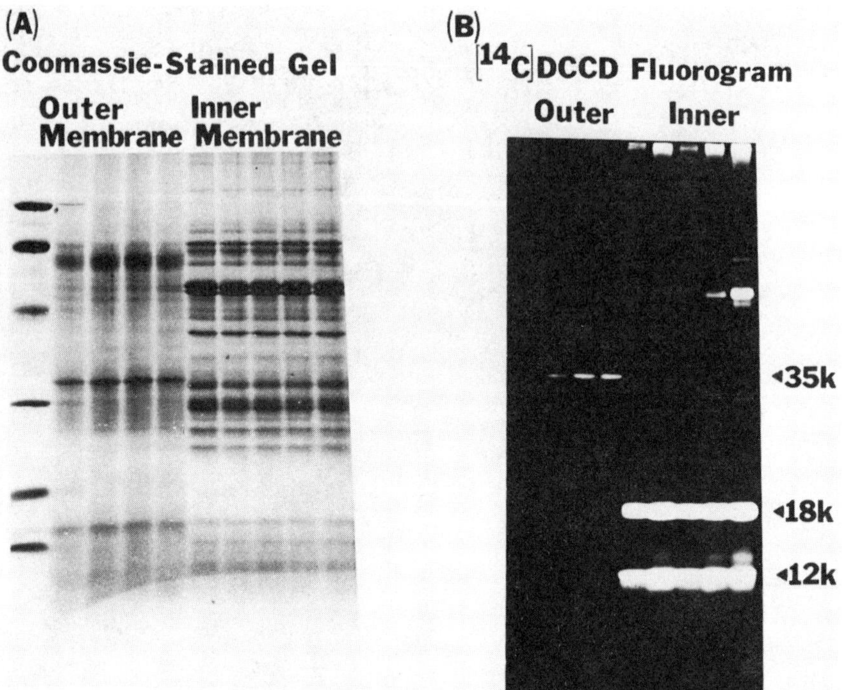

FIGURE 4. Submitochondrial localization of DCCD binding peptides. Inner and outer membrane fractions were prepared from rat liver mitochondria by the technique of Wojtczak and Sottocasa.[47] Analysis of marker enzyme activities (monoamine oxidase for outer membrane and succinate dehydrogenase for inner membrane) indicated that the final fractions were greatly enriched in inner and outer mitochondrial membranes. The purified membranes were treated with levels of [^{14}C]DCCD ranging from 2 to 20 nmol/mg protein for outer membrane (2.5 mg protein/ml) and 2 to 40 nmol/mg for inner membrane (5 mg protein/ml). DCCD-pretreated membranes were extracted with 2% Triton X-100 and analyzed by SDS-PAGE in a 14 to 20% acrylamide gradient gel. Concentrations of DCCD increase from left to right in the figure. (A) The peptide compositions of the detergent extracts were visualized by staining with Coomassie brilliant blue R-250 (Bio-Rad). Molecular weight markers obtained from Bio-Rad (lane 1) were of 92,500, 66,000, 45,000, 31,000, 21,500, and 14,400 daltons. (B) The stained gel was then impregnated with En^{3}Hance (New England Nuclear) and the DCCD-labeled peptides were identified by fluorography. As can be seen in (B) the 35,000 dalton DCCD-binding peptide was present in the outer membrane fraction, but not in the inner membrane. Conversely, the two lower molecular mass peptides (12,000 and 18,000 daltons in this gel system) were present in the inner membrane fraction but not in the outer membrane fraction. Whole rat liver mitochondria contained all three DCCD-labeled peptides (data not shown). At higher concentrations of DCCD, several additional peptides were observed to be labeled, including the beta subunit of the F_0F_1 ATPase. However, at dosages of DCCD that inhibit hexokinase binding to intact mitochondria (FIG. 2) only three DCCD-labeled peptides were present in either inner or outer membrane fractions, with apparent electrophoretic mobilities of 12,000, 18,000, and 35,000 M_r. Thus, the results obtained with purified inner and outer membrane fractions from rat liver mitochondria are in agreement with those previously described for whole AS-30D hepatoma mitochondria.[35] In addition, they establish that of the three mitochondrial peptides that are labeled at low concentrations of DCCD, only the 35,000 dalton peptide is present in the outer mitochondrial membrane. (From Nakashima & Pedersen.[34] With permission from Elsevier Science Publishers.).

appears that the channel-forming characteristics of the AS-30D pore protein are indistinguishable from those of the normal rat liver protein.

Analysis of the two proteins by SDS-polyacrylamide gel electrophoresis indicates that they possess equal mobilities on SDS-PAGE (data not shown). A comparison of the two proteins by analysis of amino acid compositions shows that they are very similar, but not identical.[35]

Although we have demonstrated hexokinase binding to partially purified hepatoma pore protein we have not been able to show binding of hexokinase to the purified pore protein reconstituted into lipid vesicles. We consider these results to be consistent with the hypothesis that the pore-forming protein by itself cannot effectively bind hexokinase. Experiments are currently underway to determine if other components of the outer mitochondrial membrane (either lipid or protein) are required for hexokinase binding activity.

INHIBITION OF MITOCHONDRIAL HEXOKINASE BINDING BY DCCD

An independent series of experiments indicates that the outer membrane pore protein of hepatoma mitochondria is in fact a part of the hexokinase receptor complex. In addition, we have presented the first evidence for a highly selective covalent inhibitor of mitochondrial hexokinase binding.[35] Treatment of either AS-30D hepatoma mitochondria or normal rat liver mitochondria with the carboxyl-specific reagent DCCD (N,N′-dicyclohexylcarbodiimide) resulted in almost complete inhibition of mitochondrial hexokinase binding activity (FIG. 2). Fifty percent of inhibition is reached at a DCCD concentration of 10 μM (2 nmol DCCD/mg of mitochondrial protein). At this concentration of [^{14}C]DCCD only three mitochondrial peptides were found to be covalently labeled (FIG. 3) of apparent M_r 12,000, 18,000, and 35,000 by SDS-polyacrylamide gel electrophoresis. One of these proteins (of 35,000 M_r) was found to be identical to the hepatoma outer mitochondrial membrane pore protein.[35] Previous results from this laboratory have shown conclusively that the hexokinase receptor site of tumor mitochondria is located on the outer mitochondrial membrane.[21] Submitochondrial localization of the three DCCD-binding proteins demonstrates that only the 35,000 dalton (pore-forming) protein is present in the outer mitochondrial membrane where the hexokinase receptor site is located (FIG. 4). We consider these results to be strong evidence that the 35,000 M_r pore protein forms at least part of the hexokinase receptor complex in tumor mitochondria and that DCCD inhibits hexokinase binding through covalent modification of the pore protein.[35] Further studies are underway to characterize the primary structures of hexokinase and the outer membrane pore protein involved in binding and catalytic or channel-forming activity.

COMPARISON OF TUMOR HEXOKINASE WITH NORMAL RAT HEXOKINASES

Although much is known about the kinetic properties of the hexokinases, very little has been determined about the structure of any eukaryotic hexokinase above the level of yeast. One of the goals of our investigation of tumor glucose utilization is to determine what differences in amino acid composition and structure exist between tumor hexokinase and normal hexokinases and to identify the role that differences in structural composition play in the binding activity and regulatory properties of these enzymes. As an initial step in this process we have identified the amino acid composition of AS-30D hexokinase (TABLE 2). For comparative purposes we have also included data on the reported amino acid compositions for the type I, type II, and type IV (glucokinase) isozymes purified from normal rat tissues.[37–39] Tumor hexokinase was solubilized from AS-30D mitochondria by addition of ATP.[21] The solubilized enzyme was purified by sequential chromatography on DEAE-cellulose and hydroxylapatite columns according to Grossbard and Schimke.[40] The final eluate contained a substantially purified enzyme with a specific activity of 128 U/mg protein. Analysis of this fraction by SDS-PAGE indicated that the enzyme was over 95% pure. This material

TABLE 2. Comparison of Amino Acid Compositions of Hepatoma Hexokinase And Normal Rat Hexokinases

Amino Acid	Tumor HK	Rat Brain HK (Type I)[a]	Rat Muscle HK (Type II)[b]	Rat Liver HK (Type IV)[c]
ASP	8.3	11.3	10.4	10.2
GLU	11.7	10.1	11.7	13.1
SER	6.8	5.5	5.9	6.9
GLY	13.6	9.2	9.7	8.9
HIS	2.0	2.1	2.6	2.4
ARG	7.0	6.2	6.3	6.2
THR	4.7	5.7	6.1	5.3
ALA	6.4	5.5	7.4	6.7
PRO	3.2	4.1	2.9	3.1
TYR	2.9	0.9	1.8	1.8
VAL	5.6	7.4	7.4	7.1
MET	2.4	3.4	3.4	4.0
CYS	3.1	2.1	—	2.0
ILE	3.8	5.7	4.1	3.8
LEU	9.0	9.0	10.4	9.1
PHE	4.1	4.1	4.1	4.0
LYS	5.6	6.9	5.9	5.5

AS-30D hepatoma hexokinase was solubilized from mitochondria and chromatographed on columns of DEAE-cellulose and hydroxylapatite. The final eluate contained a substantially purified hexokinase with a specific activity of 128 U/mg protein. The purified fraction was analyzed for amino acid composition as described in the text. For comparison the amino acid compositions of rat brain hexokinase (HKI), rat muscle hexokinase (HKII), and rat liver glucokinase (HKIV) are also included. Data are expressed as mol % amino acid residues.

[a]From Chou & Wilson.[37]

[b]From Holroyde & Trayer.[38]

[c]From Holroyde *et al.*[3]

was subjected to overnight hydrolysis at 110°C and derivatized with Edman's reagent (phenylisothiocyanate). The derivatized amino acids were analyzed by high pressure liquid chromatography on a Waters Cl8 (PICO-TAG) reverse-phase column. Individual peaks were quantified by comparison with a known injection of derivatized amino acid standards (from Pierce). As can be seen, the amino acid composition of the tumor hexokinase is most similar to that of the high K_m type IV hexokinase (glucokinase), which is the predominant isozyme present in normal rat liver. Since the amino acid compositions of all four hexokinases from normal rat liver are not known at present it is difficult to say at this point whether the tumor enzyme represents a distinct isozymic form not found in normal rat liver. However, based upon its chromatographic and kinetic properties[21,41] it would appear that tumor hexokinase is not identical to any of the known rat liver isozymes. In its elution profile from DEAE-cellulose the tumor enzyme appears to resemble either type II or type III hexokinase, while its K_m for glucose is most consistent with the type I isozyme.

DISCUSSION

Traditional approaches to the chemotherapeutic treatment of cancer have relied on the finding that rapidly dividing tumor cells are more susceptible to damage by agents that interfere with nucleic acid metabolism than are surrounding normal cells. There are two drawbacks to this approach. The first is that rapidly dividing normal cells are also highly sensitive to these same agents. Thus, delivery of effective dosages of anticancer drugs to the growing tumor is often limited by their systemic toxicity. Second, not all tumor cell lines are rapidly growing. There has been a notable lack of success in the treatment of some forms of slowly growing tumors such as the lung carcinomas. It would seem that a rational alternative approach to cancer chemotherapy would be to identify essential metabolic pathways that contain "regulatory" enzymes of different isozymic composition than normal cells and to target these tumor-specific enzymes for chemotherapeutic inhibition.[42] Results obtained so far would indicate that tumor hexokinase may provide a model system for such an approach.

We have presented evidence that the elevated levels of mitochondrially bound hexokinase are required for the high rates of glucose utilization observed in transformed cell lines.[4,23] Although tumor cells can survive *in vitro* in the absence of glucose, provided that their requirements for anabolic precursors are met,[6] it is clear that *in vivo* some degree of glucose catabolism is required for nucleic acid synthesis. In addition, inhibition of glycolysis at the level of hexokinase would result in a substantial decrease in tumor energy production in low oxygen environments.[5] Several antineoplastic agents have been reported that apparently act through selective inhibition of tumor energy production.[42–45] One of the most promising of these compounds is the antispermatogenic agent Lonidamine (1-(2,4-dichlorobenzyl)-1H-indazole-3-carboxylic acid), which is currently in phase II clinical trials in human cancer patients. Lonidamine has been shown to be an effective antineoplastic agent in several forms of tumors resistant to standard chemotherapeutic treatment.[46] Lonidamine appears to act through disruption of tumor energy production and has been reported to be a highly selective inhibitor of mitochondrially bound hexokinase activity in cancer cells.[43–45] Significantly, this compound has no apparent effect on energy production in normal tissues which contain high levels of particulate hexokinase, such as the brain. These results demonstrate the feasibility of chemotherapeutic intervention in cancer growth that is not directed towards nucleic acid synthesis. We suggest that this line of research may prove highly productive in the development of antineoplastic agents of low toxicity to normal cell lines.

REFERENCES

1. WARBURG, O., K. POSENER & E. NEGELEIN. 1924. Biochem. Z. **152:** 309–344.
2. CORI, C. F. & G. T. CORI. 1925. J. Biol. Chem. **65:** 397–405.

3. AISENBERG, A. C. 1961. The Glycolysis and Respiration of Tumors. Academic Press, Inc. New York.
4. BUSTAMANTE, E., H. P. MORRIS & P. L. PEDERSEN. 1981. J. Biol. Chem. **256:** 8699–8704.
5. NAKASHIMA, R. A., M. G. PAGGI & P. L. PEDERSEN. 1984. Cancer Res. **44:** 5702–5706.
6. MCKEEHAN, W. L. 1982. Cell Biol. Intl. Rep. **6:** 635–650.
7. REITZER, L. J., B. M. WICE & D. KENNELL. 1979. J. Biol. Chem. **254:** 2669–2676.
8. KNOX, W. E., C. JAMDAR & P. A. DAVIS. 1970. Cancer Res. **30:** 2240–2244.
9. GUMAA, K. A. & P. MCLEAN. 1969. Biochem. Biophys. Res. Commun. **36:** 771–779.
10. PEDERSEN, P. L. 1978. Progr. Exp. Tumor Res. **22:** 190–274.
11. SINGH, M., V. N. SINGH, J. T. AUGUST & B. L. HORECKER. 1974. Arch. Biochem. Biophys. **165:** 240–246.
12. SINGH, V. N., M. SINGH, J. T. AUGUST & B. L. HORECKER. 1974. Proc. Natl. Acad. Sci. USA **71:** 4129–4132.
13. KORNBLITH, P. L., C. J. CUMMINS, B. H. SMITH, R. A. BROOKS, N. J. PATRONAS & G. DI CHIRO. 1984. Progr. Exp. Tumor Res. **27:** 170–178.
14. DI CHIRO, G., R. A. BROOKS, N. J. PATRONAS, D. BAIRAMIAN, P. L. KORNBLITH, B. H. SMITH, L. MANSI & J. BARKER. 1984. Ann. Neurol. **15**(Suppl): S138–S146.
15. GRANGER, D. L., R. R. TAINTOR, J. L. COOK & J. B. HIBBS, JR. 1980. J. Clin. Invest. **65:** 357–370.
16. WEBER, M. J., P. K. EVANS, M. A. JOHNSON, T. F. MCNAIR, K. D. NAKAMURA & D. W. SALTER. 1984. Fed. Proc. **43:** 107–112.
17. TEJWANI, G. A., S. CHAUHAN, V. DURUIBE & K. K. VASWANI. 1985. Arch. Biochem. Biophys. **239:** 462–466.
18. ARANY, I., P. RADY, F. BOJAN & P. KERTAI. 1981. Environ. Res. **26:** 335–339.
19. WEBER, G. 1983. Cancer Res. **43:** 3466–3492.
20. IBSEN, K. H. & W. H. FISHMAN. 1979. Biochim. Biophys. Acta **560:** 243–280.
21. PARRY, D. M. & P. L. PEDERSEN. 1983. J. Biol. Chem. **258:** 10904–10912.
22. ROSE, I. A. & J. V. B. WARMS. 1967. J. Biol. Chem. **242:** 1635–1645.
23. BUSTAMANTE, E. & P. L. PEDERSEN. 1977. Proc. Natl. Acad. Sci. USA **74:** 3735–3739.
24. SAITO, M. & S. SATO. 1971. Biochim. Biophys. Acta **227:** 344–353.
25. KOSOW, D. P. & I. A. ROSE. 1968. J. Biol. Chem. **243:** 3623–3630.
26. GOTS, R. E. & S. P. BESSMAN. 1974. Arch. Biochem. Biophys. **163:** 7–14.
27. INUI, M. & S. ISHIBASHI. 1979. J. Biochem. **85:** 1151–1156.
28. KUROKAWA, M., J. KIMURA, S. TOKUOKA & S. ISHIBASHI. 1979. Brain Res. **175:** 169–173.
29. FELGNER, P. L., J. L. MESSER & J. E. WILSON. 1979. J. Biol. Chem. **254:** 4946–4949.
30. LINDEN, M., P. GELLERFORS & B. D. NELSON. 1982. FEBS Lett. **141:** 189–192.
31. FIEK, C., R. BENZ, N. ROOS & D. BRDICZKA. 1982. Biochim. Biophys. Acta **688:** 429–440.
32. COLOMBINI, M. 1979. Nature (Lond.) **279:** 643–645.
33. CRANE, R. K. & A. SOLS. 1953. J. Biol. Chem. **203:** 273–292.
34. NAKASHIMA, R. A. & P. L. PEDERSEN. 1985. *In* Cell Membranes and Cancer. T. Galeotti, A. Cittadini, G. Neri, S. Papa & L. A. Smets, Eds.: 183–192. Elsevier Science Publishers. New York.
35. NAKASHIMA, R. A., P. S. MANGAN, M. COLOMBINI & P. L. PEDERSEN. 1986. Biochemistry **25:** 1015–1021.
36. COLOMBINI, M. 1983. J. Membr. Biol. **74:** 115–121.
37. CHOU, A. C. & J. E. WILSON. 1972. Arch. Biochem. Biophys. **151:** 48–55.
38. HOLROYDE, M. J. & I. P. TRAYER. 1976. FEBS Lett. **62:** 215–219.
39. HOLROYDE, M. J., M. B. ALLEN, A. C. STORER, A. S. WARSY, J. M. E. CHESHER, I. P. TRAYER, A. CORNISH-BOWDEN & D. G. WALKER. 1976. Biochem. J. **153:** 363–373.
40. GROSSBARD, L. & R. T. SCHIMKE. 1966. J. Biol. Chem. **241:** 3546–3560.
41. BUSTAMANTE, E. & P. L. PEDERSEN. 1980. Biochemistry **19:** 4972–4977.
42. HILF, R., R. S. MURANT, U. NARAYANAN & S. L. GIBSON. 1986. Cancer Res. **46:** 211–217.
43. FLORIDI, A., M. G. PAGGI, M. L. MARCANTE, B. SILVESTRINI, A. CAPUTO & C. DE MARTINO. 1981. J. Natl. Cancer Inst. **66:** 497–499.

44. FLORIDI, A., M. G. PAGGI, S. D'ATRI, C. DE MARTINO, M. L. MARCANTE, B. SILVESTRINI & A. CAPUTO. 1981. Cancer Res. **41:** 4661–4666.
45. FLORIDI, A., S. D'ATRI, M. BELLOCCI, M. L. MARCANTE, M. G. PAGGI, B. SILVESTRINI, A. CAPUTO & C. DE MARTINO. 1984. Exp. Molec. Pathol. **40:** 246–261.
46. EVANS, W. K., F. A. SHEPHERD & B. MULLIS. 1984. Oncology (Suppl.) **41:** 69–77.
47. WOJTCZAK, L. & G. L. SOTTOCASA. 1972. J. Membr. Biol. **7:** 313–324.
48. BERGMEYER, H. U. 1974. *In* Methods in Enzymatic Analysis. 2nd edit. H. U. Bergmeyer, Ed. Vol. 2. Academic Press, Inc. New York.
49. DURUIBE, V. & G. A. TEJWANI. 1981. Molec. Pharmacol. **20:** 621–630.
50. LOWRY, O. H., N. J. ROSEBROUGH, A. L. FARR & R. J. RANDALL. 1951. J. Biol. Chem. **193:** 265–275.

DISCUSSION OF THE PAPER

G. SALVATORE (*University of Naples, Naples*): I have a semantic question for you. Do you really feel it appropriate to call the porin a "receptor"? If porin is simply a binding protein of the mitochondrial membrane, perhaps it should be better to call it a "binding protein," not a receptor, which is admittedly a fashionable term, but it implies a different meaning (such as transduction of a signal, internalization of the complex, modification of either the ligand or the binding molecule, etc.).

PEDERSEN: From the point of view of definition, fashion, and functional consequences I believe it is very appropriate. First, the word "receptor" simply implies that one protein or proteolipid provides a site for some other molecule. Secondly, by referring to porin as a receptor (or partial receptor) one draws more attention to it in this era obsessed with receptorology. Finally, there are subsequent consequences of its binding. If you wish to make it analogous to hormone receptors, we can speak of a sequence of events: hexokinase binding to its receptor → loss of regulation of hexokinase plus higher affinity for ATP → → increased glucose-6-phosphate, increased phospholipids, DNA, RNA → → increased cell growth and division.

T. GALEOTTI (*Catholic University, Rome*): The increase in the glycolytic rate following the addition of mitochondria to the cytosol of poorly differentiated cells may be due not only to the increase in hexokinase but also to the decrease of the $NADH/NAD^+$ ratio owing to the activation of mitochondria shuttle mechanisms. What is your opinion?

PEDERSEN: I do not wish to imply from our work that aberrations in the activity, regulation, and cellular location of hexokinase are the only factors responsible for the increase in glycolytic rate.

R. WATTIAUX (*Laboratoire di Chimie Physiologique, Namur, Belgium*): Is there a relationship between the degree of differentiation of Morris hepatomas and the binding of hexokinase?

PEDERSEN: Yes, highly differentiated and well differentiated cancer cells have little or no hexokinase activity. Poorly differentiated tumors, of which many are human tumors, uniformly have a high glucose catabolic rate and a correspondingly high hexokinase activity, of which 40–70% has been found bound to the mitochondria.

P. COLEMAN (*New York University, New York*): Do tumor mitochondrial outer membranes possess more pore protein (porin)/mg protein than normal mitochondria do?

PEDERSEN : We don't know for sure because we have not as yet been able to make an antibody to porin. However, one might guess that there is an increased amount of porin per mg of tumor mitochondria relative to normal liver mitochondria. Thus, AS-30D and Novikoff hepatoma mitochondria have between 1,000–2,000 mU hexokinase bound as isolated whereas liver mitochondria have only about 1 mU bound/mg. Moreover, liver mitochondria, even when treated with tumor hexokinase usually never bind more than 500 mU/mg protein.

SALVATORE: Do you have any information on the purity or the homogeneity of your final preparation of porin? From the increase in the specific activity during the purification procedure, it seems that the final preparation was only 100 times more active than the initial crude membrane extract. Is porin contained in the mitochondrial membrane in rather substantial amount?

PEDERSEN: Unfortunately, we cannot measure the specific activity of porin in intact mitochondria. In the Triton X-100 extract it is already subtantially purified, therefore accounting for only a 100-fold purification in the final step relative to the Triton X-100 fraction. Finally, we do have information on purity. The final porin fraction exhibits only a single, distinct band of 35 kDa upon SDS gel electrophoresis.

E. CARAFOLI (*Swiss Federal Institute of Technology, Zurich*): Do you think the main function of porin in tumor mitochondria is the binding of hexokinase? Secondly, did you consider the possibility that, in addition to, or instead of, hexokinase, porin itself is altered in tumor mitochondria?

PEDERSEN: As porin is apparently a component of the outer membrane of all mitochondria, its main function (still unidentified) in normal cells cannot be to bind hexokinase. Most likely, its main functions or one of its functions is to allow small molecules to get in and out of the mitochondria. Because one of these small molecules is ATP, it can be suggested that hexokinase bound to porin in the tumor cell provides closer access to mitochondrially generated ATP. This may occur in such a way as to not interfere with the other normal functions of porin. With regard to the second question, it is possible that porin of tumor cells is somewhat different than that of normal cells. Amino acid composition data, however, indicate that such differences may be small.

Membrane Cholesterol and Tumor Bioenergetics[a]

PETER S. COLEMAN

Laboratory of Biochemistry
Department of Biology
New York University
New York, New York 10003

INTRODUCTION

By late 1985, three phenomenological metabolic findings relevant to some of the characteristics of cancer, cellular growth, and proliferation, had acquired overwhelming support. One of these findings, on reflection, might be considered as particularly unanticipated. These discoveries are (1) the operation of the cholesterol biosynthesis pathway *de novo* is a requirement of cell growth and cell division; (2) the normally observed negative feedback regulation of cholesterogenesis, at the level of 3-hydroxy-3-methylglutaryl CoA-reductase (HMGR), the pathway's rate-limiting enzyme, is lost or severely defective in tumors; and (3) some cholesterogenic pathway intermediate, subsequent to the formation of mevalonic acid, but prior to the generation of squalene, may act in an as-yet-inexplicable fashion to "trigger" DNA replication prior to the onset of mitosis.

Evidence for the first of these findings has accumulated for more than 30 years.[1,2] It has been clearly shown that constantly proliferating cells (epidermis, developing brain, and, of course, spontaneous tumors) generate cholesterol at relatively high rates *in vivo,* whereas cells in interphase do not.[3] Acceptance of the second fundamental discovery has been due, primarily, to the pioneering and extensive studies of Siperstein and co-workers,[4,5] and augmented significantly over the last decade by Chen, Kandutsch, and Heiniger of the Jackson Laboratory,[6] as well as by the work of Sabine.[7] At the present time, the loss of feedback control over cholesterogenesis appears to be a metabolic phenotype characteristic of every vertebrate tumor examined to date, regardless of its etiological source and tissue of origin. Yet, the precipitating event that leads to this metabolic lesion is still unknown. The third finding, that a sterologenesis intermediate appears to be implicated in eliciting DNA synthesis, is of more contemporary vintage, but once again, stems from the creative work of both Siperstein's lab[8] and that of the Jackson Laboratory group.[3] Recent work along the lines of this third theme gathers increasing support for the concept that forerunners of the isoprenoid intermediates within the cholesterogenesis pathway, in addition to providing the building blocks for sterol construction, also are involved as mediators of cellular proliferation.[9–12]

Early Correlates between Mitochondrial Membrane Cholesterol and Cancer

In contemplating the larger metabolic vista that incorporates these three phenomenological facts, we have been concerned, for several years, with questions about the relationship between deregulated cholesterogenesis in tumors and tumor cell bioener-

[a]Supported in part by The National Institutes of Health Grant NCI CA28677.

getics. The principal reason for the rise in our interest in this aspect of cancer biochemistry was that the finding of decontrolled cholesterogenesis implied an overproduction of the sterol, which could result in an enrichment of tumor cell membranes with cholesterol, particularly those of tumor cell mitochondria. It is well known that mitochondrial membranes constitute a very large component of cellular membrane mass and surface area, especially in liver and heart. In fact, early documentation reported that the mitochondria isolated from experimental hepatomas possessed higher cholesterol:phospholipid ratios than those found for normal liver mitochondria,[13-15] and alterations in some of the functional properties of these tumor organelles were reported and rationalized on the basis of their different lipid content.

Of course, it is not outwardly surprising that an alteration in the lipid composition of mitochondrial (or any other biological) membranes should lead to changes in the catalytic capacities of membrane-affiliated functions. Yet, it is important to remember that different cellular membranes catalyze specific reactions, such that the "division of labor" with regard to intermediary metabolism permits one to characterize the various cellular membranes according to relatively specific functions catalyzed by them, but not by others. Consequently, an altered membrane lipid composition, which modifies the activity of certain membrane-affiliated functions, can just as surely lead to striking changes in the patterns of integrated metabolism as can the allosteric modulation of a key enzyme's activity alter the flow of intermediates through a metabolic pathway.

One of the first studies of deliberate *in vitro* manipulation of mitochondrial membrane lipid content involved an enrichment with cholesterol. Such cholesterol-enriched organelles indeed showed a direct correlation between the extent of sterol enrichment and an increase in the rate of succinate oxidation.[16] Later, this laboratory discovered that an *in vitro* method for directly increasing the cholesterol content of normal mitochondria (based on cholesterol transfer via collision from co-incubated, cholesterol-loaded Sephadex G-10 beads) led to a parallel increment in mitochondrial ATPase activity,[17] an enzymatic characteristic that had previously been observed with mitochondria isolated from rapidly proliferating, cholesterol-rich Morris hepatomas.[18] In ensuing mitochondrial studies it was also shown that manipulating the inner membrane's cholesterol:phospholipid content *in vitro* effected alterations in the rates of electron transfer between sequential redox components of the respiratory chain.[19] Some of these redox activity studies indicated that the incorporation of cholesterol into the mitochondrial inner membrane promoted a topological clustering (rather than a random dispersion) of membrane-associated proteins, a structural phenomenon that clearly contributed to some of the altered redox activities noted between respiratory chain components.

During the past decade, the limited mechanistic correlations between mitochondrial membrane cholesterol and altered mitochondrial membrane functions appear to have acquired an enlarged sphere of importance with respect to membrane pathology, inasmuch as it is now clearly established that all tumors examined possess an apparently unregulated pathway of *de novo* cholesterol biosynthesis. Clearly, if an absence or diminishing of regulatory constraints over cholesterogenesis obtains in tumors, then the sterol enrichment of the mitochondrial membranes during the processes constituting tumorigenesis can be a most logical outcome, and we may wish to search for a correlation between these two phenomena within the speculative framework on mechanisms of tumorigenesis.

Tumor Mitochondria Preferentially Export Citrate

One of the key relationships between deregulated cholesterogenesis and mitochondrial function in tumors focuses upon the tricarboxylate (citrate-malate) exchange transporter of the inner mitochondrial membrane. This transporter provides the principal means by which the acetate fragment precursor for lipid synthesis is made accessible to the cytosol, where such synthesis occurs. Cytosolic acetyl CoA, in turn, derives from the prior formation of citrate within the mitochondria. Upon export to the cytoplasm on the tricarboxylate exchange carrier, citrate is cleaved to yield acetyl CoA and oxaloacetate via the enzyme ATP-citrate lyase (EC 4.1.3.8). Since mitochondrial citrate is thus offered a choice of metabolic routes to follow (either via export or by further intramitochondrial processing, e.g., by way of the Krebs cycle), the tricarboxylate exchange carrier could play a pivotal role in contributing to the regulation of this aspect of intermediary metabolism (FIG. 1).

Much of our work has concentrated on this problem. We have accumulated evidence that leads us to conclude that, indeed, the tricarboxylate exchange carrier of

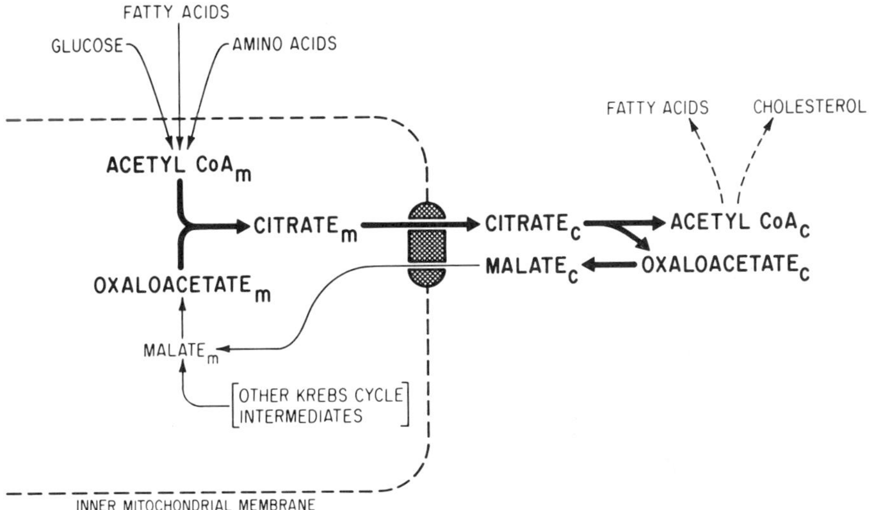

FIGURE 1. The role of the mitochondrial tricarboxylate (citrate) exchange carrier. Mitochondrially generated citrate is exported to the cytosol via the tricaboxylate carrier in exchange for the import of malate. The latter derives from the enzymatic cleavage of cytosolic citrate. Not shown are the various other anion-specific exchange carriers, particularly the glutamate-aspartate transporter, which prevent depletion of Krebs cycle intermediates and provide for efficient respiration despite the removal of acetate carbons from the scheme.

tumor mitochondria preferentially delivers mitochondrial citrate to the cytosolic compartment at a much faster rate than is evident with normal mitochondria. For example, Arrhenius plots of the kinetics of citrate export from mitochondria isolated from normal rat liver (FIG. 2, A), from host liver (the livers of rats that carry intramuscular transplants of tumors) (FIG. 2, B), and the mitochondria from Morris hepatoma 3924A (FIG. 2, C), indicate that only the tumor mitochondria fail to show any phase-transition temperature discontinuity near 12°C.[20] By extrapolation of the slopes from 12°C to 37°C, the tumor mitochondria exhibited a capacity (or potential) to export citrate at a rate at least twofold greater than that of normal mitochondria (apparent V_{max} (tumor) = 592 nmol/min/mg; apparent V_{max} (normal) = 283 nmol/min/mg).

Further studies[21] demonstrated that these isolated Morris hepatoma 3924A

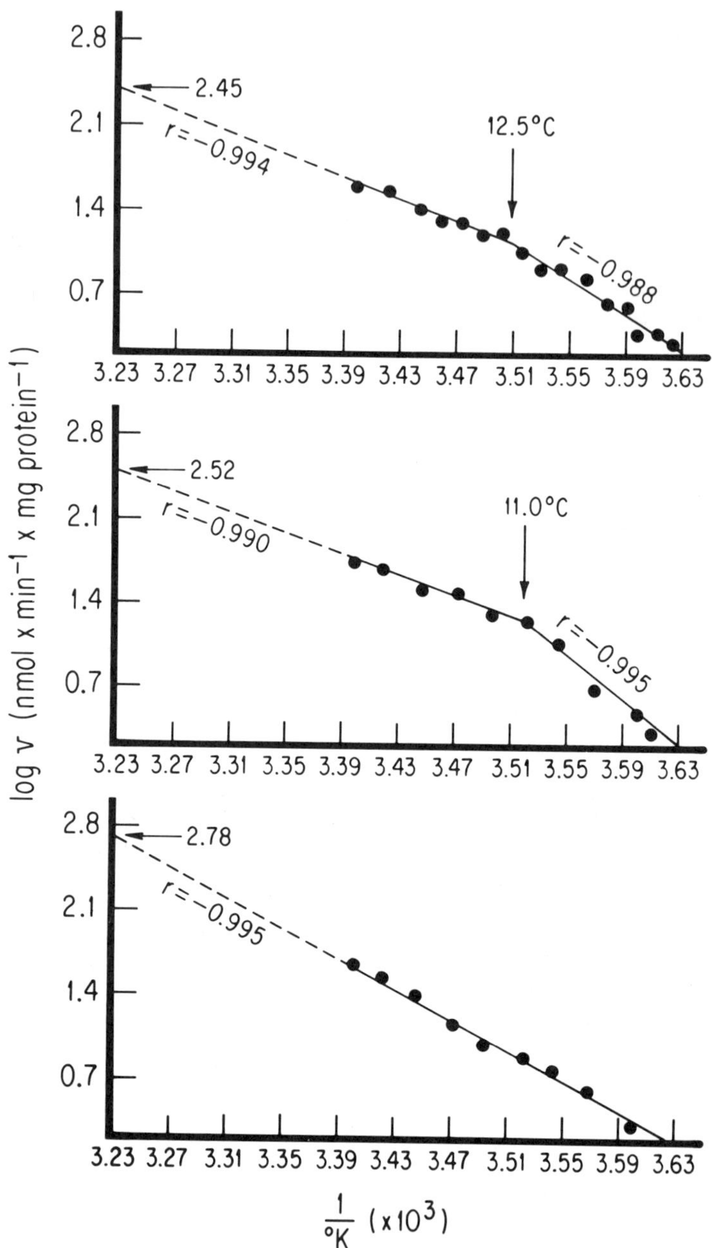

FIGURE 2. Arrhenius plots for citrate efflux from citrate-loaded, respiration-inhibited normal, host, and Morris hepatoma 3924A mitochondria. Efflux rates of [^{14}C]citrate in exchange for [^{12}C]citrate with isolated mitochondria from normal ACI rat liver (top), host liver (middle), and tumor 3924A (bottom). See text for discussion, and for detailed experimental conditions see Kaplan *et al.*[20]

mitochondria, when incubated at 25°C under conditions in which normal mitochondria display pyruvate-driven respiration, preferentially export the pyruvate-derived citrate more than fourfold faster than normal (FIG. 3). Concomitantly, the state-3 (ADP-triggered) respiratory rates with either added pyruvate or citrate were so low as to be unmeasureable (i.e., indistinguishable from state-4 rates) for the tumor organelles (TABLE 1). Furthermore, it was found that the preferential export of citrate from the tumor mitochondria could be totally blocked with the tricarboxylate carrier inhibitor

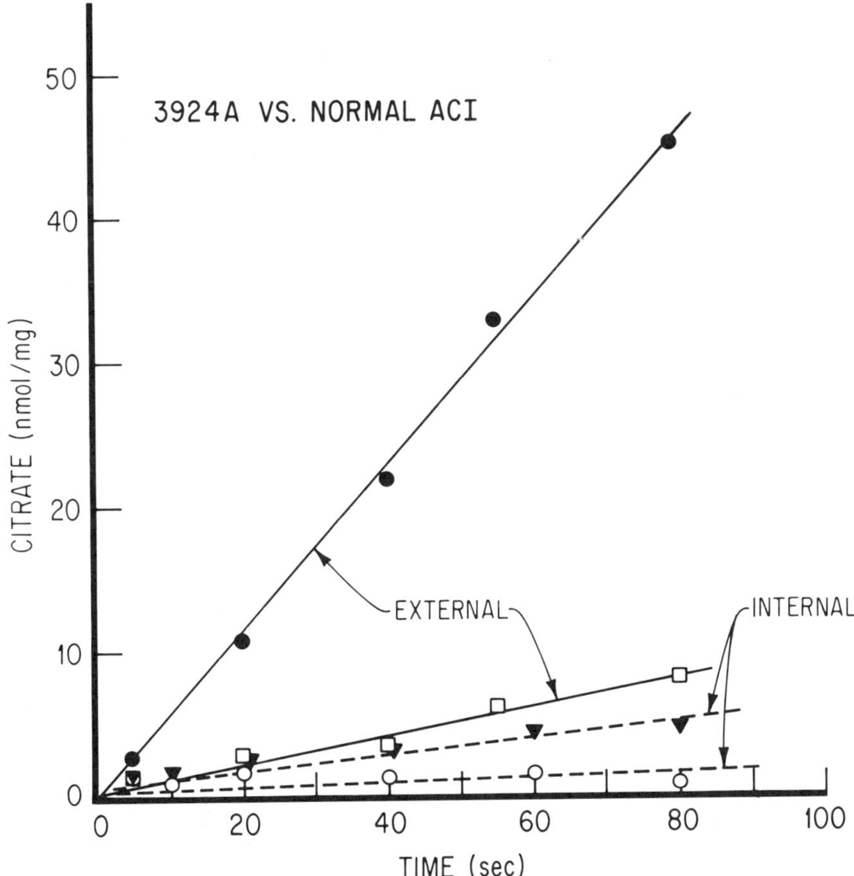

FIGURE 3. Enhanced efflux of pyruvate-derived citrate in respiring tumor 3924A versus normal liver mitochondria. Mitochondria were incubated at 25°C in a medium (pH 7.4) containing pyruvate (0.5 mM) plus malate (0.1 mM), in the presence or absence of 0.15 mM 1,2,3-benzenetricarboxylate and respiration was triggered with excess ADP. At specified time intervals, the incubation was abruptly stopped by rapid centrifugation through a silicone oil barrier into 14% $HClO_4$. The citrate content above (extramitochondrial) and below (intramitochondrial) the oil interface was assessed enzymatically, as described in Parlo and Coleman.[21] (●) Tumor 3924A (external); (▼) Tumor 3924A (internal); (□) Normal rat liver (external); and (O) Normal rat liver (internal).

TABLE 1. Mitochondrial State 3 Respiratory Rates: Morris Hepatoma 3924A and 16 versus Normal Rat Liver

	O$_2$ Consumption Rate (ng atom O/min/mg protein)			
Substrate	Hepatoma 3924A	Normal ACI	Hepatoma 16	Normal Buffalo
Pyruvate[a]	too small to measure	38.8 ± 5.1	14.4 ± 1.3	23.2 ± 3.2
Citrate[a]		43.9 ± 4.2	36.5 ± 8.0	28.6 ± 2.7

[a]Plus 1.0 mM malate.
NOTE: Experimental conditions and methods were as previously described.[21]

1,2,3-benzenetricarboxylate (BTC), thereby reversing the anomalous metabolic pattern by reestablishing both pyruvate- and citrate-driven respiration. These results imply that the tumor mitochondria carry out an almost exclusive efflux of matrix-generated citrate carbons to the extramitochondrial milieu, rather than couple citrate oxidation to electron transfer-linked phosphorylation via the Krebs cycle (TABLE 2).

To what features or characteristics peculiar to tumor mitochondria could we assign the cause of such anomalous patterns of carbon flux? Recalling the early reports of tumor mitochondrial membrane enrichment with cholesterol, together with Hackenbrock's findings on the effects of *in vitro* lipid compositional manipulation of mitoplast membranes on respiratory chain rates,[19] we suspected that the tumor mitochondrial membrane enrichment with cholesterol might alter the kinetic characteristics of the tricarboxylate-exchange carrier in tumors. We therefore employed the "solid-state" molecule transfer method,[22,17,23] to enrich exogenously isolated normal rat liver mitochondria with cholesterol. This procedure allows for a controllable titration of membrane cholesterol enrichment in a simple and rapid fashion. Normal mitochondria thus enriched with cholesterol were found to mimic the tumor mitochondria with respect to their ability to export preferentially pyruvate-generated citrate (inhibitable by BTC) (FIG. 4). By the same token, such exogenously enriched, cholesterol-loaded normal mitochondria were found less and less capable of state-3 respiration with pyruvate or citrate as they became increasingly cholesterol rich (TABLE 3). TABLE 4 indicates, as a point of interest for comparison, that isolated mitochondria of two of the Morris hepatoma series—a rapidly proliferating, highly deviated malignant tissue

TABLE 2. Mitochondrial State 3 Respiratory Rates: Hepatoma versus Normal Liver With or Without 1,2,3-Benzenetricarboxylate

	O$_2$ Consumption Rate (ng atom O/min/mg protein)			
Substrate	Hepatoma 3924A	Normal ACI	Hepatoma 16	Normal Buffalo
Pyruvate[a]	11.9 ± 2.5	29.9 ± 7.4	12.1 ± 2.2	13.4 ± 2.5
Citrate[a]	47.9 ± 9.5	48.7 ± 8.3	32.5 ± 6.5	38.9 ± 3.4

[a]Plus 1.0 mM malate.
NOTE: Experimental conditions and methods were as previously described.[21] When present, the final BTC concentration was 0.15 mM.

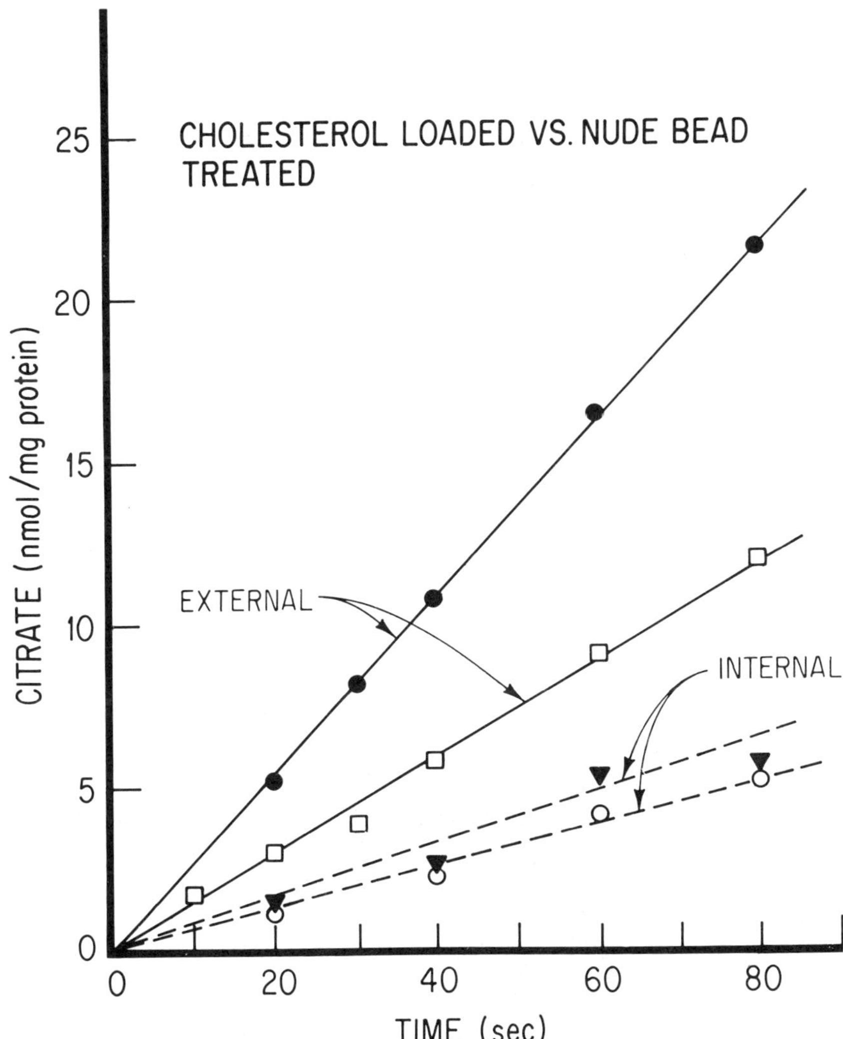

FIGURE 4. Enhanced efflux of pyruvate-derived citrate in respiring, cholesterol-enriched, normal mitochondria. Normal ACI rat liver mitochondria were first enriched exogenously 7.6-fold via the solid-state molecule transfer procedure,[17,21,23] and assessed according to methods exactly as given for FIGURE 3. As control, normal liver mitochondria were subjected to the identical solid-state molecule transfer protocol, but in the absence of exogenous cholesterol.[21] (●) Cholesterol-enriched (external); (▼) Cholesterol-enriched (internal); (□) Control (external); and (○) Control (internal).

TABLE 3. Mitochondrial State 3 Respiratory Rates: Normal Liver Mitochondria Enriched *in Vitro* With Cholesterol

	Mitochondrial Membrane Cholesterol Content (μg total cholesterol/mg protein)				
	4.9 ± 1.0[a]	4.7 ± 1.1[b]	10.1 ± 1.7	19.8 ± 2.1	37.4 ± 2.9
	O_2 Consumption Rate (ng atom O/min/mg protein)				
Pyruvate[c]	38.8 ± 5.1	50.1 ± 1.8	17.3 ± 1.9	10.9 ± 1.0	8.6 ± 2.2
Citrate[c]	43.9 ± 4.2	50.2 ± 2.8	23.6 ± 2.3	10.5 ± 1.2	6.2 ± 1.5

[a]Normal ACI rat liver mitochondria.

[b]"Nude" Sephadex bead-treated normal mitochondria; 125 mg, acetone-washed Sephadex G-10, incubated with 1 ml resuspended mitochondria (40 mg protein/ml) at 0°C for 15 min in a reciprocating shaker.[21]

[c]Plus 1.0 mM malate.

NOTE: The *in vitro* cholesterol enrichment procedure used here[21] was modified from Coleman *et al.*[17] It is clear that the cholesterol content of isolated normal mitochondria (4.9 μg/mg protein) is not affected by the "sham enrichment" procedure described here as "nude" bead treatment (4.7 μg/mg protein).

(hepatoma 3924A) and a slowly growing, minimally deviated tumor (hepatoma 16)—possess different levels of membrane cholesterol, with the tumor displaying the higher rate of proliferation *in vivo* (3924A) also containing the greater mitochondrial membrane cholesterol content. It is apparent that there are suggestive correlations between (1) the extent of tumor mitochondrial membrane cholesterol and the extent to which these mitochondria preferentially export citrate as well as (2) the level of mitochondrial membrane cholesterol and the capacity to catabolize pyruvate or citrate via Krebs cycle respiration.

Cholesterol-Rich Tumor Mitochondria Display a Truncated Krebs Cycle

The fact that cholesterol enrichment of mitochondrial membranes (whether generated during the course of tumorigenesis *in vivo* or procured *in vitro* via exogenous enrichment procedures) alters the pattern of pyruvate-generated citrate metabolism becomes much more interesting when considered together with some additional bioenergetically relevant observations. Although our data permit us to speculate that cholesterol-rich mitochondrial membranes provide a different microenvironment for the tricarboxylate-exchange carrier in the inner membrane leading to more rapid

TABLE 4. Total Membrane Cholesterol Content of Isolated Mitochondria from a Rapidly Growing (3924A) and a Slowly Growing (16) Morris Hepatoma Compared with Those from Normal Rat Liver

Total (Free *plus* Esterified) Cholesterol (μg/mg protein)			
Tumor 3924A	Normal ACI liver	Tumor 16	Normal Buffalo liver
25.7 ± 6.7 (11)	4.9 ± 0.9 (9)	13.7 ± 4.1 (8)	6.8 ± 1.6 (10)

NOTE: Experimental procedures were as previously described.[21]

citrate efflux kinetics, we also observed that all other Krebs cycle intermediates that join the catabolic cycle subsequent to citrate (i.e., isocitrate, α-ketoglutarate, glutamate, and succinate) are unaffected by the level of mitochondrial membrane cholesterol (TABLE 5). This observation appears to be valid whether such respiratory rates were examined in endogenously enriched mitochondria isolated from hepatoma 3924A or in normal organelles enriched with cholesterol *in vitro*.[21,24]

Our conclusions based on these observations can be summarized by means of a portrait for tumor cell metabolic flux (FIG. 5). The principal feature of such a metabolic pattern is the postulate of a continually operating truncated or abbreviated Krebs cycle in tumor mitochondria. With this metabolic paradigm, the generation of intramitochondrial acetyl CoA, by whatever means (carbohydrate or fatty acid catabolism, or amino acid trans- or deamination), does not contribute to the fueling of oxidative phosphorylation, but rather increases the rate of supply of extramitochondrial acetyl fragments required for lipid synthesis—particularly cholesterogenesis. Since the latter pathway is no longer as stringently supervised in tumors, it presumably occurs continuously and without the normal modes of feedback control inhibition. On

TABLE 5. Mitochondrial State 3 Respiratory Rates: Normal Liver Mitochondria Enriched *in Vitro* with Cholesterol (Some Substrates Joining the Krebs Cycle Past Citrate)

	Mitochondrial Membrane Cholesterol Content (μg total cholesterol/mg protein)				
	4.9 ± 1.0[a]	4.7 ± 1.1[b]	10.1 ± 1.7	19.8 ± 2.1	37.4 ± 2.9
	O_2 Consumption Rate (ng atom O/min/mg protein)				
Glutamate	59.8 ± 2.1	54.2 ± 4.1	48.9 ± 1.3	51.3 ± 1.9	54.8 ± 4.8
Succinate	113.0 ± 13.0	62.9 ± 2.7	55.8 ± 2.3	57.0 ± 2.2	55.4 ± 3.2

[a]Normal ACI rat liver mitochondria.

[b]"Nude" Sephadex bead-treated normal mitochondria; 125 mg, acetone-washed Sephadex G-10, incubated with 1 ml resuspended mitochondria (40 mg protein/ml) at 0°C for 15 min in a reciprocating shaker.[21]

NOTE: Experimental methods and conditions were as previously described.[21] See also TABLE 3.

the other hand, our data also indicated (TABLE 5) that those Krebs cycle substrates that fuel oxidative phosphorylation past citrate are seemingly oblivious to the presence of cholesterol-rich mitochondrial membranes, implying that the various Krebs cycle substrate anion exchange transport carriers of the mitochondrial inner membrane, except for the tricarboxylate carrier, function normally. We have proposed that glutamate may serve as the predominant respiratory fuel in tumors, and there is a great deal of evidence that supports the view that glutamate (or indirectly, glutamine) is employed in this manner.[25-27] The metabolic logic of this proposal has been discussed in some detail.[21,23]

The Role of the Mitochondrial Tricarboxylate Carrier in Deregulated Cholesterogenesis in Whole Tumor Tissue

The scheme portrayed in FIGURE 5 leads us to consider the above tumor mitochondrial data in a larger context, beyond the boundaries of classical bioenergetics

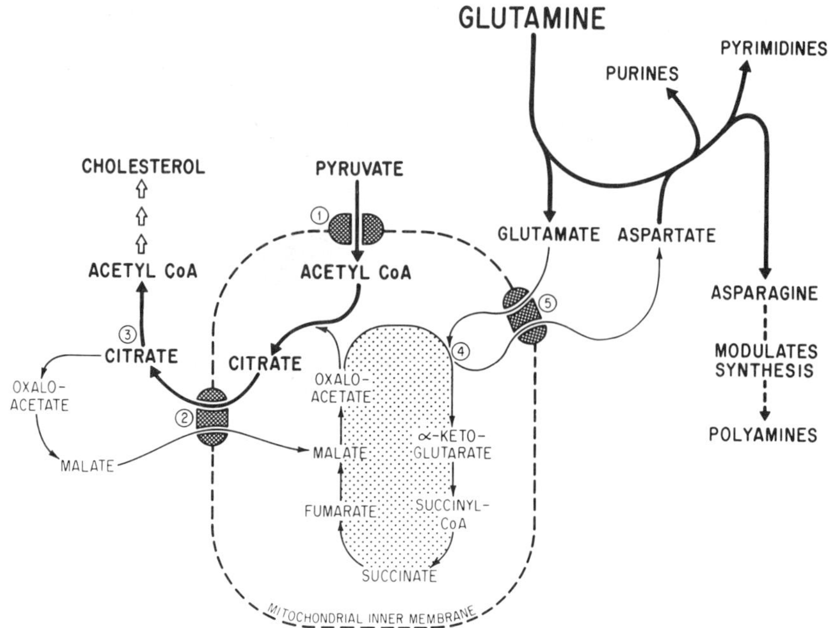

FIGURE 5. The truncated Krebs cycle of tumor mitochondria. According to this model, cholesterol-rich tumor mitochondria preferentially export the citrate generated within, yet respire efficiently, fueled by glutamate. The exported citrate provides the high levels of cytosolic acetate required for the continuously operating, deregulated cholesterogenesis pathway in tumors. The glutamate, derived via glutamine deamination ("glutaminase" activity is remarkably high in many tumors[23]) generates α-ketoglutarate (a primary respiratory substrate) and aspartate (which is exported in exchange for an incoming glutamate). The model also depicts how glutamine and aspartate serve as metabolic precursors for purine, pyrimidine, and, as has been documented for diverse tumor tissues, enhanced asparagine and polyamine synthesis.[23] The circled numbers indicate (1) the pyruvate/OH^-, (2) the tricarboxylate/malate, and (5) the glutamate/aspartate exchange transporters, respectively, while (4) is mitochondrial aspartate aminotransferase, and (3) is cytosolic ATP-citrate lyase.

per se, and focuses our attention on the cytosolic compartment where cholesterogenesis occurs. It is thus necessary to ask whether studies on whole, viable tumor cells can provide evidence that supports the concepts of a postulated truncated Krebs cycle that operates continuously in tumor tissue and is coupled with a deregulated cholesterogenic carbon flux from, say, pyruvate.

We have begun to perform some of these studies. FIGURE 6 demonstrates the flow of carbons from added [^{14}C]pyruvate into newly generated cholesterol upon incubation of tissue slices obtained from Morris hepatoma 3924A and from normal rat liver. FIGURE 7 indicates the simultaneously measured production of $^{14}CO_2$ from the added [^{14}C]pyruvate in both tissues. Finally, TABLE 6 compares, as ratios, the extent of incorporation of [^{14}C]pyruvate into cholesterol relative to CO_2 over the 4-hr incubation period for viable tissues from both the tumor and normal liver. These data disclose that in viable tumor tissue derived from the rapidly proliferating Morris hepatoma 3924A, the added pyruvate carbons are incorporated into cholesterol at a very high linear rate during the 4 hr incubation at 37°C. Concomitantly, the rate of CO_2 production from the added pyruvate is very low in the tumor. On the other hand, the pattern of pyruvate carbon flux to either cholesterol or CO_2 is precisely the inverse with normal liver tissue (FIGS. 6 and 7), where CO_2 generation is high but that of cholesterol is rather unimpressive. The relative pyruvate carbon flux to sterol versus CO_2 (TABLE 6) is nearly sevenfold greater for the tumor than for the normal liver after 4 hr. The data also disclose that while the high rate of [^{14}C]pyruvate carbon incorporation into cholesterol in tumor tissue is linear over the 4-hr time course, that for normal liver slows abruptly after 2-hr; on normalization to g tissue protein, it is practically zero for the liver between 2 and 4 hr. We suspect that this reduction in carbon flux to newly synthesized cholesterol in the normal liver tissue after 2 hr of active cholesterogenesis may be a reflection of the well-documented downshift regulatory mechanisms that have been repeatedly shown to occur in normal livers upon dietary supplementation with cholesterol.[28]

Clearly, the above results on whole cell carbon flux from pyruvate to cholesterol via the intervention of an altered processing of mitochondrial metabolites in tumor versus normal tissue appear to support the proposal of a preferential efflux of pyruvate-

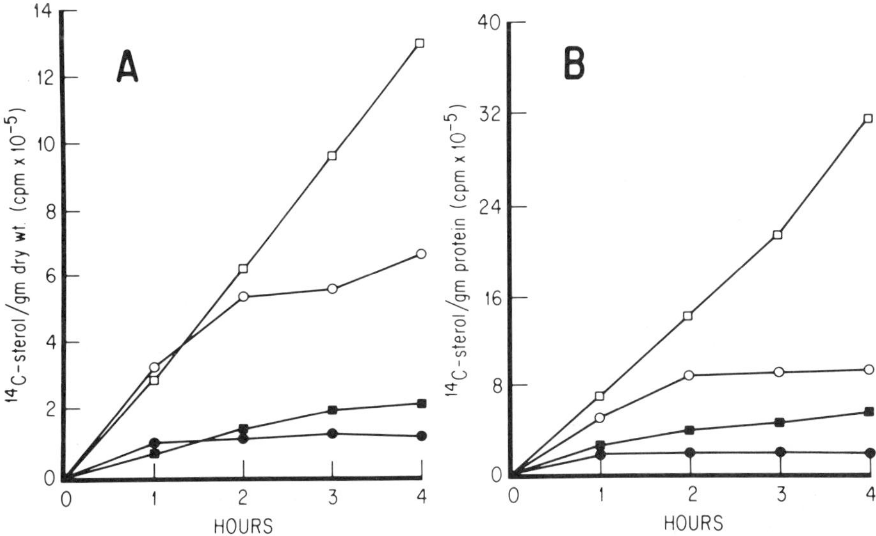

FIGURE 6. Incorporation of pyruvate-derived ^{14}C into cholesterol in viable tumor and normal liver tissue +/− benzenetricarboxylate. Thin slices (250 mg wet weight) from hepatoma 3924A and normal rat liver were incubated in Krebs-Ringer bicarbonate, pH 7.6, 37°C with and without 10 mM BTC. [U-^{14}C]pyruvate (10 mM, 0.67 μCi/ml) was added as the sole exogenously supplied substrate. Reactions were performed in O_2/CO_2 gas equilibrated, septum-covered flasks in a shaking water bath. Incubations were terminated at specified time intervals and assayed, after lipid extraction and saponification, for digitonin-precipitable ^{14}C incorporation. Metabolic viability of these tissues over the 4 hr incubation was ascertained separately by measuring their O_2 consumption rates with 5 mM glutamate. These respiratory rates were linear and similar for both tumor and normal liver for more than 4 hr, but by 5 hr had decreased substantially: for 4 hr, tumor, 2.80 mg atom O/hr/g wet weight tissue; liver, 2.73 mg atom O/hr/g wet weight tissue. Two databases were used. Data for (A) were normalized to g dry weight tissue; data for (B) were normalized to g tissue protein. (□) Tumor minus BTC; (■) Tumor plus BTC; (○) Normal liver minus BTC; and (●) Normal liver plus BTC.

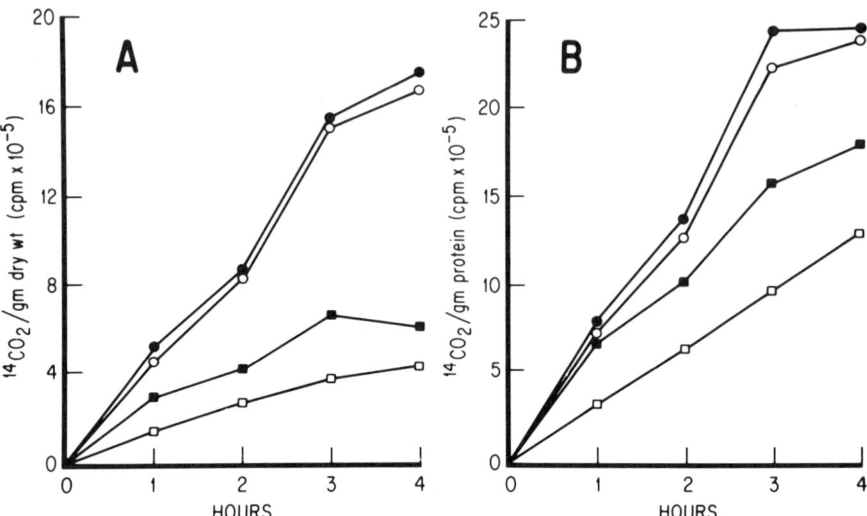

FIGURE 7. Incorporation of pyruvate-derived ^{14}C into CO_2 in viable tumor and normal liver tissue +/− benzentricarboxylate. Incubations were performed under conditions identical to those for FIGURE 6, except that a center well, containing 0.3 ml Protosol (New England Nuclear) served to absorb $^{14}CO_2$ evolved upon termination of the incubation with 0.15 ml 6 M H_2SO_4. The entire center well was counted via liquid scintillation spectrometry. Data for (A) and (B) were normalized as for FIGURE 6. (□) Tumor minus BTC; (■) Tumor plus BTC; (○) Normal liver minus BTC; and (●) Normal liver plus BTC.

derived citrate to the cytosol coupled to the continuous flow of sterol precursor carbons into cholesterol in tumors, as portrayed in FIGURE 5. We feel that the concept of a truncated Krebs cycle in mitochondria isolated from rapidly growing tumors gains credence from the above results employing viable tumor tissue. The data also signify that the mitochondrial citrate transport carrier is likely to function at a much higher rate within viable tumor tissue relative to normal liver, in a manner similar to what we had previously observed with isolated, cholesterol-rich tumor mitochondrial preparations (FIG. 3).

The most intriguing data from FIGURES 6 and 7 and TABLE 6 pertain to the response of these tissues to the presence of 1,2,3-benzenetricarboxylate (BTC), the inhibitor of citrate transport in isolated mitochondrial systems. This relatively hydrophobic structural analog of citrate was found to inhibit the incorporation of ^{14}C-pyruvate carbons into newly synthesized cholesterol in both tumor and normal tissues (FIG. 6). BTC simultaneously caused a significant (nearly 50%) increase in $^{14}CO_2$ from the added [^{14}C]pyruvate in the tumor tissue incubation, but did not affect the already high rate of $^{14}CO_2$ production in normal liver (FIG. 7).

We have drawn two conclusions from these results. First, despite the anionic nature of BTC above pH 7, the nonpolar benzene moiety appears to allow for the molecule's penetration of the plasma membrane of cells in both tumor and liver tissues, and once within these cells, the BTC can serve to block the mitochondrial tricarboxylate carrier, just as it is known to do with isolated mitochondrial systems. Thus, the preferential efflux of mitochondrial pyruvate-generated citrate is inhibited by the addition of BTC to metabolizing tumor tissue slices, and this in turn dramatically inhibits the incorporation of pyruvate carbons into newly generated cholesterol. A second and most intriguing outcome is that by blocking the more rapid efflux of citrate from the cholesterol-rich tumor mitochondria, BTC concomitantly impels the pyruvate-generated citrate within the matrix into further catabolism via the Krebs cycle, which has become the only metabolic route that now remains available to it. This promotes an increase in the rate of CO_2 evolution in the tumor tissue, a pattern more closely resembling that of normal liver tissue (FIG. 7). In fact, the presence of BTC with normal liver tissue hardly effected any change in the rate of CO_2 production, indicating that even in the absence of the tricarboxylate carrier inhibitor, mitochondrial citrate is predominantly catabolized via the Krebs cycle, and not exported to any appreciable extent under our experimental conditions. This, of course, is reflected in the relatively low rate of pyruvate carbon incorporation into sterol in the normal liver system, especially after the second hour of incubation. Thus, it may be suggested that even in viable tumor tissue, BTC can reverse the anomalous but characteristic pattern of tumor cell citrate metabolism and help to re-establish a route of citrate carbon flux to CO_2 via mitochondrial respiration, conforming to the usually observed, or more "classical," role of citrate in cellular bioenergetics that obtains with normal liver.

The Mitochondrial Tricarboxylate Carrier and Cancer: Speculations

Perhaps the most surprising information regarding the metabolic acrobatics concerned with cellular proliferation is the finding that a mevalonate derivative within the cholesterogenesis pathway may act to help "trigger" new DNA synthesis.[9–12,29] Such results recall the earlier suggestion of Chen *et al.*[6] that since cholesterol synthesis *de novo* appears to be imperative for the proliferation of cells in culture, decontrolled and continuous cholesterogenesis may be capable of inducing cell populations into anomalously enhanced proliferative activity. This as-yet speculative concept may be cautiously expanded in light of the results with the tricarboxylate carrier inhibitor BTC, described above, on viable tumor tissue. Clearly, DNA synthesis is the prelude to cell division and growth, and the most generally descriptive definition of cancer is the uncontrolled growth of cells within the organism.[30] It is certainly tempting to consider,

TABLE 6. ^{14}C Incorporation Ratios into Cholesterol and CO_2 for Tumor 3924A and Normal Liver Slices Incubated with [U-^{14}C]Pyruvate with and without BTC

	$\frac{\text{cpm } ^{14}\text{C-Sterol/g protein}}{\text{cpm } ^{14}\text{CO}_2\text{/g protein}}$				
Incubation Time (hr)	Tumor (−BTC)	Tumor (+BTC)	Normal (−BTC)	Normal (+BTC)	$\frac{\text{Tumor } (-\text{BTC})}{\text{Normal } (-\text{BTC})}$
1	2.00	0.26	0.67	0.20	2.98
2	2.20	0.34	0.66	0.12	3.33
3	2.20	0.25	0.38	0.07	5.79
4	2.50	0.28	0.37	0.06	6.76

NOTE: Normal ACI rat liver and Morris hepatoma 3924A liver slices were incubated in the presence or absence of 10 mM BTC as described for FIGURES 6 and 7. The column farthest on the right (Tumor 3924A minus BTC/Normal liver minus BTC) indicates the incremental ratios, with increasing incubation time, of pyruvate ^{14}C carbon flux into newly generated sterol relative to that into evolved $^{14}CO_2$. See text for discussion.

therefore, whether the blocking of the tumor mitochondrial tricarboxylate exchange carrier could slow or even halt the cholesterogenically linked processes deemed necessary for DNA synthesis and ensuing mitosis, by effectively "starving" the cytoplasm of the required level of citrate-derived acetyl CoA precursor for *de novo* sterol synthesis. In so doing, one might guess that the production of the putative, mevalonate-derived "trigger" for DNA synthesis would be held below its effective threshold level. With similar logic, Maltese *et al.*[31] recently reported on the successful growth suppression of a murine neuroblastoma *in vivo* by infusion of tumor-bearing animals with a specific inhibitor of hydroxymethylglutaryl CoA reductase (HMGR) called mevinolin, a synthetic competitive inhibitor of the enzyme, first described by Alberts *et al.*[32] that presumably is undergoing further testing by Merck, Sharp and Dohme (Rahway, NJ). Whether or not BTC would also prove therapeutically effective is unknown. At the very least, we feel that the data we have obtained suggest an enlarged appreciation of the important, indeed rather central, role of the mitochondrial tricarboxylate carrier in apportioning the molecular components of intermediary metabolism between the support of cellular bioenergetics and the support of one of the chief hallmarks of the wide variety of cancer cells studied—the deregulation of the cholesterol synthesis pathway in tumors.

SUMMARY

We have established that a preferential export of pyruvate-generated citrate occurs from cholesterol-rich tumor mitochondria, with both isolated mitochondrial systems as well as with viable tumor tissue slices (i.e., with whole tumors cells). Furthermore, we have demonstrated that the more rapid citrate efflux kinetics (catalyzed by the tricarboxylate exchange carrier) of isolated tumor mitochondria is completely inhibited upon addition of 1,2,3-benzenetricarboxylate (BTC) and have shown that this inhibition is apparently also obtained in viable tumor tissue when the inhibitor is added to the tissue incubation. Upon BTC inhibition of tumor mitochondrial citrate export in viable tumor tissue incubations, the incorporation of [^{14}C]pyruvate into newly synthesized cholesterol is severely inhibited as well. Among the most interesting conclusions drawn from our results, we catalog the following. (1) The preferential export of citrate from isolated tumor mitochondria appears to be coupled, functionally, to a high linear rate of incorporation of ^{14}C from pyruvate to cholesterol in viable tumor tissue slices, simultaneously supporting the postulate of a truncated Krebs cycle and corroborating the well-established deregulated and continuous cholesterogenesis pathway in tumors, especially hepatomas. (2) The extent of [^{14}C]pyruvate flux to newly generated cholesterol in either tumor or normal liver tissue is inversely related to the extent of $^{14}CO_2$ production. Despite the evolution of some CO_2 during cholesterogenesis, the predominant portion presumably arises via metabolic processing of pyruvate-generated citrate during Krebs cycle-linked respiration. (3) Isolated tumor mitochondrial systems, as well as viable tumor tissue incubations, can manifest a reversal in the pattern of enhanced mitochondrial citrate efflux coupled to increased cholesterogenesis, when BTC is added to the system. This implies that BTC, a hydrophobic but negatively charged moiety at pH 7, can indeed penetrate the plasma membrane of cells. Upon entry into the cell, BTC apparently blocks the tricarboxylate carrier of tumor tissue mitochondria, thus forcing the mitochondrial citrate into Krebs cycle-linked respiration rather than permitting it to serve as the predominant provider of an increased supply of cytosolic acetyl CoA precursor required for deregulated cholesterogenesis during the development of the tumor.

ACKNOWLEDGMENTS

The experimental data from my laboratory, some of which are reported here, were acquired painstakingly by my friends and colleagues, including: Dr. Ronald Kaplan, Dr. Risa Parlo, Dr. Alan Posner, Dr. Srinivasa Rao, and Neal Azrolan. I am indebted to them for their perseverance and I cherish their friendship. Perpetual inspiration is provided by Jane and Danny.

REFERENCES

1. SRERE, P. A., I. L. CHAIKOFF & W. C. DAUBEN. 1948. J. Biol. Chem. **176:** 829–833.
2. CHEN, H. W., H-J HEINIGER & A. A. KANDUTSCH. 1975. Proc. Natl. Acad. Sci. USA **72:** 1950–1954.
3. KANDUTSCH, A. A. & H. W. CHEN. 1977. J. Biol. Chem. **252:** 409–415.
4. SIPERSTEIN, M. D. & V. M. FAGAN. 1964. Cancer Res. **24:** 1108–1115.
5. WILEY, M. H. & M. D. SIPERSTEIN. 1976. *In* Control Mechanisms in Cancer. W. E. CRISS, T. ONO & J. R. SABINE, Eds.: 343–350. Raven Press.
6. CHEN, H. W., A. A. KANDUTSCH & H. -J. HEINIGER. 1978. Prog. Exp. Tumor Res. **22:** 275–316.
7. SABINE, J. R. 1975. Prog. Biochem. Pharmacol. **10:** 269–307.
8. QUESNEY-HUNEEUS, V., M. H. WILEY & M. D. SIPERSTEIN. 1979. Proc. Natl. Acad. Sci. USA **76:** 5056–5060.
9. FAUST, J. R., M. S. BROWN & J. L. GOLDSTEIN. 1980. J. Biol. Chem. **255:** 6546–6548.
10. FAIRBANKS, K. P., L. D. WITTE & D. S. GOODMAN. 1984. J. Biol. Chem. **259:** 1546–1551.
11. SCHMIDT, R. A., C. J. SCHNEIDER & J. A. GLOMSET. 1984. J. Biol. Chem. **259:** 10175–10180.
12. SINENSKY, M. & J. LOGEL. 1985. Proc. Natl. Acad. Sci. USA **82:** 3257–3261.
13. FEO, F., A. CANUTO, G. BERTONE, R. GARCEA & P. PANI. 1973. FEBS Lett. **33:** 229–232.
14. FEO, F., A. CANUTO, R. GARCEA & L. GABRIEL. 1975. Biochim. Biophys. Acta **413:** 116–134.
15. MORTON, R., C. CUNNINGHAM, R. JESTER, M. WAITE, N. MILLER & H. P. MORRIS. 1976. Cancer Res. **36:** 3246–3254.
16. GRAHAM, J. M. & C. GREEN. 1970. Eur. J. Biochem. **12:** 58–66.
17. COLEMAN, P. S., B. B. LAVIETES, R. BORN & A. WEG. 1978. Biochem. Biophys. Res. Commun. **84:** 202–207.
18. KASCHNITZ, R. M., Y. HATEFI & H. P. MORRIS. 1976. Biochim. Biophys. Acta **449:** 224–235.
19. HACKENBROCK, C. 1981. Trends Biochem. Sci. **6:** 151–154.
20. KAPLAN, R. S., H. P. MORRIS & P. S. COLEMAN. 1982. Cancer Res. **42:** 4399–4407.
21. PARLO, R. A. & P. S. COLEMAN. 1984. J. Biol. Chem. **259:** 9997–10003.
22. COLEMAN, P. S., A. EWELL & R. A. GOOD. 1978. Proc. Natl. Acad. Sci. USA **75:** 3766–3770.
23. COLEMAN, P. S. & B. LAVIETES. 1981. CRC Crit. Rev. Biochem. **11:** 341–393.
24. COLEMAN, P. S. & R. A. PARLO. 1984. Ann. N.Y. Acad. Sci. **435:** 129–132.
25. NYHAN, W. L. & H. BUSCH. 1958. Cancer Res. **18:** 385–393.
26. LAVIETES, B., D. H. REGAN & H. DEMOPOULOS. 1974. Proc. Natl. Acad. Sci. USA **71:** 3993–3997.
27. REITZER, L. J., B. M. WICE & D. KENNELL. 1979. J. Biol. Chem. **254:** 2669–2676.
28. RODWELL, V., J. L. NORDSTROM & J. J. MITSCHELEN. 1976. Adv. Lipid Res. **14:** 1–75.
29. QUESNEY-HUNEEUS, V., H. A. GALICK, M. D. SIPERSTEIN, S. K. ERICKSON, T. A. SPENCER & J. A. NELSON. 1983. J. Biol. Chem. **258:** 378–385.
30. PITOT, H. C. 1966. Ann. Rev. Biochem. **35:** 335–368.
31. MALTESE, W. A., R. DEFENDINI, R. A. GREEN, K. M. SHERIDAN & D. K. DONLOY. 1985. J. Clin. Invest. **76:** 1748–1754.

32. ALBERTS, W. A., J. CHEN, G. KURON, V. HUNT, J. HUFF, C. HOFFMAN, J. ROTHROCK, M. LOPEZ, H. JOSHUA, E. HARRIS, A. PATCHETT, R. MONOGHAN, S. CURRIE, E. STAPLEY, G. ALBERS-SCHONBERG, O. HENSENS, J. HIRSHFIELD, K. HOOGSTEEN, J. LIESCH & J. SPRINGER. 1980. Proc. Natl. Acad. Sci. USA **77:** 3957–3961.

DISCUSSION OF THE PAPER

P. PEDERSEN (*Johns Hopkins University, Baltimore, MD*): Would you comment on the possible advantage(s) that the loss of feedback control of cholesterol synthesis might have for the cancer cell? Perhaps it is only a nuisance that the cancer cell has to live with, and there are no advantages for growth or maintenance. I suggest this because there is *no* apparent relationship between loss of cholesterol synthesis control and malignancy/growth rate.

COLEMAN: Let me respond to your last comment first. As a matter of fact, I think that there may be a correlation between the loss of cholesterogenesis and cancer cell growth rate, but this has not been studied very carefully although it deserves to be. The data that do exist on this question are sometimes confusing, probably because of different methodologies employed (including the normalization of results to all sorts of different data bases, like cell protein, dry or wet weight tissue, tissue lipid, etc.). With regard to what role deregulated cholesterogenesis might play in the proliferation of cancer cells, I lean to the idea, not original with me, that some post-mevalonate intermediate of the sterol synthesis pathway perhaps acts as a "trigger" for new cellular DNA synthesis prior to mitosis. If this turns out to be generally verifiable, then depending on the particular tumor's carbon flux rate, a continuously operating deregulated sterol synthesis pathway would provide the cancer cell with a higher-than-normal steady-state level of such a putative trigger of DNA synthesis. My own convictions here are based on reports that show that tissues with high rates of cell division, such as intestinal mucosa as well as malignant tumors, have high rates of cholesterol synthesis, but tissues with little or no tendency to proliferate, such as muscle or kidney, have low rates.

N. SILIPRANDI (*University of Padova, Padua*): Is the activity of isocitrate dehydrogenase altered in tumor mitochondria? This information is relevant for understanding the fate of acetyl CoA in the presence of the tricarboxylate carrier inhibitor.

COLEMAN: No. I may not have described it clearly, but unlike the situation with citrate as substrate, the tumor mitochondria have no trouble respiring with added isocitrate or any other Krebs cycle intermediate past citrate, even in the presence of BTC. When we block the preferential export of citrate from tumor mitochondria, BTC "forces" citrate (and therefore isocitrate) into the cycle. We observe good state-3 respiration rates with all substrates in tumor mitochondria when BTC is added about 10 seconds after the addition of substrate. We feel this indicates that the Krebs cycle enzymes, including aconitate hydratase and isocitrate dehydrogenase, function normally in the tumor organelles, but with pyruvate and citrate this normal respiration is observed only if the preferential citrate export is blocked.

R. WATTIAUX (*Laboratoire de Chimie Physiologique, Namur, Belgium*): Have your determinations of mitochondrial cholesterol been performed on total mitochondria or on isolated mitochondrial membranes? If the determinations were done on total mitochondria, another explanation of your results is that the amount of mitochondrial matrix protein is decreased in tumors.

COLEMAN: Well yes, of course, you could be correct if your assumption about the lower mitochondrial matrix protein in tumors is valid. I'm not aware of any data that specifically point to this possibility, but I think some of our data argue against it. In one of my slides (TABLE 2) I showed that in the presence of BTC, the state-3 respiratory rate with citrate now becomes the same for tumor and normal liver mitochondria. I think that if your proposal were correct, I would have expected that a lower matrix protein level in the tumor organelle would yield a commensurately lower state-3 rate under these circumstances, but this is not the case. In these experiments we measured the total (free *plus* esterified) cholesterol content per mg total mitochondrial protein, but in other less extensive studies with sonicated submitochondrial particles, we also saw a significantly higher level of total cholesterol in the tumor preparation.

G. ESPOSITO (*University of Milan, Milan*): How do you interpret the fact that many tumor cell membranes have a very low content of cholesterol?

COLEMAN: I think that this interesting observation you refer to concerns only the plasma membrane of tumor cells, and not intracellular membranes. A while ago, in studies with a murine hepatoma maintained in suspension culture or grown as the ascites *in vivo,* we observed a lower cholesterol level in their isolated plasma membranes compared with normal mouse liver hepatocytes. However, we also found that the tumor cells actively shed or exfoliated microvesicles from their plasma membranes into the extracellular medium, and these membrane vesicles turned out to be extraordinarily enriched in cholesterol, even more so than normal liver cell plasma membranes. This shedding of cholesterol-rich microvesicles from tumor cell plasma membranes has been reported previously by others, and although we haven't followed up our own studies recently, I think that this cholesterol-rich vesicle exfoliation is quite interesting and may account for the low plasma membrane cholesterol level in tumors.

Membrane Alterations in Cancer Cells:

The Role of Oxy Radicals[a]

TOMMASO GALEOTTI, SILVIA BORRELLO, AND GIORGIO MINOTTI

Institute of General Pathology
School of Medicine
Catholic University
00168 Rome, Italy

LANFRANCO MASOTTI

Institute of Biological Chemistry
University of Parma
43100 Parma, Italy

INTRODUCTION

Radiation, chemicals, and some viruses are known to be environmental causes of cancer. The origin of tumors of different kinds has been associated for a long time to exposure to radiations of different energy.[1,2] Chemicals also can cause cancer either by a direct action on the cells or by enzymatic conversion to an active carcinogenic form. This process often involves the NADPH-cytochrome P-450 electron transport system, located mainly in the endoplasmic reticulum of the hepatocyte but also of several other tissues.[3,4] Free radicals are among the products of this cytochrome P-450–mediated reactions. Substances that give free radical intermediates are halogenoalkanes, nitro-compounds, aromatic amines, quinones, nitrosoamines, polycyclic hydrocarbons, and polyunsaturated fatty acids. These intermediates have been found to play a significant role in chemical carcinogenesis.[5,6] Free radicals are indeed responsible for a series of damaging reactions within the cell. Among their targets are DNA, nucleotide-coenzymes, thiol-dependent enzymes and membranes, whose damage is known to be induced by lipid peroxidation, responsible for structural and functional modifications as well as production of toxic intermediates.[7]

It has of course to be considered that highly reactive radicals are very short lived and cannot diffuse far from the site of generation. This consideration is particularly relevant for radicals that may have DNA as a target and are produced far from the nucleus. On the other hand, slow reacting radicals should not have too great a damaging effect even if they can diffuse more freely within cell.

Among the damaging effects of free radicals, lipoperoxidation has been studied with particular attention because of a number of important biological consequences. It consists of a chain reaction resulting in the oxidative degradation of unsaturated lipids, particularly of polyunsaturated fatty acid residues of phospholipids that constitute the bilayer of the membranes.[8,9] Several are the products of lipoperoxidation: alkanals,

[a]Supported by MPI 40% 1984 and by Consiglio Nazionale delle Ricerche, special project 'Oncology', grant no. 85.02171.44 (T.G.) and grant no. 85.02241.44 (L.M.).

alkenals, and 4-hydroxy-alkenals, ketones, alkanes, epoxy- and hydroxy-fatty acids, lipid hydroperoxides, and leukotrienes.[10]

The cellular sources of free radicals are subcellular organelles (mitochondria,[11–14] endoplasmic reticulum,[15–17] nuclear membrane,[16,18,19] peroxisomes,[20] and plasma membrane)[21] and cytoplasmic components. Species such as the reactive oxy radicals (O_2^-, H_2O_2, $OH\cdot$, singlet O_2) are generated from oxygen metabolism and potentiated by radiation, hyperoxia, and inflammation.

Lipid peroxides are byproducts of free radical chain reaction or of prostanoid metabolism. The latter occurs in the plasma membrane and has as major sources the catalytic action of lipoxygenase and prostaglandin synthetase.[22]

Defenses against the damaging action of oxy radicals are available within the cell. They are represented by the protective enzymes superoxide dismutase, glutathione peroxidase, and catalase and by antioxidants, such as α-tocopherol, β-carotene, ascorbic acid.[7]

Of special interest is the involvement of lipid peroxidation in cancer.[10,23,24] The major effects of the products of lipid peroxidation are inhibition of DNA synthesis,[10] cell division,[25–29] and tumor growth.[30] These effects and their role in cancer can be better understood by studying the regulation of lipid peroxidation in membranes from tumor cells.

It has been widely reported that in subcellular fraction, such as mitochondria and microsomes, lipid peroxidation is low.[31–36] In hepatomas, particularly, peroxidizability is much less than in hepatocytes and the rate of lipid peroxidation consistently slower.[34–36]

An interesting hypothesis to be tested is whether lipoperoxides have a role in the control of tumor growth and malignancy. Since lipoperoxides inhibit cell division, their lower concentration in hepatomas might result in a loss of control in cell proliferation. Testing this hypothesis required measurement of the levels of protective enzymes, the concentration of antioxidants, the kinetics of lipoperoxidation in different cellular membranes, and the modifications of static and dynamic properties of the membranes brought about by the degradation of the lipids.

The results of these experiments and their interpretation are reported in the following sections.

HEPATOMA MICROSOMAL AND PLASMA MEMBRANES: THEIR PEROXIDIZABILITY AND STRUCTURAL CHARACTERISTICS AS FUNCTION OF TUMOR GROWTH

Microsomes

Chemical Composition and Physical Parameters

Early work had been focused on the protective enzymes against the active oxygen species in normal and tumor cells. The most significant result had been that in general tumor cells show a decrease of enzymatic antioxidant activity.[13,37–41] It was therefore necessary to achieve wider information on the cellular production of such reactive species, on the defense mechanisms of the cell, and on the functional and structural effects on membranes of the damaging action of oxy radicals. If differences between normal and transformed cells have to be established it is desirable to employ tumors with different growth rates and degrees of differentiation: those of the Morris hepatoma series[42] afforded us such system. Indeed, it is well known that in these hepatomas the degree of differentiation is inversely proportional to the growth rate.

The first objective of our research was to characterize the membranes isolated from different tumor lines in terms of chemical composition and structural parameters. Microsomal membranes were chosen primarily because of their importance in the study of oxygen-induced peroxidative damage. The physical parameters of the membranes, i.e. the molecular order (static) and the fluidity (dynamic), were determined respectively as the order parameter and the correlation times of 1,6-diphenyl-1,3,5-hexatriene (DPH) by means of fluorescence depolarization studies. TABLE 1 shows the chemical composition and the abovementioned parameters for the different microsomal membranes. The lipid-to-protein ratio decreases because of the concomitant loss of lipids and slight increase in protein content. There is also an increase of the cholesterol-to-phospholipid ratio because the lipid loss is coupled to some increase in the cholesterol content.[45] It is important to underline the decrease in the double-bond index with increasing growth rate of the tumors. These findings can explain the observed increased molecular order of the lipids and the decreased fluidity.

Protective Enzymes and Lipid Peroxidation

TABLE 2 reports the activities of the oxy radical scavenging cytosolic enzymes and the peroxidizability of the membrane lipids, expressed as malondialdehyde and lipid hydroperoxide formation, of microsomes isolated from rat liver and hepatomas.

With respect to the enzyme activities, the most relevant result is the decrease of protective capability with increasing growth rate of the tumor. As a consequence, one would expect that in tumor membranes a higher content of lipid peroxidation byproducts should be found. Surprisingly, tumor membranes appear to be less susceptible to the damaging effect of O_2^- radicals, as indicated by their lower peroxidation rates with respect to the normal values.

In conclusion, these studies have established that in the case of hepatomas a correlation exists between the growth rate of the tumor, the chemical composition of the membrane (in terms of lipid content, cholesterol content, and degree of fatty acid unsaturation), and the molecular order of lipid bilayer and microviscosity. It was also

TABLE 1. Chemical Composition and Structural Parameters of Microsomal Membranes Isolated from Hepatocytes and Morris Hepatomas with Different Growth Rate

Microsomal Membranes	PL/Protein[a] (w/w)	Cholesterol/PL (mole/mole)	Double-Bond Index[b]	$\langle P_2 \rangle_T^c$	τ_R(ns)[d]
Rat liver	0.66	0.15	82	0.074	7.7
Hepatoma 9618A	0.60	0.18	68	0.120	10.0
Hepatoma 44	0.54	0.24	60	0.224	10.0
Hepatoma 3924A	0.32	0.33	50	0.488	9.1

[a]PL, phospholipid.

[b]The double bond index was calculated according to Waite *et al.*[43]

[c]$\langle P_2 \rangle_T$, second rank-order parameter. Temperatures were 52°C for control, 51.4°C for 9618A, 53°C for H44, and 52.2°C for 3924A.

[d]τ_R, correlation time. It is the reciprocal of the value of the component of the probe diffusion tensor perpendicular to the long axis of DPH.[44]

TABLE 2. Content of Cu-Zn Superoxide Dismutase and Activities of Cytosolic Glutathione Peroxidase and Catalase and Microsomal Lipid Peroxidation of Rat Liver and Morris Hepatomas

Tissue	Cu-Zn SOD[a] (μg/g wet weight)	GSH-Px[b] (nmol/min/mg protein)		Catalase[c] (units/mg protein)	Lipid Peroxidation[d] (nmol/10 min/mg protein)	
		Se	non Se		MDA[e]	LOOH[f]
Rat Liver	224.6	215.0	133.3	138.0	76.9	300.2
Hepatoma 9618A	136.7	49.3	166.7	101.0	64.6	216.3
Hepatoma 44	54.3	21.0	94.3	NA[g]	NA	NA
Hepatoma 3924A	11.7	27.7	4.3	6.8	15.1	51.8

[a,b]Data from Bartoli *et al.*[46,47]
[c]Data from Mochizuki *et al.*[37]
[d]Lipid peroxidation was induced by O_2^- radicals. For experimental details see legend to FIGURE 1.
[e]MDA, malondialdehyde.
[f]LOOH, lipid hydroperoxides.
[g]NA, not available.

shown that when the growth rate of the tumors increases the *in vitro* peroxidizability of their membranes decreases, while the protective activity against oxy radicals is diminished. As a consequence, the answer to this particular behavior of tumor membranes, at least in *in vitro* studies, must be looked for in directions other than the enzyme antioxidant activity.

Plasma Membranes

As pointed out in the introduction, in order to fully investigate the role of lipid peroxidation in cancer, a strategy of approach would be the investigation of more than one cellular site where there are sources of radicals as well as targets of their action. Several important metabolic processes take place in the endoplasmic reticulum, while the machinery for cell-cell interactions is located in the plasma membrane. From these considerations stems the interest in characterizing the plasmalemma, as done previously for the microsomes. The data available so far are limited to hepatocytes and hepatoma 3924A. The choice was made owing to the fact that the latter is, among those available to us, the most "deviated" tumor and therefore the comparative study of the two membrane systems could afford the most evident structural and functional differences. Characterization of plasmalemmas from slower growing tumors is in progress to get a complete picture of the possible correlation between lipid peroxidation and membrane modifications and tumor growth rate.

The results of these studies are presented in TABLE 3. It is evident that lipid peroxidation induced by superoxide radicals is much lower in the tumor plasmalemma when compared to the liver membrane. In hepatoma 3924A plasmalemmas there is a decrease of phospholipid-to-protein ratio and a remarkable increase of cholesterol-to-phospholipid ratio as compared to normal rat liver cells. However, no increase in molecular ordering has been observed in fast-growing tumor plasma membrane compared to normal. This is probably due to the fact that the molecular order is already very high in normal plasma membranes. It is worth remembering that in

TABLE 3. Chemical Composition, Structural Parameters, and Peroxidizability of Plasma Membranes from Hepatocytes and Morris Hepatoma 3924A

Plasma Membranes	PL/Protein[a] (w/w)	Cholesterol/Protein (w/w)	Cholesterol/PL (w/w)	$\langle P_2 \rangle_T^b$	τ_R(ns)	Lipid Peroxidation (nmol/10 min/mg protein)	
						MDA	LOOH
Rat liver	0.46	0.10	0.22	0.508	7.3	55.5	218.3
Hepatoma 3924A	0.28	0.12	0.43	0.502	10.5	2.5	Nil

[a]The symbols are as in TABLES 1 and 2.
[b]Temperatures were 51.5°C for rat liver and 52.6°C for hepatoma 3924A.

microsomes the order parameters range from about 0.1 to about 0.5, possibly due to the fact that these membranes are richer in phospholipids and that their cholesterol/phospolipid ratio is lower. The correlation time of DPH in the two plasma membranes tested indicates that the probe rotational motion is much more restricted in the tumor.

The data obtained for plasma membranes, therefore, concur to form a picture where the tumors seem to be characterized by a higher structural rigidity.

OXY RADICAL-MEDIATED CHANGES IN THE FUNCTION, COMPOSITION, AND STRUCTURE OF RAT LIVER AND HEPATOMA MICROSOMAL MEMBRANES

The data reported in the previous section have demonstrated that the *in vitro* susceptibility to peroxidation of microsomal and plasma membrane phospholipids is decreased in hepatomas. This is surprising, considering the finding that protective enzymes against toxic species of oxygen (e.g. superoxide dismutase and glutathione peroxidase) are also diminished and the decrease is more pronounced in those tumors that exhibit the lowest peroxidation rates. FIGURE 1 clearly shows such a behavior in a series of experiments carried out comparatively in microsomal membranes from rat liver and the "maximal deviation" hepatoma 3924A. During 40 min of incubation in the presence of O_2^- radicals, the formation of malondialdehyde and lipid hydroperoxides is strongly depressed in the tumor membranes, under experimental conditions that allow these byproducts to accumulate in large amounts from rat liver microsomal phospholipids. The low susceptibility to peroxidation of hepatoma 3924A microsomes might depend on an increase in the membrane antioxidant vitamin E, whose high concentration has been taken into account to explain the resistance of lung and heart microsomes to peroxidative agents.[48] However, previous work from our laboratory[36] has ruled out this possibility by measurements of both the vitamin E content and the lipid peroxidation activity in vitamin E–deficient hepatoma 3924A microsomes. The concentrations of vitamin E found in rat liver and hepatoma microsomes are not significantly different and no increase in lipid peroxidation is observed in vitamin E–deficient hepatoma microsomes with respect to the control.

The changes in lipid composition and in the structural organization of normal microsomal membranes exposed to O_2^- radicals may provide some explanation concerning the mechanism by which membranes of tumors with low antioxidant enzymatic defenses present the most pronounced differences in the chemical and physical parameters previously described. The induction of lipid peroxidation by O_2^- radicals in rat liver microsomes causes an increase in saturated fatty acid content, a slight increase in monoenoic and a marked decrease in polyenoic acyl residues. As a consequence, the double-bond index is reduced to about one half that of the control. In contrast, tumor membranes, which already have a very low degree of fatty acid unsaturation, do not show significant changes in the double-bond index (TABLE 4).

Therefore, the resistance to lipid peroxidation *in vitro* of hepatoma microsomes seems to depend on the low availability of polyunsaturated fatty acids rather than an increased quenching capability of the membrane towards the oxy radicals. The discrepancy between the resistance to peroxidation and the low level of oxy-radical scavenging enzymes might then be only apparent if the former characteristic is considered as the consequence of a damage that tumor membranes have already suffered *in vivo*, owing to the low enzymatic protection against toxic species of oxygen. The data, shown in FIGURE 2, on the structural changes induced in microsomal

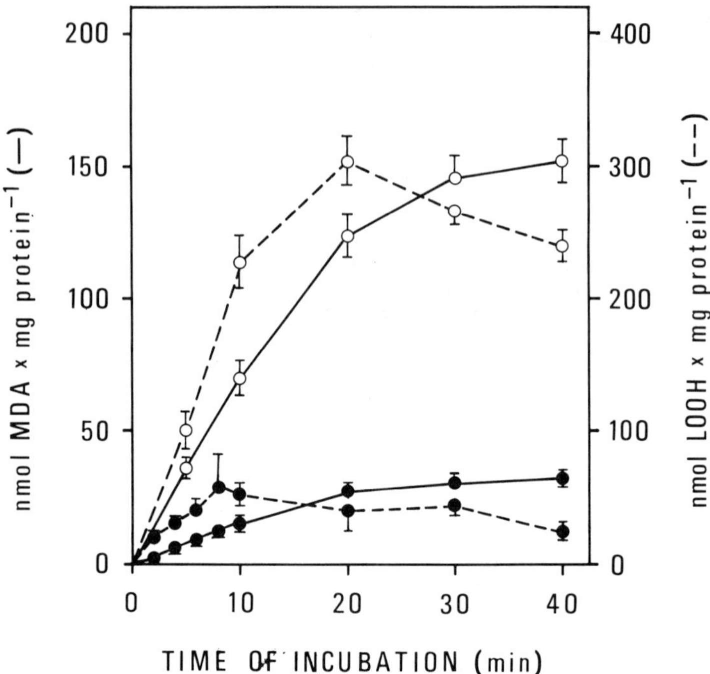

FIGURE 1. Time course of superoxide-induced lipid peroxidation of microsomal membranes isolated from rat liver (○) and Morris hepatoma 3924A (●). The formation of malondialdehyde (MDA) and lipid hydroperoxides (LOOH) was determined at 25°C during incubation, under an oxygen atmosphere, in a Dubnoff metabolic shaker. Microsomes (0.2 mg protein/ml) were suspended in 0.15 M KCl, 50 mM Tris-HCl (pH 7.5), containing 0.33 mM xanthine, 1 mM ADP, and 0.05 mM $FeCl_3$. The reaction was started by 50 μg/ml xanthine oxidase. MDA was assayed at 535 nm by the thiobarbituric acid method; LOOH were determined at 353 nm by the iodometric assay. Values are means of four to six experiments; the vertical lines correspond to twice the SE.

TABLE 4. Changes in the Chemical Composition of Fatty Acid Chains after Exposure to Oxy Radicals of Microsomal Membrane Phospholipids from Rat Liver and Morris Hepatoma 3924A

Microsomal Membranes	% Fatty Acid Classes[a]			Double-bond Index
	Saturated	Monounsaturated	Polyunsaturated	
Rat liver				
Control	41.2	5.9	35.0	102
Peroxidized, 5 min	47.3	7.8	24.0	73
Peroxidized, 10 min	54.2	7.8	19.7	55
Hepatoma 3924A				
Control	28.6	13.5	10.0	32
Peroxidized, 5 min	37.1	16.9	12.0	38
Peroxidized, 10 min	38.5	17.2	9.0	27

[a]The fatty acids reported are 16:0, 18:0, 18:1, 18:2, and 20:4.

membranes by exposure to O_2^- radicals seem to support this hypothesis. The dependence of the order parameter $\langle P_2 \rangle$ of DPH on temperature in normal and hepatoma 3924A microsomes, at different degrees of peroxidation, is illustrated. $\langle P_2 \rangle$ increases with the time of peroxidation in rat liver but not in hepatoma membranes. In the latter the value of this parameter is already at a level comparable to that of the normal membranes that have undergone peroxidation *in vitro* during 10 min of exposure to O_2^- radicals.

The fluorescence data presented so far have been analyzed by applying a mathematical approach that takes into account simultaneously static and dynamic fluorescence parameters.[44] The investigation has been extended[49] by measuring the fluorescence anisotropy decay that allows a more precise determination of such parameters. Inspection of FIGURE 3 shows that a long time plateau value (r_{∞}) different from zero is obtained, as expected, as it depends on the orientational order for the

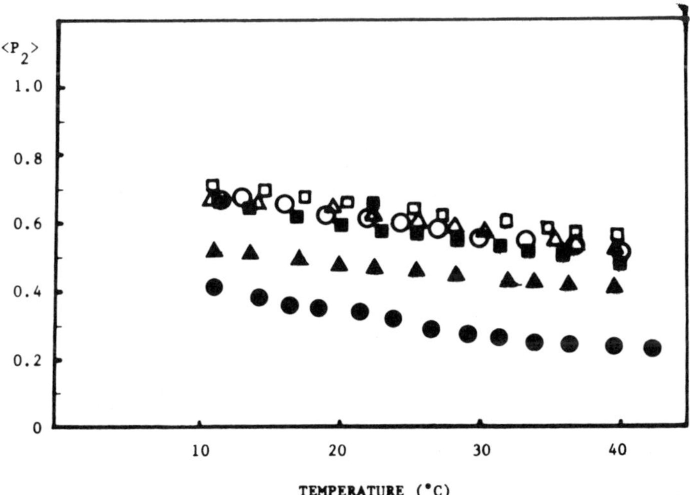

FIGURE 2. Plot of the order parameter of DPH vs. temperature of microsomal membranes. Rat liver microsomes: not peroxidized (●), after 5 min (▲), and 10 min (■) peroxidation. Hepatoma 3924A microsomes: not peroxidized (○), after 5 min (△), and 10 min (□) peroxidation.

probe. For normal membranes, $\langle P_2 \rangle$ increases from 0.52 in control to 0.53 and 0.57 in membranes peroxidized for 5 and 10 min, respectively. Tumors are more ordered as $\langle P_2 \rangle$ changes from 0.68 to 0.71 and 0.77 in the same order as normal membranes. Correlation times for normal membranes are 2.5 nsec, 4.4 nsec, and 6.8 nsec in the experimental conditions as above, indicating that peroxidation hinders the movement of the lipid chains. Tumors, with correlation times of 2.4 nsec, 3.5 nsec, and 3.4 nsec, show a much lower increase in this parameter, that is to say that the fluidity decreases with peroxidation to a much smaller extent.

An increase in the order parameter of the bilayer, dependent on the peroxidation time, has been found also in model[50] and rat liver microsomal membranes[51] by the use of the spin probes, 5-, 12-, and 16-doxylstearic acid. Moreover, our order parameters calculated from fluorescence depolarization data are in good agreement with those reported by other authors in rat liver microsomes.[52]

In conclusion, some of the chemical and physical alterations observed in cancer cell membranes seem to be due to an oxy radical–mediated mechanism, involving lipid peroxidation, under conditions of reduced enzymatic defenses. The evidence comprises (1) the loss of enzymes that scavenge superoxide, hydrogen peroxide, and lipid peroxides; (2) the similarity of certain chemical and structural properties of tumor membranes to those of corresponding normal membranes damaged by a peroxidative insult; and (3) the slight changes that peroxidative agents are able to exert *in vitro* on the same physicochemical properties in cancer cell membranes.

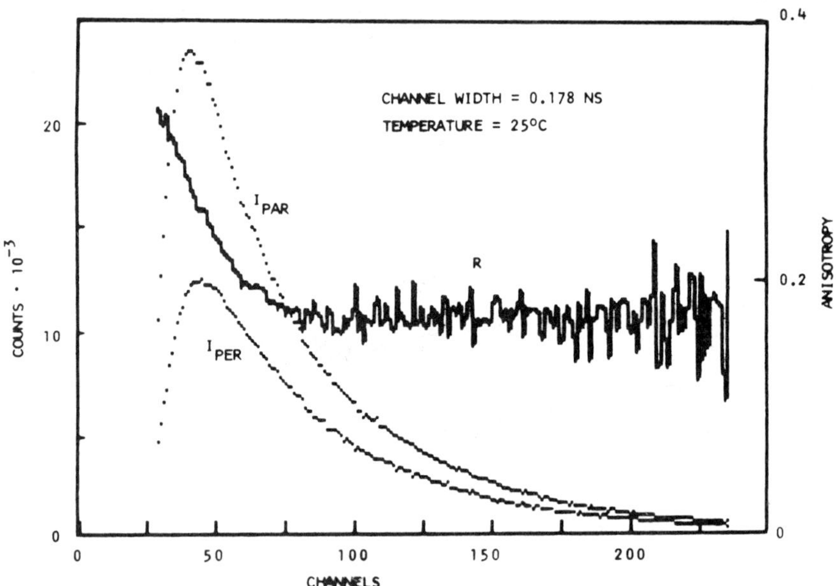

FIGURE 3. Parallel (I_{PAR}), perpendicular (I_{PER}) fluorescence intensities, and fluorescence anisotrophy (R) experimental decay course for DPH in tumor microsomal membranes. Typical decay in microsomes peroxidized 5 min.

SUMMARY

Membranes isolated from tumor cells present profound alterations in their composition, structural organization, and functional properties. In this study we have reported some of these alterations in microsomal and plasma membranes of hepatomas with different growth rate and degree of differentiation. The chemical parameters studied were the phospholipid-to-protein, the cholesterol-to-protein, and the cholesterol-to-phospholipid ratios and the fatty acid composition of the phospholipids. The physical parameters were the molecular order (static) and the fluidity (dynamic), determined, respectively, as the order parameter $\langle P_2 \rangle$ and the correlation time τ_R of the fluorescent probe 1,6-diphenyl-1,3,5-hexatriene (DPH). The functional property investigated was the ability of the membranes to undergo superoxide-induced lipid peroxidation, determined as byproduct (malondialdehyde and lipid hydroperoxides) formation and as changes in the fatty acid acyl residues. Changes in the physical state of the membrane, induced by oxy radicals, were also monitored during lipid peroxidation. A study of the antioxidant activity of the tumor cell, in terms of oxy radical enzymatic defenses (superoxide dismutase, glutathione peroxidase and catalase) was also performed.

The main results obtained are the following: (1) hepatoma membranes possess a lower phospholipid content and a lower degree of fatty acid unsaturation; on the other hand, the cholesterol-to-phospholipid ratio is increased; (2) the physical state appears characterized by an increased rigidity (increased molecular order of the lipids and decreased fluidity); (3) the membrane peroxidizability is markedly depressed and its order parameter, in contrast to liver membranes, does not increase with exposure to the action of O_2^- radicals; and (4) the oxy radical enzymatic defense mechanisms are decreased. All these alterations increase with increasing growth rate and dedifferentiation of the tumor. Considering all of the data, we are inclined to think that tumor membranes are altered structurally and functionally in part as the result of an oxy radical–induced damage that takes place *in vivo* under conditions of increased oxygen toxicity.

ACKNOWLEDGMENTS

We wish to thank Mr. Guglielmo Palombini for skillful technical assistance and Mrs. Clotilde Castellani for preparation of the manuscript.

REFERENCES

1. Upton, A. C. 1982. Physical carcinogenesis: radiation-history and sources. *In* Cancer. F. F. Becker, Ed. **1:** 551. Plenum Press. New York.
2. Storer, J. B. 1982. Radiation carcinogenesis. *In* Cancer. F. F. Becker, Ed. **1:** 629. Plenum Press. New York.
3. Miller, E. C. & J. A. Miller. 1981. Mechanisms of carcinogenesis. Cancer **47:** 1055.
4. Weisburger, J. H. & G. M. Williams. 1982. Metabolism of chemical carcinogens. *In* Cancer. F. F. Becker, Ed. **1:** 241. Plenum Press. New York.
5. Mason, R. P., W. G. Harrelson, B. Kalyanaraman, C. Mottley, F. J. Peterson & J. L. Holtzman. 1982. Free radical metabolites of chemical carcinogens. *In* Free Radicals, Lipid Peroxidation and Cancer. D. C. H. McBrien & T. F. Slater, Eds.: 378. Academic Press. London.
6. Ts'o, P. O. P., W. J. Caspary & R. J. Lorentzen. 1977. The involvement of free radicals in chemical carcinogenesis. *In* Free Radicals in Biology. W. H. Pryor, Ed. **3:** 251. Academic Press. New York.
7. Freeman, B. A. & J. D. Crapo. 1982. Biology of disease. Free radicals and tissue injury. Lab. Invest. **47:** 412.
8. Pryor, W. A. 1984. Free radicals in autoxidation and in aging. *In* Free Radicals in Molecular Biology, Aging, and Disease. D. Armstrong, R. S. Sohal, R. G. Cutler & T. F. Slater, Eds. **27:** 13. Raven Press. New York.
9. Porter, N. A. 1984. Chemistry of lipid peroxidation. Methods Enzymol. **105:** 273.
10. Slater, T. F., K. H. Cheeseman & K. Proudfoot. 1984. Free radicals, lipid peroxidation, and cancer. *In* Free Radicals in Molecular Biology, Aging, and Disease. D. Armstrong, R. S. Sohal, R. G. Cutler & T. F. Slater, Eds. **27:** 293. Raven Press. New York.
11. Boveris, A., N. Oshino & B. Chance. 1972. The cellular production of hydrogen peroxide. Biochem. J. **128:** 617.
12. Boveris, A. & E. Cadenas. 1975. Mitochondrial production of superoxide anions and its relationship to the antimycin insensitive respiration. FEBS Lett. **54:** 311.

13. DIONISI, O., T. GALEOTTI, T. TERRANOVA & A. AZZI. 1975. Superoxide radicals and hydrogen peroxide formation in mitochondria from normal and neoplastic tissues. Biochim. Biophys. Acta **403:** 292.
14. FORMAN, H. J. & A. BOVERIS. 1982. Superoxide radical and hydrogen peroxide in mitochondria. *In* Free Radicals in Biology. W. Pryor, Ed. **5:** 65. Academic Press. New York.
15. AUST, S. D., D. L. ROERIG & T. C. PEDERSON. 1972. Evidence for superoxide generation by NADPH-cytochrome c reductase of rat liver microsomes. Biochem. Biophys. Res. Commun. **47:** 1133.
16. ESTABROOK, R. W. & J. WERRINGLOER. 1976. Cytochrome P_{450}: Its role in oxygen activation for drug metabolism. *In* Drug Metabolism Concepts. M. J. Donald, R. G. Robert & D. C. Washington, Eds.: 1. American Chemical Society. Columbus, OH.
17. BARTOLI, G. M., T. GALEOTTI, G. PALOMBINI, G. PARISI & A. AZZI. 1977. Different contribution of rat liver microsomal pigments in the formation of superoxide anions and hydrogen peroxide during development. Arch. Biochem. Biophys. **181:** 276.
18. BARTOLI, G. M., T. GALEOTTI & A. AZZI. 1977. Production of superoxide anions and hydrogen peroxide in Ehrlich ascites tumour cell nuclei. Biochim. Biophys. Acta **497:** 622.
19. PESKIN, A. V., I. B. ZBARSKY & A. A. KONSTANTINOV. 1980. A novel type of superoxide generating system in nuclear membranes from hepatoma 22a ascites cells. FEBS Lett. **117:** 44.
20. MASTERS, C. & R. HOLMES. 1977. Peroxisomes: new aspects of cell physiology and biochemistry. Physiol. Rev. **57:** 816.
21. WEISS, S. J. & A. F. LO BUGLIO. 1982. Phagocyte-generated oxygen metabolites and cellular injury. Lab. Invest. **47:** 5.
22. MEAD, J. F. 1984. Free radical mechanisms in lipid peroxidation and prostaglandins. *In* Free Radicals in Molecular Biology, Aging, and Disease. D. Armstrong, R. S. Sohal, R. G. Cutler & T. F. Slater, Eds. **27:**53. Raven Press. New York.
23. MCBRIEN, D. C. H. & T. F. SLATER. 1982. Free Radicals, Lipid Peroxidation and Cancer. Academic Press. London.
24. FLOYD, R. A. 1982. Free Radicals and Cancer. Marcel Dekker. New York, N.Y.
25. WILBUR, K. M., N. WOLFSON, C. B. KENASTON, A. OTTOLENGHI, M. E. GAULDEN & F. BERNHEIM. 1957. Inhibition of cell division by ultraviolet irradiated unsaturated fatty acid. Exp. Cell Res. **13:** 503.
26. WOLFSON, N., K. M. WILBUR & F. BERNHEIM. 1956. Lipid peroxide formation in regenerating rat liver. Exp. Cell Res. **10:** 556.
27. HUTTNER, J. J., E. T. GWEBU, R. V. PANGANAMALA, G. E. MILO, D. G. CORNWELL, H. M. SHARMA & J. C. GEER. 1977. Fatty acids and their prostaglandin derivatives: inhibitors of proliferation in aortic smooth muscle cell. Science **197:** 289.
28. HUTTNER, J. J., G. E. MILO, R. V. PANGANAMALA & D. G. CORNWELL. 1978. Fatty acids and the selective alteration of *in vitro* proliferation in human fibroblast and guinea pig smooth muscle cells. In Vitro **14:** 854.
29. CORNWELL, D. G., J. J. HUTTNER, G. E. MILO, R. V. PANGANAMALA, H. M. SHARMA & J. C. GEER. 1979. Polyunsaturated fatty acids, vitamin E, and the proliferation of aortic smooth muscle cells. Lipids **14:** 194.
30. LIEPKALNS, V. A., C. ICARD-LIEPKALNS & D. G. CORNWELL. 1982. Regulation of cell division in a human glioma cell clone by arachidonic acid and α-tocopherolquinone. Cancer Lett. **15:** 173.
31. THIELE, E. H. & J. W. HUFF. 1960. Lipid peroxide production and inhibition by tumor mitochondria. Arch. Biochem. Biophys. **88:** 208.
32. UTSUMI, K., G. YAMAMOTO & K. INABA. 1965. Failure of Fe^{2+}-induced lipid peroxidation and swelling in the mitochondria isolated from ascites tumor cells. Biochim. Biophys. Acta **105:** 368.
33. PLAYER, T. Y., D. J. MILLS & A. A. HORTON. 1977. NADPH-dependent lipid peroxidation in mitochondria from livers of young and old rats and from rat hepatoma D30. Biochem. Soc. Trans. **5:** 1506.
34. BARTOLI, G. M. & T. GALEOTTI. 1979. Growth-related lipid peroxidation in tumour microsomal membranes and mitochondria. Biochim. Biophys. Acta **574:** 537.

35. GALEOTTI, T., G. M. BARTOLI, S. BARTOLI & E. BERTOLI. 1980. Superoxide radicals and lipid peroxidation in tumour microsomal membranes. *In* Biological and Clinical Aspects of Superoxide and Superoxide Dismutase. W. H. Bannister & J. V. Bannister, Eds.: 106. Elsevier/North Holland. New York.
36. BORRELLO, S., G. MINOTTI, G. PALOMBINI, A. GRATTAGLIANO & T. GALEOTTI. 1985. Superoxide-dependent lipid peroxidation and vitamin E content of microsomes from hepatomas with different growth rates. Arch. Biochem. Biophys. **238:** 588.
37. MOCHIZUKI, Y., Z. HRUBAN, H. P. MORRIS, A. SLESERS & E. L. VIGIL. 1971. Microbodies of Morris hepatomas. Cancer Res. **31:** 763.
38. BOZZI, A., I. MAVELLI, A. FINAZZI-AGRO', R. STROM, A. M. WOLF, B. MONDOVI' & G. ROTILIO. 1976. Enzyme defense against reactive oxygen derivatives. II Erythrocytes and tumor cells. Mol. Cell. Biochem. **10:** 11.
39. PESKIN, A. V., Y. M. KOEN, I. B. ZBARSKI & A. A. KONSTANTINOV. 1977. Superoxide dismutase and glutathione peroxidase activities in tumors. FEBS Lett. **78:** 41.
40. OBERLEY, L. W. & G. R. BUETTNER. 1979. Role of superoxide dismutase in cancer: a review. Cancer Res. **39:** 1141.
41. BIZE, I. B., L. W. OBERLEY & H. P. MORRIS. 1980. Superoxide dismutase and superoxide radical in Morris hepatomas. Cancer Res. **40:** 3686.
42. MORRIS, H. P. & B. P. WAGNER. 1968. Induction and transplantation of rat hepatomas with different growth rate (including "minimal deviation" hepatomas) *In* Methods in Cancer Research. H. Bush, Ed. **4:** 125. Academic Press. New York.
43. WAITE, M., B. PARCE, R. MORTON, C. CUNNINGHAM & H. P. MORRIS. 1977. The deacylation and reacylation of phosphoglyceride in microsomes of Morris hepatoma 7777 and host rat liver. Cancer Res. **37:** 2092.
44. MASOTTI, L., P. CAVATORTA, G. SARTOR, E. CASALI, A. ARCIONI, C. ZANNONI, G. M. BARTOLI & T. GALEOTTI. 1982. A fluorescence depolarization investigation of membranes from tumour cells with different growth rate. *In* Membranes in Tumour Growth. T. Galeotti, A. Cittadini, G. Neri & S. Papa, Eds.: 39. Elsevier Biomedical Press. Amsterdam.
45. HOSTETLER, K. Y. 1982. Personal communication.
46. BARTOLI, G. M., S. BARTOLI, T. GALEOTTI & E. BERTOLI. 1980. Superoxide dismutase content and microsomal lipid composition of tumors with different growth rate. Biochim. Biophys. Acta **620:** 205.
47. BARTOLI, G. M., T. GALEOTTI, S. BORRELLO & G. MINOTTI. 1982. Loss of defensive enzyme and oxygen-mediated damage of microsomal membranes in liver and hepatomas. *In* Membranes in Tumour Growth. T. Galeotti, A. Cittadini, G. Neri & S. Papa, Eds.: 461. Elsevier Biomedical Press. Amsterdam.
48. KORNBRUST, D. J. & R. D. MAVIS. 1980. Relative susceptibility of microsomes from lung, heart, liver, kidney, brain and testes to lipid peroxidation: correlation with vitamin E content. Lipids **15:** 315.
49. CAVATORTA, P., L MASOTTI, G. SARTOR, M. B. FERRARI, E. CASALI, S. BORRELLO, G. MINOTTI & T. GALEOTTI. 1985. Lipid peroxidation of microsomes from Morris hepatoma 3924A: order parameter and dynamic properties determined by fluorescence anisotropy decay. *In* Cell Membranes and Cancer. T. Galeotti, A. Cittadini, G. Neri, S. Papa & L. A. Smets, Eds.: 269. Elsevier Science Publishers. Amsterdam.
50. BRUCH, R. C. & W. S. THAYER. 1983. Differential effect of lipid peroxidation on membrane fluidity as determined by electron spin resonance probes. Biochim. Biophys. Acta **733:** 216.
51. CURTIS, M. T., D. GILFOR & J. L. FARBER. 1984. Lipid peroxidation increases the molecular order of microsomal membranes. Arch. Biochem. Biophys. **235:** 644.
52. EICHENBERGER, K., P. BÖHNI, K. H. WINTERHALTER, S. KAWATO & C. RICHTER. 1982. Microsomal lipid peroxidation causes an increase in the order of the membrane lipid domain. FEBS Lett. **142:** 59.

DISCUSSION OF THE PAPER

P. PEDERSEN (*Johns Hopkins University, Baltimore, MD*): Let me ask you the same general question I asked Peter Coleman with regard to cholesterolgenesis in tumors. Would you comment on whether the aberrations observed in "free radical" metabolism provide an advantage or a nuisance for the cancer cell? I would suspect that aberrations in "free radical" metabolism might provide an advantage as these aberrations are more profound in more malignant tumors, i.e. those that are poorly differentiated and have a high growth rate.

GALEOTTI: The observation that membranes isolated from fast-growing tumors have characteristics similar to "peroxidized" membranes, together with the decreased enzymatic defense mechanisms observable in such tumors, suggest that *in vivo* intracellular membranes, after having been subjected to an early peroxidative damage, become then unable to produce sufficient amounts of peroxidation products (lipid peroxides, epoxides, aldehydes), known to inhibit mitotic activity.

R. WATTIAUX (*University of Namur, Namur, Belgium*): You know that microsomes are heterogeneous. They are made of Golgi, endoplasmic reticulum membranes, and plasma membranes. Are you sure that the membrane composition of the rat liver microsomes is the same as the membrane composition of hepatomas? I think that such a similarity is a prerequisite to assess the comparisons you have presented.

GALEOTTI: Comparison between chemical composition and physical properties of normal and tumor microsomes has been done in many other laboratories. The membrane population may be slightly different in the two tissue preparations. However, the differences found are so marked that the exact similarity of the membrane preparation does not represent a prerequisite to assess the comparisons.

Oncogenes and Phosphatidylinositol Turnover[a]

L. C. CANTLEY,[b] M. WHITMAN,[b] S. CHAHWALA,[b]
L. FLEISCHMAN,[b] D. R. KAPLAN,[d]
B. S. SCHAFFHAUSEN,[c] AND T. M. ROBERTS[d]

[b]*Department of Physiology*

[c]*Department of Biochemistry*
Tufts University School of Medicine

[d]*Laboratory of Neoplastic Disease Mechanisms*
Department of Pathology
Dana Farber Cancer Institute
Harvard Medical School
Boston, Massachusetts 02111

INTRODUCTION

The proliferation of nontransformed vertebrate animal cells in culture requires not only nutrients but also specific growth factors. In the absence of such factors, cells arrest in G_0/G_1 and fail to synthesize DNA. Readdition of growth factors results in DNA synthesis after a lag of 8 to 12 hours. The intracellular events that lead to new DNA synthesis are obscure. However, the magnitude and diversity of cellular responses that occur as a consequence of occupation of a relatively small number of receptors by growth factors argue for an amplification pathway involving one or more second messengers.[1] The best understood second messenger pathway is cAMP generation in response to beta adrenergic agonists. This system is activated within seconds of beta-agonist binding to a specific receptor and results in pleiotropic changes in the cell, many of which can be explained by cAMP activation of a specific protein kinase. In the case of mitogen addition to quiescent cells a number of plasma membrane and cytosolic events are also activated within seconds to minutes of stimulation. However, most of these events appear not to be mediated by cAMP, suggesting the existence of other second messengers.[2]

In the past five years a number of important discoveries have been made that not only suggest candidates for second messengers of growth factor stimulation but also suggest that transformation of cells by oncogenes occurs by amplification of the same second messengers that are regulated by growth factors. A key piece of this puzzle was provided by the observation that the cellular receptor for tumor-promoting phorbol esters (agents that stimulate many of the same responses in cells as growth factors) is a calcium, phospholipid, and diacylglycerol-activated protein kinase (C-kinase).[3] Phorbol esters activated this enzyme by replacing the normal regulator, diacylglycerol. This observation suggested that the numerous cellular events that had been observed to

[a]Supported by grants GM36133 (L. C.), CA34722 (B. S.), and CA30002 (T. R.) from the National Institutes of Health. L. C. and B. S. are Established Investigators of the American Heart Association. M. W. is a National Science Foundation Predoctoral Fellow and D. K. is a predoctoral trainee supported by NSR Award GM07196.

occur in response to phorbol esters (including rapid changes in ion fluxes, phosphorylation of ribosomal proteins, induction of fos and myc genes, and growth promoting activity) are mediated by this kinase. Even more interesting was the implication that elevation of cellular diacylglycerol is the critical signal for this response. The observation that several mitogens and most notably platelet-derived growth factor (PDGF) can elevate cellular diacylglycerol levels by stimulating cleavage of phosphatidylinositol (PI) implicated the PI turnover response in growth regulation.[4] The role of PI turnover in growth regulation became even more interesting with the discovery that it is a phosphorylated form of PI that is first broken down in response to hormones and that in addition to diacylglycerol another second messenger is also generated, inositol 1,4,5-trisphosphate (IP_3).[5] IP_3 was shown to regulate cytosolic calcium levels by

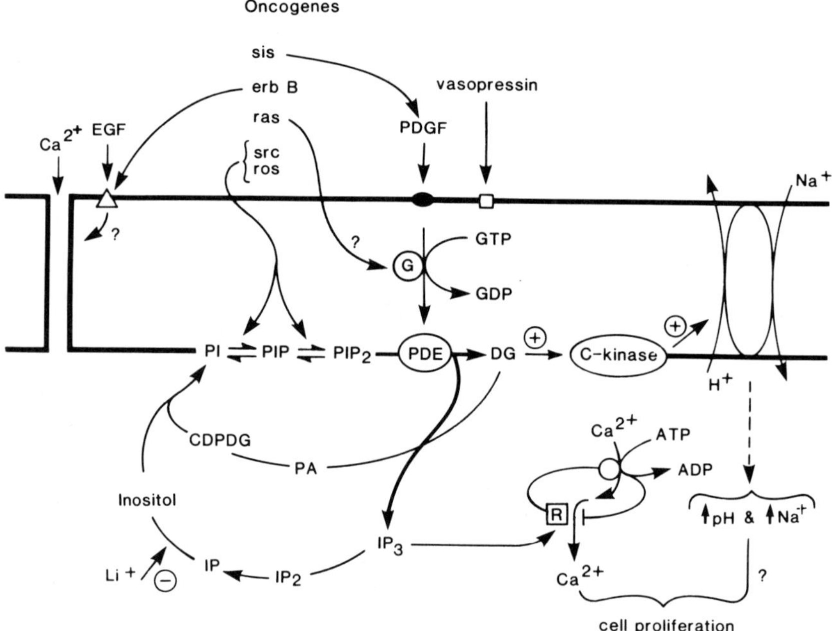

FIGURE 1. A model for potential sites of action of oncogene products on regulatory steps of phosphatidylinositol turnover and ion fluxes. See the text for discussion.

stimulating calcium release from the endoplasmic reticulum. Thus, when mitogens such as PDGF bind to target cells the diacylglycerol produced from PI turnover activates C kinase and the IP_3 elevates cytosolic calcium, which in turn activates a diverse set of responses mediated by calcium binding proteins.

Not all mitogens are capable of rapid activation of PI turnover. For example, the primary response to epidermal growth factor (EGF) appears to be calcium influx from outside the cell rather than PI turnover,[6] although a secondary PI turnover response may occur in some cells. In addition it seems likely that the various growth factor receptors activate other cellular responses independent of and in parallel with the changes in PI turnover and calcium elevation.[1] What these other responses are is unclear. The mechanism by which growth factor binding to receptor results in activation of calcium influx or PI turnover or other primary changes in the cell is still not known.

A clue to the mechanism of growth factor receptor action was provided from work on virus-encoded oncogenes. In the late seventies and early eighties a host of transforming genes was found to encode for protein tyrosine kinases.[7] Phosphorylation of proteins on tyrosine is a relatively rare event and had not been previously observed. Subsequently, Cohen's group discovered that the EGF receptor has EGF-regulated tyrosine kinase activity, suggesting a link between oncogenes and growth factor receptors.[8] This link became solid with the discovery that the virus-encoded oncogene Erb B is a truncated version of the EGF receptor.[9] A second oncogene, v-sis, was shown to be homologous to PDGF whose receptor is also a protein tyrosine kinase.[10] Thus it is now clear that many oncogenes accomplish cell transformation merely by mimicking the function of growth factors or growth factor receptors and that activation of tyrosine kinase activity is critical for the cellular response to many growth factors and oncogenes.

The major missing piece in this puzzle has been identification of the substrates for the tyrosine kinases that are relevant to the pleiotropic responses of cells to growth factors and transforming genes. To date, the identified substrates have provided no satisfactory model. An intriguing link of tyrosine kinase to the PI turnover response was proposed a couple of years ago when two research groups found that PI kinase activity copurified with two different oncogene-encoded tyrosine kinases, pp60$^{v\text{-}src}$[11] and pp68$^{v\text{-}ros}$.[12] The difficulty of separating tyrosine kinase activity from PI kinase activity suggested that these may be intrinsic activities of the same protein. However, more recent results indicate that the PI kinase is a separate enzyme that specifically binds to certain activated tyrosine kinases.[13] The possibility that this PI kinase is regulated by the tyrosine kinase activity of pp60^{src} is attractive since it could at least partially explain the increased flux of phosphate through intermediates of the PI turnover cycle observed in src and ros transformed cells.[11,12] The increased IP_3 and diacylglycerol might then explain many of the pleiotropic responses to oncogene-induced cell transformation. A model suggesting sites at which certain oncogenes may affect PI turnover and calcium influx is presented in FIGURE 1.

Our studies on polyoma virus, a DNA tumor virus, have extended previous observations on the association between transformation and both PI and PIP kinase activities, and have provided genetic, biochemical, and physiological evidence for a functional, rather than fortuitous, association between these activities.[13,14] This paper discusses some of our recent results in an attempt to understand how phosphorylation of phosphatidylinositol is regulated in transformed cells.

RESULTS AND DISCUSSION

The Transforming Gene Product of Polyoma Virus, Middle T Antigen

Middle T antigen is a 56–58 kDa phosphoprotein necessary for polyoma transformation.[15] A fraction of middle T associates at the inner surface of the cell membrane with pp60$^{c\text{-}src}$,[16] the cellular homologue of Rous sarcoma virus pp60$^{v\text{-}src}$. These two proteins immunoprecipitate as a complex from polyoma-infected or transformed cells, and middle T appears to activate pp60$^{c\text{-}src}$ as a tyrosine kinase in these immunoprecipitates.[17] Transformation-defective middle T mutants (FIG. 2) fall into two categories with respect to pp60$^{c\text{-}src}$ association: those that lack normal association with pp60$^{c\text{-}src}$

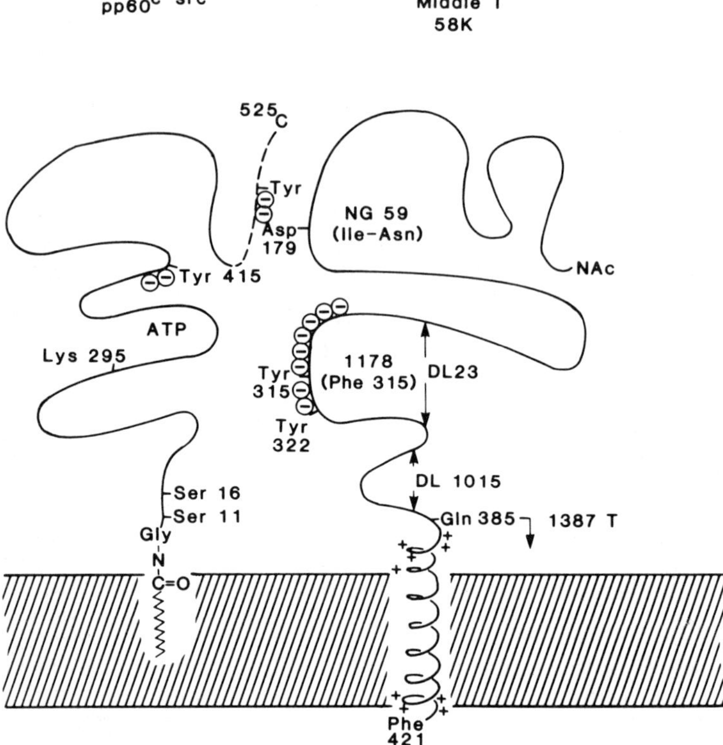

FIGURE 2. A model illustrating locations of mutation sites in the polyoma middle T gene product that affect cell transformation and association with $pp60^{c\text{-}src}$. NG 59 is a transformation-defective mutant in which Asp 179 is replaced by Ile-Asn. 1178 is a partially defective mutant in which the major site of phosphorylation by $pp60^{c\text{-}src}$, Tyr 315, is replaced by a phenylalanine. DL 23 and DL 1015 are deletion mutants that cause a decrease in transforming ability. 1387 T is a termination mutant that lacks the carboxy terminal hydrophobic domain that anchors middle T to the plasma membrane. This mutant is totally defective in transformation.

TABLE 1. Transformation Mutants in the Middle T Gene of Polyoma. Relative Amounts of Tyrosine Kinase Activity and Phosphatidylinositol Kinase Activity Associated with the Immunoprecipitated Middle T Protein

Polyoma Mutant	Transformation Competence	Tyrosine Kinase Activity	PI Kinase Activity
RA (Wild Type)	+	+	100
Mock Infected	–	–	7
NG59	–	–	10
py1387	–	–	8
d123	+/–	+	14
d11015	+/–	+	23
py1178T	+/–	+	23
d11014/py1178T	+/–	+	19

For details of the assays for PI kinase and tyrosine kinase see reference 13.

and do not activate it (Py 1387T, NG59), and those that associate normally with pp60$^{c\text{-}src}$ at the plasma membrane and activate its tyrosine kinase activity (DL23, DL 1015, 1178T, DL1014/1178T) (see Kaplan *et al.*[13] for discussion).

Association of PI Kinase with Immunoprecipitated Middle T

Using immunoprecipitates made from polyoma-infected cells with an antipolyoma T antigen antibody, we have found a phosphoinositide kinase activity associated with the middle T/pp60$^{c\text{-}src}$ complex. As shown in TABLE 1, this activity is found in immunoprecipitates made from cells infected with wild type polyoma virus, but not from cells infected with the transformation-defective middle T mutant py 1387T. Phosphatidylinositol 4-phosphate, but not diacylglycerol, is also phosphorylated by these immunoprecipitates.[14] Middle T/pp60$^{c\text{-}src}$-associated PI kinase activity can also be immunoprecipitated using antibodies to pp60$^{c\text{-}src}$; in this instance no significant PI kinase activity is associated with pp60$^{c\text{-}src}$ in the absence of transformation competent middle T.[13]

We have assayed PI kinase activity in transformation-defective middle T mutants and have found (TABLE 1) every transformation-defective mutant tested to be partially or completely defective in associated PI kinase activity. Particularly interesting are transformation-defective mutants of the second category (d123, d11015). Immunoprecipitates of these middle T mutants phosphorylate exogenously added enolase on tyrosine at rates comparable to wild type. However, these immunoprecipitates are substantially defective in their associated PI kinase activity. The critical substrates for the middle T/pp60$^{c\text{-}src}$ tyrosine kinase *in vivo* are unknown, but these results indicate that for these middle T mutants transforming activity is more closely correlated with association of the complex with PI kinase activity than with tyrosine kinase activity towards an artificial substrate.

Changes in PI Metabolism in Transformed Cells

If the middle T/pp60$^{c\text{-}src}$ complex interaction with a PI kinase activity *in vitro* indeed reflects a regulatory interaction that is physiologically significant *in vivo,* expression of middle T in cells should result in an alteration in polyphosphoinositide levels in these cells. Furthermore, if polyphosphoinositide levels are important in regulating the rate of phosphoinositide turnover, levels of the inositol phosphates should also be changed upon middle T expression. We have found that both cellular polyphosphoinositide and inositol polyphosphate levels in quiescent cells are increased by expression of wild type middle T (TABLE 2), but not by expression of the transformation-defective py1387 middle T.

It is interesting to compare these effects of middle T on PI metabolites *in vivo* with those reported for other stimuli acting on PI turnover. Most hormones are believed to act on PI breakdown by increasing the catalytic efficiency of phospholipase C.[5] Such stimuli result in a rapid decrease in polyphosphoinositide levels accompanied by a concomitant increase in the polyphosphoinositide breakdown products DAG and/or IP_3, often followed by a replenishment of polyphosphoinositides to resting levels. Persistent stimuli acting primarily through direct activation of phospholipase C to raise IP_3 and DAG levels would therefore not be expected to increase PIP_2 above resting levels. We have found that 10% serum (TABLE 2,A) strongly stimulated IP_3 levels without significantly raising PIP_2 or PIP levels. Similarly, transformation of cells by the ras oncogene, which has been postulated to activate phospholipase C,[18] raises cellular DAG levels while decreasing PIP_2 levels (TABLE 2,B). In contrast, we have found that middle T expression results in increased PIP_2 levels concomitant with an increase in IP_3, indicating that middle T acts on the PI cycle *in vivo* primarily by accelerating polyphosphoinositide formation. The oncogene products middle T and p21 ras may therefore act at distinct and experimentally distinguishable control points in the PI turnover pathway to increase levels of the critical second messengers DAG and IP_3. It is possible that serum acts on both the PI (and PIP) kinase steps and the phospholipase C step in the pathway to increase IP_3 levels while maintaining PIP_2 levels.

Although the changes in PI metabolite levels we have observed are consistent with the model for middle T action on the PI cycle we propose, it is difficult to interpret changes in steady-state levels of intermediates as changes in the kinetics of the pathway without a full understanding of the rate-limiting steps in this pathway. Our hypothesis that middle T alters PI turnover by accelerating PI and PIP phosphorylation to PIP_2 depends on the assumption that phospholipase C is not already saturated with PIP_2 in the normal cell membrane. If phospholipase C is indeed saturated with

TABLE 2. Effect of Transformation by Polyoma Virus or by Kirsten Virus on Intermediates of PI Turnover

A Polyoma-Infected Cells				
Stimulus	PIP	PIP_2	IP_3	DAG
Serum starved Mock infected NIH3T3	100	100	100	ND
Wild Type Polyoma infected	140 ± 5	133 ± 7	170 ± 12	ND
PY 1387 infected	88 ± 17	91 ± 15	102 ± 19	ND
10% serum	119 ± 32	110 ± 18	185 ± 30	ND
B Kirsten (ras)-Transformed Cells				
Cell	PIP	PIP_2	$IP_3 + IP_2$	DAG
Confluent NRK cells	100	100	100	100
Kirsten virus transformed NRK cells	95 ± 8	63 ± 4	166 ± 16	160 ± 12

The steady-state levels of intermediates were determined from cells labeled for at least 48 hours with [^{3}H]inositol or with [^{3}H]glycerol. The data are adapted from references 13 and 18.

PIP_2 under unstimulated conditions, accelerated PI or PIP kinase activity would not result in increased IP_3 or DAG production. Although the concentration of PIP_2 in the fibroblast cell membrane can be reliably determined, the K_m of plasma membrane phospholipase C for PIP_2 at physiological concentrations of calcium is not known. Therefore the question of saturation cannot yet be answered. Direct assessment of our proposal that increased PI and PIP kinase activities caused by middle T will result in increased second messenger levels will require a more complete understanding of phospholipase C as a rate-limiting step in PI turnover. Our data do not, however, exclude the possibility that middle T directly affects phospholipase C in addition to increasing PI and PIP kinase activities.

Is the PI Kinase an Intrinsic Activity of Middle T or pp60$^{c\text{-}src}$?

Genetic and biochemical analyses suggest that PI kinase activity is important for middle T action. The PI kinase itself has not yet been identified; we can, however, address the question of whether the middle T/pp60$^{c\text{-}src}$ tyrosine kinase itself functions as a PI kinase in our immunoprecipitates. Middle T itself appears to lack an ATP binding site, and therefore is unlikely to be directly responsible for the PI kinase activity we have observed. Furthermore, monoclonal antibodies that precipitate free middle T efficiently but precipitate the middle T/pp60$^{c\text{-}src}$ complex poorly have little associated PI kinase activity (B.S., D.K., M.W., unpublished observations). In view of the association of PI kinase activity with purified pp60$^{v\text{-}src}$,[11] it initially seemed plausible that pp60$^{c\text{-}src}$ might catalyze PI phosphorylation directly. We have been unable to detect significant PI kinase activity associated with pp60$^{c\text{-}src}$ purified from bovine brain or immunoprecipitated from pp60$^{c\text{-}src}$ overproducer cells lines (H. Piwinica-Worms, D. Kaplan, M. Whitman, T. Roberts, in preparation). Furthermore, phosphatidylinositol and enolase do not compete as substrates in our immunoprecipitate assay, suggesting that different catalytic sites are responsible for the tyrosine and PI kinase reactions. Finally, we have been able to separate the PI kinase activity away from the bulk of the middle T/pp60$^{c\text{-}src}$ in the immunoprecipitate (D.K., M.W. unpublished observations). This argues that a protein other than middle T or pp60$^{c\text{-}src}$ is responsible for PI phosphorylation in the complex.

The amount of PI kinase activity brought down with middle T in our immunoprecipitates is only a small fraction ($\sim$1%) of the total PI kinase present in cell lysates (unpublished observations). Too little is known about cellular PI kinases and the nature of their interaction with middle T/pp60$^{c\text{-}src}$ for us to interpret with confidence these observations. The significance of cell lysate PI kinase activity with respect to rates of phospholipid metabolism *in vivo* may itself be problematic. Recent studies of PI kinase activities in lysates or particulate fractions in normal versus transformed cells have given conflicting results with respect to changes in cellular PI kinase activity upon transformation. Sugano and Hanafusa[19] found no significant difference in PI kinase activity in detergent extracts of nontransformed versus transformed cells, while McDonald *et al.*[20] and Sugimoto and Erikson[21] reported significant increases in PI kinase activity in particulate fractions from transformed cells. Resolution of these conflicting observations may require a closer consideration of the possibly several distinct activities present in cell extracts. Harwood and Hawthorne[22] have reported distinct, differentially localized detergent-dependent and detergent-independent PI kinase activities and we have found similarly distinct activities in 3T3 fibroblasts (M. Whitman, D. Kaplan, B. Schaffhausen, unpublished observations). Transformation-sensitive PI kinase may therefore represent a subset of cellular PI kinase differing in localization or activation requirement from the bulk activity present in a cell lysate. The 1% of lysate activity present in immunoprecipitates might therefore represent a stoichiometric interaction between middle T/pp60$^{c\text{-}src}$ and this tranformation-sensitive subset of the total lysate PI kinase activity. Alternatively, and the hypothesis that we favor, is that middle T/pp60$^{c\text{-}src}$ activates one or more cellular PI kinases by tyrosine phosphorylation. The activity in immunoprecipitates reflects an enzyme substrate interaction resulting in copurification of some PI kinase with middle T/pp60$^{c\text{-}src}$. In this case, the interaction between middle T/pp60$^{c\text{-}src}$ and PI kinase is catalytic rather than stoichiometric, and therefore only a small portion of the transformation-sensitive pool of PI kinase is actually coprecipitated with middle T/pp60$^{c\text{-}src}$.

CONCLUSION

If polyoma middle T can indeed accelerate PI turnover by stimulating cellular PI and PIP kinases, the broader question that remains is: what aspects of the transformed phenotype are attributable to PI-generated second messengers? And how are the effects of these messengers mediated? Our observation that serum can stimulate IP_3 production to the same extent as expression of middle T, without producing the full range of changes characteristic of middle T expression, suggests that middle T–stimulated PI turnover is insufficient for full expression of the transformed phenotype. It therefore seems unlikely that the metabolic effects of transforming genes on cellular physiology are mediated through any single regulatory pathway and that a variety of cellular targets may be critical in transformation. Analysis of this problem will require consideration of the multiple effects of polyoma transformation on growth (e.g. serum independence, anchorage independent growth, and focus formation), and has not yet been directly addressed. How the messengers produced by PI breakdown might alter cell growth is currently under investigation in many laboratories, and a full consideration of this question is beyond the scope of this review. The immediate effects of DAG and IP_3 are presumably mediated through C-kinase and calcium binding proteins, respectively; the distal targets for these regulators, which are ultimately responsible for progression through the cell cycle, have not been identified.

SUMMARY

Products of phosphatidylinositol turnover have recently been implicated as regulators of cell growth and differentiation. Transformation of cells in culture by infection with certain viruses (Rous sarcoma virus, Kirsten sarcoma virus, and polyoma virus) or by transfection with the oncogenes carried by these viruses affect the steady-state level of intermediates in the PI turnover pathway. In addition, immunoprecipitates of the transforming gene products of Rous sarcoma virus and polyoma virus contain activities of certain enzymes in the PI turnover pathway. We have previously reported that polyoma middle T immunoprecipitates can catalyze phosphorylation of PI to phosphatidylinositol-4-phosphate (PIP). This activity is not intrinsic to middle T or c-src but is due to a cellular enzyme that specifically associates with this complex. The PI kinase is found in immunoprecipitates of the middle T protein from polyoma viruses that are capable of cell transformation but does not associate with mutants of middle T defective in transformation suggesting that this association may be important for transformation.

REFERENCES

1. WHITMAN, M., L. FLEISCHMAN, S. CHAHWALA, L. CANTLEY & P. ROSOFF. 1986. Phosphoinositides, mitogenesis and oncogenesis. *In* Phospoinositides and Receptor Mechanism. J. W. Putney, Ed.: 197–217 Alan R. Liss, Inc. New York.
2. ROZENGURT, E. 1981. Cyclic AMP: A Growth-Promoting Signal for Mouse 3T3 Cells. *In* Advances in Cyclic Nucleotide Research. J.E. Dumont, P. Greengard & G.A. Robison, Eds. **14:** 429–442. Raven Press. New York.
3. CASTAGNA, M., Y. TAKAI, K. KAIBUCHI, K. SANO, V. KIKKAWA & Y. NISHIZUKA. 1982. Direct activation of calcium activated, phospholipid-dependent protein kinase by tumor-promoting phorbol esters. J. Biol. Chem. **257:** 7847–7851.
4. BERRIDGE, M., J. HESLOP, J. P. IRVINE & K. D. BROWN. 1984. Inositoltriphosphate formation and calcium mobilization in Swiss 3T3 cells in response to platelet-derived growth factor. Biochem. J. **222:** 195–201.
5. BERRIDGE, M. J. & R. F. IRVINE. 1984. Inositoltris-phosphate, a novel second messenger in cellularsignal transduction. Nature **312:** 315–321.
6. MOOLENAR, W. H., R. J. AERTS, L. G. J. TERTOOLEN & S. W. DE LAAT. 1986. The

epidermal growth factor induced calcium signal in A 431 cells. J. Biol. Chem. **261:** 279–284.
7. HUNTER, T. & J. COOPER. 1985. Protein-tyrosine kinases. Ann. Rev. Biochem. **54:** 897–930.
8. COHEN, S., C. USHIRO, C. STOSCHECK & M. CHINKERS. 1982. A native 170,000 epidermal growth factor receptor—kinase complex from shed plasma membrane vesicles. J. Biol. Chem. **257:** 1523–1528.
9. DOWNWARD, J., Y. YARDEN, E. MAYES, G. SCRACE, N. TOTTY, P. STOCKWELL, A. ULLRICH, J. SCHLESSINGER & M. D. WATERFIELD. 1984. Close similarity of epidermal growth factor receptor and v-erb B oncogen product sequences. Nature **307:** 521–527.
10. WATERFIELD, M. D., G. T. SCRACE, N. WHITTLE, P. STROOBANT, A. JOHNSSON, A. WASTESON, D. WESTERMARK, C.-H. HELDIN, J. S. HUANG & T. F. DEUEL. 1983. Platelet derived growth factor is structurally related to the putative transforming protein p28 sis of simian sarcoma virus. Nature **304:** 35–38.
11. SUGIMOTO, Y., M. WHITMAN, L. C. CANTLEY & R. L. ERIKSON. 1984. Evidence that the Rous sarcoma transforming gene product phosphorylates phosphatidylinositol and diacylglycerol. Proc. Natl. Acad. Sci. USA **81:** 2117–2121.
12. MACARA, I. G., G. V. MARINETTI & P. C. BALDUZZI. 1984. Transforming protein of avian sarcoma virus UR2 is associated with phosphatidylinositol kinase activity: possible role in tumor igenesis. Proc. Natl. Acad. Sci. USA **81:** 2728–2732.
13. KAPLAN, D. K., M. R. WHITMAN, B. SCHAFFHAUSEN, L. RAPTIS, R. L. GARCEA, D. PALLAS, T. M. ROBERTS & L. C. CANTLEY. 1985. Polyphosphoinositide metabolism and polyoma mediated transformation. Proc. Natl. Acad. Sci. USA **83:** 3624–3629.
14. WHITMAN, M., D. R. KAPLAN, B. SCHAFFHAUSEN, L. CANTLEY & T. M. ROBERTS. 1985. Association of phosphatidylinositol kinase activity with polyoma middle-T competent for transformation. Nature **315:** 239–341.
15. SMITH, A. E. & B. K. ELY. 1983. The biochemical basis of transformation by polyoma virus. *In* Advances in Viral Oncology. G. Klein, Ed. **3:** 3–30. Raven Press. New York.
16. COURTNEIDGE, S. A. & A. E. SMITH. 1983. Polyoma virus transforming protein associates with the product of the c-src cellulargene. Nature **303:** 435–439.
17. BOLEN, J. B., C. J. THIELE, M. A. ISRAEL, W. YONEMOTO, L. A. LIPSICH & J. S. BRUGGE. 1984. Enhancement of cellular src gene product associated tyrosyl kinase activity following polyoma virus infection and transformation. Cell **38:** 767–777.
18. FLEISCHMAN, L. F., S. B. CHAHWALA & L. CANTLEY. 1986. Ras-transformed cells: Altered levels of phosphatidylinositol-4,5-bisphosphate and catabolites. Science **231:** 407–410.
19. SUGANO, S. & H. HANAFUSA. 1985. Phosphatidylinositol kinase activity in virus-transformed and nontransformed cells. Mol. Cell. Biol. **5:** 2399–2404.
20. MCDONALD, M. L., E. A. KUENZEL, J. A. GLOMSET & E. J. KREBS. 1985. Evidence from two transformed lines that the phosphorylations of peptide tyrosin and phosphatidylinositol are catalyzed by two different proteins. Proc. Natl. Acad. Sci. USA **82:** 3393–3397.
21. SUGIMOTO, Y. & R. ERIKSON. 1985. Phosphatidylinositol kinase activity in normal and rous sarcoma virus transformed cells. Mol. Cell. Biol. **5:** 3194–3198.
22. HARWOOD, J. L. & J. N. HAWTHORNE. 1969. The properties and subcellular distribution of phosphatidylinositol kinase in mammalian tissues. Biochim. Biophys. Acta **171:** 75–87.

DISCUSSION OF THE PAPER

M. BLAUSTEIN (*University of Maryland, Baltimore, MD*): As you pointed out, it is generally agreed that cellular pH goes up during cell activation, and that Na^+/H^+ exchange is accelerated. Since the Na^+/H^+ exchanger equilibrium position for H^+ distribution depends upon the Na^+ concentration gradient (i.e., at equilibrium, $\Delta\bar{\mu}_H = -\Delta\bar{\mu}_{Na}$), what is it that causes $\Delta\bar{\mu}_H$ to change when the exchanger is activated (assuming the intracellular Na^+ concentration does not change)?

CANTLEY: The Na^+/H^+ exchange system is not at equilibrium since intracellular pH is 7.1 with extracellular pH of 7.4 and the Na^+ gradient is about 10 (150 mM extracellular, 20 mM cytosol). Activation results because the apparent pK for protons on the cytosolic side shifts to a higher value.

R. DENTON (*University of Bristol, Bristol, U.K.*): Surely increased PIP_2 breakdown cannot be the sole and sufficient signal pathway involved in growth stimulation.

CANTLEY: I agree. We will probably find other targets of growth factor receptors that work cooperatively with PIP_2 breakdown to stimulate growth.

Transmembrane Signalling by Growth Factors[a]

W. H. MOOLENAAR, L. H. K. DEFIZE, B. C. TILLY, A. J. BIERMAN, AND S. W. de LAAT

Hubrecht Laboratory
Uppsalalaan 8
3584 CT Utrecht
The Netherlands

INTRODUCTION

Polypeptide growth factors regulate the proliferation of cells alone or in concert with other mitogens by inducing DNA synthesis and cell division in specific target cells. Of the known growth factors, epidermal growth factors (EGF) and platelet-derived growth factor (PDGF) are best defined; however, their molecular mechanisms of action and their *in vivo* functions are not known.

Like all polypeptide hormones, growth factors initiate their action by binding to specific receptor molecules on the cell surface. Following growth factor binding, the activated receptor triggers a cascade of rapid biochemical events in the target cell, which ultimately lead to the initiation of replicative DNA synthesis. There are at least three potential signal pathways in the action of growth factors like EGF and PDGF (FIG. 1).

One of the first consequences of growth factor–receptor interaction is the activation of a protein kinase specific for tyrosyl residues.[1–3] In fact, the receptors for growth factors like EGF and PDGF are transmembrane glycoproteins that possess intrinsic tyrosine-specific protein kinase activity.[3] Growth factor binding induces a rapid stimulation of the receptor kinase, resulting in autophosphorylation of the receptor itself as well as in the phosphorylation of various substrate proteins. The intrinsic tyrosine-specific kinase activity is shared with several viral oncogene products, such as the transforming protein of Rous sarcoma virus.[3,4] This suggests that tyrosine-specific protein phosphorylations may initiate a set of common mitogenic pathways in virus-transformed and growth factor–stimulated cells. However, it is not yet possible to relate increased tyrosine kinase activity to specific metabolic alterations in stimulated cells.

Other immediate consequences of receptor activation (FIG. 1) include the breakdown of inositol phospholipids,[5,6] a transient rise in cytoplasmic free Ca^{2+} ($[Ca^{2+}]_i$),[7,8] and the stimulation of monovalent ion transport across the plasma membrane.[9,10] Of the known ionic transport changes in growth factor–stimulated cells, the activation of electroneutral Na^+/H^+ exchange is best characterized (see next section).

ACTIVATION OF Na^+/H^+ EXCHANGE

The first direct evidence that growth factors activate an otherwise quiescent Na^+/H^+ exchanger in the plasma membrane came from studies on serum-stimulated

[a]Supported by the Netherlands Cancer Foundation (Koningin Wilhelmina Fonds) and the Organization for the Advancement of Pure Research (ZWO).

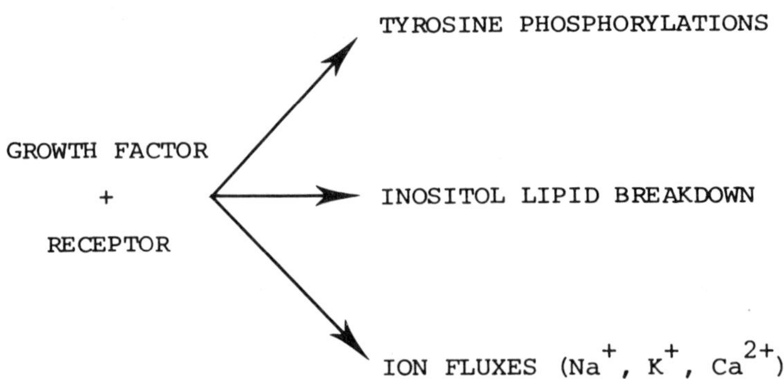

FIGURE 1. Potential signalling pathways in the action of growth factors.

neuroblastoma cells.[11,12] Re-addition of serum to growth-arrested cells leads to the rapid activation of an electrically silent Na^+ influx pathway, which is sensitive to amiloride and can be stimulated by acidifying the cytoplasm using weak acids. In addition, it was shown that amiloride-sensitive Na^+ uptake in such acid-loaded cells is coupled to the efflux of H^+ with an 1:1 stoichiometry. In a subsequent study it was shown that EGF-induced Na^+ influx in quiescent human fibroblasts is amiloride-sensitive, electroneutral, and is enhanced by cytoplasmic acid loads.[3] From these studies it was predicted that stimulation by mitogens leads to an increase in pH_i, while the accompanying entry of Na^+ results in a stimulation of the Na^+,K^+ pump. New developments in pH-monitoring techniques, particularly the synthesis of fluorescent indicators that can be trapped into the cytoplasm of small cells, have made it possible to show that the Na^+/H^+ exchanger is normally involved in the close regulation of pH_i and that activation of the Na^+/H^+ exchanger by growth factors leads to a sustained rise in pH_i of 0.2–0.3 unit.

Most cells maintain their pH at 7.0–7.4. This is well above the electrochemical equilibrium value of 6.4 that is predicted by the Nernst equation from a transmembrane potential of approximately -60mV. In vertebrate cells, the specific H^+-extruding mechanism that raises pH appears to be Na^+/H^+ exchange.[14,15] The functioning of the Na^+/H^+ exchanger in the plasma membrane and its role in pH_i homeostasis is most easily assessed by continuously monitoring the rapid recovery of pH_i to its resting level after a sudden acidification of the cytoplasm, as induced by NH_4^+ prepulse or by weak acids. In most cells, this pH_i recovery process follows an exponential time course and is entirely due to net H^+ extrusion via the Na^+/H^+ exchanger, which utilizes the energy stored in the transmembrane Na^+ gradient. The major determinant of the rate of the Na^+/H^+ exchanger is pH_i. At normal pH_i values (near 7.0) the exchanger is relatively inactive, although the steep transmembrane Na^+ gradient could theoretically raise pH_i about one unit more alkaline.

As pH_i falls below a certain "threshold," the Na^+/H^+ exchanger is increasingly stimulated. Aronson and co-workers[16] were the first to point out that the Na^+/H^+ exchanger is apparently set in motion through allosteric activation by cytoplasmic H^+ at a regulatory site that is distinct from the internal H^+ transport binding site. The relatively strong pH_i sensitivity of the exchanger is, of course, a crucial property for an H^+ extruding system to maintain pH_i at a critical level.

FIGURE 2 shows a typical example of an alkaline pH_i shift after addition of EGF to quiescent cells loaded with the pH-sensitive dye bis(carboxyethyl)carboxyfluorescein

(BCECF). The shift in pH_i is initiated within 20–30 sec and is complete by 10–15 min. The elevated pH_i persists for as long as the growth factor is present. In general, the induced alkalinizations range from 0.1–0.3 pH unit; they are inhibited by amiloride and by Na^+ removal and are accompanied by a transient increase in amiloride-sensitive $^{22}Na^+$ uptake (FIG. 1).[10] Furthermore, the rise in pH_i is converted into a fall in pH_i when the direction of the transmembrane Na gradient is reversed.[17] When taken together, these data convincingly demonstrate that the mitogen-induced pH_i rise is mediated by the Na^+/H^+ exchanger. TABLE 1 summarizes the stimuli that have been reported to raise pH_i by activating Na^+/H^+ exchange in their target cells.

Mechanism of Activation of Na^+/H^+ Exchange

How does receptor occupancy lead to activation of the Na^+/H^+ exchanger? Recent studies have shown that the activation is attributable to an alkaline shift in the pH_i sensitivity of the exchanger.[17–19] As mentioned above, this pH_i sensitivity is determined by an allosteric H^+-binding site on the cytoplasmic face of the exchanger. It thus seems plausible to assume that the altered pH_i sensitivity of the exchanger is due to some conformational change resulting in an increased pK_a of the regulatory H^+-binding site. Thus, the physiological effect of growth factors on the Na^+/H^+ exchanger is to increase its pH_i threshold, that is the level to which pH_i must rise before the exchanger virtually shuts off. Indeed, the Na^+/H^+ exchanger is only transiently stimulated by external stimuli and its activity returns to the control level once pH has attained its new stable value (FIG 2).[10]

By what biochemical steps do growth factors modify the pH_i sensitivity of the Na^+/H^+ exchanger? Current evidence indicates that protein kinase C can somehow activate the exchanger. Tumor promoting phorbol esters and synthetic diacylglycerols,

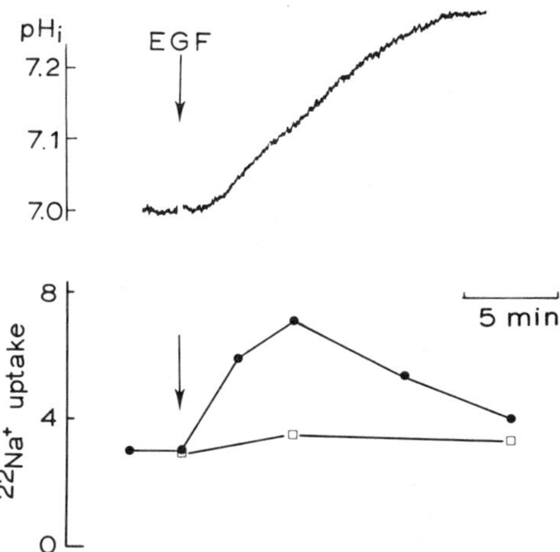

FIGURE 2. Activation of Na^+/H^+ exchange by EGF in human fibroblasts. *Upper panel.* Shift in pH_i induced by EGF (10 ng/ml). Quiescent monolayers were loaded with BCECF as described.[17] *Lower panel.* Time course of EGF-induced ^{22}Na uptake in the presence (□) and absence (●) of 1 mM amiloride. Initial rates of $^{22}Na^+$ uptake were measured over 3-min intervals in the presence of 0.1 mM ouabain to block active Na^+ efflux.

TABLE 1. Stimuli that Raise pH_i by Activating Na^+/H^+ Exchange in Their Target Cells

Stimulus	Cell Type
Serum	Fibroblasts, A431 cells
EGF	Fibroblasts, A431 cells
PDGF	Fibroblasts
Thrombin	Fibroblasts
Insulin	see footnote[b]
Vasopressin	3T3 cells
Vanadate	A431 cells
Bradykinin	A431 cells (W.H.M., unpublished data)
Lectins	T lymphocytes
Lipopolysaccharide	B-lymphoid cells
Interleukin-2	T lymphocytes
Phorbol ester/diacylglycerol	Various cell types
Hypertonicity (cell shrinking)	Various cell types

[a]Reported alkalinizations range from ~0.1–0.3 pH unit. For references see Moolenaar,[10] unless indicated otherwise.

[b]Insulin alone fails to raise pH_i in most cell types, but it often potentiates the pH_i response to other mitogens.

which bind to and directly activate kinase C, mimic growth factors in raising pH_i without producing a rise in $[Ca^{2+}]_i$.[20] Thus, a rapid increase in $[Ca^{2+}]_i$ is not essential for activation of the exchanger. Of the physiological stimuli listed in TABLE 1, many, but not all, are known stimulators of inositol lipid breakdown and, hence, of kinase C activity in their specific target cells.

Recent studies seem to indicate that there are additional pathways, not involving kinase C, by which the Na^+/H^+ exchanger can be activated. Chronic treatment of cells with phorbol esters leads to the disappearance of functional protein kinase C. Yet, kinase C–depleted 3T3 cells can still raise their pH_i in response to EGF.[21] Thus the Na^+/H^+ exchanger can be activated by at least two separate pathways, one involving the phospholipase C–protein kinase C system and the other one(s) unknown.

Ca^{2+} MOBILIZATION BY GROWTH FACTORS

In addition to activating Na^+/H^+ exchange, mitogens like EGF and PDGF induce a rapid but transient rise in $[Ca^{2+}]_i$ in their target cells, as measured by quin-2 fluorescence.[7,8,22] Addition of EGF or PDGF to responsive cells elicits a several-fold rise in $[Ca^{2+}]_i$ that is initiated without a detectable lag period and is usually complete within 30–60 sec. Thereafter, $[Ca^{2+}]_i$ gradually returns to near-basal levels over a 10-min period. FIGURE 3 schematically illustrates the time courses of both the $[Ca^{2+}]_i$ transient and the rise in pH_i as induced by EGF in responsive cells.

How does growth factor binding increase $[Ca^{2+}]_i$? The Ca^{2+} signal in response to PDGF is not prevented by removal of external Ca^{2+}, indicating that the Ca^{2+} is released from intracellular stores.[7] Indeed, PDGF provokes the rapid formation of inositol-1,4,5-trisphosphate (IP(1,4,5)), the key messenger for mobilizing Ca^{2+} from non-mitochondrial stores.[6,23] Surprisingly, the Ca^{2+} signal in response to EGF shows no contribution from intracellular stores but seems to result from net Ca^{2+} entry through a voltage-independent Ca^{2+} channel in the plasma membrane.[8] This interpretation is based on the finding that the EGF-induced $[Ca^{2+}]_i$ rise in human A431 carcinoma cells is critically dependent on the extracellular Ca^{2+} concentration, is accompanied by enhanced $^{45}Ca^{2+}$ uptake,[5] is blocked by Ca^{2+} entry blockers like La^{3+} and Mn^{2+}, while it is not accompanied by changes in transmembrane potential.

These results are intriguing because they suggest that there is a fundamental difference between the receptors for EGF and PDGF in terms of their $[Ca^{2+}]_i$-raising mechanisms; however, some caution is needed in interpreting the disappearance of the Ca^{2+}–quin-2 response to EGF when external Ca^{2+} is removed, since it is conceivable that intracellular quin-2 (a Ca^{2+} chelator) somehow interferes with the proper functioning of the EGF receptor, particularly in the absence of extracellular Ca^{2+}.

INOSITOL LIPID BREAKDOWN

Much attention has recently been focused on the role of inositol 1,4,5-trisphosphate ($IP_3(1,4,5)$) as a specific releaser of intracellular Ca^{2+} following receptor stimulation.[23] It is now becoming increasingly apparent that several additional inositol polyphosphates, with as yet unknown functions, are produced in stimulated cells. In particular, an $IP_3(1,3,4)$ isomer has been detected that seems to be formed by dephosphorylation of inositol-(1,3,4,5)-tetrakisphosphate (IP_4).[24] Furthermore, the presence of IP_5 and IP_6 has been described.[25] FIGURE 4 summarizes some of the known and proposed metabolic pathways responsible for the formation of inositol phosphates.[25] Although it seems likely that $IP_3(1,3,4)$ and IP_3 may have second messenger functions, this idea remains to be tested.

We have separated [^{3}H]inositol phosphates from EGF-treated A431 cells using an HPLC anion-exchange system. Growing A431 cells were found to contain two IP_3 isomers at roughly equal concentrations, while three prominent peaks with decreasing ionophoretic mobilities were tentatively identified as IP_4, IP_5, and IP_6, respectively (B.C. Tilly & P. van Paridon, unpublished data). Addition of EGF causes a small but significant increase in the level of $IP_3(1,4,5)$, while there is an approximate threefold increase in $IP_3(1,3,4)$ concentration within 1 min (TABLE 2). These results, although preliminary, strengthen the view that the $IP_3(1,3,4)$ isomer might have some second

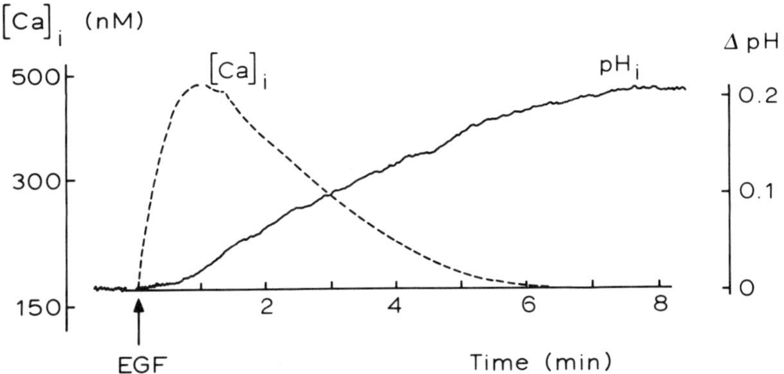

FIGURE 3. Changes in $[Ca]_i$ and pH_i in human A431 cells following the addition of EGF (100 ng/ml). For methods see Moolenaar *et al.*[8,17]

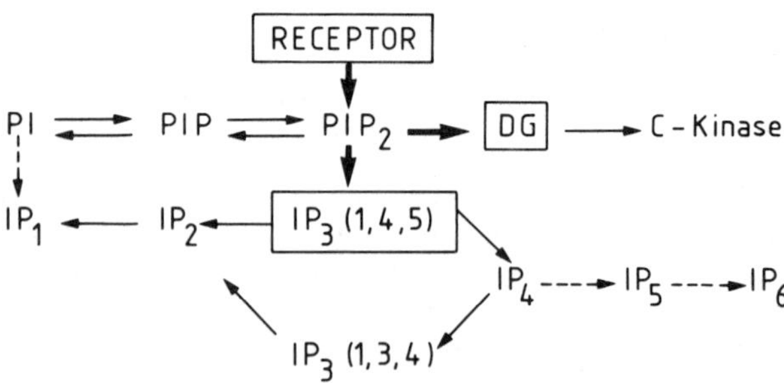

FIGURE 4. Proposed metabolic pathways responsible for the formation of inositol phosphates. PI(P) = phosphatidyl-inositol-(phosphate). IP_1 = inositol-monophosphate; IP_2 = inositol-bisphosphate. See text and Heslop *et al.*[25] for further details.

messenger role. Time-course studies and microinjection experiments should lead to the verification or disproval of this hypothesis.

DISSOCIATION OF SIGNAL PATHWAYS BY ANTI-RECEPTOR MONOCLONAL ANTIBODIES

The partial activation of post-receptor signalling pathways by phorbol esters has proved to be a fruitful approach in elucidating interrelationships among these pathways. An alternative tool for the dissociation of molecular events in the signalling cascade is provided by the availability of monoclonal antibodies to the EGF receptor. We have used three different anti-EGF receptor monoclonal IgGs, directed against distinct epitopes of the extracellular domain of the human EGF receptor,[26] to test their ability to act as partial or complete agonists of the EGF receptor. All three antibodies (named 2E9, 2D11, and 2G5, respectively) are able to immunoprecipitate a functional EGF receptor showing EGF-dependent tyrosine kinase activity. Monoclonal 2E9 is unique in that it recognizes a peptide determinant at or close to the EGF binding domain of the receptor. As a consequence, 2E9 competitively inhibits EGF receptor binding. In contrast, the other monoclonals (2D11 and 2G5) are directed to blood group A–specific carbohydrate structures on the EGF receptor and fail to affect EGF binding. We have tested these antibodies for their EGF-like properties in stimulating the receptor-mediated tyrosine phosphorylations, cytoplasmic alkalinization, Ca^{2+}

TABLE 2. [^{3}H]Inositol Phosphate Levels in A431 Cells

	d.p.m. ($\times 10^2$ per Sample)	
	Control	EGF
IP_3(1,4,5)	11	18
IP_3(1,3,4)	3.5	11
IP_4	5.3	9
$IP_5 + IP_6$	117	83

Levels of inositol phosphates were measured in nearly confluent cultures pre-labeled to isotopic equilibrium with [^{3}H]inositol (2 μCi/ml). Treatment with EGF (100 ng/ml) lasted for 60 sec. Cell extracts were processed for analysis of [^{3}H]inositol phosphates by an HPLC anion-exchange system (Tilly *et al.*, manuscript in preparation).

mobilization, and DNA synthesis. As summarized in TABLE 3, all three monoclonal IgGs can stimulate the tyrosine-specific autophosphorylation of the 170 kD EGF receptor both in isolated A431 membranes and in intact cells.[26] Interestingly, none of these antibodies is capable of triggering inositol phosphate formation and generation of ionic signals (TABLE 3), even after addition of a second crosslinking anti-IgG. Finally, the anti-receptor antibodies fail to stimulate DNA synthesis in quiescent human fibroblasts. Stimulation of the receptor's intrinsic tyrosine kinase is apparently not sufficient, by itself, to elicit a mitogenic response (for further details see Defize *et al.*[26]) Another important conclusion from those results is that stimulation of the EGF receptor kinase does not necessarily activate the post-receptor pathway that leads to phospholipase C activation and to an increase in $[Ca^{2+}]_i$ and in pH_i. These findings further support the view that the two major ionic signals are indispensable for the stimulation of DNA synthesis and cell proliferation.

TABLE 3. Comparison of the Biological Effects of EGF and Anti-EGF Receptor Monoclonal Antibodies on Human A431 Cells and Fibroblasts

	EGF	2E9	2D11
Precipitation of EGF receptor		+	+
EGF binding competition	+	+	−
Stimulation of tyrosine kinase	+	+	+
Morphological changes	+	−	+
Inositol phosphate formation	+	−	−
Rise in $[Ca^{2+}]_i$	+	−	−
Rise in pH_i	+	−	−
Stimulation of DNA synthesis	+	−	−

Stimulation of DNA synthesis was tested on quiescent fibroblasts; all other effects on A431 cells. For further details see Defize *et al.*[26]

POSSIBLE PHYSIOLOGICAL ROLE OF THE IONIC SIGNALS

Cytoplasmic Alkalinization

An early rise in pH_i appears to be a fairly common response of metabolically dormant cells to appropriate surface stimuli.

A shift in pH_i of about 0.2 unit would be expected to have considerable effects on a host of pH-sensitive processes in the cell. Growing consensus among workers in the field holds that an alkaline pH_i shift has a permissive rather than a strictly triggering role in the response of cells to mitogens. In this sense pH_i differs from the more classical second messengers such as Ca^{2+} and cAMP. Of critical importance is the question of whether cytoplasmic alkalinization, mediated by Na^+/H^+ exchange, is essential for the initiation of DNA synthesis and cell division in response to growth stimuli. Perhaps the most convincing demonstration of a signalling role for Na^+/H^+ exchange and pH_i in the initation of a mitogenic response has been made with fertilized sea urchin eggs, in which pH_i must rise by at least 0.2 unit to permit DNA synthesis to begin.[27] Using mutant fibroblasts, which lack a functional Na^+/H^+ exchanger, Pouysségur *et al.*[28,29] elegantly showed that below a certain threshold value (around 7.2) pH_i becomes limiting for cell proliferation, and furthermore that one of the critical pH_i-dependent steps in activated fibroblasts appears to be the stimulation of protein synthesis. Studies by others seem to confirm that a mitogen-induced rise in pH_i is a permissive event that is necessary but not sufficient for progression through S phase, at least in fibroblasts.[30,31] However, cytoplasmic alkalinization may have a less critical role in the mitogenic response of lymphocytes to interleukin-2 than in stimulated fibroblasts.[32]

Shifts in pH_i during the cell cycle have been observed in lower eukaryotes such as protozoa, slime molds, and yeast.[33] For example, during the *Dictyostelium* cell cycle pH_i oscillates (by an unknown mechanism) with the same period as the DNA replication cycle, with alkalinization occurring during DNA synthesis. This pH_i oscillator may have an on-off triggering function rather than a permissive role in the timing and regulation of protein and DNA synthesis.[34] It will be of considerable interest to monitor pH_i and the state of the Na^+/H^+ exchanger during the cell cycle of higher eukaryotes. Another challenge for future studies is to examine whether the ionic signals generated by growth factor receptors have their correlates in the action of certain oncogene products. One widely held idea is that oncogenes induce malignant growth at least partially via the constitutive activation of signal pathways normally involved in growth factor action. This might lead not only to permanently altered

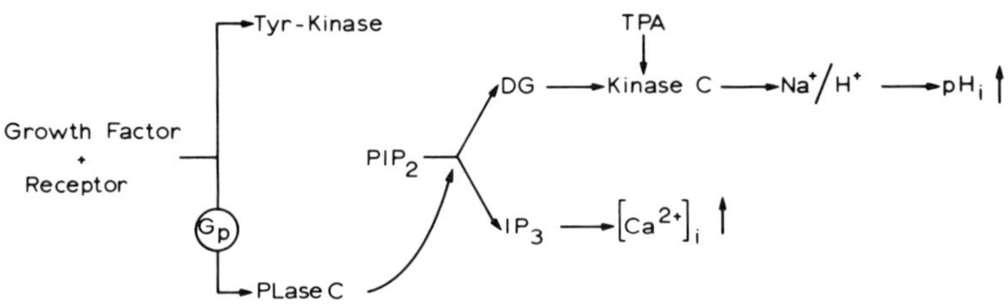

FIGURE 5. Proposed sequence of events leading from growth factor binding to ionic signals. G_p is a putative GTP-binding protein[38]; Tyr-kinase = tyrosine-specific protein kinase; PLase C = phospholipase C; PIP_2 = phosphatidylinositol 4,5-bisphosphate; DG = 1,2-diacylglycerol; IP_3 = inositol 1,4,5-trisphosphate; TPA = 12-O-tetradecanoyl phorbol-13-acetate. See text for further details.

intracellular Ca^{2+} compartments but also to an uncontrolled activation of the Na^+/H^+ exchanger and, hence, to an elevated pH_i.

Rise in $[Ca^{2+}]_i$

Another key question is what the physiological role of the $[Ca^{2+}]_i$ transient is in the action of growth factors. Numerous cellular functions are regulated by $[Ca^{2+}]_i$ and many of them through the formation of Ca^{2+}-calmodulin complexes. Of particular relevance to mitogen action is the finding that artificially raising $[Ca^{2+}]_i$ by Ca^{2+} ionophore A23187 can mimic the effects of EGF and PDGF on the rapid induction of the c-*fos* and c-*myc* proto-oncogenes.[35,36] We recently found that Ca^{2+}-mobilizing neurohormones like bradykinin and histamine similarly induce c-*fos* and *c-myc* expression as well as cell division in their target cells (W. Kruijer and W.H. Moolenaar, unpublished data). These results raise the intriguing possibility that Ca^{2+} could function as a second messenger that mediates, either directly or indirectly, the early transcriptional effects of growth factors and they provide a framework for future studies on the role of intracellular Ca^{2+} in growth control.

CONCLUDING REMARKS

In recent years much has been learned about the various molecular events that are induced by growth factor–receptor interaction. These include tyrosine-specific protein phosphorylations, inositol lipid breakdown, and the two major ionic events discussed above, namely, transient elevation of $[Ca^{2+}]_i$ and activation of Na^+/H^+ exchange resulting in a sustained rise in pH_i. There is good evidence that both ionic events are the direct consequence of the receptor-mediated hydrolysis of inositol phospholipids, while tyrosine kinase activity seems to initiate a distinct pathway, as summarized in the scheme of FIGURE 5. Although uncertainty still exists about the biological significance of each of these "early" steps in the eventual initiation of DNA synthesis, occurring many hours later, it seems likely that the various events evoked by inositol lipid breakdown act in concert with tyrosine-specific protein phosphorylations to stimulate cell proliferation. Selective pharmacological inhibitors without non-specific side effects could greatly assist in elucidating the relevance of each of the signal pathways, but such agents are lacking at present. It is obvious that mutant cells defective in Na^+/H^+ exchange activity[28] or other functions, as well as anti-receptor monoclonal antibodies that act as partial agonist[26] are of great help in analyzing the role of the various signal pathways in growth control.

REFERENCES

1. CARPENTER, G. 1984. Properties of the receptor for epidermal growth factor. Cell **37:** 357–358.
2. HELDIN, C.-H. & B. WESTERMARK. 1984. Growth factors: Mechanism of action and relation to oncogenes. Cell **37:** 9–20.
3. HUNTER, T. & J. A. COOPER. 1985. Protein-tyrosine kinases. Ann. Rev. Biochem. **54:** 897–930.
4. BISHOP, J. M. 1985. Viral oncogenes. Cell **42:** 23–38.
5. SAWYER, S. T. & S. COHEN. 1981. Enhancement of calcium uptake and phosphatidylinositol turnover by EGF in A431 cells. Biochemistry **20:** 6280–6286.
6. BERRIDGE, M. J., J. P. HESLOP, R. F. IRVINE & K. D. BROWN. 1984. Inositol trisphosphate formation and calcium mobilization in Swiss 3T3 cells in response to PDGF. Biochem. J. **222:** 195–201.
7. MOOLENAAR, W. H., L. G. J. TERTOOLEN & S. W. DE LAAT. 1984. Growth factors immediately raise cytoplasmic free Ca^{2+} in human fibroblasts. J. Biol. Chem. **259:** 8066–8069.
8. MOOLENAAR, W. H., R. J. AERTS, L. G. J. TERTOOLEN & S. W. DE LAAT. 1986. The epidermal growth factor-induced calcium signal in A431 cells. J. Biol. Chem. **261:** 279–284.
9. ROZENGURT, E. 1981. Stimulation of Na^+ influx, Na^+,K^+ pump activity and DNA synthesis in quiescent cultured cells. Adv. Enzyme Reg **19:** 61–85.
10. MOOLENAAR, W. H. 1986. Effects of growth factors on intracellular pH regulation. Ann. Rev. Physiol. **48:** 363–376.
11. MOOLENAAR, W. H., C. L. MUMMERY, P. T. VAN DER SAAG & S. W. DE LAAT. 1981. Rapid ionic events and the initiation of growth in serum-stimulated neuroblastoma cells. Cell **23:** 789–798.
12. MOOLENAAR, W. H., J. BOONSTRA, P. T. VAN DER SAAG & S. W. DE LAAT. 1981. Sodium/proton exchange in mouse neuroblastoma cells. J. Biol. Chem. **256:** 12883–12887.
13. MOOLENAAR, W. H., Y. YARDEN, S. W. DE LAAT & J. SCHLESSINGER. 1982. Epidermal growth factor induces electrically silent Na^+ influx in human fibroblasts. J. Biol. Chem. **257:** 8502–8506.

14. Roos, A. & W. Boron. 1981. Intracellular pH. Physiol. Rev. **61:** 296–434.
15. Moolenaar, W. H. 1986. Regulation of cytoplasmic pH by Na^+/H^+ exchange. Trends Biochem. Sci. **11:** 141–143.
16. Aronson, P. S., J. Nee & M. A. Suhm. 1982. Modifier role of internal H^+ in activating the Na-H exchanger in renal microvillus membrane vesicles. Nature **299:** 161–163.
17. Moolenaar, W. H., R. Y. Tsien, P. T. van der Saag & S. W. de Laat. 1983. Na^+/H^+ exchange and cytoplasmic pH in the action of growth factors in human fibroblasts. Nature **304:** 645–648.
18. Paris, S. & J. Pouysségur. 1984. Growth factors activate the Na^+/H^+ antiporter in quiescent fibroblasts by increasing its affinity for intracellular H. J. Biol. Chem. **259:** 10989–10994.
19. Grinstein, S., S. Cohen, J. D. Goetz, A. Rothstein & E. W. Gelfand. 1985. Characterization of the activation of Na^+/H^+ exchange in lymphocytes by phorbol esters: Change in cytoplasmic pH dependence of the antiport. Proc. Natl. Acad. Sci. USA **82:** 1429–1433.
20. Moolenaar, W. H., L. Tertoolen & S. W. de Laat. 1984. Phorbol esters and diacylglycerol mimic growth factors in raising cytoplasmic pH. Nature **312:** 371–374.
21. Vara, F. & E. Rozengurt. 1985. Stimulation of Na /H antiport activity by EGF and insulin occurs without activation of protein kinase C. Biochem. Biophys. Res. Commun. **130:** 646–653.
22. Hesketh, T. R., J. P. Moore, J. D. H. Morris, M. V. Taylor, J. Rogers, G. A. Smith & J. C. Metcalfe. 1985. A common sequence of calcium and pH signals in the mitogenic stimulation of eukaryotic cells. Nature **313:** 481–484.
23. Berridge, M. J. & R. F. Irvine. 1984. Inositol trisphosphate, a novel second messenger in cellular signal transduction. Nature **312:** 315–321.
24. Batty, I. R., S. R. Nahorski & R. F. Irvine. 1985. Rapid formation of IP_4 following muscarinic stimulation of rat cerebral cortical slices. Biochem. J. **232:** 211–215.
25. Heslop, J. P., R. F. Irvine, A. T. Tashjian & M. J. Berridge. 1985. Inositol tetrakis- and pentakisphosphates in GH_4 cells. J. Exp. Biol. **119:** 396–401.
26. Defize, L. H. K., W. H. Moolenaar, P. T. van der Saag & S. W. de Laat. 1986. Dissociation of cellular responses to EGF using anti-receptor monoclonal antibodies. EMBO J. **5:** 1187–1192.
27. Whitaker, M. J. & R. A. Steinhardt. 1982. Ionic regulation of egg activation. Q. Rev. Biophys. **15:** 593–666.
28. Pouysségur, J., L. C. Sardet, A. Franchi, G. l'Allemain & S. Paris. 1984. A specific mutation abolishing Na^+/H^+ antiport activity in hamster fibroblasts precludes growth at neutral and acidic pH. Proc. Natl. Acad. Sci. USA **81:** 4833–4837.
29. Pouysségur, J., A. Franchi, G. l'Allemain & S. Paris. 1985. Cytoplasmic pH, a key determinant of growth factor-induced DNA synthesis in quiescent fibroblasts. FEBS Lett. **190:** 115–118.
30. Bravo, R. & H. MacDonald-Bravo. 1986. Effect on pH on the induction of competence and progression to S-phase in mouse fibroblasts. FEBS Lett. **195:** 309–312.
31. Moolenaar, W. H., L. H. K. Defize & S. W. de Laat. 1986. Ionic signalling by growth factor receptors. J. Exp. Biol. (In press.)
32. Mills, G. B., E. J. Cragoe, E. W. Gelfand & S. Grinstein. 1985. Interleukin-2 induces a rapid increase in intracellular pH through activation of a Na^+/H^+ antiport. J. Biol. Chem. **260:** 12500–12507.
33. Busa, W. B. & R. Nuccitelli. 1984. Metabolic regulation via intracellular pH. Am. J. Physiol. **246:** R409–R438.
34. Aerts, R. J., A. J. Durston & W. H. Moolenaar. 1985. Cytoplasmic pH and the regulation of the Dictyostelium cell cycle. Cell **43:** 653–657.
35. Bravo, R., J. Burckhardt, T. Curran & R. Müller. 1985. Stimulation and inhibition of growth by EGF in different A431 cell clones is accompanied by the rapid induction of c-*fos* and c-*myc* proto-oncogenes. EMBO J. **4:** 1193–1198.
36. Tsuda, T., K. Kaibuchi, B. West & Y. Takai. 1985. Involvement of Ca^{2+} in platelet-derived growth factor-induced expression of c-myc oncogene in Swiss 3T3 fibroblasts. FEBS Lett. **187:** 43–46.

37. ROSOFF, P. H., L. F. STEIN & L. C. CANTLEY. 1984. Phorbol esters induce differentiation in a pre-B-lymphocyte cell line by enhancing Na^+/H^+ exchange. J. Biol. Chem. **259:** 7056–7060.
38. COCKROFT, S. & B. D. GOMPERTS. 1985. Role of guanine nucleotide binding protein in the activation of polyphosphoinositide phosphodiesterase. Nature **314:** 534–536.

DISCUSSION OF THE PAPER

R. MILANICK (*Yale University, New Haven, CT*): My first question concerns the maintenance of high pH_i after adding growth factors. If Na^+/H^+ exchange balances H^+ leak and H^+ production before stimulation, and if stimulation only transiently increases the activity of the Na^+/H^+ exchange, must not the fact that pH_i remains elevated imply that either the H^+ leak or the H^+ production rate decreases?

MOOLENAAR: The Na^+/H^+ exchanger activity is such that it exactly balances the acidifying effects of cellular metabolism and proton influx.

MILANICK: My second question concerns the actual importance of Na^+/H^+ exchange when HCO_3^- is present. Were all your experiments done in the absence of HCO_3^-? Would you comment on the observation by Pouysségur that cells in HCO_3^- could still be stimulated by growth factors yet he could not observe any change in pH_i?

MOOLENAAR: In some cells the Na^+- dependent Cl^--HCO_3^- exchanger is able to raise pH_i into the "permissive" range. In such conditions an additional shift in pH_i via the Na^+-H^+ exchanger is not essential anymore.

L.A. PINNA (*University of Padova, Padova*): I wonder if a recent observation made by us could be relevant to the mechanism by which PK-C is suspected to mediate the rise of pH produced by growth factors. The observation is that the effectors-independent activity of PK-C detectable with some model substrates is extremely sensitive to small variations of pH: it occurs at pH 7.3–7.4, but it becomes negligible at pH 7 or below 7.

MOOLENAAR: That is a very interesting observation that could perhaps have biological significance.

M. CANESSA (*Harvard University, Boston, MA*): Dr. Moolenaar, are there other mechanisms of action of phorbol esters to activate the Na^+/H^+ exchange that are not mediated by protein kinase C? In human red cells, an increase on cytosolic calcium activates the Na^+/H^+ exchange but not phorbol esters, despite the fact that they stimulate phosphorylation of three membrane proteins (spectrin, band 4.1 and 3).

MOOLENAAR: Perhaps, Ca^{2+}-calmodulin can activate the Na^+-H^+ exchanger. Convincing evidence for that concept is lacking, however.

P. PEDERSEN (*Johns Hopkins University, Baltimore, MD*): Does TPA increase the internal pH via the Na^+/H^+ exchanger in neuroblastoma cells even when glucose is present? I would think that the high lactic acid derived from glucose would override pH changes in the appropriate direction via the Na^+/H^+.

MOOLENAAR: Yes, glucose is always present in our experiments. The Na^+/H^+ exchanger is perfectly able to cope with any increase in metabolic acid production.

G. BIANCHI (*University of Milan, Milan*): Cell mutants, lacking a specific membrane transport system, may be useful tools to assay the precise role of these systems in the effect of G.F. (as far as known it has been done recently for Na-K co-transport). Do you know whether these studies have been carried out with mutants lacking Na/H? Do you know the influence of selective changes in pH, Na, Ca, and K on the expression of some oncogenes?

MOOLENAAR: With respect to your first question: Dr. Pouysségur in Nice has selected fibroblasts lacking a functional Na^+/H^+ exchanger. Using these mutants, he has shown that a nitrogen-induced shift in pH_i has a permissive effect on protein and DNA synthesis. As for your second question: artificially raising $[Ca^{2+}]_i$ induces the cellular *fos* and *myc* proto-oncogenes. In contrast, shifts in pH_i do not affect the induction of these genes.

R. DENTON (*University of Bristol, Bristol*): Two related questions on the monoclonal antibodies that stimulate increased phosphorylation of the EGF receptor but do not have any of the intracellular effects of EGF. (1) Do the antibodies cause increases in phosphorylation of the same sites on the receptor as EGF? (2) Do the antibodies cause the same increases in tyrosine phosphorylation of other cellular proteins?

MOOLENAAR: (1) We will analyze tryptic peptide phosphorylations in the near future. (2) Yes, the p36 protein, probably identical to lipocortin, is phosphorylated in isolated membrane preparations.

Lack of Subunit II of Cytochrome *c* Oxidase in a Patient with Mitochondrial Myopathy

MASASHI TANAKA, MORIMITSU NISHIKIMI, AND TAKAYUKI OZAWA

Department of Biomedical Chemistry
Faculty of Medicine, University of Nagoya
Nagoya, 466 Japan

SHIGEAKI MIYABAYASHI AND KEIYA TADA

Department of Pediatrics
Tohoku University School of Medicine
Sendai, 980 Japan

Mitochondrial myopathies are miscellaneous entities characterized by muscular atrophy and morphologically abnormal mitochondria. Deficiencies in the oxidative phosphorylation system have been recognized as one of the causes of mitochondrial myopathy. In order to elucidate the molecular mechanism of the deficiency of mitochondrial electron transfer chain, we have surveyed systematically the subunits of the energy-transducing complexes, Complex I (NADH-ubiquinone oxidoreductase), Complex III (ubiquinol-cytochrome *c* oxidoreductase), and Complex IV (cytochrome *c* oxidase) in a patient with cytochrome *c* oxidase deficiency limited to skeletal muscle.

A six-year-old boy had exercise intolerance for three years. He had generalized muscular atrophy and showed lacticacidemia. Muscle biopsy indicated ragged red fiber myopathy with fatty infiltration. Cytochrome *c* oxidase activity stain was negative. Cytochrome *c* oxidase activity was normal in biopsied liver, cultured fibroblasts, and leukocytes, but was markedly diminished in biopsied muscle, indicating that the defect is localized in the muscle tissue.

Rotenone-sensitive NADH-cytochrome *c* reductase and succinate-cytochrome *c* reductase activities in patient muscle mitochondria were 40% and 66% of the control values on protein base, respectively. Succinate dehydrogenase and F_1-ATPase activities were close to the normal levels, whereas cytochrome *c* oxidase activity was 3% of the control value. Spectrophotometry showed that cytochrome aa_3 was absent in patient mitochondria and that the contents of cytochromes b and $c + c_1$ were both 57% of each control value.

The immunoblotting and immunoprecipitation experiments using anti-Complex IV antibody prepared in our laboratory demonstrated that the amounts of cytoplasmically synthesized subunits IV–VII and mitochondrially synthesized subunit I in Complex IV were markedly reduced and that subunit II was absent in patient mitochondria. In contrast with the normal contents of α and β subunits of F_1-ATPase, the amounts of several subunits in Complexes I and III were also decreased in patient mitochondria. Moreover, high-resolution SDS-urea gel electrophoresis disclosed that six additional unidentified polypeptides were markedly diminished or completely missing in patient mitochondria.

The lack of subunit II in Complex IV, which is encoded by mitochondrial DNA and synthesized on mitochondrial ribosomes, might lead to impaired assembly of the complex and result in deficiencies of other subunits of Complex IV. By analogy, decreased synthesis of mitochondrially coded subunits of Complex I and III might be responsible for the deficiencies of some subunits in these complexes.

The messenger RNAs for subunits I, II, and III in Complex IV, apocytochrome *b* of Complex III, and possibly six subunits of Complex I are transcribed from H-strand of human mitochondrial DNA as a single primary transcript, cleaved precisely by putative specific endonuclease(s), polyadenylated, processed into mature messenger RNAs, and translated on mitochondrial ribosomes. The association of partial deficiencies of subunits of Complexes I and III as well as deficiencies of unidentified mitochondrial polypeptides with the lack of subunit II of Complex IV reported here might indicate that the defect in this patient lies in mitochondrial genome *per se* and/or in its expression processes.

The Relationship between the Expression of the Mitochondrial Genome and Plasma Membrane Properties in Thymocytes and Leukemic Cells in the Rat

C. VAN DEN BOGERT, S. KUZELA,[a] T. MELIS, AND A. M. KROON

Laboratory of Physiological Chemistry
State University
Groningen, The Netherlands
[a]Cancer Research Institute
Slovak Academy of Sciences
Bratislava, Czechoslovakia

We have shown before that continuous inhibition of mitochondrial (mt) protein synthesis by the tetracyclines results in proliferation arrest of several tumor systems in the rat as soon as the content of functional mitochondria is reduced to a critical limit by prolonged treatment.[1] The anti-tumor effect of the tetracyclines is not limited to cytostasis; prolonged treatment results in the disappearance of tumors. The disappearance rate of a T-leukemic tumor depends on the stage of tumor development in which tetracycline treatment is started. Once its growth is arrested the tumor disappears faster if tetracycline treatment is started in a later stage of tumor development. The concentration changes of certain proteins during tumor development might offer a clue for this phenomenon. In untreated tumor-carrying rats, these proteins (LP) are absent in the tumor cells during the early stages of tumor development, their concentration increases rapidly in the intermediate stage, and they disappear again in the terminal stages of tumor growth. Tetracycline treatment counteracts these changes. The latter observation is in line with the postulation that tetracycline treatment interferes with phenotypic changes of this rat leukemia.

The LPs that reflect these phenotypic changes are glycoproteins, only present in some lymphoid cell types and in the endothelium of lymphoid organs. In lymphoid cells their presence is restricted to the nucleus and the plasma membrane. The LPs are absent in peripheral blood lymphocytes, present in low amounts in cells of the spleen, lymph nodes, and bone marrow, and in high amounts in the lymphoid cells of the thymus (FIG. 1). Moreover, the concentration of the LPs is linked to the function of the thymus. It is therefore probable that the LPs reflect the maturation stage of various lymphoid subsets. Also in the normal thymus, tetracycline treatment appears to counteract the changes in the content of the LPs (TABLE 1). It can be concluded that the tetracyclines interfere in a complex way with the presence of some glycoproteins that reflect the maturation stage of lymphoid cells.

The only primary effect of the tetracyclines that we are aware of is inhibition of mt-protein synthesis. The tetracycline-induced effects on the LPs are found before inhibition of mt-protein synthesis leads to cytostasis and become visible when the capacity for oxidative phosphorylation is hardly reduced. It is unlikely that the tetracyclines exert a direct effect on the synthesis or presence of the LPs. This is also contradicted by the opposite effects of tetracycline treatment (maintenance of the

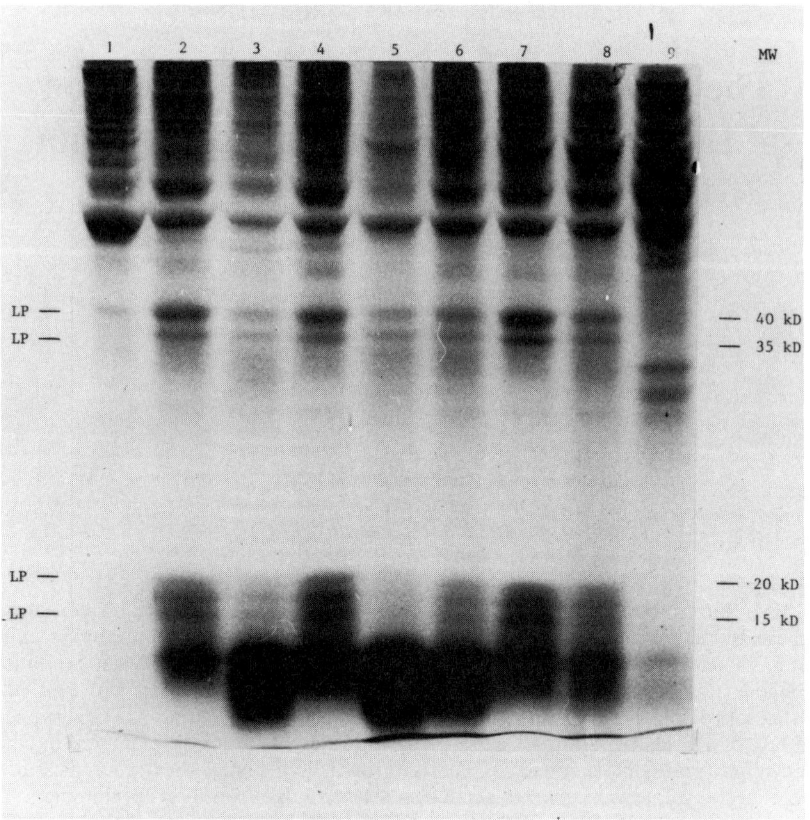

FIGURE 1. Sodium dodecyl sulphate-polyacrylamide slab gel of total cellular proteins of some rat tissues (12.5% AA). Lane 1: peripheral blood lymphocytes, Lane 2: peripheral blood leukemic cells, intermediate stage, Lane 3: normal bone marrow, Lane 4: leukemic bone marrow, Lane 5: normal spleen, Lane 6: leukemic spleen, Lane 7: normal thymus, Lane 8: leukemic thymus, Lane 9: normal liver.

TABLE 1. Effect of Tetracycline Treatment on the Concentration of Lymphoid Maturation Stage Specific Proteins in the Thymus of the Rat under Various Conditions

Condition	Tetracycline Treatment	Concentration of Proteins
Resting thymus	untreated (control)	++++
	tetracycline for 2 weeks	+++
	tetracycline for 3 weeks	++
	tetracycline for 4 weeks	++
12 days after cortisone treatment	control	+
	tetracycline treated during cortisone response	+++
4 days after induction of a T-cell dependent immune response	control	++
	tetracycline treated during immune response	++++

Oxytetracycline or doxycycline were administered to Wistar rats by means of continuous intravenous infusion. The tetracycline serum levels ranged between 4 and 8 μg/ml. Cortisone was injected intramuscularly in a dose of 2.5 mg/kg. The immune response was induced by intravenous injection of xenogeneic tumor cells. To ensure that the tetracycline serum levels were sufficiently high to impair mt-protein synthesis during the total period of the respective responses, tetracycline treatment was started 8 hours before cortisone or tumor cells were injected.

presence of the LPs in conditions that lead to their disappearance in the controls and the reverse).

A possible explanation is that mitochondria interact with nucleocytoplasmic activities of the cell by modulating the synthesis of some cell membrane–linked glycoproteins. Interference with mt-protein synthesis might result in a disturbance of this interaction and lead to changes in the maturation stage of lymphoid cells.

REFERENCES

1. KROON, A. M. & VAN DEN BOGERT. 1985. The mitochondrial genome as a target for chemotherapy of cancer. *In* Achievements and Perspectives of Mitochondrial Research. E. Quagliariello, E. C. Slater, F. Palmieri, C. Saccone & A. M. Kroon, Eds. **2:** 21–34. Elsevier Science Publishers. Amsterdam.
2. VAN DEN BOGERT, C., B. H. J. DONTJE & A. M. KROON. 1985. The antitumour effect of doxycycline on a T cell leukaemia in the rat. Leukemia Res. **9:** 617–623.

Defects of Complex I and Complex IV in Skeletal Muscle from Patients with Chronic Progressive External Ophthalmoplegia[a]

H. S. A. SHERRATT,[b] M. A. JOHNSON,[c]
AND D. M. TURNBULL[c]

Department of [b]Pharmacological Sciences and of [c]Neurology
Medical School
University of Newcastle upon Tyne
Newcastle upon Tyne NE2 4HH and NE4 6BE, England

In many mitochondrial myopathies chronic progressive external ophthalmoplegia (CPEO) is associated with a proximal myopathy and/or craniosomatic abnormalities. Partial cytochrome oxidase deficiency was found in a group of these patients.[1] Between 4–27% of skeletal muscle fibers in biopsy samples gave no cytochemical reaction for cytochrome oxidase. In two additional patients there were also defects in Complex I.[2,3] Here we report on three more patients with this clinical syndrome.

CASE HISTORIES

MR aged 62 years, had a 15 year history of bilateral ptosis, limited eye movements, and a mild proximal myopathy. RL aged 32 years, had a 20 year history of CPEO, and has peripheral retinal pigment streaks, external ophthalmoplegia, proximal muscle weakness, depressed reflexes, and cerebellar ataxia. PH aged 34 years, had a 20 year history of ptosis. There was bilateral ptosis, incomplete external ophthalmoplegia, facial diplegia, mild distal weakness, and depressed reflexes. Routine biochemistry and hematology were normal in all three patients.

METHODS

Skeletal muscle was obtained by open biopsy from patients or during hip replacement operations from controls and frozen at −150°C for cytochemistry or used for the preparation of mitochondrial fractions.[3] Other details are given in the TABLES.

RESULTS

The histochemical results are summarized in TABLE 1 and the biochemical results in TABLE 2. Biopsies from all three patients showed fibers with peripheral accumula-

[a]Supported by the Muscular Dystrophy Association of Great Britain.

TABLE 1. Cytochemical Characteristics of Fibers in Biopsies of Skeletal Muscle from Patients with CPEO

Patient	Number of RR Fibers Shown by SDH and NADH-TR Activities	Number of Cytochrome Oxidase Negative Fibers	Number of Cytochrome Oxidase Positive RR Fibers
MR	17	8	13
RL	36	60	2
PH	15	27	0

Cytochrome oxidase,[4] succinate dehydrogenase (SDH), and NADH-tetrazolium reductase (NADH-TR) activities[5] were demonstrated cytochemically in serial sections. 100 fibers were examined in each muscle sample. These changes were seen in all fiber types.

TABLE 2. Substrate Oxidations and Cytochrome Concentrations in Skeletal Muscle Mitochondrial Fractions

Mitochondrial Parameter	Patients MR	Patients RL	Patients PH	Controls (mean ± S.D.)
Oxidation Rates (nmol ferricyanide reduced/min/mg protein)				
10 mM Succinate	430	580	530	517 ± 117 (19)
10 mM Glutamate + 1 mM malate	59[a]	141	253	252 ± 81 (13)
10 mM 2-Oxoglutarate + 1 mM malate	59[a]	185	185	268 ± 91 (5)
Ratios of oxidation rates				
Glutamate + Malate / Succinate	0.14[a]	0.24	0.47	0.52 ± 0.15 (13)
2-Oxoglutarate + Malate / Succinate	0.14[a]	0.32	0.35	0.63 ± 0.20 (5)
Cytochrome Concentrations (nmol/mg protein)				
Cytochrome aa_3	0.34	0.186	0.573	0.47 ± 0.14 (14)
Cytochrome *b*	0.139	0.194	0.235	0.216 ± 0.07 (14)
Cytochrome *c*	0.385	0.349	0.758	0.499 ± 0.161 (14)
Ratios of Cytochrome Concentrations				
Cytochrome aa_3 / Cytochrome *c*	0.88	0.33[a]	0.76	0.958 ± 0.138 (14)
Cytochrome *b* / Cytochrome *c*	0.36	0.35	0.37	0.446 ± 0.117 (14)

The rates of mitochondrial oxidations were determined spectrophotometrically using 0.5 mM ferricyanide as final electron acceptor and following its reduction at 425–475 nm.[6] Rotenone was added when succinate was the substrate. This method has the advantage that the measured flux of electrons from the substrate to cytochrome *c* is not dependent on cytochrome oxidase. Cytochrome concentrations were measured by low temperature redox difference spectroscopy with succinate as reductant.[3] Protein was measured by the Lowry method. [a]Indicates that the value is more than two standard deviations from the control mean.

tion of abnormal mitochondria (ragged red fibers, RR fibers). In PH and RL large numbers of fibers had no cytochemically demonstrable cytochrome oxidase activity. There were about twice as many cytochrome oxidase negative as RR fibers, and there were very few RR fibers that showed cytochrome oxidase activity. By contrast, MR showed significantly more RR fibers than cytochrome oxidase negative fibers and a higher proportion of RR fibers had normal or near normal cytochrome oxidase activity. The morphological abnormalities in these fibers may be associated with the known deficiencies of Complex I in this patient.

MR has a severe defect in Complex I as indicated by the low rates of oxidation of NAD^+-linked substrates. There were also some cytochrome oxidase negative fibers confirming an associated defect of Complex IV. An intriguing question is whether the Complex I deficiency also occurs as a mosaic defect or whether it can occur in fibers with normal cytochrome oxidase activity. We do not know of a valid cytochemical method that specifically demonstrates NADH-ubiquinone reductase activity. The often-used method for NADH-tetrazolium reductase (NADH-TR) activity shows both NADH-ubiquinone reductase activity and NADH-cytochrome b_5 reductase activity. The latter enzyme is very active and is associated with both the outer mitochondrial and endoplastic reticulum membranes. The finding of RR fibers with mitochondria in which cytochrome oxidase is demonstrable suggests that, in these the biochemical abnormality involves another segment of the respiratory chain, which may be Complex I.

RL has a severe deficit of cytochrome oxidase activity and concentration. In addition, the rate of oxidation of NAD^+-linked substrates was in the low normal range. The oxidations by mitochondrial fractions represent the average values for the tissue and there could be severe deficiency of Complex I in some fibers.

DISCUSSION

The etiology of CPEO with proximal myopathy and/or craniosomatic abnormalities remains uncertain because of the lack of clear family histories and the variable age of onset.[1–3] These patients show characteristic morphological changes but differ in the nature of the underlying biochemical abnormalities. Clearly, the syndrome of CPEO involves several different clinical and biochemical defects and multiple lesions of the respiratory chain are likely to be common.

REFERENCES

1. Johnson, M. A., D. M. Turnbull, D. J. Dick & H. S. A. Sherratt. 1983. J. Neurol. Sci. **60:** 31–53.
2. Sherratt, H. S. A., N. E. F. Cartildge, M. A. Johnson & D. M. Turnbull. 1984. J. Inher. Metab. Dis. Suppl. **2:** 107–108.
3. Turnbull, D. M., M. A. Johnson, D. J. Dick, N. E. F. Cartlidge & H. S. A. Sherratt. 1985. J. Neurol. Sci. **70:** 93–100.
4. Seligman, A. M., M. J. Karnovsky, H. L. Wasserkrug & J. S. Hanker. 1969. J Cell Biol. **38:** 1–14.
5. Pearse, A. G. E. 1972. Histochemistry, Theroretical and Applied. 3rd edit. Churchill. London.
6. Turnbull, D. M., H. S. A. Sherratt, D. M. Davies & A. G. Sykes. 1982. Biochem. J. **206:** 511–516.

Defects in Mitochondrial Beta Oxidation[a]

H.R. SCHOLTE, I.E.M. LUYT-HOUWEN, W. BLOM,
H.F.M. BUSCH, P.C. DE JONGE, M. DE VISSER,
J.G.M. HUIJMANS, F.G.I. JENNEKENS, P.D. MOOY,
H. PRZYREMBEL, R.B.H. SCHUTGENS,
M.H.M. VAANDRAGER-VERDUIN, AND
R.N.A. VAN COSTER

Department of Biochemistry I
Erasmus University
Rotterdam, The Netherlands

In our study of the mitochondrial function in muscle disease, we assessed in isolated mitochondria the integrity of some mitochondrial dehydrogenases and respiratory chain phosphorylation with various substrates, including palmitoylcarnitine and malate.[1] We also determined the carnitine-stimulated oxidation of U-^{14}C-palmitate into CO_2 and perchloric acid–soluble intermediates.[2] In controls and most patients there was a good correlation between the oxidation of palmitoylcarnitine and radiopalmitate. However, in patients with defects in the acyl-CoA dehydrogenase complex at the level of ETF or its dehydrogenase,[3,4] the palmitoylcarnitine oxidation was decreased to 30% of controls, while the radiopalmitate oxidation was decreased to 11%. In patients with respiratory chain defects the average percentages were 24 and 78, respectively. This indicates that the control strength[5] of the various defective complexes of oxidative phosphorylation is rather small when one measures the radioactive intermediates. Only drastic reduction of the activity of these complexes was able to reduce the rate of the radiopalmitate oxidation. This method is therefore an excellent tool to study the integrity of the beta oxidation itself and also gives an indication of the presence of adequate levels of NAD^+ and CoA in the mitochondrial matrix. Applied to Duchenne patients, in whom we found defects in oxidative phosphorylation, the method showed intact beta oxidation in most patients.[6]

We found the following beta oxidation defects. In patients with low blood carnitine and abnormal urinary organic acids we found evidence of defects in ETF or ETF dehydrogenase. In a family study we detected three asymptomatic sisters, while the patient, who was wheel-chair bound with barely sufficient respiratory muscles, was completely recovered by riboflavin.[4] Another patient recovered by temporary glucose plus insulin and chronic oral carnitine and riboflavin.[3]

In patients with recurrent hypoglycemia and hypoketonemia, glycine conjugates were found of monocarboxylic and dicarboxylic acids in urine, as was systemic carnitine deficiency. In isolated muscle mitochondria of these patients we found hexanoyl-CoA dehydrogenase deficiency. Most of these patients respond well to high carbohydrate diet, without extra carnitine or riboflavin. Metabolic problems are only to be expected after fasting. In isolated mitochondria of these patients we found that an increase of carnitine from 0.5 to 5 mM (and an increase in CoA from 0.1 to 1 mM, but this increase is of minor importance) doubled the rate of U-^{14}C-palmitate oxidation. In

[a]Supported by grants from Het Prinses Beatrix Fonds, The Hague, and Het Willem H. Kröger Fonds, Rotterdam, The Netherlands.

several other patients, with all kinds of mitochondrial defects and reduced carnitine in muscle, we found a much higher activation by 5 mM carnitine than in controls. Interestingly, several of these patients had a reduced activity of medium-chain acyl-CoA dehydrogenase. It can be speculated that this enzyme is down-regulated by the primary mitochondrial defect. These persons need more muscle carnitine than controls, in which the beta oxidation is almost fully activated by 0.5 mM carnitine, which corresponds to 0.25 μmol/g muscle weight. Control muscle carnitine level is 3.96 ± 0.09, $N = 59$.

In patients with exercise-induced myoglobinuria, we found defects in carnitine palmitoyltransferase II. Only one of the three patients also had reduced CPT I activity. We detected two asymptomatic sisters with a CPT II deficiency in leukocytes. Recently we found in one of the male patients with CPT II deficiency decreased long-chain and medium-chain acyl-CoA dehydrogenase activity but no abnormal urinary organic acids.

In another patient with exercise-induced myoglobinuria with severe myalgia, we found reduced carnitine-efflux from mitochondria by addition of acetylcarnitine, palmitoylcarnitine, or carnitine. Although his palmitoylcarnitine oxidation rate was reduced, he showed normal acetylcarnitine oxidation. Later we found also in this patient decreased long-chain acyl-CoA dehydrogenase. At the moment we do not know what the primary lesion of this patient is—deficiency of the carnitine carrier or of long-chain acyl-CoA dehydrogenase. We hope that study of his cultured skin fibroblasts or muscle cells cultured by Dr. P.A. Bolhuis in Amsterdam will provide the solution, since defects in mitochondrial beta oxidation, in the patients discussed above, are expressed in cultured cells.[3,4,7,8]

REFERENCES

1. Scholte, H. R., I. E. M. Luyt-Houwen & H. F. M. Busch. 1985. Difficulties in assessing biochemical properties of abnormal muscle mitochondria. J. Inher. Metab. Dis. **8** (Suppl. 2): 149–150.
2. Van Hinsbergh, V. W. M., J. H. Veerkamp & H. Th. B. Van Moerkerk. 1978. An accurate assay of long-chain fatty acid oxidation by human muscle. Biochem. Med. **20:** 256–266.
3. Mooy, P. D., M. A. H. Giesberts, H. H. Van Gelderen, H. R. Scholte, I. E. M. Luyt-Houwen, H. Przyrembel & W. Blom. 1984. Glutaric aciduria type II: multiple defects in isolated muscle mitochondria and deficient beta-oxidation in fibroblasts. J. Inher. Metab. Dis. **7** (Suppl. 2): 101–102.
4. De Visser, M., H. R. Scholte, R. B. H. Schutgens, P. A. Bolhuis, I. E. M. Luyt-Houwen, M. H. M. Vaandrager-Verduin, H. A. Veder & P. L. Oei. 1986. Riboflavin-responsive lipid storage myopathy and glutaric aciduria type II with early adult onset. Neurology **36:** 367–372.
5. Tager, J. M., R. J. A. Wanders, A. K. Groen, W. Kunz, R. Bohnensack, U. Küster, G. Letko, G. Böhme, J. Duszynski & L. Wojtczak. 1983. Control of mitochondrial respiration. FEBS Lett. **151:** 1–9.
6. Scholte, H. R., I. E. M. Luyt-Houwen, H. F. M. Busch & F. G. I. Jennekens. 1985. Muscle mitochondria from patients with Duchenne muscular dystrophy have a normal beta oxidation, but an impaired oxidative phosphorylation. Neurology **35:** 1396–1397.
7. Huijmans, J. G. M., H. R. Scholte, W. Blom, I. E. M. Luyt-Houwen & H. Przyrembel. 1984. Enzymatic evidence for a medium-chain acyl-CoA dehydrogenase deficiency in muscle of a patient with hypoketotic hypoglycemic dicarboxylic aciduria. Pediatr. Res. **18:** 798.
8. Scholte, H. R., W. C. Hülsmann, I. E. M. Luyt-Houwen, J. T. Stinis & F. G. I. Jennekens. 1985. Carnitine palmitoyltransferase deficiencies. Biochem. Soc. Trans. **13:** 643–645.

Biochemical Criteria for NADH-CoQ Reductase Deficiency[a]

HANS R. SCHOLTE, INEZ E. M. LUYT-HOUWEN, HERMAN F. M. BUSCH, AND M. HEDWIG VAANDRAGER-VERDUIN

Departments of Biochemistry I, Neurology, and Clinical Genetics
Erasmus University
Rotterdam, The Netherlands

The main function of mitochondria is the production of ATP by the process of oxidative phosphorylation. Since oxidation and phosphorylation are tightly coupled, ATP production ceases when the oxygen supply is disturbed. Ischemia is the most important cause of failing mitochondria in man. There are also patients with an adequate oxygen supply, but decreased mitochondrial ATP production. The mitochondrial system for energy conservation is very vulnerable for genetic and environmental damage. Most of the more than 50 known mitochondrial enzyme deficiencies affect mitochondrial ATP supply. The mitochondrial disease could affect skeletal muscle only (mitochondrial myopathy/myopathy with abnormal mitochondria) or other organs as well like heart, liver, kidney, and/or brain. The myopathic patients have chronic muscle problems such as hypotony, weakness, cramps, stiffness, and/or exercise intolerance. Sporadic and familial defects are known in oxidative phosphorylation (in the respiratory chain or in the phosphorylation), in dehydrogenases, in the synthesis of oxidative substrates (non-redox reactions), and in transport processes. The most frequently discovered mitochondrial defects are listed in TABLE 1. If the ATP synthesis is severely impaired, secondary changes in mitochondrial enzyme activities and a decrease in tissue carnitine levels occur. Trends in these changes can be of help to elucidate the primary defect. Difficulties in the search for mitochondrial biochemical defects are seldom described.[1] In our experience investigation of both isolated muscle mitochondria and muscle homogenate are needed to define mitochondrial defects.

MATERIALS AND METHODS

Muscle biopsies were taken from the M. quadriceps, biceps, or gastrocnemius under local analgesia. Mitochondria were isolated from fresh tissue (0.15–2 g) and were studied within 2 hr after the isolation procedure of 1.5 hr. Homogenates (5%) were made of frozen-thawed muscle in 0.25 M sucrose, 10 mM HEPES-KOH, 1 mM EDTA (pH 7.4), stored batchwise at −70°C, and assayed immediately after thawing.

Oxidative phosphorylation was studied in intact mitochondria with fuel substrates entering the respiratory chain before NAD^+(pyruvate + malate, glutamate + malate), at NAD^+ and CoQ (palmitoylcarnitine + malate), at CoQ (succinate + rotenone), and at cytochrome *c*(ascorbate + *N*,*N*,*N′*,*N′*-tetramethyl-*p*-phenylenedi-

[a]Supported by Het Prinses Beatrix Fonds, The Hague, and Het Willem H. Kröger Fonds, Rotterdam.

TABLE 1. The Most Frequently Established Defects in Human Mitochondria

Process	Deficient Enzyme/Translocase/Compound	Features
Oxidative phosphorylation	Multiple defects	L V M
Respiratory chain	NADH-CoQ reductase	L V M
	CoQ	L V M
	Cytochrome bc_1	L V M
	Cytochrome aa_3	L
	Multiple respiratory chain defects	L V M
Phosphorylation	Loose coupling (secondary)	M
	Adenine nucleotide translocase (sec.)	L M
Dehydrogenases	Pyruvate dehydrogenase	L V
	Branched-chain ketoacid dehydrogenase	O F V
	Mid-chain acyl-CoA dehydrogenase	O F V
	Multiple acyl-CoA dehydrogenase (ETF or ETF-dehydrogenase)	L O F V
	Isovaleryl-CoA dehydrogenase	O F V
Fuel synthesis	Propionyl-CoA carboxylase	O F V
	Pyruvate carboxylase	L F V
	Multiple carboxylase (Biotinidase, holoenzyme synthetase)	L O F V
	Methyl-malonyl-CoA mutase	O F V
Transport of activated fatty acids	Carnitine (mostly secondary)	O V
	Carnitine palmitoyltransferase II	F

Features: L = Blood and urine lactate levels are increased. O = Abnormal urinary organic acids and carnitine esters. F = Deficiency is also expressed in cultured fibroblasts. V = Some of the patients respond to vitamin, carnitine, or CoQ therapy. M = Isolated mitochondria must be studied for elucidation of all defects at this level.

amine). We determined in the presence of glucose and hexokinase oxygen uptake before and after the addition of ADP, P/O ratios, and uncoupler-stimulation of ascorbate oxidation. At the same time Mg^{2+}-ATPase ± uncoupler ± oligomycin, malonyl-CoA decarboxylase ± detergents, and U-^{14}C-palmitate oxidation with 0.1 mM CoA plus 0.5 mM carnitine, plus 5 mM carnitine, 0.9 mM CoA, and plus 1 mM KCN (to measure the peroxisomal contribution) were tested. Later, total carnitine, creatine kinase, (rotenone-sensitive) NADH oxidase,[2] succinate-INT$^+$ reductase, (antimycin-sensitive) succinate-cytochrome *c* reductase and cytochrome *c* oxidase were tested in the homogenate.[1,2]

RESULTS

From a total of 214 patients investigated (most myopathic patients and controls), we found in 75 patients an oxidation rate with pyruvate + malate + ADP of less than 33 nat O_2/min/mg protein versus 82 ± 5 (SE) in 22 controls. In 47 patients we found one or more defects in oxidative phosphorylation, the others had a primary lesion in one of the mitochondrial dehydrogenases with secondary defects in oxidative phosphorylation, or not enough data were available.

We deduced that 17 patients had a defect in NADH-CoQ reductase. Twelve of them showed a decreased (to one third or less of the average control) NADH oxidase activity. Nine of them showed more defects. In the homogenates, succinate-INT^+ reductase was decreased in 4 patients, succinate-cytochrome *c* reductase in 5, and cytochrome *c* oxidase in 6. In isolated mitochondria the stimulation of ascorbate oxidation by ADP was decreased to less than 1.2 (controls 1.65 ± 0.07, $N = 22$) in 4 patients. It is possible that increased long-chain acyl-CoA had caused an inhibition of the adenine nucleotide translocase in these patients.[3] The oxidation of U-^{14}C-palmitate with 0.5 mM carnitine was decreased (to less than half of controls) in 4 patients. The relatively normal palmitate oxidation in the other patients implies normal matrix NAD^+ and CoA levels. In the patients with normal NADH oxidase, a defect at site 1 was deduced from the fact that the succinate oxidation was higher than that of the NAD^+-linked substrates. No other defects were found in this group. It is possible that the inability to oxidize NAD^+-linked substrates has been caused by a mutant NADH-CoQ reductase with a decreased affinity for NADH. Criteria for NADH-CoQ reductase deficiency are summarized in TABLE 2.

In 13 other patients we deduced a defect at the level of CoQ-cytochrome bc_1. The mitochondria oxidized succinate and duroquinol at low rates, and the latter rate was for the greater part antimycin-insensitive, in contrast to controls. A decrease in homogenate succinate-cytochrome *c* reductase was found in only three of these patients. We suspect a defect of cytochrome bc_1 in these patients (which was confirmed by cytochrome redox spectra in one of them, in the others not enough mitochondria were available). In the other 10 patients we suspect a defect at the level of CoQ or CoQ-binding protein. In 6 other patients cytochrome *c* oxidase was deficient in the homogenates. Most of them showed a low stimulation of ascorbate oxidation by ADP. In 11 other patients a defect at the level of the adenine nucleotide (or P_i) carrier was concluded from the low ADP stimulation of ascorbate oxidation, or a defect at the level of ATP synthetase from the low activity of uncoupler-stimulated Mg^{2+}-ATPase. There were no abnormal homogenate findings in this group, except for carnitine deficiency in some of the patients, which was encountered in all groups. All of the patients showed lactic acidemia.

TABLE 2. Biochemical Criteria for NADH-CoQ Reductase Deficiency

1. Intact mitochondria oxidize NAD^+-linked substrates with low velocity, while succinate and ascorbate are oxidized with higher rates.
2. Substrate import into the mitochondria is normal. (Can be concluded by the relatively normal oxidation of succinate and of U-^{14}C-palmitate+carnitine, which implies normal malate import.)
3. Adequate amounts of NAD^+ and CoA are available in the mitochondrial matrix. (Can be concluded by the normal U-^{14}C-palmitate oxidation.)
4. The activity of homogenate NADH oxidase is reduced in most patients, while those of succinate-cytochrome *c* reductase and cytochrome *c* oxidase are normal.
5. The activities of the adenine nucleotide translocase and the phosphate carrier are normal. (Can be inferred by the normal stimulation of ascorbate oxidation by ADP and by uncoupler.)
6. The activity of ATP synthetase is normal. (Can be tested by the activity of uncoupler-stimulated Mg^{2+}-ATPase.)

When other deficiencies are encountered then only in NADH-CoQ reductase, a primary deficiency of NADH-CoQ reductase cannot be concluded from the experimental data in the absence of exact knowledge of the control strength of the defective pathways on mitochondrial ATP synthesis in human muscle mitochondria.

REFERENCES

1. SCHOLTE, H. R., I. E. M. LUYT-HOUWEN & H. F. M. BUSCH. 1985. Difficulties in assessing biochemical properties of abnormal muscle mitochondria. J. Inher. Metab. Dis. **8** (Suppl. 2): 149–150.
2. FISCHER, J. C., W. RUITENBEEK, J. M. F. TRIJBELS, J. H. VEERKAMP, A. M. STADHOUDERS, R. C. A. SENGERS & A. J. M. JANSSEN. 1986. Estimation of NADH oxidation in human muscle mitochondria. Clin. Chim. Acta **155:** 263–274.
3. LAUQUIN, G. J. M., C. VILLIERS, J. W. MICHEJDA, L. V. HRNIEWIECKA & P. V. VIGNAIS. 1977. Adenine nucleotide transport in sonic submitochondrial particles: kinetic properties and binding of specific inhibitors. Biochim. Biophys. Acta **460:** 331–345.

Mitochondrial Involvement in Causing Cell Injury in Experimental Hepatic Iron Overload

A. MASINI,[a] T. TRENTI,[b] D. CECCARELLI,[a]
AND U. MUSCATELLO[a]

Istituti di [a]Patologia Generale and [b]Clinica Medica III
Universita' di Modena
41100 Modena, Italy

The biochemical mechanism of hepatocellular injury in chronic iron overload has not been experimentally established. Enhancement of peroxidative reactions by iron in the lipid membranes of cellular organelles resulting in structural and functional alterations in cell integrity appears to be the most favored hypothesis. Indeed evidence is accumulating that iron overload may result in hepatic lipid peroxidation in mitochondrial membranes not only *in vitro*[1] but also *in vivo*.[2-5] A possible involvement of liver mitochondria in causing cell damage in this pathological condition has been investigated.

METHODS

Rats were made siderotic by feeding a diet supplemented with 2.5% (w/w) carbonyl iron over an 8-week period. Iron (Fe) was measured by atomic absorption. Lipid peroxidation was detected by conjugated dienes. The mitochondrial transmembrane potential ($\Delta\psi$) was measured by tetraphenylphosphonium-selective electrode. Ca^{2+} movements were followed by a selective electrode. Reduced pyridine nucleotides were determined by enzymatic method.

RESULTS AND DISCUSSION

After 8 weeks of Fe treatment the mitochondrial Fe content increases up to 40.9 ± 8.7 nmol/mg in respect to 3.58 ± 0.74 nmol/mg of control. Enhancement of lipoperoxidative reactions in the mitochondrial membranes, as revealed by the presence of conjugated dienes, is associated with these conditions. However electron microscopy analyses do not reveal any appreciable modifications in mitochondrial ultrastructure in iron loaded rats. Accordingly these mitochondria exhibit normal oxidative metabolism, as indicated by the value of respiratory control index, ADP/O ratio, or transmembrane potential not significantly different from the control. However, as can be seen from FIGURE 1, the membrane potential of mitochondria from iron-loaded rats, which have accumulated a low pulse of Ca^{2+} (i.e. 50 nmol/mg), progressively decreases. The observation that the Ca^{2+} chelator EGTA fully restores a normal membrane potential indicates that enhancement of Ca^{2+} cycling is the energy-dissipating process responsible for the drop. This also demonstrates that the

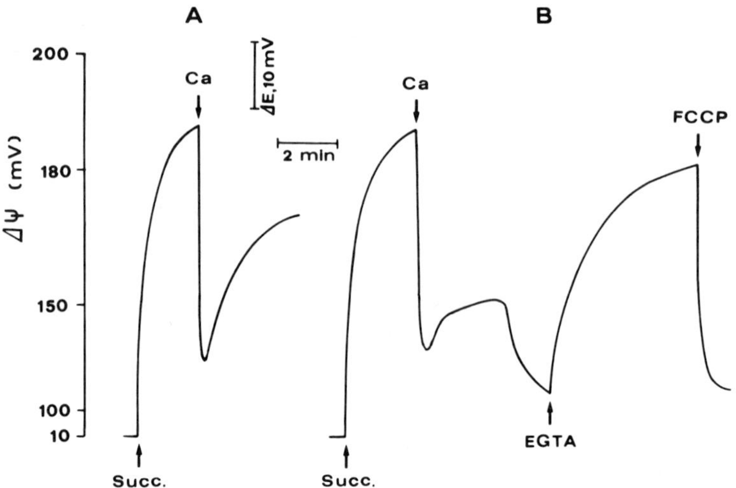

FIGURE 1. Transmembrane potential of liver mitochondria from iron-loaded rats. Arrows indicate the following additions: 2 mM succinate (Succ.); 150 μM Ca^{2+}; 0.5 mM EGTA; 0.5 μM FCCP. (A) Control mitochondria. (B) Mitochondria from 60-day iron-treated rats. The transmembrane potential ($\Delta\psi$) was measured as described in the METHODS. ΔE, electrode potential.

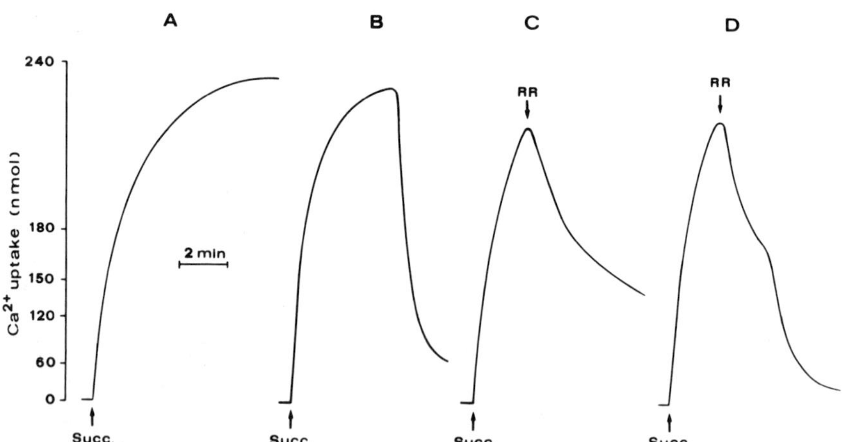

FIGURE 2. Ca^{2+} transport in liver mitochondria from iron-loaded rats. Mitochondria were incubated in the presence of 150 μM Ca^{2+}. Arrows indicate the following additions: 2 mM succinate (Succ.); 2 μM ruthenium red (RR). (A and C) Control mitochondria. (B and D) Mitochondria from 60-day treated rats. Ca^{2+} movements were followed by a selective electrode.

mitochondrial inner membrane is not irreversibly depolarized, i.e. generalized damages, under these experimental conditions.

FIGURE 2 shows that *in vivo* iron overload brings about the induction of a selective Ca^{2+} release pathway (antiport) that is not inhibitable by ruthenium red, a specific inhibitor of the electrophoretic Ca^{2+} uptake route (uniport). This event accounts for the enhancement of Ca^{2+} cycling, a process that utilizes energy to reaccumulate the released Ca^{2+} in mitochondria. Thus it may be concluded that Ca^{2+} efflux is not due to nonspecific increase in inner membrane permeability i.e. membrane potential collapse, but rather this process is the cause of membrane potential drop.

Alteration of the redox state of mitochondrial pyridine nucleotides seems to be the regulatory event in the induction of Ca^{2+} efflux from iron-loaded mitochondria.[6] Indeed, these mitochondria present a reduced content of NADH plus NAD(P)H (i.e., 2.18 ± 0.68 nmol/mg in respect to 5.72 ± 0.42 nmol/mg of the control).

The present findings indicate a strict correlation between induction of lipoperoxidative reactions in the mitochondrial membranes and the activation of a specific Ca^{2+} release route from mitochondria via the oxidation of pyridine nucleotides in *in vivo* iron overload condition. Finally this reported alteration in mitochondrial Ca^{2+} transport system may suggest a primary involvement of mitochondria in the hepatocellular damage in this pathological state.

REFERENCES

1. MASINI, A., T. TRENTI, D. CECCARELLI-STANZANI & E. VENTURA. 1985. Biochim. Biophys. Acta **810:** 20–26.
2. HANSTEIN, W. G., T. D. HEITMANN, A. D. SANDY, H. L. BIESTERFELDT, H. H. LIEM & U. MULLER-EBERHARD. 1981. Biochim. Biophys. Acta **678:** 293–299.
3. BACON, B. R., A. S. TAVILL, G. M. BRITTENHAM, C. H. PARK & R. O. RECKNAGEL. 1983. J. Clin. Invest. **71:** 429–439.
4. MASINI, A., T. TRENTI, E. VENTURA, D. CECCARELLI-STANZANI & U. MUSCATELLO. 1984. Biochem. Biophys. Res. Commun. **124:** 462–469.
5. MASINI, A., D. CECCARELLI-STANZANI, T. TRENTI & E. VENTURA. 1984. Biochim. Biophys. Acta **802:** 253–258.
6. LEHNINGER, A. L., A. VERCESI & E. A. BABABUNMI. 1978. Proc. Natl. Acad. Sci. USA **75:** 1690–1694.

On the Role of Mitochondria in Cell Injury Caused by Ca^{2+} Overload:

A Study with Vanadate and Isolated Hepatocytes

G. BELLOMO, F. MIRABELLI, P. CRINÓ,
AND G. FINARDI

Dipartimento di Medicina Interna e Terapia Medica
Clinica Medica I[a]

G. VIANI, F. ROSSI, AND P. RICHELMI

Istituto di Farmacologia II
University of Pavia
27100 Pavia, Italy

Long-term regulation of intracellular Ca^{2+} homeostasis in mammalian cells is critically dependent on active extrusion of Ca^{2+} through the plasma membrane.[1] Recently, the presence of a high-affinity Ca^{2+}-stimulated ATPase and an ATP-dependent Ca^{2+} transport have been demonstrated in the plasma membrane of a variety of cell types.[2] ATPase and the transporter have been shown to be inhibited by low concentrations of vanadate ion.[3] In this study we have analyzed the cellular alterations occurring in hepatocytes treated with vanadate in order to inhibit the plasma membrane Ca^{2+}-extruding system. As illustrated in FIGURE 1, vanadate induces a progressive accumulation of Ca^{2+} in the hepatocytes. The mechanism responsible for Ca^{2+} accumulation is likely to be due to the inhibition of the plasma membrane Ca^{2+}-extruding system rather than an increase in plasma membrane permeability or the enhanced operation of the Na^+/Ca^{2+} exchange. In fact, vanadate-treated cells fail to release the accumulated Ca^{2+} when incubated in a Ca^{2+}-free medium and do not show an inhibition of Ca^{2+} accumulation when incubated in a Na^+-depleted (choline-supplemented) medium (data not shown).

A marked decrease in intracellular glutathione concentration also occurs following vanadate treatment and precedes Ca^{2+} accumulation (FIG. 1). This decrease may be interpreted as a protective mechanism against intracellular vanadate accumulation since GSH can reduce vanadate to vanadyl and can form adducts with vanadyl itself.[4] In fact, in hepatocytes whose intracellular GSH concentration has been lowered by treatment with diethyl maleate, vanadate is much more efficient in promoting Ca^{2+} uptake (not shown).

Of the intracellularly accumulated Ca^{2+}, up to 90% is sequestered in mitochondria as revealed by the fact that it can be mobilized by treatment with uncouplers. As the mitochondria accumulate Ca^{2+}, they lose the ability to sequester further Ca^{2+} and to buffer extramitochondrial Ca^{2+} at physiological levels, and finally they release their own Ca^{2+}. Furthermore, the combined release of Ca^{2+} from mitochondrial stores, the continuous Ca^{2+} influx through the plasma membrane, and the marked inhibition of the Ca^{2+} extruding system lead to a sustained increase in cytosolic Ca^{2+} concentration and to the stimulation of Ca^{2+}-dependent enzymes such as phosphorylase. These alterations are also closely related to a depletion of intracellular ATP, suggesting a severe mitochondrial dysfunction.

Abnormalities of the cell surface morphology (blebs) occur following the derangements of intracellular Ca^{2+} homeostasis and ATP depletion induced by vanadate, probably linked to alterations occurring in the cytoskeletal structures of the cell. These alterations precede the irreversible cell injury revealed by the increased permeability to Trypan blue.

From the data presented it appears that vanadate treatment can be conveniently used as a model to further investigate the cellular and molecular mechanisms underlying Ca^{2+}-induced cell injury in hepatocytes.

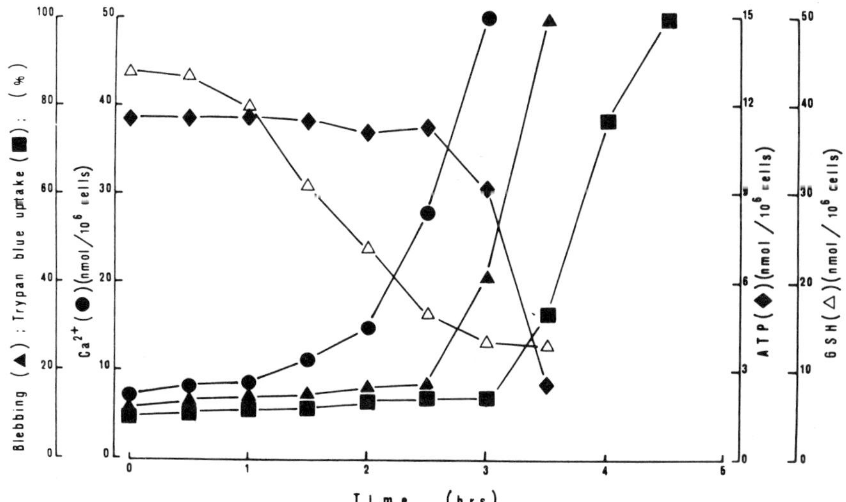

FIGURE 1. Effects of vanadate on intracellular GSH (△), ATP (◆), Ca^{2+} (●), blebbing (▲), and Trypan blue uptake (■). Hepatocytes (10^6 cells/ml) were incubated in a Krebs-Henseleit medium supplemented with 20 mM HEPES pH 7.4 in the presence of 1 mM vanadate. At the indicated times samples were taken for the measurement of the intracellular GSH, ATP, and Ca^{2+} content and analyzed for surface abnormalities and Trypan blue uptake. The methods employed are described in Bellomo & Orrenius[5] and in references therein.

CONCLUSIONS

(1) Vanadate induces Ca^{2+} accumulation in isolated rat hepatocytes in a dose- and time-dependent manner. (2) Vanadate-induced Ca^{2+} accumulation is not related to an increase in plasma membrane permeability or to the stimulation of the Na^+/Ca^{2+} exchange. Instead, it is due to the inhibition of the plasma membrane Ca^{2+}-extruding system. (3) Intracellular glutathione seems to play an important role in protecting hepatocytes against vanadate-induced Ca^{2+} accumulation. (4) Most of the Ca^{2+} accumulated in response to vanadate treatment is sequestered in the mitochondrial compartment. (5) As mitochondria progressively increase their Ca^{2+} content, they lose the ability to sequester further Ca^{2+}, to buffer extramitochondrial Ca^{2+} at physiological levels, and they release their own Ca^{2+}. (6) These alterations are associated with intracellular ATP depletion and with a sustained increase in cytosolic

Ca^{2+} concentration and are followed by plasma membrane abnormalities (blebs) and cytotoxicity.

REFERENCES

1. CAMPBELL, A. K. 1983. Intracellular calcium: its universal role as regulator. Wiley. Chichester.
2. CARAFOLI, E. 1982. Membrane Transport of Calcium. Academic Press. New York.
3. PENNISTON, J. T. 1984. *In* Calcium and Cell Function. W. Y. Cheung, Ed. **4:** 99–149. Academic Press. New York.
4. HANSEN, T. V., J. AASET & J. ALEXANDER. 1982. Arch. Toxicol. **50:** 195–202.
5. BELLOMO, G. & S. ORRENIUS. 1985. Hepatology **5:** 876–882.

Mitochondrial Myopathy due to Complex I Deficiency:

Secondary Changes[a]

H.F.M. BUSCH,[b] H.R. SCHOLTE,[c] AND
I.E.M. LUYT-HOUWEN[c]

*Departments of [b]Neurology, [b]Pathology, and [c]Biochemistry I
Erasmus University
Rotterdam, The Netherlands*

Localized lesions of the respiratory chain are increasingly recognized as the cause of myopathy or multisystem disorders. Several patients with a myopathy due to NADH-CoQ reductase have now been described.[1] Some of these patients also had a secondary carnitine deficiency syndrome.[1,2] We studied three patients with a myopathy due to complex I deficiency. Secondary changes in the function of the respiratory chain were noted and in one patient additional evidence was obtained that carnitine deficiency was secondary to the lesion at complex I.

A 15-year-old boy had fatiguable limb weakness for about four years. The second patient is his 43-year-old mother, who had similar complaints for more than 10 years.[2] Finally a 15-year-old, unrelated girl had easy fatiguability since childhood. All patients had lacticacidemia. No abnormal metabolites were identified in the urine. Muscle biopsies showed morphological and histochemical changes compatible with a myopathy with abnormal mitochondria. Isolated mitochondria had a normal coupling state and a marked reduction of the respiratory rates with NAD^+-linked substrates. In the homogenates the activity of rotenone-sensitive NADH oxidase was low.

In the boy succinate oxidation was slightly increased, ascorbate + TMPD oxidation was 53% of control values. In the homogenate the activity of succinate cytochrome *c* reductase was barely normal, that of cytochrome *c* oxidase was normal.

In his mother, succinate oxidation by isolated mitochondria was 63% and ascorbate oxidation was 44% of control values. In the homogenate, succinate cytochrome *c* reductase was reduced to 31% and cytochrome *c* oxidase to 78%, and one year later dropped to 14 and 38%, respectively. Total carnitine was only 36% of normal. Since it is reasonable to assume that the basic defect in the mother and her son is similar, these patients show that in complex I deficiency a longer duration and/or greater severity of the disease may lead to carnitine deficiency as well as additional functional derangement of the respiratory chain.

In the girl, a considerable increase in the activity of the distal part of the respiratory chain was found. Ascorbate was oxidized at twice the normal rate and oxidation of succinate + rotenone was 160% of control values. In the homogenate, succinate cytochrome *c* reductase was elevated more than fourfold, while cytochrome *c* oxidase showed a threefold increase. The nature of the defect at complex I in the girl differs from that in the first two patients in several respects. Fatiguability was more

[a]Supported in part by the Willem H. Kröger Stichting, Rotterdam and The Prinses Beatrix Fonds, The Hague.

pronounced and was riboflavin responsive, which was not the case in the other patients.

These patients show that in complex I deficiency secondary changes may occur and tend to blur or accentuate the nature of the biochemical lesion. On the one hand, carnitine deficiency and decreased activities of the distal respiratory chain may become apparent in some patients. In others the distal part of the respiratory chain may show a considerable increase in activity, accentuating the lesion at complex I.

REFERENCES

1. MORGAN-HUGHES, J. A. 1986. The mitochondrial myopathies. *In* Myology. A. G. Engel & B. Q. Banker, Eds.: 1709–1742. McGraw-Hill. New York.
2. H. F. M. BUSCH, H. R. SCHOLTE, W. F. ARTS & I. E. M. LUYT-HOUWEN. 1981. A mitochondrial myopathy with a respiratory chain defect and carnitine deficiency. *In* Mitochondria and Muscular Diseases. H.F.M. Busch, F.G.I. Jennekens & H.R. Scholte, Eds.: 207–211. Mefar b.v. Beetsterzwaag, The Netherlands.

Role of Plasma Membrane Potential in Granulocyte Activation

FRANCESCO DI VIRGILIO,[a] P. DANIEL LEW,[b]
TOMMY ANDERSSON,[c] SUSAN TREVES,[a]
AND TULLIO POZZAN[a,d]

[a]*CNR Unit for the Study of the Physiology of Mitochondria*
Institute of General Pathology
Via Loredan 16
35131, Padova

[b]*Infectious Disease Division*
Department of Medicine
University of Geneva
CH-1211 Geneva 4, Switzerland

[c]*Department of Medical Microbiology*
Linkoping University Medical School
S-581, Linkoping, Sweden

Plasma membrane potential changes associated with ligand-receptor interaction have been demonstrated in both excitable and non-excitable cells. Among the latter, hyperpolarization and/or depolarization have been described for lymphocytes,[1,2] platelets,[3] and granulocytes.[4,5] In excitable tissues such as neurons, neuroendocrine and muscle cells, the physiological role for plasma membrane depolarization has been well characterized. These cells possess plasma membrane ion channels that are sensitive to, or gated by, membrane potential.[6,7] On the other hand, the role of plasma membrane potential changes in non-excitable cells remains mysterious. In neutrophils, plasma membrane depolarization is elicited by a number of particulate and soluble agonists (e.g., formyl-methionyl oligo peptides that bind to specific cell surface receptors coupled to Ca^{2+} mobilization and phosphoinositide breakdown).

The aim of our study was to examine the relationship between plasma membrane potential changes and $[Ca^{2+}]_i$ in response to fMet-Leu-Phe in human neutrophils and in the human promyelocytic leukemia cell line HL60, after differentiation with dimethylsulfoxide (DMSO).

Our results indicate that: (1) Chemotactic peptide-dependent plasma membrane depolarization is a post-receptor event and probably depends on protein kinase C activation. In fact, receptor-activated depolarization occurred both at high and resting $[Ca^{2+}]_i$, but was inhibited at very low $[Ca^{2+}]_i$. On the contrary, phorbol myristate acetate (PMA)–induced plasma membrane depolarization was independent of $[Ca^{2+}]_i$. Threshold fMet-Leu-Phe concentrations for plasma membrane depolarization (10^{-8} M) was at least one log unit higher as compared to $[Ca^{2+}]_i$ increases (5×10^{-10} M) and coincident with the increase of NADPH-oxidase activation. Nearly maximal $[Ca^{2+}]_i$ increases were elicited by 3×10^{-9} M fMet-Leu-Phe in the absence of any significant plasma membrane potential change.

[d]Address correspondence to: Dr. Tullio Pozzan, Institute of General Pathology, Via Loredan 16, 35131 Padova, Italy.

(2) Plasma membrane potential modulates receptor-activated Ca^{2+} influx and release from intracellular stores at low fMet-Leu-Phe concentrations (10^{-9}–3×10^{-9} M), i.e. at doses that alone do not perturb plasma membrane potential. Depolarizing (gramicidin 10^{-7}–10^{-6} M or KCl 50 mM) and hyperpolarizing (valinomycin 4 μM) treatments had little influence on unstimulated $[Ca^{2+}]_i$ levels, while fMet-Leu-Phe–induced transients were significantly altered. Gramicidin and KCl decreased the fMet-Leu-Phe–induced $[Ca^{2+}]_i$ rises in Ca^{2+}-containing or Ca^{2+}-free media. However, valinomycin increased receptor-stimulated $[Ca^{2+}]_i$ rises and the effect was larger in the presence of extracellular Ca^{2+}.

(3) Plasma membrane depolarization inhibits fMet-Leu-Phe–induced $InsP_3$ formation. Yet, hyperpolarization with valinomycin has negligible effects on inositol phosphate accumulation, but strongly potentiates secretion.

It is suggested that plasma membrane depolarization is a physiological protein kinase C–activated feedback mechanism inhibiting phospholipase C stimulation and receptor-dependent $[Ca^{2+}]_i$ changes.

REFERENCES

1. Shapiro, H. M., P. J. Natale & L. A. Kamentsky. 1982. Proc. Natl. Acad. Sci. USA **76:** 5728–5730.
2. Tsien, R. Y., T. Pozzan & T. J. Rink. 1982. Nature **295:** 68–71.
3. Horne, W. C., N. E. Norman, D. B. Schwartz & E. R. Simons. 1981. Eur. J. Biochem. **120:** 295–302.
4. Korchak, H. M. & G. Weissman. 1978. Proc. Natl. Acad. Sci. USA **75:** 3818–3822.
5. Mottola, C. & D. Romeo. 1982. J. Cell. Biol. **93:** 129–134.
6. Hodkin, A. & A. F. Huxley. 1952. J. Physiol. **117:** 500–544.
7. Tsien, R. W. 1983. Ann. Rev. Physiol. **45:** 341–358.

Altered Free Cytosolic Ca^{2+} Changes in fMet Leu Phe–Stimulated Neutrophils from Patients with Bartter's Syndrome

L. CALÒ, S. CANTARO, F. DI VIRGILIO,[a] S. FAVARO, AND A. BORSATTI

Department of Internal Medicine
Postgraduate School of Nephrology
and
[a]Institute of General Pathology
University of Padova
Padova, Italy

The picture of Bartter's Syndrome (BS) includes a decreased pressor response to both Angiotensin II and noradrenaline, an anomaly in platelet aggregation, and decreased sensitivity of a distal nephron to ADH.[1–3] These events, which are not due to hypokalemia[4] or to prostaglandin overproduction[5] and are not present in similar disorders, might be in some way related to cytosolic Ca^{2+} movements, because both the contractile response of vascular smooth muscle[6] and platelet aggregation[7] and the sensitivity to ADH[8] are known to be dependent on intracellular Ca^{2+} activity. In order to verify the presence of an inherent cellular defect that alters cellular Ca^{2+} handling, we measured in BS resting and stimulated $[Ca^{2+}]_i$ levels in polymorphonuclear leukocytes. Eight patients, four males and four females, affected by BS were studied (age range 11–54). In six cases a family history of BS was present. All patients exhibited a full expression of clinical and laboratory features of the syndrome. Healthy volunteers were used as controls. Blood samples anticoagulated with citrate were taken after overnight fast; neutrophils were prepared as previously described[9] and resuspended in medium containing 138 mM NaCl, 6 mM KCl, 1 mM P_i, 0.5 mM $CaCl_2$, 5.6 mM glucose, and 20 mM HEPES (pH 7.4). $[Ca^{2+}]_i$ was monitored with quin 2[10] and the synthetic chemotactic peptide formyl methionyl leucyl phenylalanine (fMet Leu Phe) was used as a stimulant in a molar range 10^{-10}–3×10^{-7}.

Stimulated intracellular free Ca^{2+} was significantly lower in BS at fMet Leu Phe concentration higher than 3×10^{-9} (10^{-8}, 10^{-7}, 3×10^{-7}); 10^{-8} M:575 ± 122, $N = 5$ controls (C) versus 343 ± 104, $N = 8$ BS nM Ca_i^{2+}, $p < 0.005$; 10^{-7} M:619 ± 154 $N = 4$ C versus 345 ± 91 $N = 8$ BS nM Ca_i^{2+}, $p < 0.005$; 3×10^{-7} M:587 ± 154, $N = 5$ C versus 347 ± 82, $N = 8$ BS nM Ca_i^{2+}, $p < 0.005$ while no difference could be found at lower polypeptide concentration (10^{-10} M:102 ± 34, $N = 5$ C versus 112 ± 28, $N = 8$ BS nM Ca_i^{2+}; 10^{-9} M:259 ± 74, $N = 5$ C versus 235 ± 36, $N = 8$ BS nM Ca_i^{2+}; 3×10^{-9} M:356 ± 142, $N = 5$ C versus 306 ± 93, $N = 8$ BS nM Ca_i^{2+}. Furthermore resting neutrophils cytosolic free Ca^{2+} was the same in both patients and controls (106 ± 12, $N = 10$ C versus 118 ± 30, $N = 10$ BS nM Ca_i^{2+} (TABLE 1 and FIGURE 1). Our data are consistent with the hypothesis that cellular Ca^{2+} metabolism is altered in BS and may suggest the stimulating possibility of an anomalous coupling between receptor stimulation and post receptor events; in fact the bond of fMet Leu Phe and vasopressor agents (i.e. angiotensin II) with the receptors are linked to the activation of membrane polyphosphoinositide breakdown and Ca^{2+} mobilization[11] and then to the subsequent events bringing to cell activation. Among the many mechanisms involved in intracellu-

TABLE 1. fMet Leu Phe–Stimulated Ca^{2+} Changes in Neutrophils from Bartter's Patients and Controls

fMet Leu Phe M	Controls $[Ca_i^{2+}]$nM	Bartter $[Ca_i^{2+}]$nM	*t* Test
Basal	106 ± 12 (*N* = 10)	118 ± 30 (*N* = 10)	$t = 1.21$; n.s.
10^{-9}	259 ± 74 (*N* = 5)	235 ± 36 (*N* = 7)	$t = 0.75$; n.s.
3×10^{-9}	356 ± 142 (*N* = 5)	306 ± 93 (*N* = 6)	$t = 0.70$; n.s.
10^{-8}	575 ± 122 (*N* = 5)	342 ± 104 (*N* = 8)	$t = 3.68$; $p < 0.005$
10^{-7}	619 ± 154 (*N* = 4)	341 ± 91 (*N* = 8)	$t = 3.99$; $p < 0.005$
3×10^{-7}	587 ± 154 (*N* = 5)	347 ± 82 (*N* = 8)	$t = 3.70$; $p < 0.005$

lar Ca^{2+} movements, the blunted response that we have demonstrated in BS could be explained either by a defect in coupling of the receptor to polyphosphoinositide-specific phospholipase C or by the recent finding that cAMP inhibits polyphosphoinositide turnover in human neutrophils.[12] Since platelet cAMP content has been shown to be increased in BS,[2] a cAMP-mediated inhibition of polyphosphoinositide turnover could also be present in neutrophils of such patients.

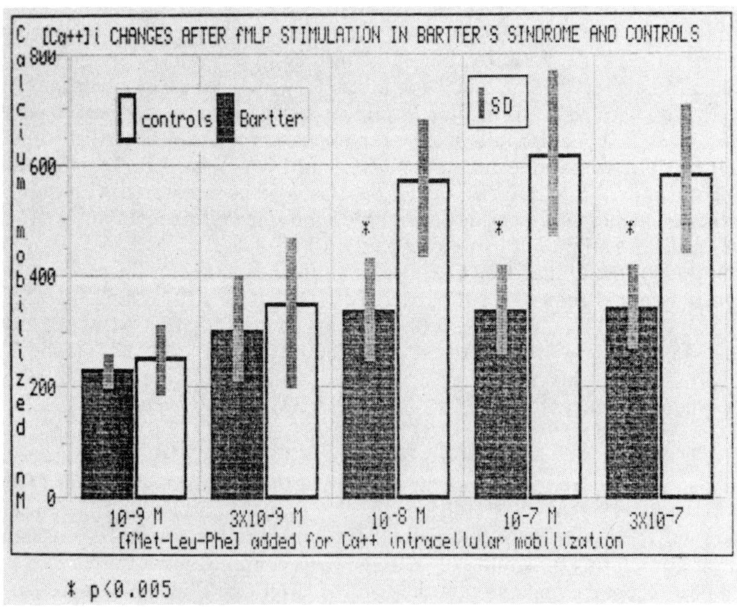

FIGURE 1. Plot of data presented in table.

REFERENCES

1. BARTTER, F. C. 1981. Hypertension **3** (Suppl. 1): I-69–I-73.
2. STOSS, J. S., M. STEMERMAN, M. STEER, E. SALTZMANN & R. BROWN. 1980. Am. J. Med. **68:** 171–180.
3. BARTTER, F. C., P. PRONOVE, J. R. GILL & R. C. MACCARDLE, JR. 1962. Am. J. Med. **33:** 811–828.
4. ZIPSER, R. D., R. K. RUDE, P. K. ZIA & M. P. FICHMAN. 1979. Am. J. Med. **67:** 263–267.
5. DELANEY, V. B., J. F. OLIVER, M. SIMMS, J. COSTELLO & E. BOURKE. 1981. Q. J. Med. **198:** 213–232.
6. CIEDEL, C. L. 1984. Fed. Proc. **43:** 2390–2398.
7. RASMUSSEN, H. & P. Q. BARRET. 1984. Physiol. Rev. **64**(3): 938–984.
8. LEVINE, S. D., W. A. KACHADORIAN, D. N. LEVIN & D. SCHLONDORFF. 1981. J. Clin. Invest. **67:** 662–672.
9. BOYUM, A. 1976. Scand. J. Immunol. **1:** 9–15.
10. TSIEN, R. Y., T. POZZAN & T. J. RINK. 1982. Nature **265:** 68–71.
11. DI VIRGILIO, F., D. P. LEW & T. POZZAN. 1984. Nature **310:** 691–693.
12. DELLA BIANCA, V., P. DE TOGNI, R. GRZESKOWIAK, L. M. VIENTINI & F. DI VIRGILIO. 1986. Biochim. Biophys. Acta. (In press.)

Renal Function in Cystic Fibrosis

B.M. ASSAEL,[a] G. MARRA, S.A. TIRELLI, G. CAVANNA, A. CLARIS APPIANI, A. GIUNTA, M. AMORETTI, AND S. MILANI

Departments of Pediatrics and Biostatistics
Medical School
University of Milan
20122 Milan, Italy

INTRODUCTION

Cystic fibrosis (CF) is an inherited disorder of generalized dysfunction of exocrine glands. The basic defect of this disease is still unknown but a disturbance of NaCl and bicarbonate transport through the cells of various epithelia has been described. Some studies have indicated that renal NaCl reabsorption and glomerular filtration rate are increased,[1,2] although this was not fully supported by other investigators.[3,4]

In this study we wish to describe abnormal tubular NaCl reabsorption and its relation with alterations in renal hemodynamics in response to an acute extracellular volume expansion with 0.9% saline.

PATIENTS AND METHODS

Hemodynamic values [Polyfructosan S clearance for glomerular filtrate (GFR), p-aminohippuric clearance for renal plasma flow (RPF)], plasma renin activity (PRA), plasma aldosterone concentration (PA), and renal handling of NaCl were studied before and after 0.9% saline load (0.4 ml/min/kg b.w. × 60 min) in 9 normal subjects and 7 CF patients. This was done under maximal water diuresis (urinary osmolality <70 mOsm/kg) to calculate the Na or Cl distal delivery according to the following definitions: free water clearance (C_{H_2O}) + sodium clearance (C_{Na}) = sodium distal delivery; (GFR × Na_p) − sodium distal delivery = sodium proximal reabsorption; sodium excretion was expressed as Fe Na (%):(C_{Na}/GFR × 100). The statistical significance of results was evaluated by analysis of variance and Tukey's HSD test. Characteristics of the patients are given in TABLE 1.

RESULTS

TABLE 2 shows the results obtained before and after saline expansion in controls and CF patients. A similar extracellular volume expansion was induced in both groups of subjects as shown by hematocrit, PRA and PA fall.

A significant increase in GFR and RPF was found only in CF group. The saline load failed to induce a depression of proximal sodium reabsorption (Na DD and Fe Na

[a]Address correspondence to: Baroukh Maurice Assael, 2a Clinica Pediatrica, Via Commenda 9, 20122 Milan, Italy.

TABLE 1. Patient Data

	Controls	Patients
Age (yr)	17 ± 2	16 ± 3
Sex	6 M + 3 F	4 M + 3 F
Body weight (Kg)	64 ± 6	50 ± 10
Schwachman score	—	84 + 11

remained unchanged throughout the test). A significant increase in natriuresis was only found in the control subjects. No differences were found between groups at the distal tubular level.

DISCUSSION

An extracellular volume expansion is expected to depress NaCl reabsorption in proximal and distal tubule.[5] This was the case in our control subjects who also exhibited no change in GFR and RPF. Under the same conditions CF subjects showed an increased proximal tubular reabsorption of NaCl and increased GFR and RPF. It is unlikely that in our conditions hyperfiltration was primary with a secondary increase in tubular reabsorption as a result of glomerulotubular balance since it is well known that

TABLE 2. Effect of Extracellular Volume Expansion with Saline (0.4 ml/min/kg × 60 min) on the Renal Function of 7 Patients with Cystic Fibrosis and 9 Healthy Control Subjects. The Measurements Were Performed under Maximal Water Diuresis (U Osm < 70 mOsm/kg)

		Basal	Saline
GFR	Control	117 ± 18[a]	120 ± 19[e]
ml/min/1.73 sq.m.	Patients	131 ± 9.2	166 ± 52[c]
RPF	Control	582 ± 113	643 ± 103
ml/min/1.73 sq. m.	Patients	621 ± 68	768 ± 201[c]
Hematocrit	Control	44 ± 3.4	41 ± 4[c]
%	Patients	43 ± 1.5	41 ± 1.7[c]
PRA	Control	1.9 ± 1.4	1.5 ± 11.5[b]
ng/ml/hr	Patients	0.9 ± 0.5	0.5 ± 0.4[b]
Distal delivery	Control	11 ± 1.2	12 ± 1.9
of Na %	Patients	9.8 ± 1.3	9.7 ± 1.8[d]
Distal delivery	Control	11.3 ± 1.3	13 ± 1.6[b]
of Cl %	Patients	10.2 ± 1.5	10 ± 2[d]
FE Na %	Control	1.4 ± 0.4	2.2 ± 0.8[c]
	Patients	1.4 ± 0.5	1.8 ± 0.7
FE Cl %	Control	1.7 ± 0.5	2.9 ± 0.8[c]
	Patients	1.7 ± 0.7	2.1 ± 0.8[d]

[a] Mean values ± SD.
[b,c] Significantly different from basal (within group) at the .05 or .01 level.
[d,e] Significant difference between groups at the .05; .01 level. For statistical analysis see METHODS.

under volume expansion this mechanism is abolished.[6] Thus it can be suggested that NaCl hyperreabsorption occurs primarily as an alteration of proximal tubular cells. This may attenuate the tubular glomerular feedback mechanism and cause an increase in RPF and GFR.[7]

REFERENCES

1. Berg, U., E. Kussofsky & B. Stranvik. 1982. Acta Paedatr. Scand. **71:** 833–838.
2. Robson, A. M., S. Tateishi, J. R. Ingelfinger, D. B. Strominger & S. Klahr. 1971. J. Pediatr. **79:** 42–46.
3. Spino, M., R. Chai & A. F. Isles. 1985. J. Pediatr. **107:** 64–68.
4. Aladjem, M., D. Lotan, H. Boichis, S. Orda & D. Katznelson. 1983. Nephron **34:** 84–87.
5. Danovitvh, G. M. & N. S. Bricker. 1976. Kidney Int. **10:** 229–234.
6. Habrle, D. A. & H. Von Bayer. 1983. Am. J. Physiol. **244:** F.
7. Blants, C. R. & J. C. Pelayo. 1984. Kidney Int. **25:** 739–746.

Phosphorylation of Band 3 Protein in Nephrolithiasis

G. CLARI,[a] B. BAGGIO,[b,c] G. MARZARO,[b] G. GAMBARO,[b]
A. BORSATTI,[b] AND V. MORET[d]

[a]*Institute of Biological Chemistry*
University of Verona

[b]*Institute of Internal Medicine*
University of Padova

[d]*Institute of Biological Chemistry*
University of Padova
35100 Padova, Italy

Previous studies[1] had shown an inheritable anomaly of red blood cell (RBC) oxalate exchange in primary calcium oxalate nephrolithiasis. Moreover, the ghosts isolated from these RBC and incubated in the presence of [γ^{32}P]ATP exhibited a faster phosphorylation of transmembrane band 3 protein (anion transporter protein) and of the underlying cytoskeletal spectrin, as compared to healthy controls.[2]

These observations prompted us to study the effect displayed on such abnormal endogenous phosphorylation by amiloride, 5,5′-dithiobis(2-nitrobenzoic acid) (DTNB), and 4,4′-diisothiocyano-2,2′-stilbene disulfonate (DIDS), three agents that have been found to reduce the oxalate exchange.[1,3]

The results reported here show that these oxalate transport inhibitors reduce the ghost endogenous phosphorylation of band 3 and of spectrin, although through a different mechanism.

Precisely, the dose-dependent inhibition by amiloride (FIG. 1 gels 1–4) appears to be competitive with respect to [γ^{32}P]ATP, being reversed by an increase of [γ^{32}P]ATP concentration (FIG. 1, gels 5–7). This indicates that amiloride brings about its inhibitory effect by competing with ATP for the same catalytic binding sites of protein kinase responsible for the phosphorylation.[4]

By contrast, DTNB and DIDS display their inhibition on the membrane protein phosphorylation (FIG. 2, gels 1–5) through a mechanism that is not reversed by increasing ATP concentrations (FIG. 2, gels 6–9). This indicates that the inhibitory effect of these agents may be due to their binding to the protein kinase or to the phosphorylatable protein substrate. The latter is more likely since it is indicated by the finding that DIDS[5,6] and DTNB[7] inhibit anion transport by binding to band 3 protein. Such a binding may induce a change of band 3 protein conformation to one less favorable for its phosphorylation as well as for its anion transport activity.

In conclusion, the above results suggest a functional link between phosphorylation of band 3 protein and transmembrane oxalate transport. It is tempting to speculate that band 3 phosphorylation might play a role in modulating the transmembrane transport of anions other than oxalate.

[c]Address correspondence to: Bruno Baggio, Institute of Internal Medicine, University of Padova, Via Giustiniani, 2, 35100 Padova, Italy.

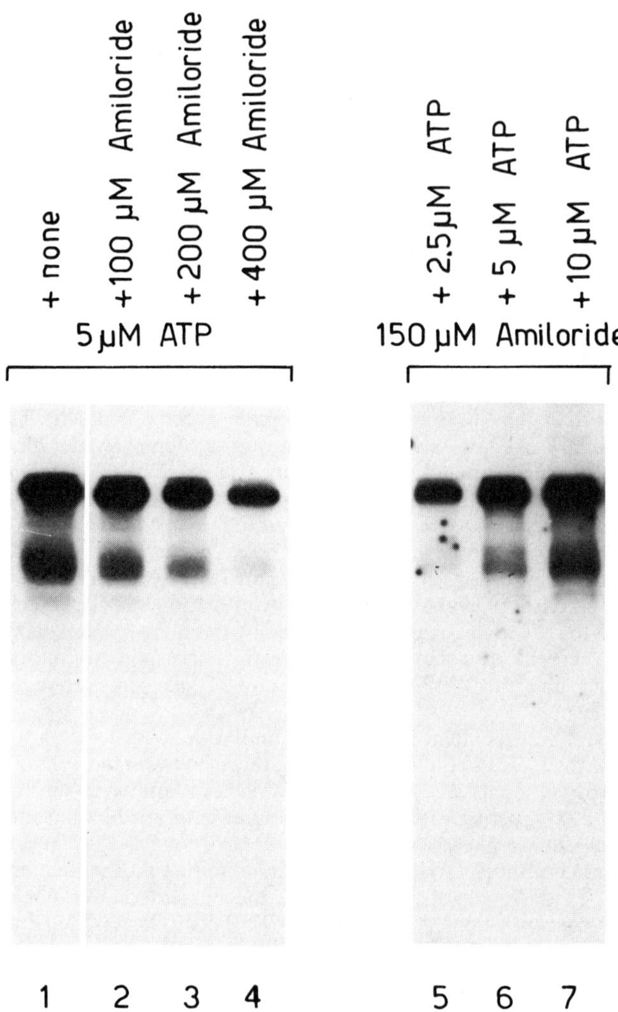

FIGURE 1. Inhibitory effect of amiloride on the RBC membrane protein phosphorylation. The ghosts were prepared from erythrocytes of stone former patients as described previously.[2] The endogenous phosphorylation was carried out at 30°C for 5 min in an incubation medium (125 μl) containing 60 mM Tris-HCl buffer pH 7.5, 10 mM $MgCl_2$, various concentrations of [γ^{32}P]ATP (about 3×10^6 cpm/nmol) and of amiloride, as indicated in the figure. Prior to addition of [γ^{32}P]ATP to initiate the phosphorylation, the ghosts were exposed to the agent at 0°C for 5 min. Experimental conditions for SDS-PAGE electrophoresis as in Baggio *et al.*[2] Autoradiograms were exposed for 60 hr at -80°C.

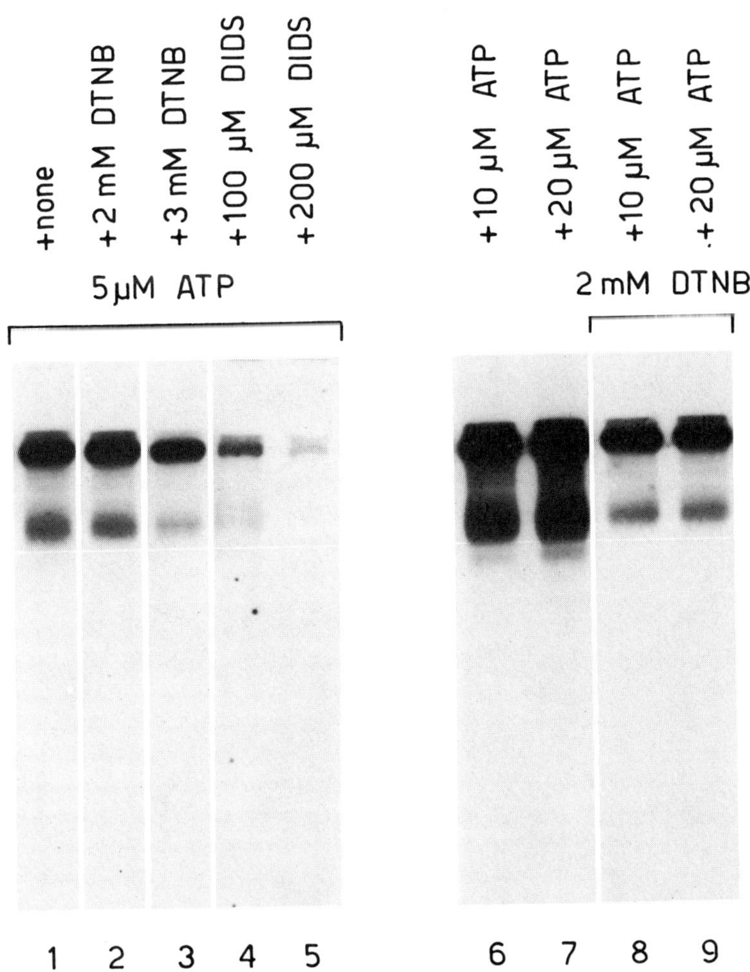

FIGURE 2. Inhibitory effect of DTNB and DIDS on RBC membrane protein phosphorylation. Experimental conditions for phosphorylation as described in FIGURE 1, except that amiloride was replaced by various concentrations of DTNB or DIDS, as indicated in figure. Autoradiograms were exposed for 50 hr at $-80°C$.

REFERENCES

1. Baggio, B., G. Gambaro, F. Marchini, E. Cicerello, R. Tenconi, M. Clementi & A. Borsatti. 1986. New Engl. J. Med. **314:** 599–604.
2. Baggio, B., G. Clari, G. Marzaro, G. Gambaro, A. Borsatti & V. Moret. 1986. IRCS Med. Sci. **14:** 368–369.
3. Baggio, B., G. Gambaro, A. Borsatti, G. Clari & V. Moret. 1984. Lancet **ii:** 223–224.
4. Holland, R., J. R. Woodgett & D. G. Hardie. 1983. FEBS Lett. **154:** 269–273.
5. Cabantchik, Z. I., P. A. Knauf & A. Rothstein. 1978. Biochim. Biophys. Acta **515:** 239–302.
6. Grinstein, S., S. Ship & A. Rothstein. 1978. Biochim. Biophys. Acta **507:** 294–304.
7. Reithmeier, R. A. F. 1983. Biochim. Biophys. Acta **732:** 122–125.

β-Adrenergic Receptors and Adenylate Cyclase Activity in Erythrocyte Membranes from Spontaneously Hypertensive Rats

P. CHATELAIN

Sanofi
Labaz-Sanofi Research Center
1120 Brussels, Belgium

P. ROBBERECHT AND J. CHRISTOPHE

Department of Biological Chemistry and Nutrition
Medical School
Free University of Brussels
1000 Brussels, Belgium

As changes of phosphoinositide content[1] and in the physicochemical properties of the lipids of the erythrocyte membranes from SHR[2] have been reported, it is of interest to compare in erythrocyte membranes from SHR and their normotensive controls (WKY) the β-adrenergic receptors and their coupling to the membrane adenylate cyclase.

TABLE 1. Dissociation Constants of β-Adrenergic Agonists and Antagonists for Erythrocyte Membranes from SHR and WKY Rats

	WKY (μM)	SHR (μM)
(±) Isoproterenol	0.91 ± 0.03	0.90 ± 0.03
(±) Adrenaline	2.5 ± 0.3	2.6 ± 0.3
(±) Noradrenaline	34.1 ± 4.5	32.3 ± 3.8
(±) Procaterol	0.32 ± 0.002	0.33 ± 0.003
(−) Propranolol	0.00035 ± 0.00005	0.00031 ± 0.00007
(+) Propranolol	0.045 ± 0.002	0.046 ± 0.003

The IC_{50} values were determined graphically and the K_i was calculated from the equation $K_i = IC_{50}/(1 + A/K_d)$ where A is the concentration of [^{125}I]HYP used in the assay (30–50 fmoles) and K_d is the dissociation constant for [^{125}I]HYP in the erythrocyte membranes. The values given are the means ± SEM of three determinations performed in duplicate.

β-adrenergic receptors present on erythrocyte membranes from spontaneously hypertensive rats and age-matched (16 weeks old) normotensive Wistar Kyoto rats were characterized by direct binding studies and measurements of adenylate cyclase activity under the same experimental conditions of buffer and temperature.

The dissociation constants (K_d) of [^{125}I]hydroxybenzylpindolol as well as the receptor densities (B_{max}) are identical in both membranes ($K_d \simeq 13 \pm 1$ pM, $B_{max} \simeq 95 + 7$ fmoles/mg proteins). Analysis of competition curves performed with various β-adrenergic compounds is indicative of a homogeneous population of β-adrenergic receptors possessing the same characteristics of affinity and stereospecificity (TABLE 1). Basal and GppNHp-, NaF-, and isoproterenol (in presence of GTP 10^{-5}M)-stimulated adenylate cyclase activity is identical in the erythrocyte ghosts from both strains (TABLE 2). For a given stimulus K_{act} is also unchanged (not shown).

These identical responses contrasted with the decrease in the total number of β-adrenergic receptors, the decrease in adenylate cyclase activity, and the lower level of cyclic AMP in SHR cardiovascular system.[3-6] In this system, the impaired β-adrenergic responsiveness was consistent with changes reported in the turnover of norepinephrine in central and peripheral neurons[7] and in the concentration of norepinephrine in plasma,[8] brainstem, and hypothalamus.[9] The normal response of SHR erythrocyte membranes to β-adrenergic agents could be due to the low affinity of the β_2-adrenergic receptors for norepinephrine. The normal response of adenylate cyclase, and the normal coupling between β-adrenergic receptors and adenylate cyclase in SHR erythrocyte membranes was surprising considering the multiple

TABLE 2. Characteristics of Basal and Stimulated Adenylate Cyclase Activity in Erythrocyte Membranes from SHR and WKY Rats

Stimulus Tested	WKY	SHR
None	0.98 ± 0.21	1.38 ± 0.28
Gpp(NH)p 10^{-4} M	10.42 ± 0.53	10.56 ± 2.58
NaF 10^{-2} M	55.80 ± 1.27	54.68 ± 2.22
GTP 10^{-5} M	1.64 ± 0.35	1.71 ± 0.34
(±) Isoproterenol 10^{-4} M + GTP 10^{-5} M	26.68 ± 1.3	27.71 ± 1.15

Adenylate cyclase activity is expressed as pmoles cAMP produced · min^{-1} · mg $protein^{-1}$. The values given are the means ± SEM of five determinations made in duplicate.

membrane alterations so far described.[1,2] Adenylate cyclase is a multiprotein complex regulated by its lipidic environment[10] and this environment is obviously modified in erythrocytes from hypertensive animals.[2] Our present results suggest that the amplitude of the variations of the bulk lipid dynamic[8] is not sufficient to alter adenylate cyclase activity or that the intrinsic proteins constituting the adenylate cyclase appear to be less sensitive to a perturbation of the lipid dynamics than calcium-binding proteins or ionic transporter. In addition to the possible control of adenylate cyclase activity by the lipid fluidity[10] discussed above, specific phospholipids have been implicated in the control of the enzyme activity. Among these phospholipids, phosphatidylinositol has a specific effect[11] and may have a dual role in the regulation of adenylate cyclase activity; when present in the outer leaflet of the membrane, phosphatidylinositol inhibits adenylate cyclase activation, but its presence in the inner leaflet may be required for enzyme activation.[11,12] The phosphoinositide content is modified in SHR erythrocyte as compared to normotensive controls.[1] The adenylate cyclase activity is identical in both erythrocyte membranes. Thus either this modification does not seem to affect adenylate cyclase in rat erythrocyte or the amplitude of the modification is not large enough to modify the enzyme activity.

REFERENCES

1. KISELEV, G., A. MINENKO, V. MORITZ & P. OEHME. 1981. Biochem. Pharmac. **30:** 883–887.
2. MONTENAY-GARESTIER, TH., I. ARAGON, M. A. DEVYNCK, PH. MEYER & C. HELENE. 1981. Biochem. Biophys. Res. Commun. **100:** 660–665.
3. LIMAS, C. & C. J. LIMAS. 1978. Biochem. Biophys. Res. Commun. **83:** 710–714.
4. DHALLA, R. C., R. V. SHARMA & T. ASHLEY. 1978. Biochem. Biophys. Res. Commun. **82:** 273–280.
5. AMER, M. S., A. W. GOWALL, J. L. PERHACK JR., H. C. FERGUSON & G. R. MCKINNEY. 1974. Proc. Natl. Acad. Sci. USA **71:** 4930–4934.
6. CHATELAIN, P., P. ROBBERECHT, P. DE NEEF, M. CLAEYS & J. CHRISTOPHE. 1979. FEBS Lett. **107:** 86–90.
7. NAKAMURA, K., M. GERALD & H. THOENEN. 1971. Naunyn Schmiedeberg's Archiv. Pharmacol. **268:** 125–139.
8. NAGOAKA, A. & W. LOVENBERG. 1976. Life Sci. **19:** 29–34.
9. YAMORI, Y., W. LOVENBERG & A. SJOERDSMA. 1970. Science **170:** 544–546.
10. ORLY, J. & M. SCHRAMM. 1975. Proc. Natl. Acad. Sci. USA **72:** 3433–3437.
11. MCOSKER, C. C., G. A. WEILAND & D. B. ZILVERSMITT. 1983. J. Biol. Chem. **258:** 13017–13026.
12. PANAGIA, V., D. F. MICHIEL, K. S. DHALLA, M. S. NIJJAR & N. S. DHALLA. 1981. Biochim. Biophys. Acta **676:** 395–400.

Hypothalamic Factor Regulates Sodium Pump Activity in Cultured Renal Tubular Epithelial Cells[a]

GARNER T. HAUPERT, JR., EDWARD CHEN, SWAPNA RAY, AND HORACIO F. CANTIELLO

Renal Unit
Massachusetts General Hospital
Harvard Medical School
Boston, Massachusetts 02114

Bovine hypothalamus contains a low molecular weight, non-peptidic molecule that has some of the characteristics and biological properties of the cardiac glycosides.[1] Principal among these is inhibition of the plasma membrane Na,K-ATPase, the enzymatic expression of the Na^+ pump. Such an inhibitor has been implicated in a number of important physiological and pathophysiological processes including the natriuresis of volume expansion[2] and the genesis of certain forms of human essential hypertension.[3] The endogenous Na transport inhibitor from hypothalamus (hypothalamic factor, HF) has been shown to meet several criteria essential for physiological relevance. It inhibits the Na,K-ATPase reversibly, with high affinity (K_d = 1.4 nM).[4] It is specific for the Na,K-ATPase and acts only from the extracellular surface, consistent with the concept of a circulating inhibitor of the sodium pump.[5] Like ouabain, the half-time for dissociation from purified membrane preparations of Na,K-ATPase was relatively long (90 min).[4] Since HF has been proposed as a putative natriuretic hormone it was of interest to test binding and dissociation characteristics to the sodium pump in intact renal epithelial cells to assess potential physiological significance.

The cultured porcine epithelial cell line, LLC-PK_1, was chosen since these cells are of renal origin, have a large number of Na pump sites (10^6/cell), and 80 to 90 percent of K^+ transport is through the Na,K-ATPase. Pump activity was measured as ouabain-sensitive $^{86}Rb^+$ influx (J_{Rb}), a marker for K^+ transport. One unit of HF activity was defined as that amount required to inhibit radioactive rubidium influx into human erythrocytes by 50% following a 3-hour incubation.[5]

FIGURE 1 shows the effects of 10^{-5} molar ouabain and saturating HF on K^+ uptake into the renal tubular cells. Uptakes are linear between 0 and 10 minutes and pump activity is inhibited by approximately 90% in both HF and ouabain-treated cells. HF did not affect ouabain-resistant uptake indicating that the inhibition is specific for K^+ flux through the Na,K-ATPase (data not shown).

To determine binding characteristics in intact cells, uptake of Rb^+ was studied after short periods of incubation with HF and after a 1 to 2 minute washout period. In contrast to membranes, no prior incubation with HF was needed for inhibition of Na,K-ATPase activity in LLC-PK_1. Saturating HF produced reduction of J_{Rb} by 33% within 10 minutes (FIG. 2). A 60-min incubation with cells followed by washout showed rapid reversal of inhibition and provoked a doubling of K^+ influx (FIG. 2).

[a]Supported by Grant HL33536 from the National Heart, Lung, and Blood Institute.

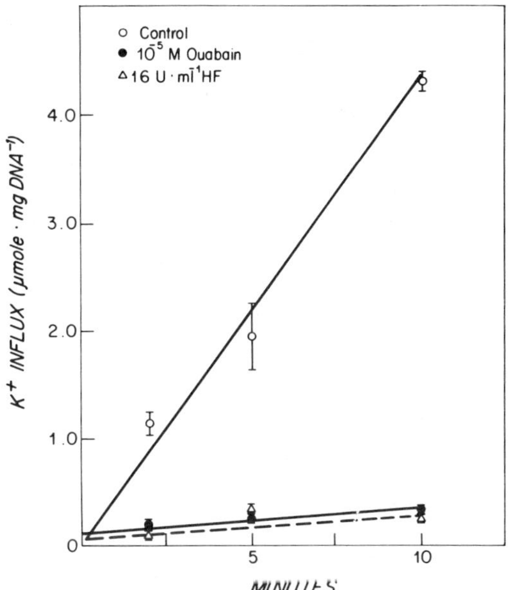

FIGURE 1. Inhibition of K^+ uptake into LLC-PK_1 cells by ouabain and hypothalamic factor (HF). Pump activity was measured as $^{86}Rb^+$ influx. 5–10 × 10^6 cells were incubated in buffer containing glucose for varying times was either HF or ouabain, and then cooled to create conditions for maximal pump activation. On removal from the cold, $^{86}Rb^+$ was added as a tracer for K^+ transport. Fluxes were run for 10 min at 37°C. Unbound counts were removed by spinning cells through 1:1 silicone:phthalate oil. Trapped counts in cell pellets were less than 1%. One unit of HF is defined in the text.

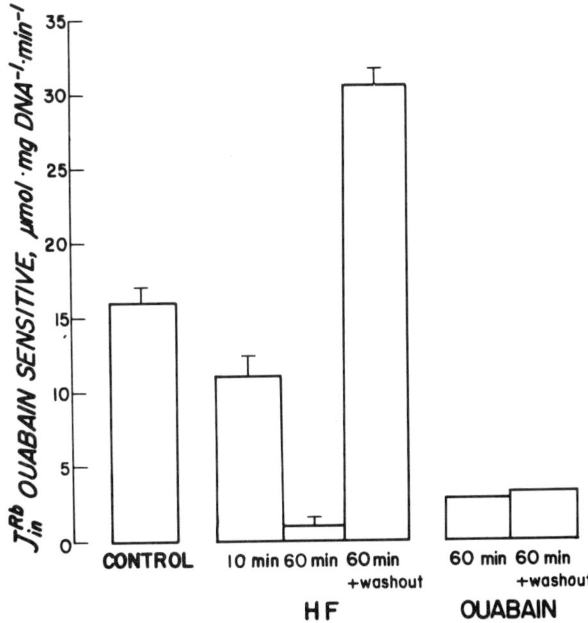

FIGURE 2. Effects of HF and ouabain on Na^+ pump activity in LLC-PK_1 measured as ouabain-sensitive $^{86}Rb^+$ influx (J_{Rb}), as a function of time of incubation with the inhibitors and following washout. Method as described in FIGURE 1. Uptake in the presence of 0.5 mM ouabain was subtracted to determine ouabain-sensitive influx. In "washout" experiments cells were washed rapidly twice in cold buffer following the incubation with HF or ouabain, with subsequent addition of the $^{86}Rb^+$ to run the flux.

Ouabain binding studies in cells before and after washout of HF showed this pump stimulation was due to increase in pump velocity, not to recruitment of additional pump sites into the membrane. The pump stimulation is probably due to the accumulation of intracellular Na^+ during the period of pump inhibition by HF. Ouabain does not demonstrate this rapid reversal of binding in LLC-PK_1 cells (FIG. 2).

The dose-response curve for HF binding to LLC-PK_1 showed a sigmoidal rather than a hyperbolic shape. While this could mean that there is more than one binding site for HF per molecule of enzyme, this would be inconsistent with our previous binding data in purified membrane preparations where the shape of the binding curves was consistent with a one-to-one enzyme-to-HF interaction.[4] An alternative explanation is that in intact cells there is a positive cooperativity in HF binding, perhaps due to allosteric changes in enzyme conformation that occur as the binding reaction proceeds. The experimental points were in fact found to fit very well to theoretical curves that describe such an allosteric type of binding reaction.

These studies represent the first demonstration that HF acts on renal cells as a potent regulator of Na,K-ATPase with binding and dissociation reactions quite different from those in isolated membranes and consistent with physiological significance. These effects could play a major role in the transepithelial movement of cations in the kidney that would account for a natriuretic effect *in vivo.*

REFERENCES

1. HAUPERT, G. T. Jr. & J. SANCHO. 1979. Proc. Natl. Acad. Sci. USA **76:** 4658–4660.
2. HILLYARD, S. D., E. LU & H. C. GONICK. 1976. Circ. Res. **38:** 250–255.
3. HADDY, F. J. & M. B. PAMNANI. 1983. Fed. Proc. **42:** 2673–2680.
4. HAUPERT, G. T. Jr., C. T. CARILLI & L. C. CANTLEY. 1984. Am. J. Physiol. **247:** F919–F924.
5. CARILLI, C. T., M. BERNE, L. C. CANTLEY & G. T. HAUPERT, Jr. 1985. J. Biol. Chem. **260:** 1027–1031.

Inhibition of Ca^{2+} Influx-Dependent Contraction of Vascular Smooth Muscle by Amiloride

S. BOVA, G. CARGNELLI, AND S. LUCIANI

Department of Pharmacology
University of Padova
35131 Padova, Italy

It is well known that the hallmark of essential hypertension is the elevated peripheral vascular resistance.[1] The relationship between the membrane control of intracellular electrolyte concentrations and the vascular smooth muscle tension is well established. An increase in the intracellular Ca^{2+} concentration has been postulated to be "the final common pathway" by which hypertension is produced.[2] Much evidence exists that the Na^+ gradient across the vascular smooth muscle cell membrane plays an important role in regulating the intracellular Ca^{2+} concentration. It is generally accepted that the Na^+ gradient controls the internal Ca^{2+} via a Na^+/Ca^{2+} exchange mechanism. Many experimental data indicate that a reduction of the Na^+ gradient across the membrane will reduce Ca^{2+} efflux and increase Ca^{2+} influx by reversed Na^+/Ca^{2+} exchange. The correlation between Na^+ metabolism and hypertension has been reassessed by Blaustein, who proposed a model of the cellular events leading to hypertension.[2] Since the potassium-sparing diuretic amiloride has been recently found to inhibit the Na^+/Ca^{2+} exchange mechanism in various tissues, including cardiac sarcolemma,[3] we considered the effect of amiloride on the contraction of guinea pig aortic strips induced by low external K^+, low external Na^+, or ouabain worth investigating. These experimental manipulations are considered to produce a contraction of vascular smooth muscle as a result of Ca^{2+} influx through the Na^+/Ca^{2+} exchange.[4] The response of aortic strips to the addition of ouabain (25 μM), KCl-free, or NaCl-free solutions was a sustained contraction reaching the maximum within 150–180 min and completely reversible by washing the preparations with the physiological salt solution. The contractions induced by ouabain, NaCl-free, or KCl-free solutions were inhibited by amiloride in a dose-dependent manner (5 μM–500 μM).[5] The calcium antagonist diltiazem had no effect under these conditions. In FIGURE 1 the inhibitory effect of 50 μM amiloride on the contractions induced by the above-described three conditions is shown. Furthermore, the effect of amiloride was tested on the time course of relaxation of KCl-free contracted aortic strips induced by the readmission of the physiological salt solution. The relaxation of the smooth muscle is ascribed to Ca^{2+} efflux from the cells through a Ca^{2+} ATPase and the Na^+/Ca^{2+} exchange.[6] As shown in FIGURE 2, amiloride (50 μM–500 μM) significantly reduced the rate of relaxation of the aortic strips.

The results reported in this paper indicate that in vascular smooth muscle amiloride inhibits both the Ca^{2+} influx-dependent contraction when the entrance of Ca^{2+} is mediated by reversed Na^+/Ca^{2+} exchange and the relaxation linked to Ca^{2+} efflux via Na^+/Ca^{2+} exchange. In conclusion, the reported observations are in agreement with Blaustein's model of the cellular events resulting in hypertension and may help to explain the antihypertensive effect of amiloride.[7]

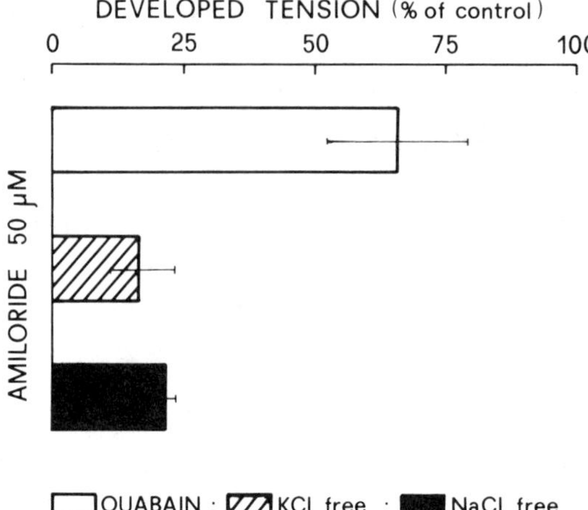

FIGURE 1. Inhibitory effect of amiloride on the ouabain (25 μM), KCl-free, NaCl-free induced contraction of guinea pig aortic strips. The mean ± SEM of six experiments is shown.

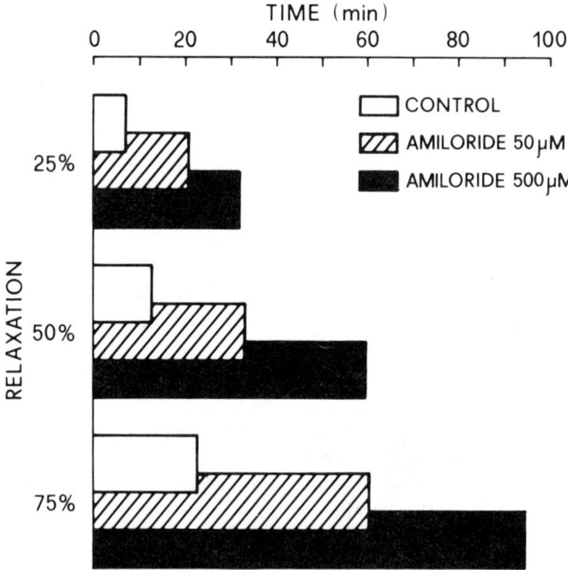

FIGURE 2. Effect of amiloride on the time course of relaxation from the KCl-free induced contraction of guinea pig aortic strips. When the contraction reached the maximum, amiloride was added. After 40 min the KCl-free solution was changed with the physiological salt solution (containing amiloride). The times at which the strips relaxed of 25%, 50%, 75% were considered. Six strips for each amiloride concentration were used.

REFERENCES

1. Conway, J. 1984. Physiol. Rev. **64:** 617–660.
2. Blaustein, M. D. 1977. Am. J. Physiol. **232**(3): C165–C173.
3. Floreani, M. & S. Luciani. 1984. Eur. J. Pharmacol. **105:** 317–322.
4. Ozaki, H. & N. Urakawa. 1979. Naunyn-Schmiedeberg's Arch. Pharmacol. **309:** 171–178.
5. Bova, S., P. Debetto, G. Cargnelli & S. Luciani. 1985. Proceedings of the British Pharmacological Society, Edinburgh 9th–11th September, C. 105.
6. Borle, A. B. 1981. Rev. Physiol. Biochem. Pharmacol. **90:** 13–153.
7. Paterson, J. W., C. T. Dollery & R. M. Haslam. 1968. Br. Med. J. **1:** 422–423.

Magnesium Sulphate Does Not Prevent the Malignant Hyperthermia Episode[a]

J. R. LÓPEZ, V. SÁNCHEZ AND M. MENDOZA

Centro de Biofisica y Bioquimica
Instituto Venezolano de Investigaciones Cientificas
Apartado 1827, Caracas, Venezuela

F. SRETER

Department of Muscle Research
Boston Biomedical Research Institute
Boston, Massachusetts 02115

Malignant hyperthermia (HM) is a pharmacogenetic syndrome of skeletal muscle that is triggered when susceptible patients or animals are exposed to certain anesthetic volatile agents and/or muscle relaxants that depolarize the post-synaptic membrane.[1] It is now well established that the pathophysiology of MH syndrome is related to a malfunction of the intracellular calcium homeostasis.[2] The aim of this study was to explore the effect of magnesium sulphate on the intracellular free calcium concentration in MH-susceptible swine and correlate the possible modifications of $[Ca^{2+}]_i$ with the effectiveness of this drug in the treatment of MH episodes.

The experiments were carried out in four MH-susceptible crossbred swine (Poland China × Pietran). Anesthesia was induced with intravenous sodium thiopental (10–15 mg/kg) and maintained with fentanyl (25 μg/kg) and N_2O/O_2 (66:34), while spontaneous movement was prevented with pancuronium (0.15 mg/kg).

The microelectrodes sensitive to calcium ions were prepared and calibrated as previously described.[3] Only those microelectrodes that showed a linear response between pCa 3 and pCa 7 (30.5 mV per decade at 37°C) were used (FIG. 1). After induction of anesthesia, a 5-cm incision was made over the peroneous longus muscle of the right hind leg. Muscle fibers were impaled with the conventional 3 M KCl microelectrode (8–10 M tip resistance) to measure the resting membrane potential (V_m) and an adjacent fiber was impaled with the calcium-selective microelectrode to measure the $[Ca^{2+}]_i$.

The measurements of $[Ca^{2+}]_i$ show a high resting myoplasmic free calcium concentration 0.19 ± 0.01 μM (M ± SEM, $N = 16$) that was not modified by the pretreatment with magnesium sulphate 0.20 ± 0.01 μM (M ± SEM, $N = 16$). The exposure to halothane without reversal of the pancuronium effect did not induce any change on $[Ca^{2+}]_i$ 0.20 ± 0.01 μM (M ± SEM, $N = 10$). However, when this effect was reverted by neostigmine there was an increase in the $[Ca^{2+}]_i$ of 2.45 times (0.49 ± 0.05; $N = 11$). This increase on $[Ca^{2+}]_i$ was associated with changes in PCO_2 (TABLE 1). However, we did not observe muscle contraction, this latter can be explained by the fact that this increase in $[Ca^{2+}]_i$ was not enough to reach the mechanical threshold. When the animals were treated with a dose of 0.25 mg/kg of

[a]Supported by grants from CONICIT of Venezuela S1-1277, Muscular Dystrophy Association, and National Institutes of Health Center Grant GM 15904-19.

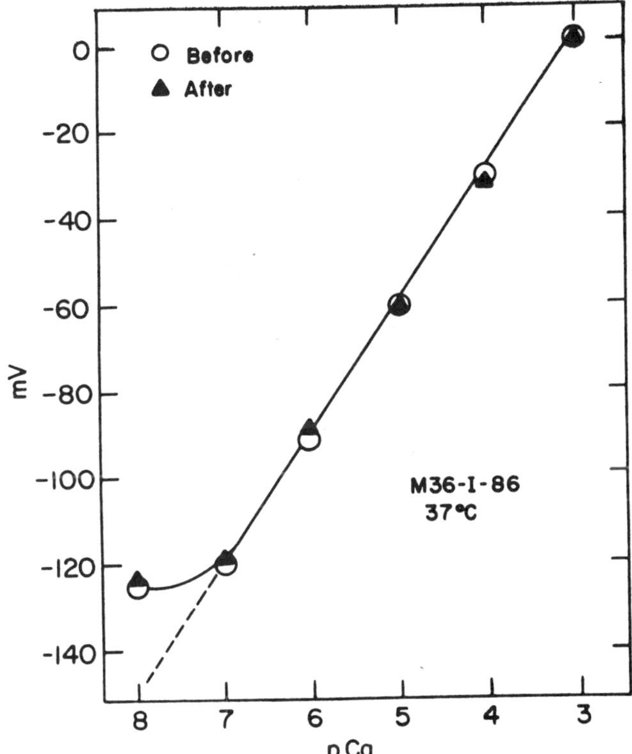

FIGURE 1. Calibration curve of Ca^{2+} selective microelectrode. The open circles represent the microelectrode response before the measurement of $[Ca^{2+}]_i$ in a muscle fiber. The filled triangle indicates the response after such measurement. The dotted line represents the Nernstian slope for an ideal Ca^{2+} response (30.5 mV/p Ca unit).

dantrolene sodium intravenously to reverse the episode, the mean value of $[Ca^{2+}]_i$ decreased to 0.13 ± 0.06 μM (M ± SEM, $N = 4$).

Magnesium sulphate has been proposed for both prophylaxis and treatment of malignant hyperthermia and it has been used with varying degrees of success.[4] Magnesium in the administered dose in the present study had no effect in either reducing the myoplasmic free $[Ca^{2+}]$ or preventing the malignant hyperthermia

TABLE 1.

	Resting Membrane Potential (mV)	$[Ca^{2+}]_i$ (μM)	PCO_2 (mm Hg)
Before $MgSO_4$	82 ± 1	0.19 ± 0.01	34 ± 2
After $MgSO_4$	81 ± 1	0.20 ± 0.01	36 ± 1
After halothane	78 ± 2	0.49 ± 0.05	69 ± 8
After dantrolene	84 ± 1	0.13 ± 0.01	38 ± 2

Values represent the M ± SEM.

syndrome. However, it was able to reduce but not to block the increment in $[Ca^{2+}]_i$ associated to the clinical manifestation of the syndrome, and therefore the changes observed in relation to CO_2 production. This attenuation in the increment of $[Ca^{2+}]_i$ might be related to the effect of Mg to inhibit Ca^{2+} release from the sarcoplasmic reticulum in skeletal muscle.[5] In conclusion, magnesium sulphate is not a satisfactory choice of drug to be used as a prophylactic agent in MH-susceptible subjects.

REFERENCES

1. GRONERT, G. A. 1980. Anesthesiology **53:** 395–423.
2. LÓPEZ, J. R., L. ALAMO, C. CAPUTO, J. WIKINSKI & D. LEDEZMA. 1985. Muscle Nerve **8:** 355–358.
3. LÓPEZ, J. R., L. ALAMO, C. CAPUTO, R. DIPOLO & J. VERGARA. 1983. Biophys. J. **43:** 1–4.
4. HALL, G. M., J. N. LUCKE & D. LISTER. 1975. Anaesthesia **30:** 308–317.
5. STEPHENSON, E. W. & R. J. PODOLSKY. 1977. J. Gen. Physiol. **69:** 1–16.

Human Glomerular Basement Membrane: Altered Binding Characteristics following *in Vitro* Non-enzymatic Glycosylation[a]

M. SENSI, P. TANZI, M. R. BRUNO, M. MANCUSO, AND D. ANDREANI

Endocrinologia I
Clinica Medica 2
Policlinico Umberto I
University of Rome
Rome, Italy

INTRODUCTION

During the last decade, the process of non-enzymatic glycosylation of body proteins associated with the hyperglycemia of diabetes mellitus has grown very much in importance. It is now believed that the structural and functional protein alterations induced by the formation of reactive glycosylation end products are partly implicated in the development of diabetic cataracts, neuropathy, and collagen alterations.[1] In the kidney, one of the organs most affected by diabetic vascular complications, non-enzymatic glycosylation could also play an important pathogenetic role. For example, the well known deposition of albumin, IgG, fibrin, and other plasma proteins along the glomerular basement membrane (GBM) of diabetic kidneys[2] could be mediated via an altered chemistry of GBM collagen induced by excessive glycosylation. Indirect support for this hypothesis is given by the reports of increased GBM glycosylation found in animals[3] and humans[4] with diabetes. In this study we have investigated whether the reactivity of normal human GBM is altered after its enhanced *in vitro* non-enzymatic glycosylation.

MATERIALS AND METHODS

Renal glomeruli were isolated[5] from the kidneys of subjects who died from causes unrelated to diabetes or renal diseases. GBM was then purified by removing all blood, endothelial, and mesangial cell components.[6] *In vitro* glycosylation was obtained by incubating GBM in phosphate buffer (pH 7.3) containing 5, 50, or 500 mM D-glucose for 10 days at 37°C. The high non-physiological sugar concentration was chosen to maximize its attachment rate to GBM at a physiological temperature with a relatively short reaction time. The extent of glycosylation was quantified separately by adding tracer doses of ^{14}C-labeled D-glucose at constant specific activity to the cold glucose solutions. The binding capacity of GBM, previously exposed to 5 mM or 500 mM

[a]Supported by grants from the National Research Council (CNR) (Target project "Preventive Medicine and Rehabilitation," Subproject "Degenerative Diseases," no. 84.02171.56) and from the Anglo-Italian Fund for research into Diabetic Vascular Disease.

D-glucose, towards ^{125}I-labeled human insulin, albumin, IgG, fibrinogen, and alpha-2-macroglobulin was tested by adding given amounts of each protein to 200 μg aliquots of GBM and incubating for ten days at 37°C in phosphate buffer. Formation of immune complexes was also studied by pre-incubating glycosylated GBM for ten days with human albumin or IgG followed by a 30-min incubation with ^{125}I-labeled rabbit anti-human albumin or IgG antibodies.

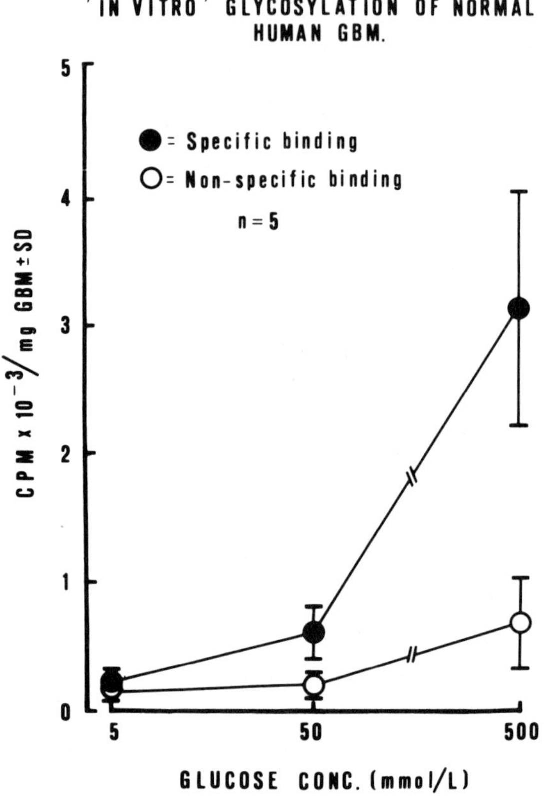

FIGURE 1. The amount of ^{14}C-labeled D-glucose bound to the GBM increases with increasing concentrations of the sugar. At 5 mM there is no difference between specific and non specific binding. This suggests very little *in vitro* glycosylation at this glucose concentration.

RESULTS AND CONCLUSIONS

The results of the *in vitro* glycosylation of GBM are shown in FIGURE 1. The graph indicates that the amount of glucose bound to the membrane increases with the increasing concentration of the sugar in the medium.

The affinity or binding capacity of glycosylated GBM towards plasma proteins is reported in FIGURE 2. Except for alpha-2-macroglobulin, all the proteins show enhanced binding over GBM pre-incubated with 500 mM D-glucose. Similarly,

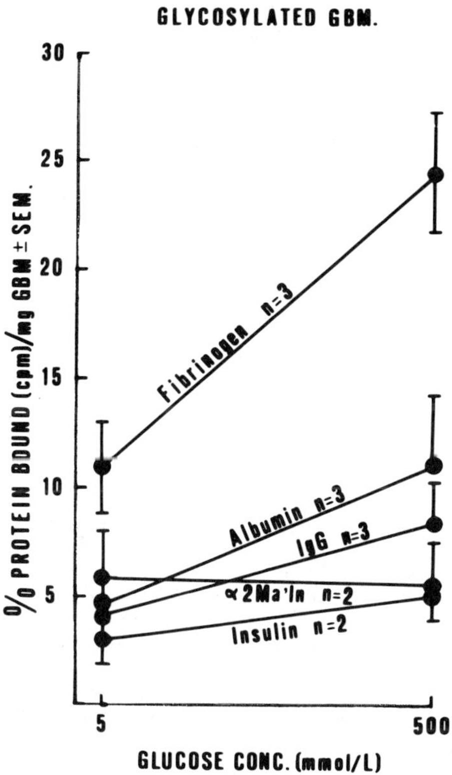

FIGURE 2. The results are expressed in cpm protein bound to GBM as a percentage of the total cpm protein added initially.

increased amounts of immune complexes are formed over glycosylated GBM (TABLE 1).

These results support the hypothesis that non-enzymatic glycosylation of structural proteins may alter their chemical characteristics. Thus, isolated normal human GBM, non-enzymatically glycosylated *in vitro,* shows an increased binding reactivity towards some of the most common plasma proteins. These experimental data could explain the

TABLE 1. Formation of Immune Complexes on Glycosylated GBM

Immune Complex	Glucose Concentration (mmol/L) 5	500	Ratio
Albumin+	19.4	30.6	1.6
Anti-Albumin	3.7	7.1	1.9
IgG+ Anti-IgG	2.7	5.2	1.9

The results are expressed as explained in the legend to FIGURE 2. The last column shows the ratio between the binding measured with glycosylated GBM and that measured with unmodified GBM.

pathomorphological findings of linear plasma protein deposition along the GBM of diabetic patients. *In vivo,* such an abnormal trapping of plasma proteins by a chemically altered GBM could make them behave as planted antibodies or antigens, favoring *in situ* deposition of immune complexes and thus initiating glomerular tissue damage. In addition, persistent accumulation of trapped plasma proteins over glycosylated diabetic GBM could, in time, contribute to its abnormal thickening: a pathomorphological state well known in the diabetic kidney.[7]

In conclusion, excessive non-enzymatic glycosylation of GBM could explain some of the pathomorphological alterations typical of the diabetic kidney. Early and continued good metabolic control, possibly coupled with an early and safe chemical inhibition of glycosylation could reduce those processes that might eventually contribute to the development of late renal diabetic complications.

REFERENCES

1. Brownlee, M., H. Vlassara & A. Cerami. 1984. Nonenzymatic glycosylation and the pathogenesis of diabetic complications. Ann. Intern. Med. **101:** 527–537.
2. Westberg, G. 1980. Diabetic nephropathy. Pathogenesis and prevention. Acta Endocrinol. **94** (Suppl.) **238:** 85–101.
3. Cohen, M. P., E. Urdanivia, M. Surma & V. Y. Wu. 1980. Increased glycosylation of glomerular basement membrane collagen in diabetes. Biochem. Biophys. Res. Commun. **95:** 765–769.
4. Uitto, J., G. A. Grant, A. J. Perejda, E. Rowald & J. R. Williamson. 1980. Glycosylation of human glomerular basement membrane collagen (GBMC): increased non-enzymatic glycosylation in diabetes. Fed. Proc. **39:** 1972–1975.
5. Spiro, G. R. 1967. Studies on the renal glomerular basement membrane. Preparation and chemical composition. J. Biol. Chem. **242:** 1915–1922.
6. Carlson, E. C., K. Brendel, J. T. Hjelle & E. Meezan. 1978. Ultrastructural and biochemical analyses of isolated basement membranes from kidney glomeruli and tubules and brain and retinal microvessels. J. Ultrastruct. Res. **62:** 26–53.
7. Osterby, R. 1972. Morphometric studies of the peripheral glomerular basement membrane in early juvenile diabetes. I. Diabetologia **8:** 84–92.

Hemodynamic Mechanisms of the Blood Pressure Response to Atrial Natriuretic Factor in the Dog:

Effects of Vagotomy

M. VOLPE, A. CUOCOLO, F. VECCHIONE,
M. CONDORELLI, AND B. TRIMARCO

1^ Clinica Medica
2^ Facoltà di Medicina
Università di Napoli
Naples, Italy

Atrial natriuretic factor (ANF) is a peptide or group of peptides having both natriuretic and vascular actions. In their initial studies on ANF, de Bold and co-workers[1] observed that, in addition to the characteristic natriuretic effect, the injection of atrial extracts in rats induced a slight but consistent reduction in blood pressure. This effect has been later confirmed with synthetic ANF peptides in both conscious and anesthetized normotensive and hypertensive animals,[2-4] and in man[5,6] in the absence of associated tachycardia. The blood pressure lowering effect of ANF was originally attributed to its vasodilator properties, but recent studies seem to indicate that more complex hemodynamic mechanisms are involved in the responses to ANF administration. In particular, it has been reported that bolus or sustained infusion of ANF can lower cardiac output.[7] The intrinsic mechanisms that lead to a reduction in cardiac output in responses to ANF have not yet been defined. In this regard, Ackermann *et al.*[8] showed that the hypotensive effect induced by the administration of crude atrial extracts in rats is mediated by a failure of cardiac output to increase in compensation for peripheral vasodilation and is attenuated by vagotomy, thus suggesting that vagal stimulation partially mediates the blood pressure reduction caused by atrial extract.

The purpose of the present study was to investigate the hemodynamic mechanisms underlying the blood pressure response to synthetic ANF and to determine whether vagal afferent stimulation also plays a role in the hemodynamic response to synthetic ANF.

The study was performed in six mongrel dogs, averaging 18 kg in weight, anesthetized with alpha-chloralose (80 mg/kg i.v.). The animals were instrumented with intravascular catheters. Mean systemic blood pressure, heart rate, hindlimb perfusion pressure, cardiac output (evaluated by thermodilution), stroke volume, calculated total peripheral resistance, and right atrial pressure were measured in control conditions, at the end of ANF infusion (Auriculin A) (1 μg/kg bolus followed by 0.1 μg/kg/min for 15 min), and after 30 min of recovery, before and after bilateral vagotomy. The results are presented in TABLE 1.

As shown in TABLE 1, the blood pressure lowering effect caused by a sustained infusion of synthetic ANF is mostly accounted for by a reduction in stroke volume and then in cardiac output. The ANF-induced fall in stroke volume is not associated with changes in right atrial pressure. Vagotomy is able to prevent the ANF-induced reduction both in stroke volume and in blood pressure.

TABLE 1. Hemodynamic Effects of a Sustained Infusion of ANF (Auriculin A)[a]

	A			V		
	C	ANF	R	C	ANF	R
MBP (mmHg)	98 ± 3	87 ± 4[b]	96 ± 3	85 ± 5	86 ± 4	86 ± 4
HR (b/min)	147 ± 8	145 ± 6	146 ± 6	149 ± 6	150 ± 6	150 ± 6
HPP (mmHg)	97 ± 3	93 ± 2	95 ± 3	94 ± 4	93 ± 5	93 ± 5
CO (1/min)	2.7 ± .2	2.4 ± .2[b]	2.8 ± .3	2.4 ± .1	2.4 ± .1	2.4 ± .2
SV (ml)	19 ± 2	15 ± 1[b]	19 ± 2	17 ± 2	17 ± 2	16 ± 3
TPR (HRU)	37.8 ± 4	35.5 ± 4	34.3 ± 4	35.3 ± 2	35.6 ± 1	35.8 ± 2
RAP (mmHg)	7.4 ± 2	6.8 ± 2	6.9 ± 2	7.5 ± 2	6.7 ± 2	6.6 ± 2

[a] 1 μg/kg bolus followed by 0.1 μg/kg/min i.v. for 15 min.

[b] Indicates $p < 0.05$.

MBP = Mean systemic blood pressure; HR = Heart rate; HPP = Hindlimb perfusion pressure; CO = Cardiac output; SV = Stroke Volume; TPR = Total peripheral resistance; RAP = Right atrial pressure. A = Control conditions; V = Vagotomy; C = control; and R = recovery. Data are presented as mean ± SEM.

The ANF-induced fall in blood pressure seems to be caused by a reduction in cardiac output; the lack of any significant effect of the peptide on heart rate and right atrial pressure supports the possibility of a negative inotropic effect of ANF. Although other possibilities cannot be ruled out by the present data, our observation that vagotomy prevents both blood pressure and stroke volume reduction suggests that vagal stimulation is involved in the systemic depressor effect of ANF.

REFERENCES

1. deBold, A. J., H. B. Borenstein, A. T. Veress & H. Sonnenberg. 1981. Life Sci. **28:** 89–94.
2. Pegram, B. L., N. C. Trippodo, F. E. Cole & A. A. MacPhee. 1984. Fed. Proc. **43:** 453 (Abst.).
3. Seymour, A. A., E. A. Marsh, E. K. Mazack, I. I. Stabilito & E. H. Blaine. 1985. Hypertension **7:** 35–42.
4. Garcia, R., G. Thibault, J. Gutkowaska, P. Hamet, M. Cantin & J. Genest. 1985. Proc. Soc. Exp. Biol. Med **178:** 155–159.
5. Cody, R. J., A. B. Covit, J. H. Laragh & S. A. Atlas. 1985. Hypertension **7:** 845 (Abst.).
6. Richards, A. M., H. Ikram, T. G. Yandle, M. G. Nicholls, M. W. I. Webster & E. A. Espiner. 1985. Lancet **1:** 545–549.
7. Lappe, R. W., J. F. M. Smiths, J. A. Todt, J. J. M. Debets & R. L. Wendt. 1985. Circ. Res. **56:** 606–612.
8. Ackermann, U., T. J. Irizawa, S. Milojevic & H. Sonnenberg. 1984. Can. J. Physiol. Pharmacol. **62:** 819–826.

Catecholaminergic Receptors in Kidney of Milano Hypertensive Rats

A. S. TIRELLI, B. M. ASSAEL, G. CAVANNA, R. ROSSI, P. FERRARI[a] AND F. SERENI

Istituti Clinici di Perfezionamento
Department of Pediatrics
University of Milan
20122 Milano
and
[a]Laboratorio Ricerche Carlo Erba Farmitalia
Nerviano, Italy

The Milano hypertensive rats develop a genetically determined hypertension where renal abnormalities in sodium handling have been shown to play a major role. It is well accepted that the catecholaminergic system plays a part in the control of tubular sodium handling—sympathetic fibers have been shown to innervate both proximal and distal tubular cells, denervation of the kidney leads to increased diuresis and natriuesis, while stimulation of renal nerves increases sodium reabsorption.[1] However, the role of each subtype of catecholaminergic receptor is controversial since renal location and their specific functions have not been fully characterized. In our study we adopted the working hypothesis that in the kidney alpha-2 receptors primarily control sodium reabsorption,[2] beta adrenoceptors are associated with glomeruli and hemodynamics,[3] while dopamine receptors are mainly located on the vasculature.[4]

It seems important to analyze the properties of these receptors in the kidney of Milano hypertensive rats (MHS) in comparison with Milano normotensive strain (MNS).

Seven-week-old rats were studied. A crude membrane preparation was obtained from each of three kidney areas: outer cortex, inner cortex, and outer medulla. Alpha-2, beta, and dopamine receptors were analyzed by specific radiolabeled binding of respectively [^{3}H]rauwolscine (79 Ci/mmol), [^{3}H]dihydroalprenolol (40.6 Ci/mmol), and [^{3}H]haloperidol (18.7 Ci/mmol) displaced by 10 μM phentolamine, 1 μM (−)-propranolol, and 10 μM cis-flupentixol. Saturation curves were transformed by Scatchard analysis[5] to obtain the maximum apparent number of binding sites (B_{max}) and the dissociation constant (K_d). Membrane proteins were determined by Lowry's method.

FIGURE 1 shows the Scatchard plots for alpha-2 receptors in MHS and MNS rats. A statistically significant decrease in B_{max} was seen in the inner cortex of MHS rats with no changes in K_d values (TABLE 1). On the other hand, B_{max} and K_d for beta receptors were identical in the kidney areas of both strains. In the outer medulla we were unable to detect specific binding of dihydroalprenolol.

A saturation curve for dopaminergic receptors could not be identified in our crude membrane preparations within a range of concentrations of [^{3}H]haloperidol from 0.5 to 10 nM.

Our study supports the view that the renal adrenergic system is involved in the development of hypertension in the MHS rats. This is shown by a decreased number of alpha-2 binding sites in the inner cortex compared to the MNS. This involvement

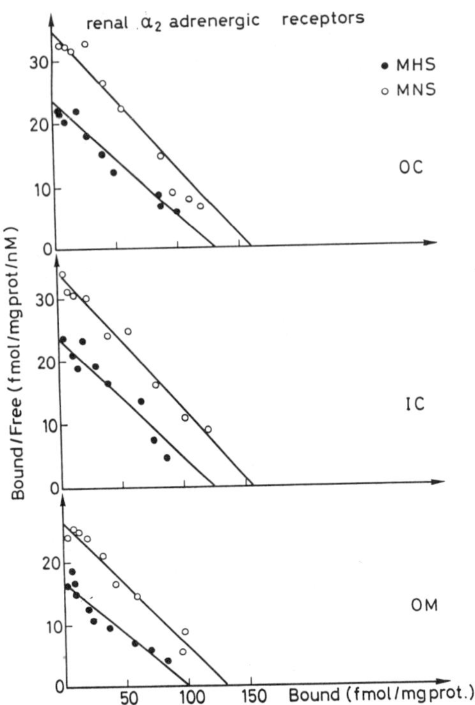

FIGURE 1. Scatchard plots of saturation curves obtained for [^{3}H]rauwolscine binding at concentrations from 0.2 to 20 nM. OC = outer cortex, IC = inner cortex, and OM = outer medulla.

TABLE 1. Maximum Number of Binding Sites (B_{max}) and Dissociation Constant (K_d) for Adrenergic Receptors in Distinct Kidney Areas of Milano Hypertensive and Normotensive Rats

		B_{max} (fmoles/mg proteins)		K_d (nM)	
		alpha$_2$	beta	alpha$_2$	beta
OC	MNS	149 ± 17.4	18.9 ± 6.5	4.35 ± 1.21	1.0 ± 0.25
	MHS	130 ± 20	12.6 ± 1.95	5.84 ± 1.76	0.79 ± 0.17
IC	MNS	155 ± 8.7	14.8 ± 2.1	4.94 ± 1.48	0.63 ± 0.14
	MHS	118 ± 24**	18.3 ± 5.7	5.14 ± 1.4	0.92 ± 0.39
OM	MNS	131 ± 21.5	n.d.	5.14 ± 0.67	n.d.
	MHS	100 ± 6.8	n.d.	6.02 ± 0.35	n.d.

** = $p < 0.05$ by Student's t test. Values are means ± S.D. of four experiments.

NOTE: MHS = Milano Hypertensive Strain, MNS = Milano Normotensive Strain, OC = outer cortex, IC = inner cortex, and OM = outer medulla, n.d. = not detectable.

seems specific since no difference was seen for beta or alpha-2 receptors in other kidney areas.

This may implicate either a down-regulation of the alpha-2 receptors, balancing a primary increase in sodium reabsorption, or a primary reduction in the presynaptic receptors number leading to increased sodium reabsorption in tubular cells.

REFERENCES

1. KIM, J. K., S. L. LINAS & R. W. SCHRIER. 1980. Pharmacol. Rev. **31:** 169–178.
2. INSEL, P. A., M. D. SNAVELY, D. P. HEALY, P. A. MUNZEL, C. L. POTENZA & E. P. NORD. 1985. J. Cardiovasc. Pharmacol. **7**(Suppl. 8): S9–S17.
3. SUMMERS, R. J. & M. J. KUHAR. 1983. Eur. J. Pharmacol. **91:** 305–310.
4. GOLDBERG, L. I. 1972. Pharmacol. Rev. **24:** 1–29.
5. SCATCHARD, G. 1949. Ann. N.Y. Acad. Sci. **51:** 660–672.

Sodium Transport in Membrane Vesicles from Kidney of the Milan Hypertensive Rat Strain

P. PARENTI, G. CASPANI, AND G. M. HANOZET

Department of General Physiology and Biochemistry
University of Milan
20133 Milan, Italy

Sodium uptake was evaluated in brush border membrane vesicles prepared from kidney cortex of rats of the Milan hypertensive strain (MHS)[1] during the prehypertensive (70 g animals) and hypertensive (200 g animals) stages. Control measurements were carried out with the corresponding Milan normotensive strain (MNS). The vesicles were prepared by differential centrifugation following precipitation of contaminating membranes with $MgCl_2$.[2] Uptake of ^{22}Na was measured by a rapid filtration technique[3] in the following intravesicular (in) and extravesicular (out) pH conditions: (1) $7_{in}/7_{out}$; (2) $5.5_{in}/7.2_{out}$; and (3) $5.5_{in}/7.2_{out}$ +100 μM carbonyl cyanide *p*-trifluoromethoxyphenyl hydrazone (FCCP). In this latter condition a stable electrical potential difference ($\Delta\psi$) occurs across the vesicle membrane.

Sodium transport was stimulated by ΔpH, indicating that Na/H exchanger was operative. Moreover a further increase of sodium uptake was obtained when $\Delta\psi$ was present, indicating the presence of a potential-sensitive sodium pathway. In all conditions, at high sodium concentration, sodium uptake was significantly higher in prehypertensive MHS animals than in normotensive controls. The difference increased in the presence of ΔpH + $\Delta\psi$ (TABLE 1). Bumetanide showed no effect in all experimental conditions, therefore the observed sodium transport cannot be accounted for by a NaCl symport. In order to evaluate the contribution of the Na/H exchanger on the overall sodium transport, the effect of amiloride was tested. The measured inhibition by 1 mM amiloride depended on the sodium concentration, however it was the same either in the presence or in the absence of FCCP. Therefore, the observed potential-sensitive sodium pathway, higher in MHS, if different from Na/H antiporter, seems to be sensitive to amiloride.

Kinetic analysis in the presence of ΔpH ± FCCP showed simple saturation curves with similar $K_{0.5}$ values, but with J_{max} values significantly higher in the prehypertensive animals than in normotensive controls. A similar pattern was found in adult animals (200 g). In the absence of ΔpH, kinetic analysis showed a more complex behavior, with almost two components: a high affinity–low capacity component and a low affinity–high capacity component. Both components showed different J_{max} values higher in MHS (TABLE 2). These data indicate the presence of a potential-sensitive sodium pathway endowed with a higher capacity in MHS, in agreement with the increase in tubular sodium reabsorption already described in MHS in studies on renal physiology.[4,5]

TABLE 1. Effect of ΔpH and Transmembrane Electrical Potential Difference on the Sodium Uptake by Brush-Border Membrane Vesicles from Kidney Cortex of MNS and Prehypertensive MHS Rats[a]

Condition	1 mM NaCl		10 mM NaCl		50 mM NaCl	
	MNS	MHS	MNS	MHS	MNS	MHS
1. pH $7.0_{in}/7.0_{out}$	.218 ± .021	.319 ± .012[b]	1.465 ± .132	1.695 ± .109	2.230 ± .324	4.560 ± .352[b]
2. pH $5.5_{in}/7.2_{out}$	.453 ± .029	.475 ± .024	2.519 ± .099	3.598 ± .160[b]	5.377 ± .181	7.582 ± .055[b]
3. pH $5.5_{in}/7.2_{out}$ + 100 μM FCCP	.621 ± .034	.666 ± .071	4.785 ± .120	8.271 ± .110[b]	11.259 ± .280	17.944 ± 1.26[b]

[a]Sodium uptake was measured: (1) in the absence of gradients, in 200 mM mannitol, 10 mM HEPES-Tris, pH 7.0; (2) in the presence of a pH gradient ($5.5_{in}/7.2_{out}$); (3) in the presence of a pH gradient ($5.5_{in}/7.2_{out}$) + 100 μM FCCP. The pH gradient was obtained by diluting 1 volume of brush-border membrane vesicles suspended in 193 mM mannitol, 90 mM MES, 17 mM Tris, pH 5.5 with 4 volumes of 160 mM mannitol, 90 mM HEPES, 45 mM Tris, pH 7.5; final pH was 7.2. Uptake is expressed as nmol/3 sec/mg protein. Values are means ± SE of four measurements. The comparison between MNS and MHS rats was carried out with the *t*-test.

[b]$p < 0.05$ compared with respective MNS controls.

TABLE 2. Kinetic Constants of the Sodium Uptake in Brush-Border Membrane Vesicles from Kidney Cortex of MNS and MHS Rats[a]

	$K_{0.5}$ MNS	MHS	J_{max} MNS	MHS
A. 70 g rats				
pH $7.0_{in}/7.0_{out}$				
high affinity	3.62 ± 0.14	8.48 ± 0.59[b]	1.00 ± 0.28	3.04 ± 0.15[b]
low affinity	30.01 ± 6.59	100.0 ± 20.0[b]	4.06 ± 0.52	14.47 ± 2.16[b]
pH $5.5_{in}/7.2_{out}$	19.31 ± 3.15	18.08 ± 1.10	7.74 ± 0.82	10.68 ± 0.38[b]
pH $5.5_{in}/7.2_{out}$ + 100 μM FCCP	22.14 ± 1.34	18.13 ± 1.64	16.62 ± 0.73	24.11 ± 1.52[b]
B. 200 g rats				
pH $5.5_{in}/7.2_{out}$	20.35 ± 3.20	30.71 ± 5.29	20.15 ± 1.85	31.21 ± 3.57[b]
pH $5.5_{in}/7.2_{out}$ + 100 μM FCCP	24.56 ± 3.73	27.57 ± 3.16	35.11 ± 3.31	46.19 ± 3.42[b]

[a]Sodium uptake initial rate (3 sec) were measured at 8–10 NaCl concentrations varying between 0.5 and 80 mM. Experimental conditions as in TABLE 1. The constants were calculated by regression analysis in Eadie Hofstee plots. $K_{0.5}$ is expressed as mM ± SE. J_{max} is expressed as nmol/3 sec/mg protein ± SE. The comparison between MNS and MHS rats was carried out with the *t*-test.

[b]$p < 0.05$ compared with respective MNS controls.

REFERENCES

1. BIANCHI, G., P. FERRARI & B. R. BARBER. 1984. *In* Handbook of Hypertension. W. de Jong, Ed. **4:** 328–349. Elsevier Science Publishers B.V. Amsterdam.
2. PARENTI P., G. M. HANOZET & G. BIANCHI. 1986. Hypertension. **8:** 932–939.
3. HANOZET, G. M., P. PARENTI & P. SALVATI. 1985. Biochim. Biophys. Acta **819:** 179–186.
4. BIANCHI, G., P. G. BAER, U. FOX, L. DUZZI, D. PAGETTI & A. M. GIOVANNETTI. 1975. Circ. Res. **36, 37** (Suppl. 1): I153–161.
5. SALVATI, P., G. P. PINCIROLI & G. BIANCHI. 1984. J. Hypertension **2** (suppl. 3): 351–352.

Volumes and Na Transports in Intact Red Blood Cells, Resealed Ghosts, and Inside-Out Vesicles of Milan Hypertensive Rats

P. FERRARI, L. TORIELLI, M. FERRANDI, AND G. BIANCHI

Istituto Ricerche Farmitalia Carlo Erba
Nerviano
and
Istituto di Scienze
Mediche dell'Università degli Studi di Milano
Milan, Italy

Many blood cell alterations have been found in essential hypertensives. However, because of the influence on cell functions of environmental factors, genetic polymorphism, aging, and experimental conditions, the pathophysiological meaning of these abnormalities is not yet clear. An animal model genetically selected for developing hypertension, such as the Milan rat strain, is a useful tool for studying the role of these factors on the cellular functions in controlled conditions. In this study we have evaluated the influence of age and blood pressure on red blood cell (RBC) volume, Na^+ content, and Na^+ transport in hypertensive (MHS) and normotensive (MNS) rats. Moreover, because membrane cytoskeleton is involved in cell volume and Na^+ transport regulation, we have evaluated the role that this component has on the MHS RBC abnormalities by measuring in different conditions the volume of RG and the Na^+ cotransport in inside-out vesicles (IOV), which are deprived of cytoskeleton in both strains.

MATERIALS AND METHODS

Male MHS and MNS of different ages were chronically cannulated for blood pressure measurement and blood removal. RBC and resealed ghosts (RG) volumes were measured by Coulter Counter+C 1000 equipment. RBC Na^+ transports were measured as described elsewhere.[1] RG and IOV were prepared according to Wood[2] and Steck,[3] respectively.

RESULTS

FIGURE 1 shows the MBP levels and the characteristics of RBC of MHS and MNS at different ages. RBC volume and Na^+ concentration are significantly lower in MHS than in MNS at any age. Na^+-K^+ cotransport is faster in MHS up to 70 days of age, then it becomes equal in both strains at older ages. The Na^+-K^+ pump, which is similar in young MHS and MNS, becomes lower in adult MHS. To evaluate the relationship

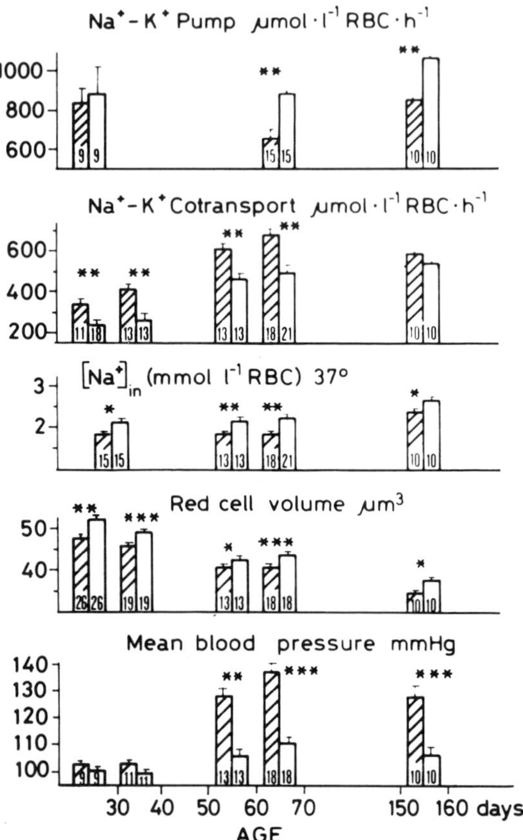

FIGURE 1. RBC Na-K pump, Na-K cotransport, Na content, cell volume, and mean blood pressure of MHS (▨) and MNS (□) at different ages. RBC ion fluxes were measured, after four washes at 37°C in 85 mM choline-chloride, 85 mM sucrose, 10 mM glucose, 1 mM $MgCl_2$, and 10 mM MOPS-Tris pH 7.4. Intracellular Na^+ content was measued in the washed packed RBC, RBC were incubated at 2% of hematocrit at 37°C divided in three aliquots containing, respectively: 2 mM KCl, 5 mM ouabain, and 5 mM ouabain + 0.2 mM bumetanide. Na^+ efflux was measured on samples taken at increasing times and the Na concentration in the supernatant of the centrifuged samples was measured by atomic absorption. The Na-K pump was estimated as ouabain-sensitive Na efflux, Na-K cotransport as ouabain-resistant, bumetanide-sensitive Na efflux, and Na passive permeability as ouabain-resistant and bumetanide-resistant Na efflux. Number of rats shown in the bars. Values are mean ± SEM. $^*p < 0.05$, $^{**}p < 0.01$, $^{***}p < 0.001$.

between smaller RBC volume and faster Na^+ cotransport in MHS, we have measured the cotransport of both strains in RBC whose volumes were made similar by hypotonicity. At 170 mOsm both the RBC volume and the Na^+ cotransport of MHS were similar to those of MNS (RBC volume μm^3: MHS 65.3 ± 2.1 and MNS = 63.4 ± 2.7; Na^+ cotransport $\mu mol \cdot hr^{-1} \cdot l^{-1}$ RBC: MHS = 112 ± 23 and MNS = 157 ± 38). RG, as fresh RBC, show a lower volume in MHS than in MNS (TABLE 1) and maintain this characteristic also after swelling in hypotonic medium up to 130 mOsm. In IOV of both strains we measured a Na^+ influx bumetanide-sensitive and chloride-dependent comparable with the Na^+-K^+ cotransport of intact RBC. This system works in IOV at the same rate in both strains (TABLE 1).

DISCUSSION

In this study we have demonstrated the following differences between MHS and MNS RBC: a lower cell volume and Na^+ concentration and a faster Na^+-K^+ cotransport in MHS. These characteristics seem linked to the integrity of cell membrane because (1) RBC volume differences between MHS and MNS are maintained in RG and (2) after removal of cytoskeleton in IOV no differences in Na^+ cotransport are detectable between the two strains. Moreover, we demonstrated that these MHS RBC abnormalities are primarily determined in the stem cell and genetically related to the hypertension[1] and are also present in MHS proximal tubular cells that are more involved in the pathogenesis of MHS hypertension.[4,5] These findings are consistent with the hypothesis that a primary alteration in the membrane cytoskeleton may be responsible for both RBC and tubular cell abnormalities.

TABLE 1. Volume and Na^+-K^+ Cotransport of RBC, RG, and IOV from 2-month-old MHS and MNS

	Volume		Na^+-K^+ cotransport	
	RBC (μm^3)	RG (μm^3)	RBC ($\mu mol \cdot hr^{-1} \cdot l^{-1}$)	IOV ($nmol \cdot min^{-1} \cdot mg^{-1}$)
MHS	44.6 ± 0.56	28.9 ± 0.81	612 ± 62	9.47 ± 1.25
	$p < 0.001$	$p < 0.05$	$p < 0.01$	
MNS	48.08 ± 0.45	32.7 ± 1.11	470 ± 29	12.7 ± 2.09
N (MHS-MNS)	6–6	6–6	13–11	8–8

RBC and RG volumes were measured with Coulter Counter + C 1000, suspending the cells in Isoton[R] II at 295 mOsm.

Na^+ transports were measured in IOV at room temperature, as $^{22}NaCl$ uptake, in presence of 1 mM ouabain on both sides of IOV, with and without 0.2 mM bumetanide outside.

The composition of flux medium was (mM): 5 NaCl, 80 KCl, 1 $MgCl_2$, and 50 Tris-Glycine pH 7.4. 5–10 μl of IOV were mixed with 20 μl of flux medium and filtered at increasing times (3, 6, and 9 sec) on Millipore filters. The filters were washed three times with cold flux medium, without ^{22}Na, and counted in a γ counter. The rate of Na uptake was expressed as $nmol \cdot min^{-1} \cdot mg^{-1}$ IOV. Protein concentration was measured by the method of Lowry.

REFERENCES

1. BIANCHI, G., P. FERRARI, D. TRIZIO, M. FERRANDI, L. TORELLI, B. R. BARBER & E. POLLI. 1985. Hypertension **7:** 319–325.

2. Wood, P. G. & H. Passow. 1981. Techniques Cell. Physiol. **P112:** 1–43.
3. Steck, T. & J. A. Kant. 1974. Methods Enzymol. **31:** 172–180.
4. Beck, F., G. Bianchi, A. Dorge, R. Rick, M. Schramm & K. Thurau. 1983. J. Hypertension **1** (Suppl. 2): 38–39.
5. Hanozet, G., P. Parenti & P. Salvati. 1985. Biochim. Biophys. Acta **819:** 179–186.

Abnormalities of Red Cell Membrane Calcium Handling in the Milan Hypertensive Strain of Rat

M. CIRILLO,[a] F. GALLETTI, AND P. STRAZZULLO

Institute of Internal Medicine and Metabolic Diseases
2nd Medical School
University of Naples
Naples, Italy

Abnormalities of extracellular calcium (Ca) metabolism and of cellular Ca handling were described in human essential hypertension as well as in rat genetic hypertension.[1] In this study we aimed to investigate some aspects of Ca handling in red blood cells (RBC) of the highly inbred strain of Milan hypertensive rats (MHS) and their normotensive control rats (MNS).

By previously used techniques[2] we measured: (a) the ^{45}Ca binding to RBC membranes and (b) the active ATP-dependent ^{45}Ca uptake in RBC inside-out vesicles (IOV), a parameter that reflects the activity of the membrane (Ca,Mg)ATPase.[3] Ca binding and uptake were analyzed in the presence of free Ca concentrations typical of the intracellular environment (10^{-8}–10^{-6} mol/l) fixed by using Ca-EGTA/EGTA buffers.[2]

TABLE 1. Apparent Affinity Constant (K_d) for Ca and Number (N) of Ca Binding Sites in Erythrocyte Membranes of MHS and MNS Rats

	MHS	MNS
K_d (nmol Ca/l)	51 ± 8	42 ± 7
N (nmol Ca/mg prot)	3.24 ± 0.30[a]	4.22 ± 0.37

[a] $p < 0.01$, M ± SEM.

A high affinity Ca binding site was detected in both MHS and MNS membranes. The Scatchard analysis of the data indicated that the number of the high affinity sites was reduced in MHS membranes while their apparent affinity constant for Ca ions did not differ between hypertensive and control rats (TABLE 1).

The Lineweaver-Burk analysis of the Ca uptake data showed that the maximal rate of the intravesicular Ca accumulation was significantly reduced in RBC IOV of MHS rats. The apparent affinity for Ca of the (Ca,Mg)ATPase was comparable in MHS and MNS rats (TABLE 2). The percent of IOV in the membrane homogenate was similar between the two strains of rats in each experiment.

These data provide further evidence of cell membrane abnormalities in this model of rat hypertension. In the RBC membranes of the Milan strain of hypertensive rats,

[a] Address correspondence to: Dr. Massimo Cirillo, Clinica Medica 2, II Facoltà di Medicina, Nuovo Policlinico, via Sergio Pansini, 5, 80131 Napoli, Italy.

TABLE 2. Apparent Affinity (K_m) for Ca and Maximal Rate (V_{max}) of Active Ca Uptake by Erythrocyte Inside-Out Vesicles of MHS and MNS Rats

	MHS	MNS
K_m (nmol Ca/1)	40 ± 4	37 ± 3
V_{max} (nmol Ca/mg prot/min)	9.03 ± 0.81[a]	13.54 ± 1.21

[a] $p < 0.01$, M ± SEM.

two systems involved in buffering the intracellular free Ca level were significantly altered. It is not clear whether the abnormalities of RBC Ca homeostasis are related to the ones observed in sodium metabolism[4] or represent a separate aspect of a membrane dysfunction.

Renal Ca clearance studies in MHS rats showed reduced tubular Ca reabsorption,[5] a finding that suggests the possibility that the renal tubular cell of these rats is characterized by a similarly defective Ca handling. Should other cell types, namely the smooth muscle cells of the vascular wall, share the same membrane abnormalities, it would be conceivable to hypothesize that such abnormalities might be relevant to the pathogenesis of hypertension given the key role of cytoplasmic Ca levels in the contraction process of smooth muscle cells.[6]

REFERENCES

1. ROBINSON, B. F. 1984. J. Hypertension **2:** 453–460.
2. CIRILLO, M., M. DAVID-DUFILHO & M. A. DEVYNCK. 1984. J. Hypertension **2**(Suppl 3): 485–487.
3. SARKADI, B. 1980. Biochim. Biophys. Acta **604:** 159–190.
4. BIANCHI, G., P. FERRARI & B. BARBER. 1984. The Milan hypertensive strain. *In* Handbook of Hypertension. W. De Jong, Ed. Vol. **4:** 328–349. Elsevier Science Publishers. Amsterdam.
5. CIRILLO, M., F. GALLETTI, M. F. CORRADO & P. STRAZZULLO. 1985. J. Hypertension **4:** 443–449.
6. BOHR, D. F. 1973. Circ. Res. **32:** 665–672.

The Relationship between Myoplasmic Calcium and Force Developed during Stimulation of Arterial Smooth Muscle:

A Study with Fura 2

GIACOMO BRUSCHI, MARIA E. BRUSCHI, ANGELO CAVATORTA, AND ALBERICO BORGHETTI

Istituto di Clinica Medica e Nefrologia
Università di Parma
43100 Parma, Italy

Calcium is believed to play a primary role in vascular contraction, but quantitative simultaneous measurements of free intracellular calcium (hereafter: Ca_i^{2+}) and force have not been reported. Novel intracellular calcium indicators, like Fura 2, provide this opportunity.[1]

Rat (Wistar-Kyoto, 8–9 weeks old) aortic medial rings were prepared by removal of adventitia and intima. The smooth muscle rings (1.5 mm wide, about 50 μm thick) were light-microscopically intact and developed about 100% of force developed by intact whole aortic rings. They were mounted between two parallel pins in a 20 ml chamber with quartz windows. The lower pin was immovable, while the upper one was connected to a very rigid (0.02 mm displacement per gram force) tension transducer. The rigidity of the transducer ensured that little geometrical changes took place during stimulation and contraction of smooth muscle samples. The chamber was held in the sample compartment of a Perkin-Elmer MPF44A spectrofluorimeter in the position normally occupied by ordinary cuvettes. The smooth muscle tissue was loaded with 10–20 μM Fura2-AM (resulting in 100–200 μM intracellular free dye, with signals 8–10 times higher than basal autofluorescence). Preparations were excited alternately at 340–380 nm by motorizing the shaft of the excitation monochromator. The emitted fluorescence was recorded at 510 nm. The 340/380 ratios were used to calculate Ca_i^{2+} as previously described,[1] using a calibration curve obtained by exposing the samples to 10 μM ionomycin in the presence of 1 nM–1 mM extracellular calcium (Ca_o^{2+}). The results of this study suggest the following conclusions: (1) the level of myoplasmic calcium is crucial for contraction, but (2) factors that increase the force:Ca_i^{2+} ratio are equally important. Conclusion (1) is suggested by the fact that under any condition (including norepinephrine (NE) and high K stimulation) the level of force was proportional to Ca_i^{2+} (FIGS. 1 and 2) and that raising Ca_i^{2+} "forcefully" with a ionophore (ionomycin) evoked Ca_i^{2+}-dependent contractions. On the other hand conclusion (2) is supported by the findings that in the initial, phasic portion of high K and NE contraction, one observes a rapid increase in force without any major increase in Ca_i^{2+}; the phasic onset of stimulation is characterized by an increase of the force:Ca_i^{2+} ratio, which is maintained as long as stimulation is conserved; that a wash (which removes NE and high K) and EGTA addition (in the presence of the above agents) produce similar effects on force (i.e. relaxation), but at very different Ca_i^{2+} levels (FIGS. 1 and 2); and that ionophore-induced contractions require much higher levels of Ca_i^{2+} than NE and high K to produce the same amount of force.

NE-induced contractions in Ca-free media have been attributed to intracellular

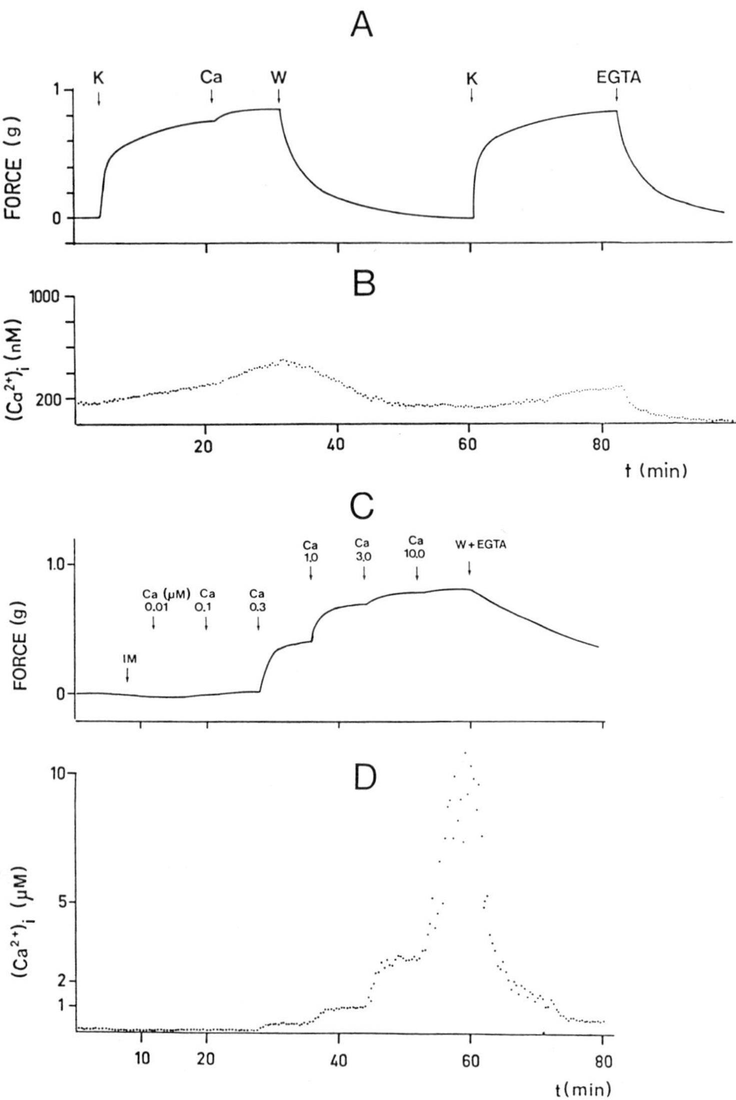

FIGURE 1. *Upper panel.* Simultaneous recordings of tension(a) and Ca_i^{2+} (b) during stimulation of aortic medial rings by equimolar substitution of NaCl with 80 mM KCl (high K stimulation). Arrows denote the onset of stimulation (K), the addition of 3 mM EGTA at pH 7.4 (EGTA), and a wash with normal medium containing 1.5 mM $CaCl_2$. *Lower panel.* Simultaneous recordings of tension (c) and Ca_i^{2+} (d) in the presence of 10 μM ionomycin (added to Ca-free medium at IM) and different extracellular calcium concentrations (0.01–10 μM) established as described previously,[1] with slight modifications. W+EGTA indicates a wash with Ca-free medium containing 0.3 mM EGTA.

calcium release. FIGURE 2a shows that although a Ca_i^{2+} transient is evoked in this condition, this is insufficient to account fully for the phasic transient of NE contraction. Most of the effect is due to an increase of the force:Ca_i^{2+} ratio. This is further illustrated by the experiments in FIGURE 2 (b and c). A 30-min stay in Ca-free medium abolishes the Ca_i^{2+} transient attributable to intracellular release, but not the phasic contraction. Similarly, in samples loaded with 1 mM Fura 2 (FIG. 2d) NE stimulation

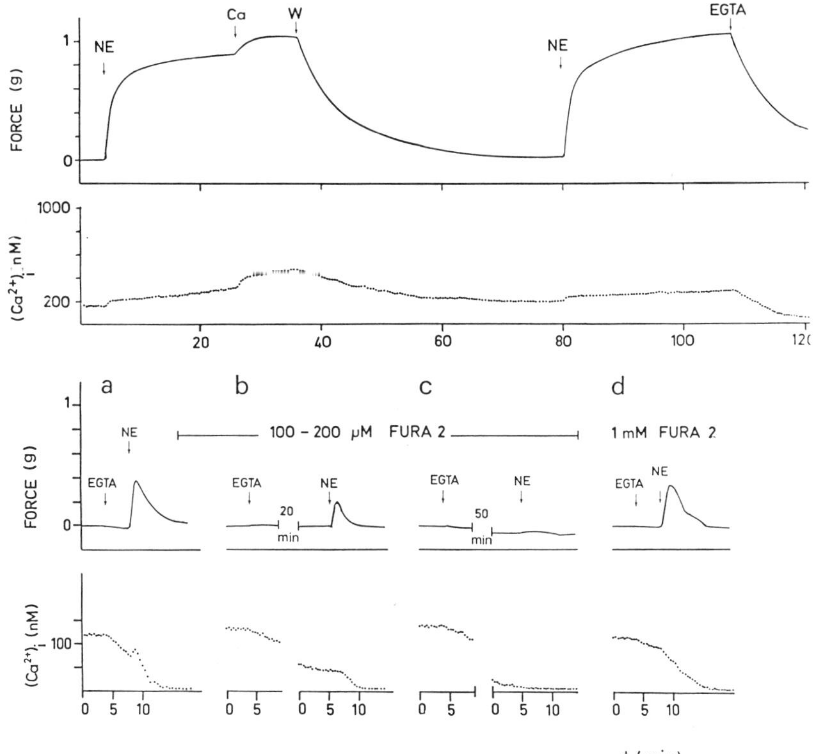

FIGURE 2. *Upper panel.* Simultaneous measurements of force and Ca_i^{2+} during stimulation with 10^{-6} norepinephrine (NE, denoted by arrow). The other symbols are as in FIGURE 1 (upper panel). *Lower panel.* (a, b, c) NE stimulation (10^{-6} M) in Ca-free medium (corresponding to 3 mM EGTA added to medium containing 1.5 mM $CaCl_2$) in preparations loaded with 100–200 μM intracellular Fura 2. (d) Procedures were the same but tissue was loaded with 1.1 mM intracellular indicator in order to buffer intracellular release. Note that a Ca_i^{2+} transient is observed only in (a), though a tension peak is present in (a), (b), (d).

evokes contraction in spite of no change in Ca_i^{2+}. The latter effect is presumably due to buffering by this unusual dye loading.

REFERENCE

1. WILLIAMS, D. A., K. E. FOGARTY, R. Y. TSIEN & F. S. FAY. 1985. Nature **318:** 558.

Endogenous Digitalis-like Factor(s) in Human Plasma:

Binding to Specific Antiserum, Red Blood Cells, and Plasma Proteins

S. BALZAN,[a] S. GHIONE, A. CLERICO, AND U. MONTALI

Consiglio Nazionale delle Ricerche
Institute of Clinical Physiology
and
Institute of Biological Chemistry
University of Pisa
56100 Pisa, Italy

Several studies[1,2] reported the presence of an endogenous substance(s) with digoxin-like immunoreactivity (DLIS) in blood of experimental animals and humans using digoxin radioimmunological (RIA) or enzyme-immunoassay (EIA) methods. Experimental studies and theoretical considerations suggest that this substance in addition to binding to antidigoxin antibodies, might also bind to the same cellular receptor (Na/K ATPase) and displace labeled cardiac glycosides from the specific binding sites. Therefore, it has been suggested that DLIS is an endogenous modulator of the membrane sodium-potassium pump.[1,2]

Detectable DLIS concentrations were previously found by us and others in blood and urine of adults (normal healthy controls, hypertensive patients, and salt-loaded healthy subjects), while higher levels were generally observed in pregnant women.[1,2] The highest levels of digoxin-like immunoreactivity were found in blood and urine of healthy newborns in the first week of extrauterine life (about 5–10-fold higher than those previously observed by us in adults). Several studies using extraction procedures and chromatographic separations suggested that DLIS is not a large charged molecule, salt, or fatty acid and also demonstrated that DLIS circulates in blood bound to plasma protein(s), from which it is detached by boiling or by methanol extraction.[1,2] More recently, Cloix *et al.*[3] suggested that DLIS could be an amino-glyco-steroid containing a carboxyl acid function and three methyl groups with a molecular weight of 431.

We have measured the DLIS concentrations with a sensitive RIA method, previously described,[4] and labeled ouabain binding activity with a radioreceptor assay (RRA) in plasma of 7 normal adults, 10 newborns, and 5 pregnant women (second or third trimester) after purification on Sep Pak C18 cartridges. RRA consisted in the incubation (at 37°C for 4 hours) of intact red blood cells (RBC) in a medium containing plasma extracts and increasing concentrations of [^{3}H]ouabain (from 2.7 nmol/l to 19.7 nmol/l). The radioactivity bound to RBC was determined after separation by filtration on GFC Whatman filters and the specific binding was calculated by subtracting the counts due to non-specific binding obtained by adding an excess of unlabeled ouabain (always less than 10%). Newborn and pregnant women plasma extracts showed significantly higher levels ($p < 0.05$) than adult plasma

[a]Address correspondence to: Dr. Balzan Silvana, C.N.R. Institute of Clinical Physiology, University of Pisa, Via Savi 8, 56100 Pisa, Italy.

TABLE 1. DLIS, Inhibitory Activity (1%), Number of Sites, and K_d Values (mean ± SD) in Normal Adults, Newborns, and Pregnant Women

Subjects	DLIS (pg/ml d.e.)	1%	Number Sites/Cell	K_d (nMol/L)
Adults (N = 7)	10 ± 7	7 ± 7	482 ± 60	7.1 ± 1.5
Newborns (N = 10)	40 ± 14	24 ± 14	418 ± 118	9.5 ± 2.9
Pregnancy (N = 5)	28 ± 17	21 ± 4	424 ± 62	9.0 ± 0.9

samples both for DLIS and for displacement capacity of [^{3}H]ouabain from the RBC membrane receptor (TABLE 1). Significant differences between the groups for K_d ($p < 0.02$) were observed, but not for the number sites (TABLE 1). Moreover, a significant correlation was found between DLIS and displacement capacity (r = 0.72, $N = 22, p < 0.001$). These data confirm that elevated concentrations of an endogenous substance (of a group of substances) with immunological and biological activity similar to cardiac glycosides are present in plasma of newborns and pregnant women.

We also performed some preliminary chromatographic studies to investigate the binding of DLIS to plasma proteins. Newborn plasma (8 ml) was eluted on Sephadex G200 column (83 × 25 cm) with 50 mM ammonium acetate. The eluted fractions were assayed with the RIA method. The chromatograhic profile of digoxin-like immunoreactivity showed a major peak in correspondence with albumin elution and a smaller peak eluted after the salt region (FIG. 1). These results indicate that DLIS circulates bound to a plasma protein (M.W. about 60,000) with a chromatographic pattern similar to albumin.

In conclusion, the present study confirms the presence of elevated levels of immunological and receptor binding activity similar to cardiac glycosides drugs in blood of newborns and pregnant women, but it is not certain whether the same

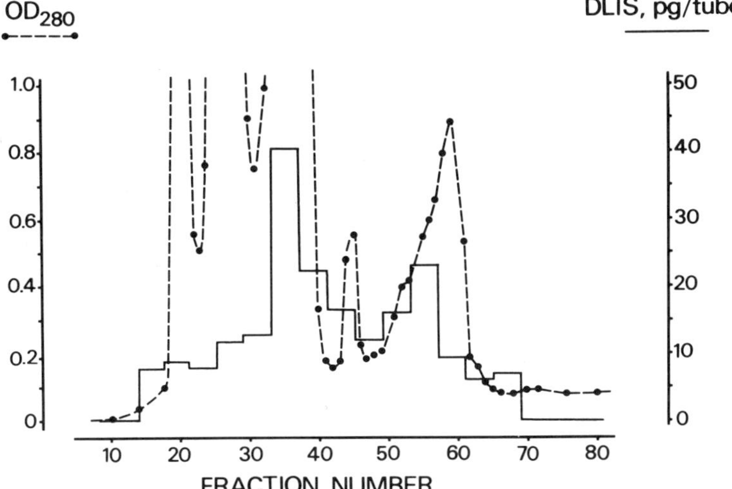

FIGURE 1. Chromatographic elution of newborn plasma on a Sephadex G 200 column. The optical activity (at 280 nm) and the immunological activity are, respectively, indicated by a dotted line and continuous line. The fraction volume was 8 ml.

endogenous substance has both properties. Further studies are necessary to characterize definitively the biochemical and physiological properties of this endogenous factor (or group of substances) and its possible role in the regulation of fluids and electrolytes in animals and humans.

REFERENCES

1. Clerico, A., S. Ghione, S. Balzan & M. G. Del Chicca. 1984. Digoxin-like immunoreactive substance: non-specific interference or a new hormone? J. Nucl. Med. All. Sci. **28**(4): 297–301.
2. Valdes, R. 1985. Endogenous digoxin-like immunoreactive factors: impact on digoxin measurements and potential physiological implications. Clin. Chem. **31**(9): 1525–1532.
3. Cloix, J. F., M. Crabos, I. W. Wainer, D. Ruegg, M. Seiler & F. Meyer. 1985. High yield-purification of a urinary Na^+-pump inhibitor. Biochem. Biophys. Res. Commun. **131**(3): 1234–1240.
4. Balzan, S., A. Clerico, M. G. Del Chicca, U. Montali & S. Ghione. 1984. Digoxin-like immunoreactivity in normal human plasma and urine, as detected by a solid-phase radioimmunoassay. Clin. Chem. **30**(3): 450–451.

Ion Transport in Red Blood Cells and Antihypertensive Therapy

E. NIUTTA, M. G. TRIPODI, C. PATI, D. CUSI,
C. BARLASSINA, F. DOSSI, AND G. BIANCHI[a]

Insitute of Medical Sciences
University of Milan
Milan, Italy

Many abnormalities in cell membrane transport of ions have been reported to occur in blood cells from patients with essential hypertension.[1-3] It is not clear whether these abnormalities are genetically linked to hypertension development or if they are secondary changes mediated by humoral factors or if they are an epiphenomenon without any relationship with hypertension.

In order to investigate the effects of the antihypertensive therapy on cation fluxes, 23 hypertensive patients have been treated with Atenolol (AT) 100 mg, Captopril (CP) 75 mg, and Canrenoate K (CK) 200 mg. Each patient received the three drugs

TABLE 1. Effects of Captopril, Atenolol, and Canrenoate K on Blood Pressure and Heart Rate

	Basal	Captopril	Atenolol	Canrenoate K
S. B. P. ortho	151.2 (±22.5)	130.8 (±19.9)[a]	131.2 (±21.4)[a]	127.8 (±17.2)[a]
D. B. P. ortho	105.3 (±10.1)	93.7 (±11.3)[a]	89.1 (±11.9)[a]	94.5 (±10.2)[a]
Heart rate ortho	84.8 (±9.1)	82.9 (±9.6)	64.9 (±7.8)[a]	88.4 (±10.3)
S. B. P. clino	157.3 (±18.5)	141.1 (±18.9)[a]	138.9 (±20.9)[a]	138.2 (±16.2)[a]
D. B. P. clino	105.1 (±7.6)	95.2 (±12.6)[a]	90.4 (±9.3)[a]	96.2 (±7.6)[a]
Heart rate clino	76.0 (±10.2)	72.6 (±8.5)[b]	60.9 (7.6)[a]	75.9 (±8.8)

Number of patients: 23 (14M, 9F) and age: 44.4 (16–57).

Systolic and diastolic blood pressure are expressed in mm Hg. Heart rate is expressed in beats/min.

[a]Indicates $p < 0.001$.

[b]Indicates $p < 0.02$.

for three months in a randomized sequence. Before and after every treatment period we measured red blood cell Na-K pump (Na-K P), Na-Li countertransport (Na-Li CNT), and Na-K cotransport (Na-K COT) activities, together with intraerythrocyte Na content (Na_{int}) and passive permeability to Na (Na pp) according to the original methods.[3]

Particular care was given in admitting only patients who suspended any previous treatment for at least three months. All three drugs significantly lowered blood pressure (TABLE 1). Na_{int} significantly decreased and Na-Li CNT significantly increased after CP and CK. CK also induced a significant increase in Na-K P activity.

[a]To whom all correspondence should be addressed.

TABLE 2. Effects of Captopril, Atenolol, and Canrenoate K on Cation Fluxes across Red Cell Membrane and on Intraerythrocyte Na Concentration

	Basal	Captopril	Atenolol	Canrenoate K
Na-K Pump	5602.0 (±1061)	5620.5 (±1036)	5574.7 (±1221)	6213.1 (±1069)[a]
Na-K cotransport	368.4 (±160)	381.4 (±131)	368.4 (±160)	440.9 (±162)
Na passive permeability	16.4 (±3.5)	17.4 (±3.0)	17.3 (±4.3)	16.8 (±4.3)
Na-Li Countertransport	279.8 (±93.4)	308.9 (±92.5)[b]	295.2 (±197)	316.7 (±115)[c]
Internal Na content	8.5 (±1.6)	7.9 (±1.4)[b]	8.3 (±1.7)	7.2 (±1.6)[d]

Number of patients = 23 (14M, 9F) and age = 44.4 (16–57) years.

Na-K pump, Na-K cotransport, and Na-Li countertransport are expressed in micromole per liter of red blood cells^{-1} · hr^{-1}. Na passive permeability is expressed in 1/hr. Internal Na content is expressed in mmol · lRBC^{-1}.

[a]Indicates $p < 0.01$. [b]Indicates $p < 0.05$. [c]Indicates $p < 0.02$. [d]Indicates $p < 0.001$.

Atenolol did not induce any detectable change in cation transport activities nor in internal Na content (TABLE 2).

The patients were than divided, according to their greatest decrease in blood pressure, in three groups: CP (6 patients), AT (10 patients), and CK (7 patients) "responders." The 7 "CK responders" had, in basal conditions, a Na_{int} significantly lower than the "CP responders" (8.1 ± 0.9 vs. 9.3 ± 1.2 $p < 0.05$).

These preliminary results indicate that some antihypertensive treatments affect ion transport across red blood cell membrane and that intracellular Na may be relevant in predicting the antihypertensive effect of some drugs.

REFERENCES

1. CANESSA, M., M. ADRAGNA, H. SOLOMON, T. M. CONNOLLY & D. C. TOSTESON. 1980. Increased Na-Li countertransport in red cells of patients with essential hypertension. New Engl. J Med. **302:** 772–777.
2. AMBROSIONI, E., F. V. COSTA, L. MONTEBUGNOLI, F. TARTAGNI & B. MAGNANI. 1981. Increased intralymphocytic sodium content in essential hypertension: an index of impaired cellular Na metabolism. Clin. Sci. **6**: 181–183.
3. CUSI D., C. BARLASSINA, P. FERRANDI, P. LUPI, P. FERRARI & G. BIANCHI. 1981. Familial aggregation of cation transport abnormalities and essential hypertension. Clin. Exp. Hypertens. **3(4)**: 871–887.

Heritability of Sodium Transport Systems and Hypertension[a]

DANIELE CUSI, GRAZIA TRIPODI, ELENA ALBERGHINI, ENRICO NIUTTA, CRISTINA BARLASSINA, EMILIO FOSSALI, FIORELLA DOSSI, AND GIUSEPPE BIANCHI

Istituto di Scienze Mediche
University of Milan
20122 Milan, Italy

INTRODUCTION

In recent years several studies have been performed on the biochemical alterations of cell membrane function in essential hypertension. However, the relations between erythrocyte (RBC), intracellular sodium ($[Na]_i$), and Na transport alterations and heredity are not clear. This paper presents a large population survey on $[Na]_i$ and the correlation between the different RBC transport systems in families in which neither, one, or both parents were hypertensive.

MATERIALS AND METHODS

$[Na]_i$, Na pump (Na P), Na-K cotransport (Na COT), and Li-Na countertransport (Li CNT) were measured according to previously described methods.[1] $[Na]_i$ was measured in 130 hypertensive adults and 148 matched normotensive controls. $[Na]_i$ and all the abovementioned Na transport systems were measured in 24 complete families (parents and young offspring, $N = 84$) with at least one hypertensive parent (in six families both parents were hypertensive) and in 18 complete families (parents and young offspring, $N = 64$) with both normotensive parents and no history of hypertension in the grandparents. None of the subjects studied were on any pharmacological treatment (antihypertensive or other).

RESULTS AND DISCUSSION

The mean values for $[Na]_i$ for the transport systems measured are summarized in TABLE 1. In agreement with previous findings of others and ours,[1-3] $[Na]_i$ was the same in hypertensive and normotensive adults. In this sample the mean value of Na COT was higher in the hypertensive than in the normotensive parents. We found that Na COT is bimodally distributed in the hypertensive subjects, with values higher and lower than in normotensive subjects.[2] Indeed this bimodality may reflect different pathogenetic mechanisms. However, independent of blood pressure, Na P correlated in husband and wife, as did $[Na]_i$ ($r = 0.3$; $p < 0.005$ for Na P and $r = 0.48$; $p < 0.001$ for $[Na]_i$), thus suggesting strong environmental influences.

[a]Supported in part by C.N.R. grants 84.02201.56.115.05593 and 85.00456.56.115.05593.

TABLE 1.

	General Population (Adults)		
	Hypertensives (N = 130)		Normotensives (N = 145)
$[Na]_i$	8.3 ± 1.8		8.3 ± 1.7
	Families		
	Hypertensive parents (N = 30)		Normotensive parents (N = 54)
$[Na]_i$	8.1 ± 1.8		8.4 ± 1.9
Na P	5991 ± 1447		5724 ± 1449
Na COT	425 ± 149	$p < 0.01$	324 ± 168
Li CNT	315 ± 139		289 ± 117
	Offspring of hypertensives (N = 44)		Offspring of normotensives (N = 21)
$[Na]_i$	7.4 ± 1	$p < 0.05$	8.9 ± 1.4
Na P	6277 ± 1861	$p < 0.02$	5128 ± 1869
Na COT	302 ± 147		303 ± 139
Li CNT	226 ± 65		267 ± 108

NOTE: Mean values for the intraerythrocyte Na concentration ($[Na]_i$), in mmol/l RBC and the Na pump (Na P), Na-K cotransport (Na COT), and Li-Na countertransport (Li CNT), in μmol/l RBC/hr, in our sample. Only $[Na]_i$ was measured in the sample of adult general population. The data are expressed as mean values ± S.D.

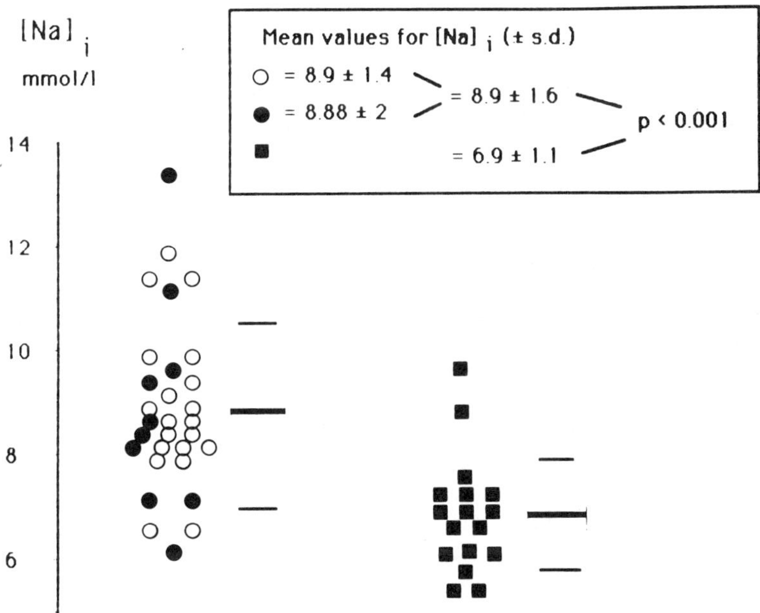

FIGURE 1. Intraerythrocyte Na concentration ($[Na]_i$) of the young offspring of both normotensive (○) or hypertensive parents divided according to the Na-K cotransport value of the hypertensive parent (<350 μmol/l RBC/hr, "low cotransport," ●; >450 μmOm/l RBC/hr, "high cotransport," ■)

No correlation was found for Na COT and Li CNT. Conversely, the mean value of Na COT of the parents correlated with the values of their offspring ($r = 0.44$; $p < 0.005$). Similar results were found for Li CNT ($r = 0.71$, $p < 0.001$), suggesting some genetic control of these transport systems. The mean value of $[Na]_i$ of the parents correlated with the values of their offspring as well as Na P, with better correlation coefficients than the correlation between husband and wife ($r = 0.73$, $p < 0.001$ for $[Na]_i$ and $r = 0.53$, $p < 0.001$ for Na P). The better correlation coefficients may indicate that, although strongly influenced by the environment, $[Na]_i$ and Na P may also be under genetic control. In spite of the strong environmental influences, $[Na]_i$ was abnormally low in the young offspring of hypertensive parents. This difference was independent of Na consumption, which was the same in the two groups (urinary Na/creatinine ratio: 1.7 ± 0.9 vs. 1.5 ± 0.6, $p =$ n.s.). The offspring of hypertensives were divided in two groups according to the Na COT of their hypertensive parent (we defined low Na COT values <350 μmol/l RBC hr and high Na COT values >450). The offspring of high Na COT hypertensive parents have significantly lower $[Na]_i$ than the offspring of low Na COT hypertensive parents (6.87 ± 1.1 vs. 8.9 ± 1.6, $p < 0.001$), the $[Na]_i$ of the offspring of low Na COT hypertensive parents is equal to that of the controls (FIG. 1).

We previously demonstrated that RBC alterations in the Milan hypertensive strain (high Na COT and low $[Na]_i$) are primarily determined in the stem cells and are genetically associated with hypertension.[4] The present results in humans suggest that in a subgroup of normotensive subjects at risk of developing essential hypertension it is possible to find the RBC alterations primarily described in the Milan hypertensive strain.

REFERENCES

1. CUSI, D., C. BARLASSINA, M. FERRANDI, P. LUPI, P. FERRARI & G. BIANCHI. 1981. Familial aggregation of cation transport abnormalities and essential hypertension. Clin. Exp. Hypertension **3**(4): 871–884.
2. CUSI, D., C. BARLASSINA, M. FERRANDI, P. PALAZZI, E. CELEGA & G. BIANCHI. 1982. Relationship between altered Na-K cotransport and Na-Li countertransport in the erythrocytes of essential hypertensive patients. Clin. Sci. **63:** 184s–187s.
3. CANESSA, M., N. ADRAGNA, H. S. SOLOMON, T. M. CONNOLLY & D. C. TOSTESON. 1980. N. Engl. J. Med. **302:** 772–777.
4. BIANCHI, G., P. FERRARI, D. TRIZIO, M. FERRANDI, L. TORIELLI, B. BARBER & E. POLLI. 1985. Red blood cell abnormalities and spontaneous hypertension in the rat—A genetically determined link. Hypertension **7:** 319–324.

Membrane Properties of the Human Colon Tumor Cell Line HT-29[a]

G. ESPOSITO,[b] E. BOMBARDIERI,[c] M. G. COCCIOLO,[c]
C. LINDI,[b] M. VALTOLINA,[c] P. MARCIANI,[c]
AND M. R. GIORIA[d]

[b]*Istituto di Fisiologia Generale e Chimica Biologica*
Facoltà di Farmacia
Università di Milano
Milan, Italy

[c]*Istituto Nazionale per lo Studio e la Cura dei Tumori*
Milan, Italy

[d]*Dipartimento di Fisiologia e Biochimica Generale*
Università di Milano
Milan, Italy

In order to modulate membrane properties of HT 29 cells, we grew these cells in media containing different concentrations of cholesterol (25, 50, or 100 μg/ml) or α_2 recombinant interferon (rIFNα-2A, 100–200 U/ml). In fact, it was demonstrated that many tumor cells present a more fluid membrane when cholesterol content is reduced.[1] Interferon displays an antiproliferative activity and affects membrane properties as well as the cytoskeleton structure.[2]

The morphology of HT-29 cells in the presence of cholesterol seemed to show an increased number of mitochondria and a modification of their aspect. The number of lipid droplets also increased (FIG. 1).

The presence of 100 μg/ml cholesterol caused an initial cell loss and a reduced growth rate until confluence; moreover, carcinoembryonic antigen (CEA) release expressed as ng/ml was constantly higher in controls, while CEA production given in ng per 10^{-6} cells was much higher in cholesterol-treated cells during the first four days of culture. The following days showed a similar production of CEA in both control and treated cells (FIG. 2). These phenomena displayed a dose-dependent effect.

The presence of rIFNα-2A determined a cell loss during the first days of culture; then the growth rate became similar to that of the control cells. CEA release was higher only in the second and third day of culture, while CEA production was higher during the first three days of culture.

The uptake of tritiated 3-O-methyl-D-glucose into HT-29 cells showed the presence of a transport system of low affinity (K_m = 20 mM) and high capacity (V_{max} = 12.9 nmoles/mg cell protein × 1.5 sec). In low Na^+ medium, V_{max} decreased (V_{max} = 6.3 nmoles/mg cell protein × 1.5 sec) while K_m did not change (K_m = 17 mM). That seemed to indicate that Na^+ affects the mobility of the carrier but not the affinity for the substrate.[3]

[a]Supported by C.N.R. grants C.T. 85.02146.44 and 85.02056.44 (Finalized-Project "Oncology").

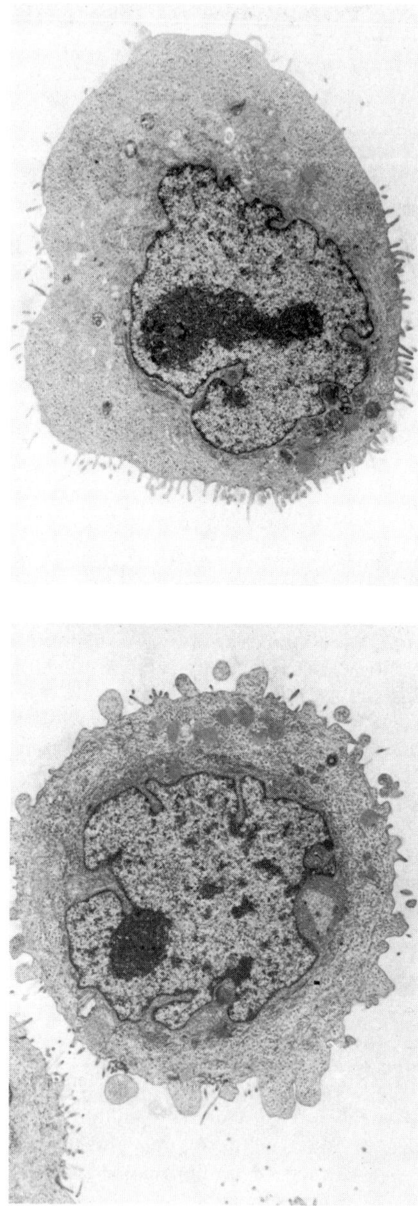

FIGURE 1. Electron micrographs of HT-29 cells. (a) Cells grown in RPMI + 10% fetal bovine serum (×8,000). (b) Same as in (a) but in medium containing 100 μg/ml cholesterol (×8,000). Figure reduction of 60%.

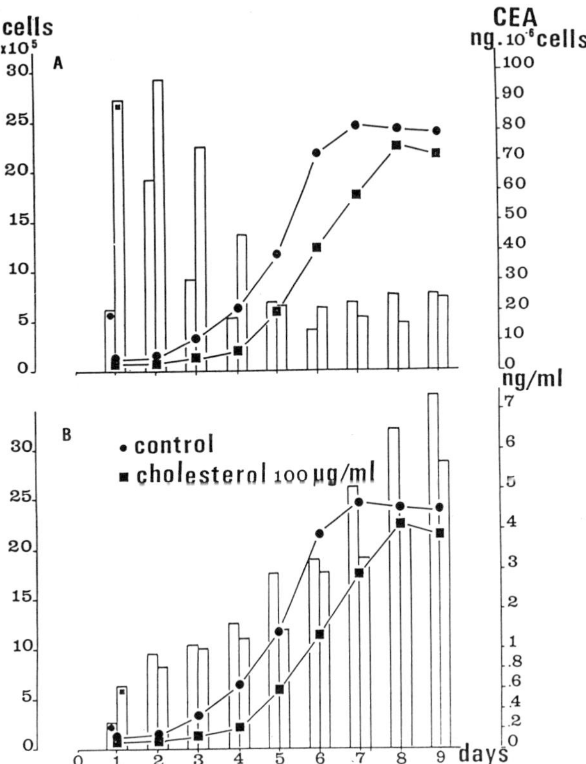

FIGURE 2. Effect of 100 μg/ml cholesterol on cell growth rate, (a) on CEA production, and (b) on CEA release into the medium.

ACKNOWLEDGMENT

The authors wish to thank Hoffmann La Roche (Basel) for supplying rIFNα-2A free of charge.

REFERENCES

1. SHINITZKY, M. & Y. BARENHOLZ. 1978. Fluidity parameters of lipid regions descrimined by fluorescence polarization. Biochim. Biophys. Acta **515:** 367–384.
2. GRESSER, I. & M. G. TOVEY. 1978. Antitumor effect of interferon. Biochim. Biophys. Acta **516:** 231–239.
3. ESPOSITO, G., V. F. SACCHI, C. LINDI, E. BOMBARDIERI, M. G. COCCIOLO, M. VALTOLINA & P. MARCIANI. 1985. Sugar uptake into HT-29 cells, a cell line of adenocarcinoma of human colon. *In* Cell Membranes and Cancer. T. Galeotti *et al.*, Eds.: 283–287. Elsevier Scientific Publishers. Amsterdam.

Altered Membrane Bound Protein Kinase Activities in Lymphoid Cells Transformed by Moloney and Abelson Leukemia Viruses[a]

LORENZO A. PINNA,[b] ANNA MARIA BRUNATI,[b]
DANIELA SAGGIORO,[c] AND LUIGI CHIECO-BIANCHI[c]

[b]Istituto di Chimica Biologica
and
[c]Cattedra di Oncologia
University of Padova
Padova, Italy

The phosphorylation of definite seryl, threonyl, and tyrosyl residues, catalyzed by specific protein kinases and reversed by protein phosphatases is a frequent short-term regulatory device affecting, among others, many membrane proteins. Considering that the oncogene of the acute transforming Abelson Leukemia virus (A-MuLV) codes for a tyrosine protein kinase that is largely integrated into the membranes of the host cell,[1] we focused our attention on the membrane-bound protein kinase activities of a murine cell line established from A-MuLV–induced thymic lymphomas. Such activities were compared with those of normal thymocytes and of a murine T-cell lymphoma line induced by Moloney leukemia virus (M-MuLV), a slow transforming retrovirus whose genome is partially overlapping that of A-MuLV but lacks oncogene sequences coding for tyrosine protein kinase. The particulate fraction from both normal and transformed cells was obtained as in Brunati *et al.*[2] and was extracted with 1% Nonidet P40 (NP40). Both the NP40-soluble and NP40-insoluble fractions were assayed for the following protein kinases: cAMP-dependent protein kinase (tested on the synthetic peptide RRASVA); "protamine kinase," providing a rough estimate of Ca^{2+}, phospholipid-dependent protein kinase C together with its spontaneously active proteolytic derivative protein kinase M;[3] casein kinase 1 (tested on casein in the presence of 1 μg/ml heparin); casein kinase 2 (evaluated as the difference between activities toward casein in the absence and presence of 1 μg/ml heparin); tyrosine protein kinase (tested on poly(Glu,Tyr)(4:1).

As shown in FIGURE 1 neither cAMP-dependent protein kinase nor CK-1 undergoes any remarkable variations upon transformation. CK-2 on the other hand is hardly detectable in the particulate fraction of normal cells while it becomes quite evident in membrane extracts of cells transformed by either M-MuLV or A-MuLV. This result is in good agreement with a report indicating a rise of CK-2 as a general feature of malignancy.[4] In addition, our data suggest that this kinase may mediate some effects of transformation by binding to cell membranes.

FIGURE 1 also shows the activities toward protamine, and poly(Glu,Tyr)4:1 are almost unchanged in M-MuLV transformed cells, while undergoing considerable increments in cells transformed by A-MuLV. That the increment of protamine kinase

[a]Supported by Italian Consiglio Nazionale delle Ricerche, Progetto Finalizzato Oncologia.

activity in A-MuLV transformed cell membranes is due to PK-C plus PK-M, was confirmed by DEAE-sepharose chromatography: about 70% eluted as a peak of Ca^{2+}, phospholipid-stimulatable kinase, while the remaining was recovered in a more retarded peak corresponding to PK-M (not shown). The rise of PK-C upon transformation by A-MuLV, observed also with B-lymphocytes (not shown), is a finding of special interest considering that infection by A-MuLV increases the phosphorylation at seryl

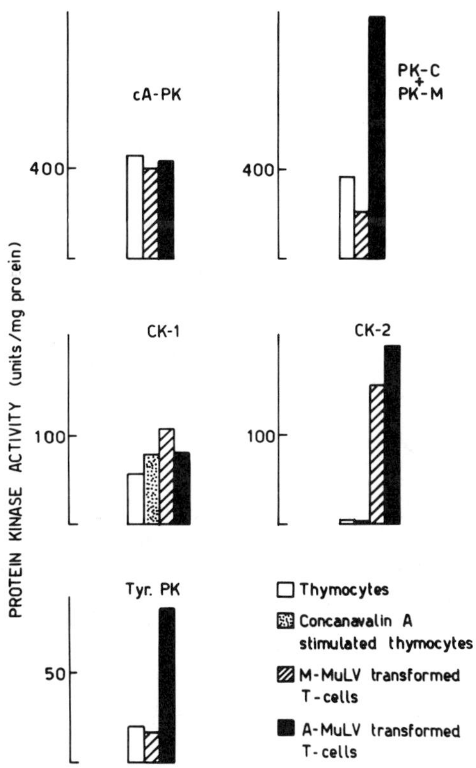

FIGURE 1. Protein kinase activities of the particulate fraction from normal thymocytes and from thymic lymphomas induced by either Moloney or Abelson leukemia viruses. Nonidet P40 extracts of particulate fractions were assayed for the following activities: cAMP-dependent protein kinase (cA-PK); protein kinase C and its proteolytic derivative protein kinase-M (PK-C + PK-M); casein kinase 1 (CK-1), casein kinase 2 (CK-2), and tyrosine protein kinase (Tyr-PK). One unit of protein kinase is defined as the amount of enzyme transferring 1 pmol P per min to the peptide or protein substrate.

residues of ribosomal protein S6,[5] which proved a good target for PK-C.[6] Our data support the concept that the tyrosine protein kinase expressed by A-MuLV promotes a direct or mediated activation of PK-C.

The impressive increment of tyrosine protein kinase in NP40 extracts of membranes from A-MuLV transformed cells cannot merely be accounted for by the tyrosine kinase expressed by the virus itself, which is not extractable from the

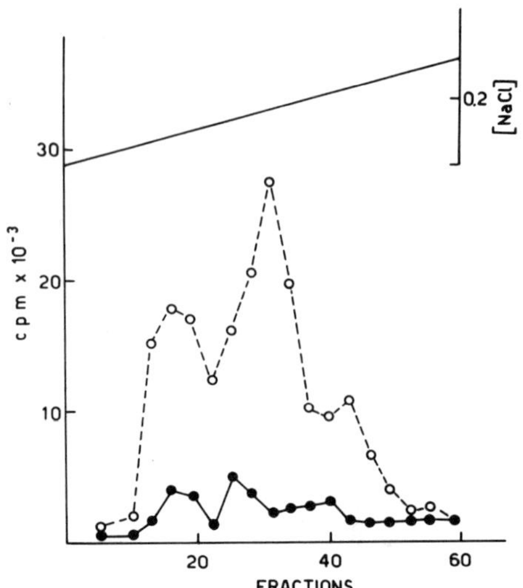

FIGURE 2. DEAE sepharose column chromatography of tyrosine protein kinase extracted with 1% Nonidet P40 from the particulate fractions of either thymus (●———●) or A-MuLV transformed thymocytes (○———○). Experimental conditions are the same as in Brunati *et al.*[2] except for the replacement of 0.1% Triton X-100 with Nonidet P40.

membrane by NP40.[7] Actually, as shown in FIGURE 2, the whole NP40-extracted tyrosine kinase activity of A-MuLV transformed cells is due to at least three distinct enzymes eluting from DEAE-sepharose with the same NaCl concentrations as three tyrosine protein kinases from spleen.[2] Therefore the oncogene product of Abelson virus apparently promotes the activation of some cellular tyrosine protein kinases, besides that of PK-C. Whether these multiple stimulatory effects occur independently or are part of a single cascade mechanism still remains an open question.

REFERENCES

1. KONOPKA, J. B. & O. N. WITTE. 1985. Biochim. Biophys. Acta **823:** 1–17.
2. BRUNATI, A. M., F. MARCHIORI & L. A. PINNA. 1985. FEBS Lett. **188:** 321–325.
3. KIKKAWA, U., R. MINAKUCHI, Y. TAKAI & Y. NISHIZUKA. 1983. Methods Enzymol. **98:** 288–298.
4. PROWALD, K., H. FISHER & O. G. ISSINGER. 1984. FEBS Lett. **176:** 479–483.
5. MALLER, J. L., J. G. FOULKES, E. ERIKSON & D. BALTIMORE. 1985. Proc. Natl. Acad. Sci. USA **82:** 272–276.
6. PARKER, P. J., M. KATAN, M. D. WATERFIELD & D. P. LEADER. 1985. Eur. J. Biochem. **148:** 579–586.
7. BOSS, M. A., G. DREYFUSS & D. BALTIMORE. 1981. J. Virol. **40:** 472–481.

Index of Contributors

(Italicized page numbers indicate discussions)